Fundamental Concepts of

Inorganic Chemistry

Volume 5

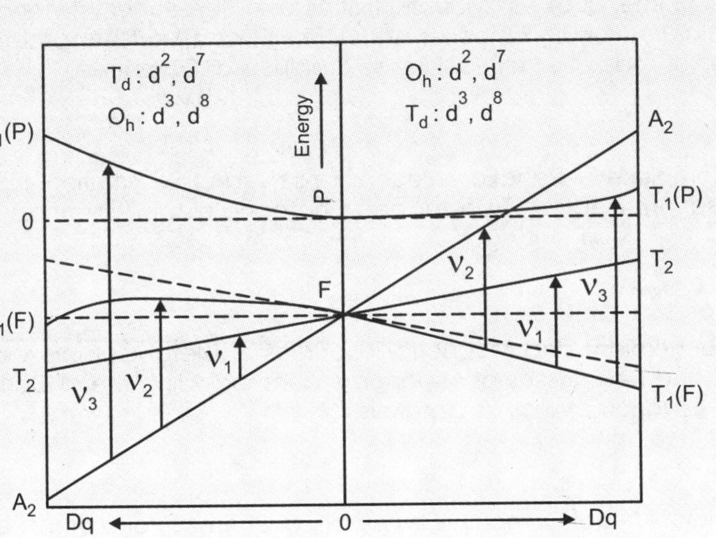

Fundamental Concepts of **Inorganic Chemistry**

Contents of Volumes 1-7

Fundamental Concepts of

Inorganic Chemistry

Volume 5

Asim K Das
MSc (Gold Medalist, CU), PhD (CU), DSc (Visva Bharati)
Professor of Chemistry
Visva Bharati University, Santiniketan 731235
West Bengal (India)

Mahua Das
MSc (CU), PhD (Visva Bharati)
Former Research Associate, Department of Chemistry
Visva Bharati University, Santiniketan 731235
West Bengal (India)

CBS

CBS Publishers & Distributors Pvt Ltd

New Delhi • Bengaluru • Chennai • Kochi • Kolkata • Mumbai
Hyderabad • Jharkhand • Nagpur • Patna • Puna • Uttarakhand

Disclaimer
Science and technology are constantly changing fields. New research and experience broaden the scope of information and knowledge. The authors have tried their best in giving information available to them while preparing the material for this book. Although, all efforts have been made to ensure optimum accuracy of the material, yet it is quite possible some errors might have been left uncorrected. The publisher, printer and authors will not be held responsible for any inadvertent errors or inaccuracies.

Fundamental Concepts of
Inorganic Chemistry
Volume 5

ISBN: 978-81-239-2352-9

First Edition: 2014
Reprint: 2015, 2016, 2018, 2019, 2021

Published by Satish Kumar Jain and produced by Varun Jain for

CBS Publishers & Distributors Pvt Ltd
4819/XI Prahlad Street, 24 Ansari Road, Daryaganj, New Delhi 110 002, India
Ph: 011-23289259, 23266861, 23266867
Fax: 011-23243014
Website: www.cbspd.com
e-mail: delhi@cbspd.com; cbspubs@airtelmail.in

Corporate Office: 204 FIE, Industrial Area, Patparganj, Delhi 110 092
Ph: 011-4934 4934 Fax: 011-4934 4935 e-mail: publishing@cbspd.com; publicity@cbspd.com

Branches

- **Bengaluru:** Seema House 2975, 17th Cross, K.R. Road, Banasankari 2nd Stage, Bengaluru 560 070, Karnataka
  Ph: +91-80-26771678/79 Fax: +91-80-26771680 e-mail: bangalore@cbspd.com
- **Chennai:** 7, Subbaraya Street, Shenoy Nagar, Chennai 600 030, Tamil Nadu
  Ph: +91-44-26680620, 26681266 Fax: +91-44-42032115 e-mail: chennai@cbspd.com
- **Kochi:** 42/1325, 1326 Power House Road, Opp. KSEB, Ernakulum, Kochi 682018, Kerala
  Ph: +91-484-4059061-65 Fax: +91-484-4059065 e-mail: kochi@cbspd.com
- **Kolkata:** 6/B, Ground Floor, Rameswar Shaw Road, Kolkata 700 014, West Bengal
  Ph: +91-33-22891126, 22891127, 22891128 e-mail: kolkata@cbspd.com
- **Mumbai:** PWD Shed, Gala No. 25/26, Ramchandra Bhatt Marg, Next to JJ Hospital, Gate No. 2, Opposite Union Bank of India, Noorbaug, Mumbai 400 009, Maharashtra, India
  Ph: +91-22-24902340/41 Fax: +91-22-24902342 e-mail: mumbai@cbspd.com

Representatives

- **Hyderabad** 0-9885175004
- **Jharkhand** 0-9811541605
- **Nagpur** 0-9421945513
- **Patna** 0-9334159340
- **Pune** 0-9623451994
- **Uttarakhand** 0-9716462459

Printed at India Binding House, Noida, UP, India

A tribute to

Gurudev Rabindranath Tagore (1861-1941)
on the occassion of his 151st Birth Anniversary

श्रद्धावान् लभते ज्ञानम्

One who has shraddha acquires knowledge

Preface

We do endeavour. He gives the strength and patience.

The present volumes 4–7 are in continuation of the existing title ***Fundamental Concepts of Inorganic Chemistry.*** The classnotes and valuable suggestions received from the esteemed readers have been shaped in these volumes.

These volumes cover the structure and bonding (VBT, CFT, and MOT), stability and reactivity, spectral and magnetic properties of metal complexes in depth. Kinetics and reaction mechanisms of ligand substitution, electron transfer and photochemical reactions have been included. Magnetochemistry and organometallic chemistry have been covered. Applications of different spectroscopic techniques (Raman, IR, NMR, ESR, Mossbauer, UV-VIS, UV-PES, etc.) have been discussed to widen the utility of the series. In developing the present extension, we have taken all the measures to retain the basic features of the existing title.

In preparing the manuscript, we have freely consulted the books and reviews of the earlier authors and have borrowed their ideas whenever it has been required. These sources are listed and acknowledged at the end of the text. We are grateful and indebted to these authors. In reality, we have picked up flowers from these gardens to prepare the garland to worship the goddess of learning.

We are extremely thankful and grateful to Mr SK Jain, Managing Director, CBS Publishers & Distributors, for his continued support. We are thankful to Mr YN Arjuna, Senior Director, Publishing, Editorial and Publicity, and the DTP staff for taking the trouble in processing the manuscript.

In spite of our best efforts, some mistakes and misconceptions might have crept in for which we beg to be excused. Constructive criticism and suggestions are always welcome to better the presentation.

Asim K. Das

Mahua Das

Contents

Contents of Other Volumes

Volume 6

Volume 7

CHAPTER 5

Reaction Mechanism: Ligand Substitution, Isomerisation and Racemisation Reactions of Metal Complexes

5.1 DIFFERENT TYPES OF REACTIONS OF METAL COMPLEXES

Different types of reactions of metal complexes are as follows:

- ligand substitution reactions
- metal substitution reactions
- isomerisation and recemisation reactions
- redox (*i.e.* electron transfer) reactions
- reactions of coordinated ligands (*i.e.* activation of ligands).

5.2 CLASSIFICATION OF SUBSTITUTION REACTION

These are classified as follows:

- nucleophilic substitution (S_N)
- electrophilic substitution (S_E)

In **nucleophilic substitution reactions, a nucleophile** (*i.e.* Lewis base) donates its electron cloud to a positively charged centre, *i.e.* nucleus (*i.e.* Lewis acid) through the replacement of another nucleophile.

$$M - X + Y \rightarrow M - Y + X \ (S_N)$$

(Charges not shown)

Here both X and Y are the nucleophiles (*i.e.* ligands). Y is the *entering* or *incoming nucleophile* while X is the *leaving group* or *leaving nucleophile.* The metal centre (*i.e.* M) is the Lewis acid. Thus in a nucleophilic substitution reaction, one Lewis base (*i.e.* ligand) displaces another Lewis base from a Lewis acid (*i.e.* metal centre).

Thus the ligand substitution reaction is a case of nucleophilic substitution reaction. In ligand substitution reactions, other ligands (*i.e. nonleaving groups)* are called the **spectator ligands**.

$$L_5M - X + Y \rightarrow L_5M - Y + X$$

Here the ligands denoted by L_5 are called the *spectator ligands that may influence the rate and process of the substitution reaction through the steric and electronic effects.*

In an electrophilic substitution reaction, the metal centres (*i.e.* electron seeking, Lewis acids) are replaced, *i.e.* one metal centre replaces another one. In metal complexes, the *ligand substitution* (*i.e.*

nucleophilic substitution) reactions are abundant while the *metal substitution* (*i.e.* electrophilic substitution) reactions are relatively rare.

5.3 DIFFERENT TYPES OF LIGAND SUBSTITUTION REACTIONS

These are illustrated with some specific examples.

Acid hydrolysis: The leaving group is replaced by H_2O (*i.e.* solvent in aqueous media) in acidic condition.

$$[Co(NH_3)_5X]^{2+} + H_2O \rightarrow [Co(NH_3)_5(OH_2)]^{3+} + X^-$$

This is also called the *aquation* or simply the *dissociation* reaction.

Base hydrolysis: The leaving group is replaced by OH^- group in aqueous media. This is also a case of dissociation reaction.

$$[Co(NH_3)_5X]^{2+} + OH^- \rightarrow [Co(NH_3)_5(OH)]^{2+} + X^-$$

Anation: The coordinated solvent molecule (*i.e.* H_2O in aqueous media) is replaced by an anion (X^-).

$$[Co(NH_3)_5(OH_2)]^{3+} + X^- \rightarrow [Co(NH_3)_5X]^{2+} + H_2O$$

This is also described as the *formation reaction.*

Solvent exchange: This is illustrated below.

$$[M(OH_2)_n] + H_2O^* \rightarrow [M(OH_2)_{n-1}(^*OH_2)] + H_2O, \text{ (in aqueous solvent)}$$

Ligand exchange: This is illustrated below.

$$[PtCl_2(py)_2] + 2^*Cl^- \rightarrow [Pt(^*Cl)_2(py)_2] + 2Cl^-$$

Formation reaction: Replacement of a coordinated solvent molecule by a ligand other that the solvent itself.

$$[M(OH_2)_n] + L \rightarrow [M(OH_2)_{n-1}(L)] + H_2O$$

Solvolysis reaction (*cf.* Hydrolysis reaction): Replacement of a ligand by the solvent itself.

$$M — L + S \text{ (solvent)} \rightarrow M — S + L$$

5.4 THERMODYNAMIC AND KINETIC STABILITY: LABILITY AND INERTNESS (*cf.* Sec. 4.1)

Thermodynamic stability of a complex is determined by β_n.

$$M + nL \rightleftharpoons ML_n, \quad \beta_n = \frac{[ML_n]}{[M][L]^n}$$

The higher value of β_n indicates its higher thermodynamic stability. Thus it gives the measure of the extent to which the equilibrium can proceed but it cannot say anything regarding the speed with which the equilibrium is attained. Thus the thermodynamic and kinetic stability are not necessarily correlated. A thermodynamically stable complex may react slowly or fast depending on the condition. This can be illustrated by considering the different cyanido complexes which are extremely stable (*i.e.* very high value of β_n) but they behave quite differently in terms of the rate of exchange of the radiocarbon labelled cyanide.

$$[Hg(CN)_4]^{2-} + 4\,^{14}CN^- \rightleftharpoons [Hg(^{14}CN)_4]^{2-} + 4CN^-, \textbf{ very fast}$$

$$[Ni(CN)_4]^{2-} + 4\,^{14}CN^- \rightleftharpoons [Ni(^{14}CN)_4]^{2-} + 4CN^-, t_{1/2} \approx 30\,s, \textbf{ very fast}$$

$$[Pt(CN)_4]^{2-} + 4\,^{14}CN^- \rightleftharpoons [Pt(^{14}CN)_4]^{2-} + 4CN^-, t_{1/2} \approx 1\text{ min}$$

$$[Mn(CN)_6]^{3-} + 6\,^{14}CN^- \rightleftharpoons [Mn(^{14}CN)_6]^{3-} + 6CN^-, t_{1/2} \approx 1\text{ h}$$

$$[Cr(CN)_6]^{3-} + 6\,^{14}CN^- \rightleftharpoons [Cr(^{14}CN)_6]^{3-} + 6CN^-,\ t_{1/2} \approx 24\text{ d},$$

$$[Fe(CN)_6]^{3-/4-} + 6\,^{14}CN^- \rightleftharpoons [Fe(^{14}CN)_6]^{3-/4-} + 6CN^-,\ \textbf{(very slow)}$$

All the above complexes are thermodynamically very stable (*cf.* $[Ni(CN)_4]^{2-}$, log $\beta_4 \approx 29$; $[Pt(CN)_4]^{2-}$, log $\beta_4 \approx 40$; $[Hg(CN)_4]^{2-}$, log $\beta_4 \approx 42$; $[Fe(CN)_6]^{3-}$, log $\beta_6 \approx 44$; $[Fe(CN)_6]^{4-}$, log $\beta_6 \approx 37$; etc.) but their kinetic stabilities differ widely. *The thermodynamically stable complexes like* $[Ni(CN)_4]^{2-}$, $[Hg(CN)_4]^{2-}$, etc. *react fast in the ligand exchange reactions but* $[Cr(CN)_6]^{3-}$ *which is also thermodynamically stable reacts slowly* and $[Fe(CN)_6]^{4-}$ also reacts very slowly.

- For some **square planar Pt(II) – complexes**, the lability and thermodynamic stability follow the same sequence, *i.e.* **the most stable complex is the most labile one.**

Complex:	$[PtCl_4]^{2-}$	$[PtBr_4]^{2-}$	$[PtI_4]^{2-}$	$[Pt(CN)_4]^{2-}$
~ logβ_4:	17	20	30	40
~ $t_{1/2}$ (min):	850	6	4	1

Increasing lability (*cf.* Secs. 5.30.5,6) **and stability** ⟶

- **Dissociation of $[Co(NH_3)_6]^{3+}$** in aqueous acidic media is thermodynamically highly favoured but the process is kinetically highly disfavoured.

$$[Co(NH_3)_6]^{3+} + 6H_3O^+ \rightleftharpoons [Co(OH_2)_6]^{3+} + 6NH_4^+,\ \log K_{eq} \approx 25.0$$

The tremendous thermodynamic driving force for the forward reaction arises from the protonation of the basic NH_3 ligands, but the complex remains unchanged in a fairly strong acid medium for weeks at room temperature. In fact, the salt $[Co(NH_3)_6]Cl_3$ can be crystallised from a hot aqueous solution of HCl without any noticeable decomposition. Thus, *stability of the complex arises not from the thermodynamic stability but from the kinetic stability.*

- **$[M(en)_3]^{3+}$ and $[M(bigH)_3]^{3+}$** (M = Cr, Co): The biguanide complex is more stable than the ethylenediamine complex but with respect to the acid catalysed dissociation reaction, $[M(bigH)_3]^{3+}$ reacts much faster than the $[M(en)_3]^{3+}$ complex.

$$[M(en)_3]^{3+} \xrightarrow{H_3O^+} [M(OH_2)_6]^{3+} + 3enH^+$$

$$[M(bigH)_3]^{3+} \xrightarrow{H_3O^+} [M(OH_2)_6]^{3+} + 3bigH_2^+$$

In fact, $[Co(bigH)_3]^{3+}$ undergoes dissociation in acid medium at a measureable rate at 30°C (*cf.* D.Banerjea et al, *J. Inorg. Nucl. Chem.*, **26,** 1233, 1964) and $[Co(en)_3]^{3+}$ remains completely unchanged in 1 M $HClO_4$ at room temperature even for months.

Note: The kinetic favour for dechelation of the biguanide ring arises from the *its possibility of protonation before ring opening* but such a protonation cannot occur prior to opening of the chelate ring for the ethylenediamine complex.

- To differentiate between the thermodynamic stability and kinetic stability, H. Taube has coined the kinetic terms: **lability** and **inertness.** The kinetically stable complexes are called the **inert complexes** while the kinetically unstable complexes are called the **labile complexes.** The simple terms — stable and unstable refer to the thermodynamic parameters.
- In ligand substitution reactions, the rate can span a very wide range of time scale — **nanosecond (ns) to years**.

$$[Cu(aq)]^{2+} + 4NH_3 \rightarrow [Cu(NH_3)_4]^{2+}$$, **(instantaneous)**
(blue purple)

$$[Co(NH_3)_6]^{3+} \text{ or } [Co(en)_3]^{3+} \rightarrow \text{Hydrolysis}$$, **(years)**

- *There is no sharp border line to distinct between the labile and inert complexes.* However, H. Taube has described the complexes as the **labile complexes** having $t_{1/2}$ (substitution half life) ≤ 30 s in a particular reaction while the complexes with $t_{1/2} > 30$ s as the **inert complexes.** Rate process for the inert complexes can be followed by the conventional techniques while for the labile complexes, it requires some special techniques (*e.g. stopped flow, P–jump, T–jump,* etc.). This is why, more information is available for the inert complexes.
- The lability or inertness depends on the **activation energy,** *i.e.* high activation energy imparts the inertness while low activation energy imparts the lability (*cf.* Fig. 4.1.1). The inertness or lability is determined by $\Delta G^{\neq}$ (**free energy of activation**).

$$\boldsymbol{\Delta G^{\neq} = \Delta H^{\neq} - T\Delta S^{\neq}}, \ \Delta H^{\neq} = \textbf{enthalpy of activation}, \ \Delta S^{\neq} = \textbf{entropy of activation.}$$

The stability of a complex is determined by the free energy change (ΔG^0) (*cf.* Fig 4.1.1) in a reaction.

$$\Delta G^0 = \Delta H^0 - T\Delta S^0$$

$\Delta G^{\neq}$ depends on the reaction pathway (*i.e.* reaction mechanism) while ΔG^0 depends on the difference in free energy between the reactant and product.

Labile and inert complexes of 3d–series

It depends on d^n–configuration.
Labile centres: d^0, d^1, d^2, d^4 (h. s.), d^5 (h. s.), d^6 (h.s.), d^7, d^8, d^9
Inert centres: d^3, d^4 (l. s.), d^5 (l. s.), d^6 (l. s.)
These can be explained by CFT. These aspects will be discussed later.

5.5 NUCLEOPHILICITY vs. BASICITY; ELECTROPHILICITY

- It has been already mentioned that the nucleophiles are basically the Lewis bases but the term **basicity** refers to the thermodynamic concept (*i.e.* equilibrium concept). The analogous kinetic term is called **nucleophilicity** (nucleophile = nucleus loving). In an associative pathway, the entering nucleophile (*i.e.* entering ligand) makes a bond with the metal centre in producing the **activated complex or transition state**. The *kinetic ease* with which a nucleophile can attack the metal centre to produce the activated complex gives the measure of its nucleophilicity. Relative nucleophilicity of a particular ligand L is measured with reference to that of a standard ligand L^0 by comparing their rates of substitution reaction (occurring through the **associative process**) at a chosen metal centre, *i.e.*

$$\text{Nucleophilicity of } L = \frac{\text{rate of substitution by } L}{\text{rate of substitution by } L^0}$$

The details of nucleophilicity scale will be discussed later (*cf.* Sec. 5.30.5)
The Lewis acids act as *electrophiles.*

- **Acidity** is a thermodynamic term while the analogous kinetic term is **electrophilicity** (electrophile = electron loving). The rate of reaction of a Lewis acid (*i.e.* metal centre in the complexes) with the Lewis bases (*i.e.* entering ligands or entering nucleophiles) gives the measure of the electrophilicity of the Lewis acid. Just like the relative nucleophilicity scale, a relative electrophilicity scale can be generated.

5.6 METHODS OF FOLLOWING KINETICS (*i.e.* RATE MEASUREMENT)

(1) **For the inert systems**, conventional techniques like: UV–Visible spectroscopic method, potentiometry, conductometry, polarography, polarimetry, titrimetry, etc. can be used to follow the rate process. In fact, reactions of the complexes of Cr(III), Co(III), Rh(III), Ir(III), Pt(II) and some complexes of Ni(II) can be studied by the conventional techniques. Some examples are given below.

● **Direct chemical analysis:** Direct chemical estimation of either the reactants or products at regular intervals during the progress of reaction is possible in many cases.

(a) $[PtCl(NH_3)_3]^+ + H_2O \rightarrow [Pt(NH_3)_3(OH_2)]^{2+} + Cl^-$
(b) $[CoCl_2(en)_2]^+ + H_2O \rightarrow [CoCl(en)_2(OH_2)]^{2+} + Cl^-$
(c) $[Co(NCS)(NH_3)_5]^{2+} + OH^- \rightarrow [Co(NH)_5(OH)]^{2+} + SCN^-$
(d) $[Co(NH_3)_5(S_2O_3)]^+ + L \rightarrow [Co(L)(NH_3)_5]^{n+} + S_2O_3^{2-}$
(L = different nucleophiles)

In the reactions (a) and (b), the released Cl^- can be estimated by *argentometry* (*i.e.* titration by $AgNO_3$) or by *potentiometry* (using an electrod reversible with respect to Cl^-, *e.g.* Ag/AgCl electrode). However, there is a possibility by the *direct attack* of Ag^+ *on the bound* Cl^- (*i.e.* overestimation of Cl^-). This can be minimised, if it is carried out at low temperature and in a mixed solvent like acetone-water mixture. In the reaction (c), released SCN^- can be also estimated similarly by titration with $AgNO_3$. In the reaction (d), the released $S_2O_3^{2-}$ can be estimated iodimetrically as usual.

● **Spectrophotometric method:** Generally, the coordination compounds are coloured and concentration of a coloured coordination compound can be determined from the optical density measurement at a suitable wavelength (A = optical density = εCl, ε = molar extinction coefficient, C = molar concentration, l = optical path length). Generally, in a particular reaction, the absorption spectra of the product and reactant are different and the most suitable wavelength for the optical density measurement is the wavelength at which absorption of the reactant and product differs most.

If the measured optical density (OD) is due to a particular compound whose formation or decay experiences a first order kinetics then the plot of $\log\dfrac{A_0 - A_\infty}{A_t - A_\infty}$ *vs.* t (time) will give the first order rate constant.

$A_t = OD$ at time t; $A_0 = OD$ at zero time; $A_\infty = OD$ at infinity (*i.e.* at the end of reaction).

Note: For a first-order reaction, we can write:

$$C_t = C_0 \exp(-kt),\ i.e.\ k = \frac{2.303}{t}\log\frac{a}{a-x}$$

$C_0 = a$ = initial concentration (*i.e.* $t = 0$), $C_t = (a - x)$ = concentration at t.

If a physical property P (like OD), directly proportional to its concentration is available, then we can write:

$$k = \frac{2.303}{t}\log\frac{P_0 - P_\infty}{P_t - P_\infty};\ \text{slope of the plot, } \log\frac{P_0 - P_\infty}{P_t - P_\infty}\ vs.\ t \text{ is } \frac{k}{2.303}$$

● **Electrometric method** (*i.e.* measurement of conductance, emf, pH, etc.): In the following aquation reactions, conductance increases remarkably during the progress of reaction.

$[CoCl_2(en)_2]^+ + H_2O \rightarrow [CoCl(en)_2(OH_2)]^{2+} + Cl^-$
$[PtCl(NH_3)_3]^+ + H_2O \rightarrow [Pt(NH_3)_3(OH_2)]^{2+} + Cl^-$

Thus conductance measurement with time can be practised to study the above reactions. The rate of Cl^- release can also be measured potentiometrically by using an electrode sensitive to Cl^-.

In the following reaction, during the progress of reaction, pH decreases due to the release of H^+ ion.

$$M^{2+} + LH \rightarrow ML^+ + H^+$$

Thus the pH measurement with time can evaluate the rate constant of the process.

● **Polarimetric method:** The measurement of optical rotation with time can be used to determine the rate constant of a racemisation processes.

$$\Delta-\left[Co(bigH)_3\right]^{3+} \rightleftharpoons \Lambda-\left[Co(bigH)_3\right]^{3+}$$

It should be taken into consideration that the light used in the optical rotation measurements may accelerate the isomerisation process in some cases. In such cases, the method becomes erratic.

● **Isotope tracer technique:** The electron exchange reactions in the couples like $[Ru(NH_3)_6]^{3+}$/$[Ru(NH_3)_6]^{2+}$, $[Fe(CN)_6]^{3-}$/$[Fe(CN)_6]^{4-}$, $[Fe(OH_2)_6]^{3+}$/$[Fe(OH_2)_6]^{2+}$, etc. may be followed by using the labelled isotope. Isotopic tracer techniques may be used to follow the metal exchange and ligand exchange reactions.

(2) For following the fast reactions of labile centres, it needs some special techniques like stopped-flow, spectrophotometry, T–jump, P–jump, NMR, etc. depending on the $t_{1/2}$ of the reactions.

● **Stopped-flow spectrophotometry:** In the stopped-flow technique, the mixing time between the reactants is about 1 ms (*i.e.* 10^{-3} s) and the reactions faster than the mixing time cannot be studied by the stopped-flow spectrophotometry. In the stopped-flow spectrophotometry, after mixing of the reactants within the reaction cell, progress of the reaction can be followed spectrophotometrically with the help of a fast recorder device.

● **Perturbation technique (*i.e.* T–jump, P–jump):** This technique is applicable for the reactions of $t_{1/2}$ in the time scale of μs (= 10^{-6} s). In this technique, the equilibrium is perturbed suddenly (fraction of μs or less) by a temperature change, *ca.* 5 – 10°C (in T–jump) or a pressure change, *ca.* a few hundred atmosphere (in P–jump). This change of temperature or pressure directs the equilibrium in a direction as demanded by *Le Chatelier's principle.* This relaxation of the system towards the new equilibrium position can be followed spectrophotometrically by using a fast recorder device. This is why, this perturbation technique is referred to as the **relaxation technique.** It may be mentioned that

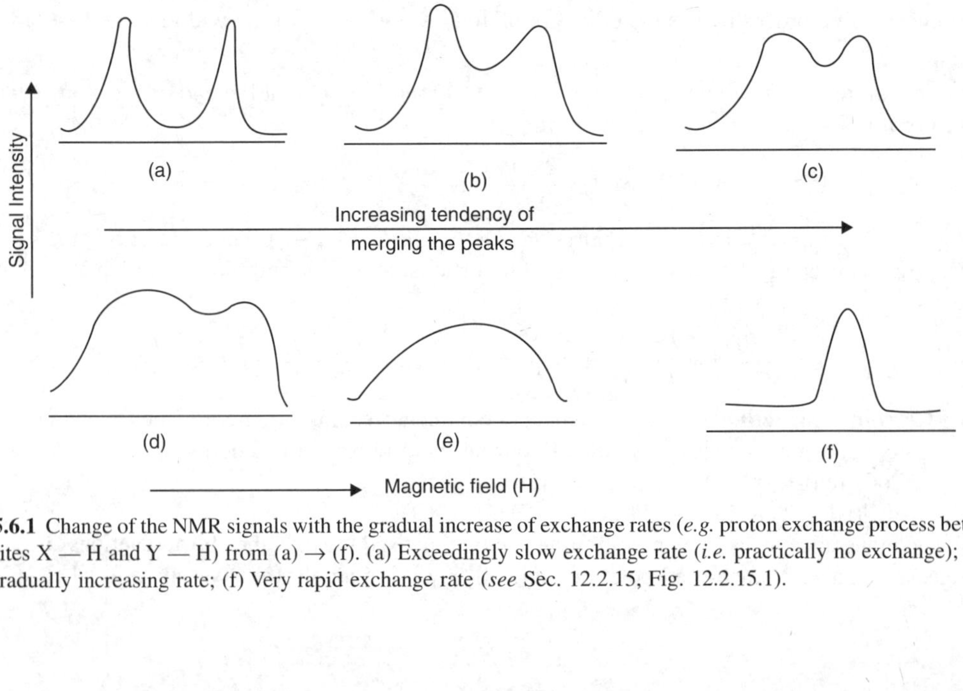

Fig. 5.6.1 Change of the NMR signals with the gradual increase of exchange rates (*e.g.* proton exchange process between two sites X — H and Y — H) from (a) → (f). (a) Exceedingly slow exchange rate (*i.e.* practically no exchange); (b) to (e) Gradually increasing rate; (f) Very rapid exchange rate (*see* Sec. 12.2.15, Fig. 12.2.15.1).

the T–jump technique can be applied to all systems but the P–jump technique can only be applied to the systems where there is a change in the number of species in the reaction (more correctly, where there is a molar volume change).

● **NMR technique** (*cf.* Sec. 12.2.15): For the studies of *fast exchange reactions* (both ligand exchange and metal exchange), the NMR method is very much important. For the studies of water exchange reactions, $^{17}OH_2$ is used. The NMR spectra of a particular NMR-active centre depends on its chemical environment. Thus the NMR spectra of ^{17}O are different for metal bound $^{17}OH_2$ and $^{17}OH_2$ present in bulk solvent. If the water exchange rate is slow, then the NMR signals of ^{17}O for two different chemical environments are quite distinct. *But with the increase of exchange rate, the signals will move to merge or overlap* (*i.e.* the peak will be broadening). If the exchange rate is extremely fast, then a single peak is noticed. Thus from the nature of merging the peaks, it is possible to determine the exchange rate constant. Merging of peaks due to an exchange reaction is qualitatively shown in Fig. 5.6.1.

5.7 MECHANISM OF LIGAND SUBSTITUTION REACTION: INTIMATE AND STOICHIOMETRIC MECHANISM

● **Stoichiometric mechanism:** It deals with the *sequence of elementary steps* leading to a chemical reaction. It looks at the *reactants, products* and *intermediates* but not at the transition states. Thus the stoichiometric mechanism looks at the species residing at the **potential minima** along the reaction coordinates. Consideration of the involved elementary steps can lead to the rate law. There are three types of stoichiometric mechanism of ligand substitution reaction. These are:

Dissociative (D), Interchange (I) and Associative (A)

Dissociative and associative reactions are the **two-step reactions** passing through an **intermediate,** while interchange reaction is a **one-step reaction** without the formation of a **true intermediate.**

● **Intimate mechanism:** It deals with the *activation process* leading to the formation of an activated complex at the rate determining step. It looks at the species residing at **the highest point** (*i.e.* activated complex or transition state) on the reaction coordinate. Thus it considers the *energetics of the formation of the activated complex* and *consequently the rate.*

If the rate of formation of the activated complex at the rate determining step (*i.e.* rate of the overall reaction) depends strongly on the nature of the entering ligand (say *L*) then it indicates that the enering ligand makes a new bond to a significant extent to generate the activated complex. Thus the *activation process* is **associative (*a*)**. On the other hand, if the reaction rate is strongly dependent on the nature of the leaving group (say *X*) and almost independent of the nature of enterging group (*L*), then it indicates that bond breaking by the leaving group is important to generate the activated complex. Thus the *activation process* is ***dissociative* (*d*)**.

Here it is worth mentioning that the **associative activation** and **dissociative activation** are denoted by **a** and **d** respectively (not by **A** and **D** which denote the stoichiometric associative and dissociative reaction mechanisms respectively).

● **Relationship between the stoichiometric and intimate mechanism: D mechanism** must involve the **dissociative (d) activation** while **A mechanism** must involve the **associative (a) activation. I mechanism** involves **both the dissociative and associative activation,** *i.e.* both bond breaking by the leaving group and bond formation by the entering group contribute at the rate determining step. If in this process of activation, bond breaking (*i.e.* dissociative) is the predominant factor then it is referred to as I_d. On the other hand, if in the interchange process, bond formation (*i.e.* associative) by the entering group is the predominant factor then it is referred to as I_a.

Thus the **combined notations** describing both the stoichiometric and intimate mechanisms are:

$$\textbf{A, I}_\textbf{a}\textbf{, I}_\textbf{d}\textbf{, D (Langford-Gray notation)}$$

● **Relationship between Langford-Gray notation and Hughes-Ingold notation used in organic chemistry:** According to Basolo and Pearson, *A mechanism* is equivalent to $\textbf{S}_\textbf{N}\textbf{2 lim}$. (*i.e.* limiting situation of S_N2) while *D mechanism* is equivalent to $\textbf{S}_\textbf{N}\textbf{1 lim}$. I_a and I_d correspond to S_N2 (not limiting) and S_N1 (not limiting) respectively.

Note: $S_N1 \Rightarrow$ substitution nucleophilic unimolecular; $S_N2 \Rightarrow$ substitution nucleophilic bimolecular.

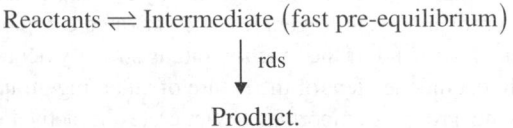

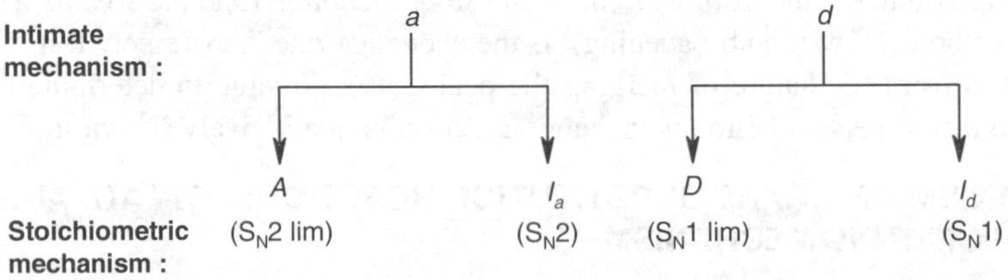

Intimate mechanism :	*a*		*d*	
	↓	↓	↓	↓
	A	I_a	*D*	I_d
Stoichiometric mechanism :	(S_N2 lim)	(S_N2)	(S_N1 lim)	(S_N1)

● For both the **D** and $\textbf{I}_\textbf{d}$ processes, the intimate mechanism is **d** (*i.e.* **d**–activation) but for the D–process, the intermediate (*i.e.* **long life time**) of lower coordination number is detectable while the intermediate is not detectable (*i.e.* **very small life time**) for the I_d process. Thus their stoichiometric mechanisms (*i.e.* sequence of elementary steps) differ.

● Similarly for both the **A** and $\textbf{I}_\textbf{a}$ processes, the intimate mechanism is **a** (*i.e.* **a**–activation). The intermediate of higher coordination number is detectable for the **A**–pathway but not for the I_a–pathway. Thus their stoichiometric mechanisms (*i.e.* sequence of elementary steps) differ.

● **Cases involving a fast pre-equilibrium step followed by the rate determing step:** The fast preequilibrium step may lead to the formation of reactive intermediates like **ion pair (IP), conjugate base (CB), conjugate acid (CA),** etc. and then these reactive intermediates participate at the rate determining steps. In such cases, notation of the reaction mechanism depends on both the nature of the reactive intermediate (formed in a rapid preequilibrium step) and the nature of rate determining step involving the reactive intermediate. This aspect is illustrated below.

$$\text{Reactants} \rightleftharpoons \text{Intermediate}\left(\text{fast pre-equilibrium}\right)$$
$$\downarrow \text{rds}$$
$$\text{Product.}$$

Notation of the mechanism: Nature of the rds (*i.e.* A or D or I) – nature of the intermediate (*i.e.* IP or CA or CB).

If the intermediate is *IP*, and the rds is a *A*–pathway, then the mechanism is designated by ***A–IP*** (*i.e.* S_N2–*IP* in old nomenclalure). Similarly we can have:

D–IP (*i.e.* S_N1 lim–IP, **in short** S_N1–*IP*) meaning *IP* as the intermediate, and rds as the *D*-pathway;

D–CA (*i.e.* S_N1 **lim–CA, in short** S_N1–*CA*) meaning *CA* as the intermediate and rds as the *D*-pathway;

D–CB (*i.e.* S_N1 **lim–CB, in short** S_N1–*CB*) meaning *CB* as the intermediate, and rds as the *D*-pathway;

Similarly, we can write: $\textbf{\textit{I}}_\textbf{\textit{d}}\textbf{–\textit{CB}, \textit{I}}_\textbf{\textit{d}}\textbf{–\textit{IP}, \textit{I}}_\textbf{\textit{a}}\textbf{–\textit{IP}}$, etc.

5.8 REACTION PROFILE FOR DIFFERENT STOICHIOMETRIC MECHANISMS

A. Associative (A) process: Two step reaction with two transition states and one intermediate

$$L_5M - X \xrightarrow[(rds)]{+Y} L_5M\begin{matrix} X \\ Y \end{matrix} \xrightarrow[\text{(fast)}]{-X} L_5M - Y$$

The entering ligand (Y) makes a new bond with the metal centre at the rate determining step to **increase the coordination number by unity.** In attaining this intermediate of higher coordination number, there is no bond breaking by the leaving group. After the rate determining step (rds), the leaving group (X) is lost from the intermediate at a faster step.

The reaction pathway involves the formation of a **single intermediate** ($Y - ML_5 - X$) and **two transition states (T.S.)** — one for the formation of the intermediate and another for the decomposition of the intermediate to the product. The energy barrier for the decomposition of the intermediate is relatively lower.

B. Dissociative (*D*) process: Two step reaction with two transition states and one intermediate

$$L_5M - X \xrightarrow[(rds)]{-X} L_5M \xrightarrow{+Y} L_5M - Y$$

The leaving group (X) is lost at the rate determining step through the rupture of the $M - X$ bond to generate an intermediate (L_5M) with **a lower coordination number** (*i.e.* coordination number decreases by unity for the unidentate leaving group). At this stage of generation of the intermediate, there is no interaction with the entering ligand. After the rate determining step, the entering group enters into the coordination sphere at a faster step to give the final product.

As in the *A* process, in the *D*–process also, there is **a single intermediate** but of lower coordination number, and **two transition states** – one leading to the formation of the intermediate and the other leading to the product from the intermediate. The energy barrier leading to the formation of the intermediate is higher.

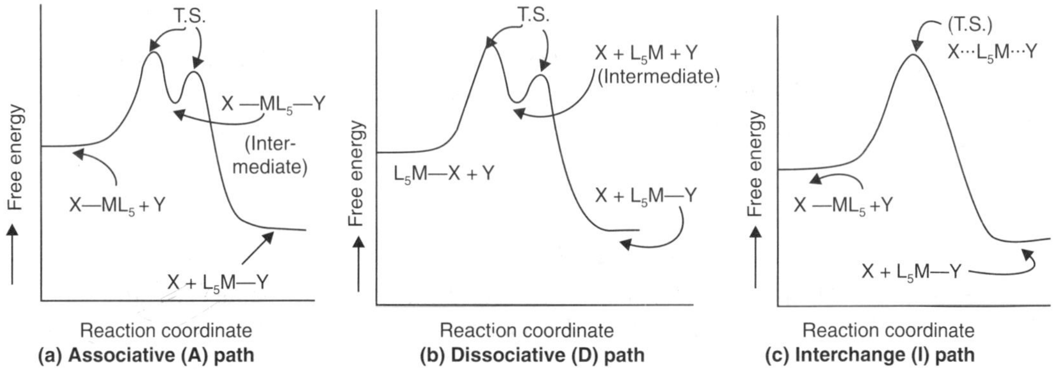

Fig. 5.8.1 Energy-profile diagram for the different mechanisms of ligand substitution reactions.

C. Interchange (I) Process: Single step reaction with a single transition state

$$L_5M - X \xrightarrow[(rds)]{+Y} L_5M \begin{matrix} \cdot\cdot X \\ \cdot\cdot Y \end{matrix} \xrightarrow[\text{(fast)}]{-X} L_5M - Y \quad (\cdots \text{ denotes breaking or forming bond})$$

Both the leaving and entering group are involved in **a single step** to generate the **activated complex (not the true intermediate because it is too short-lived to be detected).** In attaining this activated complex, both the bond breaking by the leaving group (*i.e.* **slight rate dependance on the nature of the leaving group**) and bond formation by the entering group (*i.e.* **slight rate dependance on the nature of the entering group**) are important. In other words, bond formation by the entering group and bond rupture by the leaving group go on simultaneously in a *concerted manner* and ultimately the incoming group gets completely bound with the metal centre and the leaving group is completely dislodged. The reaction profile is characterised by a **single transition state** (T.S.).

Depending on the relative importance of the bond formation by the entering group and bond dissociation by the leaving group in the interchange process, two different situations may arise giving rise to two different types of transition states.

$$L_5M - X \xrightarrow{+Y} L_5M \begin{matrix} \cdot\cdot X \\ \diagdown Y \end{matrix} \quad (I_a)$$

$$L_5M - X \xrightarrow{+Y} L_5M \begin{matrix} \diagup X \\ \cdot\cdot Y \end{matrix} \quad (I_d)$$

Bond denoted by .... makes a less contribution in the activation process.

Thus in $\mathbf{I_a}$, the new bond formation by the entering group is more important (*i.e.* rate is relatively more sensitive on the nature of the entering group) while in $\mathbf{I_d}$, bond rupture by the leaving group is more important in the activation process.

5.9 RATE LAWS FOR DIFFERENT REACTION PATHWAYS (*i.e.* STOICHIOMETRIC MECHANISMS)

5.9.1 Dissociative Mechanism (*cf.* Sec. 5.16)

In a pure **D** process (*i.e.* limiting S_N1), an intermediate of lower coordination number is produced. *Life time of the intermediate is sufficiently high* so that it can be consumed in a way that is independent of its mode of formation. The elementary steps for a *D*-process (stoichiometric mechanism) are outlined below.

$$L_5M - X \underset{k_{-1}}{\overset{k_1}{\rightleftharpoons}} L_5M + X \qquad \ldots(5.9.1.1)$$

$$L_5M + Y \xrightarrow{k_2} L_5M - Y \qquad \ldots(5.9.1.2)$$

The rate law for this mechanism is obtained by considering the **steady-state concentration** of the intermediate L_5M (*i.e. its concentration is small at any moment and it remains virtually constant during the reaction*), *i.e.*

$$\frac{d}{dt}\left[L_5M\right] = 0 = k_1\left[L_5MX\right] - k_{-1}\left[L_5M\right][X] - k_2\left[L_5M\right][Y] \quad \text{or,} \quad \left[L_5M\right] = \frac{k_1\left[L_5MX\right]}{k_{-1}[X] + k_2[Y]}$$

The rate of the overall reaction is given by:

$$-\frac{d}{dt}[L_5MX]=\frac{d}{dt}[L_5MY]=k_2[L_5M][Y]$$

$$=\frac{k_1k_2[L_5MX][Y]}{k_{-1}[X]+k_2[Y]} \qquad \ldots(5.9.1.3)$$

$$=\frac{k_1k_2[L_5MX][Y]/(k_{-1}[X])}{1+k_2[Y]/(k_{-1}[X])}$$

$$=k_{obs}[L_5MX],\ (\text{when } [Y]>>[L_5MX])$$

$$k_{obs}=\frac{k_1k_2[Y]/(k_{-1}[X])}{1+k_2[Y]/(k_{-1}[X])} \qquad \ldots(5.9.1.4)$$

- The rate equation indicates that k_{obs} (observed rate constant) shows a **complicated dependence pattern** on both **[*X*] and [*Y*]**. If *X* = solvent (*i.e.* **formation reaction**) then [*X*] = constant and the situation becomes simplified for analysis. Then we can write:

$$k_{obs}=\frac{k_1k_2[Y]/(k_{-1}[H_2O])}{1+k_2[Y]/(k_{-1}[H_2O])};\ i.e.\ k_{obs}=\frac{m[Y]}{1+n[Y]} \qquad \ldots(5.9.1.5)$$

$$m=\frac{k_1k_2}{k_{-1}[\text{solvent}]}=\frac{k_1k_2}{k'_{-1}}=k=\text{constant}$$

$$n=\frac{k_2}{k_{-1}[\text{solvent}]}=\frac{k_2}{k'_{-1}}=\text{constant};$$ (*see* Sec. 5.16 to understand the **significance of this constant**)

i.e.
$$\frac{1}{k_{obs}}=\frac{1}{m[Y]}+\frac{n}{m} \qquad \ldots(5.9.1.6)$$

The ***Burk-Lineweaver type*** **double reciprocal plot,** $\left(i.e., \frac{1}{k_{obs}}\ vs.\ \frac{1}{[Y]}\right)$ gives a straight line and from the slope and intercept of the plot, the constants *m* and *n* may be evaluated.

Thus for the **D-process of formation reaction** (*i.e.* *X* = solvent), in general we can write:

$$k_{obs}=\frac{m[Y]}{1+n[Y]} \qquad (cf.\ \text{Eqn. } 5.9.1.5)$$

or,
$$\frac{1}{k_{obs}}=\frac{1}{m[Y]}+\frac{n}{m} \qquad (cf.\ \text{Eqn. } 5.9.1.6)$$

The plots are shown in Fig. 5.9.1.1.

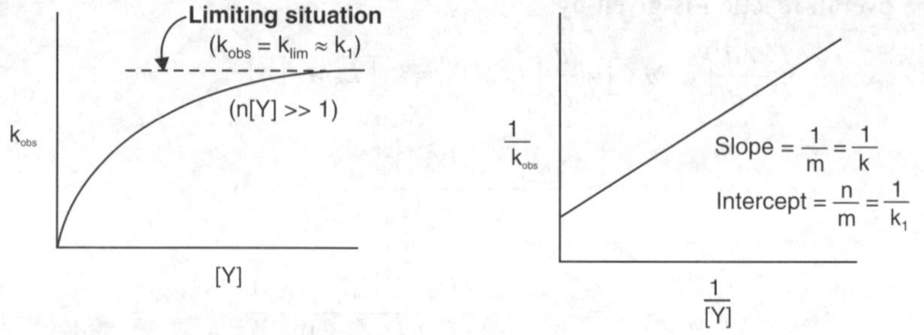

Fig. 5.9.1.1 Rate dependence on the concentration of the incoming nucleophile (*Y*) in a formation reaction occurring through the dissociative process under the condition, $[ML_5X] << [Y]$.

- For the **formation reaction,** *i.e.* X = solvent,

$$\text{rate} = \frac{k_1 k_2 [ML_5 - OH_2][Y]}{k_{-1}[H_2O] + k_2[Y]}, \text{(say, solvent = water)}$$

At **high concentration of Y**, *i.e.* under the condition, $k_{-1}[H_2O] << k_2[Y]$, we have:
rate = $k_1[ML_5 - OH_2] = k_{obs}[ML_5 - OH_2]$, (*i.e.* $k_{obs} = k_{lim} \approx k_1$ and the rate is independent of $[Y]$)
But at **low concentration of [Y],** *i.e.* under the condition, $k_{-1}[H_2O] >> k_2[Y]$, we have:

$$\text{rate} = \frac{k_1 k_2 [ML_5 - OH_2][Y]}{k_{-1}[H_2O]} = k_{obs}[ML_5 - OH_2] \text{ where } k_{obs} = k[Y], k = \frac{k_1 k_2}{k_{-1}[H_2O]} = \text{constant}$$

i.e. **first-order dependence on [*Y*]** (*cf.* Sec. 5.16).

Under the **limiting condition,** $k_{-1}[H_2O] \ll k_2[Y]$, *i.e.* at high concentration of *Y*,

rate = $k_1[ML_5 - OH_2]$, *i.e.* no dependence on $[Y]$.

5.9.2 Associative Mechanism

In the pure **A**-process (*i.e.* limiting S_N2), an intermediate of higher coordination number is produced. *The life time of the intermediate is sufficiently high and it can be detected.* The elementary steps of this stoichiometric mechanism are outlined below.

$$L_5M - X + Y \underset{k_{-a}}{\overset{k_a}{\rightleftharpoons}} L_5M\begin{matrix} X \\ Y \end{matrix} \quad (\text{i.e. } L_5MXY) \quad \ldots(5.9.2.1)$$

$$L_5M\begin{matrix} X \\ Y \end{matrix} \xrightarrow{k_b} L_5M - Y + X \quad \ldots(5.9.2.2)$$

Applying the **steady state condition** to the intermediate, L_5MXY, we can write:

$$\frac{d[L_5MXY]}{dt} = 0 = k_a[L_5MX][Y] - k_{-a}[L_5MXY] - k_b[L_5MXY]$$

i.e.

$$[L_5MXY] = \frac{k_a[L_5MX][Y]}{k_{-a} + k_b}$$

Rate of the reaction is given by:

$$-\frac{d\left[ML_5X\right]}{dt}=\frac{d\left[ML_5Y\right]}{dt}=k_b\left[L_5MXY\right]$$

$$\boxed{k_{obs}=\frac{k_a k_b[Y]}{k_{-a}+k_b}}$$

$$=\frac{k_a k_b\left[ML_5X\right][Y]}{k_{-a}+k_b}, \qquad \ldots(5.9.2.3)$$

$$=k_{obs}\left[ML_5X\right], \left(\text{when } [Y]\gg\left[ML_5X\right]\right)$$

Note: This second order kinetics of A–reaction is observed provided the second step is irreversible, *i.e.* $k_{-b}=0$. If this step is reversible, *i.e.* $k_b \neq 0$, then there will be a deviation from the second order kinetics.

5.9.3 Interchange Mechanism (Consistent with the Eigen-Wilkins Mechanism, Sec. 5.14)

In this pathway, *the intermediate is too short-lived to be detected.* Depending on the conditions, it may be I_a or I_d. This interchange mechanism operates within a *pre-formed outer-sphere (O.S.) complex.* The **outer-sphere complex** or **encounter complex** is produced in a rapid pre-equilibrium step (*cf.* **Eigen-Wilkins mechanism** *cf.* Sec. 5.14). The process can be outlined as follows:

$$L_5M—X+Y \underset{}{\overset{K_{OS}}{\rightleftharpoons}} \underset{\text{(outer-sphere complex)}}{\left(L_5MX, Y\right)} \qquad \ldots(5.9.3.1)$$

$$\underset{\text{O.S. Complex}}{\left(L_5MX, Y\right)} \xrightarrow{k_i} L_5M—Y+X \qquad \ldots(5.9.3.2)$$

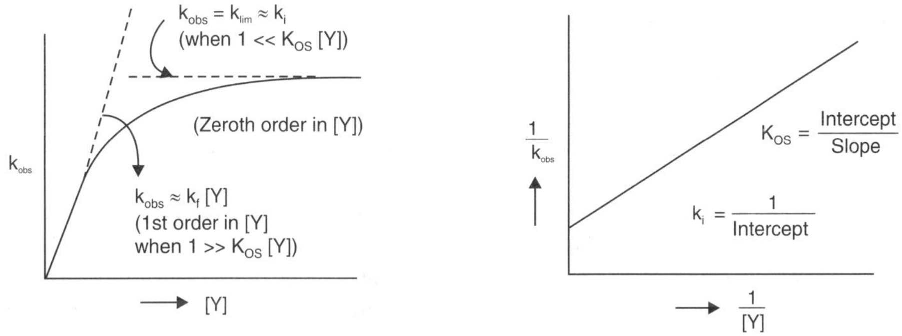

Fig. 5.9.3.1 Rate dependence on [Y] in an interchange (I) process and evaluation of k_i and K_{OS} (**double reciprocal plot,** general procedure).

The rate law can be expressed in terms of the initial concentration of the starting complex, *i.e.* $[L_5MX]_0$ and the initial concentration of *Y*, *i.e.* $[Y]_0$. In presence of *Y*, the reacting complex is also partly distributed in the outer sphere complex. Thus we can write:

$$[L_5MX]_0 = [L_5MX] + [L_5MX, Y]$$
$$= [L_5MX] + [O.S.]$$

and $$[Y]_0 = [Y] + [O.S.] \approx [Y], \text{ (when } [Y]_0 >> [L_5MX]_0)$$

$$K_{OS}=\frac{[O.S.]}{[L_5MX][Y]}$$

$$=\frac{[O.S.]}{\{[L_5MX]_0-[O.S.]\}[Y]}$$

or, $$K_{OS}[L_5MX]_0[Y]-K_{OS}[O.S.][Y]=[O.S.]$$

or, $$[O.S.]=\frac{K_{OS}[L_5MX]_0[Y]}{1+K_{OS}[Y]} \quad \ldots(5.9.3.3)$$

The rate of the reaction is given by:

$$\frac{d[L_5MY]}{dt}=k_i[O.S.]=\frac{k_iK_{OS}[L_5MX]_0[Y]}{1+K_{OS}[Y]} \quad \ldots(5.9.3.4)$$

$$=k_{obs}[L_5MX]_0,\ (\text{when } [Y]_0>>[L_5MX]_0)$$

$$k_{obs}=\frac{k_iK_{OS}[Y]}{1+K_{OS}[Y]},\ \text{or,}\ \frac{1}{k_{obs}}=\left(\frac{1}{k_iK_{OS}}\right)\frac{1}{[Y]}+\frac{1}{k_i}$$

Distinction between the I_a and I_d mechanisms (*cf.* Secs. 5.14, 5.19.9)

- Rate law: $k_{obs}\approx k_iK_{OS}[Y]$ when $1\gg K_{OS}[Y]$

$$=k_f[Y];\ k_i=\frac{k_f}{K_{OS}}$$

K_{OS} can be theoretically calculated (*cf.* Sec. 5.14). **Same form** of the rate law for both the I_d and I_a mechanisms arises.

- The **distinctions between the I_a and I_d mechanisms** are determined by the extent to which the interchange rate constant (k_i) depends upon the nature of the entering group (*Y*). The greater dependence on the nature of *Y* indicates the $\mathbf{I_a}$ mechanism while of less sensitivity on the nature of *Y* indicates the $\mathbf{I_d}$ mechanism. In the I_d mechanism, $k_{lim}\approx k_i\approx k_{ex}$ **(water exchange rate constant)**

$[Co(NH_3)_5(OH_2)]^{3+} \xrightarrow{\text{anation}} k_i$, almost independent of the nature of *Y*, *i.e.* $\mathbf{I_d}$ mechanism

$[Rh(NH_3)_5(OH_2)]^{3+} \xrightarrow{\text{anation}} k_i$, sensitive to the nature of, *Y* *i.e.* $\mathbf{I_a}$ mechanism.

Distinction among the *D*, I_d and I_a mechanisms

This aspect has been discussed in Sec. 5.16.

Plot of $\frac{1}{k_{obs}}$ *vs.* $\frac{1}{[Y]}$ gives a straight line with slope $= \frac{1}{k_i K_{OS}}$ and intercept $= \frac{1}{k_i}$, *i.e.* $K_{OS} = \frac{\text{Intercept}}{\text{Slope}}$

$$k_{obs} = \frac{k_i K_{OS}[Y]}{1 + K_{OS}[Y]}, \text{ i.e. } k_{obs} = k_{\lim} \approx k_i \text{ when } K_{OS}[Y] >> 1$$

Very often, $1 >> K_{OS}[Y]$, it leads to: $\boldsymbol{k_{obs} \approx k_i K_{OS}[Y] = k_f[Y]}$

Limiting rate laws for the I-process and rate dependence on Y: Depending on the relative value of 1 and $K_{OS}[Y]$, dfferent situations may arise.

- **Case-I (1 ≈ K_{OS}[Y]):** $k_{obs} = \frac{k_i K_{OS}[Y]}{1 + K_{OS}[Y]}$ *i.e.* **nonlinear dependence** on [*Y*] or **fractional order** in [*Y*].

- **Case-II (1 >> K_{OS} [Y]):** $k_{obs} \approx k_i K_{OS}[Y]$ *i.e.* **first order** in $[Y]$. k_{obs} depends on both the **concentration** of *Y* and its **nature** which determines both k_i and K_{OS}; but dependence of k_i on the nature of *Y* is relatively less sensitive for the I_d-path compared to that of the I_a-path.
- **Case-III (K_{OS} [Y] >> 1):** $k_{obs} = k_{\lim} \approx k_i$, *i.e.* **zeroth order** in [*Y*] but dependence of k_{obs} on the **nature** of *Y* arises for the interchange process; k_i is more sensitive on the nature of *Y* for the I_a-path.

5.9.4 General Rate Law for A, D and I Processes for the Formation Reaction (*i.e.* *X* = solvent = *S*, say)

Under the condition, $[Y] >> [L_5MX]$, we can write the general rate law for the *D* and *I* processes as follows:

$$k_{obs} = \frac{pq[Y]}{1 + q[Y]}, \quad (p, q \text{ are constants}), \qquad \ldots(5.9.4.1)$$

$$\text{Rate} = k_{obs}[Y] = \frac{pq[ML_5X][Y]}{1 + q[Y]} \quad (\textit{cf.} \textbf{ D and I processes}) \qquad \ldots(5.9.4.2)$$

- Under the condition, $\boldsymbol{q[Y] >> 1}$, Rate = $p[ML_5X]$
  i.e. first-order reaction for all types of mechanisms
- Under the condition, $\boldsymbol{1 >> q[Y]}$, Rate = $pq[ML_5X][Y]$
  i.e. second-order reaction for all types of mechanisms

A–process: $k_{obs} = \frac{k_a k_b[Y]}{k_{-a} + k_b}$, **D–process:** $k_{obs} = \frac{k_1 k_2[Y]/(k_{-1}[S])}{1 + k_2[Y]/(k_{-1}[S])}$,

I–process: $k_{obs} = \frac{k_i K_{OS}[Y]}{1 + K_{OS}[Y]}$, (*cf.* Sec. 5.14, **Eigen-Wilkins Mechanism**)

Thus for *D* and *I* (I_a and I_d) processes, the **expression is of similar type** and these processes are mainly observed for the octahedral complexes. **Pure A–process is rarely found in the octahedral complexes.**

● **It indicates that molecularity of the reaction is not necessarily indicated from the rate law.** For example, **d**–intimate mechanism does not necessarily lead to the first order kinetics. Similarly, **a**–intimate mechanism does not necessarily lead to the second-order kinetics.

Table 5.9.4.1 General rate law (*i.e.* $k_{obs} = pq[Y]/(1 + q[Y])$ for the D, I_a and I_d processes of the formation reactions through the displacement of solvent (S) (*cf.* Sec 5.16)

Parameters (p, q)	I_a	I_d	D
● $p =$	k_i	k_i	k_1
● $q =$	K_{OS}	K_{OS}	$k_2/(k_{-1}[S])$
● Dependence of p on Y	Strongly dependent	Very slightly dependent	Independent
● Dependence of p on k_{ex} (solvent exchange rate constant)	No relationship	$k_{ex} \approx p$	$k_{ex} = p$

Note: For **A–process:** $p = \left(\frac{k_a k_b}{k_{-a} + k_b}\right)\frac{1}{K_{OS}}$, $q = K_{OS}$; $1 + K_{OS}[Y] \approx 1$; $(1 >> K_{OS}[Y])$, p depends very strongly on the nature; of Y; but p does not bear any relationship with k_{ex} and p may be greater than k_{ex} and even slower than k_{ex}).

5.9.5 Implication from the Rate Laws for the D and I$_d$ Processes for the Solvent (S) Replacement Reactions

● For the ***I$_d$*–process,** $k_{obs} \approx k_i K_{OS}[Y] = k_f[Y]$, (if $1 >> K_{OS}[Y]$

For the different entering ligands, the formation rate constant (k_f) varies for the different values of K_{OS} (*outer sphere association constant*). But, k_i $\left(= \frac{k_f}{K_{OS}}\right.$, K_{OS} can be calculated theoretically) remains the same for the different types of entering ligands. Thus the calculated k_i values for the different types of ligands reacting with a particular complex remain constant and become very close to k_{ex} (solvent exchange rate constant). **This is the test for the I$_d$ mechanism and this type of mechanism is described very often as the Eigen mechanism.**

$[Ni(OH_2)_6]^{2+}$ reacting with the different types of ligands maintains a constant value of k_i (= k_f/K_{OS}) which is close to 10^4 s^{-1} at 25°C and this is comparable to k_{ex} (water exchange rate constant) of $[Ni(OH_2)_6]^{2+}$. Here it may be mentioned that though k_i is more or less constant, k_f may vary in a wide range. ($10^2 - 10^5$ dm^3 mol^{-1} s^{-1}; *cf.* Table 5.14.1).

● For the ***D*–process** (*cf.* Sec. 5.9.1) of formation reaction (*i.e.* leaving group X = solvent), we have:

$$k_{obs} = \frac{k_1 k_2 [Y]}{k_{-1}[S] + k_2[Y]} \approx \frac{k_1 k_2 [Y]}{k_{-1}[S]}; \quad \text{(under the condition, } k_{-1}[S] >> k_2[Y])$$

i.e. $$k_{obs} = k_f[Y] \text{ and } k_f = \frac{k_1 k_2}{k_{-1}[S]}$$

Here $k_1/(k_{-1}[S])$ is constant. **Thus the dependence of k_f on $[Y]$ arises from the ability of the lower coordinate intermediate (*i.e.* ML_5) to discriminate the competing nucleophiles, *i.e.* solvent and the entering nucleophile (*i.e.* Y).** Thus in a perfect D–process, the dependence of k_f on $[Y]$ arises from the sensitivity of the lower coordinate intermedite towards Y in the k_2 – step. The dislodged leaving group (X = solvent for the formation reaction) and the entering group (Y) compete for the lower coordinate intermediate and this is why, k_{obs} depends on the ratio k_2/k_{-1}.

Limiting forms of the rate laws in *d*–intimate mechanism (*cf.* Sec. 5.16)

Stoichiometric	Condition	Rate law	k_{obs}, for [Y] >> [ML₅X]	Inference
D	$k_2[Y] >> k_{-1}[X]$	$k_1[ML_5X]$	k_1	k_{obs} depends on the leaving group, but is independent of Y.
D	$k_2[Y] << k_{-1}[X]$ (*e.g.* X = solvent)	$\frac{k_1k_2[ML_5X][Y]}{k_{-1}[X]}$	$\frac{k_1k_2[Y]}{k_{-1}[X]}$	k_{obs} depends on the leaving group; dependence on Y due to the sensitivity of the k_2–step.
I_d	$K_{OS}[Y] >> 1$	$k_i[ML_5X]$	k_i	k_{obs} depends on the interchange rate constant (k_i)
I_d	$K_{OS}[Y] << 1$	$k_iK_{OS}[ML_5X]\,[Y]$	$k_iK_{OS}[Y]$	k_{obs} depends on both k_i and Y

5.9.6 Implication from the Rate Laws for the A and I_a Processes for the Solvent (S) Replacement Reactions (*cf.* Sec. 5.16)

In the ***A*-path (2 step process),** the **intermediate** of higher coordination number is **relatively long-lived** to be characterised but in the I_a–path (**single step process**), the intermediate is **too short-lived** to be characterised.

For the ***I_a*–path**, the rate law gives, $k_{obs} = k_iK_{OS}[Y]$ under the condition $1 >> K_{OS}[Y]$. Thus it follows the usual second-order kinetics as in the ***A*–process** for which the k_{obs} is given by $k_ak_b[Y]/(k_{-a} + k_b)$. Thus under these conditions, it is a difficult task to discriminate between the A and I_a paths. In fact, in the anation and formation reactions of $[Pd(OH_2)_4]^{2+}$ and $[Pt(OH_2)_4]^{2+}$, the second order kinetics is maintained to complicate the distinction between the A and I_a paths. Under the conditions, $K_{OS}[Y] >> 1$, for the I_a–mechanism, k_{obs} is given by k_i which is the *ligand interchange rate constant.* In this limiting condition, the I_a path shows the **first order kinetics** (*i.e.* rate = k_i[complex] but the A-path **always shows the second order kinetics** (*i.e.* always first order with respect to the entering ligand).

Limiting forms of rate laws for the *a*–intimate mechanisms

Stoichiometric mechanism	*Conditions*	*Rate law*	*k_{obs} for [Y] >> [ML₅X]*	*Inference*
A	$[Y]$ is very large or small	$\frac{k_ak_b[Y][ML_5X]}{(k_{-a}+k_b)}$	$\frac{k_ak_b[Y]}{(k_{-a}+k_b)}$	Always second-order kinetics
I_a	$K_{OS}[Y] >> 1$	$k_i\,[ML_5X]$	k_i	k_{obs} is the ligand interchange rate constant
I_a	$K_{OS}\,[Y] << 1$	$k_iK_{OS}[ML_5X][Y]$	$k_iK_{OS}[Y]$	k_{obs} is a composite one

A *vs.* I_a Path: Stereochemistry of the Products (*cf.* Sec. 5.30.4)

- **I_a–path (single step process,** *cf.* Fig. 5.8.1) produces the higher coordinate intermediate whose **life time is too short** (*i.e.* transient species) while **A–path (two step process,** *cf.* Fig. 5.8.1) produces an intermediate (higher coordination number) which is **relatively long-lived** to experience a rearrangement.
- **sp^3–C** can adopt the **I_a–path** (but not the A-path; *cf.* Ch. 10, Vol. 2) and the **TBP intermediate is too short-lived to experience any rearrangement** and it gives the product with **Walden inversion for** the optically active compound. On the other hand, **sp^3–Si** can adopt the **A-path** and it can stabilise the TBP intermediate by using its *d*-orbital (*cf.* sp^3d hybridisation for TBP). **The TBP intermediate is sufficiently long-lived to experience the Berry pseudorotation** to produce two different TBPs remaining in an equilibrium and from these TBPs, a racemic product is obtained.

$$\left.\begin{cases} (P)(Q)(R)C - X + Y^- \xrightarrow{I_a\text{-path}} (P)(Q)(R)C - Y + X^- \\ \qquad\qquad\qquad\qquad\qquad\qquad \text{(Inversion)} \\ (P)(Q)(R)Si - X + Y^- \xrightarrow{A\text{-path}} (P)(Q)(R)Si - Y + X^- \\ \quad \text{(Chiral compound)} \qquad\qquad\qquad \text{(Racemic product)} \end{cases}\right\} (cf.\ \text{Scheme 5.30.4.3})$$

- **Retention of stereochemistry** during the ligand substitution in **Pt(II) square planar complexes** can be explained in terms of the I_a-mechanism not by the A-mechanism (*cf.* Scheme 5.30.4.2).

In **both the I_a– and *A*–paths,** the rate is very much sensitive to the nature of the entering ligand. *The rate is faster than the solvent exchange rate if the entering ligand is a better nucleophile than the solvent. On the other hand, the rate is slower than the solvent exchange rate, if the entering ligand is a poorer nucleophile than the solvent.*

The ***A*– and I_a–paths** are well established in the ligand substitution processes of square planar complexes of Ni(II), Pd(II), Pt(II), Au(III), etc. *The pure A–path in the octahedral complexes is still called in question but the I_a–path is known to operate in many octahedral complexes.*

5.10 REPRESENTATIVE EXAMPLES OF SUBSTITUTION REACTIONS PASSING THROUGH DIFFERENT TYPES OF MECHANISMS

(A) d–reactions, *i.e.* D and I_d mechanisms: True *D*-reactions are rarely known. In fact, if any convincing evidence lacks in favour of the existence of lower coordinate intermediate, then it is better to describe the *d*–reactions as the I_d-reactions instead of the *D*–reactions.

- ***D* mechanism** (supported by high $\Delta V^{\neq}$ value which is not complicated by the charge effect) has been argued for the **anation** or **aquation** of some octahedral complexes having large **negative charges and *trans*-labilising ligands.** Anionic nature of the complex disfavours the formation of outer-sphere complex — a prerequistic condition for the I_d-path. Such examples are:

  anation of $[Co(CN)_5(OH_2)]^{2-}$; aquation of $[Fe(CN)_5L]^{3-}$ ($\Delta V^{\neq} \approx +22$ cm^3 mol^{-1}); aquation of $[Cr(NCS)_6]^{3-}$ ($\Delta V^{\neq} = +16$ cm^3 mol^{-1}); aquation of $[Co(CN)_5X]^{3-}$. Anation of $[Co(corrin)(OH_2)]$ (relevant to vitamin B_{12}) also satisfies the requirements (*cf.* Sec. 5.16) of a *D*–process.

- **I_d mechanism** operates for the substitution reactions at the aqua metal ions $[M(OH_2)_6]^{2+}$ (M = Fe, Co, Ni; *i.e.* bivalent metal ions of late members of 1st transition series) and at $[M(OH)(OH_2)_5]^{2+}$ (M = Fe, Cr). Anation of $[Co(NH_3)_5(OH_2)]^{3+}$ and hydrolysis of $[Co^{III}(NH_3)_5(X)]^{2+}$ also follow the I_d-path.

(B) a-reactions, *i.e.* A and I_a mechanisms: It has been already mentioned that when $K_{OS}[Y] << 1$, it becomes difficult to distinguish between the *A* and I_a mechanisms because in both the situations, the usual second-order kinetics is maintained. In absence of any strong evidence in favour of formation of an intermediate of higher coordination number, it is better to describe the *a*-reaction as the I_a–reaction rather than the *A*-reaction.

● In the **I_a–process**, the life time of the higher coordinate intermediate is **too short** while in the **A-process**, the life time of the higher coordinate intermediate is **relatively higher**. The TBP intermediate generated from the **sp^3–C** is **too short-lived to experience any rearrangment** and it causes the **Walden inversion** (*i.e.* I_a path) in the product while in the case of **sp^3–Si compound**, the TBP generated is stable (*cf.* role of the *d*-orbital of Si) and it can experience the **Berry pseudorotation** (*i.e.* A-path) and it gives the **racemised product** (*cf.* Sec. 5.30.4).

● Complexation (including anation) at **$[Pd(OH_2)_4]^{2+}$ and $Pt(OH_2)_4]^{2+}$** has been argued to pass through the ***A*–mechanism**. In fact, in such cases, the second order kinetics is maintained for an appreciable range of concentration of *Y* and even when the entering ligand, *i.e.* *Y* is anionic. Anionic nature of *Y* makes the high value of K_{OS} and at high concentration of *Y*, the condition $K_{OS}[Y] << 1$ is disfavoured. Thus if the process is argued to pass through the I_a–path then for the high concentration of anionic entering ligand, it is reasonable to expect the condition $K_{OS}[Y] >> 1$ leading to a first-order kinetics. In reality, it does not happen so. Thus it is reasonable to conclude that complexation at $[Pd(OH_2)_4]^{2+}$ and $[Pt(OH_2)_4]^{2+}$ occurs through the ***A*–mechanism**. It is also supported by negative values of $\Delta V^{\neq}$ for complexation with neutral ligands (where no complications occurs in the interpretation of $\Delta V^{\neq}$).

● Strong dependence on the nature of entering ligand for substitution at **$[Ti(OH_2)_6]^{3+}$** indicates the ***A*–mechanism**. The variation of rate constant covers a wide range (*ca.* $10^2 - 10^6$ dm^3 mol^{-1} s^{-1}) for different types of entering ligands.

● Highly negative value of $\Delta V^{\neq}$ for aquation of **$[RuCl(NH_3)_5]^{2+}$** and anation of **$[Ru(NH_3)_5(OH_2)]^{3+}$** indicate the ***A*–mechanism**.

Intimate *a*- and *d*-mechanisms for the octahedral complexes at different metal centres

a: Ti(III) (d^1), V(II) (d^3), V(III) (d^2), Mo(III) (d^3), Cd(II) (d^{10}), In(III) (d^{10}, post-transition element), Ir(III) (d^6), etc.

Both **a** and **d:** Cr(III) (d^3, predominantly **a**), Mn(II) (d^5), Ru(II) (d^6), Ru(III) (d^5), Rh(III) (d^6).

d: Mg(II) (d^0), Al(III) (d^0), Fe(II) (d^6), Co(III) (d^6), Ni(II) (d^8), Zn(II) (d^{10}), Ga(III) (d^{10}).

- Trivalent metal centres tend to adopt the **a–**process:
- Heavier congeners in a group tend to adopt the **a**–process;

$$\xrightarrow[\text{increasing associative character}]{\text{Al(III)} < \text{Ga(III)} < \text{In(III)}} \; ; \; \xrightarrow[\text{increasing associative character}]{\text{Co(III)} < \text{Rh(III)} < \text{Ir(III)}}$$

(C) I_a and I_d mechanisms: Depending on the sensitivity of k_i on the nature of the incoming nucleophile, the I_a and I_d mechanisms can be classified.

● Anation of $[Co(NH_3)_5(OH_2)]^{3+}$ occurs through the I_d path (*i.e.* k_i is almost independent of the nature of entering group) while anation of $[Rh(NH_3)_5(OH_2)]^{3+}$ follows the I_a path (*i.e.* k_i *is sensitive on the nature of the incoming group).*

● Anation of $[Cr(OH_2)_6]^{3+}$ occurs at different rates for different entering anions and this difference cannot be explained in terms of the difference in K_{OS}. It indicates the **associative character** in the reaction pathway. This associative character is also supported by the **LFER plot** (linear free energy relationship plot, *i.e.* ΔG^0 *vs.* $\Delta G^{\neq}$ or log K_{eq} *vs.* log *k*) for which the slope is less than unity. This aspect will be discussed later. Interestingly, $[Cr(OH)(OH_2)_5]^{2+}$ reacts through the I_d–path. Here the π–donor OH group strongly labilises the ligand H_2O bound at the *cis*–**position** (*i.e.* ***cis–effect***) to impart the dissociative character. The **good σ-donor property** of OH also labilises the coordinated water molecules to impart the dissociative activation. Besides these, the **overall less positive charge** (+2 *vs* +3) in $[Cr(OH)(OH_2)_5]^{2+}$ favours the dissociative path. ***In fact, both the σ- and π-donor properties of the nonleaving ligands reduce the positive charge on the metal centre to favour the d-activation.***

● In the same way, $[Fe(OH_2)_6]^{3+}$ reacts through the I_a path while its conjugate base, $[Fe(OH(OH_2)_5]^{2+}$ reacts through the I_d path.

● $[Cr(OH_2)_6]^{3+}$ reacts through the I_a path while $[Cr(NH_3)_5(OH_2)]^{3+}$ reacts through the I_d path. It happens so due to the better σ–donor property of NH_3 than that of H_2O.

● $[M(OH_2)_6]^{2+}$ (M = late members of 1st transition series) like $[Fe(OH_2)_6]^{2+}$, $[Ni(OH_2)_6]^{2+}$, etc. follow the I_d path in the ligand substitution process while the early members of the series, *e.g.* $[V(OH_2)_6]^{2+}$, $[Mn(OH_2)_6]^{2+}$, etc. follow the I_a path. This aspect will be discussed later.

5.11 RATE LAWS: LIMITATIONS OF THE RATE LAWS AND MECHANISMS OF LIGAND SUBSTITUTION REACTION

The rate law derived from the proposed reaction mechanism does not predict anything conclusively regarding the molecularity of the reaction. **Different reaction mechanisms may lead to the same rate law.** The rate law gives only the order of a reaction. The rate determining step may be unimolecular but the overall rate expression may be a second order one. These are the limitations of the rate law. Some of these aspects have been already discussed. This aspect is further illustrated in the following cases.

(i) **1st order kinetics and participation of solvent:** It can be shown that for two different reaction mechanisms, we obtain the same rate law.

Scheme I:

$$\left.\begin{array}{l} L_5M-X \xrightarrow[(rds)]{\text{slow }(k_1)} L_5M+X \\ L_5M+Y \xrightarrow{\text{fast}} L_5M-Y \end{array}\right\} \mathbf{D \equiv S_N1\ lim.}$$

Here, the rate determining step (rds) is a **unimolecular** one and the rate law is given by:

$$\text{rate} = \frac{[L_5MY]}{dt} = k_1[L_5M-X] \qquad \ldots(5.11.1)$$

Scheme II:

$$L_5M-X+H_2O \xrightarrow[(rds)]{\text{slow }(k_1)} L_5M-OH_2+X$$

$$L_5M-OH_2+Y \xrightarrow{\text{fast}} L_5M-Y+H_2O$$

Here the rate determining step is a **bimolecular** one and the rate law is:

$$\text{rate} = k_1[L_5MX][H_2O] = k_1'[L_5MX] \qquad \ldots(5.11.2)$$

(where $k_1' = k_1[H_2O]$ = constant, $[H_2O] \approx 55.5$ mol dm^{-3})

Thus it becomes a **pseudo first-order reaction.**

It is evident that the two different reaction mechanisms (*i.e.* **unimolecular rds** in Scheme I, **bimolecular rds** in Scheme II) are giving the same rate law. To distinguish between these possibilities, we need other observations.

(ii) **Second order kinetics for different reaction schemes:** Several reaction schemes may be proposed to explain the second order kinetics in the ligand substitution process.

Scheme I:

$$\left.\begin{aligned} L_5M - X + Y &\xrightarrow[(rds)]{\text{slow }(k_1)} L_5MXY \\ L_5MXY &\xrightarrow{\text{fast}} L_5M - Y + X \end{aligned}\right\} \mathbf{A \equiv S_N2\ lim.}$$

The above scheme leads to the following rate law:

$$\text{rate} = k_1[L_5M - X][Y] \qquad \ldots(5.11.3)$$

This *A*–mechanism leads to the second order kinetics.

Scheme II:

It involves the formation of an *outer sphere* (*O.S.*) *complex* **(remaining at steady state)** that participates at the rds.

$$\left.\begin{aligned} L_5M - X + Y &\underset{k_{-1}}{\overset{k_1}{\rightleftharpoons}} (L_5M - X)\cdot Y\ (O.S.\ complex) \\ (L_5M - X)\cdot Y &\xrightarrow[(rds)]{k_2} L_5M - Y + X \end{aligned}\right\} \textbf{Pre - equilibrium type followed by rds.}$$

Under the steady state condition of $(L_5M - X)\cdot Y$ (*O.S. complex*), we have the rate law (*cf.* Eqn. 5.9.2.3):

$$\text{rate} = \frac{k_1k_2[L_5MX][Y]}{k_{-1} + k_2} = k[L_5MX][Y] \qquad \ldots(5.11.4)$$

where $k = k_1k_2/(k_{-1} + k_2)$

Even if the **rds is a unimolecular one,** but the rate process follows *a second order kinetics.* A similar situation arises when **there is a preequilibrium step followed by a rds** (*cf.* Fig. 5.11.1).

Scheme III:

$$\left.\begin{aligned} L_5M - X + Y &\underset{k_{-1}}{\overset{k_1}{\rightleftharpoons}} L_5M\!\left\langle\begin{matrix} X \\ Y \end{matrix}\right. \\ L_5M\!\left\langle\begin{matrix} X \\ Y \end{matrix}\right. &\xrightarrow[(rds)]{k_2} L_5M - Y + X \end{aligned}\right\} \textbf{A-process}$$

Under the steady state condition of L_5MXY, an intermediate of higher coordination number, the following rate law is obtained (*cf.* Sec. 5.9.2)

$$\text{rate} = \frac{k_1k_2[L_5MX][Y]}{k_{-1} + k_2} = k[L_5MX][Y] \qquad \ldots(5.11.5)$$

where $k = k_1k_2/(k_{-1} + k_2)$

This second order kinetics is valid if the second step is a irreversible (*i.e.* $k_{-2} = 0$). If it is not irreversible (*i.e.* $k_{-2} \neq 0$), then a deviation from the second order kinetics occurs.

Scheme IV:

Let us consider the following formation reaction.

$$L_5M(OH_2) + Y \rightarrow L_5M - Y + H_2O$$

The above reaction may be considered to pass through the following mechanism where the rds is a unimolecular one.

$$\left.\begin{array}{l} L_5M - OH_2 \underset{k_{-1}}{\overset{k_1}{\rightleftharpoons}} L_5M + H_2O \\ L_5M + Y \xrightarrow{k_2} L_5M - Y \end{array}\right\} \textbf{D-process}$$

Under the steady state condition of the intermediate L_5M, we have the rate law (*cf.* Sec. 5.9.1)

$$\text{rate} = \frac{k_1k_2[L_5M - OH_2][Y]}{k_{-1}[H_2O] + k_2[Y]} = \frac{k_1k_2[L_5M - OH_2][Y]}{k'_{-1} + k_2[Y]}, \quad \ldots(5.11.6)$$

($k'_{-1} = k_{-1}[H_2O]$ = constant)

If the concentration of Y is small, *i.e.* $\mathbf{k_{-1}[H_2O] >> k_2[Y]}$, we have:

$$\text{rate} \approx k[L_5M—OH_2][Y] \text{ where } k = \frac{k_1k_2}{k'_{-1}}$$

It gives the second-order kinetics.

- At high concentration of Y, *i.e.* $\mathbf{k_2[Y] >> k_{-1}[H_2O]}$, we have:

$$\text{rate} \approx k_1[L_5M — OH_2]$$

It gives the first-order kinetics.

From the above discussed Schemes, it is evident that the ***observed second-order kinetics does not necessarily suggest the bimolecularity of the process.***

(iii) **Involvement of a rapid pre-equilibrium step:** The pre-eqilibrium step may involve the formation of an **outer-sphere complex** or **acid-base reaction.** The outer-sphere complex or the conjugate acid/base formed in the pre-equilibrium step passes through the rate determining step to give the product. The rds may involve a dissociative (**D**), an associative (**A**) or an interchange (**I**) process, but ***in all cases, the rate expression will incorporate the concentration term of the entering ligand (Y).*** These are illustrated below.

(a) **Formation of an outer-sphere (O.S.) complex at a rapid pre-equilibrium step:** The reactants may interact through electrostatically (*e.g.* ion-ion, ion-dipole) or hydrogen bonding to form the outer-sphere complex (*cf.* **Eigen Mechanism,** Secs. 5.9.3 and 5.14).

$$L_5MX + Y \overset{K_{OS}}{\rightleftharpoons} \underset{\text{(O.S. Complex)}}{L_5MX \bullet Y} \xrightarrow[\text{(slow)}]{k} L_5M — Y + X$$

It leads to:

$$K_{OS} = \frac{[O.S.]}{[L_5MX][Y]} \text{ and } [O.S.] = \frac{K_{OS}T_C[Y]}{1 + K_{OS}[Y]},$$

(T_C = total concentration of the starting complex = $[L_5MX] + [L_5MX{\cdot}Y] = [L_5MX] + [O.S.]$)

$$\text{rate} = k[O.S.] = \frac{kK_{OS}T_C[Y]}{1 + K_{OS}[Y]} \quad \ldots(5.11.7), (\textit{cf.} \text{ Eqn. } 5.9.3.4)$$

It leads to: $k_{obs} = \dfrac{kK_{OS}[Y]}{1+K_{OS}}$, when $[Y] \gg T_C$

- Under the condition, $1 >> K_{OS}[Y]$, rate $\approx kK_{OS}T_C[Y]$

  i.e. second order kinetics (*i.e.* first order dependence on both the complex and entering ligand) though the *rds* is a unimolecular one.

  This is attained at very low concentration of *Y* or for very low values of K_{OS}.

- Under the condition, $K_{OS}[Y] >> 1$, rate $\approx kT_C$

  i.e. first-order in the starting complex and zeroth order in the entering ligand.
  This condition is attained at very high concentration of *Y* or for very high values of K_{OS}.

Thus at very low concentration of *Y*, it shows a first-order dependence on [*Y*] while at very high concentration of *Y*, it shows a zeroth-order dependence on [*Y*]. This is illustrated in Fig. 5.11.1.

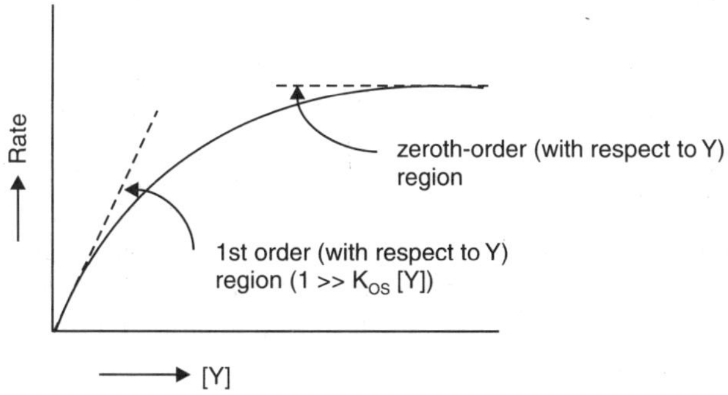

Fig. 5.11.1 Rate dependence pattern on [*Y*]

Outer-sphere complex in a steady state

If the outer-sphere complex remains in a *steady state* then the above scheme is represented as follows:

$$L_5M—X + Y \underset{k_{-1}}{\overset{k_1}{\rightleftharpoons}} \underset{complex}{O.S.} \xrightarrow[(rds)]{k_2} L_5M—Y + X$$

$$\text{rate} = \frac{k_1k_2[L_5MX][Y]}{k_{-1}+k_2} = k[L_5MX][Y],\ k = \frac{k_1k_2}{k_{-1}+k_2} \quad \ldots(5.11.8)\ \ (cf.\ \text{Eqns. } 5.11.4;\ 5.9.2.3)$$

The rate determining step is a unimolecular one, but it shows the second order kinetics.

If the outer sphere (*O.S.*) complex is an ion pair, then depending on the molecularity or nature of the rds, the reaction may be described as ***D–IP*** **(≡ S_N1 lim–*IP*, in short S_N1–*IP*)** (*i.e.* dissociative *rds*), ***A–IP*** **(≡ S_N2 lim–*IP*, in short S_N2–*IP*)** (*i.e.* associative *rds*, *Y* attacks as a nucleophile)

(b) Formation of conjugate acid or base at the pre-equilibrium step: If the leaving group (*X*) possesses suitable lone pairs to act as a base, then it may lead to the **conjugate acid** (*CA*) at the pre-equilibrium step. This happens so for the **acid catalysed aquation** of the complexes like $[Cr(OH_2)_5X]^{2+}$, $[Co(OH_2)_5X]^{2+}$, $[Co(NH_3)_5X]^{2+}$ ($X^- = N_3^-$, halide, etc.), etc.

$$\left[(H_3N)_5Co - X:\right]^{2+} + H^+ \rightleftharpoons \left[(H_3N)_5Co - XH\right]^{3+} \text{ (CA)}$$

$$\xrightarrow[\text{(rds)}]{+H_2O} \left[Co(NH_3)_5(OH_2)\right]^{3+} + XH$$

If the CA reacts through an associative pathway (*i.e.* H_2O attacks as a nucleophile) then the process is described as **A–CA** (≡ S_N2 lim–*CA*, in short **S_N2–CA).** On the other hand, if the CA participates in a dissociative process (*i.e.* XH is dislodged without the attack of H_2O), then the process is called **D–CA** (≡ S_N1 lim–*CA*, in short **S_N1–CA).**

● In the **base catalysed hydrolysis** of the complexes like $[Co(NH_3)_5X]^{2+}$, the conjugate base (CB) may be formed at the pre-equilibrium step and then the CB participates at the *rds* (*cf.* Sec. 5.25).

$$\left[Co(NH_3)_5X\right]^{2+} + OH^- \rightleftharpoons \underset{\textbf{(CB)}}{\left[Co(\ddot{N}H_2)(NH_3)_4(X)\right]^{+}} + H_2O$$

$$CB \xrightarrow[\text{(slow)}]{rds} \left[Co(NH_2)(NH_3)_4\right]^{2+} + X^-$$

$$\left[Co(NH_2)(NH_3)_4\right]^{2+} + H_2O \xrightarrow{\text{fast}} \left[Co(NH_3)_5(OH)\right]^{2+}.$$

Here in the conjugate base (CB), the good σ–donor and π–donor ligand (NH_2^-) labilises the leaving group (X). *Thus, the rds is a unimolecular one* (*i.e.* **dissociative activation**) and the mechanism is described as **D–CB** (≡ S_N1 lim–*CB*, in short **S_N1–CB).** *But it shows the second-order kinetics.*

Metal ion catalysis in aquation process

If the bound leaving group can act as a Lewis base, then it can produce a *Lewis-acid base adduct* at the pre-equilibrium.

$$L_5M - X: + M' \rightleftharpoons L_5M - X - M' \xrightarrow[(rds)]{+H_2O} L_5M - OH_2 + M'X(\text{or } M' + X).$$

The adduct (*i.e.* X–bridged binuclear complex) may participate in a dissociative or an associative (*i.e.* nucleophilic attack by H_2O) path to give the product.

(iv) **Proton ambiguity and rate law:** This aspect has been separately discussed in Sec. 5.17.

5.12 KINETICS AND MECHANISM OF SOLVENT EXCHANGE REACTIONS: DIAGNOSIS OF OTHER SUBSTITUTION REACTION MECHANISM

5.12.1 Characteristics of Solvent Exchange Reactions

The solvent exchange reaction involves the exchange between the coordinated and bulk solvent molecules.

$$[M(S)_x]^{n+} + xS^* \rightarrow [M(^*S)_x]^{n+} + xS$$

These reactions are generally followed by *isotopic exchange technique, NMR line broadening technique,* and other *relaxation* techniques. For the water exchange (*i.e.* water as the solvent) reaction, $H_2{}^{17}O$ is used and O-17 NMR line broadening technique is commonly used for measuring the exchange rate constant (k_{ex} in the range $10^3 - 10^9\ s^{-1}$) of paramagnetic ions. The water exchange process is:

$$[M(OH_2)_x]^{n+} + xH_2{}^{17}O \rightarrow [M(^{17}OH_2)_x]^{n+} + xH_2O$$

The chemical and electronic environment of free and coordinated $H_2{}^{17}O$ *are different and consequently, chemical shifts for free* and *coordinated* $H_2{}^{17}O$ *are different.* The extent of merging the NMR peaks depends on the exchange rate (*cf.* Fig. 5.6.1).

Rate laws: The kinetic behaviour of metal ions towards the solvent exchange process is of an important consideration to understand the mechanistic aspects of other substitution reactions. The water exchange process can be simply represented as:

$$M — {}^*OH_2 + H_2O \rightarrow M — OH_2 + {}^*OH_2 \quad \ldots(5.12.1)$$

(*cf.* $M — X + Y \rightarrow M — Y + X$, *i.e.* $X = {}^*OH_2$, $Y = H_2O$)

(i) If the process occurs by a **D mechanism** then the rate law is given by:

$$k_{obs} = \frac{k_1 k_2 [Y]}{k_{-1}\left[H_2O^*\right] + k_2 [Y]} = \frac{k_1 k_2 \left[H_2O\right]}{k_{-1}\left[H_2O^*\right] + k_2\left[H_2O\right]} \quad (cf. \text{ Sec. } 5.9.1)$$

$k_{obs} \approx k_1$ (under the condition, $k_2[H_2O] >> k_{-1}[H_2O^*]$)

(ii) If the process occurs through **the I process** (both I_a and I_d), the rate law is given by:

$$k_{obs} = \frac{k_i K_{OS} [Y]}{1 + K_{OS} [Y]} = \frac{k_i K_{OS}\left[H_2O\right]}{1 + K_{OS}\left[H_2O\right]}, \quad (cf. \text{ Sec. } 5.9.3)$$

$$\approx k_i, \text{ (under the condition, } K_{OS}[H_2O] >> 1)$$

Table 5.12.1.1 Water exchange rate constants (k_{ex}) (at 25° C) for $[M(OH_2)_6]^{n+}$ and their activation parameters

(a) **Bivalent metal ions**

Metal ion	V^{2+}	Cr^{2+}	Mn^{2+}	Fe^{2+}	Co^{2+}	Ni^{2+}	Cu^{2+}	Zn^{2+}
d^n	$3d^3$	$3d^4$	$3d^5$	$3d^6$	$3d^7$	$3d^8$	$3d^9$	$3d^{10}$
(~) log k_{ex}	2.0	> 8.5	7.4	6.6	6.4	4.5	9.9	7.5
$\Delta H^{\neq}$ (kJ mol^{-1})	62	–	33	41	47	57	23	–
$\Delta S^{\neq}$ (J K^{-1} mol^{-1})	0	–	6.0	21.0	37.0	32.0	25.0	–
$\Delta V^{\neq}$ (cm^3 mol^{-1})	–4.1	–	–5.4	+3.8	+6.1	+7.2	–	–

(b) **Trivalent metal ions**

Metal ion	Ti^{3+}	V^{3+}	Cr^{3+}	Fe^{3+}
d^n	$3d^1$	$3d^2$	$3d^3$	$3d^5$
(~) log k_{ex}	5.3	2.7	–5.5 (–3.7)	2.2 (5.0)
$\Delta H^{\neq}$ (kJ mol^{-1})	43	49	110 (110)	64 (42)
$\Delta S^{\neq}$ (J K^{-1} mol^{-1})	1	–27.5	16.0 (55)	12.0 (5)
$\Delta V^{\neq}$ (cm^3 mol^{-1})	–12.1	–8.9	–9.3 (+2.7)	–5.4 (7.0)

Values within the parentheses for $[M(OH)(H_2O)_5]^{2+}$

(c) **Dependence of $\Delta V^{\neq}$ ($cm^3\ mol^{-1}$) of k_{ex} on the bulkiness of solvent**

Solvent	Co(II)	Ni(II)		Solvent	Fe(III)	Ga(III)	
H_2O	+6.1	+7.2	↓ increasing d-activation	H_2O	−5.4	–	↓ increasing d-activation
MeOH	+9.0	+11.3		$OCHNMe_2$ (DMF)	−0.9	+7.9	
				$OSMe_2$ (DMSO)	−3.1	+13.1	

Refs. Adv. Chem. Sev. **49**, 55, 1965; *Coord. Chem. Rev.*, **5**, 45, 1970.

The important conclusions on water exchange rate constants (k_{ex}) are given below:

(i) **Effect of LFAE (Ligand Field Activation Energy):** The **relative inertness** for $3d^3$ (V^{2+}, Cr^{3+}) and $3d^8$ (Ni^{2+}) can be explained in terms of the relatively high **ligand field activation energy (LFAE).** This aspect of LFAE will be discussed later.

(ii) **Effect of J.T. distortion:** The **high lability** for the $3d^4$ (Cr^{2+}) and $3d^9$(Cu^{2+}) systems is due to the **Jahn-Teller distortion.** Due to this distortion, the weakly bound solvent molecules at the tetragonally elongated sites are rapidly exchanged which is almost close to the *diffusion controlled rate*. Thus, the solvent exchange preferably occurs at the tetragonally elongated sites and then a *facile vibration* leads to the rapid exchange among the six coordination sites of the octahedral geometry.

(iii) **Sign of $\Delta V^{\neq}$ value depending on the mechanistic path:** *For the solvent exchange processes $\Delta V^{\neq}$ (volume of activation) is a very reliable parameter to identify the mechanism. In such cases, there is no possibility of charge annihilation or charge creation in attaining the transition state in all possible routes of a– and d– processes.* Thus, there is no complication in $\Delta V^{\neq}$ from the electrorestriction over the solvent molecules because **in attaining the transition state** through the addition **(*a*-activation)** or removal **(*d*-activation)** of a neutral solvent H_2O molecule, there is no change in charge, *i.e.* no change in electrorestriction over the solvent molecules. ***It makes the $\Delta V^{\neq}$ parameter, a reliable parameter to diagnose the mechanistic path.***

For the solvent exchange process, the **straight-forward conclusions** are:

positive value of $\Delta V^{\neq}$ indicates **d-activation** and *negative value of $\Delta V^{\neq}$ indicates* **a-activation**.

It may be noted that molar volume of free H_2O (*i.e.* bulk solvent) is larger than that of coordinated H_2O (Sec. 5.19.5). In *d*-activation, **positive value of $\Delta V^{\neq}$** arises from the release of a coordinated H_2O molecule (*i.e.* one molecule splits into 2 molecular species). In *a*-activatiion, **negative value of $\Delta V^{\neq}$** arises because a free H_2O molecule gets fixed into the coordination sphere of the starting complex (*i.e.* 2 molecules combine into a single one).

From the $\Delta V^{\neq}$ values of k_{ex}, it is evident that for $[M(OH_2)_6]^{2+}$, the **early members** (*e.g.* V^{2+}, Mn^{2+}) of the 1st transition series adopt the **associative path** while for the **late members** (*e.g.* Fe^{2+}, Co^{2+} and Ni^{2+}), the **dissociative mechanism** is operated. The increasingly dissociative character from Mn^{2+} to Ni^{2+} is reflected in the increasingly positive $\Delta S^{\neq}$ values.

(iv) **Effect of d^n–configuration:** For the octahedral complexes $[M(OH_2)_6]^{n+}$, the associative path involves the approach of the entering solvent molecule along the three-fold axes (C_3) of the octahedral complexes. Thus, greater electron density in the t_{2g} level will disfavour the approach of the entering nucleophile electrostatically. *This is why, for the early members of M^{2+} (of the 1st transition series) having less electron density in the t_{2g} level, the associative path is favoured while for the late members of the series where the t_{2g}–level is filled in, the associative path is highly disfavoured (i.e. dissociative path is favoured).*

Thus for the water exchange process at the M^{2+} ions of Ist transition series, there is a change-over of the mechanism from a dissociative path to an associative path at Mn^{2+} in moving from right to left along the 3d-period.

(v) **Effect of the bulkiness of solvent:** With the increase in bulkiness of the solvent, the associative path leading to an increase in coordination number in attaining the transition state is disfavoured. This is why, for a particular metal centre, the $\Delta V^{\neq}$ increases in the order: DMSO (dimethyl sulfoxide) $\rangle$ DMF (dimethyl formamide) $\rangle$ H_2O. In fact, for the solvent exchange process, $\Delta V^{\neq}$ is positive (*i.e.* dissociative path) for $[Mn(DMF)_6]^{2+}$ while $\Delta V^{\neq}$ is negative for $[Mn(OH_2)_6]^{2+}$ (*i.e.* associative path). *Probably, a perfect D-mechanism operates in the solvent exchange process of* $[Ni(DMF)_6]^{2+}$.

(vi) **Effect of charge on the metal ion:** For the M^{3+} ions *e.g.* $[Ti(OH_2)_6]^{3+}$, $[V(OH_2)_6]^{3+}$, $[Cr(OH_2)_6]^{3+}$, $[Fe(OH_2)_6]^{3+}$, the associative path is operated (evidenced by the negative values of $\Delta V^{\neq}$). Higher positive charge favours the binding of the entering nucleophile.

(vii) **Hydrolysis of trivalent metal ions and difference in the reactivity of the unhydrolysed and hydrolysed species:** It is interesting to note that that though the associative path is operated for $[M(OH_2)_6]^{3+}$ (M = Fe, Cr, etc.), for their hydrolysed species $[M(OH)(OH_2)_5]^{2+}$, the dissociative path operates, *i.e.* $[M(OH)(OH_2)_5]^{2+}$ behaves like the bivalent metal ions of the late members of the 1st transition series. This difference in their mechanisms is supported by the $\Delta V^{\neq}$ values. The dissociative character (*i.e.* positive $\Delta V^{\neq}$ value) is introduced into the hydrolysed species $[M(OH)(OH_2)_5]^{2+}$ because of two grounds:

lower positive charge and *labilising effect of the good σ–donor and π–donor OH ligand.*

The labilising effect of the OH ligand leads to about 10^2-10^3 fold rate enhancement (*cf.* k_{ex} values).

(viii) **Effect of electron density in the t_{2g}–level:** In the case of $[M(OH_2)_6]^{2+}$ species, the associative character for the solvent exchange process decreases with the increase of electron density in the t_{2g} level. It is supported by the fact that there is a gradual increase of $\Delta V^{\neq}$ from Ti(III) (t_{2g}^1) to Fe(III) ($t_{2g}^3e_g^2$) for their water exchange processes.

(ix) **Unsually high k_{ex} value for $[Co(aq)]^{3+}$:** The unusually high k_{ex} value for $[Co(OH_2)_6]^{3+}$ needs an explanation. This aspect has been specially discussed in Sec. 5.13.

5.12.2 Identification of Mechanism of Ligand Substitution Reaction from the Knowledge of Solvent Exchange Rate Constant (k_{ex}) (*cf.* Secs. 5.14, 5.19.9)

When the entering ligand (Y) is other than the solvent (*i.e.* H_2O in aqueous media), the **upper limiting rate constant** (k_{lim}) can be obtained by making [Y] very large (*i.e.* $k_i \approx k_{lim} \approx k_{obs}$ at high [Y]. (*cf.* Sec. 5.9.5).

● In the process of **d–activation,** $M — OH_2$ bond breaking becomes the *rate controlling factor* for anation and other formation reactions. **Thus for such reactions, the rate cannot be greater than the water exchange rate, *i.e.*** $k_{lim}/k_{ex} \approx 1$. Obviously, the ratio k_{lim}/k_{ex} can never exceed unity. It has been observed in many cases. For the anation of $[Co(NH_3)_5(OH_2)]^{3+}$ by different types of anions (*i.e.* $Y = N_3^-$, SO_4^{2-}, Cl^-, NCS^-, etc.), the condition $k_{lim}/k_{ex} \approx 1$ has been found true ($k_{ex} \approx 10^{-4}$ s^{-1}).

● For the **I_d-type Eigen Mechanism**, the k_f/K_{OS} ratio is constant and close to k_{ex} for different types of entering nucleophiles. This aspect will be discussed later.

● For the **a–activation process,** $M — Y$ bond formation (*i.e.* bond formation by the entering ligand) is primarily important. Thus if Y is a better nucleophile than the solvent, *i.e.* water, then the situation will give $k_{lim} > k_{ex}$. On the other hand, if Y is a poorer nucleophile than the solvent, the situation will give $k_{lim} < k_{ex}$.

k_{lim}/k_{ex} for d– and a–activation (*cf.* Sec. 5.19.9)

Generally, the upper limiting rate constant (k_{lim}) is taken as the k_{obs} obtained at high concentration of the entering ligand (Y) and ***the ratio k_{lim}/k_{ex} can never exceed 1 for the d–activation but the ratio may exceed 1 for the a–activation if the entering ligand is a better nucleophile than the solvent.***

$$\left[Co(NH_3)_5(OH_2)\right]^{3+} \xrightarrow[(Y = N_3^-,\ SO_4^{2-},\ Cl^-,\ NCS^-)]{\text{anation}} \text{Product};\ k_{lim}/k_{ex} \approx 1-0.16\ (\textit{i.e.} = \textbf{d-activation})$$

(*cf.* Table 5.19.6.2.)

$$\left[Rh(NH_3)_5(OH_2)\right]^{3+} \xrightarrow[(Y = Cl^-, Br^-, SO_4^{2-})]{\text{anation}} \text{Product};\ k_{lim}/k_{ex} \approx 1-5.0\ (\textit{i.e.}\ \textbf{a-activation})$$

$$\left[Ir(NH_3)_5(OH_2)\right]^{3+} \xrightarrow{\text{anation}} \text{Product};\ k_{lim}/k_{ex} \approx 4.0\ (\textit{i.e.}\ \textbf{a-activation})$$

Thus it is the common observation that the associative character in the ligand substitution process increases for the heavier congeners.

5.13 CLASSIFICATION OF THE METAL IONS IN TERMS OF WATER EXCHANGE RATE CONSTANTS

A. Classification of the metal ions: Based on the **water exchange rate constant** (k_{ex}), Langford and Gray have divided the metal ions in four classes.

- **Class-I** ($k_{ex} \geq 10^8\ s^{-1}$): The water exchange process is exceedingly fast and the *first-order exchange rate constant* (k_{ex}) is of the order of $10^8\ s^{-1}$ (*i.e. diffusion controlled rate*). In such cases, the water molecules are mainly bound through electrostatically and the metal ions are generally of low charge and large ionic radius (*i.e.* low value of the ratio $Z^{*2}/r \leq 10 \times 10^{-28}\ C^2\ m^{-1}$).

Examples: Alkali metal ions, alkaline earth metal ions (except Be^{2+}), Cd^{2+}, Hg^{2+}, Cr^{2+}, Cu^{2+}, and several trivalent lanthanides (*e.g.* Gd^{3+}, Ho^{3+} etc.)

- **Class-II** ($k_{ex} = 10^4$ to $10^8\ s^{-1}$): The water exchange process is fast and k_{ex} lies in the range 10^4 to $10^8\ s^{-1}$. In such cases, $M—OH_2$ bond is somewhat stronger than that in the case of Class-I metal ions. To the metal-ligand bonding, cfse contribution is relatively small. The ratio Z^{*2}/r for such metal ions ranges approximately in the span $10–30 \times 10^{-28}\ C^2\ m^{-1}$.

Examples: Bivalent metal ions of 1st transition series including Zn^{2+} (except for V^{2+}, Cr^{2+} and Cu^{2+}), Ti^{3+}, In^{3+}, Mg^{2+}, some trivalent lanthanides (*e.g.* Yb^{3+}), etc.

- **Class-III** ($k_{ex} = 10^0$ to $10^4\ s^{-1}$): The water exchange process is relatively slower compared to the cases of Class I and Class II but the rate is definitely fast on an absolute scale that spans the range *ca*, 10^{-6} to $10^{10}\ s^{-1}$. In such cases, in the metal-ligand interaction, besides the electrostatic factor, cfse contributes to a significant extent. The Z^{*2}/r ratio for such metal ions is greater than about $30 \times 10^{-28}\ C^2\ m^{-1}$.

Examples: Tripositive metal ions of first transition series (*e.g.* Fe^{3+}, V^{3+}), Al^{3+}, Ga^{3+}, Be^{2+}, Pd^{2+}, etc.

- **Class-IV** (inert centres, $k_{ex} \approx 10^{-2}$ to $10^{-8}\ s^{-1}$): The water exchange process is very slow. In terms of size, such metal ions are comparable to those of Class-III, but contribution of cfse in metal-ligand interaction is more important for the metal ions of Class-IV.

Examples: Cr^{3+}, Co^{3+} (unstable as it oxidises water), Ru^{2+}, Ru^{3+} and Pt^{2+}.

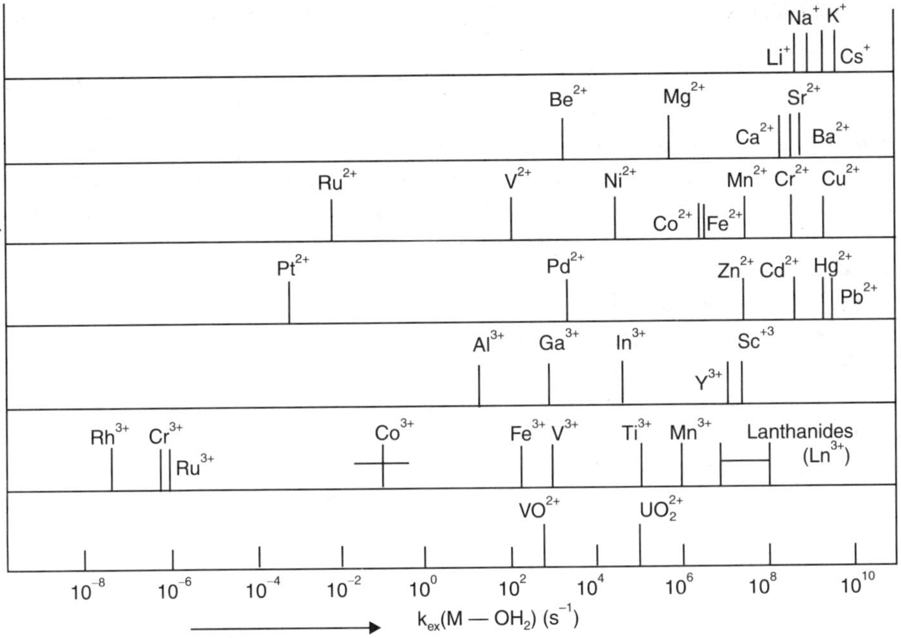

Fig. 5.13.1 Water exchange rate constants (at 25°C) for various aqua metal ions.

B. Different factors controlling the k_{ex} values: The important observations from the classification of metal ions in terms of their k_{ex} values are discussed below.

● **Electrostatic interaction in the metal-ligand bond:** The parameter, **ionic potential** (= charge/radius) or Z^{*2}/r ($Z^* = Z -$ shielding constant) gives the measure of electrostatic contribution in metal-ligand interaction. This electrostatic interaction is the main contributing factor for the metal ions of **noble gas** and **pseudo-noble gas configuration.** In fact, with the increase of ionic potential, the lability measured by k_{ex} decreases. It explains the following **lability sequences:**

(i) $Cs^+ > K^+ > Na^+ > Li^+$
(ii) $Ba^{2+} > Sr^{2+} > Ca^{2+} > Mg^{2+} > Be^{2+}$
(*cf.* Ba^{2+}, Sr^{2+}, Ca^{2+} in Class I; Mg^{2+} in Class-II; Be^{2+} in Class III)
(iii) $In^{3+} > Ga^{3+} > Al^{3+}$
(*cf.* In^{3+} in Class-I; Ga^{3+} in Class-II; Al^{3+} in Class III)
(iv) $Hg^{2+} > Cd^{2+} > Zn^{2+}$
(*cf.* Hg^{2+} and Cd^{2+} in Class-I; Zn^{2+} in Class-II)

(For the representative elements (*i.e.* s- and p-block elements), charge/radius ratio decreases in moving from top to bottom).

(v) $Na^+ > Mg^{2+} > Al^{3+}$; $Ru^{2+} > Ru^{3+}$; $Fe^{2+} > Fe^{3+}$

The parameter, charge/radius ratio indicates that stronger **Bronsted acids,** *i.e.* the $[M(OH_2)_x]^{n+}$ having the larger charge/radius ratio are possessing the lower k_{ex} values. In fact, the *nonacidic* or *feebly acidic* $[M(OH_2)_x]^{n+}$ ions are highly labile ($k_{ex} \geq 10^8\ s^{-1}$) while the acidic cations are relatively more inert.

The decrease of lability with the increase of the ratio charge/radius indicates the importance of **dissociative activation.**

For the transition metal ions, in addition to the contribution of the parameter, ionic potential (charge/radius ratio), **crystal field activation energy** (CFAE) is of an important consideration. However,

importance of the ionic potential parameter is reflected in the fact that the trivalent metal ion is less labile than the bivalent metal ions. The lability sequences: $Fe^{2+} \rangle Fe^{3+}$; $Ru^{2+} \rangle Ru^{3+}$, etc. support the fact (exception: $V^{2+} \langle V^{3+}$).

Note: To estimate the parameter charge/radius, the effective charge may be obtained from the consideration of shielding effect, *i.e.* effective nuclear charge = atomic number – shielding constant.

● **Crystal field activation energy** (CFAE) (*cf.* Sec. 5.18.2): For the *d*-block elements, CFAE is an important factor to determine the lability. Thus, increase of charge may not necessarily increase the inertness. For example, V(III) (d^2) is more labile than V(II) (d^3). In fact, for the *d*–block metal ions, charge effect is straight-forward only for the isoelectronic species like Cr^{3+} and V^{2+}; Fe^{3+} and Mn^{2+}.

● **Jahn-Teller distortion:** The exceptionally high lability of Cr^{2+} and Cu^{2+} can be explained by considering the **Jahn-Teller distortion** in the ground state. Thus in terms of *dissociative activation, the ground state geometry of such distorted complexes is very close to that of the transition state for the water exchange process.* In other words, the elongated $M — OH_2$ bond rupturs easily in the dissociative process to favour the exchange. Then the exchanged water molecule experiences a rapid exchange among the six positions through a facile vibration.

● **Lability sequence for the M^{2+} (high spin) ions of 1st transition series:** In terms of the k_{ex} values, the lability of bivalent metal ions of 1st transition series runs as:

$V^{2+}(d^3) \langle\langle Cr^{2+} (d^4) \rangle Mn^{2+} (d^5) \rangle Fe^{2+} (d^6) \rangle Co^{2+} (d^7) \rangle\rangle Ni^{2+} (d^8) \langle\langle Cu^{2+}(d^9) \rangle Zn^{2+}(d^{10})$

The actual lability sequence is:

	Cr^{2+}	≈	Cu^{2+}	⟩⟩	Mn^{2+}	≈	Zn^{2+}	⟩	Fe^{2+}	⟩	Co^{2+}	⟩⟩	Ni^{2+}	⟩⟩	V^{2+}
$\log k_{ex}$:	*ca.* 9		*ca.* 9		7.4		7.5		6.6		6.4		4.5		2.0

The above sequence can be explained by considering **both CFAE and Z^* (effective nuclear charge)** (more correctly the parameter Z^{*2}/r). The effective nuclear charge gradually increases from left to right along the period. Here it may be mentioned that for the late members of the series (approximately starting from (Cr^{2+} – Mn^{2+}), the water exchange process passes through the **dissociative path** (evidenced by +ve $\Delta V^{\neq}$ values). This dissociative path is disfavoured for increasing the Z^* values.

The abnormally high lability at Cr^{2+} and Cu^{2+} arises due to J.T. distortion. The CFAE for Mn^{2+}, Fe^{2+}, Co^{2+} and Zn^{2+} are considered to be zero (after ignoring the –ve values) for their high spin $[M(OH_2)_6]^{2+}$ complexes.

For **Ni(II) (d^8), V(II)** (d^3), the inertness arises due to high CFAE. Here, it may be mentioned that V(II) experiences an **associative path** for which CFAE is 4.26 *Dq* (assuming the pentagonal bipyramid intermediate) while Ni(II) experiences **a dissociative path** having CFAE 2 *Dq*. This explains higher inertness of V(II) than that of Ni(II).

● **Lability sequence among the congeners of a particular group of d-block metals:** For the transition metal ions, in a particular group, the lability decreases for the heavier congeners (*i.e.* heavier congeners are more inert). This can be explained by considering the *crystal field activation energy.* The lability sequences are:

(i) **$Ni^{2+} \rangle Pd^{2+} \rangle Pt^{2+}$** (*cf.* Ni^{2+} in Class-II; Pd^{2+} in Class-III; Pt^{2+} in class IV).

For Ni(II) and Pd(II), if the complexes are considered to be **square planar,** then their ligand substitution process passes through the **associative path** in which the incoming nucleophile approaches along the z-direction (*cf.* Sec. 5.30.3). In terms of MOT, it requires the participation of the **vacant antibonding d_{z^2} orbital** to accommodate the entering nucleophile. The energy of the vacant antibonding d_{z^2} orbital increases from Ni(II) to Pt(II). In terms of CFT, it happens so due to the *z*–out distortion (*i.e.* splitting of the e_g level of the corresponding octahedral complex) and crystal field splitting power increases from Ni(II) to Pt(II). In other words, in terms of MOT, the antibonding character of the $(n-1)\,d_{z^2}$ orbital increases gradually from Ni(II) to Pt(II). Thus participation of the **vacant antibonding d_{z^2} orbital** (more correctly $\boldsymbol{a_{1g}}$ in terms of MOT, see Vol. 4) to accommodate the entering nucleophile causes the loss of *lfse* (*i.e. cfse*) and this loss increases from Ni(II) to Pt(II). ***Thus the CFAE or LFAE increases from Ni(II) to Pt(II) and the lability gradually decreases from Ni(II) to Pt(II).*** Here it may be mentioned that in an alternative view, it can be argued that the incoming ligand occupies the **vacant nonbonding np_z orbital** (= a_{2u} in terms of MOT). The energy order is: $6p_z\,(Pt^{II}) > 5p_z\,(Pd^{II}) > 4p_z\,(Ni^{II})$ and it explains their lability order. Participation of the **vacant *pd*-hybrid orbital** ($p_z^0 \pm d_{z^2}^2$ combination) to receive the nucleophilic attack can also explain the lability order (see Sec. 5.30.3).

Here it is worth mentioning that Ni(II) ($3d^8$) very often adopts the octahedral geometry and its CFAE is also high (cf. other M^{2+} ions of 3*d*-series).

(ii) $Fe^{2+} > Ru^{2+}$ (cf. Fe^{2+} in Class-II; Ru^{2+} in Class-IV); (iii) $Co^{3+} > Rh^{3+} > Ir^{3+}$.

Mean residence time of a water molecule in the coordination spheres of $[M(OH_2)_6]^{3+}$ at 25°C follows the sequence: $[Cr(OH_2)_6]^{3+}$(~100 h) < $[Rh(OH_2)_6]^{3+}$ (~1.5 y) << $[Ir(OH_2)_6]^{3+}$(~300 y).

● **Unusually high lability of $[Co(OH_2)_6]^{3+}$ – Why**? The Co(III) (d^6) complexes are generally more inert than the corresponding Cr(III) (d^3) complexes (cf. CFAE values). But astonishingly, the rate of water exchange in $[Cr(OH_2)_6]^{3+}$ (*i.e.* exchange with $^{17}OH_2$) is much slower (*ca.* $k_{ex} \approx 10^{-6}\ s^{-1}$ at 25°C) than that for $[Co(OH_2)_6]^{3+}$. In fact, k_{ex} for $[Co(OH_2)_6]^{3+}$ is fairly high ($\approx 10^{-1}\ s^{-1}$). $[Co(OH_2)_6]^{3+}$ is highly oxidising ($E^0 \approx 1.8$ V) and it can oxidise water to release oxygen slowly and it is itself reduced to $Co(aq)^{2+}$. Thus, even the purest form of $Co(aq)^{3+}$ sample in aqueous solution gets invariably contaminated by **$Co(aq)^{2+}$ that may catalyse the processs**. $Co(aq)^{2+}$ is labile and its water exchange rate is fast as expected.

Overall rate benefit in the water exchange process

$$\left[Co(OH_2)_6\right]^{3+} + 6H_2O^* \rightleftharpoons \left[Co(^*OH_2)_6\right]^{3+} + 6H_2O, \text{ (Apparently rapid water exchange)}$$

$$\left[Co(OH_2)_6\right]^{2+} + 6H_2O^* \rightleftharpoons \left[Co(^*OH_2)_6\right]^{2+} + 6H_2O; \text{ (Fast, labile centre)}$$

$$\left[Co(OH_2)_6\right]^{3+} + \left[Co(^*OH_2)_6\right]^{2+} \rightleftharpoons \left[Co(OH_2)_6\right]^{2+} + \left[Co(^*OH_2)_6\right]^{3+},$$

(**Rapid electron exchange through the outer sphere process**)

The reason behind the *rapid electron exchange* in the couple $Co(aq)^{3+}/Co(aq)^{2+}$ will be discussed in the next chapter (cf. Secs. 6.3.6; 6.6).

> **Lability enchanced by coupling the favoured electron transfer reactions**
>
> Here it may be mentioned that it is the common observation:
>
> ***If the higher oxidation state of a particular metal centre is inert but its lower oxidation state is more labile, then the lower oxidation state will catalyse the ligand substitution process at the relatively inert higher oxidation state.*** Thus, Cr(II) can catalyse the ligand substitution process at Cr(III); Pt(II) can catalyse the ligand substitution process at Pt(IV) through the electron transfer process. This aspect will be discussed in the next chapter.

There is **another reason** for the enhanced k_{ex} value for $Co(aq)^{3+}$. In $[Co(OH_2)_6]^{3+}$, the $10Dq$ value provided by the H_2O ligands is a **critical 10Dq value** (*i.e.* $10Dq \approx P$). It allows the **spin isomerism** between $[Co(OH_2)_6]^{3+}$ (high spin, $t_{2g}^4\,e_g^2$) and $[Co(OH_2)_6]^{3+}$ (low spin, $t_{2g}^6\,e_g^0$). The low spin $[Co(OH_2)_6]^{3+}$ is inert because of the very high crystal field activation energy, but the high spin $[Co(OH_2O)_6]^{3+}$ is more labile because of the lower crystal field activation energy. It leads to an overall rate enhancement in the water exchange process

Overall rate benefit in the water exchange process

$$\left[Co(OH_2)_6\right]^{3+}(t_{2g}^4 e_g^2) \rightleftharpoons \left[Co(OH_2)_6\right]^{3+}(t_{2g}^6 e_g^0)$$

(labile) (Rapid water exchange process) — (inert) (Slow water exchange process)

Rapid ligand exchange in $[Co(LL)_3]^{3+}$ (LL = bpy, phen, en)

Catalysis by $Co(aq)^{2+}$ in the water exchange process of $Co(aq)^{3+}$ has been explained by coupling the electron exchange process in the $Co(aq)^{3+}/Co(aq)^{2+}$ couple. A similar mechanism can operate in the rapid exchange of the free ligand LL with those in the $[Co(LL)_3]^{3+}$ complex. Here it must be mentioned that the $[Co(LL)_3]^{3+}$ complexes (low spin, t_{2g}^6) are undoubledly inert as expected from CFAE. In fact, $[Co(phen)_3]^{3+}$ can survive from its dissociation even in 5N H_2SO_4 under refluxing for a period of three hours; $[Co(en)_3]^{3+}$ remains unchanged in 1M $HClO_4$ at 30°C even for 2 months.

Sometimes, **pseudo-substitution** reactions at Co(III) without the rupture of Co(III) – ligand bond can occur and these reactions are rapid as the process does not involve the bond rupture. In fact, such reactions occur through the **eletrophilic attack on the bound ligand.** These have been discussed separately in Sec. 5.24.

- **Effect of electron density in t_{2g} level:** It is interesting to compare the kinetic parameters for water exchange process at Sc(III), Ti(III), V(III) and Fe(III)–all of which follow the **associative path.** Higher positive charge on the metal centre favours the associative path.

		$[Sc(OH_2)_6]^{3+}$	$[Ti(OH_2)_6]^{3+}$	$[V(OH_2)_6]^{3+}$	$[Fe(OH_2)_6]^{3+}$
Electronic Config	:	t_{2g}^0	$t_{2g}^1 e_g^0$	$t_{2g}^2 e_g^2$	$t_{2g}^3 e_g^2$
~ log k_{ex} (25°C)	:	7.2	5.3	2.7	2.2
r (M^{3+}) (pm)	:	89	81	78	79
CFAE	:	0	~0	~0	0

The ionic radii of the trivalent metal ions are more or less comparable but their water exchange rate constants differ widely. It can be explained in terms of the electron population density at the t_{2g}– level to which the entering ligand approaches in an associative path. Definitely it is most favoured for $Sc^{3+}(t_{2g}^0)$ and most disfavoured for $Fe^{3+}(t_{2g}^3)$. Here it may be pointed out that CFAE for all these $[M(OH_2)_6]^{3+}$ (high spin) is zero.

Note: In terms of MOT, the t_{2g} level (*i.e.* NBMO) or e_g^* (ABMO) may act as the LUMO (in the octahedral complexes) to receive the nucleophilic attack. If there is a vacancy in the t_{2g} level to receive the nucleophilic attack it will act as the LUMO. Otherwise, e_g^* will act as the LUMO.

5.14 EIGEN-WILKINS MECHANISM (1960–65)

The stoichiometric I (interchange) mechanism for ligand substitution can be represented in general as follows:

$$L_5M - X + Y \rightarrow (Y \cdots L_5M \cdots X)^{\neq} \rightarrow L_5M - Y + X$$

The transient nature of the activated complex precludes its observation. It may be noted that the perfect *D* and *A* mechanisms operate only in some limited cases but the **I–stoichiometric mechanism is the most common one**. It may be I_a and I_d depending on the situation.

The interchange mechanism for the ligand substitution reactions of octahedral complexes is very often consistent with the **Eigen-Wilkins mechanism.** This well documented reaction mechanism of ligand substitution reaction is named after M. Eigen (who shared the **1967 Nobel Prize for Chemistry**, for his contribution in understanding the fast reactions of labile metal complexes with R.G. Norrish and G. Porter) and R.G. Wilkins. This mechanism is also described as **Eigen-Tamm** or **Eigen-Wilkins-Tamm** or simply **Eigen mechanism**.

The first step of this mechanism considers the formation of an **outer-sphere complex** or **encounter complex** in a rapid (almost diffusion controlled rate) pre-equilibrium step. In the encounter complex, the leaving group or ligand remains in the ***primary coordination sphere*** while the entering ligand (say *Y*) resides in the first solvation shell, *i.e.* in the ***secondary coordination sphere*** of the reacting complex (say L_5M—X, where X is the leaving group). This pre-equilibrium step is rapid as the solvent molecules in the secondary coordination sphere are loosely held and exchange very rapidly.

Formation of the *outer-sphere* or *encounter complex* may involve different types of binding forces, *e.g.* Coulombic attraction (for the oppositely charged reactants, *i.e.* ion-pair formation), ion-dipole interaction (when one reactant is charged and the other reactant is neutral but polar), hydrogen-bonding, etc. Then *at the second step which is the rate determining step,* the encounter complex transforms into the product complex (*i.e.* L_5M – *Y*) and *X* which resides in the secondary coordination sphere of the product complex. This rate determining step leads to the ***interchange of the X and Y ligands between the first and second coordination spheres***. Finally, the leaving group *X* diffuses away from the second coordination sphere rapidly. Here it may be mentioned that theoretically transformation of the O.S. complex into the product complex at the rate determining step may occur either by a D or an A or an *I*-process but **most commonly by an *I* (I_d or I_a)** process (*cf.* Sec. 5.9.3). It is reasonable because in the O.S. complex, the entering ligand (Y) is favourably disposed to participate in a bonding interaction with the metal centre to some extent in the transition state. Thus the whole process can be represented as follows:

$$L_5M - M + Y \overset{K_{OS}}{\rightleftharpoons} (L_5MX,\ Y)\ (\text{Outer-sphere complex}) \qquad \textbf{(Step 1)}$$

$$(L_5MX,\ Y) \xrightarrow[(rds)]{k_i} (L_5MY,\ X)\ (\text{O.S. Complex}) \qquad \textbf{(Step 2)}$$

$$(L_5MY,\ X) \xrightarrow{\text{fast}} L_5M - \text{Y} + X \qquad \textbf{(Step 3)}$$

Scheme 5.14.1 Eigen mechanism.

Step 1 and 3 are the **diffusion controlled processes** and they operate in the time scale of about ns (10^{-9}s). It indicates that formation and break up of the complex in step 1 occur very rapidly and the same thing happens for the step 3. Very often, for the sake of simplicity, step 2 and step 3 are not shown separately. These two steps are simply shown as follows:

$$(L_5MY,\ X) \xrightarrow[(rds)]{k_i} L_5M - Y + X$$

Scheme 5.14.1 leads to the following rate law (*cf.* Sec. 5.9.3).

$$\text{Rate} = \frac{k_i K_{OS}[L_5MX]_0[Y]_0}{1+K_{OS}[Y]_0};\ ([Y]_0 >> [L_5MX]_0)$$

and $$k_{obs} = \frac{k_i k_{OS}[Y]}{1+K_{OS}[Y]},\ (i.e.\ [Y]_0 = [Y]+[O.S.] \approx [Y])$$

Case-I: $k_{obs} \approx k_{\lim} \approx k_i$ when $K_{OS}[Y] >> 1$. This limiting value is obtained for the high concentration of Y.

It leads to: $$\frac{1}{k_{obs}} = \left(\frac{1}{k_i K_{OS}}\right)\frac{1}{[Y]} + \frac{1}{k_i}$$

Linear plot of $\frac{1}{k_{obs}}$ *vs.* $\frac{1}{[Y]}$ can evaluate k_i and K_{OS} (*cf.* Fig. 5.9.3.1).

$$k_i = \frac{1}{\text{intercept}};\ K_{OS} = \frac{\text{intercept}}{\text{slope}}.$$

Case II: $k_{obs} = k_i K_{OS}[Y] = k_f[Y]$, (when $1 >> K_{OS}[Y]$) *i.e.*, k_f (formation rate constant) $= k_i K_{OS}$

This is the **most common form of the rate law** for the I (I_d and I_a) process. An example of the reaction is:

$$[Ni(OH_2)_6]^{2+} + NH_3 \overset{K_{OS}}{\rightleftharpoons} \{[Ni(OH_2)_6]^{2+}, NH_3\}$$
(O.S. Complex)

$$\{[Ni(OH_2)_6]^{2+}, NH_3\} \xrightarrow{k_i} [Ni(NH_3)(OH_2)_5]^{2+} + H_2O$$

and $$k_{obs} = \frac{k_i K_{OS}[NH_3]}{1+K_{OS}[NH_3]},\ ([NH_3]_0 >> [[Ni(OH_2)_6]^{2+}]_0)$$

$$k_{obs} = k_i K_{OS}[NH_3] = k_f[NH_3],\ (\text{when } 1 >> K_{OS}[NH_3])$$

Special Case: Y = solvent; X = Solvent, *i.e.* Solvent Exchange Reactions

If the entering ligand is the solvent then a special case arises because the encounter equilibrium remains always saturated. It happens so because the encounter complex $\{L_5MX, Y\}$ is the solvated starting complex $\{L_5M(OH_2), OH_2\}$ (in aqueous solvent). ***Thus the reacting complex always exists as the encounter complex,*** because the reacting complex is always surrounded by the solvent molecules in the second coordination sphere. It leads to: $k_{obs} = k_i = k_{ex}$.

The relation, $k_i = k_f/K_{OS}$ is very important to identify a reaction mechanism. Here k_f ($dm^3\ mol^{-1}\ s^{-1}$) is called the **formation rate constant** and k_i denotes the rate constant for the transformation of an outer-sphere complex into an inner-sphere complex through the interchange process. The k_f value may be different for different ligands but if k_i remains almost constant, it occurs through the I_d pathway. ***In fact,***

outer-sphere complexation of an aqua-complex does not remarkably alter its water exchange rate constant (k_{ex}). Thus the estimated k_i values ($= k_f/K_{OS}$, k_f is an experimental quantity; K_{OS} can be computed from the theoretical models, *e.g.* Fuoss-Eigen equation for an ion-pair formation) should be very close to k_{ex} for the **dissociative transformation** (*i.e.* I_d) of the outer-sphere complex.

Distinction between I_d and I_a path in terms of Eigen-Wilkins mechanism (*cf.* Sec. 5.19.9)

For an I_d path, k_i ($= k_f/K_{OS}$) values estimated for different types of ligands differing in basicity, structure and binding sites should cover a narrow range close to k_{ex} while for an I_a path (*i.e.* associative transformation of the outer-sphere complex), the extimated k_i values should cover a huge span without having any relationship with k_{ex}. Formation reaction at the **bivalent metal ions** like $[M(OH_2)_6]^{2+}$ (M = Ni, Fe, Cr, etc.) passes through the I_d path (*i.e.* constancy in the estimated k_i values) while the **trivalent metal ions** like $[M(OH_2)_6]^{3+}$ (M = Ti, Fe, Cr, etc.) adopt the I_a path (*i.e.* the estimated k_i - values cover a wide range depending on the nature of the entering ligands). But their **hydrolysed products**, *i.e.* $[M(OH)(OH_2)_5]^{2+}$ very often adopt the I_d-path because of the lower charge (*cf.* +3 *vs.* + 2) and labilising effect of the OH group **(*cis*-effect).**

That $[Ni(OH_2)_6]^{2+}$ reacts with different types of ligands through the I_d path is evidenced by the constancy of k_i values (*cf.* Table 5.14.1) (*i.e.* $k_f = k_i K_{OS} \approx k_{ex} K_{OS}$).

Table 5.14.1 Kinetic parameters (25°C) for the formation reactions of $[Ni(OH_2)_6]^{2+}$ (*cf.* R.G. Wilkins, *Acc. Chem.* Res., **3**, 408, 1970; D.B. Rorabacher, *Inorg. Chem.*, **5**, 1891, 1966)

Entering ligand (Y)	k_f (dm³ mol⁻¹ s⁻¹)	K_{OS} (dm³ mol⁻¹)	$k_i = \left(\frac{k_f}{K_{OS}}\right)(s^{-1})$
NH_3	5×10^3	0.15	3×10^4
F^-	8×10^3	1.0	0.8×10^4
$CH_3CO_2^-$	1×10^5	3.0	3×10^4
py	4×10^3	0.13	3×10^4
SCN^-	6×10^3	1.0	0.6×10^4
Im	5×10^3	0.31	1.6×10^4
H_2O	–	–	3×10^4 ($= k_{ex}$)

Note: k_f varies in the range $10^3 - 10^5$ dm³ mol⁻¹ s⁻¹ but k_i covers a narrow range 0.6×10^4 to 3×10^4 s⁻¹ and this is close to the k_{ex} value (*i.e.* $k_f = k_i K_{OS} \approx k_{ex} K_{OS}$).

Statistical relation between k_i and k_{ex}

$$\left[M\left({}^{*}OH_2\right)_6\right] + H_2O \longrightarrow k_{ex}; \left[M\left(OH_2\right)_6\right] + Y \longrightarrow k_i$$

It can be shown: $k_i = \frac{6}{8} k_{ex} = 0.75\ k_{ex}$. This aspect has been discussed in Sec. 5.19.9.

Eigen-Fuoss equation to estimate K_{OS}: Oppositely charged reactant particles are supposed to generate the ion-pair as the encounter or outer-sphere complex. The equilibrium constant (K_{OS}) for the formation of the outer-sphere complex or encounter complex can be very often calculated by Eigen-Fuoss equation. In fact, R.M. Fuoss developed the equation by using the concept of statistical thermodynamics while it was developed by M. Eigen based on the principle of kinetics. Their result is called Eigen-Fuoss equation as given below.

$$K_{OS} = \frac{4\pi N r^3}{3000}\exp\{-U(r)/k_B T\}$$

where $U(r)$ is the **Debye-Huckel interionic potential.**

$$U(r) = \frac{Z_1 Z_2 e^2}{rD} - \frac{Z_1 Z_2 e^2 \kappa}{D(1+\kappa r)};\quad \kappa^2 = \frac{8\pi N e^2 \mu}{1000 D k_B T}$$

Thus $U(r)$ represents the sum of the attractive and repulsive potentials between the two ions bearing the charge Z_1e and Z_2e.
(N = Avogadro's number; r = distance of closest approach of the ions of charges Z_1e and Z_2e; e = charge carried by an electron in esu units; D = bulk dielectric constant, μ = ionic strength; k_B = Boltzmann's constant)

Thus the Eigen-Fuoss equation can be used to calculate the K_{OS} values for the **ion-dipole association** and **ion-pair association.** Taking, $\mu = 0.1$ mol dm^{-3} and $r = 500$ pm, the computed values of $K_{OS}/(\text{dm}^3\ \text{mol}^{-1})$ are: ~0.3, ~2, ~5, ~15 and ~100 for the charge products (Z_1Z_2) = 0, –2, –3, –4 and –6 respectively. For the ion-dipole interaction in the O.S. complex, a different equation may be used and the estimated K_{OS} value is 0.2 mol^{-1} dm^3 (Ref. J.E. Prue, *J. Chem. Soc.,* 1965, 7534).

Estimation of K_{OS} from kinetic data: Nonlinear dependence of k_{obs} on [Y] can lead to the evaluation of K_{OS}. The **double reciprocal plot** (*i.e.* Burk-Lineweaver plot), *i.e.* $\frac{1}{k_{obs}}$ *vs* $\frac{1}{[Y]}$ can evaluate K_{OS} (= intercept/slope) which may be compared with the values calculated theoretically.

5.15 MECHANISTIC STEPS FOR CHELATION BY A DIDENTATE LIGAND: STEADY-STATE MECHANISM vs. EQUILIBRIUM MECHANISM

If the entering ligand is a didentate one, then the general mechanism may be given as in Scheme 5.15.1.

$$[(H_2O)_5M\text{—}OH_2] + L\text{—}L \overset{K_{OS}}{\rightleftharpoons} \underset{(\mathrm{I})}{\{[(H_2O)_5M(OH_2)]\cdot L\text{—}L\}} \qquad \ldots(5.15.1)$$

$$\underset{(\mathrm{I})}{\{[(H_2O)_5M(OH_2)]\cdot L\text{—}L\}} \underset{k_{-1}}{\overset{k_1}{\rightleftharpoons}} \underset{(\mathrm{II})}{\left[(H_2O)_5M\diagup\begin{matrix}L\\|\\L\end{matrix}\right]} + H_2O \qquad \ldots(5.15.2)$$

$$\underset{(\mathrm{II})}{\left[(H_2O)_5M\diagup\begin{matrix}L\\|\\L\end{matrix}\right]} \underset{k_{-2}}{\overset{k_2}{\rightleftharpoons}} \underset{(\mathrm{III})}{\left[(H_2O)_4M\begin{matrix}\diagup L\\ \ \ |\\ \diagdown L\end{matrix}\right]} + H_2O,\ i.e.\ [M(OH_2)_4(L_2)] \qquad \ldots(5.15.3)$$

Scheme 5.15.1 Chelation process by a didentate ligand.

Species **I** is an outer sphere complex; species **II** is an inner-sphere complex where the didentate ligand binds as a unidentate one; species **III** is the final product.

Here k_2 and k_{-2} are the rate constants for the ring closure and ring opening step respectively. The formation (k_f) and dissociation (k_d) rate constants may be given as follows if the **steady-state condition** is maintained for the species $[(H_2O)_5 M - L - L]$.

$$\text{Rate} = k_f [M(OH_2)_6] [L — L] - k_d [M(OH_2)_4(L_2)]$$

$$k_f = K_{OS} k_1 k_2 / (k_{-1} + k_2) \quad \text{...(5.15.4)}$$

$$k_d = k_{-1} k_{-2} / (k_{-1} + k_2) \quad \text{...(5.15.5)}$$

● **If the ring closure step is faster** than the bond fission of the unidentatedly bonded ligand in the intermediate (*i.e.* $k_2 >> k_{-1}$), then the formation rate constant is given by:

$$k_f \approx K_{OS} k_1 \quad \text{...(5.15.6)}$$

$$\approx K_{OS} k_{ex}, \ (\textbf{for } \mathbf{I_d} \textbf{ path}), \quad \text{...(5.15.7)}$$

The condition $k_2 >> k_{-1}$ is attained when the ring closure (*i.e.* k_2 – step) is not sterically hindered and the first bond formed by L—L is relatively strong so that tendency of the bond to break (measured by k_{-1}) is less. The above equation is equivalent to the equation found true of complexation by the unidentate ligand through Eigen mechanism (*cf.* Sec. 5.14). This type of observation has been noted in many cases where the ring closure step is a faster one. This situation (*i.e.* faster ring closure step) leading to Eigen mechanism for the chelation process may be referred to as **normal substitution** which indicates that the didentate or polydentate ligand resembles the appropriate unidentate ligand (for which only the first step is meaningful) in terms of rate constant (k_f). The chelation by *bpy*, *phen*, *terpy*, etc. with $[Ni(OH_2)_6]^{2+}$ (*cf.* R.G. Wilkins, *Acc. Chem. Res.*, **3**, 408, 1970) follows the Eigen mechanism, *i.e.* ***normal*** I_d ***substitution***, *i.e.* $k_f \approx K_{OS} k_{ex}$ (*cf.* Table 5.15.1).

Table 5.15.1 Formation rate constants (k_f) for the formation of some representative Ni(II)-mono-chelates from $[Ni(OH_2)_6]^{2+}$ (data at 25°C) (*cf.* R.G. Wilkins, *Acc. Chem. Res.* **3**, 408, 1970).

Entering ligand:	bpy	phen	terpy	gly⁻	α–ala⁻
k_f ($dm^3 mol^{-1} s^{-1}$):	1.6×10^3	3.2×10^3	1.4×10^3	4×10^4	2×10^4

Note: The difference in k_f value arises from the difference in K_{OS} values but, k_f/K_{OS} remains approximately constant. For an approximate calculation, K_{OS} may be taken as 0.2 $dm^3 mol^{-1}$ for bpy and phen; and as 2 $dm^3 mol^{-1}$ for the anionic ligand gly⁻ and α-ala⁻.

● **Steady state mechanism *vs.* equilibrium mechanism:** If the ring closure step contributes to k_f, then k_f calculated by Eqn. 5.15.7 differs significantly from the observed k_f value. When the ring closure step is sterically hindered or the first metal-ligand bond formed is weak (*e.g.* by O–donor sites) to favour the k_{-1} step, this situation arises. If k_1, k_{-1} are sufficiently fast and $k_{-1} \rangle\rangle k_2$, then this situation arises, *i.e.* **ring closure step contributes to k_f. Under this condition, the steady-state concept will not be applicable.** Thus the **steady state mechanism** will be shifted to an **equilibrium mechanism.**

Table 5.15.1 indicates that ***phen reacts faster than bpy.*** It is believed that this difference arises due to some contribution of the ring closure step to k_f for complexation by *bpy*. In *phen*, the two binding sites are held at close proximity to the metal centre and consequently the ring closure step is easier compared to the case of *bpy* where due to the rotation around the C—C single bond, the second binding site may lie away from the coordination sphere. **Thus, for bipy, the ring closure step is more hindered.** In other words, for *bpy*, contribution of the ring closure step is more probable than the case of *phen*. Similarly, contribution of the ring closure step for *β–ala⁻* that forms a six membered ring (which is sterically more hindered) is more probable than for α–*ala*⁻. **In fact, α–ala⁻ reacts faster than β–ala⁻** (*cf.* $k_f = 1 \times 10^4$ dm^3 mol^{-1} s^{-1} for β–*ala*⁻, = 2×10^4 dm^3 mol^{-1} s^{-1} for α–*ala*⁻, in the reaction with $[Ni(OH_2)_6]^{2+}$.).

The *most favourable situation for the contribution of the ring closure step appearing as the rate determining step* (*i.e.* $k_{-1} >> k_2$) arises during the formation of ternary complexes where the primary ligand (say *A*) is tetradentate in character, bulky in size and the entering ligand binds first through the *O*-site (*i.e.* weak bond forming site) followed by ring closure through the *N*–site. (*cf.* A.K. Das, *Int. J. Chem. Kinet,* **28**, 275-282, 1996 and the references cited therein; D. Sc. Thesis by A.K. Das, Visva Bharati). This leads to the **equilibrium mechanism.**

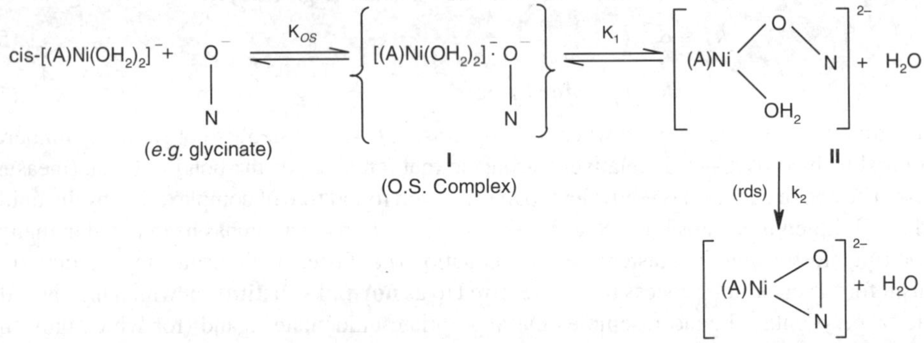

Scheme 5.15.2 Formation of ternary complex where A (primary ligand) is a tetradentate ligand (*e.g.* nta^{3-}).

Scheme 5.15.2 leads to:

$$k_f = \frac{K_{OS}K_1k_2[\text{O—N}]}{1+K_1[\text{O—N}]} \qquad \ldots(5.15.8)$$

In Scheme 5.15.2, the species **II** remains in equilibrium with the reactants. The condition $k_{-1} >> k_2$ leads to $K_1 = \dfrac{k_1}{k_{-1}}$ and *it does not allow the steady-state condition of the species* **II.** Rather, a rapid equilibrium leading to a **considerable buildup of the species II** (in which the didentate ligand binds unidentatedly through the weaker bond forming site) is established before the sluggish ring closure step (k_2).

5.16 DISTINCTION AMONG THE D, I_d and I_a PATHWAYS (*cf.* Sec. 5.19.9)

It has been already mentioned that in the ligand substitution process of octahedral metal complexes, the most commonly observed mechanisms are *D*, I_d and I_a. *They can be distinguished from the nature of rate dependence on the entering nucleophile (Y).*

A. Distinction between the D and I_d processes (*cf.* Secs. 5.9.1, 5.9.3., 5.14, 5.19.9)

Here it is worth mentioning that the **D- and I_d-mechanisms gives the similar rate law** (*cf.* Sec. 5.9.4) and it becomes a very difficult task to distinguish between the *D*- and I_d-mechanisms. If the **lower coordinate intermediate is sufficiently long-lived to discriminate the nucleophiles**, then the D-mechanism can be argued. If no convincing evidence in favour of a such long-lived intermediate is available, then it is better to propose the I_d-mechanism. For an I_d-process, we have: $k_f = k_iK_{os}$ and $k_{obs} \approx k_f[Y]$ when $1 \rangle\rangle K_{os}[Y]$ (*cf.* Sec. 5.14).

Let us try to illustrate by taking the complex $[L_5M(OH_2)]$ reacting with different types of nucleophiles in a D-process.

$$\left.\begin{array}{l} L_5M - OH_2 \underset{k_{-1}}{\overset{k_1}{\rightleftharpoons}} L_5M + H_2O \\ L_5M + Y \xrightarrow{k_2} L_5M - Y \end{array}\right\} \textbf{D-process}$$

The rate law (*cf.* Sec. 5.9.1) is given by:

$$\text{rate} = \frac{k_1k_2[L_5M(OH_2)][Y]}{k_{-1}[H_2O]+k_2[Y]} = \frac{k_1k_2[L_5M(OH_2)][Y]}{k'_{-1}+k_2[Y]} = k_f[L_5M(OH_2)][Y]$$

$$\text{where } k_f = \frac{k_1k_2}{k'_{-1}+k_2[Y]}$$

Here k_1 denotes the *first-order forward rate constant* and k'_{-1} denotes the *pseudo first-order backward rate constant* for the first step ($k_{-1}[H_2O] = k'_{-1}$ = constant). It leads to (under the condition, $[Y] \gg [L_5M(OH)_2]$):

$$\text{rate} = k_{obs}[L_5M(OH_2)] \text{ where } k_{obs} = k_f[Y] = \frac{k_1k_2[Y]}{k'_{-1}+k_2[Y]} \text{ or, } \frac{1}{k_{obs}} = \left(\frac{k'_{-1}}{k_1k_2}\right)\frac{1}{[Y]} + \frac{1}{k_1}$$

Pure dissociative process (D): Does it depend on the nature of the entering ligand?

- Yes, the rate depends nonlinearly, $k_{obs} = \dfrac{k_1k_2[Y]}{k'_{-1}+k_2[Y]}$
- Yes, the rate will depend linearly on $[Y]$, under the condition $k'_{-1} \gg k_2[Y]$
- No, under the condition, $k_2[Y] \gg k'_{-1}$, $k_{obs} = k_{lim} = k_1$ **(limiting rate law)**

Conditions favouring the D-process over the I_d process

- Anation or aquation of the **anionic complexes** disfavours the outer-sphere association due to an electrostatic repulsion. Outer-sphere association is a prerequisite condition for the I_d–path (*cf.* Eigen mechanism). Such a situaion favours the D–process; dissociation of the leaving group from the anionic complex is favoured.
- Strong *trans* or *cis*–labilising groups favour the *D*–path. Pi-donor nonleaving groups can show the ***cis–labilising effect; σ–donor ligand can labilise at the trans-position.***

Examples: Anation of $[Co(CN)_5(OH_2)]^{2-}$; anation of $[RhCl_5(OH_2)]^{2-}$; aquation of $[Fe(CN)_5L]^{3-}$; aquation of $[Cr(NCS)_6]^{3-}$; aquation of $[Co(CN)_5X]^{3-}$; etc.

Life time of the intermediate in the D- and I_d-processes

- In a D-process, the lower coordinate intermediate is sufficiently long lived (*i.e.* finite life time) while in the I_d-process, the lower coordinate intermediate is too short lived to discriminate the nucleophiles.

If the concentration of the entering ligand is sufficiently low to maintain the situation, $k'_{-1} \gg k_2[Y]$, then we have:

$$k_{obs} = \frac{k_1 k_2}{k'_{-1}}[Y] = k_f [Y]$$

Thus k_f (formation constant) depends on the ratio k_2/k'_{-1}.

What is the significance of the ratio k_2/k'_{-1} (more correctly, k_2/k_{-1})?

It means that the lower coordinate intermediate, *i.e.* L_5M is **sufficiently long lived** to react competitively with the solvent (*i.e.* k_{-1} path) and Y. The kinetic competition between H_2O and Y for the lower coordinate intermediate generates the term k_2/k'_{-1}. ***The value of this ratio depends on the relative nucleophilicities of the solvent and Y (i.e. k_2–path).*** For an **I_d-process**, the intermediate is **too short lived** to discriminate the nucleophiles. Thus in a D-process, k_f depends on the nature of Y but in the I_d-process, k_f is relatively insensitive of the nature of Y provided K_{OS} does not vary significantly (*cf.* $k_f = k_i K_{OS}$).

For an **I_d path** (Sec. 5.9.3), formation of the outer-sphere complex is the required condition and if this step can be made disfavoured then the *D*–process may operate. **Anation of a highly anionic complex** will disfavour the formation of outer sphere complex due to an electrostatic repulsion. This will disfavour the I_d path. The following anation reactions meet this condition.

$$\left[Co(CN)_5(OH_2)\right]^{2-} + Y^- \longrightarrow \left[Co(CN)_5(Y)\right]^{3-} + H_2O$$

$[RhCl_5(OH_2)_5]^{2-} + Y^- \rightarrow [RhCl_5(Y)]^{3-} + H_2O$, ($Y^- = Cl^-, Br^-, I^-, SCN^-, N_3^-$, etc.).

In fact, for these reactions, the $\frac{k'_{-1}}{k_2}$ $\left(= \frac{\text{slope}}{\text{intercept}}\right.$, slope and intercept obtained from the linear plot $1/k_{obs}$ *vs.* $1/[Y^-]$; *cf.* Fig. 5.9.1.1) values determined depend on the nature of the entering anion. $k_{-1}/k_2 = 1$ for the water exchange process, *i.e.* $Y = H_2O$ and $k_2 = k_{-1}$. It supports the operation of the *D*–mechanism.

B. Distinction between the D and I_a processes

Now the question arises to distinguish between the *D* and I_a processes. Both of which depend on the nature of the entering ligand.

Under the condition, $k_2[Y] \gg k'_{-1}$, the **limiting rate law for the D–process** is given by:

rate $\approx k_1[L_5MX]$, *i.e.* $k_{obs} \approx k_{lim} = k_1$; k_1 describes the water loss step.

Thus k_{obs} gives the rate constant for the loss of the bound H_2O molecule to produce the lower coordinate intermediate. Under this limiting condition, k_{obs} for the D–process is independent of the nature of the entering ligand. **This situation is never attained for the I_a–process** [*cf.* $k_{obs} = \frac{k_i K_{OS}[Y]}{1 + K_{OS}[Y]} \approx k_i K_{OS}[Y]$ (when $1 \gg K_{OS}[Y]$); and $k_{lim} \approx k_{obs} \approx k_i$ (when $K_{OS}[Y] \gg 1$), k_i depends on the nature of Y in the I_a path] **and it makes the difference between the D and I_a pathways.** ***Thus in a D-process, $k_{lim} \approx k_1$ is independent of the nature of Y but in an I_a-process, $k_{lim} \approx k_i$ is dependent on the nature of Y.***

C. Example of D, I_d and I_a reactions (*cf.* Sec. 5.19.9)

- $\left[Co^{III}(corrin)(OH_2)\right] \xrightarrow{\text{anation}}$ (Mimicking the vitamin B_{12})
- $\left[Co(CN)_5(OH_2)\right]^{2-} \xrightarrow{\text{anation}}$

} **D-process**, *i.e.* k_{obs} depends on $[Y]$ under the condition $k'_{-1} >> k_2[Y]$ and k_{obs} is independent of $[Y]$ under the condition, $k'_{-1} << k_2[Y]$; the lower coordinate intermediate is sufficiently **long lived**.

- **D-process:** The ratio k'_{-1}/k_2 obtained from the double reciprocal plot (*i.e.* $1/k_{obs}$ *vs.* $1/[Y^-]$, *cf.* Fig. 5.9.1.1) has been found to vary for the different entering anions. From the estimated $\frac{k'_{-1}}{k_2}$ ratios, the $\frac{k_{-1}}{k_2}$ ratios can be calculated (*cf.* $k'_{-1} = k_{-1}[H_2O]$). For the water exchange process, the ratio k_2/k_{-1} becomes unity. ***The ratio k_2/k_{-1} gives the measure of the relative nucleophilicities of H_2O and Y or Y^-.*** The values of k_2/k_{-1} are as follows for the **anation of $[Co(CN)_5(OH_2)]^{2-}$.**

  k_2/k_{-1} = 1 (H_2O), 6 (Cl^-), 11.7 (I^-), 20.4 (NCS^-), 23.9 (py).

- **D-process:** Aquation of $[Cr(NCS)_6]^{3-}$($\Delta V^{\neq}$ = +16 cm^3 mol^{-1}), $[Co(CN)_5X]^{3-}$, $[Fe(CN)_5X]^{3-}$, ($\Delta V^{\neq} \approx$ + 22 cm^3 mol^{-1}).

- **I_d - process:** $\left[Co(NH_3)_5(OH_2)\right]^{3+} \xrightarrow{\text{anation}} \Big\} I_d\,;\ k_{obs} \approx k_i K_{OS}[Y] = k_f[Y],\ k_i\left(=\frac{k_f}{K_{OS}}\right) \approx$ constant (independent of *Y*)

In fact, the lower coordinate **intermediate $[Co(NH_3)_5]^{3+}$ cannot** live longer than few picoseconds.

- **I_a - process:** $\left[Ti(OH_2)_6\right]^{3+}, \left[Cr(OH_2)_6\right]^{3+}, \left[Fe(OH_2)_6\right]^{3+} \xrightarrow{\text{anation}} \Big\} I_a\,;\ k_{obs} \approx k_i K_{OS}[Y] = k_f[Y],\ k_i\left(=\frac{k_f}{K_{OS}}\right)$ not constant and varies in a wide range depending on the nature of *Y*.

5.17 PROTON AMBIGUITY AND LIMITATION OF THE RATE LAW (*cf.* D. Seewald, *Inorg. Chem.*, **2**, 643, 1963)

If both the metal centre and ligand participate in protonation and deprotonation equilibria, then different kinetic paths may have the same type of proton dependence giving rise to **proton ambiguity** and the paths are kinetically indistinguishable. This aspect is illustrated in Scheme 5.17.1.

$$\underset{(CA)}{\left[M(OH_2)_6\right]^{m+}} + \underset{CB}{L^{n-}} \xrightarrow[\textbf{(path-I)}]{k_1} \left[M(OH_2)_5L\right]^{(m-n)+}$$

$$\left[M(OH_2)_6\right]^{m+} \underset{}{\overset{-H^+,\,K_h}{\rightleftharpoons}} \left[M(OH)(OH_2)_5\right]^{(m-1)+} \qquad L^{n-} \overset{-H^+,\,K_a}{\rightleftharpoons} LH^{(n-1)-}$$

$$\underset{CB}{\left[M(OH)(OH_2)_5\right]^{(m-1)+}} + \underset{(CA)}{LH^{(n-1)-}} \xrightarrow[\textbf{(path-II)}]{k_2} \left[M(OH_2)_5L\right]^{(m-n)+}$$

Scheme 5.17.1 Illustration of proton ambiguity and isomeric reaction paths.

The rate law for the Scheme 5.17.1 can be obtained as follows:

$$\left[M(OH_2)_6\right] \overset{K_h}{\rightleftharpoons} \left[M(OH)(OH_2)_5\right] + H^+;\ \textit{(charges not shown)}.$$

$$K_h = \frac{[H^+]\left[M(OH)(OH_2)_5\right]}{\left[M(OH_2)_6\right]};\ T_M = \left[M(OH_2)_6\right] + \left[M(OH)(OH_2)_5\right]$$

$$= \frac{[H^+][M(OH)(OH_2)_5]}{T_M - [M(OH)(OH_2)_5]}$$

i.e. $$[M(OH)(OH_2)_5] = \frac{K_h T_M}{K_h + [H^+]};$$

and $$[M(OH_2)_6] = T_M - [M(OH)(OH_2)_5] = \frac{T_M[H^+]}{K_h + [H^+]}$$

$$LH \overset{K_a}{\rightleftharpoons} L + H^+, \text{(charge of the ligand not shown)}$$

i.e. $$K_a = \frac{[L][H^+]}{[LH]},\ T_L = [L] + [LH]$$

i.e. $$K_a = \frac{[L][H^+]}{T_L - [L]}$$

i.e. $$[L] = \frac{T_L K_a}{K_a + [H^+]};\ \text{and } [LH] = T_L - [L] = \frac{T_L[H^+]}{K_a + [H^+]}$$

$$r_1\,(\textbf{rate in path-I}) = k_1[L][M(OH_2)_6]$$

$$= \frac{k_1 T_M[H^+]}{(K_h + [H^+])} \times \frac{K_a T_L}{(K_a + [H^+])}$$

$$r_2\,(\textbf{rate in path-II}) = k_2[LH][M(OH)(OH_2)_5]$$

$$= \frac{k_2 T_L[H^+]}{(K_a + [H^+])} \times \frac{K_h T_M}{(K_h + [H^+])}$$

$$\text{Total rate} = r = r_1 + r_2$$

$$= \frac{T_M T_L[H^+]}{x}\{k_1 K_a + k_2 K_h\},\ \text{where } x = (K_h + [H^+])(K_a + [H^+]),$$

$$= \frac{T_M T_L}{x}\{k_1 K_a[H^+] + k_2 K_h[H^+]\}.$$

From the above equation, it is evident that *both the k_1– and k_2–path show the same type of hydrogen ion dependence* (*i.e.* first-order) and they remain **kinetically indistinguishable in terms of [H⁺] dependence**. In fact, the plot,

$$\frac{x(\text{rate})}{T_M T_L}\ vs.\ [H^+],\ \text{slope} = k\,(say) = k_1 K_a + k_2 K_h$$

gives a straight line with slope = $k_1K_a + k_2K_h$. From this slope, it is not possible to find the contribution of the k_1–path *and* k_2–path separately.

k_1 path $\Rightarrow$ $(CA)_M + (CB)_L$: called ***k(CACB)*** **path;**

k_2 path $\Rightarrow$ $(CB)_M + (CA)_L$: called ***k(CBCA)*** **path;**

and these are called the **isomeric reaction pathways.**

● For the metal centres like Fe(III) (having log $K_h \approx -3.0$) which can provide $[Fe(OH_2)_6]^{3+}$ (*CA*) and $[Fe(OH)(OH_2)_5]^{2+}$ (*CB*) even in a fairly strong acidic condition, such a proton ambiguity prevails. From other evidences it can be shown that though the unhydrolysed species is the predominant one in an acidic condition, *the reaction mainly passes through the step involving* $[Fe(OH)(OH_2)_5]^{2+}$ (*cf.* A.K. Das, *Bull. Chem. Soc. Jpn.*, **65,** 2205 – 2210, 1992; *Transition Met. Chem.*, **17**, 484-488, 1992; and the references cited there in).

Detailed analysis indicates the following results:

$k = k_1K_a + k_2K_h \approx k_2K_h$, $k_2 \approx k_iK_{os}$ (I – path) and $k_i \approx k_{ex}$ of $[M(OH)(OH_2)_5]^{2+}$, *i.e.* $[Fe(OH)(OH_2)_5]^{2+}$ reacts in the I_d-path. If ***the k_1K_a term is argued to contribute to k, it leads to an unusually high value of k_1.***

● To solve the problem with the proton ambiguity, we are, to consider other factors because from the experimental data, we cannot determine k_1 and k_2 separately. This is illustrated in the following reaction between Fe(III) and $HCrO_4^-$ at 25°C (*cf.* J.H. Espenson et al, *Inorg. Chem.*, **8**, 1051, 1969).

$$\textbf{k(CACB) path}\begin{cases} HCrO_4^- \rightleftharpoons CrO_4^{2-} + H^+, \quad K_a \approx 3\times10^{-7}\ \text{mol dm}^{-3} \\ \left[Fe(OH_2)_6\right]^{3+} + CrO_4^{2-} \xrightarrow{k_1} \left[Fe(OCrO_3)(OH_2)_5\right]^+ + H_2O \end{cases}$$

$$\textbf{k(CBCA) path}\begin{cases} \left[Fe(OH_2)_6\right]^{3+} \rightleftharpoons \left[Fe(OH)(OH_2)_5\right]^{2+} + H^+, \quad K_h \approx 1.7\times10^{-3}\ \text{mol dm}^{-3} \\ \left[Fe(OH)(OH_2)_5\right]^{2+} + HCrO_4^- \xrightarrow{k_2} \left[Fe(OCrO_3)(OH_2)_5\right]^+ + H_2O \end{cases}$$

$$k = k_1K_a + k_2K_h = 15\ s^{-1}\ \text{(experimentally found).}$$

If we consider that the term 'k_1K_a' contributes to k (= 15 s^{-1}), it should be at least in the order of 10^0 to 10^1. Taking $k_1K_a = 10$, we get $k_1 = 10/K_a \approx 3.3 \times 10^7$ mol^{-1} dm^3 s^{-1} **which is unusually high** (*cf.* $k_{ex} \approx 10^2$ s^{-1} for $[Fe(OH_2)_6]^{3+}$). **In fact, this calculated value is not reasonable.** On the other hand, let us consider k_2K_h to contribute to k. Taking $k_2K_h \approx 10$, *i.e.* $k_2 = 10/K_h \approx 10^4$ mol^{-1} dm^3 s^{-1}. Taking $k_2 \approx k_iK_{OS}$ (I-process), the estimated k_i value is in the order of 10^4 s^{-1}.This value is quite reasonable for $[Fe(OH)(OH_2)_5]^{2+}$. In fact, the water exchange rate constant (k_{ex}) is in the order of 10^4 s^{-1} for $[Fe(OH)(OH_2)_5]^{2+}$. Thus in the present case, we may reasonably conclude that the ***k(CACB) path does not pratically contribute to k, i.e. k ≈ k_2K_h.***

It may be mentioned that besides the *k*(CACB), *k*(CBCA) paths, other possible reaction paths in Scheme 5.17.1 are as follows:

$[M(OH_2)]_6$ + LH $\Rightarrow$ *k*(CACA)

$[M(OH)(OH_2)_5]$ + L $\Rightarrow$ *k*(CBCB)

It may be mentioned that these *k(CACA) and k(CBCB) paths are not isomeric* and *they do not introduce any proton ambiguity.*

5.18 INTERPRETATION OF LABILITY AND INERTNESS IN TERMS OF THE VALENCE BOND THEORY (VBT) AND CRYSTAL FIELD THEORY (CFT) OF OCTAHEDRAL METAL COMPLEXES

5.18.1 Application of VBT in Explaining the Kinetic Behaviour of Octahedral Complexes

(a) Outer and inner orbital octahedral complexes: Valence bond theory (VBT) classifies the octahedral complexes as the **outer orbital complexes** (sp^3d^2) and **inner orbital complexes** (d^2sp^3). The involved *d*–orbitals in hybridisation are $d_{x^2-y^2}$ and d_{z^2}. Thus in the outer orbital complexes, the *nd* orbitals participate along with the *ns* and *np* orbitals for hybridisation while in the inner orbital complexes, the $(n-1)d$ orbitals participate along with the *ns* and *np* orbitals for hybridisation.

The inner orbital complexes are formed only when at least two $(n-1)d$ orbitals are available for d^2sp^3 hybridisation. Thus the situation is straight-forward for the d^0, d^1, d^2 and d^3 configurations. For d^n $(n > 3)$ configuration, it needs rearrangement of electrons through pairing and/or excitation of *d*–electrons for inner orbital complex formation.

(b) Kinetic behaviour of the outer orbital octahedral complexes: In terms of VBT, the outer orbital complexes are considered to be labile. The outer orbital complexes formed by $Mn^{2+}(3d^5)$, $Fe^{2+}(3d^6)$, $Co^{2+}(3d^7)$, $Ni^{2+}(3d^8)$, $Cu^{2+}(3d^9)$, $Zn^{2+}(3d^{10})$, etc. are labile. In the same way, the complexes of Cd^{2+}, Hg^{2+}, Al^{3+}, Ga^{3+}, In^{3+} are expected to be labile because they form the outer orbital complexes.

The nd orbitals are of higher energy than the (n–1)d orbitals. This is why, the outer orbital complexes (sp^3d^2) are **considered to be relatively less stable** than the inner orbital complexes. Such complexes follow the **dissociative path** (*i.e.* bond breaking mainly contributes to the activation energy) and the process gets kinetically favoured due to the low bond dissociation energy.

However, if the charge on the metal centre increases, the metal-ligand bond becomes stronger and consequently the dissociative path gets disfavoured. *Thus the lability of the outer orbital complexes decreases with the increase of the positive charge on the central atom.* It explains the lability sequence:

$$AlF_6^{3-} > SiF_6^{2-} > PF_6^- > SF_6$$

In fact, the outer orbital complexes (using the 3*d*, 3*s* and 3*p* orbitals in sp^3d^2 hybridisation) like SF_6, PF_6^- are very much inert because of the very high positive charge on the central atom.

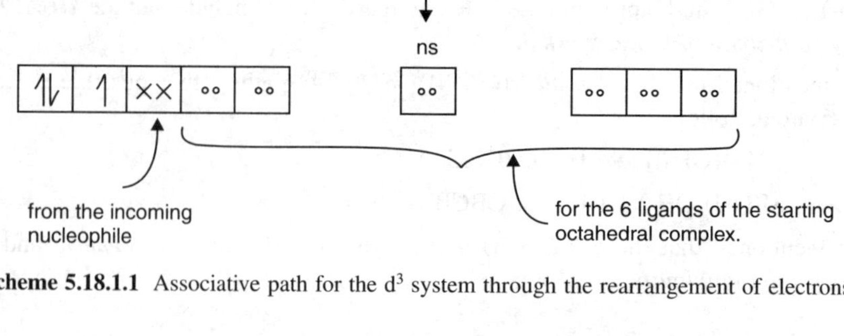

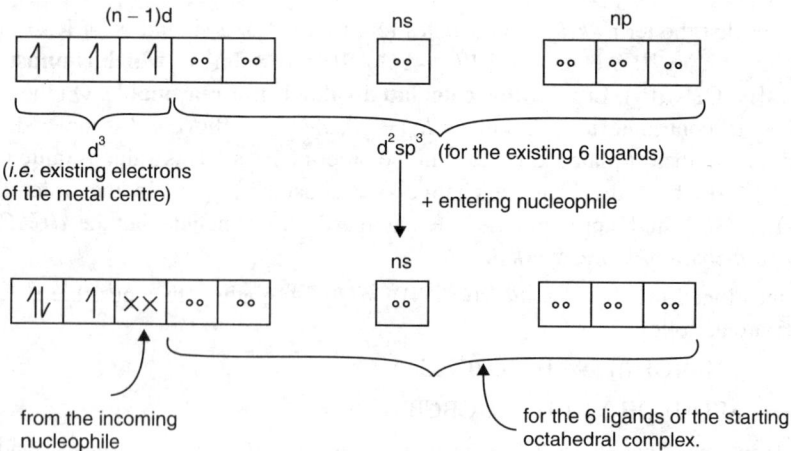

Scheme 5.18.1.1 Associative path for the d^3 system through the rearrangement of electrons.

(c) Kinetic behaviour of the inner orbital octahedral complexes: The d-orbitals d_{xy}, d_{yz} and d_{xz} (*i.e.* t_{2g} set in terms of CFT) of the $(n-1)$ shell are kept out of hybridisation. *Participation of the inner $(n-1)d$ orbitals in hybridisation is considered to make* **the relatively stronger bonds** (compared to those of the outer orbital complexes). Because of this stronger metal-ligand bond in the inner orbital complexes, *the dissociative pathway is expected to be disfavoured but the* ***associative pathway will be favoured*** *through the formation of a new metal-ligand bond by the entering nucleophile to produce an intermediate of higher coordination number.*

If the substitution process is argued to pass through an **associative path** to produce a 7-coordinate intermediate from the starting octahedral complex, then the ease of formation of the intermediate depends on the **availability of a vacant *d*–orbital.** ***It indicates that at least one of the d–orbitals d_{xy}, d_{yz}, d_{xz} which do not participate in bonding in the starting octahedral complex, must remain vacant to accommodate the nucleophilic attack by the entering nucleophile.*** This situation is prevailing for d^0, d^1 and d^2 systems.

Sometimes, the rearrangement of the electrons may be required to make one such d-orbital vacant. For example, in the inner orbital complexes of the d^3 and d^4 electronic configurations, as such there is no vacant d–orbital to accommodate the incoming nucleophile in the associative process. However, redistribution of these electrons through pairing can make one d–orbital vacant for the incoming nucleophile. It is illustrated for the d^3–system in Scheme 5.18.1.1.

Thus we can conclude as follows:

- **Labile inner-orbital complexes**

d^0, d^1, d^2: vacant d–orbital is already available to accommodate the entering nucleophile; and these are labile.

- **Inert inner-orbital complexes**

d^3, d^4: redistribution of electrons through pairing can provide a vacant d–orbital for the entering nucleophile; the process is relatively disfavoured compared to the cases of d^0, d^1, d^2 systems and consequently, the d^3, d^4 centres are less labile.

d^5, d^6: to provide a vacant d-orbital for the entering nucleophile, it needs the promotion of existing d–electrons to some higher energy orbitals, and thus the process is highly disfavoured and it introduces the inertness.

(d) Limitations of the VBT in explaining the kinetic behaviour of the octahedral complexes: The important drawbacks are given below.

(i) The **outer orbital complexes** have been suggested to participate in a **dissociative path** based on the argument that the bonds in the outer orbital complexes are weaker than in the inner orbital complexes. But the assumption of weaker bonds in the outer orbital complexes is not experimentally justified. For example, the hydration energies of the metal ions having the d^7, d^8, d^9 and d^{10} configurations (*i.e.* outer orbital hexaaqua complexes) are much higher than those of the d^0, d^1 and d^2 configurations forming the inner-orbital hexaaqua complexes. This fact weakens the argument of VBT.

(ii) In VBT, it was suggested that the inner orbital complexes participate in an **associative path** based on the argument that the bonds in the inner orbital complexs are stronger. However, the argument is not really justified.

Thus, the ***basic assumptions: dissociative process for the outer orbital complexes and associative process for the inner orbital complexes are not justified.***

(iii) The serious objection against the VBT treatment is that it cannot predict anything regarding the order of lability. Thus it is purely a qualitative treatment.
For example, the difference in lability among the inner orbital complexes of the d^0, d^1 and d^2 configurations cannot be predicted and these are lumped into a group provided they do not differ significantly in charge. Similarly, all the inner orbital complexes of d^3, d^4, d^5 and d^6 are lumped into a group described as the inert complexes.

5.18.2 Application of Crystal Field Theory (CFT) in Explaining the Kinetic Behaviour of the Octahedral Metal Complexes (*cf.* Sec. 5.30.10 for CFAE of the square planar complexes)

Activation energy controls the reaction rate. To the activation energy, loss of cfse (crystal field stabilisation energy) in attaining the activated complex from the starting complex contributes. This contribution is described as **crystal field activation energy** (CFAE), more correctly **ligand field activation energy** (LFAE). This CFAE or LFAE mainly contributes to $\Delta H^{\neq}$.

CFAE = loss of cfse in forming the activated complex
= cfse of the starting complex – cfse of the activated complex.

Calculation of cfse of the starting complex whose geometry is known is straight-forward. But calculation of cfse of the activated complex is a difficult task without the exact knowledge of the structure. Without this knowledge, some approximation may be considered as suggested by Basolo and Pearson. For the **associative path,** the activated complex may be considered to be **seven coordinate** of two

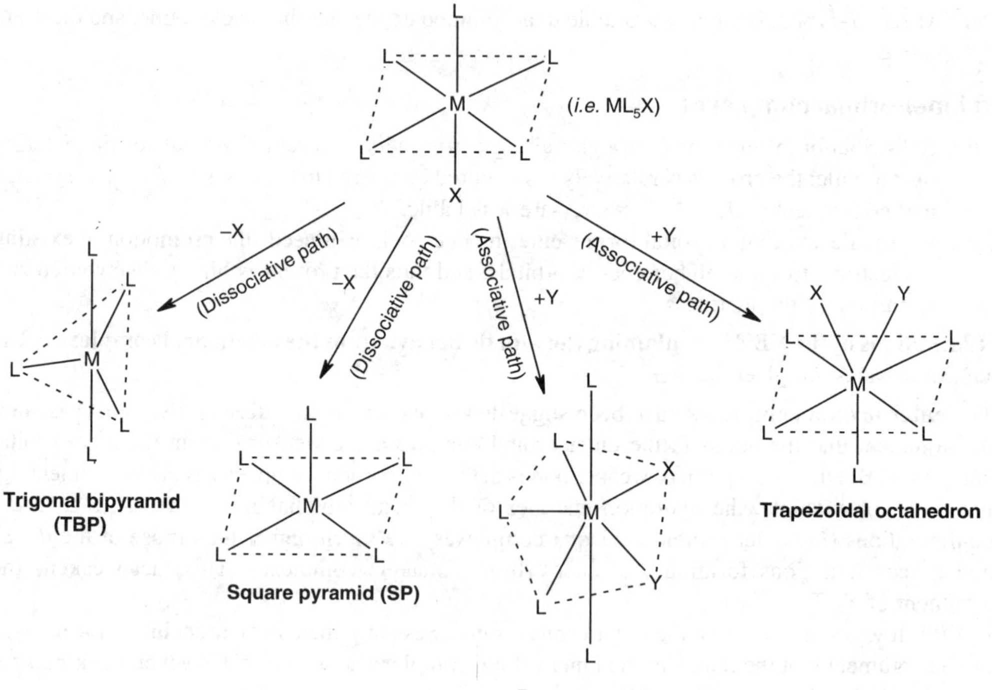

Fig. 5.18.2.1 Structures of the intermediates/activated complexes in dissociative (D) and associative (A) paths of ligand substitution in the octahedral ML_5X complexes: $L_5M — X + Y \rightarrow L_5M – Y$.

probable geometries: *pentagonal bipyramid* (*i.e.* the entering ligand approaches towards the midpoint of an edge of the octahedron) and *trapezoidal octahedron* or *capped octahedron* (*i.e.* the entering ligand approaches towards the centre of a trigonal face of the octahedron). The cfse of such geometries like pentagonal bipyramid or capped octahedron may be calculated easily. Similarly, for the **dissociative path** the activated complex may be approximately considered to be **five coordinate** of two probable geometries: *square pyramid* and *trigonal bipyramid.*

D–path:

$$\underset{(O_h)}{X{-}ML_5} \xrightarrow{(rds)} \underset{(TBP \text{ or } SP)}{ML_5} + X$$

$$\text{CFAE} = \text{cfse of } ML_5X - \text{cfse of } ML_5$$

The 5–coordinate intermediate formed from the octahedral geometry through the rupture of $M — X$ bond is a square pyramid (SP) which may change into a trigonal bipyramid (TBP). In presenting the CFAE in Table 5.18.2.1, the TBP geometry has not been considered.

A–path:

$$X - ML_5 \xrightarrow{+Y,\ (rds)} ML_5XY$$ (pentagonal bipyramid or trapezoidal octahedron or capped octahedron)

$$\text{CFAE} = \text{cfse of } ML_5X - \text{cfse of } ML_5XY$$

Based on the above arguments, Basolo and Pearson have presented the CFAE values (Table 5.18.2.1) for both the dissociative and associative paths for the strong and weak field ligands.

- Thus the **positive value of CFAE** indicates the loss of *cfse* in forming the activated complex or transition state and the rate becomes slow. The more positive value of CFAE indicates the slower reaction.

Table 5.18.2.1 Crystal field activation energy (CFAE) for reactions of octahedral complexes.

System	*Calculated values of CFAE (in Dq_o)**					
	Strong field cases			*Weak field cases*		
	D	*A*	*A*	*D*	*A*	*A*
	Square pyramid	*Pentagonal bipyramid*	*Trapezoidal octahedral*	*Square pyramid*	*Pentagonal bipyramid*	*Trapezoidal octahedral*
d^0	0	0	0	0	0	0
d^1	–0.57	–1.28	–2.08	–0.57	–1.28	–2.08
d^2	–1.14	–2.56	–0.68	–1.14	–2.56	–0.68
d^3	2.00	4.26	1.80	2.00	4.26	1.80
d^4	1.43	2.98	–0.26	–3.14	1.07	–2.79
d^5	0.86	1.70	1.14	0	0	0
d^6	4.00	8.52	3.63	–0.57	–1.28	–2.08
d^7	–1.14	5.34	–0.98	–1.14	–2.56	–0.68
d^8	2.00	4.26	1.80	2.00	4.26	1.80
d^9	–3.14	1.07	–2.79	–3.14	1.07	–2.79
d^{10}	0	0	0	0	0	0

See Sec. 5.30.10 for CFAE of the square planar complex. * The negative values of CFAE are essentially zero.

- The **negative value of CFAE** indicates that the *cfse* of the activated complex is more than that of the starting complex. It implies that the transition state is more stable (in terms of *cfse*) than the reactant but it is improbable. In fact, the negative value of CFAE arises in the cases where the additional *cfse* due to the J.T. distortion in the starting complex is ignored in estimating the CFAE. It may happen so for Cr(II) (d^4), Cu(II) (d^9), etc. For the sake of simplicity, the negative values of CFAE are considered as zero and such complexes react very fast.

Based on the CFAE, the **major conclusions** regarding the lability orders for the reactions of octahedral complexes are given below.

(a) **Dissociative (D)–pathway (square pyramid activated complex)**

(i) **Lability sequence for the strong field ligand:**

$$\underbrace{d^0, d^1, d^2, d^7, d^9, d^{10}}_{\text{labile}} >> \underbrace{d^5 > d^4 > d^3,\ d^8 > d^6}_{\text{inert}}$$

(ii) **Lability sequence for the weak field ligand:**

$$\underbrace{d^0, d^1, d^2, d^4, d^5, d^6, d^7, d^9, d^{10}}_{\text{labile}} >> \underbrace{d^3,\ d^8}_{\text{inert}}$$

(b) **Associative (A)–pathway (Pentagonal bipyramid activated complex)**

(i) **Lability sequence for the strong field ligand:**

$$\underbrace{d^0, d^1, d^2, d^{10}}_{\text{labile}} >> \underbrace{d^9 > d^5 > d^4 > d^3,\ d^8 > d^7 > d^6}_{\text{inert}}$$

(ii) **Lability sequence for the weak field ligand:** $\underbrace{d^0, d^1, d^2, d^5, d^6, d^7, d^{10}}_{\text{labile}} >> \underbrace{d^4, d^9 > d^3,\ d^8}_{\text{inert}}$

(c) **Associative (A)–pathway (Trapezoidal octahedron activated complex)**

(i) **Lability sequence for the strong field ligand:**

$$\underbrace{d^0, d^1, d^2, d^4, d^7, d^9, d^{10}}_{\text{labile}} >> \underbrace{d^5 > d^3,\ d^8 > d^6}_{\text{inert}}$$

(ii) **Lability sequence for the weak field ligand:**

$$\underbrace{d^0, d^1, d^2, d^4, d^5, d^6, d^7, d^9, d^{10}}_{\text{labile}} >> \underbrace{d^3,\ d^8}_{\text{inert}}$$

Thus it is evident that **irrespective of the reaction pathway,** the octahedral complexes react as follows:

- d^3, d^8 and d^6 (low spin) centres are **inert**;
- d^3, d^8 and low spin complexes of d^4, d^5 and d^6 are **inert** and the reactivity order is: $d^5 > d^4 > d^3$, $d^8 > d^6$
- d^0, d^1, d^2, d^7 (except strong field ligand, A-path through the pentagonal bipyramidal activated complex), d^9 and d^{10} centres are **labile**;
- high spin complexes of d^4, d^5, d^6 and d^7 are **labile.**
- d^4 (high spin) and d^9 centres are labile because of the J.T. distortion.
- the lability decreases for the **heavier congeners** because of the increase of CFAE.

d^5 *vs.* d^4: The above conclusions have been verified experimentally. Generally, the d^5 system reacts faster than the d^4 system **for the strong field ligands**. However in the case of **exchange of CN^- ligand in $[M(CN)_6]^{n-}$** (n = 3, 4), the lability order runs as: $d^4 \rangle\ d^5 \rangle\ d^3 \rangle\ d^6$. It indicates that the process goes on through the *trapazoidal octahedral activated complex since only this pathway predicts this order where* d^4 *reacts faster than* d^5.

Factors favouring the associative path

- **High effective nuclear charge** and **high overall positive charge of the complex** favouring a new metal-ligand bond formation (*cf.* higher valent metal ions *vs.* lower valent metal ions; heavier congeners *vs.* lighter congeners)
- **Large size of the central metal** to increase the coordination number (*cf.* heavier congeners *vs.* lighter congeners)
- **Low *d*–electron density** (*cf.* early members *vs.* late members of the transition series) specially **in the t_{2g} level** for the octahedral complexes; more correctly, LUMO of low energy)

Effect of charge and periodic position of the metal centre: Increase of charge is expected to reduce the lability but in terms of CFT, to predict the lability, we should consider the CFAE. With the increase of positive charge, the crystal field splitting power increases sufficiently (*i.e.* **10Dq value increases)** and sometimes it may favour the formation of low spin complexes. **This is why, generally the metal ions of higher charge react more slowly. Because of the same ground, the heavier congeners having the higher crystal field splitting power react more slowly.** It is supported by the lability orders: Co(III) 〉 Rh(III) 〉 Ir(III); Pt(II) 〈 Pd(II) 〈 Ni(II), etc. This aspect has been discussed in connection with the classification of metal ions in terms of water exchange rate constant (*cf.* Sec. 5.13).

V(III) is more labile than V(II) — why?

V^{3+} is expected to be more inert than V^{2+} in terms of the charge/radius ratio. But in reality, V(III) (d^2) is more labile than V(II) (d^3). It is expected in terms of CFAE which is much higher for V(II).

Macrocyclic effect

The so called labile centres may show the inertness in the macrocyclic complexes.

5.18.3 Some Applications of the Knowledge of CFAE and Limitations of CFAE

- **Ignorance of the contribution of $\Delta S^{\neq}$:** $\Delta G^{\neq}$ (= $\Delta H^{\neq} - T\Delta S^{\neq}$) determines the rate constant but CFAE mainly contributes to determine $\Delta H^{\neq}$. $\Delta S^{\neq}$ is an important parameter to determine the rate constant. This is why, prediction from CFAE only is not always correct.

- **Lability of the d^6, d^7 and d^8 systems:** CFAE of the low spin $d^6(t_{2g}^6)$, high spin $d^7(t_{2g}^5e_g^2)$ and $d^8(t_{2g}^6e_g^2)$ complexes runs in the sequence $d^7(\sim 0\ Dq)$ 〈 $d^8(2\ Dq)$ 〈 $d^6(4\ Dq)$ and lability also follows the reverse sequence. Thus the complexes of Fe(II), Co(II) and Ni(II) having the same oxidation state follow the lability sequence.

Co(II)(h.s.) 〉 Ni(II) 〉 Fe(II)(l.s.)

If both the d^7 and d^6 systems form the high-spin complexes, then the CFAE runs in the sequence: $d^6(t_{2g}^4e_g^2,\ 0\ Dq) \approx d^7(t_{2g}^5e_g^2 \sim 0\ Dq)$ 〉 $d^8(t_{2g}^6\ e_g^2,\ 2\ Dq)$.

It explains why the **high-spin complexes** like $[Fe(OH_2)_4(phen)]^{2+}$, $[Fe(OH_2)_3(terpy)]^{2+}$ dissociate faster than the corresponding Ni(II)–complexes.

- **Dissociation of $[M(phen)_3]^{2+}$ to its *bis*-complex (M = Fe, Ni) – failure of the prediction from CFAE:** Low spin $[Fe(phen)_3]^{2+}$ dissociates faster than $[Ni(phen)_3]^{2+}$

	Electronic Config.	*CFAE*	*k* (min^{-1}) (for *tris*- to *bis*-complex)	$\Delta S^{\neq}$ (J mol^{-1} K^{-1})	E_a (kJ mol^{-1}) ($E_a \approx \Delta H^{\neq} + RT$)
$[Fe(phen)_3]^{2+}$:	t_{2g}^6	4 *Dq*	4.8×10^{-3}	117.5	135
$[Ni(phen)_3]^{2+}$:	$t_{2g}^6 e_g^2$	2 *Dq*	1.3×10^{-4}	4.2	110

Though the CFAE is higher for the Fe(II)–complex, it dissociates faster than the corresponding Ni(II)-complex. It happens so due to the **entropic benefit,** *i.e.* higher positive value of $\Delta S^{\neq}$ for the Fe(II)–complex. It is suggested that the Fe(II) complex experiences **a change over in spin state,** *i.e.* low spin to high spin (*cf.* Sec. 4.4.4). **Thus the corresponding transition state is expanded causing elongation of the metal-ligand bond length**. Here it is worth mentioning that in the Fe(II)-bpy/phen system, the *tris*-complex is low-spin (diamagnetic) while the *bis*- and *mono*- complexes are high spin (Sec. 4.4.4). There is no change over in the spin state for the Ni(II) complex and consequently there is no such entropic benefit. Thus the its higher positive value of $\Delta S^{\neq}$ explains the higher dissociation rate of the Fe(II)–complexes inspite of its higher CFAE. The similar situation occurs for their bpy–complexes but the effect is less pronouned as the entropic benefit is relatively less for $[Fe(bpy)_3]^{2+}$ (k = 9.7 × 10^{-3} min^{-1}; E_a = 119.3 kJ mol^{-1}, $\Delta S^{\neq}$ = 71.5 JK^{-1} mol^{-1}).

It indicates that to determine the relative rates we need the activation parameters (i.e. both $\Delta H^{\neq}$ *and* $\Delta S^{\neq}$*) as in some cases the prediction from only CFAE (which mainly contribute to* $\Delta H^{\neq}$*) may not be true.*

● **Dissociation of $[M(bpy)_3]^{2+}$ to its *bis*-complex (M = Cr, Ni) – failure of the prediction from CFAE:** The rates of dissociation and their activation parameters of $[Cr(bpy)_3]^{2+}$ and $[Ni(bpy)_3]^{2+}$ are quite interesting.

	CFAE	*k* (min^{-1})	E_a (kJ mol^{-1})	$\Delta S^{\neq}$ (J mol^{-1} K^{-1})
$[Cr(bpy)_3]^{2+}$ ($t_{2g}^4 e_g^0$):	1.43 *Dq*	22.8	94.9	54.5
$[Ni(bpy)_3]^{2+}$ ($t_{2g}^6 e_g^2$):	2.0 *Dq*	0.18	93.2	8.5

$$E_a \approx \Delta H^{\neq} + RT$$

Though the CFAE values for the said complexes are comparable, low-spin Cr(II)-complex dissociates faster because of **an entropic benefit** that probably arises from the **change over in spin state in attaining the transition state** as in the case of $[Fe(phen)_3]^{2+}$ complex. It again indicates that to compare the rates, we need the knowledge of activation parameters (both $\Delta H^{\neq}$ and $\Delta S^{\neq}$) besides the CFAE.

● **Dissociation of $[V(phen)_3]^{2+}$ *vs.* $[Ni(phen)_3]^{2+}$ – failure of the prediction of CFAE:** $[V(phen)_3]^{2+}$ (d^3) dissociates fasters than $[Ni(phen)_3]^{2+}$ ($t_{2g}^6\ e_g^2$) though CFAE (= 2 *Dq*) is the same in both cases. It is probably due to the less important metal→ligand π–bonding in the V(II)–complex because **V(II) is a poorer π-door than Ni(II).** It makes the **metal-ligand bond weaker in the V(II)-complex.** Interestingly, the water exchange rate for Ni(II) ($k_{ex} \approx 10^4$ s^{-1}) is higher than that for V(II) ($k_{ex} \approx 10^2$ s^{-1}). This is due to the higher CFAE value (= 2 *Dq*) for V(II) for which $Dq \approx$ 1200 cm^{-1} while for Ni(II) $Dq \approx$ 850 cm^{-1}.

● **Lability sequence of $[M(OH_2)_6]^{3+}$ of 1st transition series:** The water exchange rates for the following $[M(OH_2)_6]^{3+}$ follow the sequence:

$$[Sc(OH_2)_6]^{3+} \rangle [Ti(OH_2)_6]^{3+} \rangle [V(OH_2)_6]^{3+} \rangle [Fe(OH_2)_6]^{3+}$$

The water exchange process for these systems goes through the ***associative path*** and CFAE for all these system is zero, but the k_{ex} values differ widely (*cf.* ~10^7 s^{-1} for Sc^{3+}, ~10^2 s^{-1} for Fe^{3+}). *Thus, CFAE cannot explain this wide variation of k_{ex}.* It can explain in terms of the **hindrance experienced by the entering ligand approaching towards the t_{2g} orbitals** (acting as the LUMO in terms of MOT) in an associative path. The higher electron density at the t_{2g} level will repell the entering ligand more to increase the activation energy. The electron density in the t_{2g} level is minimum at Sc^{3+} while it is maximum at Fe^{3+}. It explain the sequence of k_{ex}:

$$Sc^{3+}(t_{2g}^0),\ Ti^{3+}(t_{2g}^1),\ V^{3+}(t_{2g}^2),\ Fe^{3+}(t_{2g}^3e_g^2)$$

● **Inertness of $[Cr(OH_2)_6]^{3+}$:** The very low value of k_{ex} for $[Cr(OH_2)_6]^{3+}$ (t_{2g}^3) arises from two factors: high electron density at the t_{2g}–level and high CFAE.

5.18.4 Comparison of the Predictions on Lability from VBT and CFT

(i) **d^n (n = 0, 1, 2):** The lability of the systems like d^0, d^1, d^2 having at least one $(n-1)d$ orbital unoccupied has been explained in terms of VBT by considering the **associative mechanism**. CFT also predicts the lability of these systems in terms of CFAE but it does not mean that these systems will have to react in the associative pathway. Thus in terms CFT, to explain the lability of such systems having at least one empty $(n-1)d$ orbital, it is not necessarly to impose the associative path. However, it is found that such systems very often adopt the associative path (*cf.* $\Delta V^{\neq}$ values for water exchange processes, Table 5.12.1.1) because of the availability of a vacant t_{2g} orbital to accommodate the nucleophile.

(ii) **d^0, d^{10} and high spin d^n (n = 4, 5, 6):** Both the theories predict the lability for the d^0 systems (*e.g.* nontransition elements, lanthanides), d^{10} systems (*e.g.* posttransition elements, Cu(I), Zn(II), Cd(II), Hg(II), Ga(III). etc.), d^9 system, and high-spin (*i.e.* outer orbital) complexes of d^4, d^5 and d^6 but the arguments are different.

(iii) **d^3 and d^8:** There is a **disagreement between the theories** in connection with the lability of d^8–system. VBT predicts that the d^8 system forms an outer orbital complex and it is labile but CFT predicts it to be inert in terms of CFAE. In fact, CFAE predicts the comparable inertness of the d^3 and d^8 systems.

Here it may be mentioned that the octahedral complexes of Ni(II) (d^8) react fast generally. However, the Ni(II)–centre is less labile than the other bivalent metal centres, like Mn(II), Fe(II), Co(II) and Cu(II). But the octahedral Cr(III) (d^3) complexes react slowly. Thus apparently, the predictions for the d^3 and d^8 systems in terms of CFAE are incorrect. But as such this comparison is not meaningful where one centre is trivalent and the other centre is divalent. *Comparison in terms of CFAE is meaningful only when the 10Dq values (which depend on the charge of the metal, periodic position of the metal and nature of the ligand) are comparable.* Definitely, for the octahedral complexes, $10Dq$ value for Cr^{3+} is higher than that for Ni^{2+}. Thus the comparison becomes more appropriate for Cr^{3+} (d^3) and Cu^{3+} (d^8) bearing the same charge and having the position in the same period.

5.19 DIFFERENT FACTORS GOVERNING THE REACTION RATE AND REACTION MECHANISMS: DIAGNOSIS OF REACTION MECHANISM OF LIGAND SUBSTITUTION PROCESSES

5.19.1 Effect of Charge and Size of the Metal Centre

(a) **Dissociative and associative activation**

(i) Higher positive charge and lower ionic radius (*i.e.* **higher ionic potential**) will increase the metal-ligand bond strength. This will favour the **associative path** and disfavour the **dissociative path.**

In fact, the metal ions having the high ionic potentials tend to adopt the associative path. For the metal ions bearing the low ionic potential values, the reverse is true (*i.e.* dissociative path is favoured and associative path is disfavoured).

- **The trivalent metal centres** show the more associative character than the divalent metal centres.
- **The heavier congeners** in a group possess more Z^* to favour the associative path more, *e.g.* Ir(III) adopts the **a**–activation while Co(III) adopts the **d**–activation.
- Effect of the **charge/radius ratio** on the **water exchange process** has been illustrated (*cf.* Sec. 5.12.1). $[M(OH_2)_6]^{3+}$(M = Cr, Fe) participates in the I_a process while $[M(OH)(OH_2)_5]^{2+}$ participates in the I_d process.
- **Larger size of the metal ion** will favour the accommodation of the approaching nucleophile to favour the associative path (*cf.* heavier congeners *vs.* lighter congeners).

(ii) For the complexes, the **effect of overall charge** on the rate process is very much important. The already present ligands can neutralise the charge to different extents. The σ– and π–bonding properties of these ligands can significantly monitor the electronic environment of the metal centre.

A significant increase in rate with the decrease of the overall charge of the complex supports the **dissociative path.** It is observed in the **aquation** of Cl^- ligand in the chlorido-ammine complexes of Co(III). On the other hand, for the corresponding Cr(III)–complexes, the rate is less sensitive with the decrease of overall positive charge. It supports the **different mechanisms** for the said Co(III) and Cr(III) complexes.

Complex:	$[CoCl(NH_3)_5]^{2+}$	$[CoCl_2(NH_3)_4]^+$	$[CoCl(en)_2(NH_3)]^{2+}$	$[CoCl_2(en)(NH_3)_2]^+$
k_{aq} (s^{-1}, 25°C):	~10^{-6}	~10^{-2} to 10^{-3}	~10^{-7}	~10^{-5}

Complex:	$[CrCl(NH_3)_5]^{2+}$	$[CrCl_2(NH_3)_4]^+$	$[CrCl_2(en)_2]^+$
k_{aq} (s^{-1}, 25°C):	9.7×10^{-6}	4×10^{-5}	~10^{-4}–10^{-5}

(For Co(III), the *cis*-complex reacts faster (Sec. 5.2.1). Here for the sake of simplicity, the *cis– trans* isomers are not specified).

It is evident that for the Co(III)–complexes, the rate significantly increases with the decrease of the overall positive charge. It supports the **dissociative activation.** For the Cr(III) complexes, the rate increases less significantly with the decrease of overall positive charge. It indicates that for the Cr(III)–complexes, bond breaking leading to the transition state is not so important compared to that for the Co(III)–complexes. ***In fact, the Cr(III) complexes experience the more associative activation.***

(b) Interchange process

It is quite expected that for an associative (A) mechanism, bond formation by the entering ligand leading to the activated complex will be favoured with the increase of positive charge on the complex. It will enhance the rate. But for an **interchange (I) process,** the expectation is not so straight-forward. In the interchange process, both the bond formation by the entering nucleophile and bond dissociation by the leaving group are important. With the increase of positive charge of the complex, bond formation by the entering nucleophile will be favoured but bond dissociation by the leaving group will be disfavoured. These **two opposing factors** will operate and consequently the effect of charge on the interchange process cannot be concluded so easily. Even in the I_a process, effect of charge does not produce any straight-forward conclusion. To identify the I_a process, we should rely on the effect of other parameters like sensitivity on the nature of the entering ligand rather than the effect of charge.

(c) Effect of charge — complication in interpretation due to the change in electronic environment by the introduced ligand

To change the overall charge of a complex, it is essentially required to introduce the differently charged ligands which may vary significantly in their σ– and π– bonding properties that may influence the rate and even the reaction mechanism. For example, $[M(OH)(OH_2)_5]^{2+}$ is more reactive than $[M(OH_2)_6]^{3+}$ (M = Fe, Cr) and $[M(OH)(OH_2)_5]^{2+}$ adopts the I_d path due to the presence of the better π–donor ligand (OH^-) that shows the ***cis*–labilising effect** (*i.e.* ***cis*–effect**). On the other hand, $[M(OH_2)_6]^{3+}$ adopts the I_a path. Obviously, the **less overall positive charge** on $[M(OH)(OH_2)_5]^{2+}$ (compared to that of $[M(OH_2)_6]^{3+}$) favours the dissociative action of the hydrolysed species which may behave kinetically like the bivalent metal ions, *i.e.*, $[M(OH_2)_6]^{2+}$.

5.19.2 Effect of *d*–Electronic Configuration

This aspect has been discussed in Secs. 5.18.1–2 in terms of CFAE to estimate the relative lability.

The trivalent metal centres adopt the associative character more (effect of more positive charge) than the divalent metal centres. In a period, the **heavier congeners** having the larger size and more Z^* (effective nuclear charge) adopt the more associative character.

5.19.3 Effect of Temperature: Arrhenius Equation and Eyring Equation: Activation Parameters

The temperature dependence of rate constants (k) in terms of Arrhenius relation is given by:

$$k = A \exp(-E_a/RT) \text{ or, } \ln k = \ln A - E_a/RT = \text{constant} - E_a/RT \quad \ldots(5.19.3.1)$$

A is called the **Arrhenius frequency factor;** E_a is called the **Activation energy** which can be determined from the linear plot of ln$k$ *vs.* $1/T$.

In terms of **collision theory of reaction,** $A = PZ$ where A is called **the probability factor** or **steric factor** and Z is the **collision number**. The factor P cannot be related with any physical quantity, specially when the reactions are studied in solution. This is why, **transition state theory** is preferred. In terms of transition state theory, the reactants are in equilibrium with an **activated complex** which leads to the product formation. This activated complex lies in the region of highest energy along the reaction coordinate in the energy-profile diagram (*cf.* Fig. 5.8.1). ***This region of highest energy state along the reaction coordinate in the process of conversion of the reactant(s) to the product(s) represents the transition state (T.S.)*** T.S. theory of a bimolecular reaction may be represented as:

$$A + B \underset{}{\overset{K^{\neq}}{\rightleftharpoons}} \underset{\text{(Activated Complex)}}{X^{\neq}} \xrightarrow{k^{\neq}} \text{product} \quad \ldots(5.19.3.2)$$

$$K^{\neq} = \frac{[X^{\neq}]}{[A][B]}, \text{ or, } [X^{\neq}] = K^{\neq}[A][B] \quad \ldots(5.19.3.3)$$

Rate of the process is determined by the rate at which the activated complex decomposes to the product, *i.e.*

$$\text{rate} = k^{\neq}[X^{\neq}]$$

$k^{\neq}$ is related to the of passage of the activated complex to the product.

Passage of the activated complex to the product occurs through a *crucial vibration of frequency* ($\nu^{\neq}$) that will carry the activated complex to the product through the transition state along the reaction coordinate. It leads to:

$$\text{rate} = k^{\neq}[X^{\neq}] = \kappa\nu^{\neq}[X^{\neq}] \quad \ldots(5.19.3.4)$$

Here κ is the proportionality constant which is described as the **transmission coefficient** that may vary in the range 0.5 to 1.0. The transmission coefficient actually represents the probability of passing of the activated complex into the product. **If κ is considered to be unity** then we have:

$$\text{rate} = \nu^{\neq}[X^{\neq}]$$

The frequency ($\nu^{\neq}$) of this crucial vibration of the activated complex is related to the energy quanta $h\nu^{\neq}$. The mean energy $(\overline{\varepsilon})$ of an oscillator is:

$$\overline{\varepsilon} = \frac{h\nu}{\exp(h\nu/k_BT)-1} \approx \frac{h\nu}{(1+h\nu/k_BT)-1} \approx k_BT,$$

(when $h\nu/k_BT$ is small, at high temperature, *i.e.* $\frac{h\nu}{k_BT} \to 0$)

Under the condition we have:

$$\overline{\varepsilon} = h\nu^{\neq} = k_BT, \text{ or, } \nu^{\neq} = k_BT/h \quad \ldots(5.19.3.6)$$

Approximation in Conventional Eyring Equation

- **Transmission coefficient κ:** It has been considered to be unity. But the activated state may experience two possibilities: it may move to the product and it may also back to the reactant. If κ is taken to be unity then it ignores the probability of returning the activated state to the reactant. In reality, it may not happen.

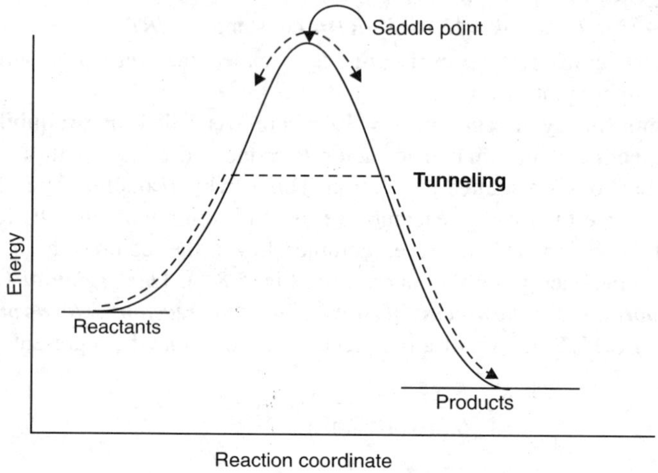

Fig. 5.19.3.1 Tunneling mechanism in some particular reactions.

- **Tunneling effect:** Tunneling is not possible in classical mechanics, but it is possible quantum mechanically. In the conventional Eyring equation, tunneling is ignored. The tunneling is favoured when the potential energy barrier is sufficiently thin and the particle to tunnel must be light **Thus tunneling effect is important for electron transfer; H–atom transfer, H^+ transfer and H^- transfer processes.** Tunneling is not favoured for the heavy isotope deuterium. *In fact, the high kinetic isotope ratio (H vs. D) is an indication of tunneling effect.*

At very low temperature, the rate process over the energy barrier is less important than the rate process through the tunneling. It indicates that the activation energy at low temperature is less.

It leads to:

$$\text{rate} = \left(\frac{k_B T}{h}\right)[X^{\neq}] = \left(\frac{k_B T}{h}\right) K^{\neq}[A][B] \qquad \ldots(5.19.3.7)$$

Thus assuming transmission coefficient to be unity, we can write:

$$\text{Rate} = \left(\frac{k_B T}{h}\right)[X^{\neq}] = \left(\frac{k_B T}{h}\right) K^{\neq}[A][B] = k_2[A][B] \qquad \ldots(5.19.3.8)$$

k_B = Boltzmann's constant; h = Planck's constant;

k_2 = second-order rate constant (in $dm^3\ mol^{-1}\ s^{-1}$).

$K^{\neq}$ can be expressed with the thermodynamic function $\Delta G^{\neq}$.

$$\Delta G^{\neq} = -RT\ln K^{\neq}\ ;\ -\frac{\Delta G^{\neq}}{RT} = \ln K^{\neq}\ ;\ \text{or } K^{\neq} = \exp(-\Delta G^{\neq}/RT)$$

It leads to:

$$\text{rate} = \left(\frac{k_B T}{h}\right)\exp\left(-\frac{\Delta G^{\neq}}{RT}\right)[A][B];$$

$$= \left(\frac{k_B T}{h}\right)\exp\left(-\frac{\Delta H^{\neq}}{RT}\right)\exp\left(\frac{\Delta S^{\neq}}{R}\right)[A][B];\ \Delta G^{\neq} = \Delta H^{\neq} - T\Delta S^{\neq} \qquad \ldots(5.19.3.9)$$

$\Delta G^{\neq}$ = Standard free energy of activation
$\Delta H^{\neq}$ = Standard enthalpy of activation
$\Delta S^{\neq}$ = Standard entropy of activation
} The superscript zero to indicate the standard state is dropped for the sake of simplicity.

It gives:

$$k_2 = \left(\frac{k_B T}{h}\right)\exp\left(-\frac{\Delta H^{\neq}}{RT}\right)\exp\left(\frac{\Delta S^{\neq}}{R}\right), \text{ (Called \textbf{Eyring Equation})} \qquad \ldots(5.19.3.10)$$

$$\text{or, } \frac{k_2 h}{k_B T} = \exp\left(-\frac{\Delta H^{\neq}}{RT}\right)\exp\left(\frac{\Delta S^{\neq}}{R}\right) \text{ or, } \ln\left(\frac{k_2 h}{k_B T}\right) = -\frac{\Delta H^{\neq}}{RT} + \frac{\Delta S^{\neq}}{R}$$

$$\text{or, } -\ln\left(\frac{k_2 h}{k_B T}\right) = \frac{\Delta H^{\neq}}{RT} - \frac{\Delta S^{\neq}}{R}, \text{ (another form of \textbf{Eyring Equation})} \qquad \ldots(5.19.3.11)$$

(A) Evaluation of $\Delta H^{\neq}$ and $\Delta S^{\neq}$ from Eyring equation:

Thus from the linear plot of $-\ln\left(\frac{k_2 h}{k_B T}\right)$ *vs.* $\frac{1}{T}$, $\Delta H^{\neq}$ can be obtained from the slope, *i.e.* slope $= \frac{\Delta H^{\neq}}{R}$

The temperature is generally varied within a span of 30°C. **The intercept** $\left(= -\frac{\Delta S^{\neq}}{R}\right)$ at $\frac{1}{T} = 0$ (*i.e.* $T = \infty$) **cannot be experimentally attained.** To determine $\Delta S^{\neq}$, the estimated $\Delta H^{\neq}$ value is used in the above Eyring equation for a known value of k_2.

It may be shown that uncertainty in the evaluation of $\Delta S^{\neq}$ is more than that in $\Delta H^{\neq}$. In fact, uncertainty in $\Delta S^{\neq}$ is usually at least ± 2.5 J K^{-1} mol^{-1}.

(**Note:** It may be noted that the linear plot, $\ln k_2$ vs. $\frac{1}{T}$ may deviate under the conditions when tunneling effect contributes to the rate process significantly).

$\Delta S^{\neq}$ can be correlated with the pre-exponential factor A of Arrhenius equation while $\Delta H^{\neq}$ is correlated with E_a as follows:

$$E_a = \Delta H^{\neq} - P\Delta V^{\neq} + RT = \Delta H^{\neq} - \Delta nRT + RT = \Delta H^{\neq} - (\Delta n - 1)RT$$

Δn = change in the number of molecules in forming the activated complex, *i.e.* $\Delta n = 0$ (unimolecular process), $\Delta n = -1$ (bimolecular process). Thus, $E_a = \Delta H^{\neq} + RT$ for the **unimolecular gas phase reaction** and for the **reactions in solutions where $\Delta V^{\neq}$ is small.** In general, we have:

$$\Delta S^{\neq} = R\left[\ln A + \ln\left(\frac{h}{k_B T}\right) - 1\right]$$

$$\Delta H^{\neq} = E_a - xRT,\ (x = 1 \text{ for unimolecular reaction; } = 2 \text{ for bimolecular reaction,})$$

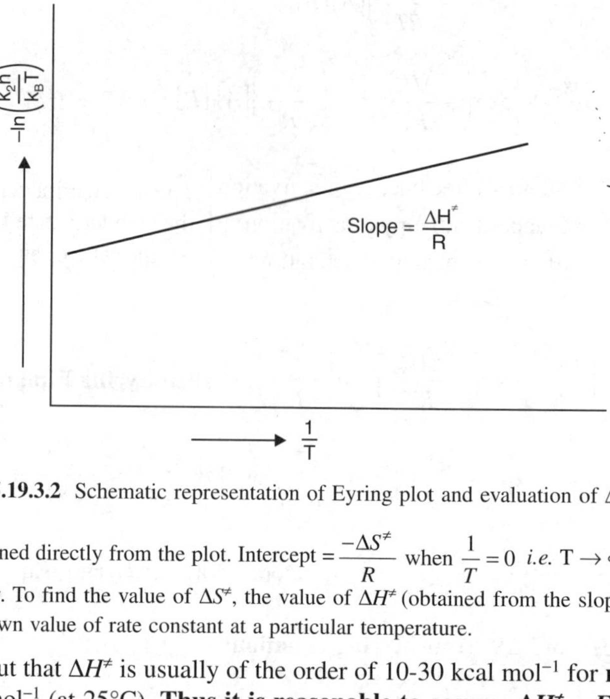

Fig. 5.19.3.2 Schematic representation of Eyring plot and evaluation of $\Delta H^{\neq}$.

Note: $\Delta S^{\neq}$ cannot be obtained directly from the plot. Intercept = $\frac{-\Delta S^{\neq}}{R}$ when $\frac{1}{T} = 0$ *i.e.* T → ∞. This condition cannot be realised experimentally. To find the value of $\Delta S^{\neq}$, the value of $\Delta H^{\neq}$ (obtained from the slope of the plot) is used in Eyring equation fcr a known value of rate constant at a particular temperature.

It may be pointed out that $\Delta H^{\neq}$ is usually of the order of 10-30 kcal mol^{-1} for **reactions in solution** while RT is 0.6 kcal mol^{-1} (at 25°C). **Thus it is reasonable to assume $\Delta H^{\neq} \approx E_a$.**

(B) Significance of $\Delta H^{\neq}$ and $\Delta S^{\neq}$ in understanding the reaction mechanism:

(a) Factors controlling $\Delta H^{\neq}$: $\Delta H^{\neq}$ may include **different energy factors** like electrostatic interaction, bond breaking, bond formation, internal reorganisation within the reactant molecules, solvation and desolvation, etc. involved in producing the transition state from the starting reactants. If a preequilibrium is involved in the process then the ΔH change in the preequilibrium process is also included in $\Delta H^{\neq}$.

(b) Enthalpy of transition (ΔH_T): The parameter called **enthalpy of transition** (ΔH_T) is given by the change of enthalpy in the formation of the product from the transition state. Thus ΔH_T is given by the energy difference between the transition state and product, *i.e.* $\Delta H - \Delta H^{\neq}$ where ΔH gives the enthalpy change in the formation of the product from the reactants.

● ***Different reactants but same product and same $\Delta H_T \Rightarrow$ same transition state.***

● **Identification of the mechanistic path of aquation of $[Co(NH_3)_5X]^{2+}$ by considering $\Delta H^{\neq}$ and ΔH_T parameters:** The knowledge of ΔH_T and $\Delta H^{\neq}$ may be helpful in understanding the reaction mechanism. **In aquation of $[Co(NH_3)_5X]^{2+}$** (X = different univalent anions like Cl^-, Br^-, NO_3^-, NO_2^-, NCS^-, etc.) leading to the same product $[Co(NH_3)_5(OH_2)]^{3+}$, **ΔH_T is more** or **less the same.** It indicates the **common transition state** attained in a ***dissociative path***. For aquation of $[Co(NH_3)_5X]^{2+}$, $\Delta H^{\neq}$ value increases with the increase of the Co(III)—X bond strength because it experiences a dissociative (*d*) activation. In *fact,* $\Delta H^{\neq}$ *bears a linear relationship with* $\bar{\nu}_1$ (wave number in cm^{-1} for the first absorption band) *which is related with* $10Dq_o$ (average value). All these support the *dissociative activation* in the aquation of X^- of $[Co(NH_3)_5X]^{2+}$.

(c) Factors controlling $\Delta S^{\neq}$: The magnitude of $\Delta S^{\neq}$ depends mainly on two factors:

(i) change in the number of particles in forming the transition state from the reactants;

(ii) relative **electrorestriction** over the solvents by the reactants and transition state.

● In a **dissociative path,** in attaining the transition state, the number of particles increases to make **$\Delta S^{\neq}$ positive**. On the other hand, for an **associative path,** the number of particles decreases in producing the transition state to **make $\Delta S^{\neq}$ negative.** These straight-forward conclusions operate in the reactions where in attaining the transition state **there is no creation** or **annihilation of charge, so that electrorestriction over the solvent does not change significantly, in producing the transition state.** This may happen for the solvent exchange reactions, formation reactions where the metal bound solvent molecule is replaced by the *neutral ligands* (*e.g.* $[Ni(OH_2)_6]^{2+} + NH_3 \rightarrow$ product).

● When two reactants of same charge combine to generate the transition state, the T.S. will be more solvated than the reactants. Similarly, if a reactant dissociates (*e.g.* aquation $[Co(NH_3)_5X]^{2+}$) to attain the T.S. which is nothing but an ion pair, the T.S. will be more solvated. *Thus it leads to separation of charge in producing the T.S.* These are the examples of the processes **leading to creation of charges in attaining the T.S. Such processes tend to make $\Delta S^{\neq}$ negative because the T.S. is more solvated than the reactant.**

● When two reactants of opposite charges combine to generate the T.S., the **charge neutralisation** (*i.e.* **charge annihilation**) will slacken the eletrorestriction over the solvent molecules. **This will tend to make the $\Delta S^{\neq}$ positive.**

If there is any preequilibrium, the ΔS value of the preequilibrium step will also contribute to $\Delta S^{\neq}$.

(C) $\Delta S^{\neq}$ and diagnosis of reaction mechanism:

● **$\Delta S^{\neq}$ for the dissociative and associative process:** We conclude that ***after correction for the electrorestriction effect,*** a large negative value of $\Delta S^{\neq}$ is an indication of the associative mechanism, and a large positive value of $\Delta S^{\neq}$ is an indication of the dissociative mechanism.

● **Mechanism of aquation of $[M(NH_3)_5X]^{2+}$ and $\Delta S^{\neq}$:** In the equation of $[M(NH_3)_5X]^{2+}$ (X = uninegative anion), if the **dissociative process** goes on, then the departed X^- in the T.S. undergoes solvation. In such cases, $\Delta S^{\neq}$ bears a linear relationship with the ΔS values of solvation of X^-. For $X^- = Cl^-$, Br^- and I^-, the extent of hydration (*i.e.* solvation) varies as: $Cl^- > Br^- > I^-$, *i.e.* Cl^- is most extensively solvated, *i.e.* electrorestriction effect causes the **highest loss of entropy in the solvation of Cl^-**. Consequently ΔS for solvation of these halides follows the sequence: $I^- > Br^- > Cl$. In fact, $\Delta S^{\neq}$ values also follow the same sequence to support the dissociative mechanism and such mechanisms may be described as **solvent assisted dissociation** (SAD).

Aquation of $[Cr(NH_3)_5X]^{2+}$:

$X^- = Cl^-$	Br^-	I^-
$\Delta S^{\neq} = -9.0$	-6.0	-2

- **Dissociation of $[M(LL)_3]^{2+}$(M = Fe, Ni; Cr; LL = bpy, phen):** In these cases, importance of $\Delta S^{\neq}$ has been illustrated in Sec. 5.18.3.
- **Solvent exchange reaction and $\Delta S^{\neq}$:** For the **solvent exchange reactions** where there is no complication from the standpoint of charge creation or annihilation, the $\Delta S^{\neq}$ value is a reliable parameter to diagnose the reaction mechanism (*cf.* Table 5.12.1.1).
- **Ring closure rate determining step and $\Delta S^{\neq}$:** Ring closure causes the loss of entropy. In the chelation process, if the **ring closure** step contributes significantly to the rate determining step, *the* $\Delta S^{\neq}$ *becomes negative.*
- **Reliability of the $\Delta S^{\neq}$ values to diagnose a reaction mechanism:** It has been already mentioned that in the experimental procedure of evaluation of $\Delta S^{\neq}$, the uncertainty is relatively large (at least ± 2.5 J K^{-1} mol^{-1}). **This is why, too much reliance should not be put on the small values of $\Delta S^{\neq}$ to diagnose a reaction mechanism.**

(D) Effect of $\Delta H^{\neq}$ and $\Delta S^{\neq}$ on the rate constant:

More the positive value of $\Delta H^{\neq}$, more the rate retardation; more the positive value of $\Delta S^{\neq}$, more the rate acceleration.

(E) Isokinetic trend and reaction mechanism:

In comparable systems, different reactions passing through a **similar reaction mechanism,** a linear relationship (called **isokinetic trend**) between $\Delta H^{\neq}$ and $\Delta S^{\neq}$ is maintained. It may be mentioned that on the rate, ***$\Delta H^{\neq}$ and $\Delta S^{\neq}$ can balance each other's effect***. Thus similar rates may be obtained for different $\Delta H^{\neq}$ and $\Delta S^{\neq}$ values. The isokinetic trend is expressed as:

$$\Delta H^{\neq} = \alpha + \beta \Delta S^{\neq}$$

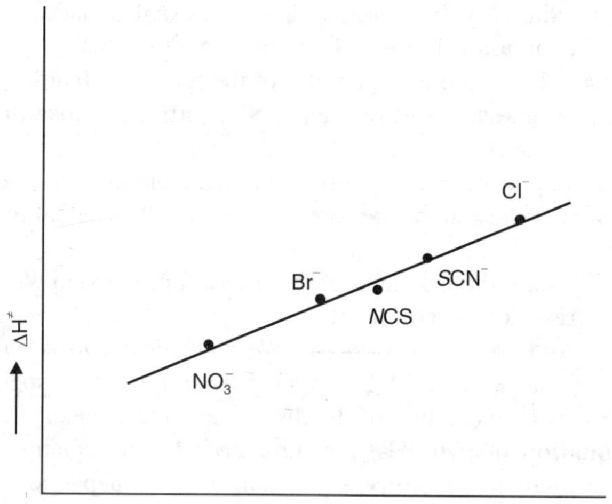

Fig. 5.19.3.3 Qualitative representation of the isokinetic plot for the anation of $[Cr(OH)(OH_2)_5]^{2+}$ by different ligands like Cl^-, Br^-, SCN^-, NO_3^-, etc. (*cf.* D. Thusius, *Inorg. Chem.*, **10**, 1106, 1971).

Here, α is actually $\Delta G^{\neq}$ (= $\Delta H^{\neq} - T\Delta S^{\neq}$) and β is called the **isokinetic temperature (= slope of the isokinetic plot)** at which all the related reactions occur at the same rate. Above this temperature, the rates are controlled by the $\Delta S^{\neq}$ values and below this temperature, the rates are controlled by the $\Delta H^{\neq}$ values. The isokinetic trend is also described as **compensation effect** where the effect of $\Delta H^{\neq}$ is balanced by $\Delta S^{\neq}$. It may be mentioned that the conclusion regarding the reaction mechanism from the isokinetic relationship has been criticised by may workers.

5.19.4 The Linear Free-Energy Relationship (LFER) and Leaving Group Effect and Reaction Mechanism

The kinetic and thermodynamic aspects of a reaction are distinctly different. Rate constant (k) is the kinetic parameter while equilibrium constant (K_{eq}) is the thermodynamic parameter.

$$\Delta G^{\neq} = -RT \ln\left(\frac{kh}{k_B T}\right), \ i.e.\ \log k = -\frac{\Delta G^{\neq}}{2.303RT} - \log\left(\frac{h}{k_B T}\right)$$

and,
$$\Delta G^0 = -RT \ln K_{eq}, \ i.e.\ \log K_{eq} = -\frac{\Delta G^0}{2.303\ RT}$$

However, in some cases $\Delta G^{\neq}$ and ΔG^0 (*i.e.* log k and log K_{eq}) bear a linear relationship which is called the **Linear free energy relationship** (LFER) having the form.

$$\Delta G^{\neq} = a\Delta G^0 + b, \text{ or } \log k = A \log K_{eq} + B \qquad \ldots(5.19.4.1)$$

To understand the origin of *LFER*, let us consider the aquation of $L_5M — X$, *i.e.*

$$L_5M — X + H_2O \underset{k_{-1}}{\overset{k_1}{\rightleftharpoons}} L_5M — OH_2 + X^-, \qquad \ldots(5.19.4.2)$$

Taking $K_{eq} = \frac{k_1}{k_{-1}}$, we can write:

$$\log k_1 = \log K_{eq} + \log k_{-1} \qquad \ldots(5.19.4.3)$$

If the aquation process (*i.e.* forward reaction of the above equilibrium Eqn. 5.19.4.2) goes on through a dissociative path, **then the activated complex and products resemble,** *i.e.* X^- is substantially dissociated in the transition state. Thus in the transition state, fate of X^- is almost comparable to that in the product. In fact, free X^- remains highly solvated in both the transition state and product.

If the anation reaction (*i.e.* backward reaction of the above equilibrium) passes through the *d*–activation (*i.e.* $M—OH_2$ bond breaking is the main contributing factor to the rate determining step), then the nature of X^- is less important to describe the nature of the transition state. Thus it is reasonable to conclude that the ***value of log k$_{-1}$ is approximately constant for the d–mechanism of anation at a particular metal centre***. Thus for the *d*–activation of both the forward (*i.e.* aquation) and backward (*i.e.* anation) processes of the above equilibrium, we can write (*cf.* Eqns. 5.19.4.1 and 5.19.4.3).

$$A = 1.0 \text{ and } B = \log k_{-1} = \text{constant.}$$

Then it gives a linear plot of log k *vs.* log K_{eq}.

Thus we can conclude that for the said aquation reactions, the variation of ΔG^0 values with the variation of X^- arises mainly from the variation of the Co(III)–X bond strength and solvation of X^-. These are quantitatively reflected in the same way in $\Delta G^{\neq}$. Thus LFER demonstrates the **leaving group effect** (*i.e.* **nucleofugality**).

The unit slope of the plot supports the **d**–activation. Some workers have proposed that the slope less than unity indicates the **a**-activation. However, this prediction of *a*-activation has been criticised by many workers.

● **LFER for aquation of $[Co(NH_3)_5X]^{2+}$ and $[Rh(NH_3)_5X]^{2+}$:** For aquation of $[Co(NH_3)X]^{2+}$ (X^- = monovalent anions like NCS^-, Cl^-, Br^-, I^-, NO_3^-, etc.), the linear plot of log k_{aq} *vs.* log K_{aq} with slope = 1 supports the **dissociative process.** For the bivalent anions (X^{2-}) like SO_4^{2-}, $C_2O_4^{2-}$, etc. the LFER line is parallel to the LFER line for the monovalent anions.

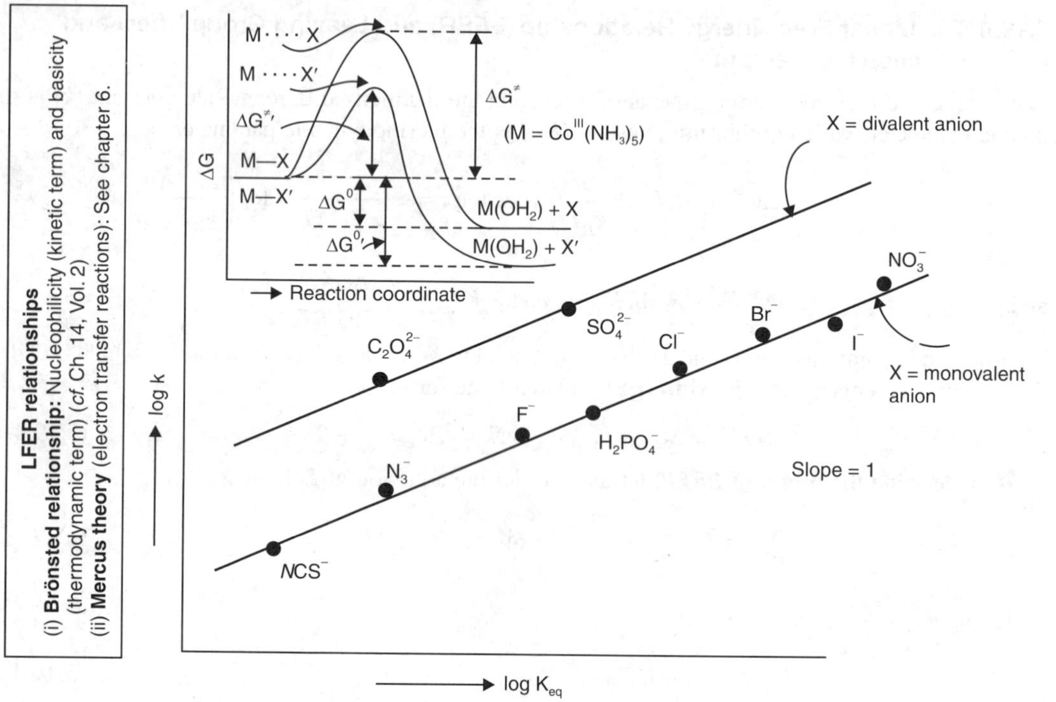

Fig. 5.19.4.1 Qualitative representation of LFER plot for aquation of $[Co(NH_3)_5X]^{n+}$, *i.e.* $[Co(NH_3)X]^{n+} + H_2O \rightarrow [Co(NH_3)_5(OH_2)]^{3+} + X^{-/2-}$. (*cf.* C.H. Langford, *Inorg. Chem.*, **4**, 265. 1965; T.W. Swaddle et.al, *Inorg. Chem.*, **8**, 1604, 2504, 1969; A. Haim, *Inorg. Chem.*, **9**, 426, 1970.)

For the aquation of $[Rh(NH_3)_5X]^{2+}$ complexes, the LFER line gives a slope slightly less than unity. It indicates some associative character in the rate process. Interestingly, the reactivity for the halides as the leaving groups is different for the Co(III) and Rh(III) centres.

$[Co(NH_3)_5X]^{2+}$: k_{aq}: $I^- \rangle Br^- \rangle Cl^-$; Bond strength: Co – Cl ⟩ Co – Br ⟩ Co – I

$[Rh(NH_3)_5X]^{2+}$: k_{aq}: $Cl^- \rangle Br^- \rangle I^-$; Bond strength: Rh – I ⟩ Rh – Br ⟩ Rh – Cl

Rh(III) being softer than Co(III), prefers the halides in the sequence: $I^- \rangle Br^- \rangle Cl^-$. It indicates that for the aquation of $[Rh(NH_3)_5X]^{2+}$, ***the bond breaking still contributes significantly.***

● **Aquation of $[Cr(OH_2)_5X]^{2+}$ vs. $[Cr(NH_3)_5X]^{2+}$ and LFER plots:** For the aquation of $[Cr(OH_2)_5X]^{2+}$, the slope of the LFER plot is only 0.56 while for the aquation of $[Cr(NH_3)_5X]^{2+}$, the corresponding slope is ca. 0.91. It indicates the more associative character in the aquation process of $[Cr(OH_2)_5X]^{2+}$. The more dissociative character for $[Cr(NH_3)_5X]^{2+}$ arises due to the presence of the better σ–donor ligands (*cf.* NH_3 is a better σ–donor ligand than H_2O) (*see* Sec. 5.19.6 for explanation in terms of energy of the LUMO).

● **Co(III) centre *vs.* Cr(III) centre:** From the slope of the LFER plots, we can have some idea regarding the extent of *d*– and *a*–activation. In the above mentioned aquation processes, the entering ligand, *i.e.* H_2O makes a stronger bond (*i.e.* I_a path) with the Cr(III) centre in the transition state (slope less than unity) while H_2O makes a very weak interaction with the Co(III) centre (*i.e.* I_d path) in the transition state.

In the associative process, the approaching nucleophile towards the t_{2g}-orbital experiences relatively less repulsion for the Cr(III)-complexes (t_{2g}^3) than for the Co(III)-complexes (t_{2g}^6). **Thus the associative path is relatively less favoured for the Co(III)-complexes in terms of the t_{2g}-electron density (t_{2g}^6 *vs.* t_{2g}^3).** In terms of MOT, the LUMO to receive the nucleophilic attack is either the partially filled NBMO, *i.e.* SOMOs (*i.e.* t_{2g}^3) or the vacant ABMO (*i.e.* e_g^*) for the Cr(III) complexes while for the Co(III)-complexes, the **high energy vacant ABMO** (*i.e.* e_g^*) **can only act as the LUMO to receive the nucleophilic attack** because the low energy NBMO (t_{2g}^6) is completely filled in. It indicates that the LUMO (*i.e.* e_g^*) of the Co(III)-complexes is of much higher energy than the LUMO (which may be the partially filled t_{2g}^3-orbitals, *i.e.* SOMOs or the vacant e_g^*) of the Cr(III)-complexes (*cf.* Sec. 5.19.6 for $[Cr(NH_3)_5X]^{2+}$ *vs.* $[Cr(OH_2)_5X]^{2+}$). Thus, in this line of argument (*i.e.* **energy of the LUMO**) also, the associative process is relatively less favoured for the low-spin Co(III) complexes (t_{2g}^6).

5.19.5 Effect of Pressure and Volume of Activation ($\Delta V^{\neq}$)

The pressure dependence of rate constant (k) can be described as follows:

$$\ln(k_2/k_1)_T = \frac{-\Delta V^{\neq}(P_2 - P_1)}{RT}, \; i.e., \; k \propto \exp\left(-\frac{P\Delta V^{\neq}}{RT}\right)$$

Rate dependence on pressure and $\Delta V^{\neq}$

We have the relationship: $\Delta G = -RT \ln K_{eq}$, **(thermodynamic parameters at standard condition, the zero superscript is dropped for the sake of simplicity)**

$$V = \left[\frac{\partial G}{\partial P}\right]_T \quad i.e. \quad \left(\frac{\partial \Delta G}{\partial P}\right)_T = \Delta V, \quad (P = \text{hydrostatic pressure})$$

It leads to: $$\left(\frac{\partial \ln K_{eq}}{\partial P}\right)_T = -\frac{\Delta V}{RT}$$

From the conventional transition state theory, we have:

i.e. $$k = \frac{k_B T}{h} K^{\neq} \quad i.e. \quad \ln k = \ln K^{\neq} + \ln\left(\frac{k_B T}{h}\right)$$

i.e. $$\left(\frac{\partial \ln k}{\partial P}\right)_T = \left(\frac{\partial \ln K^{\neq}}{\partial P}\right)_T, \quad \left(\frac{k_B T}{h} \text{ is a constant}\right)$$

The variation of equilibrium constant ($K^{\neq}$) with pressure is given by:

$$\left(\frac{\partial \ln K^{\neq}}{\partial P}\right)_T = -\frac{\Delta V^{\neq}}{RT}, \quad \Delta V^{\neq} = \text{volume of activation}$$

Thus we can write:

$$\left(\frac{\partial \ln k}{\partial P}\right)_T = \left(\frac{\partial \ln K^{\neq}}{\partial P}\right)_T = -\frac{\Delta V^{\neq}}{RT} \quad i.e. \quad \ln\left[\frac{(k_2)_{P_2}}{(k_1)_{P_1}}\right] = -\frac{\Delta V^{\neq}}{RT}(P_2 - P_1)$$

or, $\ln k = A - \frac{P\Delta V^{\neq}}{RT}$, *i.e.* $\log k = \frac{A}{2.303} - \frac{P\Delta V^{\neq}}{2.303RT}$, A = some constant.

where k_1 and k_2 denote the rate constants at pressure P_1 and P_2 respectively at constant temperature T and $\Delta V^{\neq}$ is the **volume of activation** measuring the molar volume change that occurs on going to the transition state from the reactants. The linear plot of ln k *vs.* P gives a straight line of slope = $-\Delta V^{\neq}/RT$.

(a) Factors controlling $\Delta V^{\neq}$: As in the **case of $\Delta S^{\neq}$**, the following two factors play the important roles to determine the $\Delta V^{\neq}$ values.

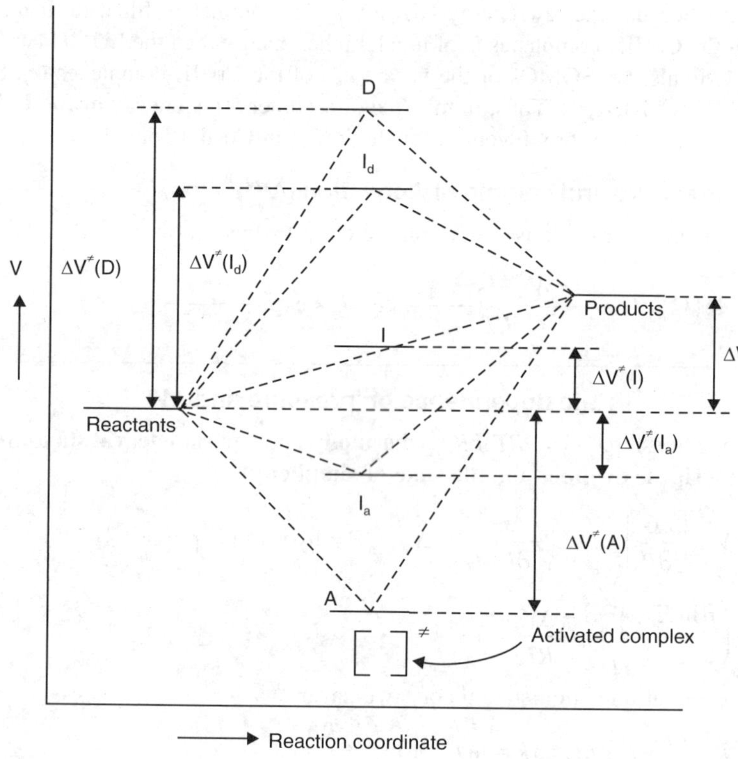

Fig. 5.19.5.1 Volume profile for different possible mechanisms of a substitution reaction (*e.g.* $[M(OH_2)_6]^{n+} + L \rightleftharpoons [M(OH_2)_5(L)]^{n+} + H_2O$) where the molar volume (V) of the products is larger than that of reactants, *i.e.* ΔV = molar volume of the reaction = +ve)

Note: Here the volume change due to outer-sphere complex formation is not shown. In fact, if the entering ligand is neutral, then the $\Delta V_{OS,}$ *i.e.* molar volume change due to outer-sphere complex formation is negligible. In the present diagram, along the reaction coordinate position of outer-sphere complex is not separately shown.'

(i) **Change in the number of free particles in attaining the transition state:** In a dissociative path, the leaving group is at least partially dislodged and released to the bulk solvent in attaining the transition state. It should make $\Delta V^{\neq}$ positive. On the other hand, in an associative path, the entering ligand formerly present in the bulk solvent gets coordinated to the metal centre. It should make $\Delta V^{\neq}$ negative. *These predictions are made without considering any other factor like change of electrorestriction over the solvent molecules.* Based on the above argument, *i.e.* consideration of ***intrinsic bond angles and bond strength,*** we may conclude as follows:

	D – path	I_d – path	I_a – path	*A* – path
$\Delta V^{\neq}$:	more positive	less positive	less negative	more negative

These are illustrated in Fig. 5.19.5.1.

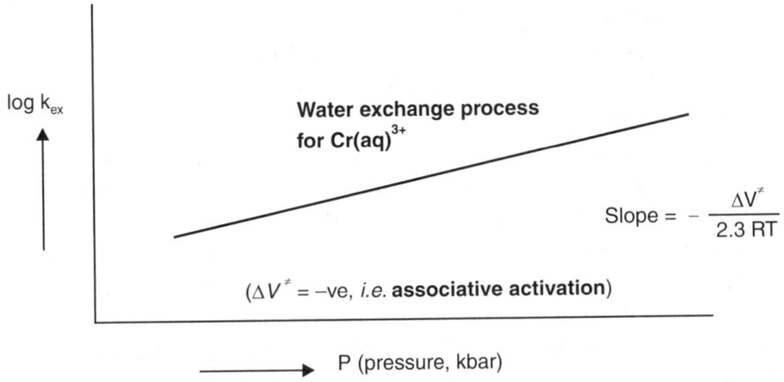

Fig. 5.19.5.2 Qualitative representation of the effect of pressure on k_{ex} for $[Cr(H_2O)_6]^{3+} + H_2O^* \rightarrow [Cr(OH_2)_5(^*OH_2)]^{3+} + H_2O$.

(ii) **Change of electrorestriction over the solvent in attaining the transition state:** In attaining the transition state, if charge separation or charge creation occurs (*as in the case of dissociative aquation of* $[M(NH_3)_5X]^{2+}$) then the T.S. exercises a more electrorestriction over the solvent than the starting reactants, *i.e.* the T.S. is more solvated than the reactants. In other words, in attaining the T.S., some relatively freer water molecules (*i.e.* bulk water molecules) are **immobilised in the hydration sphere of the T.S.** ***It may be noted that molar volume of bulk water is more than the molar volume of water present in the hydration sphere. It makes $\Delta V^{\neq}$ negative.*** In attaining the T.S., if charge neutralisation or charge annihilation occurs (*as in the case of associative anation of* $[M(OH_2)_6]^{3+}$), the reactants exercise a more electrorestriction than the T.S. over the solvent, *i.e.* the reactants are more solvated than the T.S. It makes $\Delta V^{\neq}$ positive.

Thus $\Delta V^{\neq}$ is arising from two factors:

$$\Delta V^{\neq} = \Delta V^{\neq}_{\text{intrinsic}} + \Delta V^{\neq}_{\text{solvent}}$$

Table 5.19.5.1 Rate constants and activation parameters for aquation of *L* of $[M(NH_3)_5L]^{3+}$.

L		$10^5 k_{aq}$ (s^{-1})	$\Delta S^{\neq}$ ($J\,K^{-1}\,mol^{-1}$)	$\Delta V^{\neq}$ ($cm^3\,mol^{-1}$)	
M = Cr	H_2O^*	5.2	0	–5.8	associative path
	$OSMe_2$	1.95	–15.1	–3.2	
	$OCHNH_2$	5.1	–12.2	–4.8	
	$OC(NH_2)_2$	2.0	–22.3	–8.2	
M = Co	H_2O^*	0.59	+28.1	+1.2	dissociative path
	$OSMe_2$	1.8	+10.1	+2.0	
	$OCHNH_2$	0.58	+12.2	+1.1	
	$OC(NH_2)_2$	5.5	–10.1	+1.3	

(*cf.* N.J. Curtis, *et al*, *Inorg. Chem.*, **28,** 329, 1989) * water exchange process

(b) **Cases where the electrorestriction factor** (*i.e.* $\Delta V^{\neq}_{solvent}$) **absent:** If $\Delta V^{\neq}_{solvent}$ (*i.e.* electrorestriction factor) remains absent as in the case of ***solvent exchange reaction, aquation of neutral ligand, formation of complexes by the neutral ligands***, then $\Delta V^{\neq}$ ($\approx \Delta V^{\neq}_{intrinsic}$) parameter appears as a good parameter to diagnose the reaction mechanism (*i.e.* ***+ve value of*** $\Delta V^{\neq} \Rightarrow$ ***d–character; –ve value of*** $\Delta V^{\neq} \Rightarrow$ ***a–character***). These are illustrated in the following cases:

(i) **Solvent exchange reaction:** Diagnosis of the mechanism of solvent exchange reaction in terms of $\Delta V^{\neq}$ (Table 5.12.1.1) has been discussed in detail in Sec. 5.12.1.

(ii) **Aquation of neutral ligands (*L*) in** $[M(NH_3)_5L]^{3+}$ (*cf.* N.J. Curtis, G.A. Lawrance, and R. van Eldik, *Inorg. Chem.*, **28**, 329, 1989): The $\Delta V^{\neq}$ values are given in Table 5.19.5.1.

From the $\Delta V^{\neq}$ values for aquation of uncharged ligands (*L*), it appears that the Co(III)–complexes participate in ***d*–activation** while the corresponding Cr(III)–complexes participate in **a–activation.**

(c) **Cases where the electrorestriction factor (*i.e.*** $\Delta V^{\neq}_{solvent}$**) present:** If $\Delta V^{\neq}_{solvent}$ (*i.e.* electrorestriction factor) makes a significant contribution to $\Delta V^{\neq}$, then the diagnosis of reaction mechanism from $\Delta V^{\neq}$ is not so straight-forward. It is illustrated in the following examples.

(i) **For the aquation of anionic ligand (*X*) in** $[Co(NH_3)_5X]^{x+}$**:** In $[Co(NH_3)_5(SO_4)]^+$ (*i.e.* $X^{2-} = SO_4^{2-}$) and $[Co(NH_3)_5Cl]^{2+}$, the $\Delta V^{\neq}$ values are –18.3 and –10.6 cm³ mol⁻¹ respectively. *But there are other evidences that the processes pass through the dissociative activation.* It may be mentioned that the $\Delta V^{\neq}$ value for the water exchange process in $[Co(NH_3)_5(OH_2)]^{3+}$ is +1.2 $cm^3\ mol^{-1}$ and it is an indication of the *dissociative character* in the water exchange process. In fact, in the dissociative pathway of aquation of the above Co(III)–ammine complexes, the charge creation in the T.S. makes a powerful electrorestriction over the solvent. ***Thus, in such cases*** $\Delta V^{\neq}_{solvent}$ ***dominates over*** $\Delta V^{\neq}_{intrinsic}$ ***to make*** $\Delta V^{\neq}$ ***negative.***

(ii) **Aquation of** $[M(NH_3)_5X]^{2+}$ **(M = Co, Cr):** Aquation of $[M(NH_3)_5X]^{2+}$ (M = Co, Cr) experiences –ve value of $\Delta V^{\neq}$ in spite of the dissociative mechanism due to the more solvated nature of the T.S., a consequence of charge separation.

	$[Co(NH_3)_5X]^{2+}$			$[Cr(NH_3)_5X]^{2+}$		
X^- :	Cl^-	Br^-	NCS^-	Cl^-	Br^-	NCS^-
$\Delta V^{\neq}$: ($cm^3\ mol^{-1}$)	–10.6	–9.2	–4.0	–10.8	–10.2	–8.6

Relation between $\Delta V^{\neq}$ **and** $\Delta S^{\neq}$

From our discussions, it has been pointed out that the same factors contribute to the both parameters. ***A negative*** $\Delta V^{\neq}$ ***value corresponds to a negative*** $\Delta S^{\neq}$ ***and a positive*** $\Delta V^{\neq}$ ***value corresponds to a positive*** $\Delta S^{\neq}$ ***value. In fact, a good correlation between the*** $\Delta V^{\neq}$ ***and*** $\Delta S^{\neq}$ ***values prevails for the reactions where the electrorestriction effects are important.*** But the uncertainty in the estimated $\Delta S^{\neq}$ value is relatively more. Thus $\Delta V^{\neq}$ (obtained from the slope of the plot, ln*k vs. P*, *cf.* Fig. 5.19.5.2) appears a better parameter, specially when there is no complication to diagnose the reaction mechanism. But determination of $\Delta S^{\neq}$ is very easy (*i.e.* temperature dependence of the rate constant) while determination of $\Delta V^{\neq}$ needs a very costly equipment which can allow us to follow the pressure dependence of rate constant. A quantitative relationship between $\Delta V^{\neq}$ and $\Delta S^{\neq}$ will help us to estimate $\Delta V^{\neq}$ from the $\Delta S^{\neq}$ values. A relationship has been found (*cf.* M.V. Twigg, *Inorg. Chim. Acta.*, **24**, L84, 1977) for the aquation and water exchange processes of the complexes of d^3 and d^6 (low spin) of the type $[M(NH_3)_5X]^{n+}$ (X = H_2O, halides and other anions).

$$\Delta V^{\neq} = (1.04 \pm 0.10)\Delta S^{\neq} - (4.4 \pm 0.3); \ (\Delta V^{\neq} \text{ in cm}^3\text{ mol}^{-1}, \Delta S^{\neq} \text{ in cal K}^{-1}\text{ mol}^{-1})$$

It may be noted that ΔV of the aquation reaction is also negative.

$$[M(NH_3)_5X]^{2+} + H_2O \rightarrow [M(NH_3)_5(OH_2)]^{3+} + X^-, \Delta V = -ve.$$

The –ve value of ΔV supports the –ve of $\Delta V^{\neq}$, even if the process passes through the dissociative path.

(iii) **Aquation of $[RuCl(NH_3)_5]^{2+}$ and anation (*i.e.* reverse reaction) of $[Ru(NH_3)_5(OH_2)]^{3+}$:**

$$\left[RuCl(NH_3)_5\right]^{2+} \xrightarrow{\text{(aquation)}} \left[Ru(NH_3)_5(OH_2)\right]^{3+}, \Delta V^{\neq} = -30 \text{ cm}^3 \text{ mol}^{-1}$$

$$\left[Ru(NH_3)_5(OH_2)\right]^{3+} + Cl^- \xrightarrow{\text{(anation)}} \left[RuCl(NH_3)_5\right]^{2+}, \Delta V^{\neq} = -20 \text{ cm}^3 \text{ mol}^{-1}$$

The $\Delta V^{\neq}$ values for the above mentioned aquation and anation reactions are in **conformity with the associative path.** Aquation through the associative path does not lead to any charge creation or charge annihilation in attaining the T.S. Thus the highly –ve value of $\Delta V^{\neq}$ for aquation is a strong evidence in favour of the associative path. Aquation through the dissociative path will generate two opposing contributions of $\Delta V_{\text{intrinsic}}$ (+ve) and $\Delta V^{\neq}_{\text{solvent}}$ (-ve due to the charge creation at the T.S.). Thus the resultant $\Delta V^{\neq}$ can never be so highly negative.

$\Delta V^{\neq}$ for anation through the dissociative path will not get any contribution from $\Delta V^{\neq}_{\text{solvent}}$ (*i.e.* electrorestriction factor) and $\Delta V^{\neq}$ will be positive as usual. Anation through the associative path will tend to make $\Delta V^{\neq}$ positive due to charge neutralisation at the transition state. ***Thus the –ve value of $\Delta V^{\neq}$ is a convincing evidence in favour of the associative path.*** The *relatively less negative value for the anation* reaction compared to that for the aquation reaction is reasonable in terms of relaxation of electrorestriction in generating the associative transition state, *i.e.* positive contribution of $\Delta V^{\neq}_{\text{solvent}}$ (electrorestriction factor) to $\Delta V^{\neq}$ is opposed by $\Delta V^{\neq}_{\text{intrinsic}}$ which makes a negative contribution in an associative path. Here $\Delta V^{\neq}_{\text{intrinsic}}$ predominates over $\Delta V^{\neq}_{\text{solvent}}$.

5.19.6 Effect of the Nature of Leaving and Entering Group

(a) **Dissociative activation:** For the **dissociative activation mechanism,** the nature of the **nucleofugal,** *i.e.* leaving group (X) is very much important because it needs the rupture of the M—X bond to attain the transition state. Then the entering group (Y) enters into the coordination sphere in a fast step. Thus the dissociative process is more or less independent (under the limiting condition; *cf.* Secs. 5.9.5, 5.16) of the nature of entering ligand. This aspect is illustrated in some cases.

- **Aquation of $[Co(NH_3)_5L]^{n+}$:** The aquation rates for $[Co(NH_3)_5(L)]^{n+}$ (L = NO_3^-, I^-, Cl^-, SO_4^{2-}, F^-, N_3^-, NCS^-, etc.) cover a huge range *ca.* 10^{-5} to 10^{-10} s^{-1} (*cf.* Table 5.19.6.1) *i.e.* the aquation rate is highly sensitive to the nature of the leaving group (L). This suggest the **dissociative activation.**

Table 5.19.6.1 Aquation rates of $[Co(NH_3)_5(L)]^{n+}$ complexes (at 25°C) (*cf.* M.L. Tobe, *Adv. Inorg. Bionorg. Mech.*, **2**, 1, 1984).

L	k_{aq} (s^{-1})	*L*	k_{aq} (s^{-1})
NO_3^-	2.4×10^{-5}	SO_4^{2-}	8.9×10^{-7}
I^-	8.3×10^{-6}	F^-	8.6×10^{-8}
Cl^-	1.7×10^{-6}	N_3^-	2.1×10^{-9}
Br^-	6.3×10^{-6}	NCS^-	3.7×10^{-10}
		$H_2PO_4^-$	2.6×10^{-7}

$k_{ex}(H_2O)$ for $[Co(NH_3)_5(OH_2)]^{2+} = 5.8 \times 10^{-6}$ s^{-1}

Rate dependence on the nature of entering ligand (Y) in d-activation

Here it should be cautioned that the rate is independent of the entering ligand in the *d*-activation process under the *limiting conditions only* [*i.e.* $k_2[Y] \gg k_{-1}[X]$ for the *D*-process, $K_{OS}[Y] \gg 1$ for the I_d -process, *cf.* Secs. 5.9.5, 5.16).

The dissociative activation for the aquation of $[Co(NH_3)_5X]^{2+}$ ($X^- = NO_3^-$, halides, etc.) has also been supported by an almost constant value (~105 kJ mol^{-1}; Sec. 5.19.3b) of ΔH_T (*enthalpy* of transition). ΔH_T gives the energy difference between the transition state and product.

● **Anation of $[Co(NH_3)_5(OH_2)]^{3+}$:** Anation rates for $[Co(NH_3)_5(OH_2)]^{3+}$ for $Y^{n-} = N_3^-$, SO_4^{2-}, Cl^-, NCS^- cover a relatively a narrow range (1.6 – 10.0) × 10^{-5} s^{-1} (*cf.* Table 5.19.6.2)

Table 5.19.6.2 Limiting rate constants (k_a) for anation of $[Co(NH_3)_5(OH_2)]^{3+}$ by Y^{n-} (at 45°C).

Y^{n-}	k_a (s^{-1})	k_{ex}/k_a	Y^{n-}	k_a (s^{-1})	k_{ex}/k_a
N_3^-	10^{-4}	1.0	Cl^-	2.1×10^{-5}	4.76
SO_4^{2-}	2.4×10^{-5}	4.2	NCS^-	1.6×10^{-5}	6.25

k_{ex} (45°) ≈ 10^{-4} s^{-1}

The almost *insensitivity of anation rate constants* (*cf.* Table 5.19.6.2) towards the nature of entering ligand indicates the dissociative activation. It is again further supported by the fact that the *anation rate is almost close to the water exchange rate* (*cf.* k_{ex}/k_a varies in the range 1–6).

● **Substitution at $[M(OH_2)_6]^{2+}$ (M = late members of the 1st transition series, *e.g.* Co, Ni):** In the formation of $[M(H_2O)_5Y]^{n+}$ from $[M(OH_2)_6]^{2+}$ (M = Co, Ni), the second order formation rate constant is also fairly insensitive for different types of entering ligand *Y* (both charged and anionic) provided the K_{OS} values do not differ significantly. This supports the dissociative activation.

Dependence on the nature of the entering ligand in D, I_d and I_a process (*cf.* Sec. 5.16): In an I_d path, the formation rate constant does not show any strong dependence on the nature of the entering ligand provided the K_{OS} values for the different entering ligands do not vary significantly (*cf.* Secs. 5.9.3, 5.14, 5.16). ***In a perfect D process, there is a slight dependence on the nature of the entering ligand because for the long-lived 5-coordinate intermediate there will be a competition between the solvent and the entering ligand.*** However, it differs from the I_a process. This aspect has been discussed in detail in Sec. 5.16.

(b) **Associative activation:** For the **associative activation mechanism,** the rate constant is highly sensitive to the nature of the entering nucleophile. This is illustrated in the following cases.

● **Substitution at $[Ti(OH_2)_6]^{3+}$:** The rate of ligand substitution process at $[Ti(OH_2)_6]^{3+}$ is very much sensitive to the nature of the entering ligand (*cf.* Table 5.19.6.3). This is an indication of the associative character.

● **Substitution at $[V(OH_2)_6]^{2+}$:** $[V(OH_2)_6]^{2+}$ (d^3) participates in **an associative path** in the anation process. It is evidenced from the rate dependence on the nature of the entering ligand. The second order rate constant (k_f in mol^{-1}dm^3 s^{-1}) for the different anions follows the following sequence:

$$k_{f(Cl^-)} : k_{f(NCS)^-} : k_{f(N_3^-)} = 1 : 2 : 10$$

Table 5.19.6.3 Rate constants for anation of $[Ti(OH_2)_6]^{3+}$ (at 13°C).

Y^{n-}	k ($dm^3\ mol^{-1}\ s^{-1}$)
NCS^-	8×10^3 (at ca. 9°C)
$CH_3CO_2^-$	1.8×10^6
$ClCH_2CO_2^-$	2.1×10^5

k_{ex} (H_2O) = $4.8 \times 10^5\ s^{-1}$, *i.e.* k_{ex} (in $dm^3\ mol^{-1}\ s^{-1}$) = $4.8 \times 10^5/55.5 = 8.7 \times 10^3$; *cf.* H. Diebler, *Z. Phys. Chem.*, **68**, 64, 1969.

In terms of interchange associative path (*i.e.* I_a-path), under the condition $1 >> K_{OS}[Y]$, we have k_f (second order formation rate constant) = $k_i K_{OS}$ (*cf.* Sec. 5.9.3). Here the entering (Y^-) ligands are monovalent anions and they are of comparable size. Thus it is reasonable to conclude that K_{OS} (*i.e.* ion pair association constant) remains more or less the same for Cl^-, NCS^- and N_3^-. Thus the difference in k_f arises from the difference in k_i which depends on the nature of the entering nucleophile.

● **Anation of $[Cr(OH_2)_6]^{3+}$ *vs.* $[Cr(NH_3)_5(OH_2)]^{3+}$:** Anation of $[Cr(OH_2)_6]^{3+}$ passes through the associative process and the rate constant is highly sensitive to the nature of the entering ligand (Y). But anation of $[Cr(NH_3)_5(OH_2)]^{3+}$ passes through the dissociative process. For the anation of $[Cr(OH_2)_6]^{3+}$ by the anions like Cl^-, Br^-, I^-, NO_3^-, NCS^- and SCN^-, the formation rate constants (k_f) cover a wide range, 7.3×10^{-7} to $8 \times 10^{-10}\ dm^3\ mol^{-1}\ s^{-1}$. On the other hand, for the anation of $[Cr(NH_3)_5(OH_2)]^{3+}$, the rate constants (k_f) cover relatively a narrower range $(0.7 - 4.2) \times 10^{-4}\ dm^3\ mol^{-1}\ s^{-1}$ for the anions like Br^-, Cl^-, NCS^-.

Table 5.19.6.4 Rate constants for the anation of $[Cr(OH_2)_6]^{3+}$ and $[Cr(NH_3)_5(OH_2)]^{3+}$

Y^{n-}	$[Cr(OH_2)_6]^{3+}$ k ($dm^3\ mol^{-1}\ s^{-1}$)		$[Cr(NH_3)_5(OH_2)]^{3+}$ k ($dm^3\ mol^{-1}\ s^{-1}$)	
Cl^-	3×10^{-8}	Associative character	0.7×10^{-4}	Dissociative character
Br^-	1×10^{-8}		3.7×10^{-4}	
NCS^-	1.8×10^{-6}		4.2×10^{-4}	

It is evident that not only the mechanisms of ligand substitution process at $[Cr(OH_2)_6]^{3+}$ and $[Cr(NH_3)_5(OH_2)]^{3+}$ differ but their reactivities also differ significantly. The reactivity of the ammine complex is increased by a factor about 10^4. The NH_3 ligands are much better σ–donor ligands than the H_2O ligands. This **better σ–donor properties of the NH_3 ligands** labilise the sixth aqua ligand through the dissociative process. These good σ-donor ligands reduces the positive charge on the metal centre to favour the bond dissociation at the rds. The better σ–donor NH_3 ligands can also stabilise the activated complex of lower coordination number while the aqua complex requires the **nucleophilic assistance** to stabilise the activated complex.

● **Possibility of the nucleophilic attack on Cr(III) depending on the energy of vacant orbital (*i.e.* LUMO):** In the associative path, the incoming nucleophile will occupy the LUMO which may be the **nonbonding t_{2g} orbitals** or the e_g^* orbitals depending on the d^n-configuration of the metal centre. For the $d^{0,1,2}$ systems, the NBMO (*i.e.* t_{2g} orbitals) may act as the LUMO to receive the nucleophilic attack. For the d^3 system (t_{2g}^3), the NBMOs can act as the **SOMOs** (singly occupied MOs) and they may receive the nucleophilic attack through pairing of the electron in the t_{2g}-level. This is an energy requiring process. To avoid this, the incoming nucleophile (in an associative path) may occupy the empty e_g^* orbital (in terms of MOT) of Cr(III). In the **aqua complex**, because of the weak σ–donor property of the aqua ligands, the empty e_g^* orbital is not raised much (*cf.* Fig. 3.17.4.1). ***Thus, this low lying*** e_g^* ***orbital may be occupied by the incoming nucleophile in the associative path.*** In the corresponding ammine complex, energy of the e_g^* orbital is raised higher because of the better σ–donor property of the NH_3 ligands. Thus in the ammine complex, the associative path becomes relatively more disfavoured.

● **Formation reactions at $[M(OH_2)_6]^{3+}$ and its hydrolysed product, *i.e.* $[M(OH)(OH_2)_5]^{2+}$ (M = Fe, Cr):** The unhydrolysed species having a higher positive charge adopt the I_a-path while the hydrolysed species adopt the I_d-path because of the factors: ***lower positive charge favouring the d-activation and disfavouring the a-activation; good σ-donor and π-donor property of the OH group labilising the coordinated water molecules through the d-activation.***

Difference in the pathways of reactions of the Co(III) (low spin) and Cr(III) Complexes (*cf.* Secs. 5.19.4, 5.21-22, 24, 5.25.3)

- Generally, the Co(III) complexes adopt the dissociative path like **I_d path** for their ligand substitution reactions.
- The Cr(III) complexes generally adopt the associative path like the **I_a path** for their ligand substitution reactions like aquation of $[Cr(OH_2)_5X]^{2+}$, formation reactions at $[Cr(OH_2)_6]^{3+}$, solvent exchange reactions at Cr(III), etc. However, $[Cr(OH)(OH_2)_5]^{2+}$ adopts the **I_d–path** (due to the ***cis*–effect**, *i.e.* labilising effect of the good σ- and π-donor property of the OH group); anation of $[Cr(NH_3)_5(OH_2)]^{3+}$ and aquation of $[Cr(NH_3)_5(X)]^{2+}$ adopt also the **I_d-path.** The reason behind the changeover from the I_a path to the I_d path has been explained already (*cf.* Sec. 5.19.6).
- The six electrons (*i.e.* t_{2g}^6) in the Co(III)–complexes offer a more repulsion to the approaching nucleophile in the associative path compared to the Cr(III) complexes bearing three electrons (*i.e.* t_{2g}^3).
- The **vacant antibonding orbital (*i.e.* LUMO) to accommodate the incoming nucleophile** in the associative path is energetically raised more in the Co(III) (low-spin) complexes compared to that in the Cr(III)–complexes. For the Cr(III) complexes, even the partially filled NBMO (*i.e.* t_{2g}^3) can act as the LUMO to receive the nucleophilic attack through the rearrangement of electrons in the t_{2g}-level. Thus the associative path is less favoured for the Co(III) complexes compared to that for the Cr(III) complexes (*cf.* Sec. 5.19.6).
- In terms of **CFAE**, the Co(III) (low spin) complexes are more inert than the Cr(III) complexes.

5.19.7 Steric Effects of the Spectator (Nonleaving) Ligands

Steric acceleration and steric retardation: If the starting complex is sterically crowded then in the dissociative process, the rate experience a **steric acceleration** while in the associative process the rate experiences a **steric retardation.** In a dissociative process, the sterically crowded complex experiences **a steric relaxation** in attaining the activated complex through the dissociation of the leaving group. On the other hand, in an associative process, in the activated complex, the **steric crowding is enhanced** to disfavour the process. These aspects are illustrated in the following cases.

(i) **Amino acid exchange reaction at the square planar complexes of Cu(II):** The amino acid exchange rate at the square planar complexes of Cu(II), decreases with the increase of steric crowding (*i.e.* **steric retardation**). The rate sequence for the amino acids is:

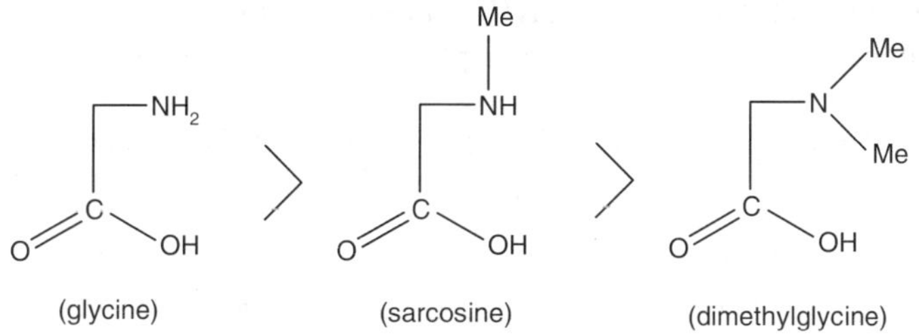

The above rate sequence supports the **associative activation** in the ligand exchange process of the amino acids at the square planar complexes of Cu(II).

(ii) **Amino acid exchange reaction at the octahedral complexes of Co(II) and Ni(II):** On the other hand, for the octahedral complexes of Co(II) and Ni(II), the opposite rate sequence, *i.e.* dimethyl glycine ⟩ sarcosine (*i.e.* N-methyl glycine) ⟩ glycine supports the **dissociative activation** experiencing the **steric acceleration.**

(iii) **Steric acceleration in the aquation of Co(III)-amine complexes:** The aquation reactions of *trans*–$[CoCl_2(N — N)_2]^+$ (where N — N denotes the didentate ethylene diamine ligand bearing different substituents) experience a **steric acceleration** (*cf.* Sec. 5.21, Table 5.21.1).

trans–$[CoCl_2(N - N)_2]^+ + H_2O \rightarrow [CoCl(N - N)_2(OH_2)]^{2+} + Cl^-$

Substitution on the ethylenediamine skeleton may occur on both the 'N' and 'C' site but *the rate is more sensitive to the substitution on N.* The steric acceleration is in conformity with the dissociative activation.

(iv) **Aquation of *trans*-$[CoCl_2(N_4)]^+$ and *trans*-$[CoCl(N_4)X]^{n+}$:** The aquation rates (for the loss of Cl^-) of *trans*–$[CoCl(N_4)(X)]^{n+}$ and *trans*–$[CoCl_2(N_4)]^+$ increase from *cyclam* (N_4 = cyclam, *i.e.* tetraazacyclotetradecane) complex to *tet–b* (N_4 = hexamethyl derivative of cyclam) complex. This *steric acceleration (cf.* Sec. 5.21, Table 5.21.2) is in conformity with the dissociative activation.

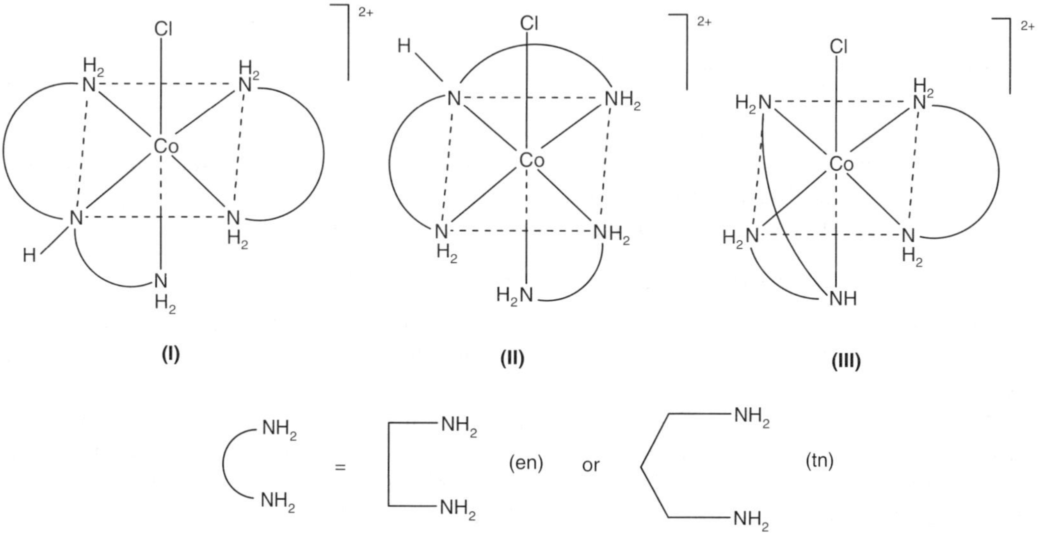

Fig. 5.19.7.1 Three geometrical isomers of $[CoCl(dien)(N — N)]^{2+}$ where N — N = en or tn.

(v) **Aquation reactions of Co(III)- *vs.* Cr(III)-ammine complexes:** Steric effects on the aquation rates of Co(III)–ammine and Cr(III)–aminine complexes are of different types.

	$[CoCl(NH_3)_5]^{2+}$	$[CoCl(MeNH_2)_5]^{2+}$	$[CrCl(NH_3)_5]^{2+}$	$[CoCl(MeNH_2)_5]^{2+}$
$k_{aq}(s^{-1})$ (50°C)	3.6×10^{-5}	6.5×10^{-4}	1.8×10^{-4}	1.4×10^{-5}
	(Steric acceleration)		(Steric retardation)	

It indicates that the Co(III)–complexes experiences the **dissociative activation** while the corresponding Cr(III)–complexes experience the **associative activation.**

(vi) **Aquation of [MCl(dien)(en)]$^{2+}$ and [MCl(dien)(tn)]$^{2+}$ (M = Co *vs.* Cr):** [CoCl(dien)(tn)]$^{2+}$ and [CoCl(dien)(en)]$^{2+}$ complexes can produce three geometrical isomers (Fig. 5.19.7.1).

The *tn–complex aquates much faster than the corresponding en–complex* (*e.g.* for the **II** isomer, $k_{tn}/k_{en} \approx 32$ at 32°C). **Interestingly, for the corresponding Cr(III)–complexes, both the tn– and en–complexes react at comparable rates.**

In the dissociative path of aquation, the five coordinate activated complex adopts the trigonal bipyramidal geometry where the diamine (N — N) ring is to be stretched considerably from 90° to 120° (for the isomer **II**) in going from the octahedral to the trigonal bipyramidal geometry. *This angle expansion is more constrained for the 5–membered diamine ring (en) than for the 6–membered diamine (tn) ring.*

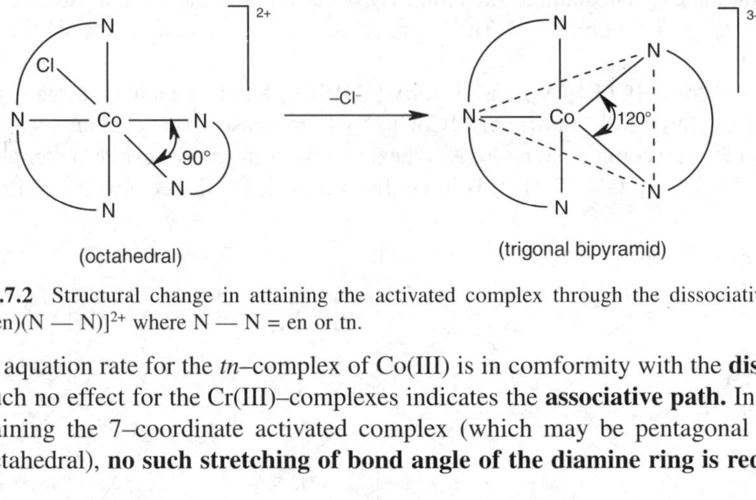

Fig. 5.19.7.2 Structural change in attaining the activated complex through the dissociative path for [CoCl(dien)(N — N)]$^{2+}$ where N — N = en or tn.

The higher aquation rate for the *tn*–complex of Co(III) is in comformity with the **dissociative path.** Absence of such no effect for the Cr(III)–complexes indicates the **associative path.** In the **associative path**, for attaining the 7–coordinate activated complex (which may be pentagonal bipyramidal or trapezoidal octahedral), **no such stretching of bond angle of the diamine ring is required.**

5.19.8 Electronic Effects of the Spectator or Nonleaving Group

Electronic effects (*i.e.* electron pushing and withdrawing) exerted by the nonleaving groups can control the reaction rate and mechanism of the ligand substitution process. This aspect is illustrated in the following cases.

- **Base hydrolysis of the Co(III)-ammine complexes:** It passes through the **D—CB or I_d—CB mechanism** in which the CB (generated at the pre-equilibrium step) experiences the **dissociative activation by the labilising effect of** the NH_2 group (which is a good σ- and π-donor group). The different aspects of this D–CB mechanism have been discussed in Sec. 5.25.

- **Aquation of [CoCl(en)$_2$(L)]$^{n+}$** (*cf.* Sec. 5.21): If the nonleaving π–donor L–group occupies the *cis*–position then the aquation rate is faster than the corresponding *trans*–complex. The higher reactivity of the *cis*–isomer arises through the ***cis*–effect** by the nonleaving L-group. It indicates the **dissociative pathway.** In fact, to stabilise the 5–coordinate intermediate by the nonleaving π–donor ligand (L), the corresponding square pyramid for the *cis*–complex is in the right configuration but the square pyramid for the corresponding *trans*–complex is not in the right configuration. In fact, the square pyramid obtained from the *trans*–complex is to be changed into the trigonal bipyramidal geometry *so that the vacant orbital (available due to the departure of the leaving group) of the metal centre can participate in π-bonding with the nonleaving group (L) (cf. Fig. 5.21.1).* It raises the activation energy.

The *cis*–complexes having the π–donor nonleaving ligand (L) react faster than the complexes where L is a π–acceptor ligand (*e.g.* CN^-) or a non π–bonding ligand like NH_3. It can be explained by the dissociative mechanism.

If L is a *π–withdrawing group (e.g. NO_2^-, CN^-, etc.) then the trans–isomer* is more reactive than the *cis*–isomer. It can be explained by considering the **associative mechanism.**

Effect of L (nonleaving group) on the aquation of $[CoCl(en)_2L]^{n+}$

Thus in the aquation of $[CoCl(en)_2L]^{n+}$, if the nonleaving group **L is a π donor** ligand, then the process passes through the **dissociative path** and the *cis*–complex reacts faster than the *trans*–isomer. On the other hand, if **L is a π acceptor ligand,** then it adopts the **associative path** where the *trans*–isomer reacts faster. These aspects have been discussed in detail in Sec. 5.21.

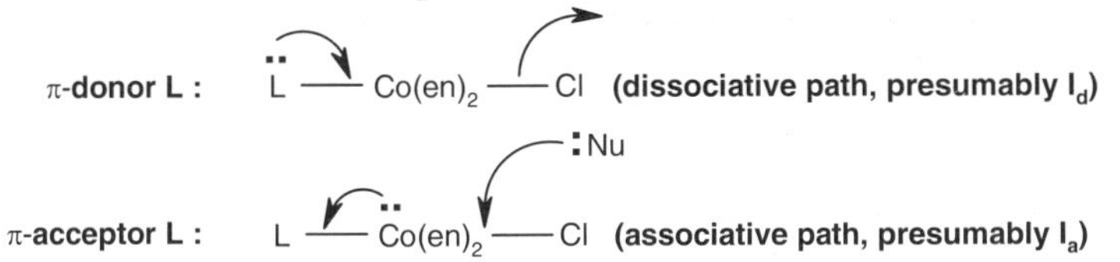

● **Reactivity of $[M(OH_2)_6]^{3+}$ *vs.* $[M(OH)(OH_2)_5]^{2+}$ (M = Fe, Cr, *cf.* Sec. 5.19.6):** The higher reactivity and preferred I_d mechanism for $[M(OH)(OH_2)_5]^{2+}$ while the lower reactivity and preferred I_a mechanism for $[M(OH_2)_6]^{3+}$ can be explained by considering the electronic effect of the nonleaving OH group. In fact, the nonleaving π–donor group OH shows the ***cis*–effect** to impart the dissociative path. Here it may be mentioned that H_2O is a poor π–donor ligand. It may be mentioned that the better σ-donor property of OH (compared to that of OH_2) also labilises the coordinated H_2O molecule and imparts the dissociative activation.

● **Reactivity of $[Cr(OH_2)_6]^{3+}$ *vs.* $[Cr(NH_3)_5(OH_2)]^{3+}$ (*cf.* Sec. 5.19.6):** $[Cr(OH_2)_6]^{3+}$ adopts the I_a path while $[Cr(NH_3)_5(OH_2)]^{3+}$ adopts the I_d path. It is due to the better σ–donor property of NH_3 compared to that of H_2O.

5.19.9 Comparison between the Rate of Formation Reaction (*i.e.* Replacement of Metal Bound Water) and Water Exchange Reaction (*cf.* Sec. 5.12.2)

Comparison of the formation rate constants with the water exchange rate constants (k_{ex}) is helpful in many cases to diagonse the reaction pathway. This aspect has been illustrated in Sec. 5.12.2 and some aspects are discussed below.

(a) **Interchange (I) process (Secs. 5.9.3; 5.14) :** It has been already pointed out that compared to the perfect *D* and *A* processes, the interchange processes (*i.e.* I_d and I_a) are more well documented in the ligand substitution reactions of octahedral complexes.

Under the limiting condition, $K_{OS}[Y] << 1$, we have $k_f = K_{OS}\, k_i$ (*cf.* Secs. 5.9.3, 5.14) and the ratio k_f/K_{OS} is more or less insensitive to the nature of the entering ligand for the I_d process. K_{OS} (outer sphere association constant) depends on the nature of the entering group (Y) for a particular metal complex and consequently k_f (in dm^3 mot^{-1} s^{-1}) depends on the nature of Y but k_i ($= k_f/K_{OS}$) remains more or less constant and close to k_{ex}. This aspect has been illustrated (*cf.* Table 5.14.1) for the following reactions where Y varies in structure and basicity widely.

$$\left[Ni(OH_2)_6\right]^{2+} + Y^{x-} \longrightarrow \left[Ni(OH_2)_5(Y)\right]^{n+} + H_2O$$

Statistical Relationship between k_i and k_{ex}

k_i = rate constant for the ligand penetration leading to the inner-sphere complex from the outer-sphere complex in the interchange process (*cf.* Secs. 5.9.3, 5.14).
k_{ex} = first order rate constant for the water exchange process.

$$[M(OH_2)_6] + L : k_i \ ; \ [M(^*OH_2)_6] + H_2O : k_{ex}$$

k_i and k_{ex} are statictically related (*cf.* B. Permutter – Hayman et. al, *J. Coord. Chem.*, **6**, 31, 1976) as: the ligand L residing in the outer coordination sphere can exchange with any one of the six water molecules residing in the inner coordination sphere but the ligand penetration competes with the water molecules present in the outer coordination sphere. Consequently the chance of ligand penetration is lowered by a factor of eight (8 faces of octahedral structure, where the entering ligand can approach; ligands approach along the C_3-axis passing through the centres of opposite trigonal faces; t_{2g}-orbitals are projected along the C_3-axes). Thus:

$$k_i = \frac{6}{8} k_{ex} = 0.75\ k_{ex}$$

(b) **Identification of the I_d and I_a paths in terms of the limiting ratio k_{lim}/k_{ex} (*cf.* Secs. 5.9.5-6, 5.12.2):** Under the upper limiting condition, $K_{OS}[Y] \rangle\rangle 1$, we have: $k_{obs} = k_{lim} \approx k_i$; and under the limiting condition, $1 >> K_{OS}[Y]$, we have: $k_{obs} \approx k_i K_{OS}[Y]$, $k_f \approx k_i K_{OS}$.

In fact, in the I_d pathway, the rate determining step is the M — OH_2 bond rupture for both the formation reaction and water exchange reaction. ***Thus in the I_d path, the ratio k_f/K_{OS} ($= k_i$) is constant and close to the k_{ex} value. This is the diagnostic test for the I_d path.***

Relationship between k_i or k_{lim} and k_{ex} for the Interchange Processes

- Under the condition, $K_{OS}[Y] << 1$, $k_{obs} = k_i K_{OS}[Y]$ and $k_f = k_i K_{OS}$; k_f (in dm^3 mol^{-1} s^{-1}):

$$k_i = \frac{k_f}{K_{OS}} = \text{constant and close to } k_{ex} \text{ for the different entering ligands } (\mathbf{I_d\ path}).$$

$$k_i = \frac{k_f}{K_{OS}} = \text{dependent on the nature of the entering ligand and no relationship with } k_{ex}\ (\mathbf{I_a\ path})$$

- Under the condition, $K_{OS}[Y] \rangle\rangle 1$, $k_{obs} = k_{lim} \approx k_i$ (in s^{-1}); generally the upper limiting rate constant (k_{lim}) is taken as the k_{obs} at high $[Y]$.
- k_{lim}/k_{ex} can be less than or greater than unity (***I_a* path**).
- For the **I_d-path**, $k_i = k_{lim} \approx k_{ex}$.
- k_{lim}/k_{ex} can never exceed unity and the ratio is close to unity (***I_d* path**).

For the **associative activation** (*i.e.* I_a path), under the condition $K_{OS}[Y] \langle\langle 1$, we can apply the condition, $k_f = K_{OS}k_i$ but in such cases the ratio k_f/K_{OS} (= k_i) is not constant and it depends on the nature of the entering ligand. There is no relationship between $k_i = k_f/K_{OS}$ or = $k_{\lim}$ (under the condition $K_{OS}[Y] >> 1$) and k_{ex}. In the I_a pathway, the rate determining step is mainly controlled by the bond formation by the entering ligand during its interchange with the metal bound solvent molecules. If the entering ligand is a better nucleophile than the solvent, *i.e.* H_2O, then k_i or $k_{\lim}$ is greater than k_{ex}. On the other hand, if the entering ligand is a poorer nucleophile than the solvent then k_i or $k_{\lim}$ is less than k_{ex}.

Table 5.19.9.1 Identification of reaction mechanism from the knowledge of $k_{\lim}/k_{ex}$ ratio.

(a) **$[Co(NH_3)(OH)_2]^{3+} + Y^{n-}$**

Y^{n-}	$k_{\lim}$ (s^{-1}), 45°C	$k_{\lim}/k_{ex}$	
Cl^-	2.1×10^{-5}	0.21	
NCS^-	1.6×10^{-5}	0.16	**Dissociative activation, *i.e.* I_d path.**
N_3^-	1.0×10^{-4}	1.0	
SO_4^{2-}	2.4×10^{-5}	0.24	

$k_{ex} = 1 \times 10^{-4}\ s^{-1}$, ($k_{\lim} = k_{obs}$ (obtained at high [Y])

(b) **$[Rh(NH_3)_5(OH_2)]^{3+} + Y^{n-}$** (*cf.* F. Monacelli, *Inorg. Chim. Acta.*, **2,** 263, 1968)

Y^{n-}	$k_{\lim}$ (s^{-1}), 65°C	$k_{\lim}/k_{ex}$	
Br^-	7.9×10^{-3}	4.9	
Cl^-	4.2×10^{-3}	2.6	**Associative activation, *i.e.* I_a path.**
SO_4^{2-}	1.7×10^{-3}	~1.0	

$k_{ex} = 1.6 \times 10^{-3}\ s^{-1}$, ($k_{\lim} = k_{obs}$ (obtained at high [Y])

(c) **$[Ir(NH_3)_5(OH_2)]^{3+} + Y^{n-}$** (*cf.* E. Borghi, *et al*, *Inorg. Nucl. Chem. Lett.*, **6,** 667, 1970)

Y^{n-}	$k_{\lim}$ (s^{-1}), 85°C	$k_{\lim}/k_{ex}$	
Cl^-	9.2×10^{-4}	4.2	**Associative activation, *i.e.* I_a path.**

$k_{ex} = 2.2 \times 10^{-4}\ s^{-1}$, ($k_{\lim} \approx k_{obs}$ (obtained at high concentration of [Y])

Table 5.19.9.2 Kinetic parameters for the formation reactions at $[Fe(OH_2)_6]^{3+}$ and $[Fe(OH)(OH_2)_5]^{2+}$ (Refs. M. Grant and R.B. Jordan, *Inorg. Chem.*, **20,** 55, 1981; S. Gouger and J. Stuehr, *Inorg. Chem.*, **13,** 379, 1974).

Entering ligand	$[Fe(OH_2)_6]^{3+}$		$[Fe(OH)(OH_2)_5]^{2+}$	
(Y)	k_f	k_f/K_{OS}	k_f	k_f/K_{OS}
Cl^-	9.4	1.88	1.1×10^4	0.55×10^4
Br^-	50	10.0	4.1×10^4	2×10^4
SCN^-	127	25.4	1×10^4	0.5×10^4
$CH_2ClCO_2^-$	4.9×10^3	10^3	2.8×10^4	1.4×10^4

(k_f in $dm^3\ mol^{-1}\ s^{-1}$, $k_i = k_f/K_{OS}$, k_i in s^{-1}), $k_{ex} \approx 10^2\ s^{-1}$ for $\left[Fe(OH_2)_6\right]^{3+}$, $\approx 10^4\ s^{-1}$ for $\left[Fe(OH)(OH_2)_5\right]^{2+}$

$K_{OS} = 2\ dm^3\ mol^{-1}$ (for $Z_AZ_B = -2$), = $5\ dm^3\ mol^{-1}$ (for $Z_AZ_B = -3$).

Table 5.19.9.3 Kinetic parameters for the formation reactions at $[Cr(OH_2)_6]^{3+}$ and $[Cr(OH)(OH_2)_5]^{2+}$ centres

Entering ligand (Y)	$[Cr(OH_2)_6]^{3+}$ k_f	$[Cr(OH_2)_6]^{3+}$ k_f/K_{OS}	$[Cr(OH)(OH_2)_5]^{2+}$ k_f	$[Cr(OH)(OH_2)_5]^{2+}$ k_f/K_{OS}
Cl^-	2.9×10^{-8}	5.8×10^{-9}	2.8×10^{-5}	1.4×10^{-5}
Br^-	9.0×10^{-9}	4.5×10^{-9}	1.7×10^{-5}	0.85×10^{-5}
NCS^-	1.8×10^{-6}	3.6×10^{-7}	4.9×10^{-5}	2.5×10^{-5}
k_{ex} (s^{-1})	2.4×10^{-6}		1.8×10^{-4}	

k_f in dm^3 mol^{-1} s^-; $k_i = k_f/K_{OS}$; k_i in s^{-1}; $K_{OS} = 2$ (for $Z_AZ_B = -2$), $= 5$ (for $Z_AZ_B = -3$); *cf.* J.H. Espenson, *Inorg. Chem.*, **8,** 1534, 1969.

Thus in the I_a path, k_{lim}/k_{ex} may be greater or less than unity depending on the relative nucleophilicities of the entering ligand and water. But in the I_d pathway, $k_i = k_{lim} \approx k_{ex}$ and the ratio k_{lim}/k_{ex} can never exceed unity and it is close to unity.

The said criteria for the I_a path are well documented for the substitution reactions at $[Ti(OH_2)_6]^{3+}$, $[Cr(OH_2)_6]^{3+}$, $[Rh(NH_3)_5(OH_2)]^{3+}$, $[Ir(NH_3)_5(OH)_2]^{3+}$.

(c) **Reactivity of $[M(OH_2)_6]^{3+}$ *vs.* $[M(OH)(OH_2)_5]^{2+}$:** For $[M(H_2O)_6]^{3+}$, (M = Fe, Cr), variation of the ratio k_f/K_{OS} in a wide range for the different ligands indicates the **associative mechanism** (*i.e.* I_a). On the other hand, the ratio k_f/K_{OS} is more or less constant for the different ligands entering at the $[M(OH)(OH_2)_5]^{2+}$ centre. This suggests the **dissociative path** (*i.e.* I_d) at $[M(OH)(OH_2)_5]^{2+}$. In $[M(OH)(OH_2)_5]^{2+}$, the *cis-effect of the π–donor OH group introduces the higher reactivity and dissociative character*. The better σ-donor property of OH (compared to that of OH_2) also labilises the remaining coordinated water molecules and it imparts a dissociative character towards the substitution reaction. The overall lower charge (+2 *vs.* +3) also favours the dissociative path at $[M(OH)(OH_2)_5]^{2+}$.

(d) **Rate dependance on the nature of the entering ligand for a pure D-process (*cf.* Secs. 5.9.1, 5.16):** For the *D*-process (where the I_d path is disfavoured because of the disfavour in the formation **of the outer-sphere complex**, *e.g.* anation or aquation of highly negatively charged complex) in the anation of the complexes like $[Co(CN)_5(OH_2)]^{2-}$, $[RhCl_5(OH_2)]^{2-}$, etc. the ratio k_2/k_{-1} obtained from the double reciprocal plot, $1/k_{obs}$ *vs.* $1/[Y]$ depends on the nature of the entering ligand *Y*.

$$\left.\begin{array}{l} L_5M - OH_2 \underset{k_{-1}}{\overset{k_1}{\rightleftharpoons}} L_5M + H_2O \\ L_5M + Y \xrightarrow{k_2} L_5M - Y \end{array}\right\}$$

D-process, Y = labelled water or any other ligand (*cf*. Secs. 5.9.1 and 5.16).

Note: $(k_2/k_{-1})_{\text{exchange}} = 1$ **(Water exchange reaction)**

I_a *vs.* D (*cf.* Sec. 5.16): What is the difference?

Rate depends on the nature of the entering ligand (*Y*) for both the *D* and I_a process. But under the **limiting condition** (*i.e.* high [*Y*]):

$k_{obs} \approx k_{lim} = k_1$ (***D*-process**; $k_2[Y] >> k_{-1}[H_2O]$ and k_1, *i.e.* water loss process is independent of Y); while $k_{obs} \approx k_{lim} = k_i$ (**I_a-process**; $K_{OS}[Y] >> 1$) which depends on the nature of *Y*.

- $k_{lim} \approx k_1$ (independent of Y) ⇒ **D-process:** $k_{lim} \approx k_i$ (dependent of Y) ⇒ **I_a process**

The variation of the ratio (k_2/k_{-1}) with the nature of Y arises from the ***competition between the solvent and Y for the long lived 5–coordinate intermediate.***

Table 5.19.9.4 (k_2/k_{-1}) ratio for the formation reaction at $[RhCl_5(OH_2)]^{2-}$ through the *D*–process.

Y	Cl^-	Br^-	SCN^-	N_3^-	H_2O *(exchange)*
k_2/k_{-1}:	2.1×10^{-2}	1.6×10^{-2}	7.8×10^{-2}	0.14	1.0

(*cf.* D. Robb *et al*, *Inorg. Chim. Acta.*, **3**, 383, 1969)

Table 5.19.9.5 (k_2/k_{-1}) ratio for anation of $[Co(CN)_5(OH_2)]^{2-}$ through the *D*-process.

Y	Cl^-	I^-	NCS^-	*py*	H_2O *(exchange)*
k_2/k_{-1}	6.0	11.7	20.4	23.9	1

(*cf.* A. Haim, *et al.*, *Adv. Chem. Ser.*, **49,** 31, 1965)

Table 5.19.9.6 (k_2/k_{-1}) ratio for the anation of $[Co^{III}(corrin)(OH_2)]$ through the *D*-process.

Y	SCN^-	I^-	Br^-	N_3^-	HSO_3^-	H_2O *(exchange)*
k_2/k_{-1}	1.64×10^{-2}	1×10^{-2}	7.1×10^{-3}	8.6×10^{-3}	1.2×10^{-3}	1 ($k_{ex} = 14\ s^{-1}$)

(*cf.* D. Thusius, *J. Am. Chem. Soc.*, **93,** 2629, 1971)

5.19.10 Effect of Ionic Strength on Rate Constant

From the **Debye-Huckel theory,** the rate dependence of a bimolecular reaction on ionic strength (I) may be expressed as follows:

$$\log k_I \approx \log k_0 + 2QZ_AZ_B\sqrt{I}\ ,\ \log\left(\frac{k_I}{k_0}\right) = 2QZ_AZ_B\sqrt{I} \qquad \ldots(5.19.10.1)$$

k_I = rate constant at ionic strength I; k_0 = rate constant at zero ionic strength;

$Q \propto \varepsilon^{-3/2}$ where ε = dielectric constant of the solvent;

Z_A and Z_B are the ionic charges of the reactant A and B.

I = ionic strength $= \frac{1}{2}\Sigma c_i Z_i^2$ where c_i is the molar concentration of the ion i having the charge Z_i.

It is evident that the plot of log k_I *vs.* $\sqrt{I}$ gives a straight line with slope = $2QZ_AZ_B$.

In aqueous solvent, Q = 0.509 (at 25°C) and thus the slope becomes $1.02Z_AZ_B$.

● Thus it is evident that if the product Z_AZ_B (*i.e.* **reacting species are of opposite charges**) is negative, the rate will linearly decrease with $\sqrt{I}$.

Examples: $[CoCl(NH_3)_5]^{2+} + OH^- \rightarrow$; $[Co(C_2O_4)_3]^{3-} + Fe^{2+} \rightarrow$; $[Cr(OH_2)_6]^{3+} + SCN^- \rightarrow$ etc.

● If the **reacting species are of the same type of charge** (*i.e.* Z_AZ_B = +ve), then the rate increases with $\sqrt{I}$.

Example: $[CoBr(NH_3)_5]^{2+} + [Hg]^{2+} \rightarrow$; etc.

Origin of the Effect of Ionic Strength on Rate Constant

$$A + B \xrightleftharpoons{K^{\neq}} X^{\neq} \xrightarrow{k^{\neq}} \text{Product}, \left(cf.\ \text{Transition State Theory}\right)$$

The rate is expressed in terms of the concentration term not in terms of the activity term. It gives:

$$\text{rate} \propto [X^{\neq}],\ i.e.\ \text{rate} = k^{\neq}[X^{\neq}];\ K^{\neq} = \frac{a_{X^{\neq}}}{a_A a_B} = \frac{[X^{\neq}]}{[A][B]}\frac{f_{\neq}}{f_A f_B}$$

It leads to:

$$\text{rate} = k^{\neq}[X^{\neq}] = k^{\neq}K^{\neq}[A][B]f_A\, f_B/f_{\neq}$$

or, $$\text{rate} = k_0[A][B]\frac{f_A f_B}{f_{\neq}} = k[A][B],\ k_0 = k^{\neq}K^{\neq}$$

Here, $k = k_0\dfrac{f_A f_B}{f_{\neq}}$; or $\log k = \log k_0 + \log\dfrac{f_A f_B}{f_{\neq}}$

According to **Debye-Huckel theory:**

$$\log f = -QZ^2\sqrt{I}$$

Using this relationship, we get:

$$\log k = \log k_0 + \log f_A + \log f_B - \log f_{\neq}$$

$$= \log k_0 - \sqrt{I}Q\left[(Z_A)^2 + (Z_B)^2 - (Z_A + Z_B)^2\right] = \log k_0 + 2QZ_A Z_B\sqrt{I}$$

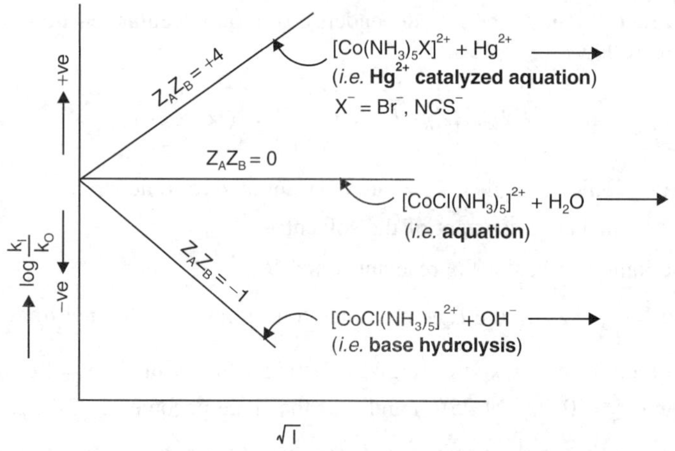

Fig. 5.19.10.1 Qualitative representation of the effect of ionic strength on the rate constant (*i.e.* linear plot of $\log(k_I/k_0)$ *vs.* $\sqrt{I}$) ignoring the **secondary salt effect.**

This is why, the kinetic studies are carried out at a *fixed* and *fairly high ionic strength* (*ca.* 0.1–1 mol dm^{-3}) maintained by the *inert electrolytes.* The inert electrolytes used for maintaining the ionic strength are expected not to interact with the reactants. Here it may be mentioned, that the Eqn. 5.19.10.1 is valid at low I-values.

Primary and secondary salt effect: The effect of inert electrolytes on the rate process has been expressed in Eqn. 5.19.10.1 in terms of ionic strength (I). This is called *primary salt effect.*

Sometimes, the electrolyte (used for maintaining the ionic strength) may participate in some specific interactions with the reactants. **Such electrolytes show the secondary salt effect.** In fact, when the reactants are highly charged, they may participate in **ion-pair formations** with the ions from the inert electrolytes. Reactivity of such ion pairs is different compared to that of the free reacting ions. For example, in the formation reactions of $[Cr(OH_2)_6]^{3+}$, it forms ion pairs with the anions (X^{n-}) of the electrolyte used for maintaining the ionic strength. Reactivities of $[Cr(OH_2)_6]^{3+}$ and ion pair like $\{[Cr(OH_2)_6]^{3+} . X^{n-}\}$ are different. **It introduces the secondary salt effect.** Similarly, in the Hg^{2+} catalysed aquation of $[CoBr(NH_3)_5]^{2+}$, *i.e.*

$$\left[CoBr\left(NH_3\right)_5\right]^{2+} + Hg^{2+} \xrightarrow{H_2O} \left[Co\left(NH_3\right)_5\left(OH_2\right)\right]^{3+} + HgBr^+,$$

the ion associations like $\{[CoBr(NH_3)_5]^{2+} \cdot X^{n-}, Hg^{2+} \cdot X^{n-}\}$ (X^{n-} represents the anion coming from the electrolyte) may also participate in the reaction with different reactivities. *Existence of the secondary salt-effect is indicated by the deviation from the linear plot of logk vs.* $\sqrt{I}$.

To avoid the secondary salt effect (*e.g.* ion-pair formation), generally **perchlorate and nitrate salts** are used. Because of the large size of the monovalent anions like ClO_4^-, NO_3^- (*i.e.* ***low charge density***), possibility of ion-pair formation is reduced. Moreover, NO_3^- and ClO_4^- are of very **poor complexing anions.** Consequently, untoward complexation by NO_3^- or ClO_4^- may be avoided.

5.19.11 Hammett Relationship: Linear Gibbs Energy Relationship

It expresses the relation of reactivities of a series of *meta–* and *para–* substituted aromatic compounds with a common reactant.

$$\log\left(\frac{k}{k_0}\right) = \log\left(\frac{K}{K^0}\right) = \rho\log\left(\frac{K_a}{K_a^0}\right) = \rho\sigma,\ i.e.\ \log k = \log k_0 + \rho\sigma \qquad ...(5.19.11.1)$$

k and k_0 are the rate constants for the reactions of the *meta–* or *para–* R–substituted and *unsubstituted aromatic compounds* respectively; K and K^0 are the equilibrium constants of the corresponding reactions; K_a and K_a^0 are the acid dissociation constants of the substituted and unsubstituted benzoic acid respectively; σ called *substituent constant* depends only on the substituent but not on the reaction series; ρ called *reaction constant* depends on the reaction under consideration.

It is evident that for the **ionisation reactions,** by definition, ρ **is taken as unity**, *i.e.*

$$\log\left(\frac{K_a}{K_a^0}\right) = \rho_a\sigma = \sigma,\ \left(\rho_a = 1\right)$$

The above equation may be expressed as follows:

$$\mathbf{\log k = \log k_0 + \rho\sigma \text{ and } \log K = \log K^0 + \rho\sigma} \qquad ...(5.19.11.1)$$

where $\sigma = \log K_a - \log K_a^0 = pK_a^0 - pK_a$ ...(5.19.11.3)

The linear plot of log k *vs.* σ or log$K$ vs. σ gives the slope = *reaction constant* (ρ). The *substituent constants* (σ) of different groups for *meta–* and *para–* positions are available in literature.

The above relations indicate that **for two homologus series of reactions,** the rate constants (say k_1 and k_2) and the equilibrium constants (K_1 and K_2) are related as follows:

$$\log k_1 = \text{constant} + \rho\log k_2 \text{ and } \log K_1 = \text{constant} + \rho\log K_2 \text{ where } \rho = \rho_1/\rho_2$$

The rate constant (k) is related with the standard free energy of activation ($\Delta G^{\neq}$; the zero subscript to indicate the standard state is dropped for the sake of simplicity) and the equilibrium constant (K) is related with the standard free energy (ΔG_0) of reaction.

$$k = \frac{(k_B T)}{h}\exp\left(-\Delta G^{\neq}/RT\right) \text{ i.e. } \log k = \log\left(\frac{k_B T}{h}\right) - \frac{\Delta G^{\neq}}{2.303RT};\ \log K = -\frac{\Delta G_0}{2.303\,RT}$$

Thus the Hammett relationships indicate the linear relationship between the Gibbs free energy, of activation or reaction, for different reaction series. Thus for two series of homologus reactions, a linear relationship between $\Delta G^{\neq}_{(1)}$ and $\Delta G^{\neq}_{(2)}$ or $\Delta G_{0(1)}$ and $\Delta G_{0(2)}$ exists.

i.e. $\Delta G_1^{\neq} = \text{constant} + \rho\Delta G_2^{\neq}$; or $\Delta G_{0(1)} = \text{constant} + \rho\Delta G_{0(2)}$, $\rho = \rho_1/\rho_2$

● ***Positive σ–value*** *indicates the electron withdrawing property of the group* (*e.g.* $-NO_2$, –Cl, etc.) *compared to that of hydrogon.* It means that the substituted benzoic acid is stronger than the unsubstituted benzoic acid, *i.e.* $pK_a^0 \rangle\, pK_{a,}$

● ***Negative σ–value*** *indicates the electron pushing effect of the group* (*e.g.* $-CH_3$, $-C_2H_5$). It means that the substituted benzoic acid is weaker than the unsubstituted one, *i.e.* $pK_a^0 < pK_a$.

It may be mentioned that for a particular substituent R, the σ–value may be different for the *meta*– and *para*– positions because of the different electronic effect arising from *meta*– and *para*–positions. The group R can show only the **inductive effect** at the *meta*–position but at the para-position, it can show both the **inductive effect** and **mesomeric effect** (if possible, as in the case of OH group).

Hammett relationships for two homologous series of reactions

$$\left.\begin{array}{l}\log k_1 = \log k_{0(1)} + \rho_1\sigma \\ \text{and } \log k_2 = \log k_{0(2)} + \rho_2\sigma\end{array}\right\}$$

$$\left.\begin{array}{l}\log K_1 = \log K_{0(1)} + \rho_1\sigma \\ \text{and } \log K_2 = \log K_{0(2)} + \rho_2\sigma\end{array}\right\}$$

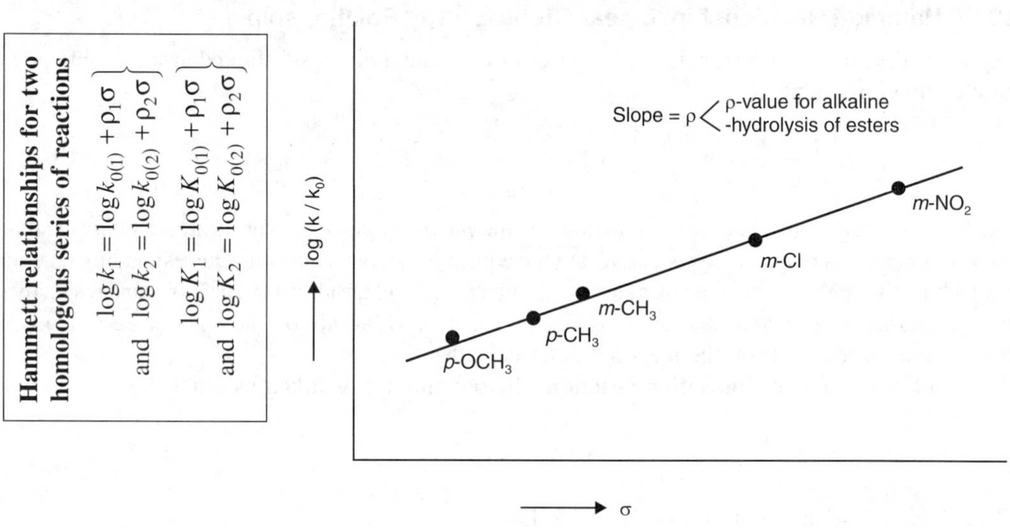

Fig. 5.19.11.1 Qualitative representation of Hammett plot for base hydrolysis of $[(H_3N)_5Co—O—C(=O)—C_6H_4R]^{2+}$ 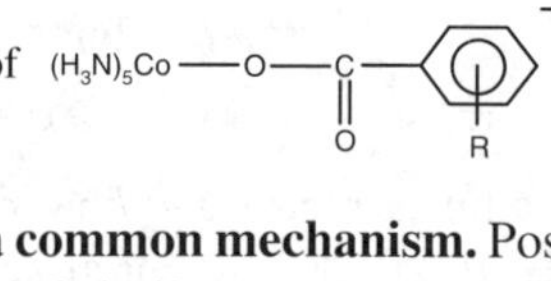 in aqueous methanol (*cf.* F. Aprile *et al, J. Inorg. Nucl. Chem.*, **21**, 325, 1961).

The linear plot indicates that the series of reaction pass through a common mechanism. Positive value of the *reaction constant* (which is constant for a particular series of similar reactions), indicates that the k-value increases as σ becomes more positive.

Determination reaction constant (ρ) helps us establish the nature of reaction. It is illustrated below.

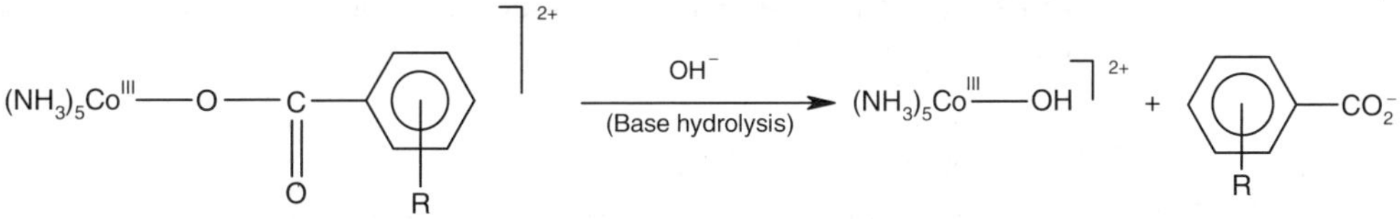

(R at the *m*– or *p*–position)

The reaction can occur either through the 'Co—O' or 'C—O' bond fission. The ρ-value for 'C—O' bond fission (*i.e.* alkaline hydrolysis of ester) is in the range 1.8 – 2.5. But, *the* ρ*–value* (*ca.* 0.75 – 0.9) *experimentally determined for the* above mentioned *base hydrolysis reactions is much smaller than the above range.* This smaller value of ρ has been argued for the Co—O bond cleavage. The ρ-value indicates the **sensitivity of the reaction series to the ring substitution.** In alkaline hydrolysis of ester, the C—O bond which ruptures is closer to the aromatic ring and it makes the ρ-value larger. On the other hand, in the said base hydrolysis, the Co—O bond which ruptures lies relatively further away (compared to the C—O bond) from the ring and it makes the ρ value relatively less (*cf.* 0.75 – 0.9 *vs.* 1.8 – 2.5). Thus in the base hydrolysis of complex, the reaction is relatively less sensitive to the ring substitution.

5.19.12 Taft Relationship

Hammett equation was found inadequate for the reactions of aliphatic compounds and *ortho*–substituted compounds. To overcome these limitations, Hammett equation was modified by Taft as follows:

$$\log(k/k_0) = \sigma^* \rho^* \qquad \ldots(5.19.12.1)$$

σ^* called *polar substituent constant;* ρ^* called *reaction constant* bearing the same significance as in Hammett equation. σ^* is defined as:

$$\sigma^* = 0.4\left[\log(k/k_0)_{\text{basic}} - \log(k/k_0)_{\text{acidic}}\right]$$

The factor 0.4 is arbitrarily chosen to bring the σ^* values to nearly the same scale of Hammet σ-values.

k denotes the rate constant for acid or base hydrolysis of an ester having a substituent (R) alpha to the carbonyl group and k_0 denotes the corresponding value for the Me-substituted ester. Thus, $\sigma^* = 0$ for Me-substitution by definition.

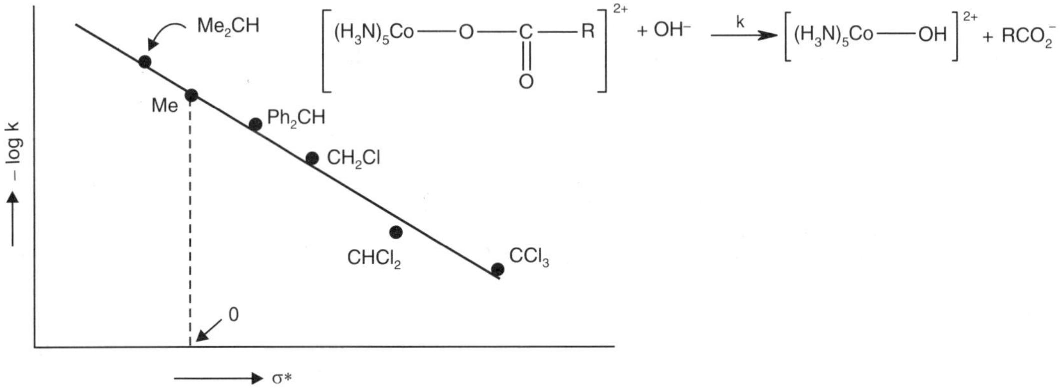

Fig. 5.19.12.1 Qualitative representation of Taft plot for the given reaction (*cf.* W.E. Jones, *et al, J. Chem. Soc., A,* 1481, 1966).

Determination of ρ* (**Taft reaction constant**) for the following reaction can decide whether the 'Co — O' or 'C — O' bond ruptures.

$$\left[(NH_3)_5Co-O-\overset{\overset{\displaystyle O}{\|}}{C}-R\right]^{2+} \xrightarrow[\text{(Base hydrolysis)}]{OH^-} \left[(NH_3)_5Co-OH\right]^{2+} + RCO_2^-$$

$R = CH_3, C_2H_5, (CH_3)_2CH$, etc.

5.19.13 Effect of Solvent on the Rate Process

If the two reacting species bear the charges Z_A and Z_B, then effect of the **dielectric constant** (ε) of the medium on the rate constant is expressed as follows:

$$\log k_\varepsilon = \log k_\infty - \frac{Z_A Z_B e^2}{2.303\varepsilon r_{AB} k_B T} \qquad \ldots(5.19.13.1)$$

e = electronic charge; r_{AB} = distance between the species in the transition state; k_ε = rate constant in a medium of dielectric constant ε; k_∞ = rate constant in a medium of dielectric constant ∞; k_B = Boltzmann constant; T = temperature in Kelvin scale.

For infinite dielectric constant, the last term of the equation becomes zero, *i.e.* when there is no electrostatic force between the reacting species.

The above relation indicates that the plot of log k_ε *vs.* 1/ε is linear and sign of the slope depends on the sign of the product Z_AZ_B. The calculated value of r_{AB} from the slope of the plot is quite reasonable.

It is evident that if the product Z_AZ_B is positive (*i.e.* reacting species are of the same charge) then the rate decreases with the increase of 1/ε (*i.e.* reciprocal of the dielectric constant). If the product Z_AZ_B is negative, then the rate constant increases with the increase of 1/ε. If the product Z_AZ_B is zero (*i.e.* one of the reacting species is neutral), then the rate constant does not depend on ε.

5.20 HYDROLYSIS OF OCTAHEDRAL COBALT(III)-AMMINE COMPLEXES

Hydrolysis of the cobalt(III)-ammine complexes (say, $[Co(N_5)X]$) may be represented simply as:

$$[Co(N_5)X] \xrightarrow{\text{water}} [Co(N_5)(OH_2)] + X^-$$

$$[Co(N_5)(OH_2)] \underset{+H^+}{\overset{-H^+}{\rightleftharpoons}} [Co(N_5)(OH)]$$

(N_5 stands for the non-replaceable amine/ammine ligands occupying the five coordination sites)

In acidic condition, the predominant product is $[Co(N_5)(OH_2)]$ while in relatively basic condition, the predominant product is $[Co(N_5)(OH)]$. The reaction pathways also depend on the pH and the rate law is given by two terms as follows:

$$-\frac{d[Co(N_5)X]}{dt} = k_A[Co(N_5)X] + k_B[Co(N_5)X][OH^-]$$

k_A and k_B denote the acid and base hydrolysis rate constants respectively.

- If $k_A \gg k_B[OH^-]$, the second term, *i.e.* base hydrolysis path is less important but when $k_A \ll k_B[OH^-]$, the first term, *i.e.* aquation or acid hydrolysis path becomes less important.
- Generally, $k_B \approx 10^6 k_A$, hence the acid hydrolysis becomes important at pH < 8 (*i.e.* $[OH^-] = 10^{-6}$ mol dm^{-3}) and the base hydrolysis becomes important at pH greater than 8.

Co(III) and Cr(III) complexes used as the model complexes for understanding the mechanistic aspects of ligand substitution processes of octahedral complexes—why?

In fact, coordination chemistry started to develop with mainly the Co(III)–complexes. The important factors behind this choice are:

● they can be easily prepared; ● they are both kinetically and thermodynamically stable; ● the compounds are not costly; ● because of their inertness, they can be used for kinetic studies by using the conventional techniques; ● for the kinetic studies of labile centres, etc. ● it needs special techniques which are costly.

5.21 AQUATION OR HYDROLYSIS UNDER ACIDIC CONDITION OF COBALT(III) AMMINE/AMINE COMPLEXES

In terms of the principle of microscopic reversibility, aquation or acid hydrolysis reactions are the reverse of anation reactions. The acid hydrolysis (pH < 5) of acidopentaamminecobalt(III) salt is shown below.

$$\left[Co(NH_3)_5X\right]^{2+} + H_2O \underset{\text{Anation}}{\overset{\text{Hydrolysis}}{\rightleftharpoons}} \left[Co(NH_3)_5(OH_2)\right]^{3+} + X^-$$

The rate law for aquation of $[Co(N_5)X]$ shows a first order dependence on the starting complex. There are **different possibilities** to explain the observed rate law. The simplified schemes are given below. For details of the D, A and I mechanisms, Secs. 5.9.1-3 are to be considered.

(i) $$\left.\begin{aligned}&\left[Co(N_5)X\right]^{2+} \xrightarrow[(rds)]{k} \left[Co(N_5)\right]^{3+} + X^- \\ &\left[Co(N_5)\right]^{3+} + H_2O \xrightarrow{\text{fast}} \left[Co(N_5)(OH_2)\right]^{3+}\end{aligned}\right\}\ \textbf{Dissociative path}$$

i.e. rate = $k[Co(N_5)X^{2+}] = k$ [Complex] ...(5.21.1)

(ii) $$\left.\begin{aligned}&\left[Co(N_5)X\right]^{2+} + H_2O \xrightarrow[(rds)]{k} \left[Co(N_5)(OH_2)(X)\right]^{2+} \\ &\left[Co(N_5)(OH_2)X\right]^{3+} \xrightarrow{\text{fast}} \left[Co(N_5)(OH_2)\right]^{3+} + X^-\end{aligned}\right\}\ \textbf{Associative path}$$

i.e. rate = $k[Co(N_5)X^{2+}][OH_2]$

$= k \times 55.5 \times [Co(N_5)X^{2+}]$

$= k'[Co(N_5)X^{2+}] = k'$ [Complex] ...(5.21.2)

(iii) $$\left[Co(N_5)X\right]^{2+} + H_2O \xrightarrow[(rds)]{k} \left[Co(N_5)(OH_2)\right]^{3+} + X^- \Big\}\ \textbf{Interchange path}$$

rate = $k[Co(N_5)X^{2+}][H_2O]$

$= k'[Co(N_5)X^{2+}] = k'$ [Complex] ...(5.21.2)

Note: The above rate laws are the simplified forms (under some conditions). The actual schemes and rate laws for the D, A and I processes have been discussed in detail earlier.

$Y = H_2O$ (in the present case)

$$\textbf{\textit{D}-process (Sec. 5.9.1)}: \text{rate} = \frac{k_1 k_2 [\text{Complex}][Y]}{k_{-1}[X] + k_2[Y]} \approx k_1[\text{Complex}], \quad \text{when } k_2[Y] >> k_{-1}[X]$$

$$\textbf{\textit{A}-process (Sec. 5.9.2)}: \text{rate} = \frac{k_a k_b [\text{Complex}][Y]}{k_{-a} + k_b} = k[\text{Complex}][Y],\ k = \frac{k_a k_b}{k_a + k_b}$$

$$\textbf{\textit{I}-process (Sec. 5.9.3)}: \text{rate} = \frac{k_i K_{OS} [\text{Complex}][Y]}{1 + K_{OS}[Y]} \approx k_f [\text{Complex}][Y], \quad \text{when } 1 >> K_{OS}[Y],\ k_f = k_i K_{OS}$$

The above associative and interchange paths will lead to the *pseudo-first order conditions*. ***Thus the three possible pathways lead to the same rate law.*** However, these possibilities can be distinguished by considering the effects of different factors on the rate process. These are discussed below. *The evidences support the dissociative (more correctly I_d path) path for the said aquation reactions.*

(A) Effect of Charge (*cf.* Sec. 5.19.1): The aquation rate of $[CoCl(NH_3)_5]^{2+}$ is about *ca.* 300 times slower than that of *trans*–$[CoCl_2(NH_3)_4]^+$; and *cis*–$[CoCl_2(NH_3)_4]^+$ aquates *ca.* 3 times faster than the corresponding *trans*–isomer.

Aquation rate: *cis*-$[CoCl_2(NH_3)_4]^+ \rangle$ *trans*-$[CoCl_2(NH_3)_4]^+ \rangle\rangle [CoCl(NH_3)_5]^{2+}$.

Aquation rate of other complexes follow the sequences:

$$[CoCl_2(NH_4)_4]^+ \rangle\rangle [CoCl(NH_3)_4(OH_2)]^{2+};$$

$$cis\text{–}[CoCl_2(en)_2]^+ \rangle\rangle cis\text{–}[CoCl(en)_2(NH_3)]^{2+}$$

Thus, it is evident that the aquation rate remarkably increases with the decrease of the overall charge of the complex. *This nature of dependence supports the dissociative activation (i.e. bond breaking is primarily important in attaining the transition state).* It is justified because removal of an anionic ligand like Cl^- from a more positively charged complex is more difficult.

If **associative activation** is argued, then the more positively charged complex should react faster.

(B) Effect of steric crowding (causing steric accleration) (*cf.* Sec. 5.19.7):

● **Aquation of *trans*-$[CoCl_2(N—N)]^+$ (N — N = en or substituted en) (*cf.* Sec. 5.19.7):** Aquation of the complexes of the type *trans*–$[CoCl_2(N—N)_2]^+$ where N—N denotes ethylenediamine or substituted ethylenediamine experiences a *steric acceleration* (*i.e.* with the increase of steric crowding, the rate of aquation increases).

$$trans\text{-}\left[CoCl_2(N—N)_2\right]^+ + H_2O \longrightarrow trans\text{-}\left[CoCl(N—N)_2(OH_2)\right]^{2+} + Cl^-$$

This **steric acceleration** is an indication of the **dissociative activation**.

● **Aquation of *trans*-$[CoCl_2(N_4)]^+$ and $[CoCl(N_4)X]^+$ (N_4 = cyclam or its derivative):** Similarly, Co(III)–complexes of the type $[CoCl_2(N_4)]^+$ and $[CoCl(N_4)(X)]^+$ where N_4 denotes the macrocyclic ligand *cyclam* or its hexamethyl derivative, *tet–b* (which provides more steric crowding than cyclam) experience a **steric acceleration** in the aquation process.

$$trans\text{–}[CoCl(N_4)X]^+ + H_2O \rightarrow trans\text{–}[Co(N_4)(OH_2)X]^{2+} + Cl^-$$

This **steric acceleration** is in conformity with the **dissociative activation**.

Table 5.2.1.1 Steric effect of nonleaving ligands (N—N) on the aquation rate of *trans*–$[CoCl_2(N — N)_2]^+$ (at 25°C) (*cf.* R.G. Pearson, *et al, J. Am. Chem. Soc.*, **75**, 3089, 1953).

N — N	10^5k (s^{-1})	*N — N*	10^5k (s^{-1})
(i) 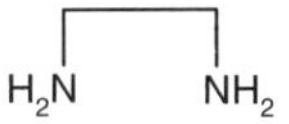 (en)	3.2	(iv) 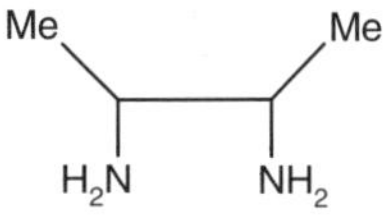 (*meso*-bn)	42.0
(ii) 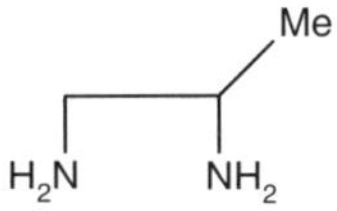 (pn)	6.2	(v) 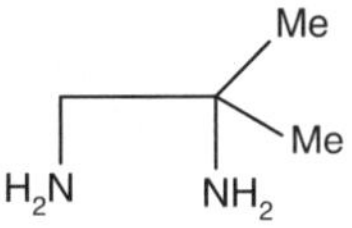	22.0
(iii) 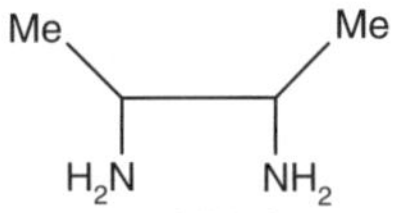 (*dl*-bn)	15.0	(vi) 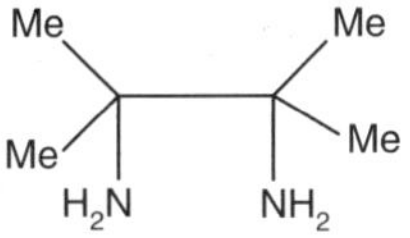	very fast (instantaneous)

N — N	10^5k (s^{-1})
(i) 	1.7
(ii) 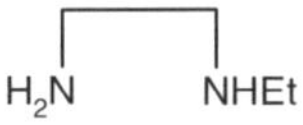	6.0
(iii) 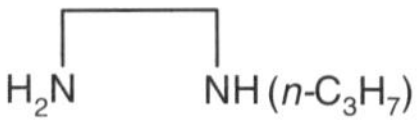	11.8

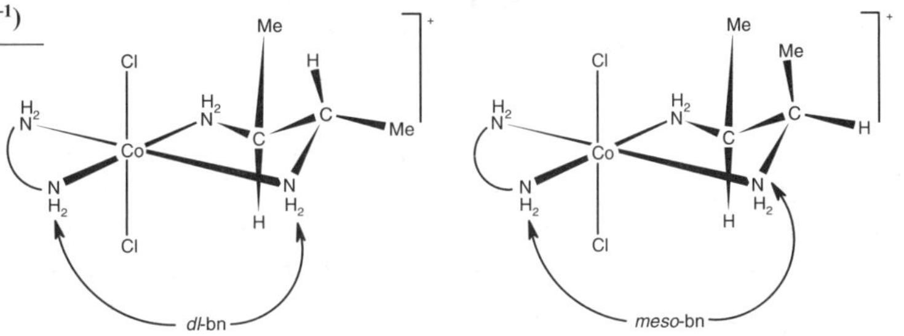

Chiral-bn sterically less crowded; Me-groups on the opposite sites of the chelate ring

Achiral-bn sterically more crowded (Aquation rate faster)

Table 5.21.2 Aquation rates of $[CoCl(N_4)X]^+$ to illustrate steric acceleration.

Complex	*k* (s^{-1}) *25°C (for loss of* Cl^-*)*
trans–$[CoCl_2(cyclam)]^+$	1.1×10^{-6}
trans–$[CoCl_2(tet–b)]^+$	9.3×10^{-4}
trans–$[CoCl(cyclam)(NCS)]^+$	1.1×10^{-9}
trans–$[CoCl(NCS)(tet–b)]^+$	7.0×10^{-7}
trans–$[Co(CN)Cl(cyclam)]^+$	4.8×10^{-7}
trans–$[Co(CN)Cl(tet–b)]^+$	3.4×10^{-4}

Cyclam = 1, 4, 8, 11 – tetraazacyclotetradecane
tet–b = *dl*–1, 4, 8, 11 – tetraaza–5, 5, 7, 12, 12, 14–hexamethylcyclotetradecane (the corresponding *meso*–isomer is called **tet-a**)

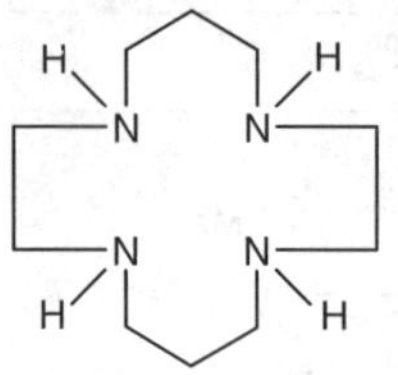

(cyclam)

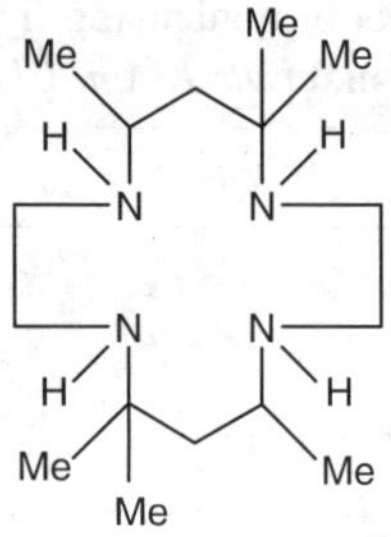

(*tet-b*)

- **Aquation of $[CoCl(MeNH_2)_5]^{2+}$ *vs.* $[CoCl(NH_3)_5)]^{2+}$:** $[CoCl(MeNH_2)_5]^{2+}$ aquates faster than $[CoCl(NH_3)_5]^{2+}$ by a factor *ca.* 20 at 50°C suggesting the **dissociative activation (steric acceleration).** *It may be noted that the corresponding Cr(III)–complex experiences steric retardation in aquation supporting the associative activation* (*cf.* Sec. 5.19.7) .
- **Aquation of $[CoCl(dien)(tn)]^{2+}$ *vs.* $[CoCl(dien)(en)]^{2+}$:** $[CoCl(dien)(tn)]^{2+}$ aquates much faster than $[CoCl(dien)(en)]^{2+}$. This rate enhancement (*i.e.* **steric acceleration**) can be explained by considering the dissociative activation. The corresponding 5–coordinate intermediate of trigonal bipyramidal geometry is more stable for $[CoCl(dien)(tn)]^{2+}$. This aspect has been discussed in Sec. 5.19.7 (*cf.* Figs. 5.19.7.1-2).

(C) Electronic effect (*e.g. cis*–effect and *trans*–effect) of the nonleaving group (L) (*cf.* Sec. 5.19.8): The observations for aquation of $[CoCl(en)_2(L)]^{n+}$ (*i.e.* replacement of Cl^- by H_2O) are:

(i) **Pi–donor nonleaving groups** like Cl^-, OH^-, NCS^- at the *cis*–positions can accelerate the aquation process. The *cis*– complex aquates faster than the corresponding *trans*–isomer, *i.e.* $k_{cis} \rangle k_{trans}$.

(ii) If the **pi–donor nonleaving group** resides at the ***cis*–position**, there is a **100% retention of configuration,** *i.e.* the product is of *cis*–configuration (*cf.* Figs. 5.21.3-4).

(iii) If the **pi-donor nonleaving group** resides at the *trans–position,* the aquated product is a mixture of *cis*– and *trans*–configuration (*i.e.* **it leads to isomerisation**; *cf.* Fig. 5.21.3-4).

(iv) If the nonleaving group is a **weak π–donor** (*e.g.* H_2O) or having no π–bonding property (*i.e.* neither π–acceptance nor π–donation as for the ligand NH_3), then both the *cis*– and *trans*– isomers react at comparable rates, *i.e.* $k_{cis} \approx k_{trans}$.

(v) If the nonleaving group is a **strong π–acceptor ligand** (*e.g.* CN^-, NO_2^-) then the corresponding *trans*– isomer aquates faster, *i.e.* $k_{trans} \rangle k_{cis}$. However, this effect (*i.e.* ***trans*–effect**) is less pronounced than the *cis*– effect (shown by the π–donor ligand residing at the *cis*–position.

Table 5.21.3 Aquation rate constants of $[CoCl(en)_2(L)]^{n+}$ at 25°C showing the effect of the nonleaving group (L). $[CoCl(en)_2(L)]^{n+} + H_2O \rightarrow [Co(en)_2(L)(OH_2)]^{(n+1)+} + Cl^-$.

L	*Electronic effect of L*	*cis*–$[CoCl(en)_2(L)]^{n+}$		*trans*–$[CoCl(en)_2(L)]^{n+}$		$\frac{k_{cis}}{k_{trans}}$
		k (s^{-1})	*% of cis-product*	*k* (s^{-1})	*% of trans-product*	
OH^-	**good π–donor**	1.2×10^{-2}	100	1.6×10^{-3}	25	7.5
Cl^-		2.4×10^{-4}	100	3.5×10^{-5}	65	6.85
NCS^-		1.1×10^{-5}	100	5.0×10^{-8}	30–50	220

(Contd.)

(Table 5.21.3 contd.)

L	*Electronic effect of L*	*cis–[CoCl(en)$_2$(L)]$^{n+}$*		*trans–[CoCl(en)$_2$(L)]$^{n+}$*		$\frac{k_{cis}}{k_{trans}}$
		k (s^{-1})	*% of cis-product*	*k* (s^{-1})	*% of trans-product*	
H_2O	**weak π–**	1.6×10^{-6}		2.5×10^{-6}		0.64
*N_3^-	**donor**	2.5×10^{-4}	100	2.4×10^{-4}	80	1.04
NH_3	**non π–bonding**	5×10^{-7}	100	3.4×10^{-7}	100	1.47
NO_2^-	**strong π–**	1.1×10^{-4}	100	1.0×10^{-3}	100	0.11
CN^-	**acceptor**	6.2×20^{-7}	100	8.2×10^{-5}	100	7.5×10^{-3}

* N_3^- (azide ion) can act either as a pi–donor or as a pi–acceptor. However, in the present case, it does not act strongly in either capacity. (*cf.* M.L. Tobe, *Inorg. Chem.*, **7**, 1260, 1968; W.G. Jackson *et al*, *Inorg. Chim. Acta.*, **60**, 115, 1982; C.K. Poon *et al*, *J. Chem. Soc. A*, 1549, 1968).

● **Mechanistic path for the π–donor ligand (L):** When the leaving group is dissociated, it produces a square pyramidal geometry (C.N. = 5). The starting octahedral complex uses six sp^3d^2 or d^2sp^3 hybrid orbitals to bind the 6 ligands. Thus dissociation of a Cl^- ligand from the octahedral complex gives a vacant d^2sp^3 or sp^3d^2 orbital. If the π–donor ligand L resides at the *cis*– position of the leaving group, then the lone pair of L can be donated to the vacant orbital to stabilise the 5–coordinate square pyramidal intermediate or activated complex. In other words, if the reaction experiences a **dissociative activation,** the 5-coordinate activated complex or the transition state is stabilised. Thus the pi–donor ligand (L) favours the dissociation of the leaving group at the *cis*–position. This phenomenon is described as the ***cis*-effect in dissociative activation.**

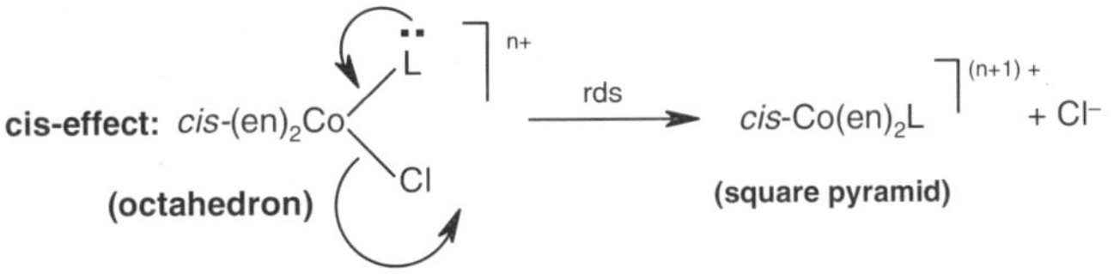

The π-donor properties of some representative ligands can be represented as follows:

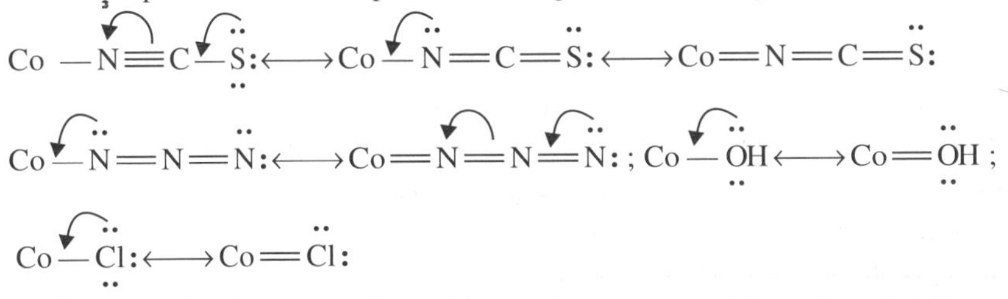

If the π–donor nonleaving group L resides at the *trans*– position, then the 5–coordinate square pyramidal geometry (obtained due to the dissociation of Cl^-) cannot be stabilised by L as such. But transformation of the **square pyramidal (SP)** geometry into a **trigonal bipyramidal (TBP) geometry** allows the stabilisation of the intermediate by L. In other words, if the reaction is supposed to pass through the dissociative activation, the activated complex or transition state is to be stabilised by the π-donor ligand L through a geometrical transformation. ***This required transformation requires energy and it raises the activation energy to slow down the reaction, i.e.*** $k_{trans} < k_{cis}$.

$$trans\text{-}\left[CoCl(en)_2(L)\right]^{n+} \xrightarrow{-Cl^-} \underset{(SP)}{\left[Co(en)_2L\right]^{(n+1)^+}} \rightleftharpoons \underset{(TBP)}{\left[Co(en)_2L\right]^{(n+1)^+}} \xrightarrow{+H_2O} \underset{(cis+trans)}{\text{Product}}$$

This transformation (SP to TBP) leads to **isomerisation.**

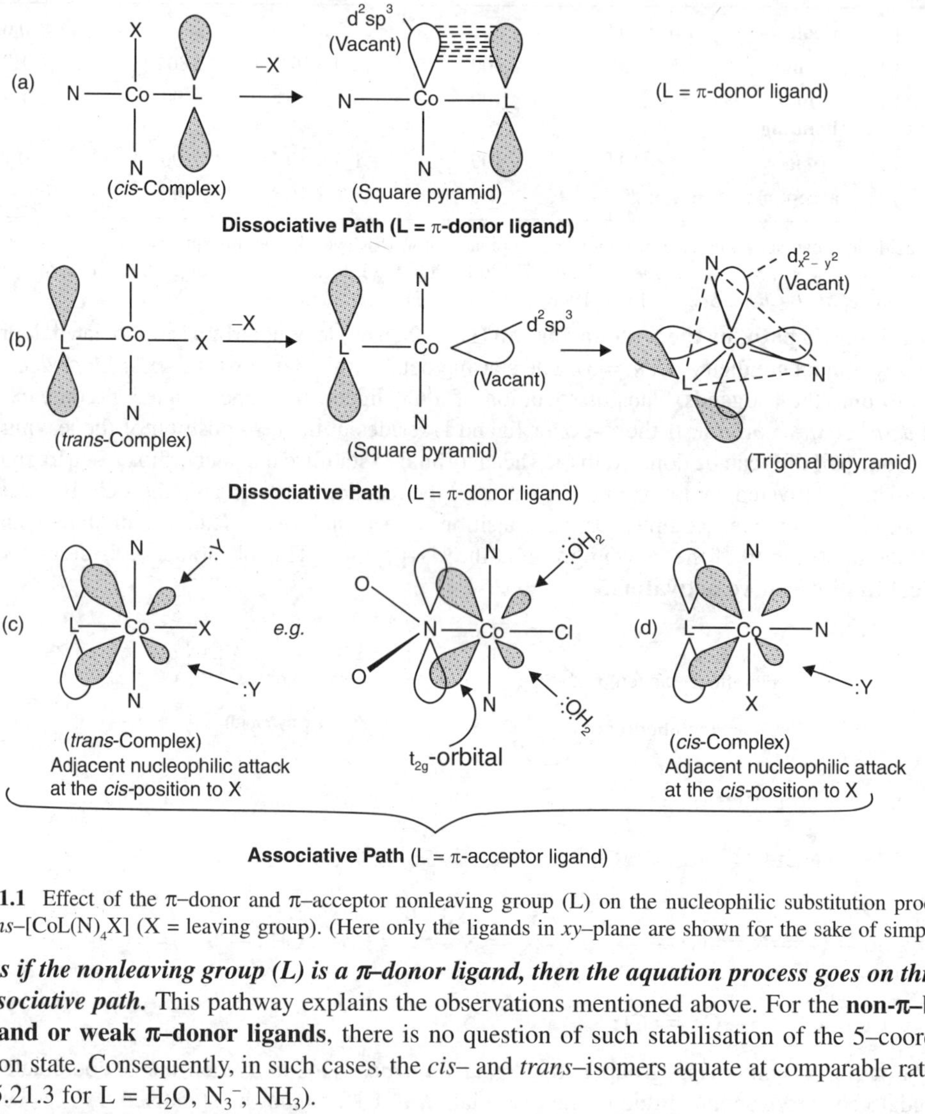

Fig. 5.21.1 Effect of the π–donor and π–acceptor nonleaving group (L) on the nucleophilic substitution process of *cis–/trans–*$[CoL(N)_4X]$ (X = leaving group). (Here only the ligands in *xy*–plane are shown for the sake of simplicity).

Thus if the nonleaving group (L) is a π–donor ligand, then the aquation process goes on through the dissociative path. This pathway explains the observations mentioned above. For the **non-π–bonding ligand or weak π–donor ligands**, there is no question of such stabilisation of the 5–coordinate transition state. Consequently, in such cases, the *cis–* and *trans–*isomers aquate at comparable rates (*cf.* Table 5.21.3 for L = H_2O, N_3^-, NH_3).

The said ***cis*–effect** also explains why ***cis*–$[CoCl_2(NH_3)_4]^+$** aquates faster than ***trans*–$[CoCl_2(NH_3)_4]^+$**.

● **Mechanistic path for the π–acceptor nonleaving group (L):** The π–acceptor nonleaving ligands (L) withdraw the electron cloud from the metal centre and *thus it favours the nucleophilic attack at the metal centre at the trans position where the electron cloud is depleted.* In fact, the metal → ligand π-bonding effectively increases positive charge on the metal centre and *it favours the associative path.*

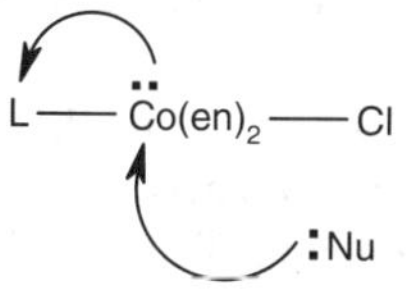

(associative path)

The π-acceptor properties of some representative ligands can be represented as follows:

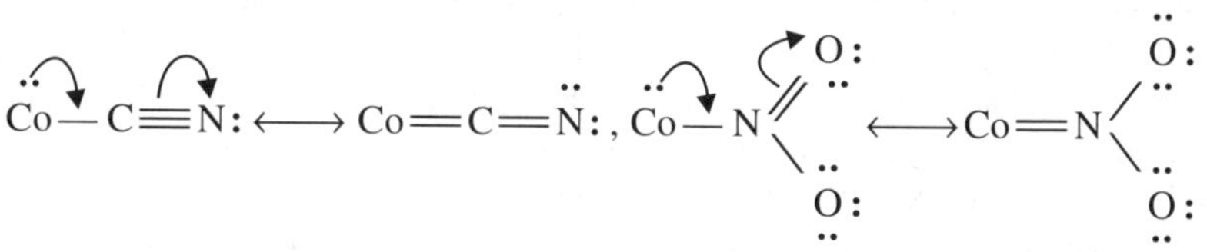

For the nonleaving π–acceptor ligands like NO_2^-, CN^-, etc., the electrons from the nonbonding *d*–orbitals (*i.e.* filled t_{2g}-orbitals) of the metal centre are pushed into the vacant orbitals of NO_2^- or CN^- (*cf.* Fig. 5.21.1c, d). For the *trans*– isomer, the **electron density is lowered in both the octahedral faces** adjacent to the leaving group (*i.e.* Cl^-). On the other hand, for the *cis*–isomers, the electron withdrawal causes the **electron depletion in one of the octahedral faces** adjacent to the leaving group. In other words, electron depletion surrounding the leaving group is more for the *trans*-isomer. ***Thus, the trans– isomer favours better than the cis–isomer for the approach of the incoming nucleophile*** (*i.e.* H_2O). In other words, the *trans*–isomer stabilises the transition state (attained through the associative activation) better than the *cis*–isomer.

Difference in the mechanistic paths of ligand substitution process in the octahedral Co(III) complexes depending on the nature of the π–bonding properties of the nonleaving group (L)

(i) If L is a strong **pi–donor ligand** (*e.g.* OH^-, Cl^-, etc.), it favours the **dissociative activation** (*i.e.* bond breaking by the leaving group is primarily important in producing the transition state). The *cis*– isomer (*i.e.* L and leaving group at the *cis*– positions) react faster than the corresponding *trans*–isomer. The *cis*– isomer tends to retain the configuration in the product while the *trans*–isomer leads to isomerisation (*i.e.* mixed products of *cis*– and *trans*– isomers).

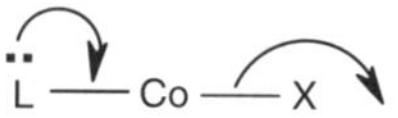

(i.e. **dissociative activation**) } $k_{cis} > k_{trans}$; *cis*-: 100% retention of configuration; *trans*- : isomerisation.

(ii) If L is a strong **pi–acceptor ligand** (*e.g.* NO_2^-, CN^-, etc.), it favours the **associative activation** (*i.e.* bond formation by the incoming nucleophile is primarily important in attaining the transition state). The *trans*– isomer (*i.e.* L and leaving group at the *trans*– positions) reacts faster than the *cis*–isomer. There is a 100% retention of stereochemical configuration for both the *cis*– and *trans*– isomers.

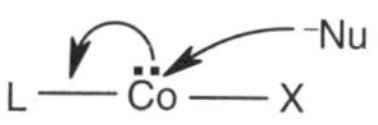

(Adjacent attack)

(*i.e.* **associative activation**) } $k_{cis} < k_{trans}$; 100% retention of configuration in both the *cis*- and *trans*-isomers.

For the π–acceptor nonleaving ligand (L), the nucleophile (*i.e.* H_2O in the aquation process) attacks at the *trans*–positions of L, *i.e.* at the *cis*-positions to X for the *trans*– isomer and at the *cis*– position of X for the corresponding *cis*– isomer (*cf.* Fig. 5.21.1c, d). **Thus this associative activation** (*i.e.* bond making by the incoming nucleophile) **is totally stereoretentive** (*i.e. cis* → *cis*, *trans* → *trans*) **and it becomes important for the π–acceptor nonleaving ligand.** This associative activation is well understood for the substitution reactions of square planar complexes of Pt(II) in the presence of strongly *trans*–directing π–acceptor ligands.

● **Mechanistic path for the non-π-bonding nonleaving group (L):** The *non–π–bonding nonleaving ligands* like NH_3 may also lead to a such associative path but they cannot show any special effect to stabilise the transition state or activated complex in the aquation process in both the *cis*– and *trans*–isomers. In fact, such non-π-bonding nonleaving ligands fail to favour both the *d*- and *a*- mechanisms consequently, both these isomers react at comparable rates.

(D) Electronic effect (inductive effect) of the nonleaving group: The bond strength between the Co(III) centre and leaving group is partially controlled by the electronic effect such as **inductive effect** of the nonleaving group. This is illustrated in the following aquation process.

$$[CoCl(en)_2(R—py)]^{2+} + H_2O \rightarrow [Co(en)_2(OH_2)(R—py)]^{3+} + Cl^-$$

With the increase of basicity of the pyridine ligand (R–py), the electron pushing inductive effect of the ligand increases. This enhanced electron density on the metal centre weakens the 'Co—Cl' bond 'strength'. If the aquation process experiences the **dissociative activation,** then with the increase of basicity of the pyridine ligand (R–py), the rate should increase. In fact, it happens so to support the dissociative activation.

Basicity order of R–py: pyridine 〈 3–methylpyridine 〈 4–methylpyridine 〈 4–methoxypyridine

10^5 k (s^{-1}) at 50°C: 1.1 1.3 1.4 1.5

Note: It is the general observation that the **good σ-donor ligands** (*i.e.* good basic ligands) impart the favourable **dissociative character** at the metal centre, *e.g.* $[Cr(OH_2)_6]^{3+}$ adopts the associative activation while $[Cr(NH_3)_5(OH_2)]^{3+}$ adopts the dissociative activation (*cf.* σ-basicity: NH_3 〉 OH_2). It may be noted that the π-donor ligands like OH^-, Cl^- are also the good σ-donor ligands and they favour the dissociative activation.

(E) Effect of the leaving group (X) in $[Co(NH_3)_5X]^{n+}$: The aquation rate decreases with the increase of the thermodynamic bond strength of the 'Co—X' bond. The aquation rate follows the sequence:
X^-: NO_3^- 〉 I^- 〉 Br^- 〉 H_2O 〉 Cl^- 〉 SO_4^{2-} 〉 F^- 〉 CH_3COO^- 〉 NH_3 〉 NCS^- (N–bonded) 〉 N_3^- 〉 NO_2^- (*N*–bonded)

Thus with the increase of thermodynamic affinity of Co(III) for X, the aquation rate decreases. It has been noted that the activation energy (E_a measuring the rate constant) increases linearly with the average ligand field strength measured by $\bar{\nu}$ (wave number in cm^{-1}) of the first d – d absorption band of the complex.

$$E_a : \xrightarrow[\text{Increasing ligand field strength}]{Br^- < NCS^- \text{ (N-bonded)} < NO_2^- \text{ (N-bonded)}}$$

All these indicate that rupture of the 'Co—X' bond is playing an important role in attaining the transition state. However, it cannot tell us the role of H_2O (*i.e.* entering nucleophile) in the transition state.

(F) Linear free energy relationship (LFER) for aquation of $[Co(NH_3)_5X]^{n+}$ (*cf.* Sec. 5.19.4): For this aquation process, the LFER plot, *i.e.* plot of log k_{aq} *vs.* log K_{eq} gives a good straight line having the slope = 1. This supports the *dissociative activation.* This aspect has been discussed in Sec. 5.19.4.

(G) Knowledge of enthapy of transition ($\Delta H_T = \Delta H - \Delta H^{\neq}$) (*cf.* Sec. 5.19.3b): ΔH_T (difference in enthalpy between the transition state and product) has been found to be more or less constant (*ca.* 25 kcal mol^{-1}) for aquation of $[Co(NH_3)_5X]^{2+}$ (X^-= Cl^-, Br^- and NO_3^-) and it is also very much comparable to ΔH_T for the water exchange process at $[Co(NH_3)_5(OH_2)]^{3+}$. *This supports the existence of a* ***common transition state*** *for both the aquation process and water exchange process.* Thus independence of ΔH_T on the nature of the leaving group supports the **dissociative activation.**

(H) Electronic effect of the leaving group (X = acetate group) indicating some associative activation: In the aquation of $[Co(NH_3)_5X]^{2+}$ (X = acetate or substituted acetate), the basicity of the acetates was varied by a factor *ca.* 10^4 in moving from $CF_3CO_2^-$ to $Me_3CCO_2^-$. Thus it is reasonable to assume that the 'Co—X' bond strength should vary by a such large factor. If the 'Co—X' bond rupture is supposed to determine the activation energy solely then the rate constants for the aquation process should cover a large span. Interestingly, the aquation rate constants cover only a narrow span of 10. *This wide span of basicity of X^- against the narrow span of aquation rate constants indicates that the process is not a pure dissociative process.* Thus nucleophilic attack by H_2O is also partly important to produce the transition state.

If R—CO_2^- (*i.e.* X^-) is a weaker base, it will make a weaker Co—X bond (*favouring the dissociative activation*). But at the same time, it will create a lesser amount of electron density on the Co(III)–centre and it will facilitate the nucleophilic attack by H_2O (*i.e. associative activation*).

R — CO — O — Co ⟵ :Nu

If the associative activation is considered to operate then the aquation will be favoured most for X^- = CF_3COO^- causing the maximum electron depletion at the Co(III) centre and will be disfavoured most for X^- = Me_3CCOO^-. It happens so for the aquation process. *But the polar nonionic H_2O is not a very good nucleophile to make a huge difference in rate constants for the aquation process.*

Table 5.21.4 Aquation rate constants of some $[Co(NH_3)_5X]^{2+}$ (X^- = acetate and substituted acetates) complexes.

X^-		*pK_a for HX*		k_{aq} (min^{-1}) *(70°C)*	
$CF_3CO_2^-$	↑ Increasing electron withdrawing inductive effect	0.3	↓ Increasing basicity of X^-	3.3×10^{-3}	↑ Increasing aquation rate
$CCl_3CO_2^-$		0.7		3.2×10^{-3}	
$CHCl_2CO_2^-$		1.3		9.6×10^{-4}	
$CH_2ClCO_2^-$		2.85		3.5×10^{-4}	
$CH_3CO_2^-$		4.74		4.9×10^{-4}	
$(CH_3)_2CHCO_2^-$		4.82		1.6×10^{-4}	
$(CH_3)_3CCO_2^-$		5.0		2.6×10^{-4}	

(*cf.* F. Basolo, *et al., J. Phys. Chem.,* **56,** 22, 1952).

It may be noted that in presence of a better nucleophile like OH^- (which is a better nucleophile than H_2O), the span of variation of rate constants becomes larger.

The narrow span of aquation rate constants (k_{aq}) for the said series of complexes may be explained by considering the nucleophilic attack by the poor nucleophile H_2O to produce the transition state. ***It only rules out the perfect D–process but not the interchange process where both the bond rupture and bond formation by H_2O (as a nucleophile) contribute to determine the activation energy.***

(I) Role of the solvent (in favour of the SAD mechanism): Aquation rates for *trans*– $[Co(en)_2(NH_3)X]^{2+}$ follow the following sequences (X^- = Cl^-, Br^-, NO_3^-).

Aquation rate:	$NO_3^- \rangle Br^- \rangle Cl^-$	**$\Delta S^{\neq}$ controlled process.** 'Co—X' bond breaking energy is partly compensated by solvation of X^- in attaining the T.S. It makes $\Delta H^{\neq}$ relatively less important for controlling the rate sequence.
$\Delta H^{\neq}$:	$NO_3^- \rangle Br^- \rangle Cl^-$	
$\Delta S^{\neq}$:	$NO_3^- \rangle Br^- \rangle Cl^-$	

The above observation can be explained by considering the **solvent assisted dissociation** (SAD) **mechanism**. H–bonding interaction with the leaving group by the solvent will favour the rupture of the Co—X bond. Because of this H–bonding interaction, the 'Co—X' bond is elongated in the transition state and finally breaks down completely and H_2O enters into the vacated coordination site. The H–bonding interaction between the leaving group and water (solvent) follows the sequence:

$Cl^- \rangle Br^- \rangle NO_3^-$

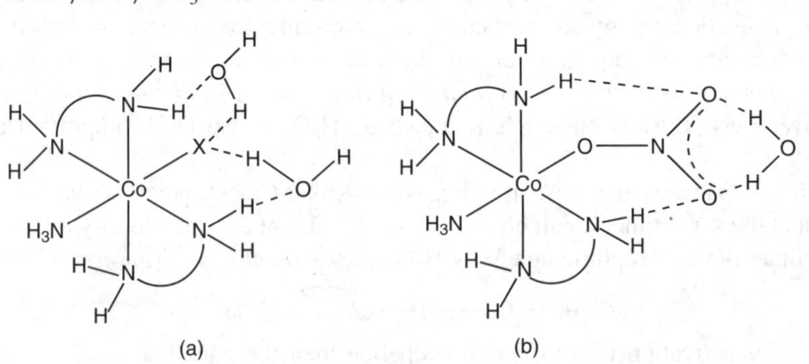

Fig. 5.21.2 Solvent assisted dissociation (SAD) of X^- in aquation of *trans*–$[Co(en)_2(NH_3)X]^{2+}$(X^- = halide, NO_3^-). (a) Hydration of leaving group for X^- = halide; (b) Intramolecular H–bonding of the leaving group, $X^- = NO_3^-$ leading to a less hydration of the leaving group NO_3^- at the stage of bond rupture. (*cf.* M.L. Tobe, *J. Chem. Soc.*, 3776, 1959; F. Basolo et al., *J. Am., Chem. Soc.*, **82,** 1077, 1960).

The order between Cl^- and Br^- is quite reasonable in terms of charge density on X^- (*i.e.* **higher charge denstiy causes higher solvation). The very low solvation for NO_3^- is believed to be due to the fact that the O–atoms of the ligated nitrate participate in an intramolecular H–bonding interaction with the —NH_2 groups of the coordinated amine ligand** (*i.e.* en). Consequently, before the complete dislodgement of the NO_3^- ligand, H–bonding interaction with the solvent is insignificant. However, NO_3^- once dislodged is solvated more than Cl^- or Br^-. This is why, the equilibrium constant (K_{aq}) for aquation follows the sequence:

K_{aq} : $NO_3^- \rangle Cl^- \rangle Br^-$ (trend of *H*–bonding interaction of the dislodged X^- with water).

The aquation rate (k_{aq}) sequence is governed by the $\Delta S^{\neq}$ sequence. More solvation of the departing ligand in the transition state will lower both the $\Delta H^{\neq}$ value (*i.e.* **enthalpic favour**) and $\Delta S^{\neq}$ value (*i.e.* **entropic disfavour;** lowering of $\Delta S^{\neq}$ arises from the electrorestriction over the solvent by H–bonding with the departing ligand). **Thus the loss of entropy in attaining the transition state is minimum for $X^- = NO_3^-$ which is least solvated in the transition state.**

(**Note:** SAD mechanism for the aquation of $[Cr(NH_3)_5X]^{2+}$ (X^- = Cl^-, Br^-, I^-) has also been argued. The aquation rate constants follow the sequence:

$$\begin{cases} k_{aq} : \ I^- > Br^- > Cl^- \\ \Delta S^{\neq} : \ I^- > Br^- > Cl^- \end{cases}$$ (*i.e.* least solvation of the departing ligand in the transition gives the maximum entropic favour).

Interestingly, **$\Delta H^{\neq}$ values remain more or less the same** for X^- = Cl^-, Br^- and I^- (*i.e.* the rate difference is controlled by $\Delta S^{\neq}$). The energy required to rupture the Co—X bond varies as: Co—Cl >

Co—Br > Co—I but the energy released through the solvation of departing halide follows the opposite order. It makes $\Delta H^{\neq}$ more or less the same. Thus in the SAD mechanism, the bond breaking energy is partially compensated by the solvation energy of the leaving group.

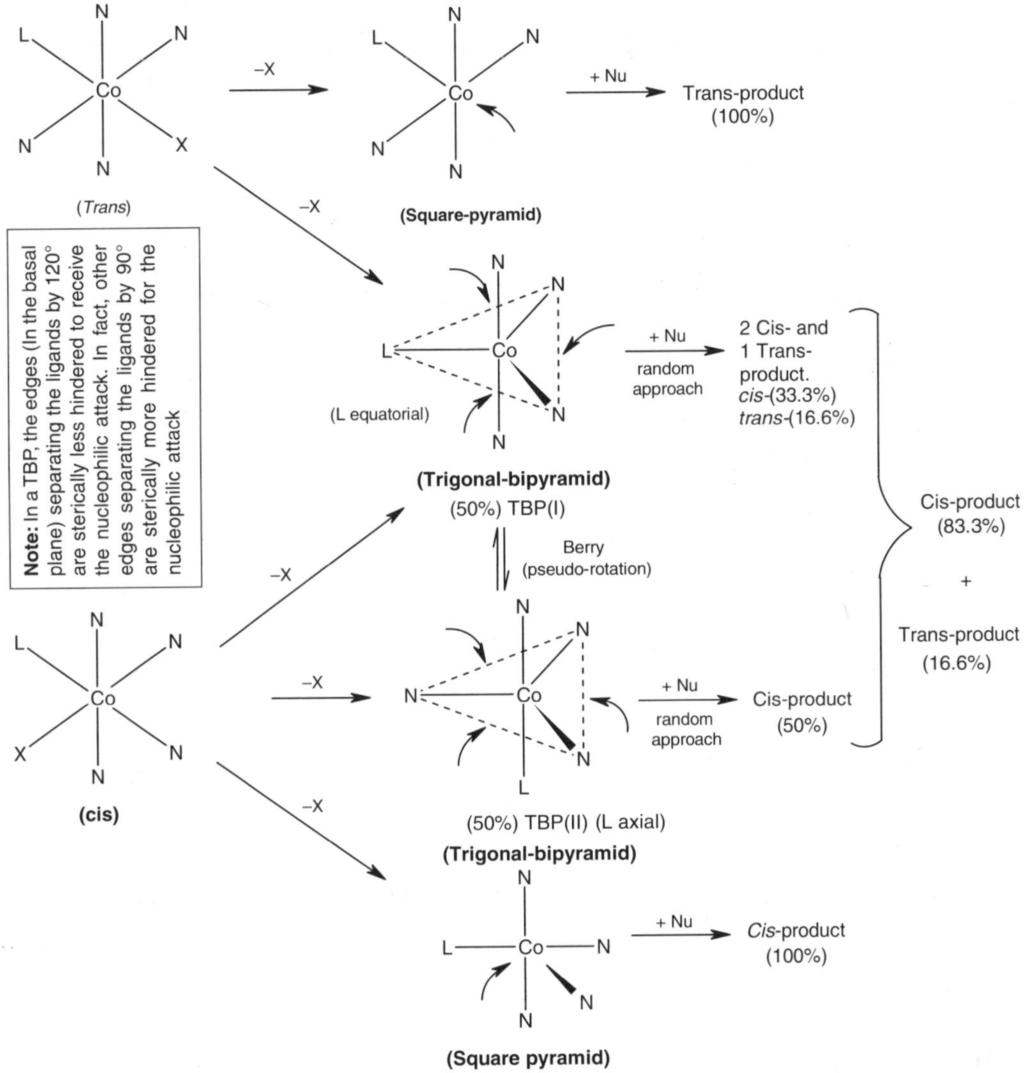

Fig. 5.21.3 Stereochemistry of the products of *cis–/trans–* $[Co(N)_4LX]^{n+}$ reacting through the dissociative (D) path. **Note:** N = nonleaving groups like NH_3; X = leaving group like Cl^-; L = Nonleaving group. **TBP(I)** (L at equatorial position) and **TBP(II)** (L at axial position) are supposed to be equally probable (*i.e.* 50% each) and they are mutually interconvertible through Berry's pseudorotation (*cf.* Fig. 5.28.2.2).

(J) Effect of pressure on the aquation process (*i.e.* interpretation of $\Delta V^{\neq}$) (*cf.* Sec. 5.19.5): In the aquation process for the *anionic leaving group,*

$$[Co(NH_3)_5X]^{n+} + H_2O \rightarrow [Co(NH_3)_5(OH_2)]^{(n+1)+} + X^-,$$

if the **dissociative path** operates, **separation of charge in the transition state** will tend to make $\Delta V^{\neq}$ negative because of the enhanced electrorestriction over the solvent imposed by the transition state. Thus here $\Delta V^{\neq}$ value will be less positive compared to that of the other dissociative processes (*e.g.* water exchange process, aquation of neutral leaving group) where there is no change of charge in attaining the **dissociative transition state.**

If the aquation process passes through the **associative path** (*i.e.* nucleophilic attack by H_2O), there will be ***no change of overall charge in attaining the transition state. Consequently, $\Delta V^{\neq}$ will be negative as usual.***

Thus, in the aquation process of an anionic leaving group, it is complicated to conclude firmly the mechanistic path in terms of the $\Delta V^{\neq}$ value. These aspects are discussed below.

$$[Co(NH_3)_5X]^{2+} + H_2O \rightarrow [Co(NH_3)_5(OH_2)]^{3+} + X^-$$

$$\begin{matrix} X^- = & Cl^- & Br^- & NCS^- \\ \Delta V^{\neq}\left(cm^3\ mol^{-1}\right): & -10.6 & -9.2 & -8.6 \end{matrix} \Bigg\} \begin{matrix} \Delta V^{\neq} \text{ for water exchange} \\ \text{process} = +1.2\ cm^3 mol^{-1} \end{matrix}$$

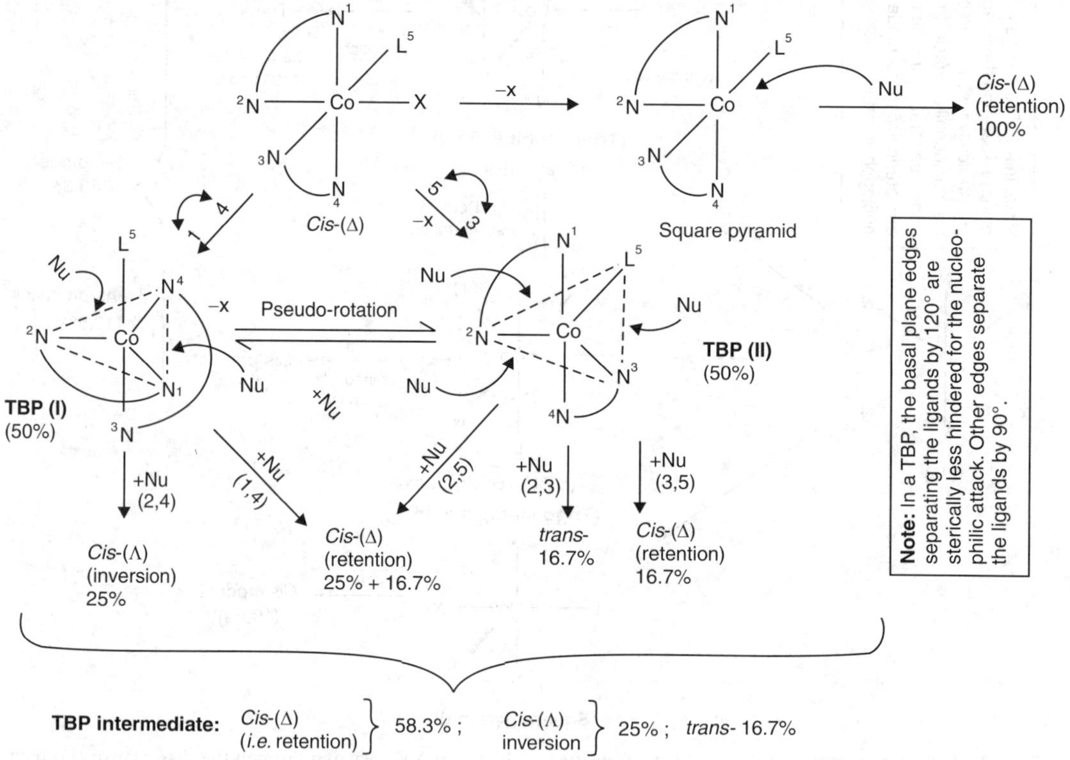

Fig. 5.21.4 Stereochemistry of the products of *cis*–$[Co(N—N)_2(L)X]^{n+}$ acting through the dissociative process.

Note:
- **TBP(I)** (L in axial position) and TBP(II) (L in equatorial position) are considered to be equally probable (*i.e.* 50% each) and they can be mutually interconverted through Berry's pseudo-rotation taking N(2) as the pivotal site.
- TBP(I) is generated, if the N(1) and N(4) sites move towards each other. TBP(II) is generated, if the N(3) and L(5) sites move towards each other.
- For TBP(I), nucleophilic attacks along the (1, 4) edge and (2, 4) edge are supposed to be equally probable. For TBP(II), nucleophilic attacks along the three edges (3, 5; 2, 3; 2, 5) are supposed to be equally probable.

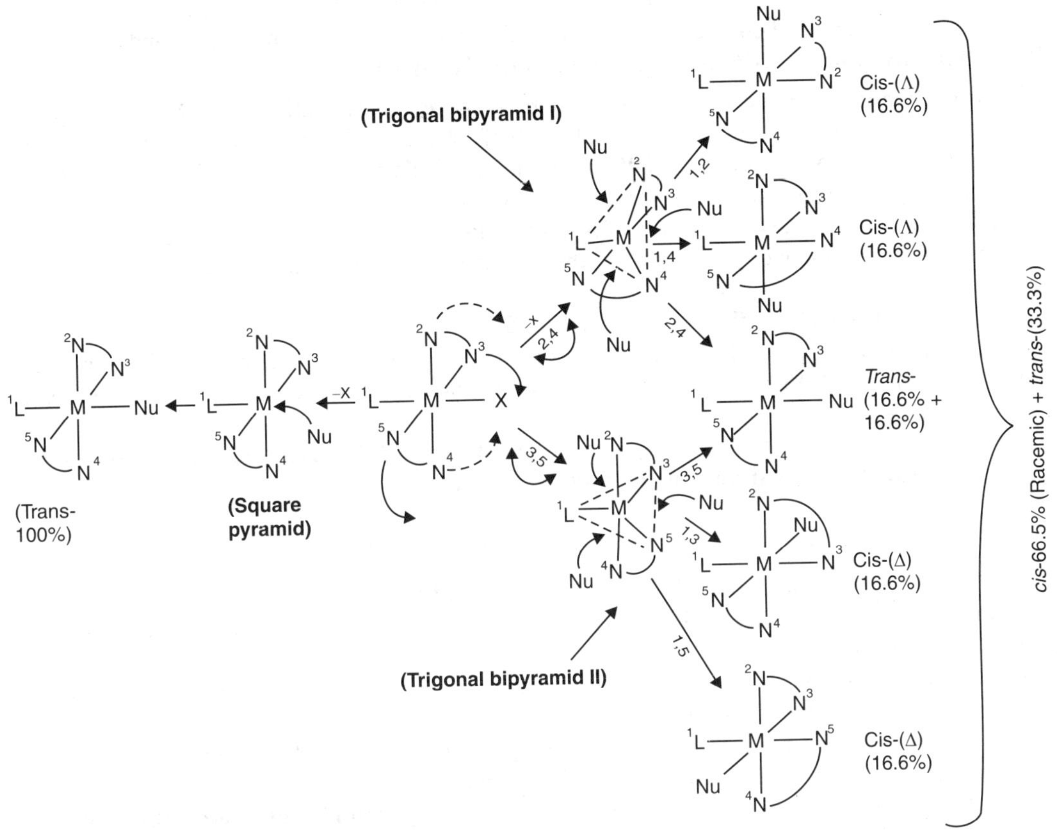

Fig. 5.21.5 Stereochemistry of products of *trans*–$[Co(N—N)_2(L)(X)]^{n+}$ reacting through the dissociative path.

Note: • Dissociation of X can lead two TBP–geometries *i.e.* TBP(I) and TBP(II). The TBP(I) is generated, if the N(2) and N(4) sites move towards each other. The TBP(II) is generated, if the N(3) and N(5) sites move towards each other. In TBP(I), N(3) and N(5) are kept at the axial positions while in TBP(II), N(2) and N(4) are kept at the axial positions.

• In calculating the product distribution, TBP(I) and TBP(II) are considered to be equally probable and each nucleophilic attack has been considered to be equally probable (*e.g.* for TBP (II), all the possible three nucleophilic attacks are equally probable).

The positive value of $\Delta V^{\neq}$ for the water exchange process conclusively supports the dissociative path. The negative value of $\Delta V^{\neq}$ for the said aquation processes is roughly the **same as the molar volume change (ΔV)** for the completed aquation reaction. *It indicates that the transition state resembles the final products.* Thus it can be concluded that in spite of the negative value of $\Delta V^{\neq}$, the said aquation processes pass through the dissociative path.

Similarly the negative value of $\Delta V^{\neq}$ (= –18.3 $cm^3\ mol^{-1}$) for the aquation of $[Co(NH_3)_5(SO_4)]^+$ can be argued for the dissociative path.

• Let us consider the **aquation of the neutral leaving group** in the following reaction.

$$[Co(NH_3)_5(L)]^{3+} + H_2O \rightarrow [Co(NH_3)_5(OH_2)]^{3+} + L,\ \Delta V^{\neq} = \text{positive value},$$

(leaving group is neutral).

$$L = OSMe_2,\ OCHNH_2,\ OC(NH_2)_2,\ \text{etc.}$$

In the above cases, there is no complication in interpreting the $\Delta V^{\neq}$ value as there is no possibility of creation or destruction of charge in attaining the transition state through the both *d*– and *a*–activation. ***Thus the positive value of*** $\Delta V^{\neq}$ ***conclusively supports the dissociative path.***

(K) Stereochemical change in the aquation process (*cf.* Secs. 5.25.3, 5.28.2): The important observations of aquation of *cis*–/*trans*– $[CoCl(en)_2(L)]^{n+}$ are given below (L = non-leaving group).

● For the ***cis*–complex**, there is a **100% retention of configuration irrespective of the nature of** L (*i.e.* there is no isomerisation). There are some exceptions.

● For the ***trans*–complex**, the degree of **isomerisation** depends on the π–bonding properties of L. With the increase π–donor property of L, the tendency to isomerise increases.
It follows the order.

$$L = OH^- \rangle NCS^- \rangle Cl^-, \text{ (cf. Tables 5.21.3, 5-6)}$$

● If the non-leaving group L is a *π–acceptor* ligand (*e.g.* NO_2^-, CN^-) or a non–π–bonding ligand (*e.g.* NH_3), there is **100% retention of configuration for both the *cis*– and *trans*–complexes** (*cf.* Tables 21.5.3, 5–6).

● For the rigid tetradentate ligand (N_4) like cyclam, there is a **100% retention of configuration for the complex which can exist only as the *trans*–complex,** *i.e.*

$$\left[CoCl(cyclam)(L)\right]^{n+} \xrightarrow{+H_2O} trans\text{-}\left[Co(cyclam)\cdot(OH_2)\right]^{(n+1)^+}$$

(100% retention of configuration irrespective of the nature of L).

(a) Prediction of Stereochemical Results in Terms of the Dissociative Path: The stereochemistry of the products is outlined in Figs. 5.21.3-5.

The important features of the dissociative path (*i.e.* formation of the 5–coordinate intermediate or activated complex) are given below:

(i) If the 5–coordinate intermediate is a **square pyramid** (*i.e.* removal of one octahedral position) **then there will a 100% retention of configuration.**

(ii) If the 5–coordinate intermediate adopts the **trigonal bipyramidal geometry**, **then there is a stereochemical change.**

(iii) If the nonleaving group (L) is a **π–donor ligand** (*e.g.* OH^-, NCS^-, Cl^-, etc.) then the square pyramidal intermediate obtained from the ***cis*-complex** is stabilised as the vacant orbital can receive back the electron cloud from the adjacent π–donor ligand L. The square pyramidal intermediate obtained from the ***trans*–isomer** cannot be stabilised by the π–donor ligand L unless it is converted into a trigonal bipyramidal geometry. ***This is the driving force for the conversion of the square pyramidal geometry into the trigonal bipyramidal geometry.*** It explains: better the π–donor property of L, more the chance for conversion into the trigonal bipyramidal geometry; *i.e.* more the isomerisation for the *trans*-complex.

(iv) In the **trigonal bipyramidal intermediate**, there are **two types of edges** — edges separating the ligands by 90° and edges separating the ligands by 120°. ***Obviously, the three edges at the basal or equatorial plane are the better positions to receive the nucleophilic attack because other edges separating the ligands by 90° are sterically more hindered.***
Thus in a trigonal bipyramidal intermediate obtained from *trans*– $[CoCl(N)_4(L)]^{n+}$, *e.g.* *trans*– $[CoCl(NH_3)_4(L)]^{n+}$, there are three possible ways to receive the nucleophilic attack. **There are two possible trigonal bipyramidal geometries remaining in an equilibrium through Berry's pseudorotation** (*cf.* Figs. 5.21.3-4).
There are also two possible trigonal bipyramidal geometries for *trans*– $[CoCl(N — N)_2(L)]^{n+}$, *e.g.* *trans*-$[CoCl(en)_2(L)]^{n+}$.

(v) If the nonleaving group (L) is a non–π–bonding ligand (*e.g.* NH_3) or a π–acceptor ligand (*e.g.* CN^-, NO_2^-, etc.), the square pyramidal geometry **(if formed)** does not have any driving force for its conversion into a trigonal bipyramidal geometry. Consequently there will be no isomerisation even if the aquation process goes on through the dissociative path. In fact, other evidences indicate the **associative path,** specially for the π–accepter nonleaving ligand. However, this associative path also explains the 100% retention of configuration (*cf.* Fig. 5.21.1c, d).

(vi) **Isomerisation (*cf.* Tables 5.21.3, 5-6) arises due to the dissociative path when the 5–coordinate intermediate (activated complex) adopts the trigonal bipyramidal structure.** Thus isomerisation, a consequence of the conversion of square pyramidal intermediate to the trigonal bipyramidal intermediate, leads to **positive $\Delta S^{\neq}$ value** (M.L. Tobe, 1968). On the other hand, for the square pyramidal intermediate (*i.e.* retention of configuration), **$\Delta S^{\neq}$ value is less and negative.**

Table 5.21.5 Stereochemical change in aquation of $[CoCl(N)_4(L)]^{n+}$, $[CoCl(N-N)_2(L)]^{n+}$ and $[CoCl(N_4)(L)]^{n+}$. (leaving group = Cl^-, *i.e.* aquation of Cl^-) (*cf.* M.L. Tobe, *Inorg. Chem.*, **7,** 1260, 1968; C.K. Poon, *et al.*, *J. Chem., Soc. A,* 1549, 1968).

***cis*–complex**	***Retention of Configuration (%)***	**$\Delta S^{\neq}$ (J K^{-1} mol^{-1})**
cis–$[CoCl(en)_2(CN)]^+$	100	–20
cis–$[CoCl(en)_2(OH)]^+$	100	–42
cis–$[CoCl_2(en)_2]^+$	100	–20
cis–$[CoCl(en)_2(NCS)]^+$	100	–59
cis–$[CoCl(en)_2(NO_2)]^+$	100	–13
cis–$[CoCl(en)_2(NH_3)]^{2+}$	100	–46
cis–$[CoCl(en)_2(N_3)]^+$	100	–17
cis–$[CoCl_2(cyclam)]^+$	100	–25
***trans*–complex**		
trans–$[CoCl_2(NH_3)_4]^+$	50 ± 10	+36
trans–$[CoCl(en)_2(CN)]^+$	100	–8
trans–$[CoCl(en)_2(NO_2)]^+$	100	–8.5
trans–$[CoCl(en)_2(OH)]^+$	25	+84
trans–$[CoCl_2(en)_2]^+$	65	+60
trans–$[CoCl(en)_2(N_3)]^+$	80	0
trans–$[CoCl(en)_2(NH_3)]^{2+}$	100	–46
trans–$[CoCl_2(cyclam)]^+$	100	–13
trans–$[CoCl(cyclam)(OH)]^+$	100	–29

Note: ● It has been mentioned that for the *cis*–$[CoCl(N—N)_2(L)]^{n+}$ complex, there is a complete retention of configuration, but in some cases, 15 – 30% conversion of the *cis*– into the *trans*–product has been noted. **This finding contradicts the earlier observation.** *It indicates that the 5–coordinate square pyramid intermediate is sufficiently long-lived in such cases to convert it into a trigonal bipyramidal geometry in some cases.*

● If the *cis*–product retains the **same enantiomeric configuration** of the starting *cis*–complex, then it can be assumed that the *cis*–product originates from the square pyramid intermediate and trigonal bipyramidal intermediate where **L–occupies the equatorial position** (*cf.* Fig. 5.21.4). Thus the loss of optical activity is due to only isomerisation to the *trans*–product and not due to recemisation.

- It may be noted that the trigonal bipyramidal intermediate bearing **L at the axial position** will lead to a racemic mixture (*cf.* Fig. 5.21.4).

Table 5.21.6 Prediction of stereochemical results in terms of dissociative path (*cf.* Figs. 5.21.3-5).

Complex	*Square pyramid intermediate*		*Trigonal bipyramidal intermediate**	
	cis–product	*trans–product*	*cis–product*	*trans–product*
cis–$[CoCl(N)_4(L)]^{n+}$	100%	0%	83.3%	16.7%
trans–$[CoCl(N)_4(L)]^{n+}$	0%	100%	83.3%	16.7%
Λ-*cis*–[CoCl(N—N)$_2$(L)]$^{n+}$	100%	0%	Λ = 58.3% Δ = 25%	16.7%
trans–[CoCl(N—N)$_2$(L)]$^{n+}$	0	100%	66.6% (Racemic)	33.3%

*Considering the equal stabilities of both the trigonal bipyramidal intermediates (*cf.* **Figs. 5.21.3–5**).

(b) Prediction of Stereochemical Results in Terms of Associative Path: The observation is:

100% retention of configuration for both the cis– and and trans– complexes (*cf.* Fig. 5.21.1 c, d). In fact, this has been experimentally verified for Cr(III), Rh(III), Ir(III) and Ru(III) complexes experiencing the associative activation. Even for the Co(III)–complexes like $[Co(en)_2(L)X]^{2+}$ where **L is a strong π–acceptor ligand,** the associative activation is quite reasonable.

$$\left.\begin{array}{l} cis\text{-}\left[RuCl_2(NH_3)_4\right]^+ \xrightarrow{\text{aquation}} 100\%\ \text{retention} \\ trans\text{-}\left[RuCl_2(NH_3)_4\right]^+ \xrightarrow{\text{aquation}} 100\%\ \text{retention} \end{array}\right\} (\text{aquation of } Cl^-)$$

The 100% retention of configuration can be explained by considering the adjacent attack (i.e. adjacent to the leaving group) by the incoming nucleophile (*cf.* Fig. 5.21.1c, d).

(L) Conclusion: The evidences given above (*A* – *K*) support the fact that the Co(III)–complexes aquate mainly through the dissociative path (most probably I_d path). However an associative activation has been argued in some special cases when the nonleaving group is a good π-acceptor ligand (*cf.* **K(b)** of this section) and the leaving group is a powerful electron withdrawing group (*cf.* **H,** Table 5.21.4 of this section).

5.22 AQUATION OR ACID HYDROLYSIS OF THE OCTAHEDRAL CHROMIUM(III) COMPLEXES (*cf.* Secs. 5.19.5-6)

(A) Solvent assisted dissociation (SAD) mechanism for $[Cr(NH_3)_5X]^{2+}$ (X^- = Cl^-, Br^-, I^-): The characteristic features of SAD mechanism for aquation of $[Co(NH_3)_5X]^{2+}$ have been discussed. The same features are prevailing here also.

$$k_{aq}: \quad I^- > Br^- > Cl^-$$

$$\Delta S^{\neq}: \quad I^- > Br^- > Cl^-$$

In the SAD path, the departing ligand is solvated through H-bonding and *the entering nucleophile* H_2O *simultaneously slips into the position vocated by* X^-. The rate sequence can be explained in terms of $\Delta S^{\neq}$. This aspect of SAD mechanism has been discussed in Sec. 5.21.1.

(B) Group replacement factor (GRF): In aquation of $[Cr(NH_3)_5X]^{2+}$ (X^- = halide), the GRF (= $k_{R-X})/k_{R-Cl}$) values run in the sequence: $I^- \rangle Br^- \rangle Cl^- \rangle\rangle F$. It rules out the pure associative path for which GRF values should run in the sequence: $F^- \rangle\rangle Cl^- \rangle Br^- \rangle I^-$. F^- (most electron withdrawing) should make the metal centre most suitable for the nucleophilic attack. The GRF values indicate the importance of bond breaking in the activation process (*cf.* the conclusion of **SAD mechanism**).

(C) LFER plot: The slopes of LFER plots (*i.e.* log k_{aq} *vs.* log K_{eq}) for the aquation of $[Cr(NH_3)_5X]^{2+}$ and $[Cr(OH_2)_5X]^{2+}$ are 0.91 and 0.56 respectively. It indicates that separation of X^- is about 90% complete in the transition state for the aquation of $[Cr(NH_3)_5X]^{2+}$ while the separation is only 50–60% complete for the aquation of $[Cr(OH_2)_5X]^{2+}$. It indicates the more dissociative character in the aquation of $[Cr(NH_3)_5X]^{2+}$ (*cf.* Secs. 5.19.4, 6 for explanation in terms of energy of the LUMO).

(D) Steric retardation and activation parameter: $[CrCl(NH_2Me)_5]^{2+}$ aquates slower than $[CrCl(NH_3)_5]^{2+}$ because of the **steric retardation** as expected in an associative activation. Higher $\Delta H^{\neq}$ and $\Delta S^{\neq}$ values for the sterically more crowded complex may be due to the **weaker bond formation by H_2O at the associative T.S.** However, the higher $\Delta H^{\neq}$ value may indicate the ***importance of bond breaking in attaining the transition state.***

	10^6k_{aq} (s^{-1})	$\Delta H^{\neq}$ (kJ mol^{-1})	$\Delta S^{\neq}$ (J K^{-1} mol^{-1})
$[CrCl(NH_3)_5]^{2+}$:	8.7	93	−29
$[CrCl(NH_2Me)_5]^{2+}$:	0.25	110 ↓	−2 ↓
		Enthalpic disfavour	**Entropic favour**

Here it is worth mentioning that $[CoCl(NH_2Me)_5]^{2+}$ aquates faster than $[CoCl(NH_3)_5]^{2+}$ because of the **dissociative activation experiencing the steric acceleration** (*cf.* Sec. 5.21B)

(E) Volume of activation: Aquation of $[Cr(NH_3)_5L]^{3+}$ (L = neutral ligands like DMSO, DMF, urea, etc.) experiences the negative $\Delta V^{\neq}$ values (*cf.* analogous Co-compounds experience positive $\Delta V^{\neq}$ values; *cf.* Sec. 5.21J). It indicates the associative activation for such Cr(III) complexes.

A plot of $\Delta V^{\neq}$ values of aquation of $[Co(NH_3)_5X]^{2+}$ versus the $\Delta V^{\neq}$ values of aquation of $[Cr(NH_3)_5X]^{2+}$ gives the slope far away from unity. It indicates that $[Cr(NH_3)_5X]^{2+}$ complexes aquate in a different mechanism compared to the analogous $[Co(NH_3)_5X]^{2+}$ complexes which adopt the dissociative path.

(F) Effect of charge: The aquation rates follow the following sequences.

$[MCl(NH_3)_5]^{2+} \langle [MCl_2(NH_3)_4]^+$; rate enhancement by a factor about 10^3–10^4 (for M = Co), and by a factor about 10^1 only (for M = Cr).

Thus for the Co(III)-complex, the rate is very much sensitive to the overall charge of the complex while for the Cr(III)-complex, the effect is less significant. ***Thus for the Co(III)-complexes, dissociative activation is primarily important but for the Cr(III) complex, it is not so important.***

(G) Comparison between calculated and experimentally found activation energy (E_a): For the aquation of $[M(NH_3)_5X]^{2+}$ (M = Co, Cr), activation energy has been calculated by considering possible the intermediates (*i.e.* activated complex) like *square pyramid* (*i.e.* dissociative path), *pentagonal bipyramid* (*i.e.* associative path), *trapezodial octahedron* (*i.e.* associative path). For this calculation of activation energy, change of cfse, and interelectronic repulsion energy in the formation of the transition state were mainly considered. For the Co(III)-complexes, the observed E_a values were found in good agreement with those calculated for the square pyramidal intermediate. On the other hand, for the Cr(III) complexes, observed E_a values were in good agreement with those calculated for those predicted for the pentagonal bipyramid intermediate. These support the **dissociative path for the Co(III)-complexes** and the **associative path for the Cr(III) complexes.**

Conclusion: Aquation of the Cr(III)-complexes generally experiences the associative activation activation (most probably the I_a-path) while the analogous Co(III) complexes generally experience the dissociative activation. The reason behind this difference has been discussed in Sec. 5.19.6.

5.23 ELECTROPHILIC CATALYZED AQUATION OF THE OCTAHEDRAL COMPLEXES

Very often the leaving groups (X) are the conjugate bases of weak acids. They can bind with the **electrophiles** (E) (*e.g.* H^+, M^{n+}) even when coordinated in a metal complex. Sometimes, the polydentate leaving groups can bind with such electrophiles either in partially dechelated condition (*e.g. en, bpy*) or in fully coordinated condition (*e.g.* ox^{2-}, $acac^-$, etc.). This binding interaction with the electrophile favours the removal of the group (X). The process can be illustrated as follows in a general way.

(i) $M—X+E \underset{}{\overset{K_E}{\rightleftharpoons}} M—X—E \xrightarrow[(+H_2O)]{k_E} M—OH_2+E—X,$

$$[M—X—E]=\frac{K_E[E][M—X]}{1+K_E[E]}, \text{ (cf. Sec. 5.9.3)}$$

(ii) $M—X \xrightarrow[(+H_2O)]{k_0} M—OH_2+X^-$

$$\text{Rate of aquation}=k_0[M—X]+k_E[M—X—E]=k_0[M—X]+\frac{k_EK_E[E][M—X]}{1+K_E[E]}$$

$$=\left\{k_0+\frac{k_EK_E[E]}{1+K_E[E]}\right\}[M—X].$$

The second term involving the k_E term represents the contribution of the catalyzed path and the first term involving the k_0 term represents the contribution of the uncatalyzed path to the aquation process. Depending on the condition, the above rate equation may be reduced as follows:

$$\text{rate} \approx \{k_0 + k_EK_E[E]\}[\text{complex}];\ 1 \gg K_E[E]$$

$$\approx \{k_0 + k_E\}\ [\text{complex}];\ K_E[E] \gg 1$$

Very often k_E increases with K_E (measuring the stability of interaction between the electrophile and leaving group). This stability interaction measured by K_E can be easily understood by HSAB principle. If the leaving group provides the hard donor sites (*e.g.* O, F, N, etc.) then the hard electrophiles, *i.e.* hard acids (*e.g.* H^+, Be^{2+}, Al^{3+}, etc.) are suitable. On the other hand, if the leaving group is a softer one (*e.g.* Cl, Br, I, S, etc.), then the soft acids (*e.g.* Hg^{2+}, Ag^+, Tl^+, etc.) are more suitable.

5.23.1 Brönsted Acid Catalyzed Aquation Reactions

The catalyzed path is:

$$M—X+H^+ \rightleftharpoons \underset{\text{CA}}{\left[M—X—H\right]^+} \xrightarrow{+H_2O} M—OH_2+XH$$

If the protonated species $[M—XH]^+$ aquates through the *dissociative path,* then the process is referred to as the **D–CA mechanism** (≡ **S_N1–CA** in old nomenclature); if the protonated species aquates through the associative path, then the process is referred to as the **A–CA** (≡ **S_N2–CA**) mechanism.

Some examples of the acid catalyzed aquation reactions are discussed below.

- **Aquation of $[M(NH_3)_5X]^{n+}$, $[M(NH_3)_4X_2]^{n+}$ (X** having the free basic sites): Aquation of the metal complexes $[M(NH_3)_5X]^{n+}$ (M = Co, Cr; X = F^-, NO_2^-, N_3^-, CN^-, CO_3^{2-}, SO_4^{2-}, oxalate, acetylacetonate, etc.) is acid catalyzed. Here the metal bound leaving group X undergoes protonation because of the presence free basic sites.

- **Aquation of $[M(NH_3)_5X]^{2+}$ (X = Cl, Br, I) *vs.* $[M(NH_3)_5F]^{2+}$:** $[M(NH_3)_5X]^{2+}$ (X = Cl, Br, I) does not experience the acid catalyzed aquation but $[Co(NH_3)_5F]^{2+}$ experiences the H^+ catalyzed aquation. Protonation of the bound Cl, Br or I is less likely because of two reasons: *these softer sites do not like* H^+ *for interaction;* HCl, HBr *and* HI *are much stronger acids than* HF (*i.e.* basicity order: $F^- >> Cl^- > Br^- > I^-$) *and consequently protonation of the heavier halides is less likely.*
- **Aquation of $[M(L—L)_3]^{n+}$:** For some ligands like en, bpy, phen, etc. there is **no free basic** site for protonation when they remain coordinated. In such cases, acid catalysis is possible, if the ligand undergoes **partial dechelation** to provide the protonation sites. On the other hand, the ligands like oxalate, acetylacetonate, biguanide, etc. may undergo protonation even when they remain coordinated because of the presence of **free basic site.** Acid catalyzed aquation of such ligands becomes relatively easier. In fact, the biguanide complexes of Co(III) or Cr(III) are thermodynamically more stable than their corresponding ethylenediamine complexes. But in terms of the acid catalyzed aquation, ***the biguanide complexes are more labile than the ethylenediamine complexes***.

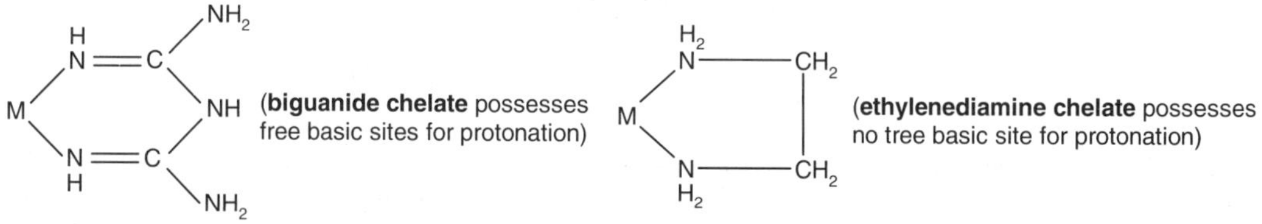

It is reasonable to expect that the more basic ligands will favour the acid catalyzed aquation path more. For the biguanide complexes, the ease of acid catalyzed aquation follows the sequence (*cf.* D. Banerjee, *et al., J. Inorg. Nucl. Chem.,* **26,** 1233, 1964): biguanide ⟩ *n*-hexylbiguanide ⟩ phenylbiguanide

Interestingly, the thermodynamic stability of the complexes also follow the same sequence.

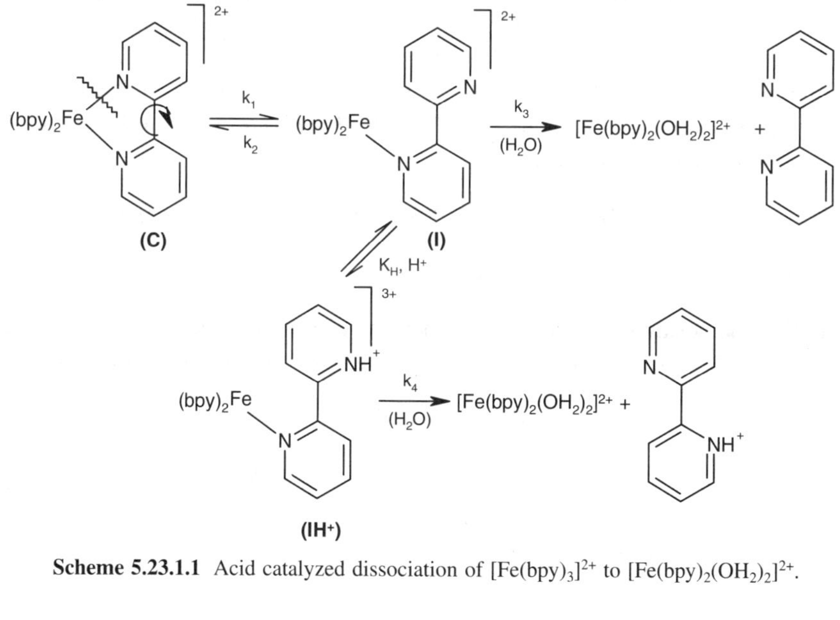

Scheme 5.23.1.1 Acid catalyzed dissociation of $[Fe(bpy)_3]^{2+}$ to $[Fe(bpy)_2(OH_2)_2]^{2+}$.

- **Acid catalysed aquation of $[M(bpy)_3]^{2+}$ *vs.* $[M(phen)_3]^{2+}$ (M = Fe, Ni):** Acid catalyzed aquations of $[M(bpy)_3]^{2+}$, $[M(phen)_3]^{2+}$ (M = Fe, Ni) are quite intersecting in terms of acid dependence. ***The bpy-complex shows the acid dependence while the phen-complex does not show any acid dependent path*** (*cf.* J.H. Baxendale *et al., Trans. Faraday Soc.*, **46,** 736, 1950; J.E. Dickens *et al., J. Am. Chem. Soc.*, **75,** 1286, 1957).

In the *bpy* and *phen* complexes, for protonation of their complexes, partial dechelation (*i.e. one ended dissociation*) of the coordinated ligand is required. The ***bpy* complex** may experience such a **one ended dissociation** accompanied by protonation of the released end of the ligand. Then it experiences a fast rupture of the remaining metal-ligand bond leading to the complete loss of the ligand. This pathway is illustrated in scheme 5.24.1.1. ***It may be noted that the intermediate having a one ended dissociated ligand experiences a rotation around the C–C single bond to take away the dislodged N-site from the coordination sphere.*** This intermediate may also experience the uncatalyzed aquation.

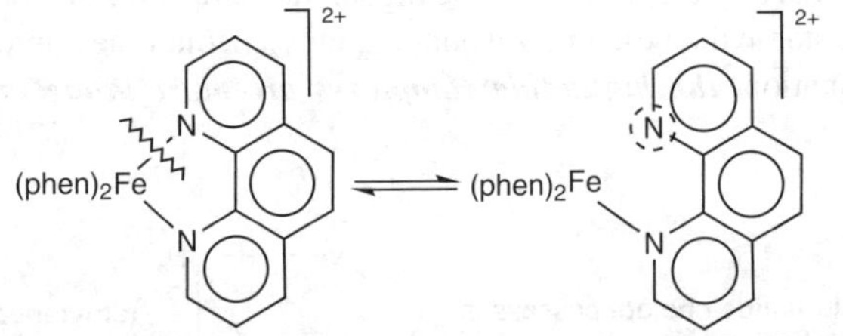

Scheme 5.23.1.2 Possibility of one-ended dissociation of the phen-chelate ring in $[Fe(phen)_3]^{2+}$. The dislodged N-site denoted by N is still within the coordination sphere as the rigidity of the structure prevents this dislodged site to go away from the coordination sphere. ***Thus, there is no real existence of the one-ended opening of the phen-chelate ring.***

In the case of phen-complexes, one ended dissociation pathway is not possible. *In phen, the rotation around the C–C bond is prevented due to its rigid structure where the third ring remains fused to the bpy-skeleton.* Thus, even if the *phen*-chelate ring experiences the said one ended dissociation, the dislodged N-site still remains wthin the coordination sphere. Thus, practically there is no meaning of such one ended dissociation of the *phen* ring. *For the loss of the phen ligand, both the 'M–N' bonds must rupture simultaneously at the rate determining step.* Then the completely dislodged *phen* ligand undergoes protonation rapidly and this prevents its reentry into the coordination sphere.

Thus presence of acid is required to prevent the overall backward reaction but not to catalyze the forward dechelation reaction kinetically.

$$\left[M(N-N)_3\right]^{2+} \rightleftharpoons \left[M(N-N)_2\right]^{2+} + N-N, \textbf{(slow)}\ (N-N = phen)$$

$$N-N+H^+ \rightleftharpoons N-NH^+, \textbf{(fast)}, \left(pK_a \ of \ N-NH^+ \approx 4\right)$$

By considering the steady-state condition to the species **(I)** (*cf.* Scheme 5.24.1.1) having a **half-bonded bpy ligand** (*i.e.* partially dissociated bpy ring), the observed rate constant of aquation is given by:

$$\frac{d[I]}{dt} = 0 = k_1[C] - k_2[I] - k_3[I] - k_4\left[IH^+\right];\ K_H = \frac{\left[IH^+\right]}{[I]\left[H^+\right]} \text{ or, } \left[IH^+\right] = K_H[I]\left[H^+\right]$$

$$= k_1[C] - k_2[I] - k_3[I] - k_4K_H[I]\left[H^+\right]$$

or, $$[I]=\frac{k_1[C]}{(k_2+k_3)+k_4K_H[H^+]}$$

$$\text{Rate}=k_3[I]+k_4[IH^+]=\frac{k_1k_3[C]}{(k_2+k_3)+k_4K_H[H^+]}+\frac{k_1k_4K_H[H^+][C]}{(k_2+k_3)+k_4K_H[H^+]}$$

$$=\frac{k_1(k_3+k_4K_H[H^+])[C]}{(k_2+k_3)+k_4K_H[H^+]}$$

$$i.e.\ k_{obs}=\left\{\frac{k_3+K_Hk_4[\mathrm{H}^+]}{(k_2+k_3)+K_Hk_4[\mathrm{H}^+]}\right\}k_1 \text{ when, } [\mathrm{H}^+]\rangle\rangle[C]$$

- At high acidity, $K_Hk_4[H^+] \rangle\rangle k_3$ and $K_Hk_4[H^+] \rangle\rangle (k_2 + k_3)$ and it leads to:

  $k_{obs} = k_1$, (*i.e.* **acid independent**)
- At very low acidity, $k_3 \rangle\rangle K_Hk_4[H^+]$ and $(k_2 + k_3) \rangle\rangle K_Hk_4[H^+]$ and it leads to:

$$k_{obs}=\frac{k_1k_3}{k_2+k_3},\ (i.e.\ \textbf{acid independent}).$$

- At the intermediate range of $[H^+]$, k_{obs} shows a **complicated pattern of acid dependence** (*cf.* Fig. 5.24.1.1).

The **limiting values** of k_{obs} are:

$$\boldsymbol{k_{obs}}\ (\textbf{lower limit})=\frac{k_1k_3}{k_2+k_3}(\text{at very low acidity})$$

$$\boldsymbol{k_{obs}}\ (\textbf{upper limit})=k_1\ (\text{at very high acidity})$$

$$\frac{k_{obs}\ (\text{lower limit})}{k_{obs}\ (\text{upper limit})}=\frac{k_3}{k_2+k_3}\approx\frac{k_3}{k_2}\approx 0.16$$

(experimentally found)

It signifies that at very low acidity, from the 5-coordinate intermediate (I), approximately 84% chances are for reforming the bond to give back the *tris*-chelate and approximately 16% chances are for complete dissociation to give the *bis*-complex.

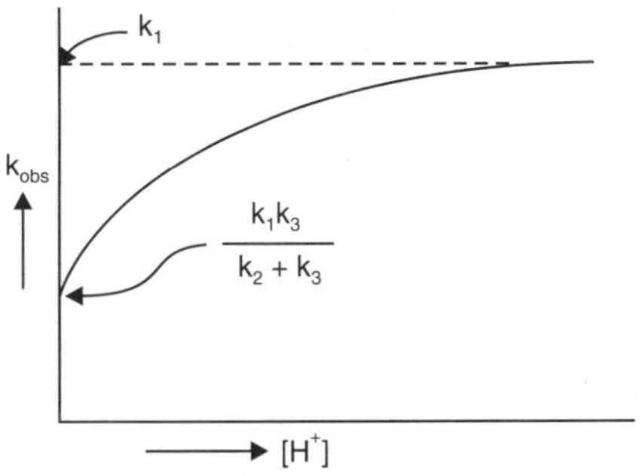

Fig. 5.24.1.1 Acid dependence of the aquation rate of $[Fe(bpy)_3]^{2+}$.

- **Aquation of $[Ni(N—N)_3]^{2+}$ *vs.* $[Fe(N—N)_3]^{2+}$:** (i) For dechelation of the *tris*-chelate of Ni(II) to the *bis*-chelate, it follows the same behaviour as found for the corresponding Fe(II) complexes.

(ii) The Ni(II) complexes differ from the Fe(II) complexes in the following aspect.

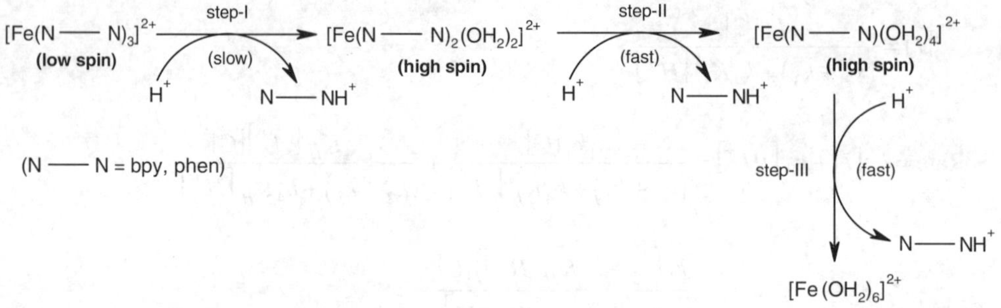

Here the lability sequence runs as:

Step I ⟨⟨ Step-II, Step-III

i.e. ***the bis– and mono-chelates are more labile than the tris-chelate.*** *It happens so because the tris-chelate is a low-spin complex having a very high crystal field activation energy while the lower complexes are the high-spin complexes having a relatively smaller crystal field activation energy.* In the case of corresponding Ni(II)-complexes, it does not occur so because there is no such spin-state change in moving from the higher complexes to lower complexes.

5.23.2 Metal Ion Catalyzed Aquation Reactions (*cf.* D. Banerjea, *Transition Met. Chem.*, **12,** 97, 1987)

The didentate and the multidentate ligands like oxalate, malonate, acetylacetonate, aminopolycarboxylate may bind with the catalyzing metal ions to produce the **binuclear bridged intermediates** which subsequently undergo aquation. It is illustrated below for the **oxalato complex**.

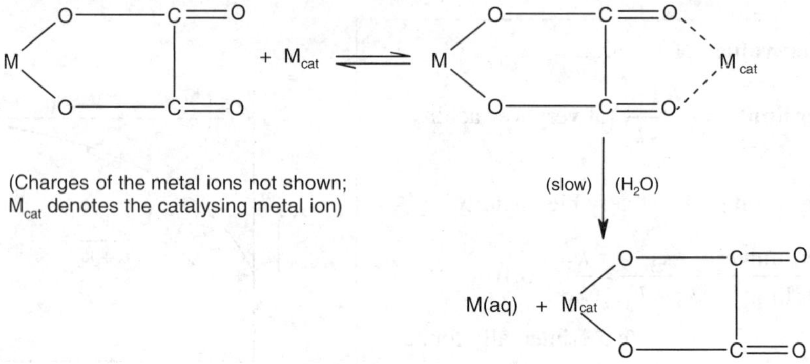

For the **biguanide complex,** the intermediate binuclear complex may be represented as follows:

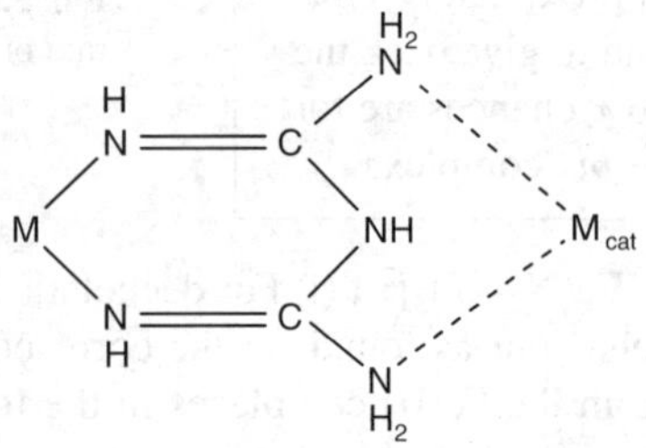

The catalyzing power of M_{cat} depends on HSAB matching and complexing power that can be measured by Z^{*2}/r where $Z^* = Z$ – shielding constant.

Effect of the nature of the catalyzing metal ion on the reaction pathway: The metal ion catalyzed aquation of $[Co(NH_3)_5X]^{2+}$ (X = Cl, Br, I) furnishes some information regarding the transition state. In presence Hg^{2+} as a catalyst, the aqua product $[Co(NH_3)_5(OH_2)]^{3+}$ bears the same ratio of $^{16}O/^{18}O$ for different halides when the aquation reaction is carried out in water enriched in ^{18}O content. This observation supports the **perfect D-process,** *i.e.* in attaining the transition state, the leaving group X^- is completely dislodged so that it cannot influence the aquation process.

Interestingly, in presence of the catalyzing metal ions like Ag^+ or Tl^+, the ratio of $^{16}O/^{18}O$ content in the aqua product was found to depend on the nature of the leaving group X^-. ***It indicates that though a common intermediate is produced for the Hg^{2+} catalyzed aquation of $[Co(NH_3)_5X]^{2+}$, no such common intermediate is produced for the Ag^+ or Tl^{3+} catalyzed aquation process.*** In fact, for the Ag^+ or Tl^+ catalyzed aquation reaction, in the transition state, both the leaving group and entering nucleophile (H_2O) remain bound as in the **interchange process** where the nucleophilic attack by H_2O is relatively less important than the bond breaking by the leaving group. Thus the suggested mechanism is I_d in which the leaving group is not completely dislodged at the transition state. ***This is why, the partially dislodged leaving group at the transition state influences the entry of the nucleophile (i.e. H_2O).*** This extent of dislodgement (catalyzed by the metal ion) of the leaving group depends on the nature of the leaving group and catalyzing metal ion. For this reason, for the reaction passing through the I_d-path, the ratio of ^{16}O and ^{18}O in the product depends on the nature of the leaving group for a particular metal ion catalysing the process.

5.24 PSEUDO-SUBSTITUTION REACTION WITHOUT THE METAL-LIGAND BOND BREAKING THROUGH THE ELECTROPHILIC ATTACK ON THE BOUND LIGAND

The ligand substitution process has been found in many cases to proceed fast apparently even for the inert metal centres like Co(III), Rh(III), Ir(III). These are explained by considering the **pseudo-substitution** reaction where the actual metal-ligand bond does not rupture. Some representative examples are shown below:

(i) **Acid catalyzed aquation of $[M(CO_3)(NH_3)_5]^+$ (M = Co, Rh, Ir)** (*cf.* A.B. Lamb *et al.*, *J. Am. Chem. Soc.*, **67,** 468, 1945): The said carbonato complexes experience a rapid aquation (*i.e.* acid hydrolysis) to form the aqua-complex, $[M(NH_3)_5(OH_2)]^{3+}$.

$$\left[M(CO_3)(NH_3)_5\right]^+ \xrightarrow{H_3O^+} \left[M(NH_3)_5(OH_2)\right]^{3+} + CO_2(aq).$$

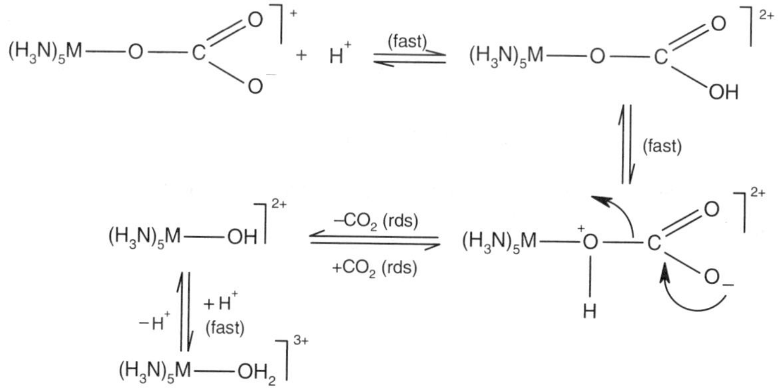

When the said aquation reactions are carried out in presence of H_2O^*, no O^* is found in the aqua product. It supports that the M—O bond remains intact and cleavage of the 'C—O' bond gives off CO_2. The probable mechanism involves the *electrophilic attack* by the proton (*i.e.* H_3O^+) on the O-atom of the bound carbonate ligand.

In fact, formation and aquation of the bicarbonato/carbonato complexes can be shown as follows:

$$\left[(H_3N)_5M-{}^*OH_2\right]^{3+} + HCO_3^- \underset{k_d}{\overset{k_f}{\rightleftharpoons}} \left[(H_3N)_5M-O^*-CO_2^-\right]^+ + H^+ + H_2O$$

The k_f (in s^{-1}) and k_d (in $dm^3\ mol^{-1}\ s^{-1}$) values at 25°C for the said complexes are as follows:

M(III) =	**Co(III)**	**Rh(III)**	**Ir(III)**
k_f (s^{-1}):	1.10	1.13	1.45
k_d ($dm^3\ mol^{-1}\ s^{-1}$):	220	470	590

It is evident that the k_f and k_d values are very much comparable for Co(III), Rh(III) and Ir(III). The lability order is: Co(III) ⟩⟩ Rh(III) ⟩⟩ Ir(III). *Thus, from the comparable values of k_f and k_d, it is reasonable to conclude that for the said processes, the metal-ligand bond replacement does not occur.* For the aquation of $[MBr(NH_3)_5]^{2+}$, the aquation rates follow the sequence.

$$k_{aq}\ (s^{-1}),\ 25°C: 6.3 \times 10^{-6}\ (Co),\ 1 \times 10^{-8}\ (Rh)\ \text{and}\ 2 \times 10^{-10}\ (Ir).$$

In these bromido-complexes, aquation goes through the M—Br bond replacement. Similarly, aquations of $[Co(NH_3)_5(SO_4)]^+$, $[Co(NH_3)_5(NO_3)]^{2+}$ and $[Co(CH_3COO)(NH_3)_5]^{2+}$ proceed through the 'Co—O' bond replacement and the processes are slow as usual.

(ii) **Formation of $[M(NH_3)_5(-ONO)]^{2+}$ from $[M(NH_3)_5(OH_2)]^{3+}$ by HNO_2–NO_2^- buffer:** The nitrito-isomer, $[M(NH_3)_5(ONO)]^{2+}$ is thermodynamically less stable than the corresponding nitro or nitrito(κN)-complex (*i.e.* linkage isomer). In the said reaction, the nitrito(κO) isomer (*kinetically controlled product*) is produced at a fairly fast reaction through the **pseudo-substitution process** as follows:

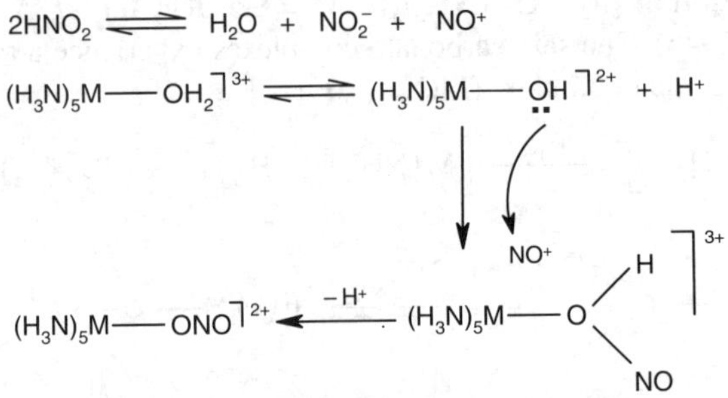

(M = Cr, Co, Rh, Ir, etc.)

Formation of the said nitrito(κO) complex may be also explained by taking the reactive species N_2O_3 as follows:

$$2HNO_2 \rightleftharpoons N_2O_3 + H_2O$$

$$[(H_3N)_5Co-{}^*OH]^{2+} + N_2O_3 \rightleftharpoons \left[(H_3N)_5Co-{}^*O---H \atop O=N-\cdot\cdot O-N=O \right]^{2+}$$

$$\downarrow rds$$

$$[H_3N)_5Co-{}^*ONO]^{2+} + HONO \ (i.e. \ HNO_2)$$

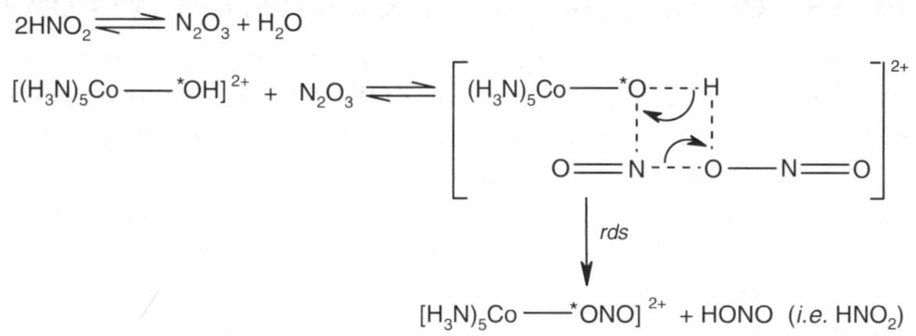

Thus the real substitution reaction outlined below is ruled out.

$$\left[(H_3N)_5Co-{}^*OH \right]^{2+} + NO_2^- \longrightarrow \left[(H_3N)_5Co-NO_2 \right]^{2+}$$

$$\text{or} \ \left[(H_3N)_5Co-ONO \right]^{2+} + {}^*OH^-$$

For Co(III), the nitrito(κO) isomer is **red coloured** and the nitro isomer is **yellow coloured.** In this pseudosubstitution process, the M—O bond does not break down. The proposed mechanism has been supported by the fact that when the reaction is carried out in water enriched with H_2O^*, there is no incorporation of O* in the product.

Here it may be mentioned that formation of the nitrito(κO) isomer occurs fairly fast even at very low temperature but on heating, the nitrito(κO) isomer **slowly** (as expected from the inertness of low-spin d^6 metal centres) isomerizes to the nitro or nitrito(κN) isomer through the M—O bond rupture.

$$(H_3N)_5M-OH_2 \rceil^{3+} \xrightarrow[\text{fast, (low temp.)}]{HNO_2-NO_2^-} (H_3N)_5M-ONO \rceil^{2+}, \textbf{(Kinetically controlled product)}$$
(Red)

$$\downarrow \text{(slow)} \ \text{Heating}$$

$$(H_3N)_5M-NO_2 \rceil^{2+}, \textbf{(Thermodynamically controlled product)}$$
(Yellow)

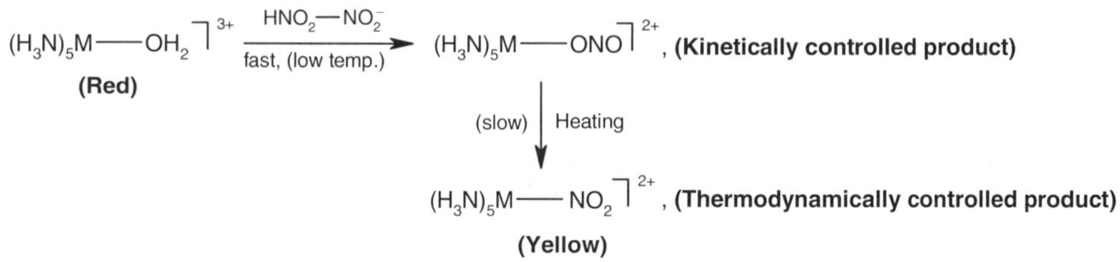

(iii) Catalyzed aquation of $[(H_3N)_5M(-ONO)]^{2+}$: The aquation process can be catalyzed by Cl^-, Br^- in acidic condition. It goes on through the pseudo-substitution process as outlined below.

$$(H_3N)_5M-{}^*ONO \rceil^{2+} \xrightarrow{H^+, X^-} (H_3N)_5M-{}^*O-N{\overset{X}{\underset{O}{\big\langle}}} \rceil^{2+}$$

(M = Co, Cr)

$$\downarrow -NOX$$

$$(H_3N)_5M-{}^*OH_2 \rceil^{3+} \xleftarrow{H_3O^+} (H_3N)_5M-{}^*OH \rceil^{2+}$$

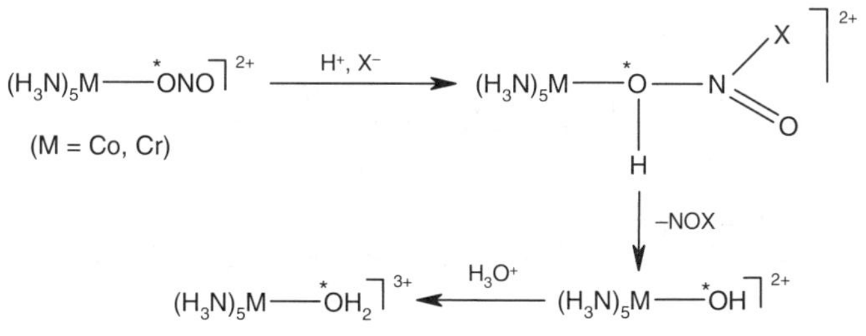

NOX is rapidly hydrolyzed to HNO_2 and HX. In this aquation process, there is no metal-ligand bond rupture.

5.25 BASE HYDROLYSIS OF THE OCTAHEDRAL COBALT(III) AMMINE COMPLEXES

We have already mentioned (*cf.* Sec. 5.20) that in the rate law for the hydrolysis of the complex like $[Co(NH_3)X]^{2+}$, the base hydrolysis term is: $\boldsymbol{k_B}$ **[complex][OH⁻]**. The base hydrolysis rate constant is about 10^5–10^6 times higher than the acid hydrolysis rate constant (*i.e.* aquation process).

For $[CoCl(NH_3)_5]^{2+}$, $k_{aq} \approx 10^{-6}\ s^{-1}$, $k_B \approx 0.25\ dm^3\ mol^{-1}\ s^{-1}$ at 25° C.

The term k_B[complex][OH⁻] may arise either due to a **genuine S_N^2 (*i.e.* A) rate determining step or due to the existence of a preequilibrium involving OH⁻ followed by a S_N^1 (*i.e.* D) type rate determining step.** In the preequilibrium, proton abstraction from the coordinated NH_3 leads to the formation of *a* ***conjugate base (CB)*** $[Co(NH_2)(NH_3)_4X]^+$ which subsequently participates in the rate determining dissociative step leading to the expulsion of the leaving group X^-. Then the 5-coordinate intermediate experiences a nucleophilic attack at a faster step. This idea was first proposed by *Garrick* and then developed and popularized by *Basolo and Pearson*. This process of mechanism of base hydrolysis is called **S_N^1–CB** (*i.e.* **D–CB** in modern nomenclature). The mechanistic steps are outlined below:

X^- = halide, N_3^-, $OSO_2CF_3^-$ etc.

$$\underset{\text{(Complex)}}{\left[Co(NH_3)_5X\right]^{2+}} + OH^- \underset{}{\overset{K_{CB}}{\rightleftharpoons}} \underset{\textbf{(CB)}}{\left[Co(NH_2)(NH_3)_4X\right]^{+}} + H_2O$$

$$\left[Co(NH_2)(NH_3)_4X\right]^{+} \xrightarrow[(rds)]{k} \underset{\textbf{(5 coordinate intermediate)}}{\left[Co(NH_2)(NH_3)_4\right]^{2+}} + X^-$$

$$\left[Co(NH_2)(NH_3)_4\right]^{2+} \xrightarrow[\text{(fast)}]{+H_2O} \left[Co(NH_2)(NH_3)_4(OH_2)\right]^{2+} \xrightarrow{\text{fast}} \left[Co(NH_3)_5(OH)\right]^{2+}$$

$$K_{CB} = \frac{K_a}{K_w},\ K_a = \text{Corresponding acid dissociation constant of the coordinated } NH_3 \text{ ligand}$$

The rate law of the above process is given below.

$$\text{Rate} = k[CB]; \qquad K_{CB} = \frac{[CB]}{[\text{complex}][OH^-]}$$

$$[\text{complex}]_T = [\text{complex}] + [CB]$$

$$= \frac{[CB]}{K_{CB}[OH^-]} + [CB] = [CB]\left\{\frac{1 + K_{CB}[OH^-]}{K_{CB}[OH^-]}\right\}$$

i.e.
$$[CB] = \frac{[\text{complex}]_T K_{CB}[OH^-]}{1 + K_{CB}[OH^-]},\ [\text{complex}]_T << [OH^-]\ \textit{i.e.}\ [OH^-]_T \approx [OH^-]$$

i.e.
$$\text{Rate} = \frac{kK_{CB}[\text{complex}]_T[OH^-]}{1 + K_{CB}[OH^-]} = k_B[\text{complex}]_T[OH].$$

Free NH_3 is a very weak Brönsted acid, *i.e.* its proton releasing power is negligibly small. But the metal coordinated NH_3 is a relatively a stronger acid. It happens so for the metal coordinated H_2O. Obviously, this acid strength increases for the higher valent metal, metal centre (*cf.* pK_a (free water) = 15.74, pK_a of $[Fe(OH_2)_6]^{3+} = 2.2$).
pK_a of the ammine proton in the complex is *ca.* 15-16 (*cf.* pK_a of free $NH_3 \ggg 15$). Consequently, K_{CB} **will have a very low value** (*ca.* $10^{-1} - 10^{-2}$ mol^{-1} dm^3) and the condition $1 \gg K_{CB}[OH^-]$ will be maintained even for the reasonably high concentration of $[OH^-]$. Thus we can write:

$$\text{Rate} = -\frac{d[\text{complex}]}{dt} = kK_{CB}[\text{complex}]_T[\text{OH}^-]$$

$$= k_B[\text{complex}][\text{OH}^-]; \left(\text{when } K_{CB} \text{ is very small, } [\text{complex}]_T \approx [\text{complex}]\right)$$

(**Note:** If the conjugate base (*CB*) is supposed to remain at a steady state $\left(i.e. \ \frac{d[CB]}{dt} = 0\right)$ having $K_{CB} = \frac{k_1}{k_{-1}}$, we get the following rate expression as follows:

$$\frac{d[\text{CB}]}{dt} = 0 = k_1[\text{complex}][\text{OH}^-] - k_{-1}[\text{CB}][\text{H}_2\text{O}] - k[\text{CB}]$$

i.e. $$[\text{CB}] = \frac{k_1[\text{complex}][\text{OH}^-]}{k_{-1}[\text{H}_2\text{O}] + k}; \text{Rate} = k[\text{CB}].$$

It gives: $$\text{Rate} = \frac{k_1 k[\text{complex}][\text{OH}^-]}{k_{-1}[\text{H}_2\text{O}] + k}$$

$$= k_B[\text{complex}][\text{OH}^-], \text{ where } k_B = \frac{kk_1}{k + k_{-1}[\text{H}_2\text{O}]}$$

5.25.1 Characteristics of the D–CB Process

- **Strict first-order in $[OH^-]$:** Here it should be mentioned that in the said D–CB mechanism, K_{CB} value is exceedingly small (*ca.* 10^{-1}–10^{-2} mol^{-1} dm^3). Consequently, **saturation kinetics** will never be attained, *i.e.* it will maintain strictly a first-order dependence on $[OH^-]$ for an appreciable range of OH^-.
- **Deprotonation site to generate the CB:** The experimental evidences indicate that deprotonation occurs from the NH_3 group present in the *trans*-position to the leaving group *X*. But there are some evidences for proton abstraction from the *cis*-NH_3 group (with respect to the leaving group).
- **Role of the $-\ddot{N}H_2$ group in CB for rate enhancement:** Formation of the CB ***reduces the overall cationic charge*** (*cf.* $[CoCl(NH_3)_5]^{2+}$ *vs.* $[CoCl(NH_2)(NH_3)_4]^+$) and it favours the **dissociative expulsion** of the leaving group (*i.e.* **charge effect**). Besides this charge effect, the 'NH_2' group shows the **electronic effects** to favour the rate process. The 'NH_2' group is a strong **σ-donor** and also a strong **pi-donor.** This strong σ-donor property labilizes the *trans*–position **(trans-effect)** and the π-donor property labilize the *cis*-position (*i.e.* **cis-effect**). This labilizing effect of the NH_2 group present in the CB favours the **dissociative activation** and explains the unusually high rate of the base hydrolysis process. The *trans*-effect (*i.e.* σ-donor property) is probably more important

than the *cis*-effect (*i.e.* π-donor property) in this labilisation process in the CB. ***This explains why deprotonation preferably occurs at the trans-prosition***. However, deprotonation from the *cis*-position (with respect to *X*) is not completely ruled out.

- **Geometry of the five coordinate intermediate:** Through the expulsion of the leaving group from the CB at the rate determining step, it adopts the *trigonal bipyramidal geometry* where the NH_2 group occupies one equatorial position (unless the other nonleaving ligands are too rigid to rearrange the square pyramid initially formed to the trigonal bipyramid). The trigonal bipyramidal (TBP) geometry is **sufficiently long-lived** to allow the Berry's pseudorotation. ***The trigonal bipyramidal intermediate is stabilized by the π-donor NH_2 group present at the equatorial site (see Fig. 5.21.1b).*** In a TBP geometry (*cf. sp^2d + pd; sp^2d*-hybrid orbitals towards the equatorial ligands; ***pd***-hybrid orbitals towards the axial ligands); ***the metal centre can receive electrons from the equatorial ligands better than from the axial ligands through the σ-bond (cf. Bent's rule). Thus the σ-donor property of the NH_2 group residing at the equatorial position of the TBP also stabilises the TBP.*** Formation of this long-lived five-coordinate intermediate (trigonal bipyramidal geometry) explains the **isomerization that occurs for both the *cis*– and *trans*–complex.** [*cf.* In the equation process, isomerization is noted generally in the *trans*-complexes like *trans*-$[Co(NH_3)_4LX]^{n+}$ (L = π-donor nonleaving group). For the corresponding *cis*– complex, 100% retention of configuration is noted in the aquation process].
- **High $\Delta S^{\neq}$:** The initially generated square pyramidal intermediate rearranges to the long-lived trigonal bipyramidal intermediate (stabilised by the NH_2 group) which may experience the Berry pseudorotation. Formation of such rearranged TBPs is indicated by high $\Delta S^{\neq}$.
- **Comparable reactivity of both *cis*– and *trans*– $[Co(en)_2\,LX]^{n+}$:** If L is a nonleaving π-donor ligand, then the *cis*-complex reacts much faster than the *trans*– isomer in the aquation process. *But in contrast to the aquation process* (*i.e.* acid hydrolysis), *both the cis– and trans– isomers react at comparable rates in base hydrolysis.* A good π-donor ligand at the *cis*-position to the leaving group can stabilize the 5-coordinate square pyramid without rearrangement (*cf.* acid hydrolysis) to a trigonal bipyramid. ***In the base hydrolysis, the NH_2 group is generated generally at the trans-position to the leaving group.*** The square pyramid generated through the expulsion of the leaving group (*trans*– to NH_2) rearranges to the trigonal bipyramid which is stabilized by the **σ- and π-donor** property of the NH_2 group at the equatorial site (Fig. 5.21.1b).

  The trigonal bipyramidal intermediate (**stabilised by the σ- and π-donor property of the NH_2 group residing at the equatorial position, Fig. 5.21.1b**) is sufficiently **long lived** to experience the Berry's pseudorotation. Thus in the rearranged TBP intermediate (rearrangement supported by high $\Delta S^{\neq}$), ligands both at *cis*- and *trans*- to the leaving group will occupy the equivalent position. It **explains the comparable reactivity of both the *cis*- and *trans*- isomers in base hydrolysis** (*cf.* difference in reactivity in aquation or acid hydrolysis).

  Thus the D–CB mechanism can be outlined as in Scheme 5.25.1.1.
- **Steric acceleration:** Base hydrolysis of $[CoCl(RNH_2)_5]^{2+}$ is faster than that of $[CoCl(NH_3)_5]^{2+}$. This steric acceleration supports the dissociative activation of the CB.
- **LFER plot:** The linear plot **$\log k_{OH}$ *vs.* $\log K_{OH}$** with the slope close to unity supports the dissociative activation. The linear plot, **$\log k_{OH}$ *vs* $\log k_{aq}$** indicates the common mechanism (*d*-activation) for both the base hydrolysis and aquation.
- **Entering ligand is H_2O not OH^-:** In the base hydrolysis product, OH–group is present but it is not the entering group. In fact, H_2O as a nucleophile attacks at the 5-coordinated intermediate at a faster step.

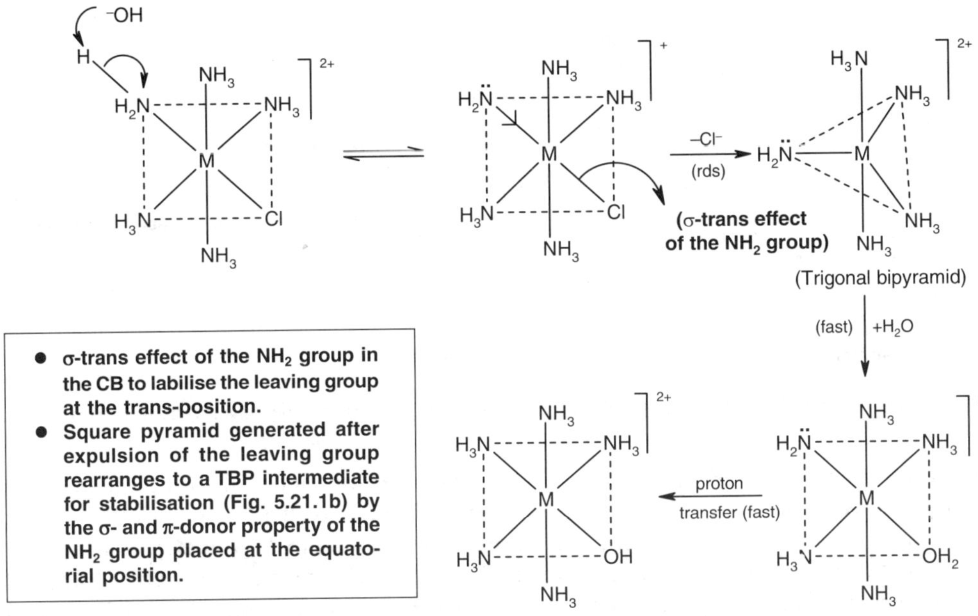

−OH
−Cl−
(rds)
(σ-trans effect
of the NH2 group)
(Trigonal bipyramid)
(fast)
+H2O
proton
transfer (fast)
• σ-trans effect of the NH2 group in the CB to labilise the leaving group at the trans-position.
• Square pyramid generated after expulsion of the leaving group rearranges to a TBP intermediate for stabilisation (Fig. 5.21.1b) by the σ- and π-donor property of the NH2 group placed at the equatorial position.

Scheme 5.25.1.1 Schematic representation of D–CB mechanism for the base hydrolysis of $[CoCl(NH_3)_5]^{2+}$.

That the OH group incorporated into the product comes from the nucleophilic attack (at a faster step) by H_2O not by ^{-}OH is evidenced from the use of ^{18}O labelled water. The isotope exchange equilibrium is:

$$H_2{}^{16}O + {}^{18}OH^- \rightleftharpoons H_2{}^{18}O + {}^{16}OH^-,\ K_{ex} = 1.04$$

The equilibrium constant ($K_{ex} \rangle 1$) indicates that $H_2{}^{18}O$ is less acidic than $H_2{}^{16}O$**,** so hydroxide contains the less ^{18}O content, *i.e.*

$$\left(\frac{^{18}O}{^{16}O}\right)_{\text{hydroxide}} \left\langle \left(\frac{^{18}O}{^{16}O}\right)_{\text{water}}\right.$$

The $^{18}O/^{16}O$ ratio incorporated into the product complex is in conformity with the ratio found in water not with the ratio found in hydroxide. It supports the fact that water (not hydroxide) enters as the nucleophile.

● **Requirement of the ammine groups (with dissociable protons) *trans* to the leaving group and the base hydrolysis of $[CoCl(NH_3)(tren)]^{2+}$:** In most of the cases, CB is formed through the abstration 'N—H' proton at the *trans*-position of the leaving group. In fact, the **red isomer** of $[CoCl(NH_3)(tren)]^{2+}$ having the $-NH_2$ group at the *trans*-position of 'Cl' **hydrolyses much faster** than the **purple isomer** lacking in such dissociable proton(s) at the *trans*-position.

For the **red isomer,** the CB generated from the removal of a proton at the *trans*-position (to Cl^-) can experience the **strong *trans*-labilising effect** due to the strong σ-donor property of the NH group. This favoured dissociative expulsion of Cl^- (*trans*-effect of the NH group) readily produces the trigonal bipyramidal (TBP) geometry (**a favourable geometry for the tripodal tren ligand,** *cf.* Fig. 2.11.7.8) which is stabilised by the both the σ- and π-donor property of the NH group when placed at the equatorial position of the TBP (*cf.* Fig. 5.21.1b).

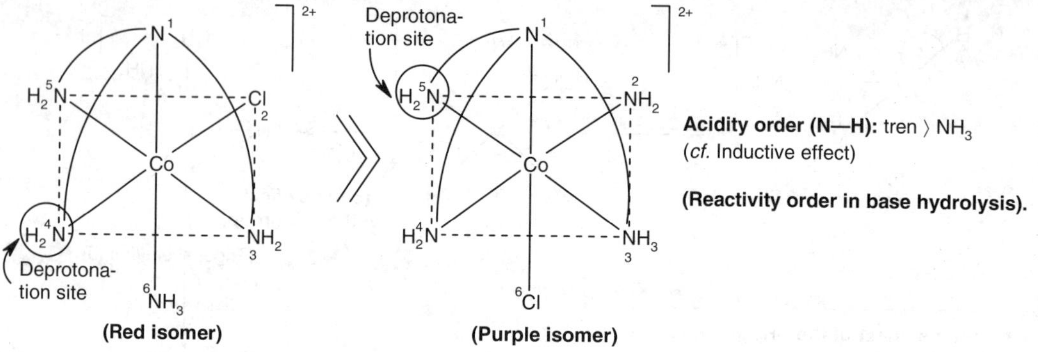

Fig. 5.25.1.1 Red and purple isomers of $[CoCl(NH_3)(tren)]^{2+}$ and their reactivity order in base hydrolysis. (*cf.* Fig. 2.5.3.11 for structures) (*cf.* D.A. Buckingham *et al.*, *Inorg. Chem.*, **14**, 1485, 1975).

For the **purple isomer,** the CB generated from the removal of a proton at the *cis*-position to the leaving group can experience the *cis*-**labilising effect** (*i.e. cis*-effect of the NH group) to cause an expulsion of the leaving group (*i.e.* Cl^-) generating a **square pyramidal (SP) geometry** which can be stabilised by the π-donor property of the NH group (*cf.* Fig. 5.21.1b). To generate the **more stable TBP intermediate** where both the σ- and π-donor property of the NH group can operate, the initially formed SP is to experience an rearrangement (Fig. 5.21.1b). **This will raise the activation energy.** Thus from the reactivity order of the red and purple isomer, we can conclude:

(i) *trans*-effect (due to the σ-donor property) of the NH group is **more important** than its *cis*-effect (due to the π-donor property) in the CB to cause a **dissociative expulsion of the leaving group**.

(ii) the 5-coordinate TBP is **more stabilised** than the SP by the electronic effect of the NH group.

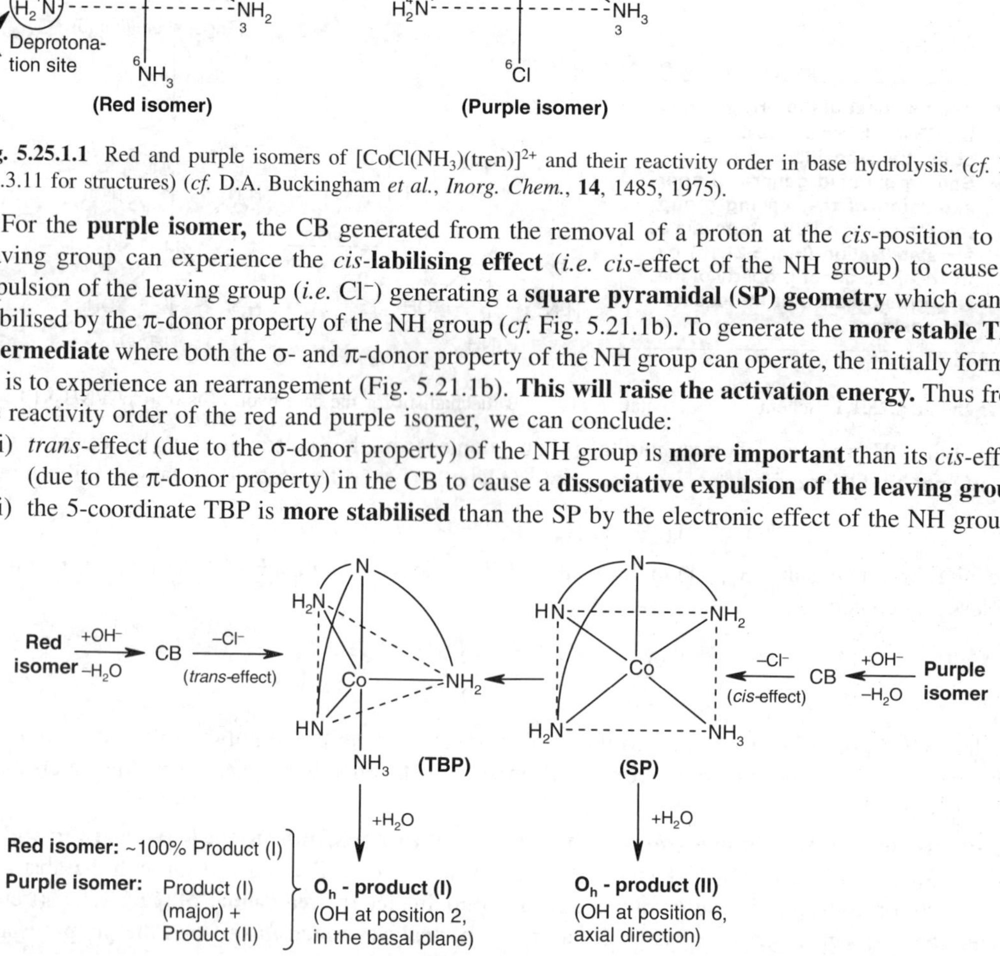

Fig. 5.25.1.2 Base hydrolysis products of the red and purple isomers of $[CoCl(NH_3)(tren)]^{2+}$ (*cf.* Fig. 5.25.1.1 for numbering of the positions in the product).

The hydrolysed product, *i.e.* $[Co(NH_3)(OH)(tren)]^{2+}$ (*i.e.* product **I**) obtained from the red isomer **retains the configuration completely.** The product is obtained by the nucleophilic attack of OH_2 along any side of the three equatorial edges of the TBP intermediate. The purple isomer gives **two products: major product** (*i.e.* product **I**) (same as in the case of red isomer; OH at the basal plane) obtained from the TBP intermediate, and the **minor product** (*i.e.* product **II**) (OH at the axial position-6, *trans* to the tertiary-N) obtained from the SP intermediate.

- **Entry of the nucleophiles at the faster steps:** Other nucleophiles (say Y^-) besides H_2O (in aqueous media) may also invade the 5-coordinate intermediate at the faster steps to give the product containing Y. However, their concentration terms are not involved in the rate expression. Thus these nucleophiles are the *spectators* of the five coordinate intermediate complex and they are not involved in the generation of this intermediate. *However, OH$^-$ or any other base (B) must be present to generate the CB.*

$$[Co(NH_3)_5X]^{2+} + B \rightleftharpoons [Co(NH_2)(NH_3)_4X]^+ \textbf{(CB)} + BH^+$$

$$\downarrow -X^- \quad \text{rds}$$

$$[Co(NH_2)(NH_3)_4Y]^+ \xleftarrow[\text{(fast)}]{Y^-} [Co(NH_2)(NH_3)_4]^{2+} \xrightarrow[\text{(fast)}]{+H_2O} [Co(NH_3)_5(OH)]^{2+}$$

$$\downarrow \text{(fast)} \quad +H_2O$$

$$[Co(NH_3)_5(Y)]^{2+}$$

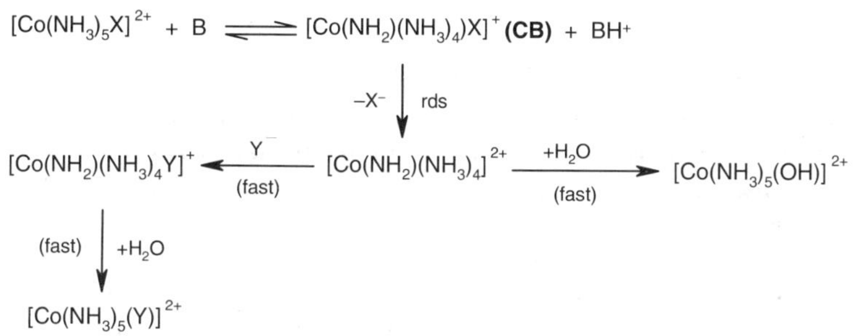

- **Presence of an acidic proton in the starting complex:** To produce the CB, the parent complex must possess a dissociable proton to **generate a good σ-donor as well as a π-donor group** (*e.g.* $\overset{..}{N}H_2$, $\overset{..}{N}H$). Otherwise, the proposed base hydrolysis mechanism will not operate. This is why, the complexes like $[CoCl_2(py)_4]^+$, $[CoCl(CN)_5]^{3-}$, etc. **lacking in such dissociable protons** do not respond to the said D–CB mechanism. However, their hydrolysis simply occurs through an interchange process slowly without any special rate benefit.

  In fact, such complexes experience aquation slowly through the I_d-**path followed by the proton abstraction by OH$^-$ from the aqua product in a fast step.** Consequently, the rate does not depend on [OH$^-$].

$$\left[CoCl(py)_5\right]^{2+} \xrightarrow[(I_d\text{-path})]{\text{aquation}} \left[Co(OH_2)(py)_5\right]^{3+} \xrightarrow[-H_2O, \text{ (fast)}]{OH^-} \left[Co(OH)(py)_5\right]^{2+}$$

$$\text{Rate} = k_A[\text{complex}][OH^-]^0 = k_A[\text{complex}]$$

- **Base hydrolysis, Co(III)- vs. Cr(III)-ammine complexes:** In the base hydrolysis process, compared to the Co(III)-ammine complexes, the corresponding Cr(III)-ammine complexes react *ca.* 3×10^3 times slower. *Probably, the lower electron affinity of Cr(III) (compared to that of Co(III)) makes the electron donating $-NH_2$ group (of the CB) less efficient to labilise the leaving group in the CB at the rate determining step.* The higher electron affinity of Co(III) compared to that of Cr(III) is evidenced by the fact that under comparable condition, it is easier to produce Co(II) from Co(III) while it is more difficult to produce Cr(II) from Cr(III). The E^θ values also support this fact. **This is why, the electron pushing effect (both by σ- and π-bond) of the NH$_2$ group towards M(III) is expressed better in the Co(III) complex** and this effect plays the most important role in the **favoured dissociative activation of the CB.** Here it is worth mentioning that for the heavier members like Ru(III), the A—CB mechanism rather than the D—CB mechanism operates (*i.e.* base hydrolysis: $[CoCl(NH_3)_5]^{2+}$ (**D—CB**); $[RuCl(NH_3)_5]^{2+}$ (**A—CB**)).

- **Base hydrolysis of the complexes having no acidic proton**

  The complexes like $[CoCl_2(py)_4]^+$, $[Co(bpy)_2(NO_2)_2]^+$, $[CoBr(CN)_5]^{3-}$, etc. experience the base hydrolysis very slowly without showing any dependence on $[OH^-]$ in the rate process. In fact, such complexes simply experience **the aquation through the I_d path followed by the rapid transformation into the hydroxo species through the acid-base reaction.** It is illustrated below:

$$\left[CoBr(CN)_5\right]^{3-} \xrightarrow[(I_d\ \text{path})]{\text{aquation}} \left[Co(CN)_5(OH_2)\right]^{2-} \xrightarrow[\substack{\text{(deprotonation of}\\ \text{the aqua ligand)}}]{OH^-\ \text{(fast)}} \left[Co(CN)_5(OH)\right]^{3-}$$

- **Reactivity order: Co(III) *vs.* Cr(III) complexes**

  In terms of CFAE, the Co(III)-complexes (low-spin) should be more inert than the Cr(III)-complexes. **The reversal order of reactivity in base hydrolysis can be probably explained by considering** the fact that higher electron affinity of Co(III) makes the NH_2 (*i.e.* amido) group **quite active** in terms of electron donation to Co(III) (by both σ- and π-bond). In fact, because of the much lower electron affinity of Cr(III), the NH_2 group in the CB of Cr(III)-complexes becomes **quite ineffective** to **labilise the leaving group**. The relative electron affinities of Co(III) and Cr(III) may be understood from their relative ease to form Co(II) and Cr(II) respectively in solution. It is also supported by their E^0 values of the aqua-complexes, *i.e.* Co(III) is more easily reduced to Co(II).

$$Co^{3+}/Co^{2+}\left(E^0 = 1.8\ \text{V}\right),\ Cr^{3+}/Cr^{2+}\left(E^0 = -0.41\ \text{V}\right)$$

- **D–CB *vs.* I_d–CB:** Stereochemical results indicate that the leaving group shows some sort of influence on the stereochemical change (*cf.* Table 5.25.3.1). It indicates that the leaving group is not completely dislodged (as expected in a pure *D*-process) in the transition state, *i.e.* the leaving group is still partly bonded with the metal centre to influence the entry of the nucleophile. Thus the process may be considered as a concerted process which may be better described as the *I_d-CB* rather than the *D–CB*. Other evidences in favour of I_d-CB have been discussed in Sec. 5.25.2.

5.25.2 Relative Importance of the σ-Donor and π-Donor Property of the NH_2 Group to Stabilise the 5-Coordinate T.S. in D-CB Mechanism: Deprotonation Site — *Cis*- or *Trans* Position?

- The **π-donor property** of the NH_2 group can stabilise both the 5-coordinate geometries—square pyramid (SP) and trigonal bipyramid (TBP) in the T.S or intermediate. To stabilise the SP, the **NH_2 group should reside at the *cis*-position of the leaving group** (*cf.* Sec. 5.21 and Fig. 5.21.1a) but to stabilise the TBP, **the NH_2 group must reside at the equatorial position** (*cf.* Fig. 5.21.1b) of the TBP.

- If the **π-donor property** of the NH_2 group were the main factor to stabilise the 5-coordinate T.S. or intermediate, **then the NH_2 group would be preferably generated** (through proton abstraction) **at the *cis*-position of the leaving group** (*cf.* aquation process) and the T.S. or intermediate would be preferably SP. But the experimental findings do not support this prediction.

- If the **π-donor property** of the NH_2 group were the main factor to stabilise the 5-coordinate T.S. or intermediate, then the 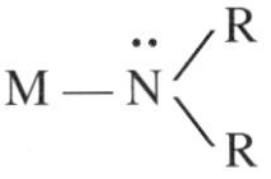 **planar moiety** (*i.e. sp*^{2}–N) **would be the most efficient one** but the experimental findings do not support this expectation.
- All these findings indicate the necessity of **an alternative explanation** which suggests that **the σ-donor (*i.e.* σ-basicity) property** of the amido group of the CB plays the main role for the dissociative expulsion of the leaving group from the CB and to stabilise the TBP T.S. or intermediate. **In fact, a very good σ-donor ligand like the NH_2-group can stabilise the TBP structure** and it happens also so for the substitution reactions of **Pt(II)-square planar complexes** where the T.S. or intermediate adopts the TBP structure.
- In the octahedral CB, the NH_2 (*i.e.* amido) group can strongly labilise the *trans*-group (*i.e.* **trans effect**) by dint of its σ-donor property while it can labilise the *cis*-position (*i.e.* **cis-effect**) by its π-donor property. In most of the cases, the NH_2 group is generated through proton removal at the *trans*-position to the leaving group. **It indicates that the σ-*trans*-effect of the NH_2 group (*i.e.* σ-donor property) is more important than its *cis*-effect (*i.e.* π-donor property) in the D-CB mechanism.**
- The TBP T.S. or intermediate can be stabilised by both the σ- and π-donor property of the NH_2 group (when placed at the equatorial position of the TBP) while the SP T.S. or intermediate is stabilised by the π-donor property only of the NH_2 group. **Thus the NH_2 group stabilises the TBP T.S. or intermediate more compared to the SP T.S. or intermediate.**
- If the SP structure is produced from the CB where the NH_2 group is generated at the *cis*-position to the leaving group, then the SP structure will have **to rearrange to attain the more stable TBP structure** (*cf.* electronic property of the NH_2 group stabilises the TBP structure more than the SP structure). **This energy requiring structural rearrangement will raise the activation energy.** Alternatively, if the NH_2 group is generated at the *trans*-position to the leaving group, then **the CB can readily attain the TBP structure** through the dissociative expulsion of the leaving group (*cf. trans*-labilising effect of the NH_2 group). This is why, proton abstraction preferably occurs at the *trans*-position of the leaving group to produce the CB.
- All these propositions are experimentally verified by comparing the base hydrolysis of the **red and purple isomers of [CoCl(NH_3)(tren)]$^{2+}$** (Figs. 5.25.1.1–2).

5.25.3 Evidences in Favour of the D–CB or I_d–CB Pathway Rather than the A-Pathway in Base Hydrolysis of Cobalt(III)-Ammine Complexes.

The observed rate law (= k_B[complex][OH^-]) can be attained either from the *associative (A) path* or from the *D–CB* or *I_d–CB* path discussed above. There are evidences to support the *D–CB* or *I_d–CB* path and to reject the associative path. These are discussed below.

(i) **Rate enhancement – charge effect and electronic effect of the –NH_2 group of CB:** Just by considering the better nucleophilic character of OH^-, compared to that of H_2O, the tremendous rate acceleration by a factor *ca.* 10^6 (compared to that of the acid hydrolysis) cannot be explained. This rate benefit arises from the **charge effect** (+2 charge of the starting complex becomes +1 in the CB) favouring the dissociative activation and **the electronic activation effect of the NH_2–group** (generated in the *CB*) *that shows the trans-labilizing effect* in the CB and stabilising effect on the TBP T.S. (by both the good σ- and π-donor property).

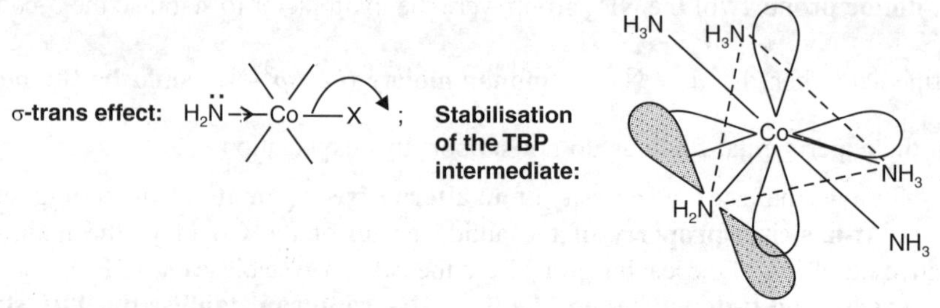

(*cf.* Fig. 5.21.1.1b)

(ii) **No rate enhancement in the base hydrolysis of $[Co(CN)_5X]^{3-}$ or *trans*–$[CoCl_2(py)_2]^+$:** If the simple associative path involving the nucleophilic attack by OH^- were operative, then these complexes would also react at comparable rates with those of $[Co(NH_3)_5X]^{2+}$. *In fact, the complexes lacking in dissociable protons cannot participate in the D–CB mechanism and consequently they do not gain any rate benefit.*

In fact, such complexes lacking in dissociable protons experience the base hydrolysis very slowly and the processes are also independent of $[OH^-]$ (*i.e.* **simple dissociative path**). It suggests that the complexes like $[Co(CN)_5X]^{3-}$, *trans*–$[CoCl_2(py)_2]^+$ experience the simple dissociative path (without any additional rate enhancement) while the complexes like $[Co(NH_3)_5X]^{2+}$ experience the *D–CB* path (with a tremendous rate benefit) showing a rate dependence on $[OH^-]$ in the base hydrolysis process.

(iii) **Entry of other nucleophiles:** In the D-CB path, OH^- acts in the proton abstraction process to produce the CB in a rapid pre-equilibrium step. The CB **generates the 5-coordinate intermediate through the expulsion of the leaving at the rate determining step.** Then the nucleophiles present in the media will attack the 5-coordinate intermediate. Thus it is reasonable to expect that if the base hydrolysis of $[Co(NH_3)_5X]^{2+}$ is carried out in presence of other nucleophiles Y^- (*e.g.* SCN^-, N_3^-, NO_2^-, etc.), the product will contain the Y-group but the rate will not depend on such nucleophiles. It has been experimentally observed. *It may be pointed out that without any base, entry of such nucleophiles under the conditions is exceedingly slow.*

In nonaqueous solvents like dimethylsulfoxide (DMSO), the replacement of X^- by Y^- (*e.g.* N_3^-, SCN^-, etc.) occurs very slowly in absence of any base capable of abstracting the N–H protons. But in presence of a trace amount (*i.e.* catalytic amount) of base *B* (*e.g.* OH^- or pipyridine), the process is tremendously catalyzed.

$$\left[Co(NH_3)_5X\right]^{2+} + Y^- \xrightarrow{\text{(in } DMSO)} \left[Co(NH_3)_5Y\right]^{2+} + X^- \; (t_{1/2} \text{ in hr.})$$

$$\left[Co(NH_3)_5X\right]^{2+} + Y^- \xrightarrow[\left(\substack{\textbf{catalytic amount} \\ \textbf{of base B}}\right)]{\text{(in DMSO)}} \left[Co(NH_3)_5Y\right]^{2+} + X^- \; (t_{1/2} \text{ in min.})$$

The rate of the catalyzed reaction depends on [B] but not on $[Y^-]$. These findings can be explained in terms of the *D-CB* mechanism.

$$\left[Co(NH_3)_5X\right]^{2+} + B \rightleftharpoons \left[Co(NH_2)(NH_3)_4X\right]^{+} \textbf{(CB)} + BH^+$$

$$CB \xrightarrow{rds} \left[Co(NH_2)(NH_3)_4\right]^{2+} + X^-$$

$$\downarrow \text{(fast)} \; +Y^-$$

$$\left[Co(NH_2)(NH_3)_4Y\right]^{+} \xrightarrow[\text{(fast)}]{+BH^+} \left[Co(NH_3)_5Y\right]^{2+} + B$$

The proportion of the 5 coordinate intermediate captured by Y^- depends on its nucleophilicity and concentration.

(iv) **Competition results in the products in the base hydrolysis of $[Co(NH_3)_5X]^{n+}$ in presence of other nucleophiles and D–CB *vs.* I_d–CB mechanism:** When the base hydrolysis of $[Co(NH_3)_5X]^{n+}$ (X = neutral group like $OSMe_2$; uninegative anion like I^-, $OSO_2CF_3^-$; dinegative anion like OSO_3^{2-}) is carried out in presence of SCN^-, then the mixed product, *i.e.* $[Co(NH_3)_5(OH)]^{2+}$ and $[Co(NH_3)_5Y]^-$ (Y = NCS^- which may bind through the *N*– or *S*– end) is obtained. The proportion of NCS^- entry (as NCS^- and SCN^-) depends on the charge of the leaving group X^{m-}. **With the increase of the anionic charge of the leaving group, the proportion of $[Co(NH_3)_5(OH)]^{2+}$ increases and the proportion of NCS-content in the product decreases.** For the neutral leaving group, the NCS^- content product is *ca.* 18%, for the anionic leaving group, *i.e.* X^- it is *ca.* 13% and for X^{2-} it is decreased to 7%.

Anation in acidic and in basic solution

$$\left[Co(NH_3)_5X\right]^{2+} + Y^- \longrightarrow \left[Co(NH_3)_5Y\right]^{2+} + X^-$$

- **Acidic condition:** $\left[Co(NH_3)_5X\right]^{2+} \xrightarrow[\text{(dissociative process)}]{\text{aquation}} \left[Co(NH_3)_5(OH_2)\right]^{3+}$

$$\downarrow \text{(anation)} \; +Y^- \text{ (dissociative process)}$$

$$\left[Co(NH_3)_5Y\right]^{2+}$$

(*i.e.* aquation followed by anation; **both steps dissociative**)

- **Basic condition:** $\left[Co(NH_3)_5X\right]^{2+} \underset{}{\overset{+OH^-}{\rightleftharpoons}} CB \xrightarrow[rds]{-X^-} \left[Co(NH_2)(NH_3)_4\right]^{2+}$

(*i.e.* through the CB formation)

$$\downarrow \text{(fast)} \; +Y^-, H_2O$$

i.e. **D-CB** mechanism.

$$\left[Co(NH_3)_5Y\right]^{2+}$$

These observations can be explained by the D-CB or I_d – *CB mechanism not by associative mechanism.* The leaving group (X^{m-}) remains in the secondary coordination sphere of the long-lived 5– coordinate intermediate. Consequently, more negative X^{m-} will discourage the anionic nucleophiles like NCS^- to attack the 5-coordinate intermediate. In other words, it may be said that the leaving group X^{m-} remains partly bonded in the transition state to prevent/discourage the incoming approaching the anionic nucleophile Y^- compared to the approaching neutral nucleophile H_2O. **This second view supports the *I_d-CB* rather than *D-CB*.** However, the competition results in the product can be explained by the *D-CB* or *I_d-CB* path not by the associative path.

(v) **Comparable reactivities of both *cis*– and *trans*– $[Co(en)_2LX]^{n+}$ (X = leaving group):** In the D-CB mechanism, it has been suggested that after expulsion of the leaving group, the 5-coordinate square pyramidal intermediate rearranges to the ***long-lived trigonal bipyramidal geometry*** *where the* 'NH_2' *group occupies the equatorial position to show its both σ- and π-donor property. The trigonal bipyramid is sufficiently long-lived to allow the* **Berry's pseudorotation**. Thus the ligands both *cis*– and *trans*– to the leaving group (present in the starting complex) occupy the equivalent positions in the trigonal bipyramidal geometry. This is why, both the *cis*– and *trans*– *isomers react at the comparable rates in base hydrolysis* (*cf.* aquation rates very often differ for the *cis*– and *trans*– isomers).

Here it may be pointed out that if the nonleaving group (L) is a π-donor ligand, then the trigonal bipyramid will be stabilized more compared to the system having the π-acceptor nonleaving group (*e.g.* L = CN^-). **This is why, the complexes bearing the π-acceptor nonleaving groups show the low reactivity.**

(vi) **Steric acceleration:** In the base hydrolysis, $[M(NH_2Me)_5(OSO_2CF_3)]^{2+}$ (M = Co, Rh, Ir, Cr) reacts much faster than $[M(NH_3)_5(OSO_2CF_3)]^{2+}$. This **steric acceleration** supports the **dissociative activation.** It may be pointed out that in aquation (*i.e.* acid hydrolysis), the Co(III)-complexes show also the **steric acceleration** but the Cr(III), Ir(III)-complexes show the **steric retardation** suggesting the **associative activation.**

(vii) **Nucleophilicity *vs.* basicity – importance of the N–H proton abstraction:** That the N–H proton abstraction is an important step in the base hydrolysis reaction is established from the observation in the reaction between $[Co(NH_3)_5X]^{2+}$ and H_2O_2/OH^- mixture. In the presence of H_2O_2, the following *acid-base interaction* (*i.e.* proton transfer) in alkaline condition occurs:

$$H_2O_2\,(A_1) + OH^-\,(B_2) \rightleftharpoons HO_2^-\,(B_1) + H_2O\,(A_2)$$

HO_2^- *(hydroperoxide ion) is a better nucleophile than* OH^- *but is a weaker Brönsted base than* OH^-. Thus with the addition of H_2O_2, the better nucleophile HO_2^- generates at the cost of the better base, *i.e.* OH^- (which can abstract the N–H proton).

If the reaction proceeds through the **associative path**, the rate should increase with the generation of the better nucleophile. On the other hand, if the reaction passes through the **D-CB or I_d-CB mechanism,** then the rate should decrease with the decrease of the better base (*i.e.* proton abstracting agent). Thus the observation supports the D-CB or I_d-CB mechanism not the associative mechanism.

(viii) **Activation parameters:** The proposed mechanism involving the **dissociative transformation into the 5-coordinate trigonal bipyramid** has been supported by the *large positive values of* $\Delta S^{\neq}$ *and* $\Delta V^{\neq}$. The preequilibrium step leading to the CB through the **charge destruction** makes a positive contribution to both $\Delta S^{\neq}$ and $\Delta V^{\neq}$; the **dissociative transformation** of the CB to a square pyramid geometry followed by its transformation into a trigonal bipyramid makes also a positive contribution to $\Delta V^{\neq}$. But this dissociative transformation step leading to the **separation of an anionic ligand (X^{n-})** creates charge which makes a negative contribution to $\Delta V^{\neq}$ from the electrorestriction imposed on the solvent. Thus the resultant $\Delta V^{\neq}$ value is less positive.

The factors to contribute $\Delta V^{\neq}$ in the D-CB mechanism are: (i) charge destruction in the formation of CB (preequilibrium step) causing a positive contribution; (ii) dissociative transformation of the CB to TBP causing a positive contribution; (iii) charge separation in the transformation of the CB into a TBP causing a negative contribution.

Complex:	$[CoCl(NH_2Me)_5]^{2+}$	$[CoCl(NH_2Et)_5]^{2+}$	*cis*–$[CoCl_2(en)_2]^+$
$\Delta V^{\neq}$ ($cm^3\ mol^{-1}$): (for base hydrolysis)	+32.7	+31.0	+28

Relationship between the $\Delta G^{\neq}$ values for acid hydrolysis and base hydrolysis

- For the **Co(III)-ammine complexes**, there is a linear relationship between $\ln k_H$ and $\ln k_{OH}$. It indicates the linear relationship between $\Delta G_H^{\neq}$ and $\Delta G_{OH}^{\neq}$. It suggests a **common mechanistic path** (which is the dissociative activation) for both the acid hydrolysis (k_H) and base hydrolysis (k_{OH}) path.
- There is no such linear relationship between k_H and k_{OH} for the Cr(III)-ammine complexes. It indicates that the acid hydrolysis and base hydrolysis for the **Cr(III)-complexes** occur through the **different mechanisms**. In fact, their acid hydrolysis mainly passes through the I_a path while the base hydrolysis occurs through the dissociative path.

(ix) **Stereochemical change in base hydrolysis and D–CB *vs.* I_d–*CB* mechanisms:** Stereochemical change in the base hydrolysis of the complexes like $[Co(en)_2(NH_3)X]^{n+}$, $[Co(en)_2(L)X]^{m+}$ (L and X are anionic, X^- is replaced by OH^-) has been extensively studied. *It has been found that isomerization occurs for both the cis– and trans– isomers.* This can be explained by considering the formation of a **long-lived 5-coordinate** trigonal bipyramid intermediate that experiences the pseudo-rotation to produce two possible trigonal bipyramids remaining in an equilibrium (*cf.* Figs. 5.21.3-5).

However, the extent of retention of configuration depends partly on the nature of the leaving group X. *It indicates that the leaving group probably remains partly bonded to the transition state to influence the stereochemistry of the product* (*i.e.* nucleophilic attack by H_2O is influenced by the leaving group). It indicates some sort of interchange process. **Thus the *D-CB* mechanism should be better described as the I_d-*CB* mechanism.**

Table 5.25.3.1 Stereochemical change in the base hydrolysis of $[Co(en)_2(NH_3)X]^{n+}$ at 25° C (*cf.* Figs. 5.21.3-5) (*cf.* D.A. Buckingham, *et al. Inorg. Chem.*, **18**, 1985, 1979; *J. Am. Chem. Soc.*, **90,** 6654, 1968).

	(a) cis– $[Co(en)_2(NH_3)X]^{n+}$		*cis-product*	
X	*%-cis product*	*%-trans product*	*% (Retention)*	*% (Inversion)*
$OSMe_2$	77	23	84.5	15.5
Cl^-	78	22	81	19
Br^-	78	22	77	23
NO_3^-	77	23	80	20

	(b) trans– $[Co(en)_2(NH_3)X]^{n+}$	
X	*%-cis product*	*%-trans product*
Cl^-	76	24
Br^-	81	19

Stereochemical Change for $[Co^{III}(en)_2LX]$: Acid Hydrolysis *vs.* Base Hydrolysis

(A) Acid Hydrolysis (*cf.* Sec. 5.21): 100% retention of configuration for the *cis*-compounds irrespective of the electronic nature of the **non-leaving group** L. 100% retention of configuration for the *trans*-compound when L is a non-π-bonding ligand (*e.g.* NH_3) or a π-acceptor ligand (*e.g.* CN^-). The *trans*-compounds experience a geometrical isomerisation for L = π-donor ligands (*e.g.* OH^-).

- For L = π-acceptor ligand, an **associative path** leads to stereoretentivity for both the *cis*- and *trans*- compounds.
- For L = π-donor ligand, stereoretentivity is attained through the formation of a **square pyramidal transition state** (dissociative path) as in the case of *cis*-compounds. Geometrical isomerisation occurs for the *trans*-compounds producing a **trigonal bipyramidal transition state/intermediate from the square pyramidal structure.**
- For L = non-π-bonding ligand, stereoretentivity can be attained in a dissociative path passing through the **square pyramidal transition state;** and associative path also leads to stereoretentivety.

(B) Base Hydrolysis (*cf.* Sec. 5.25): Both the *cis*- and *trans*-compounds experience the geometrical isomerisation due to the formation of a **long-lived trigonal bipyramidal intermediate** which is stabilised by the good σ-donor and π-donor properties of the NH_2 group when placed at the equatorial position.

Reactivity of $[Co^{III}(en)_2LX]$: Acid Hydrolysis *vs.* Base Hydrolysis

(A) Acid Hydrolysis: L = π-donor ligand, $k_{cis} \rangle k_{trans}$ **(dissociative path)**; L = weakly π-bonding ligand or non-π-bonding ligand, $k_{cis} \approx k_{trans}$; L = π-acceptor ligand, $k_{trans} \rangle k_{cis}$ **(associative path)**.
(B) Base Hydrolysis: Irrespective of the nature of L, $k_{cis} \approx k_{trans}$.

D-CB *vs.* I_d–CB in Base Hydrolysis of the Co(III) Complexes

- Distribution of product in presence of different nucleophiles depends on the nature of the leaving group. It suggests the I_d–CB path rather than the D-CB path.
- Distribution of stereochemical configuration in the product depends on the nature of the leaving group. It suggests the I_d–CB path rather than the D–DB path.

Comparison of the Reactivity of Co(III) (low-spin) and Cr(III) (*cf.* Secs. 5.19.5–6, 5.22)

In terms of CFAE, the Co(III) (low-spin) complexes are more inert than the Cr(III) complexes. It is experimentally verified in many reactions such as:

Ligand exchange: $\left[M(CN)_6\right]^{3-} + 6^*CN^- \rightleftharpoons \left[M(^*CN)_6\right]^{3-} + 6CN^-$

Aquation: $\left[M(NH_3)_5X\right]^{2+} + H_2O \longrightarrow \left[M(NH_3)_5(OH_2)\right]^{3+} + X^-, \left(X^- = Cl^-, Br^-, NCS^-\right).$

Dissociation: $\left[M(AA)_3\right]^{3+} + H_3O^+ + H_2O \longrightarrow \left[M(AA)_2(OH_2)_2\right]^{3+} + AAH^+$
(acid catalyzed) (AA = en, bigH, etc.)

However, there are **some exceptions** where the reactivity of Co(III) is found more than that of Cr(III). These are:

- **Water exchange reaction:** It occurs through a special mechanism discussed earlier.
- **Base hydrolysis:** It does not occur through the simple associative or dissociative path. It has been already discussed.
- **Pseudo-substitution:** Sometimes, due to **pseudo-substitution process**, apparently the Co(III) complexes may appear more labile. These aspects have been discussed already (*cf.* Sec. 5.24).

5.25.4 D-CB *vs.* D-IP Pathway in Base Hydrolysis

The rate law of base hydrolysis (*D-CB* path) of $[M(NH_3)_5X]^{n+}$ is:

$$\text{Rate} = \frac{kK_{CB}[\text{complex}][OH^-]}{1 + K_{CB}[OH^-]}$$

It was argued that the value of K_{CB} is very small and consequently the condition $1 \rangle\rangle K_{CB}[OH]$ is always maintained to show the strict first-order dependence on $[OH^-]$ throughout the study. *In other words, for the D-CB path, there would be never saturation kinetics or mixed order behaviour.* But, for base hydrolysis of $[Cr(NCS)(NH_3)_5]^{2+}$, the rate *vs.* $[OH^-]$ plot deviates from the linearity and the plot levels off to a constant value at the relatively higher value of $[OH^-]$. *This mixed order or saturation kinetics cannot be explained in terms of the D-CB mechanism.*

To explain the mixed order kinetics of base hydrolysis of $[Cr(NCS)(NH_3)_5]^{2+}$ the following *D-IP* mechanism has been suggested (D. Banerjea *et al*, *Z. Anorg. Allgm. Chem.*, **99,** 361, 1968).

$$\left[Cr(NCS)(NH_3)_5\right]^{+} + OH^- \underset{}{\overset{K_{OS}}{\rightleftharpoons}} \underset{\text{(Ion pair, } IP)}{\left[Cr(NCS)(NH_3)_5\right]^{2+} \cdot OH^-}$$

$$\left[Cr(NCS)(NH_3)_5\right]^{2+} \cdot OH^- \xrightarrow{k} \left[Cr(NH_3)_5(OH)\right]^{2+} + NCS^-$$

$$\text{rate} = \frac{kK_{OS}[\text{complex}][OH^-]}{1 + K_{OS}[OH^-]} = \frac{kK_{IP}[\text{complex}][OH^-]}{1 + K_{IP}[OH^-]} \qquad \left(\text{Taking } K_{OS} = K_{IP}\right)$$

where, $$k_{obs} = \frac{kK_{IP}[OH^-]}{1 + K_{IP}[OH^-]}, \text{ or } \frac{1}{k_{obs}} = \frac{1}{k} + \frac{1}{\left(kK_{IP}[OH^-]\right)}, \left(\text{when } [OH^-] >> [\text{complex}]\right)$$

The plot of $\frac{1}{k_{obs}}$ *vs.* $\frac{1}{[OH^-]}$ can estimate K_{IP} ($\approx$ 10 mol^{-1} dm^3). The estimated value is in good agreement with the theoretically calculated value. *The relatively high value of K_{OS} (i.e. K_{IP}) is responsible for the observed* ***mixed order kinetics.***

Here it is interesting to note that though $[Cr(NCS)(NH_3)_5]^{2+}$ adopts the *D-IP* path, $[Cr(NH_3)_5(N_3)]^{2+}$ adopts the *D-CB* path. **It is suggested the conjugate base (CB) and ion pair (IP) are the two limiting situations of the following system.**

$$M — NH_2^- \text{----} H^+ \text{----} OH^-$$

If the proton is primarily under the control of the —NH_2^- group, it leads to the IP. On the other hand, if the proton is primarily under the control of OH^-, it leads to the CB.

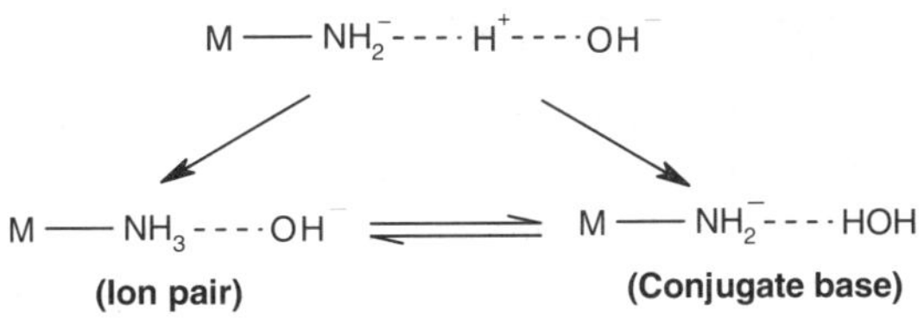

- If the protonic character of the 'N–H' proton is sufficiently high then the CB is favoured over the IP.
- If the protonic character of the N–H proton is relatively less then the IP formation is favoured over the CB.

In $[M(NH_3)_5X]^{2+}$, the protonic character of the 'N—H' proton depends on the electron withdrawing/pushing character of 'X'. *More electron pushing character of X will reduce the protonic character.* ***NCS^- is more electron pushing than N_3^-***. Consequently, in $[M(NH_3)_5(N_3)]^{2+}$, the acidic or protonic character of the N—H proton is more compared to that in $[M(NCS)(NH_3)_5]^{2+}$ to favour the *D-CB* path for the azido complex.

The difference in electron pushing character of NCS^- and N_3^- can be understood by considering their resonating structures.

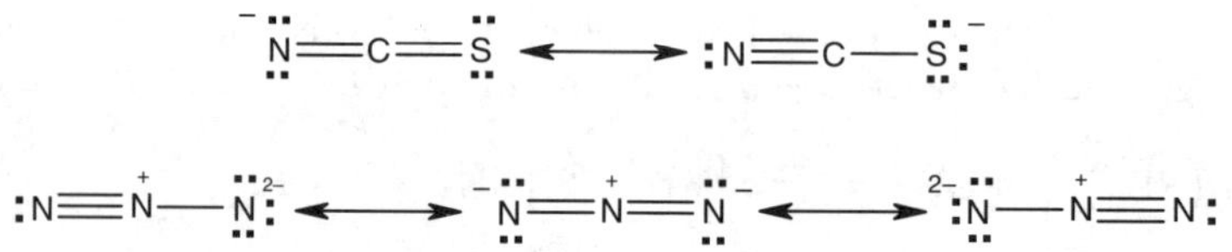

In N_3^-, because of the positively charged central N-atom, the terminal N-atoms are the weaker donors. This is supported by the fact that N_3^- is more electronegative than NCS^-; ***and N_3^- is a poorer nucleophile than NCS^-***. It makes the following difference.

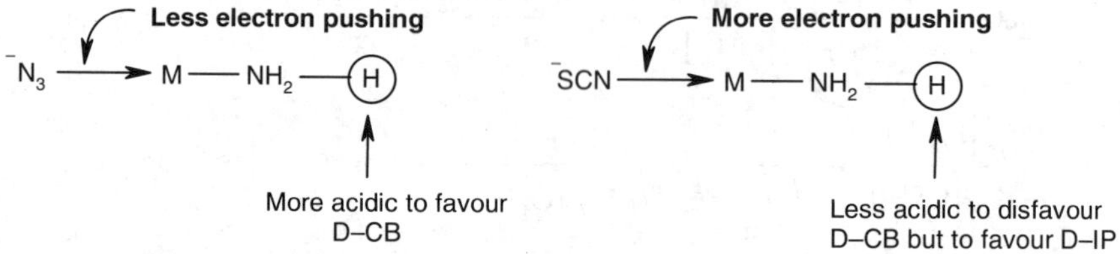

Thus D-CB *vs.* D-IP mechanistic profile depends on the electronegativity or nucleophilicity of X^-. It has been experimentally observed in the base hydrolysis of *trans*– $[Rh(en)_2X_2]^+$ where the electronegativity sequence of X^- runs as: $I^- < SCN^- < N_3^- < Br^- < Cl^-$. The change-over from D-IP to D-CB is noted between SCN^- (*N*-bonded) and N_3^-, *i.e.*

$$X^- : \quad \xleftarrow{\text{favouring } D\text{–}IP} \quad I^- \;\; SCN^- \;\; N_3^- \;\; Br^- \;\; Cl^- \quad \xrightarrow[\text{increasing electronegativity and favouring } D\text{–}CB]{}$$

5.26 KINETIC ASPECTS OF FORMATION REACTIONS AT IRON(III)

$[Fe(OH_2)_6]^{3+}$ shows a very strong tendency to undergo hydrolysis even in a fairly strong acidic condition.

$$\left[Fe(OH_2)_6\right]^{3+} + H_2O \rightleftharpoons \left[Fe(OH)(OH_2)_5\right]^{2+} + H_3O^+, \; pK_a \approx 2.0$$

Thus the observed reactivity is the sum of activity of $[Fe(OH_2)_6]^{3+}$ (*i.e.* unhydrolyzed species) and $[Fe(OH)(OH_2)_5]^{2+}$ (*i.e.* hydrolyzed species). In acidic condition, $[Fe(OH_2)_6]^{3+}$ is definitely more abundant than $[Fe(OH)(OH_2)_5]^{2+}$, *but reactivity of the hydrolyzed species is about 10^3 times faster than that of the unhydrolyzed species (cf.* $k_{ex} \sim 10^2\ s^{-1}$ for $[Fe(OH_2)_6]^{3+}$; $k_{ex} \sim 10^5\ s^{-1}$ for $[Fe(OH)(OH_2)_5]^{2+}$). *Thus very* *often, contribution to the observed total reactivity is mainly coming from the less abundant species,* $[Fe(OH)(OH_2)_5]^{2+}$. Here it may be mentioned that though **$[Fe(OH_2)_6]^{3+}$ reacts through the I_a-path, $[Fe(OH)(OH_2)_5]^{2+}$ reacts through the I_d-path.** (*cf.* A.K. Das, *Bull. Chem. Soc. Jpn.*, **65**, 2205, 1992; *Transition Met. Chem.*, **17**, 484, 1992 and the references cited therein).

$$\left[Fe(OH_2)_6\right]^{3+} \xrightarrow[(I_a\text{-path})]{+\text{ligand}} \text{Product}$$

$$-H^+ \Big\downarrow\Big\uparrow$$

$$\left[Fe(OH)(OH_2)_5\right]^{2+} \xrightarrow[(I_d\text{-path})]{+\text{ligand}} \text{Product}$$

High cationic charge of $[Fe(OH_2)_6]^{3+}$ imparts the **associative activation** to it while the overall **less cationic charge** of $[Fe(OH)(OH_2)_5]^{2+}$ favours it to adopt the **dissociative activation** (*cf.* effect of charge, Sec. 5.19.1). Besides this, the **better σ-donor and π-donor property of the $^-$OH group** (compared to those of OH_2) introduce the dissociative character in $[Fe(OH)(OH_2)_5]^{2+}$. In fact, the *cis*-effect (*cf.* Sec. 5.21) (π-donor property of the $^-$OH group) labilises the H_2O molecule at the *cis*-position.

If the ligand (say LH) remains in the following equilibrium under the experimental conditions,

$$LH \rightleftharpoons L^- + H^+$$

then complication may arise in analysing the rate constants due to **proton ambiguity** that has been discussed separately (*cf.* Sec. 5.17).

Other metal ions like $[Cr(OH_2)_6]^{3+}$, $[Ce(OH_2)_6]^{4+}$, etc. having a strong tendency for hydrolysis show the similar behaviour observed for Fe(III).

5.27 CATALYSIS BY ELECTRON TRANSFER REACTIONS IN THE LIGAND SUBSTITUTION REACTIONS OF OCTAHEDRAL COMPLEXES

If a particular metal centre can exist in two different oxidation states where the higher oxidation state is relatively inert and the lower oxidation state is relatively labile, then the lower oxidation state of the metal can catalyze the ligand substitution process at the relatively inert centre of higher oxidation state. This catalysis originates from the electron transfer reactions (through the inner-sphere mechanism) between the two different oxidation states of the metal centre. Thus, Co(II) can catalyze the substitution reactions at Co(III); Cr(II) can catalyze the substitution reactions at Cr(III); Pt(II) can catalyze the substitution reactions of Pt(IV); etc.

Catalysis by Co(II) in the water exchange process of Co(III) has been already discussed (Sec. 5.13). The other examples will be discussed in the next chapter.

5.28 ISOMERIZATION REACTIONS IN THE OCTAHEDRAL COMPLEXES

5.28.1 Linkage Isomerization

- **Nitrito ⇌ Nitro (intramolecular path):** The nitrito complex, $[Co(NH_3)_5(ONO)]^{2+}$ (**Red**) undergoes isomerization even in solid state to the thermodynamically more stable (*cf.* **symbiotic effect**, Chapter 2) nitro complex, $[Co(NH_3)_5(NO_2)]^{2+}$ (**Yellow**)

$$\left[Co(NH_3)_5(^*ONO)\right]^{2+}(\text{red}) \xrightarrow[(\text{slow})]{\text{heating}} \left[Co(NH_3)_5(NO_2)\right]^{2+}(\text{yellow}),\ i.e.\ [Co(NH_3)_5(NO^*O)]^{2+}$$

The rearrangement occurs through an **intramolecular path**. It is evidenced by the fact that there is no uptake of labelled nitrite added to the solution during the isomerisation process. The intramolecular path may be outlined as follows:

$$\left[(H_3N)_5Co\text{—}{}^*ONO\right]^{2+} \rightleftharpoons \left[(H_3N)_5Co\overset{{}^*O}{\underset{N=O}{\diagdown}}\right]^{2+} \rightleftharpoons \left[(H_3N)_5Co\text{—}NO{}^*O\right]^{2+}$$

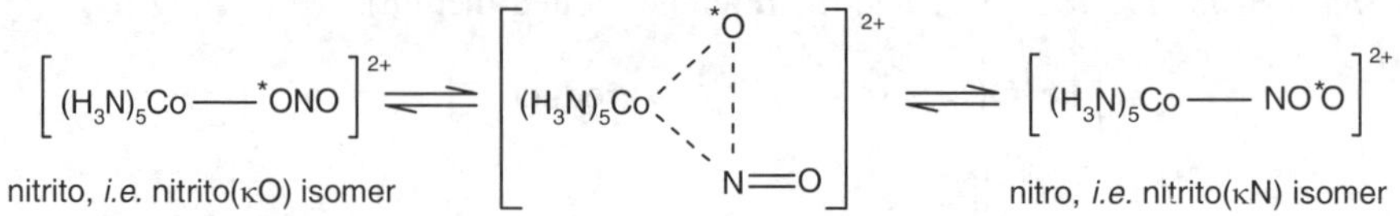

nitrito, *i.e.* nitrito(κO) isomer   nitro, *i.e.* nitrito(κN) isomer

The above mechanism has been described as S_Ni (*substitution, nucleophilic, internal displacement*). The similar mechanism has been suggested for the isomerisation of nitrito to nitro complexes of Rh(III) and Ir(III).

For the isomerisation of nitrito to nitro complex of Co(III), the $\Delta V^{\ne}$ value is negative (*ca.* −5 to − 7 cm^3 mol^{-1} for such isomerisation reactions of Co(III), Rh(III) and Ir(III) complexes). In fact, all complexes experiencing the nitrito to nitro isomerisation through the S_Ni path shows the **negative value of $\Delta V^{\ne}$**. In the **cyclic transition state** both the N– and O– sites remain bonded to the metal centre. It may be mentioned that the molar volume of the nitro complex is less than that of the nitrito isomer.

$$\left[(H_3N)_5Co\text{—}N\overset{{}^*O}{\underset{O}{\diagup}}\right]^{2+} \qquad \left[(H_3N)_5Co\text{—}{}^*O\text{—}N{\diagdown}_O\right]^{2+}$$

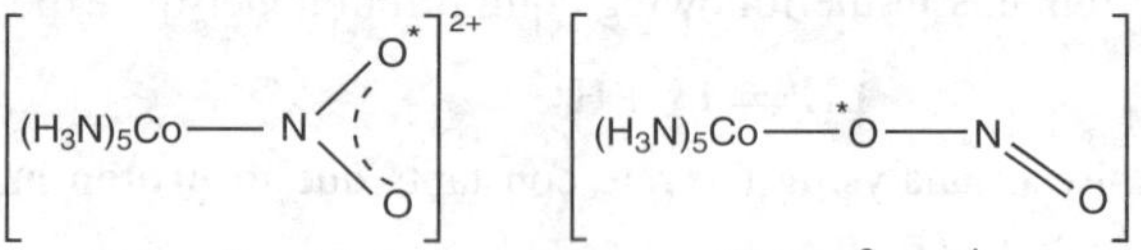

69.4 cm^3 mol^{-1} 82.3 cm^3 mol^{-1}

● **Thiocyanato $\rightleftharpoons$ Isothiocyanato (intramolecular path):** Isomerisation of $[Co(NH_3)_5(SCN)]^{2+}$ to the more stable (*cf.* symbiotic effect) isomer $[Co(NH_3)_5(NCS)]^{2+}$ occurs through the **intramolecular path.**

$$\left[(H_3N)_5Co\text{—}S{\diagdown}_{C\equiv N}\right]^{2+} \longrightarrow \left[(H_3N)_5Co\text{- - -}S{\diagdown}_{C\equiv N}\right]^{2+} \longrightarrow \left[(H_3N)_5Co\text{—}N{=}C{=}S\right]^{2+}$$

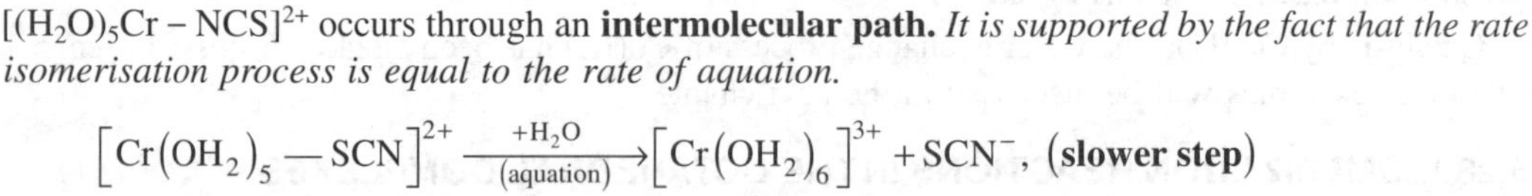

thiocyanato *i.e.* thiocyanato(κS) isomer   isothiocyanto *i.e.* thiocyanato(κN) isomer

This intramolecular path is again supported by the **small negative value of $\Delta V^{\ne}$** (= −5.3 cm^3 mol^{-1}).

● **Thiocyanato $\rightleftharpoons$ Isothiocyanato (intermolecular path):** Isomerisation of $[(H_2O)_5Cr - SCN]^{2+}$ to $[(H_2O)_5Cr - NCS]^{2+}$ occurs through an **intermolecular path.** *It is supported by the fact that the rate of isomerisation process is equal to the rate of aquation.*

$$\left[Cr(OH_2)_5 - SCN\right]^{2+} \xrightarrow[\text{(aquation)}]{+H_2O} \left[Cr(OH_2)_6\right]^{3+} + SCN^-, \textbf{(slower step)}$$

$$\left[Cr(OH_2)_6\right]^{3+} + SCN^- \xrightarrow{\text{fast}} \left[Cr(OH_2)_6\right]^{3+} \cdot SCN^- \text{ and } \left[Cr(OH_2)_6\right]^{3+} \cdot NCS^-$$

$$\left.\downarrow\right._{\text{anation}}^{-H_2O} \qquad\qquad \left.\downarrow\right._{\text{anation}}^{-H_2O}$$

$$\left[Cr(OH_2)_5 - SCN\right]^{2+} \qquad \left[Cr(OH_2)_5 - NCS\right]^{2+}$$

It is assumed that the anation process (that leads to two different linkage isomers) is relatively faster than the aquation process. *It makes the isomerisation rate equal to the aquation rate.* Aquation releases SCN^- whose reentry through anation leads to the linkage isomers.

Intermolecular vs. Intramolecular Path for
M—SCN ⇌ M—NCS and M—ONO ⇌ M—NO_2

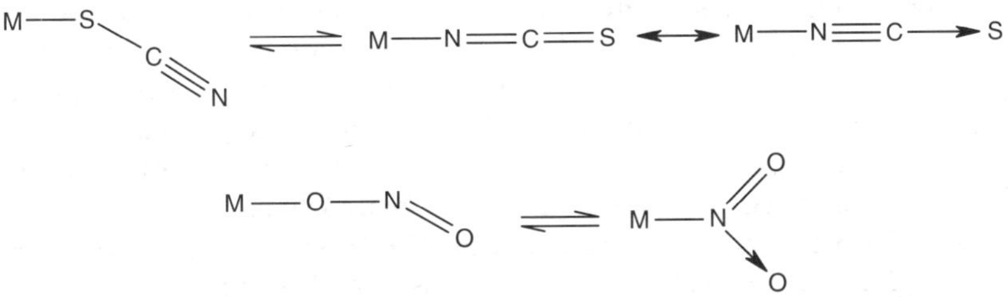

It is obvious that for the **intramolecular path** (*i.e.* S_Ni), the **cyclic transition state** experiences a less strain for the interconversion between the nitro and nitrito isomers compared to that for the interconversion between the thiocyanto and isothiocyanto isomers. This is why, the S_Ni process is more likely in M—ONO ⇌ M—NO_2 than in M—SCN ⇌ M—NCS.

5.28.2 Geometrical Isomerisation and Racemisation for the $[M(A—A)_2X_2]$ Type Complexes

(a) ***Cis–trans* isomerisation of the octahedral complexes of the type $[M(A—A)_2X_2]^+$:** It is illustrated for $[CoCl_2(en)_2]^+$.

$[Co(CO_3)(en)_2]^+ + HCl(aq) \rightarrow$ mixture of *cis*– and *trans*– $[CoCl_2(en)_2]Cl$.

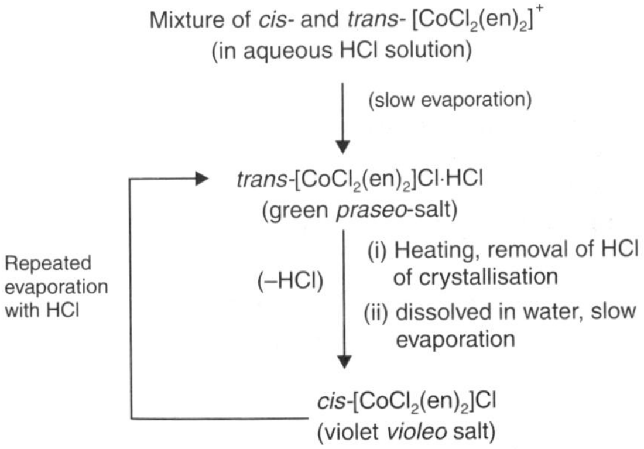

A particular isomer that will separate out depends on its solubility. The *less soluble isomer, cis*–$[CoCl_2(en)_2]Cl$ separates from an aqueous solution. On the other hand, the *trans*–isomer, *trans*–$[CoCl_2(en)_2]Cl{\cdot}HCl$ separates from an aqueous solution of HCl. It may be mentioned that from an aqueous solution of *cis*– and *trans*– isomers of $[CoCl_2(en)_2]NO_3$, the *trans*– isomer being less soluble separates first. Thus, the solubility of an isomer depends on the ion residing in outer-sphere.

The *cis–trans* isomerisation of $[CoCl_2(en)_2]^+$ is believed to pass through the aquation process.

$$cis\text{-}\left[CoCl_2(en)_2\right]^+ + H_2O \rightleftharpoons cis\text{-}\left[CoCl(en)_2(OH_2)\right]^{2+} + Cl^-$$

$$\updownarrows$$

$$trans\text{-}\left[CoCl_2(en)_2\right]^+ + H_2O \rightleftharpoons trans\text{-}\left[CoCl(en)_2(OH_2)\right]^{2+} + Cl^-$$

It may be noted that aquation of *cis*–$[CoCl_2(en)_2]^+$ produces 100% *cis*–$[CoCl(en)_2(OH_2)]^{2+}$ as the initial product while aquation of *trans*–$[CoCl_2(en)_2]^+$ gives a mixture of *cis*– and *trans*– product (*cf.* Sec. 5.21). The above **intermolecular path** is supported by the fact that when the said isomerisation is carried out in presence of labelled chloride, *i.e.* $^*Cl^-$, there is a random distribution of $^*Cl^-$ in the complex. *It may be mentioned that replacement of one anion by another anion occurs through the aquation (i.e. direct replacement does not occur).* When the isomerisation is carried out at a relatively higher temperature, further aquation of $[CoCl(en)_2(OH_2)]^{2+}$ to $[Co(en)_2(OH_2)_2]^{3+}$ may occur. Then the successive anation of $[Co(en)_2(H_2O)_2]^{3+}$ may lead to an isomerisation.

(b) **Mechanism of geometrical isomerisation of the $[M(A—A)_2X_2]$ type complexes through the TBP intermediate (*i.e.* activated complex):** In such cases, a **dissociative mechanism** *leading to a trigonal bipyramid* causes isomerisation. This trigonal bipyramidal geometry intermediate having X at an equatorial site provides the path of isomerisation (Fig. 5.28.2.1).

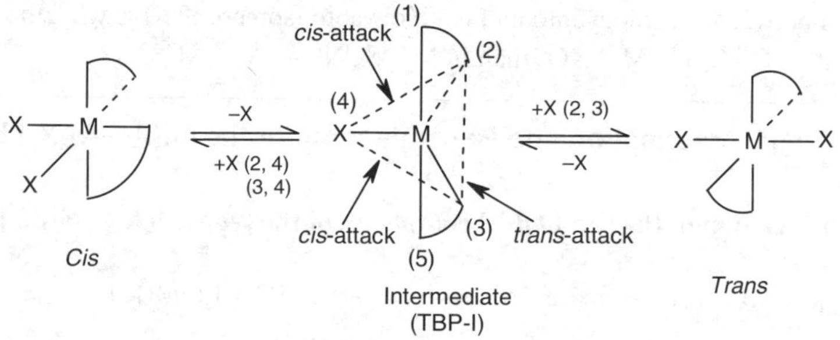

Fig. 5.28.2.1 Dissociative mechanism (intermolecular path for *cis– trans* isomerisation of the complexes like $[Co(en)_2X_2]^+$ (*cf.* Fig. 5.21.4 and 5.28.2.2).

Note: The TBP given in Fig. 5.28.2.1 gives the *cis*-product with the same optical activity (*i.e.* Δ gives the Δ isomer). There is another possible trigonal bypyramid (having X at one axial site) that will produce a *cis*–**racemic mixture** without geometrical isomerisation. In fact, these two TBPs may be interconverted through Berry pseudorotation (*cf.* Figs. 5.21.4 and 5.28.2.2).

In the trigonal bipyramid (*cf.* Fig. 5.28.2.1), the nucleophile enters along the edges of the basal triangle where the ligands are separated by 120° (*cf.* other edges separate the ligands by 90°; *i.e.* **edges of the basal triangle are sterically less hindered towards the nucleophilic attack**). Entry along the edge (2 – 3) produces the *trans*– isomer while entry along the edges (2 – 4, 3 – 4) produces the *cis*–isomers. If all these sites of nucleophilic attack are equally accessible then the entry of X will give the *cis*– and *trans*– isomers in 2:1 ratio.

It is expected, if optically pure (*i.e.* Δ or Λ) *cis*–$[CoCl_2(en)_2]^+$ is allowed to isomerise as in Fig. 5.28.2.1, then it will be accompanied with the **loss of optical activity** (*cf.* Figs. 5.21.4, 5.28.2.2). In fact, the rate of *cis*– to *trans*– isomerisation is almost comparable to the rate of racemisation (*i.e.* **loss of optical activity**) and is only slightly less than the rate of chloride exchange (using $^*Cl^-$).

Thus the findings are:

(i) isomerisation (*i.e. cis -trans*) is accompanied with the loss of optical activity.

(ii) k (*cis*– to *trans*– isomerisation) = k (racemisation) ≈ k (Cl^- exchange).

These observations indicate that there is a **common rate determining step** for the geometrical isomerisation, and Cl^- exchange processes. This common rate determining step is the dissociative transformation (through the expulsion of one X^-) of the octahedral complex to a trigonal bipyramidal intermediate **keeping preferably one X (which is a π–donor ligand) at an equatorial position**. These findings are in agreement with the mechanism outlined in Fig. 5.28.2.1.

(c) **Mechanism of isomerisation accompanied with the racemisation for *cis*-$[M(A—A)_2X_2]^+$ type complexes:** Here it may be mentioned that if the trigonal bipyramid intermediate is supposed to experience **Berry's pseudorotation** (*i.e.* the unidentate ligand X may occupy both the equatorial and axial positions), then two different trigonal bipyramids (**TBP(I)** and **TBP(II)**) may be formed; *cf.* Fig. 5.28.2.2). The Berry's pseudorotation leading to two different trigonal bipyramidal geometries is favoured if the said 5–coordinate intermediate is **long-lived and X^- (unidentate ligand) does not have any strong preference to occupy an equatorial or axial position.** In fact, if X^- is a good π–donor ligand, it will have a preference to occupy the equatorial position (*i.e.* TBP (I)).

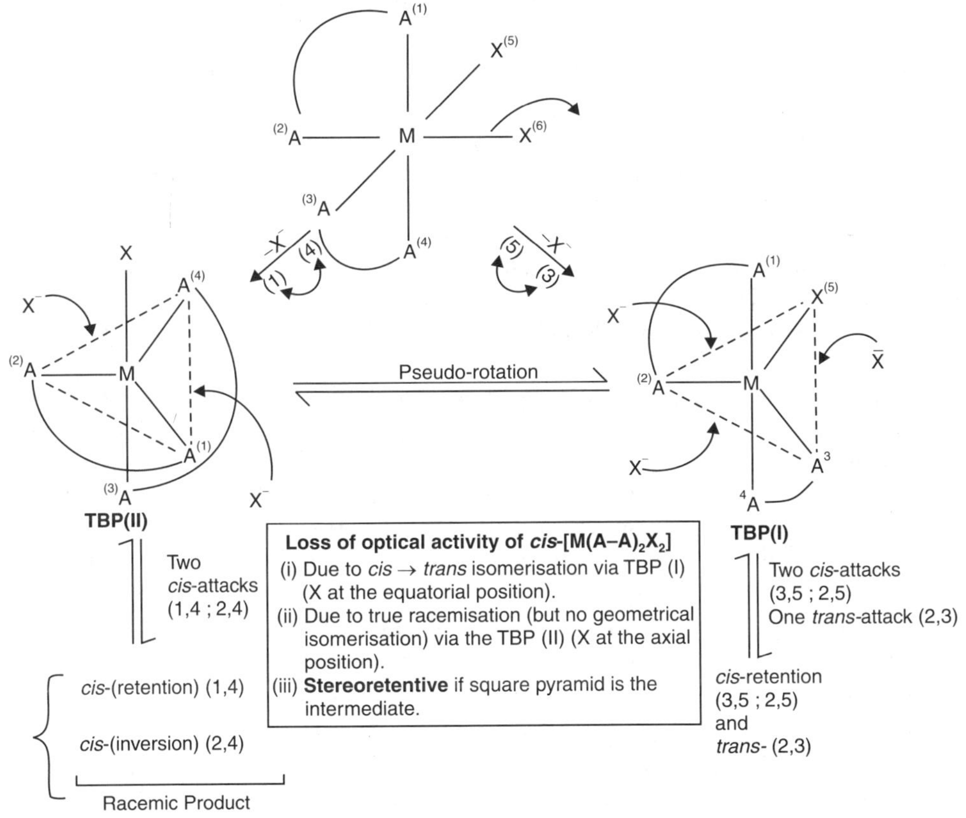

Fig. 5.28.2.2 Trigonal bipyramids as the intermediates for *cis–trans* isomerisation of the $[M(A—A)_2X_2]^+$ type complexes along with the racemisation of the optically active *cis*–complex, (*cf.* Fig. 5.21.4).

Regeneration of $[M(A—A)_2X_2]^+$ from the five coordinate TBP intermediate **structure II** having X in the axial position (Fig. 5.28.2.2) will give only the *cis*–product which is a **racemic mixture.** On the other hand, the TBP intermediate, **structure I** keeping X at one equatorial position can lead to *cis*–(which is optically pure and **retention of the same chirality**) and *trans*– product in 2:1 ratio.

The important conclusions are as follows:

- If X is a **good π-donor ligand** then TBP (I) (**X at the equatorial position**) will the preferred over the TBP(II) **(X at the axial position).**
- TBP(I) leads to the **geometrical isomerisation** (*cis* ⟶ *trans*) and the *cis*-product with the retention of the same optical activity. The loss of optical activity along this route is due to the isomerisation to the *trans*-complex **which is optically inactive.**
- TBP(II) causes the **true racemisation without any geometrical isomerisation.**
- Measured **loss of optical activity** (some times called **apparent** or **measured racemisation**) can occur due to both the geometrical isomerisation (via TBP(I)) and true racemisation (via TBP (II)).
- Rate (loss of optical activity) *i.e.* Rate (measured racemisation) ≈ Rate (*cis* ⟶ *trans*) indicates TBP(I) to be the predominant species (when X is a good π-donor ligand).
- Rate (loss of optical activity) *i.e.* Rate (measured racemisation) ⟩ Rate (*cis* ⟶ *trans*) (≠ 0) indicates both TBP(I) and TBP(II) to contribute.
- Rate (loss of optical activity) *i.e.* Rate (measured racemisation) > > Rate (*cis* ⟶ *trans*) (≈ 0, *i.e.* practically no geometrical isomerisation) indicates TBP(II) to be the predominant species.

(d) **Isomerisation and racemisation of *cis*-$[Cr(C_2O_4)_2(OH_2)_2]^-$ in neutral solution:** Here it may be pointed out that both the intermediate TBP structures I and II are not equally stable. Thus the product distribution depends on the relative proportions of these two structures. In fact, for *cis*– $[Cr(C_2O_4)_2(OH_2)_2]^-$ (optically active), the rate of racemisation is much faster than the rate of *cis*– to *trans*– isomerisation. It suggests that **the loss of optical activity of the starting *cis*-complex is not only due to its isomerisation to the *trans*-complex via TBP(I).** In other words, both the TBP(I) and TBP(II) (causing the true racemisation but no geometrical isomerisation) are involved and the TBP(II) may predominate over the TBP(I).

Here it is worth mentioning that in **neutral or slightly acidic condition** (pH ≈ 3 – 7), *cis*-$[Cr(C_2O_4)_2(OH_2)_2]^-$ experiences isomerisation and recemisation through the TBP intermediate generated from the **Cr—OH$_2$ bond cleavage** ($\Delta V^{\neq}$ = positive) but in **acidic condition** (pH ≈ 0 – 2), the TBP intermediate is generated by **acid catalysed one-ended dissociation (*cf.* Fig. 5.29.2) of one oxalate chelate ring** making a free —CO_2^- group which imposes an electrorestriction over the solven to make $\Delta V^{\neq}$ negative. Subsequently, ring closing by the protonated open end of the oxalate gives back the octahedral product with racemisation and isomerisation.

(e) **Isomerisation and racemisation for *cis*-$[Co(en)_2(OH_2)X]^{2+}$ (X = OH^-, Cl^-, Br^-, N_3^-):** The experimental observations are (recemisation is measured by the **loss of optical activity):**

k(*cis* ⟶ *trans*) ≈ k(recemisation); $\Delta V^{\neq}$ ≈ +7 cm^3 mol^{-1} (*i.e.* dissociative path)

It indicates the dissociative process (cleavage of the Co—OH$_2$ bond) predominantly passes through the TBP(I) where X occupies an equatorial position of the TBP structure. It is reasonable because of the π-donor property of the given X-ligands.

(f) **Trans ⟶ cis isomerisation of $[Cr(OH_2)_2(ox)_2]^-$ *vs.* $[Cr(mal)_2(OH_2)_2]^-$:** The *trans*– ⟶ *cis*– isomerisation of $[Cr(OH_2)_2(ox)_2]^-$ and $[Cr(mal)_2(OH_2)_2]^-$ (ox = oxalate, mal = malonate) occurs through the dissociative transformation into a 5–coordinate intermediate. *But in the case of oxalato complex, the chelate ring opens* (*i.e.* **one ended dissociation of the chelate ring**) to give a five coordinate TBP intermediate while for the malonato complex, the aqua ligand dissociates to give the 5–coordinate TBP intermediate.

$$[Cr(C_2O_4)_2(OH_2)_2]^- \xrightarrow{\text{rds}} \left[Cr(C_2O_4)(OH_2)_2 \right.$$

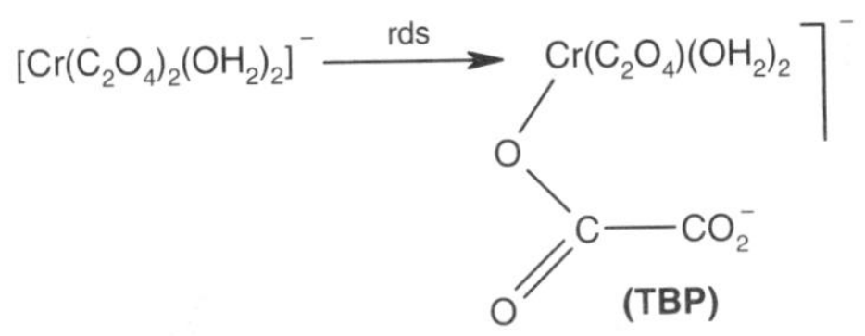

(one ended dissociation of the oxalate chelate ring)

(TBP)

$$[Cr(C_3H_2O_4)_2(OH_2)_2]^- \xrightarrow{\text{rds}} \left[Cr(C_3H_2O_4)_2(H_2O) \right]^- + H_2O$$
(TBP)

It is suggested that the six membered chelate ring formed by the malonate is more stable than the five membered chelate ring formed by the oxalate in the TBP intermediate where the chelate ring spans the equatorial and axial position. *This is why, the oxalato chelate ring experiences the one–ended dissociation while the malonato complex experiences the dissociation of one aqua ligand.*

The above propositions are experimentally supported by the $\Delta V^{\neq}$ values. For the oxalato complex, $\Delta V^{\neq}$ is **highly negative** ($= -16.6$ cm^3 mol^{-1}) while for the malonato complex it is **positive** ($= + 8.9$ cm^3 mol^{-1}) as expected for the dissociative process. For the oxalato complex, one ended dissociation of the chelate ring generates a free —CO$_2^-$ group which can impose an electrorestriction over the solvent molecules to make $\Delta V^{\neq}$ negative. On the other hand, dissociation of one neutral ligand (*i.e.* H$_2$O) neither creates nor destructs any charge. Thus $\Delta V^{\neq}$ is positive as usual.

The isomerisation of *trans* → *cis* for [Cr(C$_2$O$_4$)$_2$(OH$_2$)$_2$]$^-$ can be catalysed by the bivalent metal ions (*e.g.* Cu^{2+}, Ni^{2+}, Co^{2+}, Ca^{2+}, etc.) which favour **the one-ended dissociation of the oxalate ring** through complexation with the partially dissociated oxalate moiety. Thus the TBP generated gives back the octahedral complex through ring closure and simultaneous release of the metal ion.

(g) *Trans* → *cis* isomerisation of [Co(CH$_3$COO)(en)$_2$(OH$_2$)]$^{2+}$: The *trans–cis* isomerisation of the complex is associated with the positive $\Delta V^{\neq}$ value ($= +7.9$ cm^3 mol^{-1}). It suggests the generation of a 5–coordinate TBP intermediate through the dissociation of the neutral aqua ligand.

5.29 RACEMISATION OF THE OCTAHEDRAL COMPLEXES (*cf.* Sec. 5.33 for tetrahedral complexes)

(1) *cis*–[M(L — L)$_2$X$_2$]: Racemisation of such optically active complexes can be explained in terms of the generation of trigonal bipyramid. These aspects have been discussed in Sec. 5.28.2 (*cf.* Figs. 5.21.4; 5.28.2.2).

(2) [M(L — L)$_3$]: Racemisation of such *tris*–chelates can occur through two possible routes:

(A) intermolecular mechanism and (B) intramolecular mechanism

(A) **Intermolecular dissociative mechanism:** This process involves the reversible dissociation of a ligand giving rise to a symmetrical *bis*–chelate.

Reentry of L—L to the **symmetrical intermediate** [M(L — L)$_2$] can give both the Δ and Λ isomers with an equal probability. This leads to racemisation. It is evident that this process will also lead to the ligand exchange with the free ligand L—L present in solution.

In this process, we can reasonably expect that the rate of racemisation should be comparable (within the experimental error limit) to the rate of dissociation of the *tris*– to *bis*– complex and also the rate of ligand exchange of the [M(L—L)$_3$] complex with the free ligand L — L.

$$k_{rac} \approx k_{diss} (\text{tris} \to \text{bis}) \approx k_{ex}$$

$$[M(L-L)_3] \rightleftharpoons [M(L-L)_2] + L-L \rightleftharpoons [M(L-L)_3]$$
$$\Delta \qquad\qquad\qquad\qquad\qquad\qquad \Lambda$$

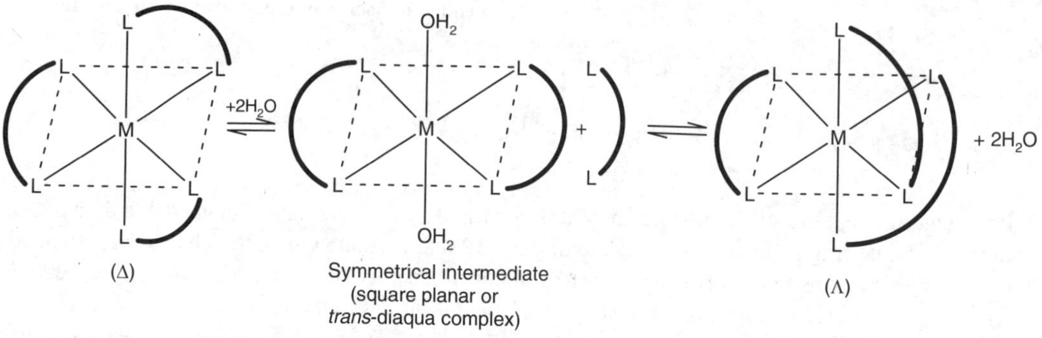

Fig. 5.29.1 Intermolecular dissociative path of racemisation of $[M(L—L)_3]$ through the formation of a symmetrical intermediate, $[M(L—L)_2]$ (square planar) or *trans*-$[M(L—L)_2(OH_2)_2]$ (*cf.* Sec. 5.33; **optically active tetrahedral** compounds may racemise through the dissociative path giving rise to a trigonal planar intermediate).

- **Racemisation of $[M(phen)_3]^{2+}$:** Among the $[M(phen)_3]^{2+}$ complexes (M = Cr, Fe, Co, Ni), **for the $[Ni(phen)_3]^{2+}$ complex, the dissociative intermolecular path is the only path for racemisation and it is supported by the fact** k_{diss} (tris → bis) ≈ k_{rac}.

Table 5.29.1 Comparison of rate constants for dissociation (k_{diss}) of *tris*– to *bis*– complex and racemisation (k_{rac}), (at 25°C) of $[M(phen)_3]^{2+}$ complexes (*cf.* E.L. Blinn *et al.*, *Inorg. Chem.*, **15,** 2952, 1976; R.G. Wilkins *et al., J. Chem. Soc. (London),* 1763, 1957).

M	k_{diss} (s^{-1})	k_{rac} (s^{-1})
Cr	1.7×10^{-2}	0.12
Fe	7.0×10^{-5}	6.5×10^{-4}
Co	0.16	20.0
Ni	7.5×10^{-6}	9.4×10^{-6}

For all other $[M(phen)_3]^{2+}$ complexes (M = Cr, Fe, Co), $k_{rac} > k_{diss}$ *indicates that the racemisation process probably passes through the different path, i.e.* **intramolecular path.** Here it may be mentioned that in such cases *both the inter– and intramolecular paths may also go on simultaneously* to make the racemisation rate faster than the dissociation rate; or the racemisation through the intramolecular path is faster than the intermolecular dissociative path. In general, we may write:

$$k_{rac} = k_{intra} + k_{inter} \ (= k_{diss})$$
$$= k_{diss}, \ (\text{when } k_{intra} = 0)$$

- **Racemisation of $[Ni(bpy)_3]^{2+}$ *vs.* $[Ni(phen)_3]^{2+}$:** Both these complexes undergo racemisation through the **dissociative intermolecular pathway.** In fact, for both the complexes, the kinetic parameters (*i.e. k*, $\Delta H^{\neq}$ and $\Delta S^{\neq}$) are identical (within the experimental error limit) for dissociation (*tris*– to *bis*– complex) and racemisation in acid media under the comparable conditions, *i.e.* $k_{rac} \approx k_{diss}$.

However, acid dependence of rates is of different types for $[Ni(bpy)_3]^{2+}$ and $[Ni(phen)_3]^{2+}$. The rates are independent of acid concentration for $[Ni(phen)_3]^{2+}$ while the rates are dependent on acid concentration for $[Ni(bpy)_3]^{2+}$ which can experience the **one ended dissociation of the chelate ring** accompanied by a rotation around the C—C bond and protonation of the released end of the ligand followed by the complete loss of the ligand. In the phen-complex, because of the rigid structure of the ligand, no such one-ended dissociation process is possible. This makes the difference in acid dependence pattern of their rates. This aspect has been discussed in detail in Sec. 5.24.1.

In acidic media, the dissociated chelating ligand (*i.e.* **phen**) remains protonated and it cannot recombine with the *bis*–chelate to regenerate the *tris*–chelate, *i.e.* **racemisation is not noticeable in a strongly acidic condition.** In neutral water, the dissociated ligand does not remain protonated, and recombination of the dissociated ligand to produce the *tris*–chelate is quite feasible. ***In fact, in neutral media, dissociation (i.e. tris–complex to bis–complex) is negligible and racemisation is predominant.*** *In acidic media where the dissociated ligand remains in a protonated condition, the reverse is true.* As, the rate determining step is the same for both the raceimisation and dissociation processes, **the rate of recemisation in neutral media is the same as the rate of dissociation in acidic media for $[Ni(phen)_3]^{2+}$.**

In general, in acidic condition, racemisation of $[M(N—N)_3]^{2+}$ (N—N = bpy, phen) through the chelate ring opening (both one end and two end opening) stops due to protonation of the dislodged ligand. This aspect has been discussed later.

(B) **Intramolecular mechanism:** This can occur in three possible ways.

(i) Formation of a **symmetrical or stereochemically nonrigid intermediate of lower coordination number** (*e.g.* 5, 4) through the one ended dissociation of the chelate ring(s);

(ii) Formation of a **symmetrical intermediate of higher coordination number** (*e.g.* 7, 8) through the ligation by monodentate auxiliary ligand (s);

(iii) Formation of a **symmetrical intermediate without any change of coordination number** through the **distortion or twisting of the octahedral structure**.

(a) Formation of an intermediate of lower coordination number through the one–ended opening of a chelate right: One-ended dissociation of a chelate ring can generate a **stereomobile** (*i.e.* fluxional or stereochemically nonrigid) **5–coordinate** (*e.g.* trigonal bipyramid, square pyramid) intermediate. These 5-coordinate intermediates (*i.e.* SP and TBP) can undergo Berry pseudorotation with scrambling of ligand sites. Religation by this free end of the partially dissociated ligand can lead to both Δ and Λ isomers. One-ended dissociation in two chelate rings produces a **4-coordinate symmetrical intermediate** from which both Δ- and Λ isomers can be generated through the religation by the free ends. These are illustrated in Fig. 5.29.2.

● **Racemisation of $[M(C_2O_4)]^{3-}$ (M = Co, Cr, Rh) is solution:** $[M(C_2O_4)_3]^{3-}$ (M = Co, Cr, Rh) experiences racemisation in this path way. In fact, for such complexes, the racemisation rate is much faster than the ligand exchange rate with the labelled oxalate, $*C_2O_4^{2-}$. It indicates that the racemisation for such complexes does not occur through the **dissociative intermolecular path.** However, the acid-catalysed exchange of ^{18}O in $H_2{}^{18}O$ enriched water with all the oxygens in $[Cr(C_2O_4)_3]^{3-}$ – $^{18}OH_2$ system occurs almost in the same rate of racemisation under the comparable conditions. Thus the observation is:

$$k(\text{racemisation}) \approx k(\mathbf{O}\text{-exchange}) >> k(\text{oxalate exchange}), \text{ (for } \mathbf{[Cr(ox)_3]^{3-}})$$

In $[Co(C_2O_4)_3]^{3-}$ and $[Rh(C_2O_4)_3]^{3-}$, the six outer oxygens (*i.e.* noncoordinating carbonyl oxygens) exchange with $^{18}OH_2$ much faster than the remaining six oxygens (*i.e.* coordinated O–atoms) and than the racemisation process.

The outer noncoordinating carbonyl O-atoms may exchange with ^{18}O of $H_2{}^{18}O$ (through the nucleophilic attack by $H_2{}^{18}O$ on the carbonyl-C) without any M—O bond cleavage. But the said O-exchange for the metal coordinated O-centres can occur only when the M—O bond cleaves. Co(III) and Rh(III) are kinetically more inert than Cr(III) and this is why, the Co(III) and Rh(III) complexes

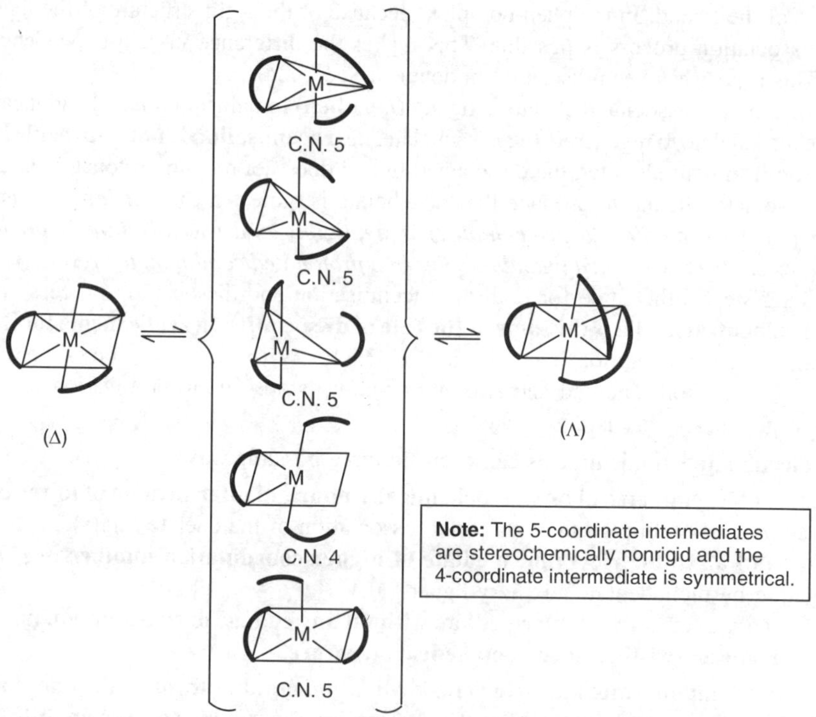

Fig. 5.29.2 Intramolecular path of racemisation of the $[M(A-A)_3]$ type chelate through one-ended dissociation of the chelate ring.

experience the O-exchange in **both ways**: in fully coordinated oxalate and in one end open oxalate. In Cr(III), the exchange process mainly goes through the one-ended dissociation of the oxalate ring.

The acid catalysed or M^{2+}-catalysed recemisation of *cis*-$[Cr(C_2O_4)_2(OH_2)_2]^-$ also occurs through the TBP intermediate produced from the one-ended dissociation of the oxalate ring (*cf.* Sec. 5.28.2). The highly **negative value of $\Delta V^{\neq}$** (= −16.3 cm³ mol⁻¹) for racemisation of $[Cr(ox)_3]^{3-}$ is due to the electrorestriction over the solvent imposed by the free —CO_2^- group generated through the one-ended dissociation of the oxalate ring in the transition state.

All these finding support that the racemisation of $[M(ox)_3]^{3-}$ occurs through the one-ended dissociation of a chelate ring giving rise to a stereochemically nonrigid 5-coordinate intermediate. **Such a one-ended dissociation of a chelate ring will not lead to the exchange of oxalate.**

One-ended dissociation of the chelate ring generates the free —CO_2^- group at which oxygen exchange can occur. Oxygen exchange rate for the six oxygens (coordinated to the metal centre) in $[M(C_2O_4)_3]^{3-}$ should be identical with the racemisation rate provided the racemisation process goes on through the intermediate having the one-ended dissociated chelate ring. *The noncoordinated six carbonyl oxygens may experience the oxygen exchange without any ring cleavage.* However, random ring opening and ring closure will allow the exchange of all the 12 oxygens of the three bound oxalate ligands at a comparable rate in $[M(C_2O_4)_3]^{3-}$ as in $[Cr(C_2O_4)_3]^{3-}$.

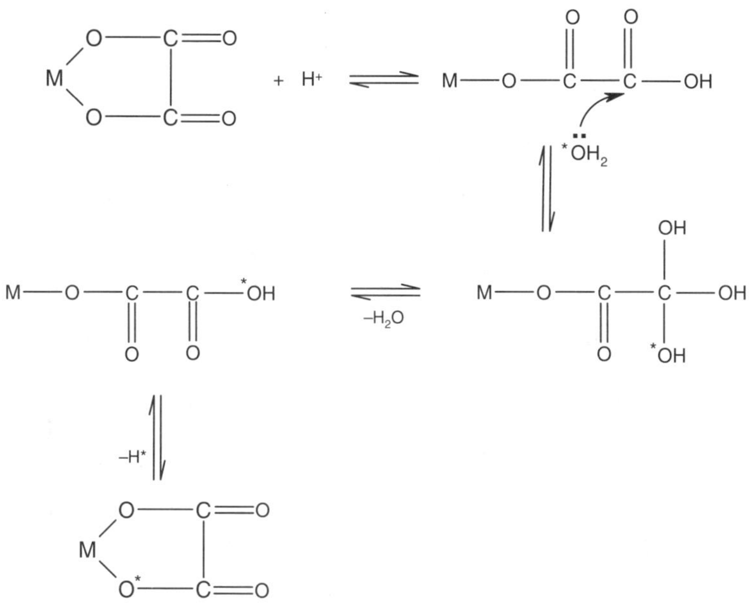

Fig. 5.29.3 O-exchange in $[M(ox)_3]^{3-}$ through the one ended dissociation of the oxalate ring.

● **Racemisation and linkage isomerisation in a Co(III) complex, *tris*(acetylacetylacetonato)-cobalt(III) where one CH_3 group is substituted by CD_3:** Dissociative pathway (*i.e.* **one-ended dissociation of the chelate ring**) has been argued for racemisation of the following compound.

The complex (in chlorobenzene) undergoes also linkage isomerisation but the *the rate of racemisation is faster than the rate of linkage isomerisation.* It is proposed that the 5–coordinate intermediate is common for the both processes. However, for the linkage isomerisation, it needs a rotation around the 'C—C' bond of the one-ended dissociated chelate. This additional operation makes the linkage isomerisation process slower. The process is outlined in the scheme (Fig. 5.29.4).

A → B → C: (linkage isomerisation + racemisation) where a rotation around the C—C bond is required.

A → B → D: (racemisation without linkage isomerisation) where the rotation around the C—C bond is not required.

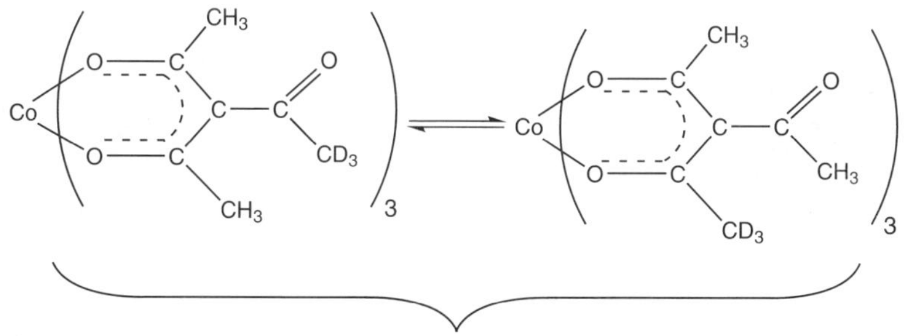

Linkage isomers

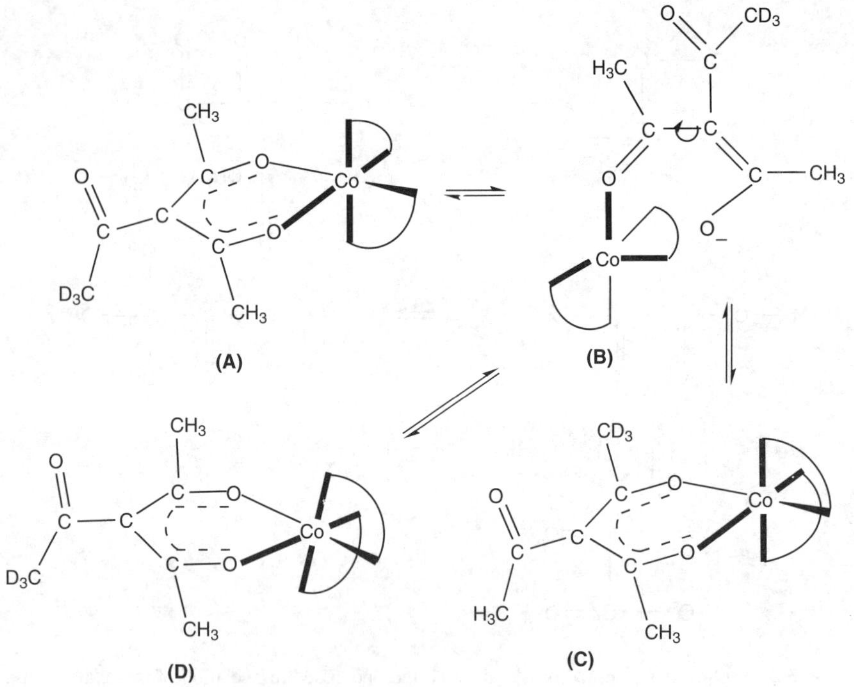

Fig. 5.29.4 Linkage isomerisation and racemisation through the 5-coordinate intermediate (*cf.* A.Y. Girgis, *et al., J. Am. Chem. Soc.,* **92,** 7061, 1970).

(b) **Formation of an intermediate of higher coordination number:** Formation of a symmetrical intermediate like **cubic structure** (C.N. 8; Charonnat, 1931) (Fig. 5.29.5) by the ligation of two

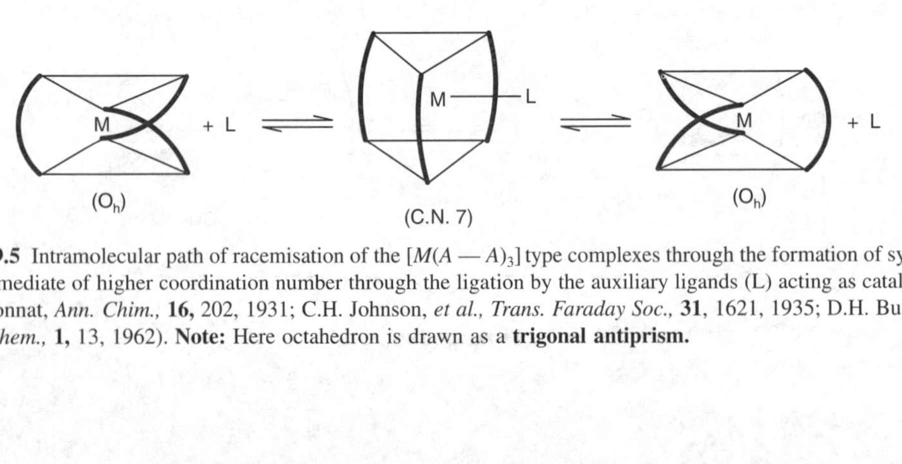

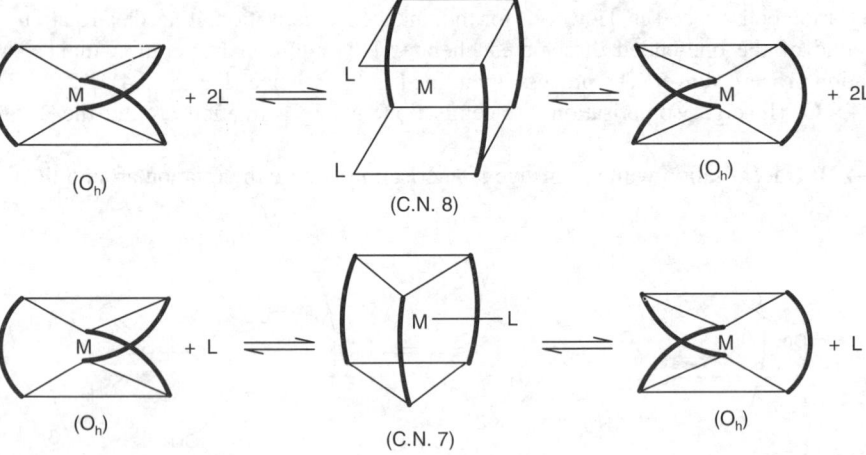

Fig. 5.29.5 Intramolecular path of racemisation of the $[M(A - A)_3]$ type complexes through the formation of symmetrical intermediate of higher coordination number through the ligation by the auxiliary ligands (L) acting as catalysts. (*cf.* R. Charonnat, *Ann. Chim.,* **16,** 202, 1931; C.H. Johnson, *et al., Trans. Faraday Soc.,* **31**, 1621, 1935; D.H. Bush *et al., Inorg. Chem.,* **1,** 13, 1962). **Note:** Here octahedron is drawn as a **trigonal antiprism.**

monodentate **auxiliary ligands** (which may be the coordinating solvents) can lead to both Δ and Λ isomers through the expulsion of these auxiliary ligands from the cubic structure. Similarly, a symmetrical intermediate like **capped octahedron** (C.N. 7; Bush *et al.*, 1962) may be also formed for racemisation (Fig. 5.29.5).

● **Racemisation of $K_3[Cr(C_2O_4)_3] \cdot 2H_2O$ *vs.* $K_3[Cr(C_2O_4)_3]$ in solid state:** Optically active, $K_3[Cr(C_2O_4)_3] \cdot 2H_2O$ undergoes racemisation on heating (in solid state) at about 120°C in an evacuated sealed tube. It is believed that the molecules of water of crystallisation ligate as the auxiliary ligands to give the intermediate of cubic structure that leads to racemisation. In fact, the anhydrous complex, $K_3[Cr(C_2O_4)_3]$ does not experience racemisation under this condition. For the anhydrous complex, there is no auxiliary ligand for expansion of coordination number to pave the way for racemisation.

● **Base catalysed racemisation of $[Co(edta)]^-$:** Base catalysed racemisation of $[Co(edta)]^-$ is believed to pass through the formation of a 7–coordinate intermediate (**capped trigonal prism**). The activation parameters ($\Delta H^{\neq}$, $\Delta S^{\neq}$) can support this proposed mechanism.

$$\Delta H^{\neq} = 134 \text{ kJ mol}^{-1}; \ \Delta S^{\neq} = 230 \text{ J K}^{-1} \text{ mol}^{-1}$$

The high enthalpy of activation is quite reasonable because formation of the proposed 7–coordinate intermediate needs considerable stretching of the metal-ligand bonds. Fixing of OH^- (which is highly solvated in free condition) requires desolvation that will make a large positive value of $\Delta S^{\neq}$.

(c) Twist mechanism: Distortion of the octahedral complex can generate a ***symmetrical intermediate*** that can give both the Δ and Λ isomers.

Ray and Dutt first proposed (1943) the twist mechanism (which is described as the **tetragonal or rhombic twist)** for the racemisation of *tris*-(biguanide)cobalt(III) ion, $[Co(bigh)_3]^{3+}$. The complex is exceedingly stable both kinetically and thermodynamically. This is why, Ray and Dutt argued that racemisation through the dissociation of the chelate ring is not possible. Ligation by the auxiliary ligands to produce a symmetrical intermediate of higher coordination number is also quite unlikely in terms of the inherent inertness of Co(III). Rather twisting is more probable for the racemisation.

In Ray-Dutt twist, keeping one chelate ring fixed, other two chelate rings rotate in opposite directions along their own plane by 45° to give a distorted octahedral geometry which may be described as a transition state (C_{2v}) for the [M(L — L)$_3$] type chelate. In fact, Ray-Dutt twist is described as the **rhombic twist because of this C_{2v} symmetry of the transition state.** This transition state may be also attained by rotating (through 60°) a trigonal face which is not associated with a three fold axis (C_3). Actually, this rotation is carried out about the **imaginary C_3 axis.** From this intermediate, retracing its step will give the starting optical isomer or continuation of rotation by the same magnitude will give the isomer of opposite chirality. Ray-Dutt twist is illustrated in Fig. 5.29.6-7.

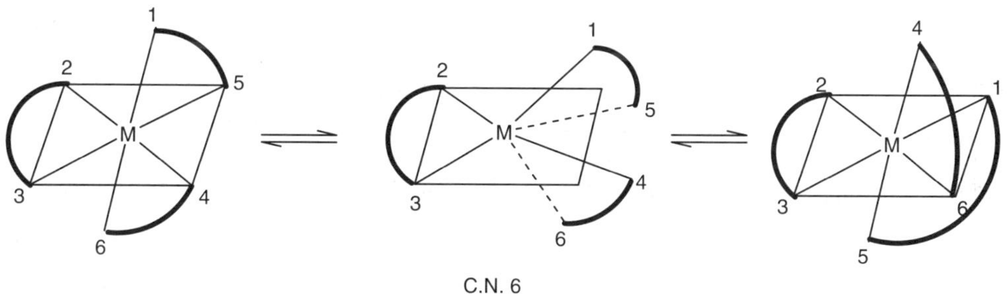

Fig. 5.29.6 Ray-Dutt twist (*i.e.* rhombic twist) mechanism for racemisation of $[M(A - A)_3]$ type chelates (*cf.* P. Ray and N.K. Dutt, *J. Indian Chem. Soc.*, **20**, 81, 1943). **Note:** One chelate ring denoted by 2, 3 remains unchanged while between the other two chelate rings, one chelate ring moves up while the other ring moves down simultaneously.

Racemisation in [M(AB)$_2$] type tetrahedral complexes (*cf.* Sec. 5.33). It passes through the **intramolecular rearrangement** giving rise to a symmetrical square planar intermediate.

$$\underset{(T_d,\Delta)}{[M(AB)_2]} \rightleftharpoons \underset{(planar)}{[M(AB)_2]} \rightleftharpoons \underset{(T_d,\Lambda)}{[M(AB)_2]}$$

This path is similar to twisting mechanism of racemisation.

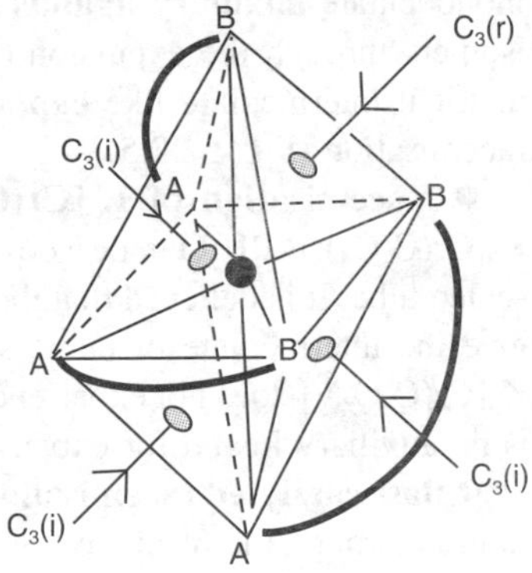

(a)

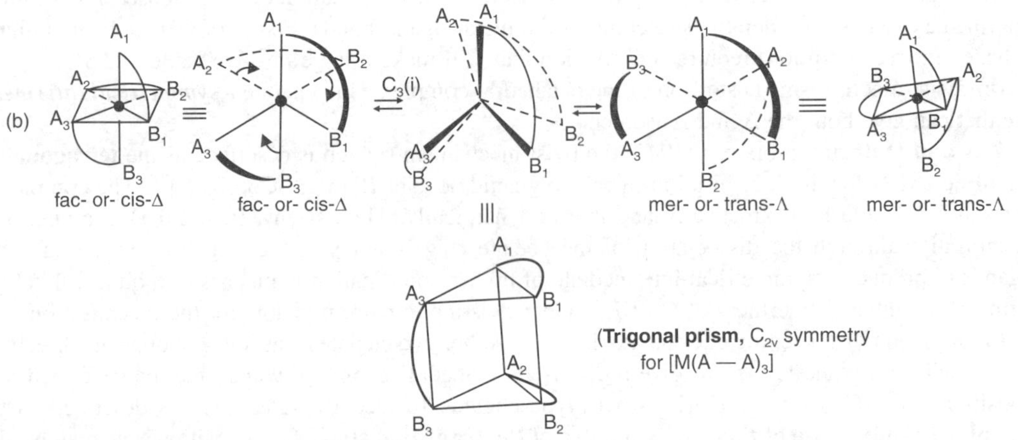

(● stands for the metal centre)

Fig. 5.29.7 (a) Position of imaginary C_3 axis [*i.e.* C_3(i)] and real C_3 axis [*i.e.* C_3(r)] in *fac*– or *cis*– Δ isomer [M(AB)$_3$]. (b) Schematic operative stages of Ray-Dutt twist in [M(A – B)$_3$] type chelate (*cf.* Sec. 10.9.3, Vol. 2).

Note: Here racemisation is coupled with geometrical isomerisation. **If A = B, the intermediate possesses C_{2v} symmetry.**

Bailar later proposed (*cf.* J.C. Bailar, *J. Inorg. Nucl. Chem.*, **8,** 165, 1958) another type of twist mechanism (called the **trigonal twist**) involving the rotation of a trigonal face with respect to the opposite face by 60° about a C_3–axis leading to a trigonal **prismatic transition state** having the $\boldsymbol{D_{3h}}$ **symmetry (for the [M(A—A)$_3$] type complex).** The Bailar twist may be also rationalised in another way by considering the octahedral chelate complex as the **trigonal antiprism** complex. In fact, the octahedral chelate appears as a trigonal antiprism when viewed through a C_3–axis. The trigonal antiprism may be represented by two triangles — solid triangle indicating the upper plane and dashed triangle indicating the plane below the paper (*cf.* Fig. 5.29.8). Twisting of the upper triangle gives the trigonal prism (D_{3h} symmetry for the [M(A—A)$_3$] type complexes). Because of this $\boldsymbol{D_{3h}}$ **symmetry of the transitions state,** this twist mechanism is called the *trigonal twist.* From this transition state, retracing will give the starting optical isomer while continution by another 60° will give the optical isomer of opposite chirality. This is illustrated in Figs. 5.29.8-9.

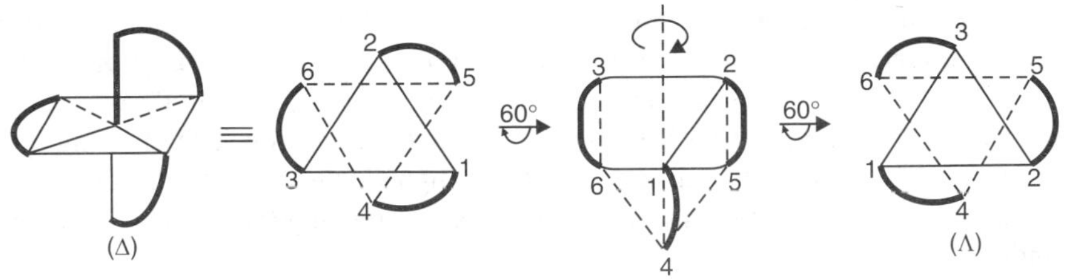

Fig. 5.29.8 Schematic representation of **Bailar twist** ('*M*' placed at the centre of the geometry, not shown).

(d) Distinction between the Bailer twist and Ray-Dutt twist: This distinction cannot be done for the [M(A — A)$_3$] type complexes having the symmetrical chelate but it can be done for the [M(A—B)$_3$] type complexes bearing the unsymmetrical chelating ligand.

The Bailar twist on [M(A—B)$_3$] will lead to **only racemisation without any geometrical isomerisation** (*i.e. cis–trans, i.e. fac–mer* isomerisation). Thus it will convert *cis*–Δ to *cis*–Λ. On the other hand, the Ray-Dutt twist on [M(A—B)$_3$] will lead to both inversion of chirality and geometrical isomerisation. The *cis*–Δ will lead to **both inversion of chirality and geometrical isomerisation.** Thus *fac*- or *cis*–Δ will lead to *mer*- or *trans*–Λ.

$$\underset{cis\text{-}\Lambda \text{ or } fac\text{-}\Lambda}{\left[M(A—B)_3\right]} \xleftarrow[\text{twist}]{\text{Bailar}} \underset{cis\text{-}\Delta \text{ or } fac\text{-}\Delta}{\left[M(A—B)_3\right]} \xrightarrow[\text{twist}]{\text{Ray-Dutt}} \underset{trans\text{-}\Lambda \text{ or } mer\text{-}\Lambda}{\left[M(A—B)_3\right]}$$

Thus we may conclude that the intramolecular twist mechanism will lead to the inversion of chirality always but not necessarily the *cis–trans, i.e. fac–mer* isomerisation.

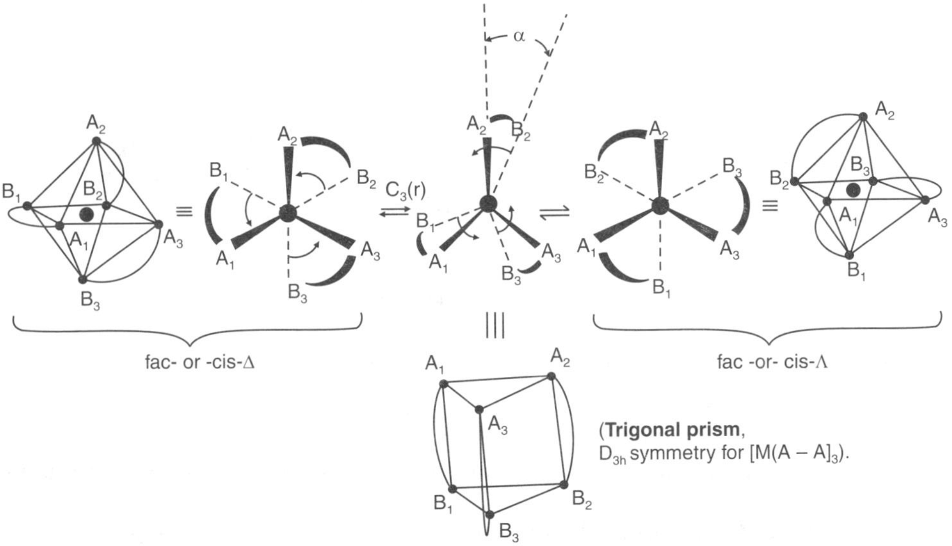

Fig. 5.29.9 Bailer twist (*i.e.* trigonal twist) and twist angle α for racemisation of a *tris*–chelate. Illustrated for the [M(A—B)$_3$] type chelates; rotation along the C_3–axis, *i.e.* C_3(r) passing through the octahedral face bearing the three ligating sites A's or three B's. (If A = B, the intermediate is a trigonal prism having the D_{3h} symmetry). (M placed at the centre of the geometry) (*cf.* Sec. 10.9.3, Vol. 2).

Note: ***For the [M(A — B)$_3$] type chelate, Bailar twist leads to racemisation without any cis–trans or fac–mer isomerisation cf. Ray-Dutt twist leads to both racemisation and fac–mer isomerisation simultaneously.***

(e) Factors favouring the Bailar twist or Ray-Dutt twist: Bailar twist leads to a **transition state (D_{3h}) of higher symmetry** while Ray-Dutt twist leads to a **transition state (C_{2v}) of relatively lower symmetry** for the $[M(A–A)_3]$ complex. In the D_{3h} transition state, the vertical distance of the trigonal prism is given by the **bite distance (*b*) of the ligand**. In the C_{2v} transition state, the length of one edge of a triangular face is given by the bite distance (*b*) of the ligand, while the lengths of the two edges of the triangular face are dictated by the distance (*l*) between the ligating (*i.e.* donor) atoms of the neighbouring ligands.

- **Chelating ligands with small bite distances (*b*) are known to stabilise the trigonal prismatic geometries** (D_{3h} symmetry). In fact, the favourable condition for the Bailar twist is attained when: ***b* is sufficiently smaller than *l*** (distance between the donor atoms of neighbouring ligands). In fact, when *b* is small (*i.e.* small twist angle, *cf.* Fig. 5.29.8), the ground state structure is already considerably distorted towards the transition state, *i.e.* trigonal prismatic configuration.
- Favourable condition for the Ray-Dutt twist is attained when:

  ***b* is sufficiently larger than *l*.**
- Under the condition, $b \approx l$, both the twist mechanisms can go on simultaneously.

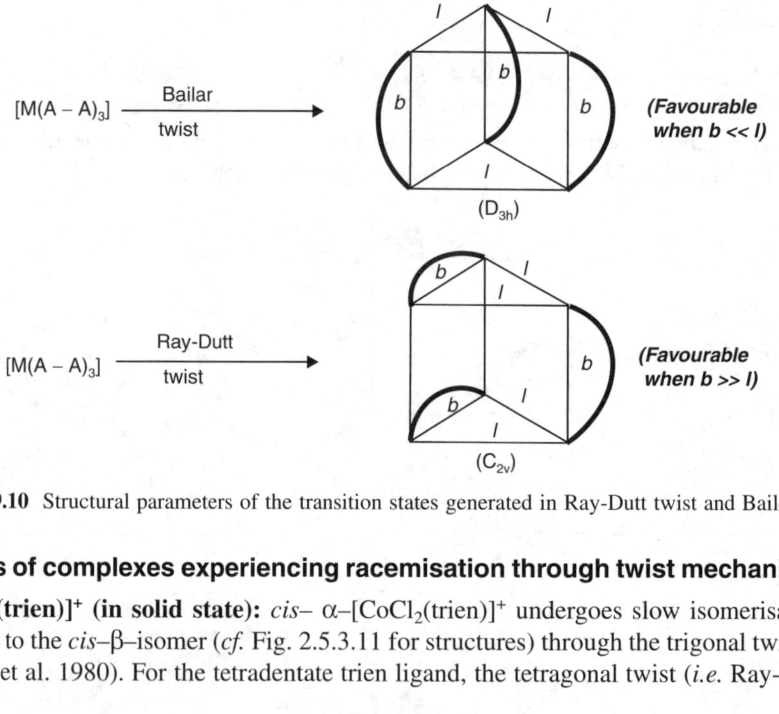

Fig. 5.29.10 Structural parameters of the transition states generated in Ray-Dutt twist and Bailer twist.

(f) Examples of complexes experiencing racemisation through twist mechanism:

• ***cis*-$[CoCl_2(trien)]^+$ (in solid state):** *cis*– α–$[CoCl_2(trien)]^+$ undergoes slow isomerisation in solid state at 157°C to the *cis*–β–isomer (*cf.* Fig. 2.5.3.11 for structures) through the trigonal twist (*i.e.* Bailar twist) (Bailar et al. 1980). For the tetradentate trien ligand, the tetragonal twist (*i.e.* Ray-Dutt twist) is not possible.

• **$[Ni(N-N)_3]^{2+}$ *vs.* $[Fe(N-N)_3]^{2+}$ (N – N = bpy, phen):** The rate of raceimisation is comparable to that of dissociation (*tris* to *bis*) for the Ni(II)–complexes while the rate of racemisation is greater than the rate of dissociation for the Fe(II) complexes. In fact, for these Fe(II)–complexes, racemisation occurs through **both the intramolecular and intermolecular paths** while for the Ni(II)–complexes, the racemisation occurs exclusively through the *intermolecular path*.

$$k_{rac} \approx k_d (tris \rightarrow bis \text{ complex}) \approx k_{ex} (\text{Ligand exchange}) \text{ for } [Ni(bpy)_3]^{3+} \text{ and } [Ni(phen)_3]^{2+} ;$$

$$k_{rac} \rangle\, k_d \approx k_{ex} \text{ for } [Fe(bpy)_3]^{2+} \text{ and } [Fe(phen)_3]^{2+}.$$

- **$[Ni(N-N)_3]^{2+}$ (Racemisation process):** Only intermolecular path.
- **$[Fe(N-N)_3]^{2+}$ (Racemisation process):** Both the intermolecular and intramolecular paths.

The intramolecular path may involve the ***one-ended dissociation of the chelate ring*** or ***twisting mechanism.***

- **$[Fe(phen)_3]^{2+}$ *vs.* $[Fe(bpy)_3]^{2+}$:** For **$[Fe(phen)_3]^{2+}$**, one ended dissociation of the chelate ring is not possible (*cf.* Sec. 5.24.1). Thus its intramolecular path involves the **twisting mechanism** only.

For **$[Fe(bpy)_3]^{2+}$**, ***both the twisting mechanism and one ended dissociation of the chelate ring are possible for the intramolecular path.*** The life-time of the 5–coordinate intermediate produced through the one ended dissociation of the chelate ring decreases with the increase of acidity. *Thus the rate of racemisation through this intramolecular path decreases with the increase of acidity and it becomes almost zero at high acidity.* At high acidity, racemisation through the dissociative intermolecular path also stops as the dissociated ligand gets protonated and **religation by the protonated ligand is not possible.** Consequently, regeneration of the *tris*–complex becomes difficult. Thus at high acidity both the intermolecular dissociative path and the intramolecular one-ended dissociation of chelate ring path fail to contribute to racemisation. **Thus at high acidity, the limiting residual racemisation observed for $[Fe(bpy)_3]^{2+}$ is due to the twisting mechanism only.**

- **Racemisation of $[Ni(bpy)_3]^{2+}$ or $[Ni(phen)_3]^{2+}$:** Through the formation of a symmetrical *bis*–complex (*i.e.* intermolecular path) *i.e.* $k_{rac} \approx k_d \approx k_{ex}$.
- **Racemisation of $[Fe(phen)_3]^{2+}$:** Through the formation of a symmetrical *bis*–complex (*i.e.* intermolecular path) + through the twisting mechanism (*i.e.* intramolecular path).
- **Racemisation of $[Fe(bpy)_3]^{2+}$:** Through the formation of a symmetrical *bis*–complex (*i.e.* intermolecular path) + through the one-ended dissociation of chelate ring (*i.e.* intramolecular path) + through the twisting mechanism (intramolecular path).
- **Twisting mechanism $[Fe(N—N)_3]^{2+}$ *vs.* $[Ni(N—N)_3]^{2+}$ (N—N = bpy, phen):** The twisting mechanism operates for the racemisation of $[Fe(N-N)_3]^{2+}$ but it remains absent for $[Ni(N-N)_3]^{2+}$. The Fe(II)–complex may remain in a spin equilibrium, $l.s.\left(t_{2g}^6 e_g^0\right) \rightleftharpoons h.s.\left(t_{2g}^4 e_g^2\right)$. The **high-spin state** (where the metal-ligands bonds are relatively weaker) probably allows the twisting mechanism. The said spin-equilibrium is quite reasonable as the corresponding *bis*–complex is known to exist in high-spin state. For the Ni(II)–complex ($t_{2g}^6 e_g^2$), such a spin-isomerism is not possible.

(**Note:** The intramolecular path of twisting mechanism definitely contributes for both the $[Fe(N—N)_3]^{2+}$ complexes. This contribution is the limiting rate found at high acidity where no dissociative path can contribute to the racemisation process. For $[Fe(phen)_3]^{2+}$, one-ended dissociation of the chelate ring (*i.e.* intramolecular path) is not possible for its racemisation. For $[Fe(bpy)_3]^{2+}$, racemisation through the intramolecular path may occur by both twisting mechanism and one-ended dissociation of the chelate ring.

- ***cis*–$[MCl_2(en)_2]Cl$ (M = Co, Cr) (in solid state):** *cis*–$[MCl_2(en)_2]Cl$ (M = Co, Cr) undergoes racemisation slowly at 150 – 180°C in solid state **without any *cis–trans* isomerisation** (D. Banerjea and J.C. Bailar, *Transition Met. Chem.*, **10,** 331, 1985). The *trans*–compound does not experience isomerisation under the similar condition. Interestingly, both the Co(III) and Cr(III) complexes show the comparable rates and activation parameters while in normal ligand substitution process, Co(III) reacts much more slowly than Cr(III).

These observations can be explained by considering the tetragonal twist (*i.e.* Ray-Dutt twist) mechanism. It may be mentioned, that operation of Bailar twist on *cis*–$[MCl_2(en)_2]^+$ will lead to **racemisation along with *cis–trans* isomerisation.** If racemisation occurs through the one-ended dissociation of a chelate ring or dissociation of one Cl^- leading to a 5–coordinate intermediate, then it will also experience the concommitant *cis–trans* isomerisation. All these support the recemisation of *cis*-$[MCl_2(en)_2]^+$ in solid state through the **Ray–Dutt twist** mechanism.

- **$[Co(bigH)_3]^{3+}$:** The robust (kinetically inert) complexes like $[Co(bigH)_3]^{3+}$, etc. undergo racemisation through the intramolecular twist mechanism (Ray and Dutt, 1943).
- **$[Cr(L—L)_3]^{3+}$ (L—L = bpy, phen) *vs.* $[Cr(ox)_3]^{3-}$:** The very **small positive values of $\Delta V^{\neq}$** (+3.3 $cm^3\ mol^{-1}$) for racemisation of $[Cr(bpy)_3]^{3+}$ and $[Cr(phen)_3]^{3+}$ support the **intramolecular twist mechanism** of racemisation because in this path, there is no appreciable change in molar volume in attaining the transition state (T.S.). The **highly negative values of $\Delta V^{\neq}$** (*ca.* −12 to −16 $cm^3\ mol^{-1}$) for racemisation of the anionic complexes like $[Cr(ox)_3]^{3-}$, $[Cr(N—N)(ox)_2]^-$ can be rationalised in terms of the **one-ended dissociation of oxalate ring** in the transition state giving rise to the TBP structure. The released —CO_2^- group (in the T.S.) puts **a strong electro-restriction over the solvent** to cause this negative value of $\Delta V^{\neq}$.

5.30 LIGAND SUBSTITUTION REACTIONS IN SQUARE PLANAR COMPLEXES

Transition metal ions of d^8–configuration (*e.g.* Ni^{II}, Pd^{II}, Rh^I, Ir^I, Au^{III}) generally form the square planar complexes. Ni^{II} forms the square planar complexes with the very strong field ligands only; otherwise it has a tendency to form the octahedral complexes. The Pd^{II} centre has a tendency to be reduced to Pd^0. Au(III) is a fairly strong oxidising agent. Relatively very few square planar complexes of Rh^I and Ir^I have been studied. In fact, understanding of ligand substitution process in the square planar complexes emerges basically from the studies of Pt(II) complexes.

5.30.1 Characteristics of the Ligand Substitution Process in Square Planar Complexes

(i) **Associative activation:** The square planar complexes are produced by using the dsp^2 hybrid orbitals of the central atom and thus there remains **the vacant p_z–orbital,** *i.e.* np_z orbital, perpendicular to the molecular plane **(in terms of VBT).** This **vacant p_z–orbital** can accommodate the incoming nucleophile to allow the *associative activation.* The **vacant antibonding d_{z^2} orbital** in terms of MOT (*xy*-plane as the molecular plane) can also receive the nucleophilic attack. In other words, the **vacant hybrid *pd*-orbital** (developed from $p_z^0 \pm d_{z^2}^2$ combination) can perform the task (see Sec. 5.30.3 for details).

The five coordinate intermediates or transtion states for the d^8–systems are not unlikely. Characterisation of the trigonal bipyramidal species like $[Pt(SnCl_3)_5]^{3-}$, $[Ni(CN)_5]^{3-}$ in the crystals of $[Cr(en)_3]^{3+}$–salt, $[Fe(CO)_5]$, $[Co(CO)_3L_2]^+$, etc. supports the possibility of expansion of C.N. 4 to 5 during the reaction. The **strong π–acceptor ligands** like CN^-, $SnCl_3^-$, etc. can stabilise the trigonal bipyramidal structure. Five coordinate compounds of the d^8 systems like Fe(0), Ru(0), Os(0), Rh(I), Ir(I), etc. with the π–acid ligands are now well known. In fact, in the reaction of $[AuCl_4]^-$ with NCS^-, the proposed five coordinate intermediate has been detected.

(ii) **Two term rate law:** The following substitution reaction follows the two-term rate law.

$$ML_3X + Y \rightarrow ML_3Y + X$$

$$-\frac{d[ML_3X]}{dt} = (k_s + k_Y[Y])[ML_3X],\quad i.e.\ \boldsymbol{k_{obs} = k_s + k_Y[Y]},\ \text{when } [Y] >> [ML_3X].$$

Experimental evidences indicate that the k_s-path (independent of [Y]) involves the solvent (S) as the first entering nucleophile. The overall reaction may be simply represented as follows to get the two term rate law.

$$L_3M—X + S \underset{k_{-1}}{\overset{k_1}{\rightleftharpoons}} L_3M—S + X;\ L_3M—S + Y \xrightarrow{k_2} L_3M—Y + S$$

$$L_3M—X+Y\xrightarrow{k_3}L_3M—Y+X$$

Under the **steady state condition** applied to L_3MS, we have:

$$\frac{d[L_3MS]}{dt}=0=k_1[L_3MX][S]-k_{-1}[L_3MS][X]-k_2[L_3MS][Y]$$

or, $$[L_3MS]=\frac{k_1[L_3MX][S]}{k_{-1}[X]+k_2[Y]}=\frac{k_1'[L_3MX]}{k_{-1}[X]+k_2[Y]},\ k_1[S]=\text{constant}=k_1'(\text{say})$$

$$\text{Rate} = k_2[L_3MS][Y] + k_3[L_3MX][Y]$$

$$=\frac{k_1'k_2[L_3MX][Y]}{k_{-1}[X]+k_2[Y]}+k_3[L_3MX][Y]$$

$$\approx k_1'[L_3MX]+k_3[L_3MX][Y];$$ (when $k_{-1}[X] << k_2[Y]$ which is most likely, if not X is added externally)

i.e. $k_{obs} = k_s + k_Y[Y]$, (under the pseudo-first order conditions of excess Y)
(where $k_1' = k_s$, $k_3 = k_Y$).

It may be noted that this rate law derivation does not tell us regarding the mechanistic paths of the substitution process.

The k_s–path does not depend on the nature of the entering ligand but **it depends on the nucleophilicity of the solvent.** On the other hand, the k_Y–path is strongly dependent on the nature of Y but not so strongly on the nature of the solvent. However, if **the nucleophilicity of Y depends on the nature of solvent, then k_Y shows its dependence on the solvent.** The two term rate law arises from two parallel pathways, both experiencing the associative activation.

(iii) **Stereospecific substitution:** In most of the cases, stereochemistry of the starting complex is retained fully in the product, *i.e.* a *trans*– reactant gives a *trans*– product and a *cis*–reactant gives a *cis*–product.

Retention of this stereochemistry can be explained by considering the formation of an intermediate of ***trigonal bipyramidal geometry*** *in which the entering group, leaving group and the group originally present to the trans-position of the leaving group occupy the equatorial positions.* Thus retention of stereochemistry again supports the associative activation.

Note: In some cases (*e.g.* square planar phosphine complexes of Pt(II)), there is a change of stereochemistry (*i.e.* case of isomerisation). In such cases, ***the proposed trigonal bipyramid intermediate is sufficiently long lived to experience the Berry pseudorotation to interchange the axial and equatorial ligands***.

Chemistry of Platinum and Russia

The abundance of platinum ores in Russia led the Russian chemists to explore the coordination chemistry of platinum extensively. In fact, many characteristic aspects (like stereospecific ligand displacement) of the reactions of square planar complexes of platinum(II) were first discovered by the Rusian chemists.

(iv) **Effect of the entering group and leaving group:** The rate ***strongly*** depends on the **nucleophilicity of the entering ligand.** It follows the following sequence.

Y: CN^-, CO $\rangle$ PR_3 $\rangle$ H^- $\rangle$ I^- $\rangle$ Cl^- $\rangle$ H_2O **(strong dependence)**

The rate also **mildly** depends on the nature of the leaving group. It follows the following sequence:

X: H_2O $\rangle$ Cl^- $\rangle$ I^- $\rangle$ H^- $\rangle$ PR_3 $\rangle$ CO, CN^- **(mild dependence)**

It is interesting to note that the above mentioned sequences follow the reverse order. ***Thus, the good entering groups (i.e. good nucleophiles) are the poor leaving groups (i.e. poor nucleofugals).***

Here it is important to compare the extent or degree of rate dependence on the nature of the entering group and leaving group. It is illustrated for the following reactions.

trans–$[PtCl_2(py)_2]$ + Y → *trans*–$[PtCl(py)_2(Y)]^{n+}$ + Cl^-, (*cf.* Table 5.30.5.1)

k_Y (dm^3 mol^{-1} s^{-1}):	4.7×10^{-4}	4.5×10^{-4}	3.7×10^{-3}	0.107	6.0
Y:	NH_3	Cl^-	Br^-	I^-	tu

Here the rate constant changes by a factor 10^4. For changing the entering nucleophile Cl^- to I^-, the rate increases by a factor of about 10^3. On the other hand, for changing the leaving group Cl^- to I^-, the rate changes by a factor only 3.5 under the comparable conditions (cf. Table 5.30.6.1). Thus the rate process is strongly sensitive to the nature of the entering group but it depends mildly on the nature of the leaving group (cf. Table 5.30.6.1). These observations are in favour of the *associative activation*.

(v) **Effect of the spectator ligands (*i.e.* nonleaving groups):** If the *trans*– ligand (to the leaving group) is a good nucleophile, then the rate is favoured. This phenomenon is described as the ***trans–effect***. The nonleaving *cis*–ligand can also mildly influence the reation rate. This phenomenon is called the ***cis–effect***. *It may be mentioned that the trans–effect is more pronounced than the cis–effect.*

Other possible ways of ligand replacement in square planar complexes of Pt (II)
(*cf.* R.J. Cross, *Chem. Soc. Rev.,* 14, 197, 1985)

In addition to nucleophilic substitution, other processes like *electrophilic substitution and oxidative addition followed by reductive elimination* can also lead to ligand replacement reactions. These are illustrated for the complex, $[Pt^{II}(Cl)(L)_2(Me)]$, (L = PMe_2Ph)

$$\left[Pt^{II}Cl_2L_2\right] \xleftarrow[-MeHgCl]{+HgCl_2} \left[Pt^{II}(Cl)L_2(Me)\right] \xrightarrow[-Cl^-]{+N_3^-} \left[Pt^{II}L_2(Me)N_3\right]$$

$$\begin{pmatrix}\text{Electrophilic}\\ \text{substitution}\end{pmatrix} \qquad\qquad \begin{pmatrix}\text{Nucleophilic}\\ \text{substitution}\end{pmatrix}$$

$$\downarrow \begin{array}{l} +Cl_2 \\ \text{(oxidative} \\ \text{addition)} \end{array}$$

$$\left[Pt^{IV}Cl_3L_2(Me)\right] \xrightarrow[\begin{pmatrix}\text{Reductive}\\ \text{elimination}\end{pmatrix}]{-MeCl} \left[Pt^{II}Cl_2L_2\right]$$

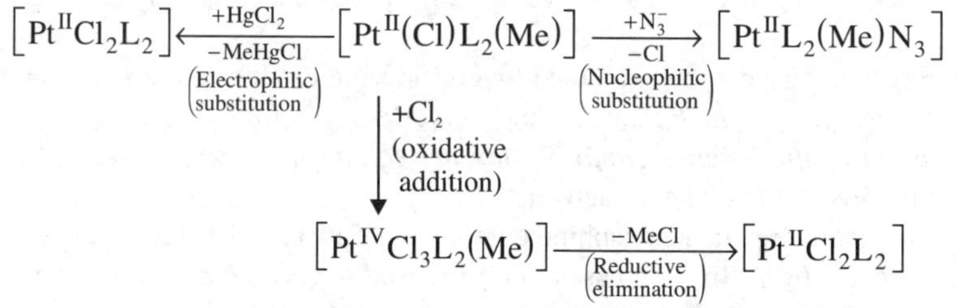

5.30.2 Characteristic Features of the Two-term Rate law

(i) **k_s and k_Y rate constants in a particular solvent:** The rate law for the reaction.

$$ML_3 - X + Y \rightarrow ML_3 - Y + X$$

is : $k_{obs} = k_s + k_Y[Y]$, when $[Y] \gg [ML_3X]$

From the plot of k_{obs} *vs.* $[Y]$, we get k_s (= intercept) and k_Y (= slope). At a particular solvent, the k_Y values are different for different entering groups in a particular reaction while k_s remains the same for all entering nucleophiles. It may be illustrated (Fig. 5.30.2.1) for the following reaction.

trans-$\left[PtCl_2(py)_2\right]$ + Y $\longrightarrow$ *trans*-$\left[PtCl(py)_2(Y)\right]$ + Cl^- (charge of the product complex not shown)

(ii) **k_s and k_Y rate constants in different solvents:** If a particular reaction is carried out in different solvents then the k_s–value (= intercept) will change depending on the coordinating behaviour of the solvent and the k_Y value may also change depending on the nature of the solvent. *The change in the k_s- value arises from the change of nucleophilicity of the solvent* (S) which will first enter as a ligand followed by the substitution of the coordinated solvent by the entering ligand in a relatively faster path.

k_s–path: $ML_3 — X + S \xrightarrow[(-X)]{\text{slow}} ML_3 — S \xrightarrow[(-S)]{+Y(\text{fast})} ML_3 - Y$

k_Y–path: $ML_3 — X + Y \xrightarrow[(-X)]{} ML_3 — Y$

In a particular solvent, for different entering ligands, the k_Y–value differs depending on the nature of the nucleophilicity of the entering ligand.

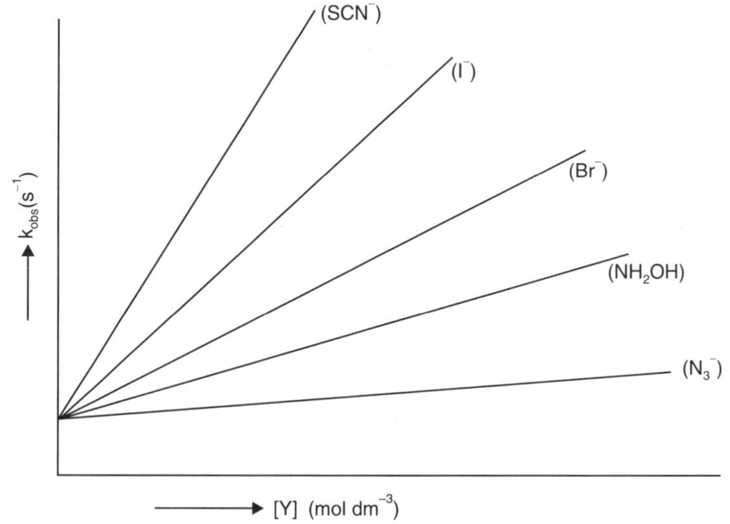

Fig. 5.30.2.1 Qualitative representation of the two term rate law for the nucleophilic substitution of Cl^- by *Y* from *trans*–$[PtCl_2(py)_2]$ in methanol solvent.

The change in the k_Y–value for a particular entering ligand in different solvents arises from the fact that the nucleophilicity of the entering nucleophile (i.e. Y) depends on the degree of solvation in a solvent. Thus, the nucleophilicity of a particular ligand is different in different solvents.

(iii) **Relative contributions of k_s and k_Y paths:** Sometimes, the k_Y path may have a negligible contribution and sometimes, the k_s–path may have also a negligible contribution depending on the situation. These are all experimentally verified.

$$[Pt(bpy)Cl_2] + py \rightarrow [Pt(bpy)Cl(py)]^+ + Cl^-$$

i.e.

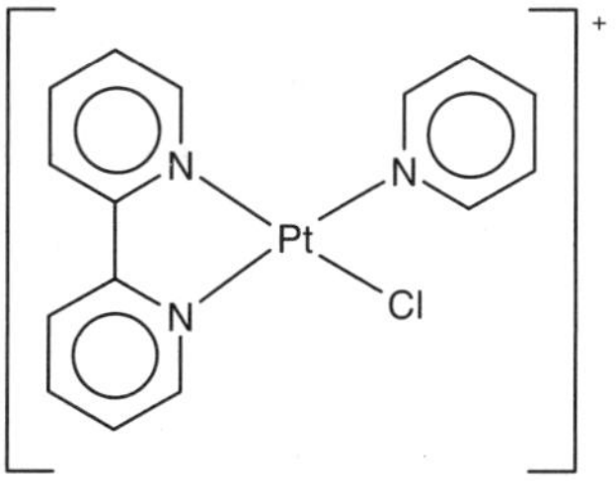

For the above reaction in methanol solvent, k_s is almost zero while k_Y is *ca.* 6×10^{-3} dm^3 mol^{-1} s^{-1}.

For the following ligand exchange reaction, k_s is almost zero in the noncoordinating solvent hexane while in the coordinating solvent methanol, k_Y is almost zero.

trans–[PtCl$_2$(N*HEt$_2$)(PPr$_3$)] + NHEt$_2$ → *trans*–[PtCl$_2$(NHEt$_2$)(PPr$_3$)] + *NHEt$_2$

Hexane (a noncoordinating solvent) cannot act as a nucleophile. Hence, absence of the k_s–path is quite reasonable. Interestingly, in the coordinating solvent methanol, the reaction mainly occurs through the solvent dependent path (*i.e.* k_s–path). *It is believed that the k_Y path which is an associative path is* ***disfavoured for the bulky entering nucleophile like diethylamine*** **(Et$_2$NH)** *and the reaction proceeds through the alternative path, i.e. k_s–path* (which is also an associative path) involving the relatively less bulky nucleophile like MeOH (*i.e.* solvent).

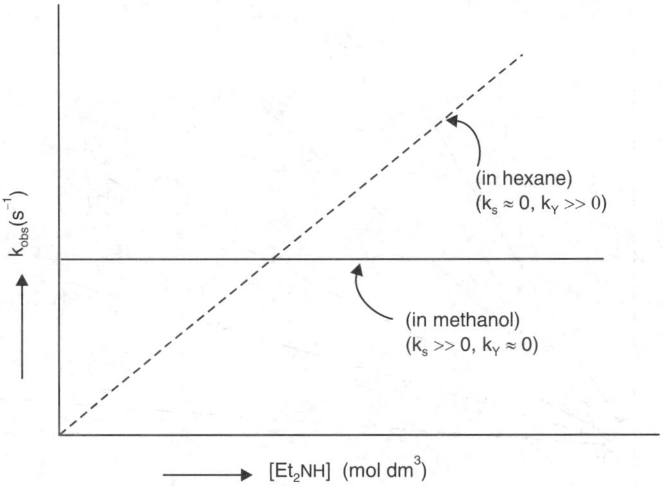

Fig. 5.30.2.2 Qualitative represention of the relative importance of k_s– and k_Y–paths in different solvents for the ligand exchange reaction: *trans*–[PtCl$_2$(N*HEt$_2$)(PPr$_3$)] + NHEt$_2$ →

(iv) **Reaction course in the k_s-path:** The k_s–path (*i.e.* solvent dependent path) involves a **double displacement process** in which each step is an associative path.

$$ML_3X + S \xrightarrow[(k_1)]{\text{slow}} ML_3(S) + X,\ S = \text{solvent} \quad \textbf{(associative path)}$$

$$ML_3(S) + Y \xrightarrow{\text{fast}} ML_3(Y) + S \quad \textbf{(associative path)}$$

The rate of this path is: k_1[complex][S] = k_s [complex] because [S] is a constant.

- The k_s–path involves the formation of the intermediate *$ML_3(S)$ that can be trapped in some cases to give the experimental evidence in favour of the proposed routes.* In fact, it has been possible to trap the aqua-intermediate in the following reaction by adding a base that can abstract a proton from the coordinated water molecule to produce a hydroxo species. ***The hydroxido group is a poor leaving group to allow the characterisation of the hydroxido complex.***

$$[\text{Pt(dien)X}]^+ \xrightarrow[(-\text{X}^-),\ \text{slow}]{\text{H}_2\text{O}} [\text{Pt(dien)(OH}_2)]^{2+} \xrightarrow[(-\text{H}_2\text{O}),\ \text{fast}]{+\text{Y}^-} [\text{Pt(dien)(Y)}]^+$$

$$[\text{Pt(dien)(OH}_2)]^{2+} + B \rightleftharpoons [\text{Pt(dien)(OH)}]^+ + \text{BH}^+ \text{ (fast)}$$

(inert to substitution by Y$^-$).

Obviously contribution of the k_s-path increases with the increase of coordinating power of the solvents having the comparable steric factors.

(v) **Relative contributions of the k_s and k_Y paths for different metal centres of d^8 configuration:** Here it should be mentioned that the relative contributions of the k_s and k_Y paths depend on the nature of the starting complex and metal centre in a particular solvent. In the Au(III)-system (which experiences more *associative activation* due to its higher charge), the k_Y-path is more predominant than the k_s-path compared to those in the analogous Pt(II)-complex. This is illustrated in the following data:

Reaction	M	$k_{H_2O}\,(s^{-1})$	$k_Y\,(dm^3\,mol^{-1}s^{-1})$	$\frac{k_Y}{k_s}\,(dm^3mol^{-1})$
$[MCl(dien)]^{n+} + Br^- \longrightarrow [MBr(dien)]^{n+} + Cl^-$	M = Au(III)	0.50	154	308
	M = Pt(II)	8.0×10^{-5}	5.3×10^{-3}	66
$[MCl_4]^{n-} + {}^*Cl^- \longrightarrow$ (exchange reaction)	M = Au(III)	6.0×10^{-3}	1.47	245
	M = Pt(II)	3.8×10^{-5}	~ 0	–

Rate constant are at 20°C for Au(III), and at 25°C for Pt(II) (cf. F. Basolo, *et al.*, *Inorg. Chem.*, **3,** 1087, 1964; **2,** 921, 1963; L.I. Elding *et al.*, *Acta. Chem. Scand. Sect. A,* **32,** 867, 1978).

5.30.3 Mechanism of the Reaction

Considering the ***dsp*² hybridisation** of the starting square planar complex (in the *xy*-plane), the **vacant** $\boldsymbol{np_z}$ (i.e. **$6p_z$ orbital for Pt(II)**) nonbonding orbital (= $\boldsymbol{a_{2u}}$ in terms of MOT) or the **vacant *pd*-hybrid orbital** (developed from the combination of vacant np_z^0 and filled $(n\text{–}1)\,d_{z^2}^2$-orbital) may receive the incoming nucleophile from the *z*-direction. In terms of MOT, the **vacant antibonding $\boldsymbol{d_{z^2}}$ orbital** (more correctly $\boldsymbol{a_{1g}}$, see Sec. 3.17.6, Vol. 4) may also receive the nucleophilic attack from the *z*-direction. Thus, the nucleophilic attack (leading to the **loss of CFSE or LFSE measuring the CEAE or LFAE**) along the *z*-direction in the **associative path** initially produces the **5-coordinate square pyramidal geometry.** Now, the entering group can move in any of the four possible directions.

The possible directions (*cf.* Scheme 5.30.3.1) are: towards L_1, towards L_2, towards *X* (which is the leaving group), towards *T* (*trans*–position with respect to the leaving group *X*). ***The most favoured direction of movement of Y is in the direction of the ligand which is least strongly bound.***

Trigonal bipyramidal transition state (I_a-path) or intermediate (A-path)

Formation of a trigonal bipyramidal transition state bearing the entering group (*Y*), leaving group (*X*) and its *trans*–group (*T*) in the trigonal equatorial plane and two *cis*–groups (L_1 and L_2, with respect to the leaving group in the starting square planar complex) in the axial positions can explain the *trans*–effect, *cis*–effect, stereochemistry of the product, etc.

This approach can explain the phenomena *trans*–effect and *trans*– influence to be discussed later. Here the leaving group *X* gradually weakens its attachment with the metal centre and bends away. This will transform the square pyramid into a trigonal bipyramid having the three ligands (*Y*, *T* and *X*), *i.e.* entering group, leaving group and the *trans*–group (with respect to the leaving group) in the equatorial plane. The other two groups (L_1 and L_2, *trans*–ligands at the starting square planar complex) will occupy the axial positions of the trigonal bipyramid. *As the leaving group X moves away from the trigonal plane of the trigonal bypyramid, the angle Y — M — T opens up to* 180° *leading to a square pyramid geometry placing the leaving group (X) at the axial position.* Then the expulsion of *X* generates the square planar geometry.

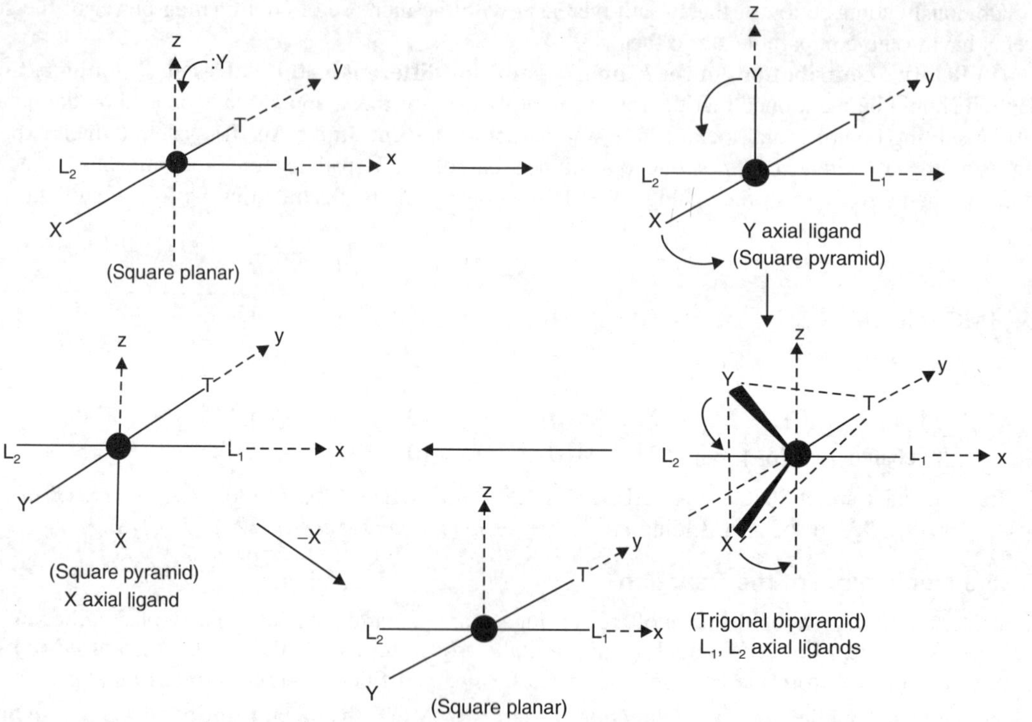

Scheme 5.30.3.1 Schematic representation of the associative mechanism of ligand substitution at the square planar complex of Pt(II).

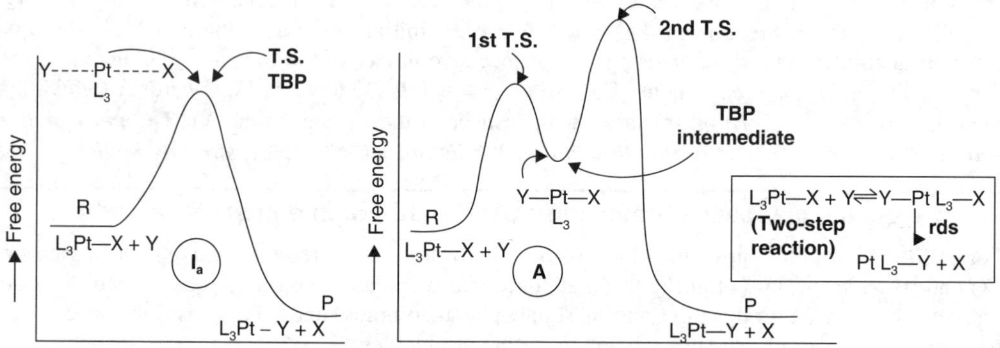

Fig. 5.30.3.1 Energy profile diagram for sustitution reaction at the square planar complexes of Pt(II) (say $L_3Pt—X$) (a) for the **one step** I_a path; (b) for the **two step** *A*–path (cf. Fig. 5.8.1) (R = Reactants; P = Products).

Note: In the **special A-path** (Fig. b) explaining the effects **of both the entering group and leaving group, reversible addition of the nucleophile** is followed by the ***rds*** **(Second T.S.) of the loss of the leaving group** accompanied with the structural rearrangement. To explain the leaving group effect, alternatively, it has been suggested that it is due to the fact: the orbital-overlap (σ and specially π) possibilities of the leaving group in the equatorial positions of TBP are substantially changed from those in the starting square planar complex. In an **ideal A-path** (Fig. 5.8.1) explaining **only the effect of the entering group,** the *rds* **(First T.S.)** is the addition of the nucleophile keeping the leaving group intact.

The proposed trigonal bipyramid species produced during the reaction may be considered as an *activated complex* or as a *true intermediate.* In the energy profile diagram, the activated complex lies at a peak and thus it refers to a *transition state.* On the other hand, the intermediate is relatively more stable and long-lived than the activated complex from which this intermediate is produced. ***In terms of life time, the activated complex is too short-lived to be detected while the intermediate is relatively long-lived and it is detectable.*** The energy profile diagram for substitution in a square planar complex involving a trigonal bipyramid activated complex or a trigonal bipyramidal intermediate is shown in Fig. 5.30.3.1. In the energy profile diagram, *presence of two maxima and one minimum in the curve suggests the existence of a true trigonal bipyramid intermediate.* This represents the **A-process** involving **two steps**. In the energy profile diagram, presence of a single maximum represents the existence of trigonal bipyramidal transition state. It represents the **I_a–process** involving a **single step**.

The two term rate law arises from the mechanistic paths of Scheme 5.30.3.2.

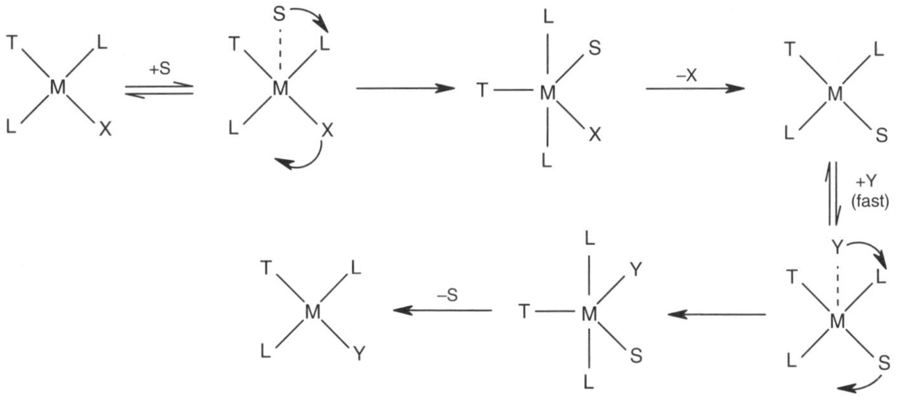

k_s-path (entry of S followed by fast replacement of S by Y)

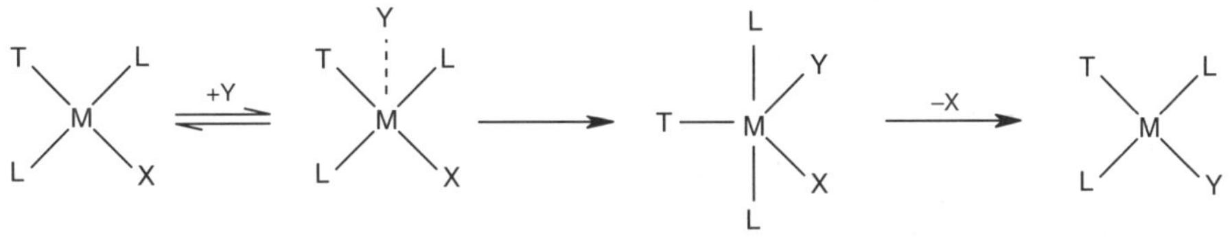

k_Y-path

Scheme 5.30.3.2 Associative pathways for both k_s and k_Y paths in ligand substitution process in square planar complexes (For details of the mechanistic steps, *see* Scheme 5.30.3.1).

5.30.4 Stereochemistry of the Product

Retention of stereochemistry is a characteristic property of the ligand substitution process in square planar complexes. Thus it leads to:

cis–complex → *cis*–product,

trans–complex → *trans*–product.

Retention of stereochemistry can be explained by considering the *trigonal bipyramidal activated complex* (*i.e.* I_a process) bearing the entering group (*Y*), leaving group (*X*) and *trans*–group (*T*) in the trigonal plane. If the trigonal bipyramid is a **long lived intermediate** (*i.e.* A–process), then it can

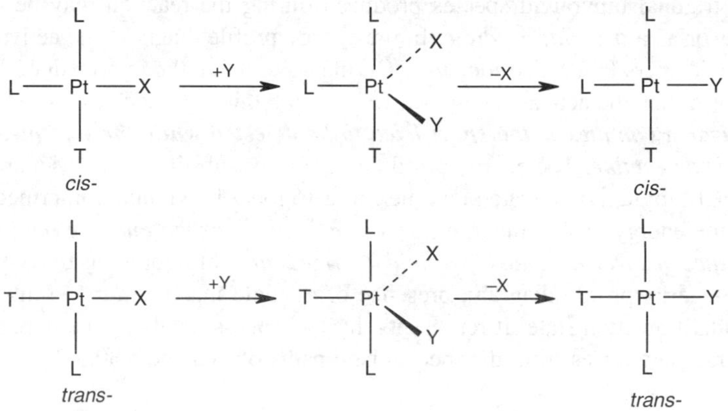

Scheme 5.30.4.1 Retention of stereochemical configuration in substitution reaction of square planar complexes of Pt(II).

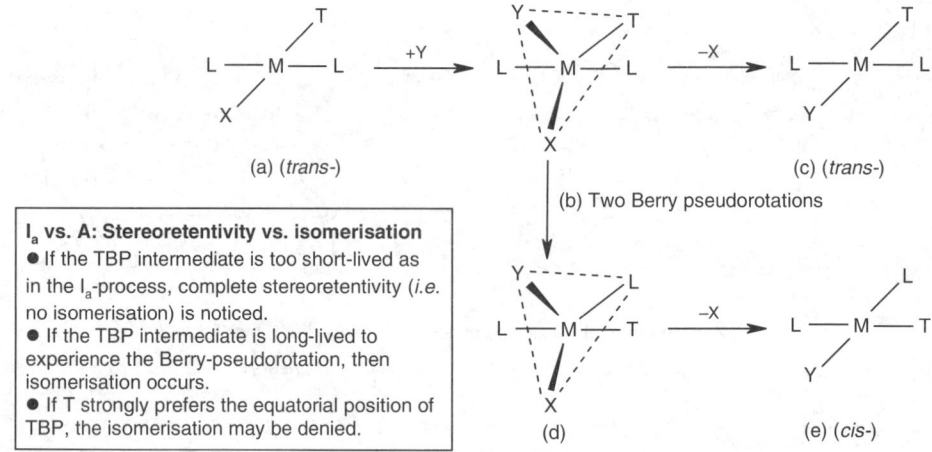

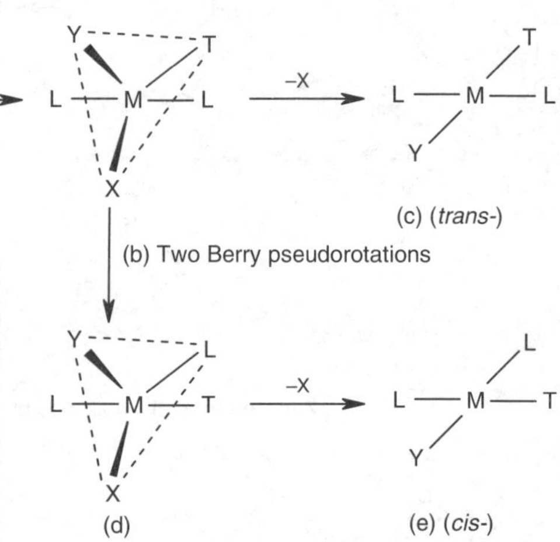

I_a vs. A: Stereoretentivity vs. isomerisation

- If the TBP intermediate is too short-lived as in the I_a-process, complete stereoretentivity (*i.e.* no isomerisation) is noticed.
- If the TBP intermediate is long-lived to experience the Berry-pseudorotation, then isomerisation occurs.
- If T strongly prefers the equatorial position of TBP, the isomerisation may be denied.

Scheme 5.30.4.2 Schematic representation of isomerisation in substitution reactions of the square planar complexes **I_a-path:** (a) → (b) → (c); **A-path:** (a) → (b) → (d) → (e) **Note:** For the conversion, (b) → (d), it needs **two successive Berry-pseudorotations** – keeping T as the pivotal group in first pseudorotation and L as the pivotal group in the second pseudorotation.

I_a *vs.* A: Tetrahedral C-compound *vs.* Tetrahedral Si-Compound

For the C-compounds, **I_a-process** is possible but **A-process is not possible;** for the Si-compounds, **A-process is possible** (*see* Vol. 2, Ch. 10). Thus for the C-compounds, in nucleophilic substitution, a short-lived TBP intermediate is produced and it leads to **Walden inversion.** By using the vacant *d*-orbital, Si can stabililise the TBP intermediate (A-process) which is sufficiently **long-lived to experience the Berry pseudorotation and it may lead to** racemisation.

experience an interchange between the equatorial and axial ligands. Thus for the **long-lived stereomobile** (*i.e.* fluxional) **trigonal bipyramid intermediate,** the *trans*–(*T*) and *cis*–ligand (L) can exchange their positions in the trigonal bipyramid. *Under the circumstances, the starting trans–complex will give a* *mixture of cis– and trans– products* (*i.e.* an isomeric mixture). This aspect is illustrated in Scheme 5.30.4.2. Two successive substitution reactions will occur to carry out the ligand catalysed isomerisation reaction (*cf.* Sec. 5.30.11) through the formation of a **stereomobile TBP intermediate.**

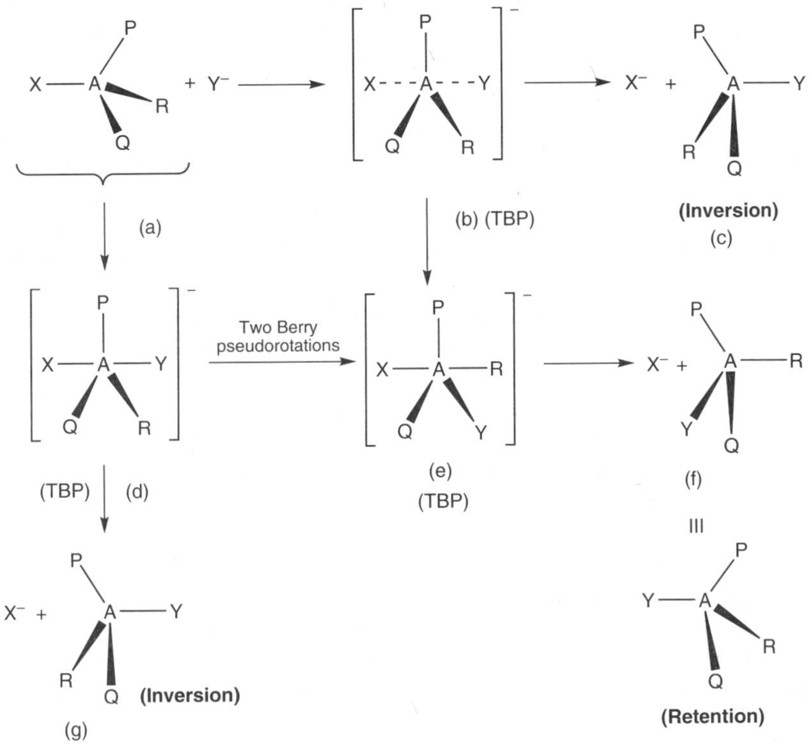

Scheme 5.30.4.3 **I_a-path** (a → b → c) for the teterahedral C-compounds leads to inversion; **A-path** (a → d → e → f, a → d → g) for the tetrahedral Si-compounds causes racemisation. **Note:** For the conversion, (d) → (e), it needs two successive Berry pseudorotations keeping R and Y as the pivotal groups respectively in these successive psuedorotations.

5.30.5 Entering Group Effects and Nucleophilicity Parameter

(a) **Nucleophilicity and basicity:** In an associative path, the rate depends significantly on the nature of the entering nucleophile (*cf.* Sec. 5.3). This rate dependence is very often measured in terms of the *nucleophilicity* of the entering group. Thus it is a ***kinetic parameter*** which determines the ability of a Lewis base to act as an entering group to influence the reaction rate.

Basicity gives the proton afinity of the entering Lewis base (Y). This is a ***thermodynamic parameter*** and it is measured by the pK_a value of the corresponding conjugate acid (YH^+).

Thus nucleophilicity is a *kinetic parameter* and basicity is a ***thermodynamic parameter***. Very often these two parameters run parallel, *i.e.* a more basic ligand is a better nucleophile (*cf.* **Brönsted relationship** suggesting LFER; *see* Vol. 3, Ch. 14) but not necessarily always.

(b) **Entering group effects in the substitution process of square planar complexes of Pt(II):** It follows the two term rate law.

$$k_{obs} = k_s + k_Y[Y],\ [Y] = \text{concentration of the entering ligand} >> [\text{complex}]$$

There are two paths: first-order (which is actually a pseudo-first order) k_s path and second order k_Y path. This k_Y–path shows the dependence on the nucleophilicity of the entering ligand. This second order path may be considered to compare the relative nucleophilicity of the entering groups.

(c) **Swain-Scott equation to the nucleophilic substitution reactions of C-compounds:** To measure the relative nucleophilicities, the earliest attempt was given by Swain-Scott equation.

$$\log\left(\frac{k_Y}{k_0}\right) = sn,\ s = 1 \text{ for } CH_3Br \quad \text{where } k_0 = \frac{k_{H_2O}}{[H_2O]}$$

where k_Y is the rate constant (for the associative process) of the following reaction.

$$CH_3 - Br + Y \longrightarrow CH_3 - Y^+ + Br^-,\ (s = 1)$$

Similarly $k_0\left(= k_{H_2O}/[H_2O]\right)$ is obtained for the reaction, CH_3Br + H_2O, where H_2O enters as the nucleophile.

Thus, Swain-Scott equation gives the relative ability of a nucleophile (say *Y*) to replace Br^- from CH_3Br compared to that of the reference nucleophile H_2O. Here, it is assumed that both *Y* and H_2O participate in an associative path to replace Br^- from CH_3Br.

The constant *n* is described as the *nucleophilic constant,* a characteristic parameter of the nucteophile *Y*. On the other hand, *s* is a *discrimination constant* that depends on the nature of the substrate (*i.e.* electrophile). Thus it varies from substrate (*i.e.* electrophile) to substrate. This is why, *s* is described as a *substrate constant.* By definition:

$$n = 0 \text{ for } H_2O \text{ (nucleophile)};\ s = 1 \text{ for } CH_3Br \text{ (electrophile)}$$

The knowledge of *s* and *n* expressing the characteristic features of electrophiles and nucleophiles respectively can help us to predict the reaction rate of an associative reaction.

(d) **Extension of Swain-Scott approach to nucleophilic substitution reactions of square planar complexes of Pt(II):** Swain-Scott approach was utilised by Basolo, Pearson and their coworkers to develop the **nucleophilicity scale (n_{Pt})** by considering the nucleophilic substitution reactions (at 30°C) at *trans*–$[PtCl_2(py)_2]$ (*i.e.* **reference substrate or electrophile**) in methanol solution where MeOH is taken as the *reference nucleophile*. It is defined as follows:

$$sn_{Pt} = \log\frac{k_Y}{k_{CH_3OH}} = \log k_Y - \log k_{CH_3OH}$$

Here k_Y is second order rate constant (having unit $dm^3\ mol^{-1}\ s^{-1}$) of the following reaction.

$$trans\text{–}[PtCl_2(py)_2] + Y \rightarrow trans\text{–}[PtCl(py)_2(Y)]^{n+} + Cl^-,\ (s = 1)$$

k_{CH_3OH} denotes the rate constant for the same reaction involving CH_3OH as the reference nucleophile where CH_3OH is the solvent. Obviously, k_{CH_3OH} is a pseudo first-order rate constant (having unit s^{-1}). This is why k_{CH_3OH} is divided by the concentration of the solvent methanol (= 25 mol dm^{-3} at 30°C), so that the **nucleophilicity parameter becomes dimensionless.** It may be noted that to make the Swain-Scott nucleophilicity parameter (*n*) dimensionless, the rate constant k_{H_2O} is to be divided by 55.5 mol dm^{-3}, concentration of water as a pure solvent. This is why, it is customary to express the nucleophilicity parameter for the Pt(II)–substrate as follows:

$$sn^0_{Pt} = \log\left[\frac{k_Y}{k_{CH_3OH}/[CH_3OH]}\right],\ s = 1 \text{ for } trans\text{-}\left[PtCl_2(py)_2\right]$$

The ***substrate constant (s)*** regarded as the ***nucleophilic discrimination factor*** is taken 1.0 for the reference substrate *trans*–$[PtCl_2(py)_2]$. The n^0_{Pt} values for various nucleophiles are given in Table 5.30.5.1.

Table 5.30.5.1 Effect of entering ligand on the rate constant (k_Y) in the reaction *trans*–$[PtCl_2(py)_2]$ + Y → *trans*–$[PtCl(py)_2(Y)]^{n+}$ + Cl^-, at 30°C (if not mentioned otherwise) in methanol and n^0_{Pt} values (*cf.* U. Belluco *et. al., J. Am. Chem. Soc.,* **87**, 241, 1965; R.G. Pearson *et al., ibid,* **90**, 319, 1968; M. Becker *et al., Inorg. Chim. Acta.,* **116**, 47, 1986).

Y	k_Y **(dm³ mol⁻¹ s⁻¹)**	n^0_{Pt}	pK_a *of YH⁺*	$n^0_{CH_3I}$
CH_3OH*	2.7×10^{-7} (25°C)	0.0	−1.7	0.0
Cl^-	4.5×10^{-4}	3.04	−5.7	4.37
NH_3	4.7×10^{-4}	3.07	9.25	5.50
pipyridine	–	3.13	11.21	7.30
pyridine	5.5×10^{-4}	3.19	5.23	5.23
NO_2^-	6.8×10^{-4}	3.22	3.37	5.35
N_3^-	1.55×10^{-3}	3.58	4.74	5.78
Br^-	3.7×10^{-3}	4.18	−7.7	5.80
I^-	1.07×10^{-1}	5.46	−10.7	7.42
CN^-	4.0 (25°C)	7.14	9.3	6.70
Ph_3P	249.0 (25°C)	8.93	2.73	7.0
Ph_3As	–	6.90	–	4.77

*The rate constant value refers to that of k_{CH_3OH}.

In Table 5.30.5.1, $n^0_{CH_3I}$ values are also given and it is defined as for n^0_{Pt}. For $n^0_{CH_3I}$, the substrate is CH_3I where the replacement of I^- by different nucleophiles is considered. There is a correlationship between n^0_{Pt} and $n^0_{CH_3I}$ in general but there are exceptions.

> n^0_{Pt} ***vs.*** $n^0_{CH_3I}$ **:** It is evident that these two nucleophilicity parameters do not bear any *close correlationship.* It is due to the fact that the centres receiving the nucleophilic attack in the associative path of reaction are different: one is sp^3-C while the other is polarisable Pt(II) (*i.e.* **soft centre**). The LUMOs to receive the nucleophilic attack are different in two cases.

It is evident from the Table 5.30.5.1, that the nucleophilicity of the entering ligands towards the soft Pt(II) centre measured by n^0_{Pt}, depends on the **softness of the ligands**. It indicates that the **soft-soft matching** is playing an important role to determine the rate constant k_Y and consequently, the nucleophilicity of Y. The rate constants cover the span about 10^9 (*i.e.* 10^2 to 10^{-7}). The rate constants and n^0_{Pt} vary in the sequence for the different donor atoms as:

$$Cl^- < Br^- < I^-;\ O < S;\ NH_3 < AsPh_3 < PPh_3.$$

It explains the order of **ligand exchange rate in $[PtX_4]^{2-}$:** k_{ex} for X^- = $CN^- > I^- > Br^- > Cl^-$.

- n^0_{Pt} ***vs.*** pK_a **(YH⁺) and polarisability of Y:** Here it is interesting to note that the nucleophilicity measured by n^0_{Pt} does not bear any correlationship with the basicity of the ligands (measured by the pK_a values of the conjugate acids). But, interestingly, n^0_{Pt} bears a good correlation with the **polarisability** (*i.e.* **softness**) the ligands. ***With the increase of softness of Y, the rate constant*** (k_Y) ***and nucleophilicity*** (n^0_{pt}) ***increases*** (*cf.* $I^- > Br^- > Cl^-$).
- n^0_{Pt} **and π-bonding property of the ligand:** Effect of the nature of the entering ligand on the rate constant can be rationalised in terms of the ***stabilisation of the trigonal bipyramidal activated complex***

in which one equatorial position is occupied by the entering ligand. The equatorial ligands can stabilise the trigonal bipyramidal activated complex through π–bonding. *Thus a good π–bonding ligand is a good entering ligand and at the same time a good polarisable ligand* (*cf.* Pt^{II}–centre is a polarisable centre) *is also a good entering ligand.* The general order of the entering ligands to favour the rate is:

$$Ph_3P \rangle tu \rangle CN^- \rangle SCN^- \rangle I^- \rangle Br^- \rangle N_3^- \rangle Cl^-$$

● **Nucleophilic discrimination factor (*s*):** The nucleophilic discrimination factor (*s*) is obtained from the following equation.

$$sn_{Pt}^0 = \log\left[\frac{k_Y}{k_{CH_3OH}/[CH_3OH]}\right] = \log\left(\frac{k_Y}{k_{CH_3OH}^0}\right) = \log k_Y - \log k_{CH_3OH}^0$$

or, $$\log k_Y = sn_{Pt}^0 + \log k_{CH_3OH}^0$$

Plot of log k_Y *vs.* n_{Pt}^0 gives a straight line having slope = *s*. For the substrate, *trans*–$[PtCl_2(py)_2]$, *s* is taken as unity. ***The larger value of s indicates that the reaction rate is more sensitive to the change of nucleophilicity.*** It has been noted that the Pt(II) complexes having the softer ligands show the higher *s*–values (*cf.* Table 5.30.3.2).

Table 5.30.5.2 Nucleophilic discrimination factors.

Complex	*s*
trans–$[PtCl_2(en)]$	0.64
$[PtCl(dien)]^+$	0.65
trans– $[PtCl_2(py)_2]$	1.00
trans–$[PtCl_2(PEt_3)_2]$	1.43

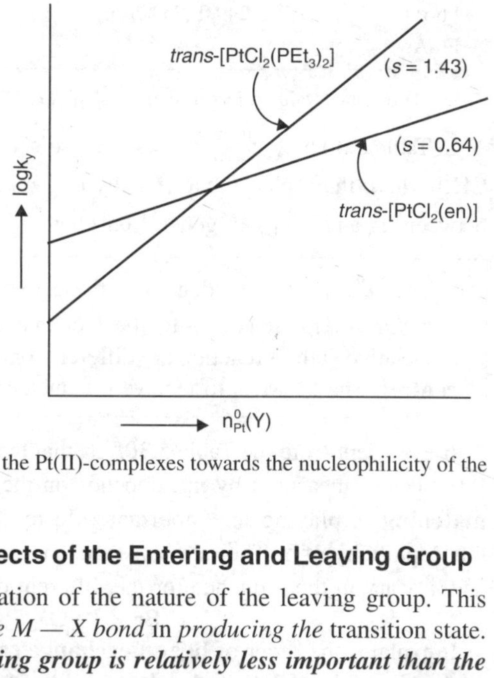

Fig. 5.30.5.1 Qualitative representation of the responsiveness of the Pt(II)-complexes towards the nucleophilicity of the entering groups.

5.30.6 Effect of the Leaving Group: Relative Effects of the Entering and Leaving Group

There is also a marked effect on the rate for the variation of the nature of the leaving group. This indicates that there is a *considerable perturbation of the M — X bond* in *producing the* transition state. ***However, this bond breaking/purturbation by the leaving group is relatively less important than the bond formation by the incoming nucleophile to produce the transition state***. This is why, the rate is more sensitive to the nature of the incoming nucleophile than the nature of the leaving group (*cf.* the relative effects of the π-bonding properties of the entering and leaving group; Sec. 5.30.15B). It supports the I_a mechanism.

Ligand exchange reactions–relative effects of the same entering and leaving group: The above arguments can explain the order of ligand exchange rate at $[PtX_4]^{2-}$ (*cf.* more **importance as the entering group):**

$$k_{ex}: \quad [Pt(CN)_4]^{2-} \rangle [PtI_4]^{2-} \rangle [PtBr_4]^{2-} \rangle [PtCl_4]^{2-};$$
$$\sim t_{1/2}\ \text{(min)}: \quad 1 \qquad 4 \qquad 6 \qquad 850$$

cf. **Nucleophilicity order:** $CN^- \rangle I^- \rangle Br^- \rangle Cl^-$; (Table 5.30.5.1)

cf. **Leaving group order:** $Cl^- \rangle Br^- \rangle I^- \rangle CN^-$

It indicates the more importance of the entering group (*i.e.* I_a or A process) to determine the rate.

Table 5.30.6.1 Effect of the nature of the incoming nucleophile and leaving group on the rate constant (25°C).

(a) $[PtBr(dien)]^+ + Y \rightarrow [Pt(dien)Y]^{n+} + Br^-$		(b) $[Pt(dien)X]^+ + py \rightarrow [Pt(dien)(py)]^{2+}$ ($[py] = 5.9 \times 10^{-3}$ mol dm^{-3})	
Y	k_Y *(dm³ mol⁻¹ s⁻¹)*	*X*	k_{obs} *(s⁻¹)*
Cl^-	8.8×10^{-4}	Cl^-	3.5×10^{-5}
I^-	2.3×10^{-1}	Br^-	2.3×10^{-5}
py	3.3×10^{-3}	I^-	1.0×10^{-5}
N_3^-	7.7×10^{-3}	N_3^-	8.3×10^{-7}
SCN^-	4.3×10^{-1}	SCN^-	3.0×10^{-7}

(*cf.* H.B. Gray *et al., Inorg. Chem.* **1,** 481, 1962).

From Table 5.30.6.1, it is evident that in changing Y (*i.e.* entering group), from Cl^- to I^-, the second order rate constant changes by a **factor about 10³** while in changing X^- (*i.e.* leaving group) from I^- to Cl^-, the observed rate constant changes by **a factor only 3.5**.

Effect of the leaving group on rate process follows the sequence as the for entering group.

$$NO_3^- \rangle H_2O \rangle Cl^- \rangle Br^- \rangle I^- \rangle, N_3^- \rangle SCN^- \rangle NO_2^- \rangle CN^-$$

It may be noted that the good entering ligands are the poor leaving groups. It can be explained by considering the formation of a trigonal bipyramidal activated complex in which the equatorial plane is occupied by the leaving group, entering group and the *trans*–directing ligand. ***The trigonal bipyramidal activated complex is stabilised by the σ- and π–bonding property of these equatorial ligands***. In fact, out of these three equatorial ligands, two ligands can simultaneously participate in π–bonding. The *trans*–activating and good entering ligands are very often the good π–bonding ligands. If the leaving group is itself a good π–bonding ligand, then it cannot be easily dislodged to give the product and consequently the rate decreases. Because of the same ground, the *trans*–activating ligands are also the poor leaving groups.

To determine the property of a leaving group, both the σ– and π–bonding properties of the ligand are to be considered. **The ligands like CN^-, SCN^-, etc. from both strong σ– and π–bonds (synergistic effect) and these are poor leaving groups.** One the other hand, halides which form only weaker bonds are the better leaving groups.

5.30.7 Effect of the Steric Crowding

It is well known that for a reaction passing through the associative path, with the increase of the steric crowding in the starting complex, there will be a rate retardation (**steric retardation**). On the other hand, for a reaction passing through the dissociative path, there will be a rate acceleration (**steric acceleration**) with the increase of steric crowding in the starting complex. Let us consider the rate constant for the hydrolysis of *cis*–$[PtCl(L)(PEt_3)_2]^+$ *i.e.*

$$\begin{bmatrix} Et_3P & & Cl \\ & Pt & \\ Et_3P & & L \end{bmatrix}^+ + H_2O \longrightarrow \begin{bmatrix} Et_3P & & OH_2 \\ & Pt & \\ Et_3P & & L \end{bmatrix}^{2+}$$

L =

(pyridine)   (2-methylpyridine)   (2, 6-dimethylpyridine)

k (s^{-1}) :   0.8   2.0 × 10^{-4} 1 × 10^{-6}
(25°C)

→ decreasing rate and increasing steric crowding

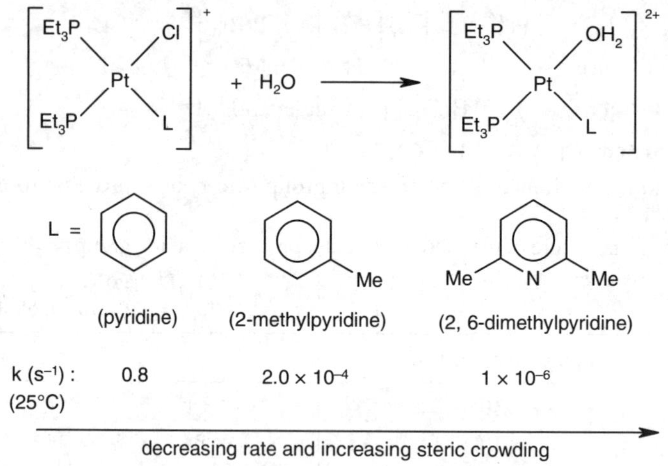

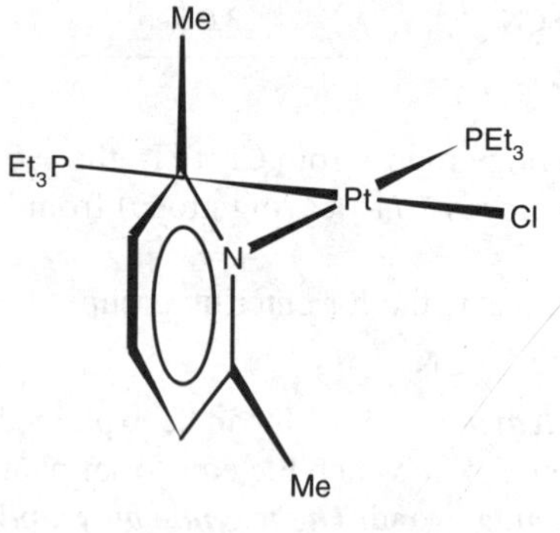

Structure of cis-[PtClL(PEt$_3$)$_2$]
L = 2,6-dimethylpyridine

Structure of the activated
complex (TBP) for
trans-[PtClL(PEt$_3$)$_2$]

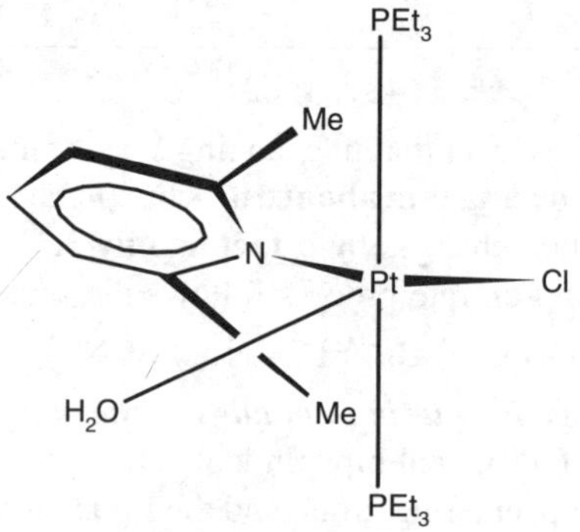

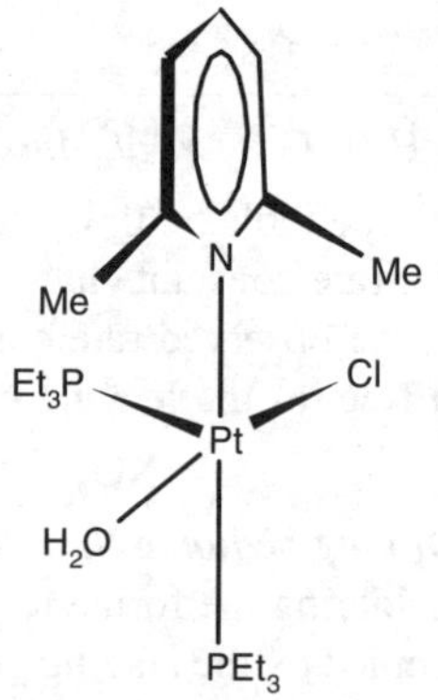

Structure of the activated
complex (TBP) for
cis-[PtClL(PEt$_3$)$_2$]

For L = 2, 6–dimethylpyridine, in the *cis*–complex, the *ortho*-Me groups can block the positions above and below the plane of the complex to hinder the approach of the entering ligand in the associative path. For 2–methylpyridine, the *ortho*–Me group can block either the above or below position to cause this hindrance. Definitely for the unsubstituted pyridine, the hindrance is minimum. ***Thus the steric retardation of rate is in conformity with the associative path.*** *Here it may be mentioned that this steric retardation is relatively less for the corresponding trans–complex (i.e. L is trans–to Cl⁻).* This can be explained by considering the formation of trigonal bipyramidal activated complex. **For the cis–complex,** the activated complex bear Cl⁻ (leaving group), PEt$_3$ (*trans*–to the leaving group) and H$_2$O (incoming nucleophile) in the trigonal plane of the trigonal bipyramid keeping L (*i.e. py* or substituted *py*) and PEt$_3$ in the axial directions. **Thus L causes the steric hindrance at 90°.** For the corresponding **trans–complex,** the activated complex places L in one equatorial position to cause the steric hindrance at 120°C. ***Obviously, the steric crowding is more pronouned in the activated complex, for the starting cis–complex,*** *i.e.* the *trans*-complex aquates faster than the *cis*-complex.

5.30.8 Activation Parameters

The activation parameters of some representative substitution reactions of square planar complexes are given in Table 5.30.8.1.

Table 5.30.8.1 Activation parameters for the reaction (in methanol). $[Pt^{II}Br(L)(PEt_3)_2] + Y \rightarrow [Pt^{II}(L)(PEt_3)_2(Y)] + Br^-$, L = 2, 4, 6–trimethylphenyl *i.e.* 2, 4, 6–$Me_3C_6H_2$ (*cf.* R. van Eldik *et al., Inorg. Chem.*, **18,** 572, 1979).

	cis–Complex + Y →				*trans–Complex + Y →*			
Y =	*I⁻*		*SC(NH₂)₂*		*I⁻*		*SC(NH₂)₂*	
	k_s	k_Y	k_s	k_Y	k_s	k_Y	k_s	k_Y
$\Delta H^{\neq}$ (kJ mol^{-1}):	84	63	79	59	80	59	71	43
$\Delta S^{\neq}$ (J K^{-1} mol^{-1}):	–59	–121	–71	–121	–52	–115	–80	–130
$\Delta V^{\neq}$ (cm^3 mol^{-1}):	–67	–63	–71	–54	–16	–16	–17	–11

k_s (s^{-1}), k_Y ($dm^3 mol^{-1} s^{-1}$)

The data in Table 5.30.8.1 indicate that for both the neutral and anionic entering ligands, both $\Delta S^{\neq}$ and $\Delta V^{\neq}$ values are significantly negative. It supports the **associative mechanism.** Here it may be noted that the starting complex is neutral and consequently entry of the nucleophile (which may be neutral or anionic) does not create or destroy any charge in the associative transition state.

The $\Delta V^{\neq}$ values for the following reactions are:

$[Pt(dien)X]^+ + Y^- \rightarrow [Pt(dien)(Y)]^+ + X^-$

$X^- = Cl^-, Br^-, I^-, N_3^-$; $Y^- = OH^-, I^-, N_3^-, NO_2^-, SCN^-$, etc.

$\Delta V^{\neq}$ (k_s–path) = –8 to –15 cm^3 mol^{-1}

$\Delta V^{\neq}$ (k_Y–path) = –6 to –12 cm^3 mol^{-1}

Here it is interesting to note that the $\Delta V^{\neq}$ values *are more negative for the k_s–path and it is in conformity with the* **interchange (I) path**. In the interchange mechanism, both the bond formation by the entering group and the bond dissociation by the leaving group occur simultaneously. In the k_s–path, the entering group is H_2O (*i.e.* solvent) and the leaving group is anionic, *i.e.* X^-. In the transition state, the leaving group X^- (which is dissociating) is solvated to cause the electrorestriction to make $\Delta V^{\neq}$ negative; and fixation of H_2O as the entering group also makes $\Delta V^{\neq}$ negative. Thus in the k_s–path (**interchange process**) both the leaving group and entering group make the $\Delta V^{\neq}$ negative. In the k_Y–path, fixing of Y^- in the transition state makes $\Delta V^{\neq}$ negative definitely but it needs desolvation to make a positive contribution to $\Delta V^{\neq}$; and the dissociating X^- (in the **interchange process**) experiences solvation in the transition state to make a negative contribution to $\Delta V^{\neq}$ through electrorestriction over the solvent. Thus $\Delta V^{\neq}$ (resultant) value becomes less negative for the k_Y–path. ***These data support the simultaneous bond rupture by the leaving group (X^-) and bond formation by the entering group (Y^-) in the transition state.***

$$Y^- \cdots\cdots Pt^{2+} \cdots\cdots X^-_{aq}, Y^- \text{ may be solvent } S.$$

If the **simple associative mechanism** is considered in the k_s-path, then fixing of H_2O as a nucleophile in the trasnition state (T.S.) makes $\Delta V^{\neq}$ negative and there is **no electrorestriction effect** (*i.e.* neither charge is created nor destroyed in attaining the T.S.). In the k_Y-path, fixing of the anionic nucleophile in the cationic complex in the associative T.S. leads to **charge neutralisation.** It makes $\Delta V^{\neq}$ more positive. Thus the simple associative path can also explain the less negative $\Delta V^{\neq}$ for the k_Y-path. But relatively **the less negative value of $\Delta V^{\neq}$** (for both k_s and k_Y) indicate the ***I_a* process rather than the A-process.**

5.30.9 Effect of Overall Charge

The ligand substitution rate is hardly influenced by the overall charge of the complex.

Complex:	$[PtCl(NH_3)_3]^+$	*trans*-$[PtCl_2(NH_3)_2]$	$[PtCl_4]^{2-}$
k_{ex} (s^{-1}) (for *Cl^-): (25°C)	1.5×10^{-5}	3.5×10^{-5}	3.8×10^{-5}

Complex:	$[PtCl_4]^{2-}$	$[PtCl_3(NH_3)]^-$	$[PtCl_2(NH_3)_2]$	$[PtCl(NH_3)_3]^+$
~ $t_{1/2}$ (aquation):	300 min	→ increasing	→	700 min.

(aquation of Cl^- for 0.1 mol dm^{-3} reactant)

For the *Cl^- exchange reaction or aquation reactions of the above complexes, the overall charge changes from +1 to −2 (*i.e.* 3 unit) or −2 to +1 (*i.e.* 3 unit), but the rate marginally changes only (**by a factor only 2-3; *cf.*** D. Banerjea, F. Basolo and R.G. Pearson, *J. Am. Chem. Soc.*, **79,** 4055, 1957). Here it may be mentioned that for aquation or anation of Co(III)–complexes, the rate constant drastically depends on the charge, *e.g.* aquation rate changes ***by a factor about 10^3*** in going from $[CoCl(NH_3)_5]^{2+}$ to $[CoCl_2(NH_3)_4]^+$ for a change of only one unit of positive charge.

The fairly insensitiveness of the rate constants with the variation of overall charge of the square planar complexes can be explained by considering the interchange mechanism where both bond breaking by the leaving group and bond formation by the entering group contribute to the activation process. If bond breaking were the predominant factor to the activation process (as in the purely dissociative process), then the rate constant would decrease remarkably with the increase of positive charge of the complex. In fact, it happens so for the octahedral Co(III)–complexes. If the bond formation by the entering nucleophile were the predominant factor to the activation process, then the rate constant would increase significantly with the increase of the overall positive charge. If both the bond rupture and bond formation simultaneously contribute to the activation process as in the **interchange (I) mechanism** then the effect of charge is mutually cancelled due to the opposing phenomena, *i.e.* bond formation and bond rupture.

5.30.10 Effect of the Central Metal and Crystal Field Activation Energy

(i) **Reactivity order for different d⁸-systems:** Reactivity of the square planar complexes depends strongly on the nature of the central metal ion. The reactivity order for Ni(II), Pd(II) and Pt(II) follows the sequence:

$$\text{Ni(II)} \gg \text{Pd(II)} \gg \text{Pt(II)}$$

It is found in many reactions:

(a)

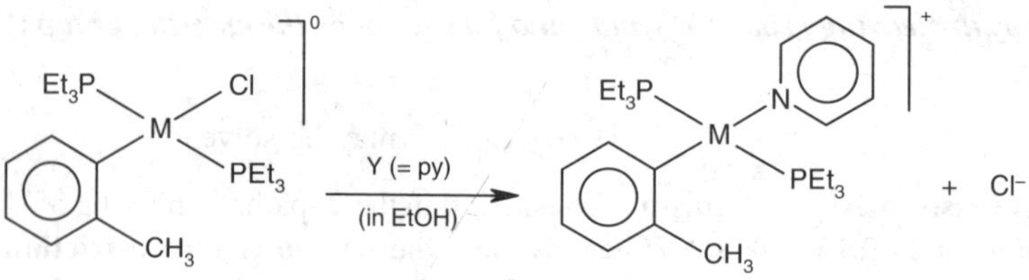

k_Y ($dm^3\ mol^{-1}\ s^{-1}$): M = Pt(II): 6.7×10^{-6}; Pd(II): 0.58; Ni(II): 33.0

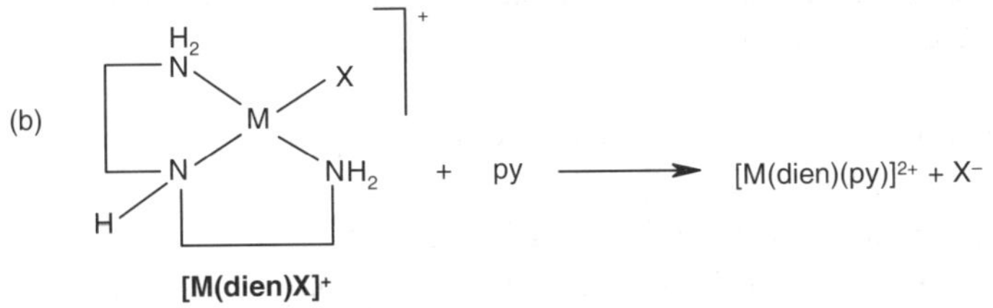

k_{py} (dm^3 mol^{-1} s^{-1}), 25°C, in aqueous media

X^-	M = Pd(II)	Pt(II)
I^-	3.5×10^{-2}	1.0×10^{-5}
SCN^-	0.31	3.0×10^{-7}
NO_2^-	6.5	5×10^{-8}

(c) The rate constant (24°C) for the exchange of $^*CN^-$ with $[M(CN)_4]^{2-}$ follows the following sequence.

	Au(III)	Pt(II)	Pd(II)	Ni(II)
$k_{ex}\left(dm^3\ mol^{-1}\ s^{-1}\right)$:	3.9×10^3	26.0	1.2×10^2	5×10^5

(ii) **Reactivity order in terms of the energy of the vacant orbital (*i.e.* LUMO) to receive the nucleophilic attack:** The observed reactivity order can be explained by considering the ease of formation of a 5–coordinate activated complex by using the vacant (non-bonding in nature) $(n + 1)p_z$–orbital in the associative pathway assuming the starting square planar complex to lie in the *xy*-plane. For Ni(II), Pd(II) and Pt(II), the said orbitals are $4p_z$, $5p_z$ and $6p_z$ respectively. Obviously, for Pt(II), it is energetically most disfavoured while it is energetically most favoured for Ni(II). For Au(III), it is $6p_z$ but its energy is decreased significantly because of higher positive charge. It causes the enhanced reactivity for Au(III) compared to that for Pt(II) which also uses the $6p_z$ orbital.

In an alternative view in terms of MOT, it may be argued that in the associative path, the incoming group will occupy the vacant d_{z^2} orbital having some antibonding character. Due to the higher crystal field splitting for the heavier congeners, this d_{z^2} orbital becomes less easily available for the incoming ligand in the case of heavier congeners.

(iii) **CFAE and reactivity order:** The reactivity of square planar complexes in ligand substitution process can be better understood in terms of *crystal field activation energy* (CFAE). In the associative

Table 5.30.10.1 Crystal Field Activation Energy (CFAE) for the ligand substitution reactions of square planar complexes (assuming the strong field ligands).

d^n	*cfse (Dq)*			*CFAE (Dq)*	
	Square planar (a)	*Trigonal bipyramidal (b)*	*Square pyramidal (c)*	*(a) – (b)*	*(a) – (c)*
d^7	26.84	13.34	19.14	13.50	7.70
d^8	24.56	14.16	18.28	10.40	6.28
d^9	12.28	7.09	9.14	5.19	3.14
d^{10}	0	0	0	0	0

path, two possible geometries for the activated complex may be considered. These are *trigonal bipyramid* and *square pyramid.* CFAE for these possible two geometries may be calculated and compared (*cf.* Table 5.30.10.1).

In terms of CFAE, the reactivity order should be (for comparable *Dq* values):

$$d^{10} \rangle\rangle\ d^9 \rangle\ d^8 \rangle\ d^7$$

It explains the **lability sequence: Zn(II) 〉 Cu(II) 〉 Ni(II)**

However, complication arises if the starting complex is in higher coordination number due to the ligation by solvents.

The *Dq*–value increases of the heavier congeners and it makes the heavier congeners more inert. If explains **lability sequence: Ni(II) 〉 Pd(II) 〉 Pt(II).**

5.30.11 Nucleophile Catalysed Cis–Trans Isomerisation in Square Planar Complexes of Platinum(II)

The geometrical isomerisation of Pt(II) complexes is catalysed by a ligand added to the solution. It passes through an **intermolecular process.** Here it may be mentioned that the substitution process in the square planar complexes of Pt(II) is **stereospecific** and *it can be concluded that the isomerisation occurs by a* **double displacement process** *(each step involves a trigonal bipyramidal activated complex).*

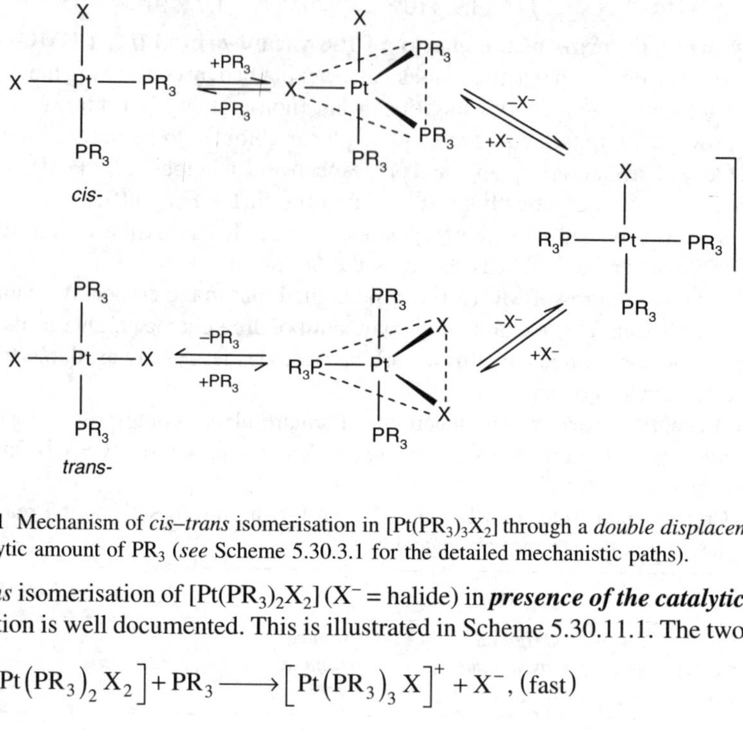

Scheme 5.30.11.1 Mechanism of *cis–trans* isomerisation in [Pt(PR_3)$_3$$X_2$] through a *double displacement process* in the presence of catalytic amount of PR_3 (*see* Scheme 5.30.3.1 for the detailed mechanistic paths).

The *cis–trans* isomerisation of [Pt(PR_3)$_2$$X_2$] ($X^-$ = halide) in ***presence of the catalytic amount of PR₃*** in CH_2Cl_2 solution is well documented. This is illustrated in Scheme 5.30.11.1. The two step process is:

1st step: $cis\text{-}\left[\mathrm{Pt}(\mathrm{PR_3})_2\mathrm{X_2}\right] + \mathrm{PR_3} \longrightarrow \left[\mathrm{Pt}(\mathrm{PR_3})_3\mathrm{X}\right]^+ + \mathrm{X}^-$, (fast)

2nd step: $\left[\mathrm{Pt}(\mathrm{PR_3})_3\mathrm{X}\right]^+ + \mathrm{X}^- \longrightarrow trans\text{-}\left[\mathrm{Pt}(\mathrm{PR_3})_2\mathrm{X_2}\right] + \mathrm{PR_3}$, (slow)

Thus in the 1st step, PR_3 acts as the entering group while X^- acts as the leaving group. In the 2nd step, the reverse event occurs.

Mechanistic Paths of *cis-trans* Isomerisation in Square Planar Complexes

The probable routes are discussed in short. (i) **Nucleophilic attack giving rise to a stereomobile TBP intermediate:** This mechanism is outlined in Scheme 5.30.11.2 and it works well for the square planar complexes of Ni(II), Pd(II) and Pt(II). For the **more inert Pt(II)-systems,** it needs a good nucleophile as a catalyst while for the **more labile Ni(II) or Pd(II) complexes, even the solvent** can act as the required nucleophile for the isomerisation. (ii) **Ligand dissociation giving rise to a symmetrical trigonal planar intermediate:** In some rare cases, this route has been found to operate. In general, this dissociative path is most unlikely for the square planar complexes which generally experience the associative activation in the ligand substitution process. (iii) **Intra-molecular rearrangement with a tetrahedral intermediate:** This route is well documented for the **Ni(II)-complexes** where stabilities of the square planar and tetrahedral structures do not differ significantly (*cf.* Sec. 2.3 conformational isomerism).

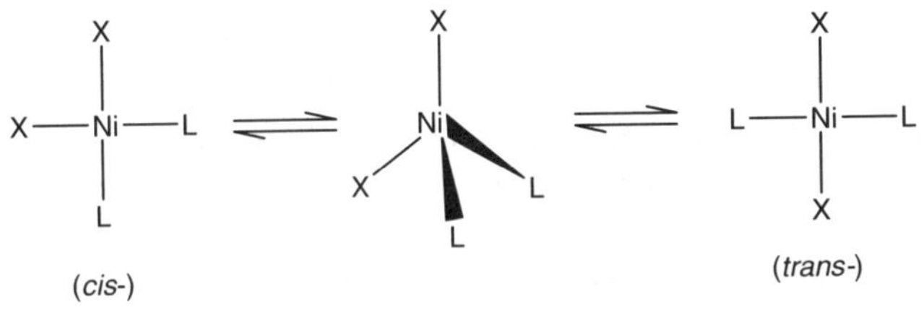

A general mechanism of the **nucleophile (Nu) catalysed *cis–trans* isomerisation** of the square planar complexes is outlined in Scheme 5.30.11.2 where **TBP-I** and **TBP-II** remain in an equilibrium through **Berry pseudorotation** which leads to an exchange between the axial and equatorial ligands. The conversion TBP-I → TBP-II needs two successive Berry pseudorotations keeping L and X as the pivotal groups respectively in these successive pseudorotations. TBP-I generates the *cis*-isomer while TBP-II generates the *trans*-isomer.

The mechanism of chloride catalysed isomerisation in the [PtCl(L)(N—O)] type square planar complex (N—O, amino acidate anion like glycinate) has been outlined in Scheme 5.30.11.3. Here, also formation of a **stereomobile trigonal bipyramidal intermediate** explains the isomerisation.

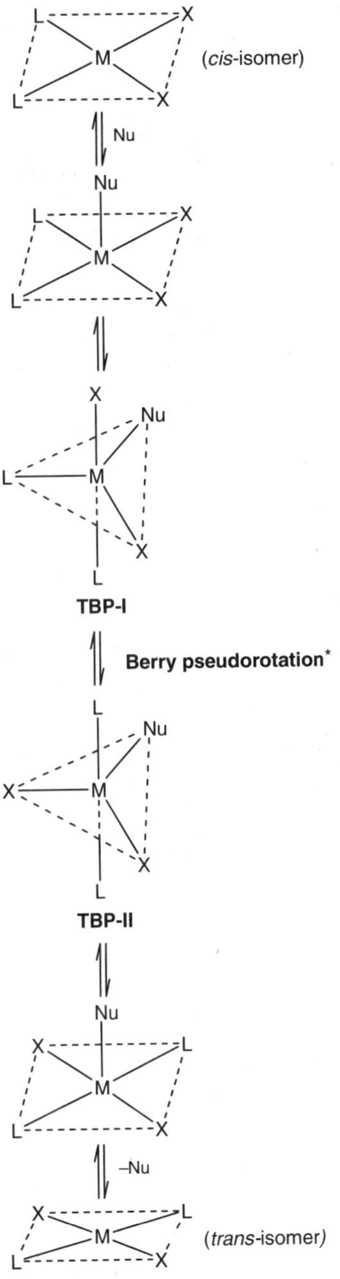

Scheme 5.30.11.2 A general mechanism for the nucleophile (Nu) catalysed *cis–trans* isomerisation of square planar complexes [ML_2X_2] (**see* Note of Scheme 5.30.4.2).

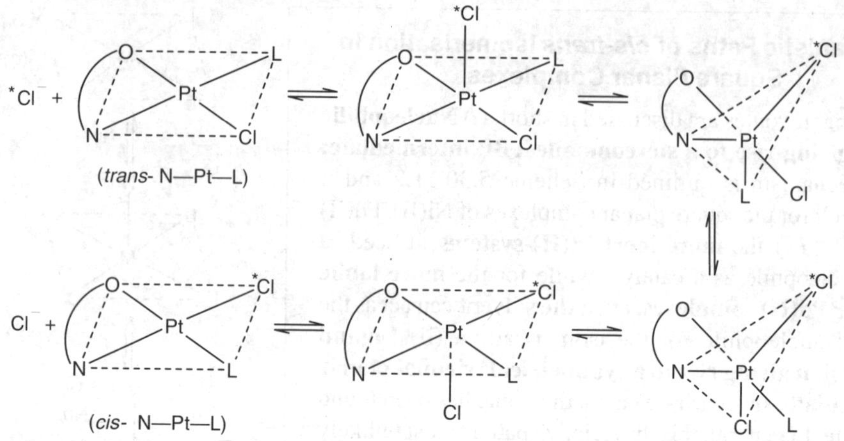

Scheme 5.30.11.3 Chloride catalysed isomerisation in [PtCl(L)(N—O)] where N—O = glycinate, sarcosinate, etc. (Charges not shown).

Trigonal bipyramid *vs.* square pyramid activated complex in the ligand substitution process of square planar complexes

Here it is interesting to note that in terms of cfse, square pyramidal geometry is favoured over the trigonal bipyramidal geometry. ***But the trigonal bipyramidal geometry is favoured because of the following reasons:***

- ***lower steric repulison*** among the ligands in the trigonal bipyramidal geometry.
- two of the equatorial ligands in the trigonal bipyramid can simultaneously participate in the π–bonding interaction; **in fact the stabilisation earned through the π–bonding can easily compensate the loss of cfse.**

Probably, because of the above mentioned two factors, the trigonal bipyramidal activated complex occurs in the ligand substitution process of the square planar complexes. However, if the equatorial ligands cannot participate in efficient π–bonding, then square pyramid geometry may also describe the activated complex.

5.30.12 Effect of the Spectator Ligand *trans–* to the Leaving Group: *Trans–*effect and *Trans–*influence

It has been already mentioned that the ligand substitution process in the square planar complexes of Pt(II) is stereospecific *i.e. the entering ligand occupies the position strictly vacated by the leaving group.* There is another factor: *spectator ligand tends to labilise its trans position.* Thus the ligand which will act as the leaving group depends on the *trans*–labilising effect of the spectator ligands.

These two important characteristic features of the ligand substitution reactions of square planar Pt(II) complexes are:

stereospecificity, and ***trans*–labilising effect**

These two factors largely control the chemistry of square planar complexes of Pt(II). These two aspects were demonstrated in classical synthesis of *cis*– and *trans*–isomers of $[PtCl_2(NH_3)_2]$.

● **Illustration of *trans*-effect in the synthesis of *cis*- and *trans*- isomers of $[PtCl_2(NH_3)_2]$:** The *cis*–isomer (called also α–isomer in old literature) was prepared by Peyrone, by treating an aqueous solution of $K_2[PtCl_4]$ with excess NH_3 + NH_4Cl buffer at low temperature (*ca.* 0°C) for a few days. The *trans*–isomer (*i.e.* β–isomer) was prepared by Reiset by heating $[Pt(NH_3)_4]Cl_2$ at about 250°C for about 10 min a in a current of dry HCl–vapour in absence of air.

In both Peyrone and Reiset reactions, the first step is a simple ligand replacement reaction where all the four groups (Cl^- or NH_3) are the same. Thus at this first-step, there is only one possible product. At the 2nd–step, in each reaction, there are two possible products but in reality, one product is produced. It is important to note that in each case, ***at the 2nd step, the entering ligand goes to the trans–position of the chloride ligand***. In each case, the other possible product through the substitution of the ligand *trans* to NH_3 doest not occur. In other words, we may conclude in the language of kinetics as follows:

the *trans*–ligand with respect to the Pt — Cl bond is more labile than the *trans*–ligand with respect to the Pt — NH_3 bond.

***cis*-$[PtCl_2(NH_3)_2]$:**
(Peyrone, 1845)

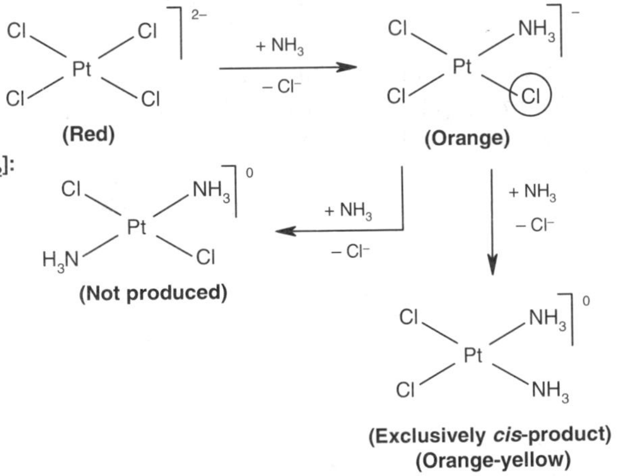

***trans*-$[PtCl_2(NH_3)_2]$:**
(Reiset, 1844)

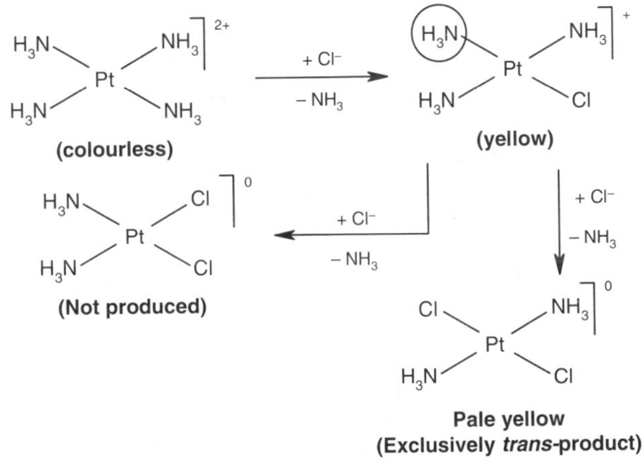

This ability of a spectator ligand to direct the incoming group to its *trans*–position is described as ***trans*–effect** (Chernyaev, 1927). By considering the Peyrone reaction and Reiset reaction, it is evident that the ***trans–effect (i.e. trans–labilising effect) of Cl^- is higher than that of NH_3.***

● **Kinetically controlled product (*trans*-effect):** *Here it must be mentioned that the above mentioned trans–effect is a kinetic phenomenon. It gives the kinetically controlled product which may not be the thermodynamically stable product.*

● **Determination of *trans*-series:** By carrying out a large number of reactions, it is possible compare the *trans*–directing or *trans*–labilising power of the ligands. It is illustrated below to determine the relative *trans*–directing power of *A* and *B*.

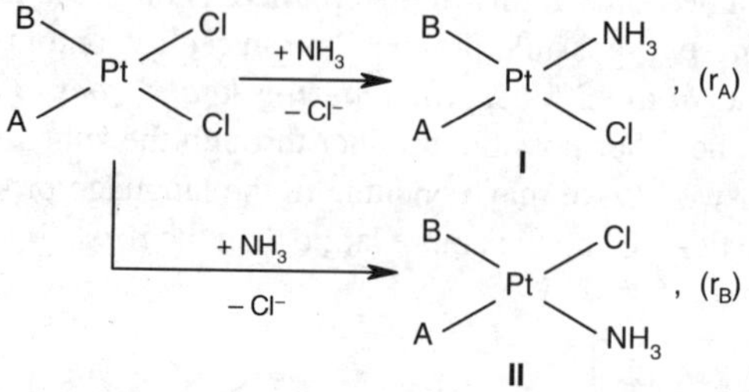

(Charges not shown)

The starting complex through the substitution of one Cl^- by NH_3 can lead to two possible products I and II. If the *reaction rate* (r_A) leading to the product I is faster than the *reaction rate* (r_B) leading to the product II, then we can conclude that the *trans*–effect of A is greater than that of B. Thus we have:

$$r_A \rangle r_B \Rightarrow A \rangle B \text{ (trans–effect).}$$
$$r_B \rangle r_A \Rightarrow B \rangle A \text{ (trans–effect).}$$

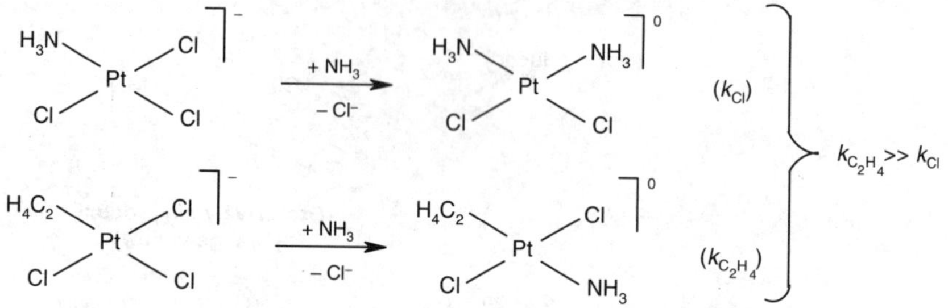

The above rate data indicate that the *trans*–effect of C_2H_4 is larger than that of Cl^-. Determination of the *trans*–effect series is illustrated in the following reaction.

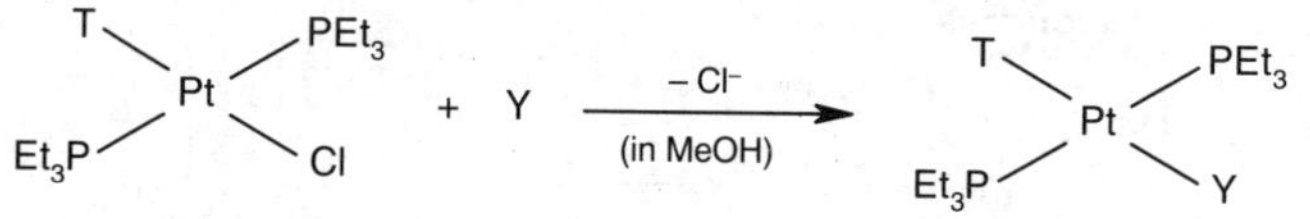

(*trans*-[$Pt^{II}Cl(PEt_3)_2$(T)]) *trans*-[$Pt^{II}(PEt_3)_2$(T)(Y)] (Charges not shown)

Y	k_Y ($dm^3\ mol^{-1}\ s^{-1}$) T = Cl^-	T = $C_6H_5^-$	T = CH_3^-
N_3^-	2×10^{-4}	8×10^{-3}	7×10^{-2}
Br^-	9.3×10^{-4}	1.8×10^{-2}	11.5×10^{-2}

The relative values of k_Y in the above reaction depends on the nature of the *trans*–ligand (T). The *trans*–directing or *trans*–labilising sequence from the above reaction is:

$$CH_3^- \rangle C_6H_5^- \rangle Cl^-$$

The same order is maintained in the above reactions for $Y = NO_2^-$, py.

Note: Here it may be mentioned that the reaction rate is jointly determined by the *trans*–ligand (T), incoming nucleophile (*Y*) and leaving group. For comparison, the leaving group may be kept the same but the effects of the groups *T* and *Y* may not be always compartmentalized and they may work in some sort of interdependent manner, specially when the incoming nucleophile (*Y*) is a very good nucleophile. Thus the *trans*–directing series based on the k_Y values may be different for the different entering nucleophiles. In fact, in the reaction between *trans*–$[Pt^{II}Cl(PEt_3)_2(T)]$ and I^-, the *trans*- directing series appears:

$$CH_3^- \rangle Cl^- \rangle C_6H_5^-$$

This is the limitation to determine the trans–directing series in terms of the k_Y values.

By considering the relative reaction rates, the ligands can be arranged in the following **order of *trans*–directing power.**

***trans*-effect series:** CN^-, CO, NO, $C_2H_4 \rangle PR_3, AsR_3, H^- \rangle CH_3^- \rangle C_6H_5^-$, $SC(NH_2)_2 \rangle NO_2^-, I^-, SCN^-, N_3^- \rangle Br^- \rangle Cl^- \rangle NH_2OH$, py $\rangle RNH_2, NH_3 \rangle OH^- \rangle H_2O$.

● **Trans–influence:** This is a *thermodynamic term* measuring the bond strength at the *trans*–position of a particular ligand. The greater bond weakening power at the *trans*–position indicates the higher *trans*–influence of the ligand T. It is illustrated below:

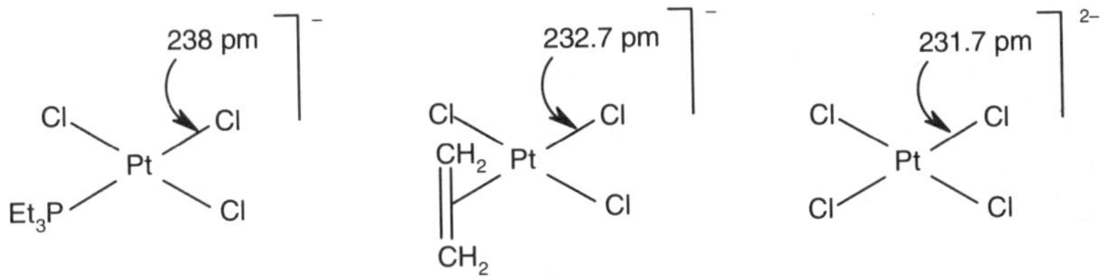

Here the Pt — Cl bond length increases for the *trans*– ligand (T) in the following sequence:

$$T = PEt_3 \rangle C_2H_4 \rangle Cl^-$$

Thus the *trans*–Pt — Cl bond strength changes in the sequence:

$$T = Cl^- \rangle C_2H_4 \rangle PEt_3$$

Trans–effect *vs.* *Trans*–influence in Square Planar Complexes

● ***Trans*– effect (a kinetic phenomenon):** It indicates the effect of a coordinated ligand upon the rate of substitution of a ligand at its *trans*–position. It is very often described as the **kinetic *trans*–effect (KTE)**.

● ***Trans*–influence (a thermodynamic phenomenon):** It describes the effect of a coordinated ligand on the bond strength at its *trans*–position in the ground state. It is also described as the ***static trans effect*** **(STE).**

Trans–effect and *Trans*–influence in Octahedral Complexes

These phenomena are also known in the octahedral complexes. In the octahedral complexes, *trans*–$[Co(en)_2(OH_2)X]^{n+}$ (X = $S_2O_3^{2-}$, RSO_2^-, SO_3^{2-} where R = alkyl group; and X binds through the *S*–end), the *trans*– Co — O bond (with respect to the Co — S bond) experiences both *trans*–effect (*i.e.* kinetic labilisation) and *trans*–influence (*i.e.* bond lengthening) (*cf.* Fig. 5.30.15.7).

The ligand which can weaken its *trans*–bond more is considered to have its more *trans*–influence. Thus in terms of the bond weakening phenomenon, the order of common ligands with regard to their *trans*–influence is:

***trans*-influence series:** $H^- \rangle PR_3 \rangle SCN^- \rangle I^-$, CH_3, CO, C_2H_4, $CN^- \rangle Br^- \rangle Cl^- \rangle NH_3 \rangle {}^-OH$.

5.30.13 Application of Trans–Effect in the Synthesis of Isomers of Platinum(II) Complexes

Common Guidelines for the Routes of Synthesis

(i) **Synthesis of *cis*– and *trans*– $[PtL_2X_2]$:**

- If *trans*–effect, X > L,

***cis*–isomer:** $[PtX_4] \xrightarrow[-2X]{+2L} cis\text{-}[PtL_2X_2]$ (*cf*. synthesis of *cis*-$[PtCl_2(NH_3)_2]$)

***trans*–isomer:** $[PtL_4] \xrightarrow[-2L]{+2X} trans\text{-}[PtL_2X_2]$ (*cf*. synthesis of *trans*-$[PtCl_2(NH_3)_2]$)

(*trans*-effect: $Cl^- > NH_3$)

- If *trans*–effect, L > X,

***cis* - isomer :** $[PtL_4] \xrightarrow[-2L]{+2X} cis\text{-}[PtL_2X_2]$

***trans* - isomer :** $[PtX_4] \xrightarrow[-2X]{+2L} trans\text{-}[PtL_2X_2]$

cf. synthesis of *cis*– and *trans*–isomers of $[PtCl_2(PR_3)_2]$ where *trans*–effect: $PR_3 \rangle Cl^-$

(ii) **Synthesis of *cis*– and *trans*– $[PtX_2(Y)(Z)]$**

If *trans*–effect: Z ⟨ X ⟨ Y

- ***cis*–isomer:** $[PtX_4] \xrightarrow[-X]{+Z} [PtX_3(Z)] \xrightarrow[-X]{+Y} cis\text{-}[PtX_2(Y)(Z)]$
- ***trans*–isomer:** $[PtX_4] \xrightarrow[-X]{+Y} [PtX_3(Y)] \xrightarrow[-X]{+Z} trans\text{-}[PtX_2(Y)(Z)]$

cf. synthesis of *cis*– and *trans*– isomers of

$[PtCl_2(NH_3)(NO_2)]^-$ (where *trans*– effect: $NO_2^- \rangle Cl^- \rangle NH_3$),

$[Pt(C_2H_4)Cl_2(NH_3)]$ (where *trans*– effect: $C_2H_4 \rangle Cl^- \rangle NH_3$)

$[Pt(CO)Cl_2(py)]$ (where *trans*– effect: CO ⟩ Cl^- ⟩ py)

(a) **Synthesis of *cis*– and *trans*–isomers of $[Pt(C_2H_4)Cl_2(NH_3)]$ from $K_2[PtCl_4]$ by considering the *trans*–effect series:** $C_2H_4 \rangle Cl^- \rangle NH_3$

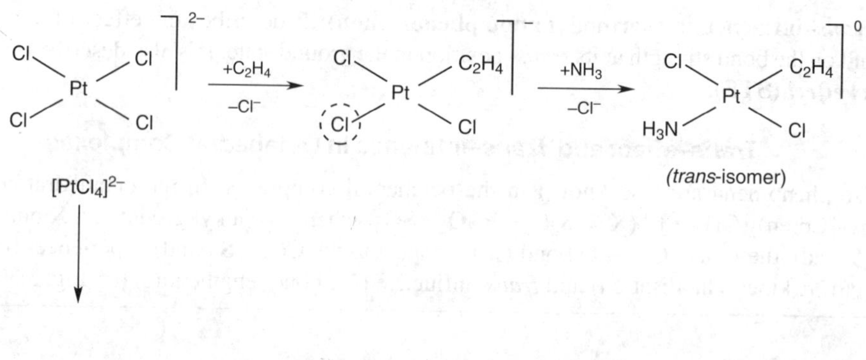

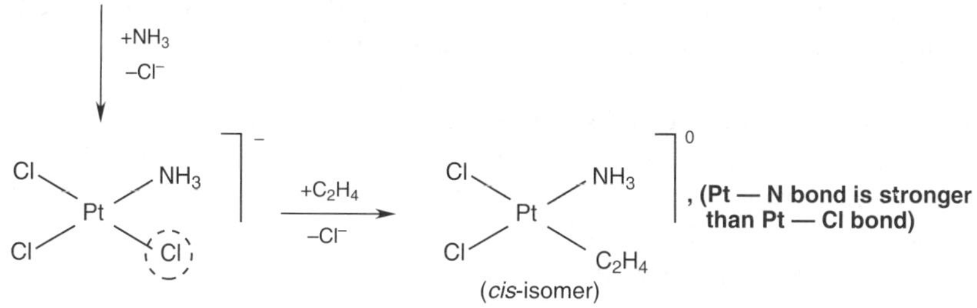

The ligand to be replaced at the second step according to the *trans*–effect series is indicated by dotted circle.

(b) **Synthesis of the *cis*– and *trans*– isomers of $[PtCl_2(NH_3)(NO_2)]^-$ from $K_2[PtCl_4]$:**
The *trans*–effect series is: $NO_2^- \rangle Cl^- \rangle NH_3$
***cis*–isomer:** NH_3 followed by NO_2^-;
***trans*–isomer:** NO_2^- followed NH_3

$$[PtCl_4]^{2-} \xrightarrow[-Cl^-]{+NH_3} [PtCl_3(NH_3)]^- \xrightarrow[-Cl^-]{+NO_2^-} cis\text{-}[PtCl_2(NH_3)(NO_2)]^-$$

$$[PtCl_4]^{2-} \xrightarrow[-Cl^-]{+NO_2^-} [PtCl_3(NO_2)]^{2-} \xrightarrow[-Cl^-]{+NH_3} trans\text{-}[PtCl_2(NH_3)(NO_2)]^-$$

(c) **Synthesis of three isomers of** $[Pt(NH_3)(NH_2OH)(NO_2)(py)]^+$ **having the *trans*–effect series:**
$NO_2^- \rangle Cl^- \rangle py \geq NH_3 \sim NH_2OH$
Bond strength: Pt — N ⟩ Pt — halogen
The possible three geometrical isomers are:

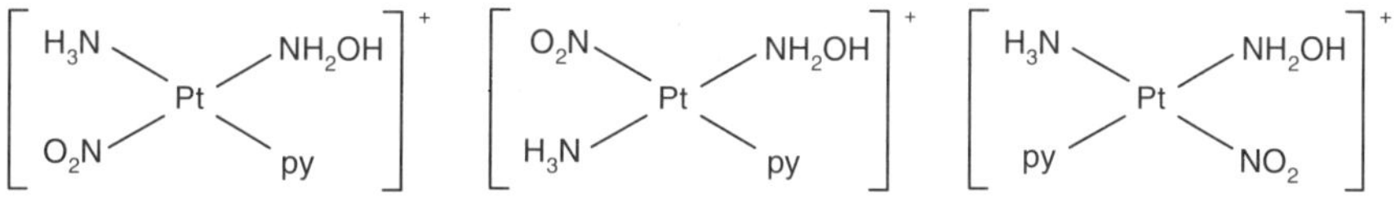

I II III

- **Synthesis of the isomer I**

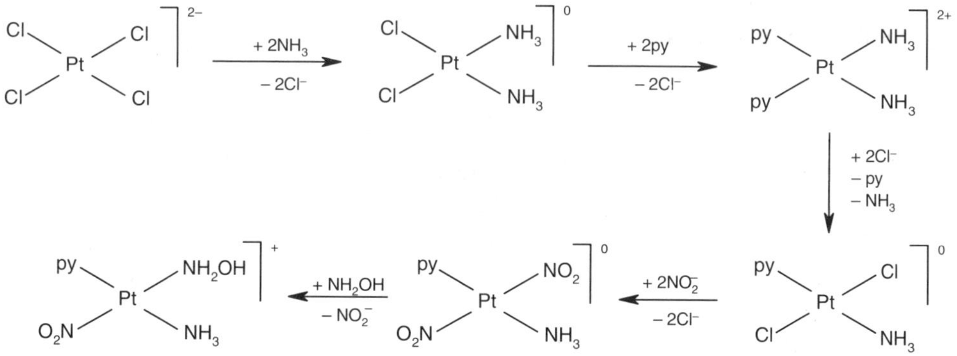

Here it is evident all the steps (except the 2nd step leading to replacement of two Cl^- ligands by *py*) are in conformity with the *trans*–effect series. At the 2nd step, because of the higher *trans*–labilising effect of Cl^-, it is expected that the coordinated NH_3 ligands (lying at the *trans*–positions of Cl^-) should be replaced by the entering ligands (*i.e. py*). But it does not happen so. To rationalise this step, it is required to consider the relative bond stength which runs in the sequence:

$$Pt — NH_3 \rangle Pt — Cl$$

It makes the Pt — Cl bond more labile than the Pt — NH_3 bond. *Thus it is very often found that when other things are equal, i.e. the Pt — X (* = Br, Cl) *bond is generally more labile than the Pt — N bond.* This lability order (originating from the bond strength effect) may oppose the lability order expected from the *trans*–effect series.

The **isomer-I** can also be synthesised in the following route.

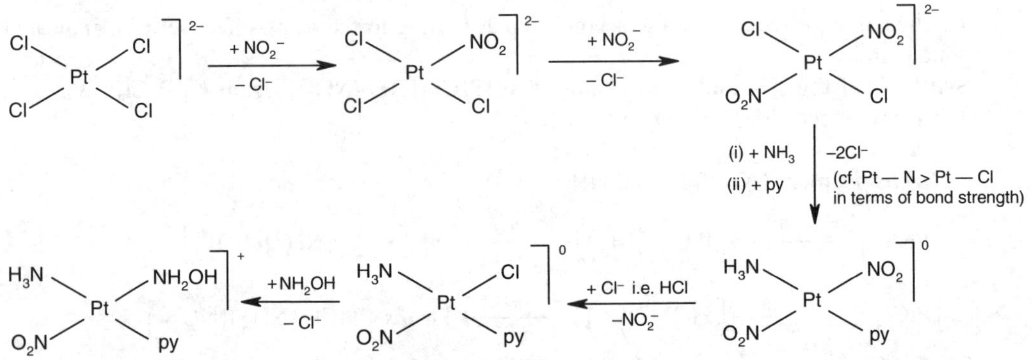

- **Synthesis of the isomer–II**

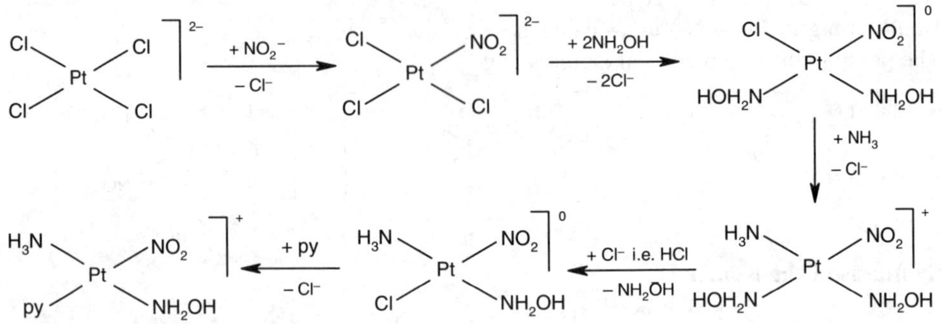

- **Synthesis of the isomer III**

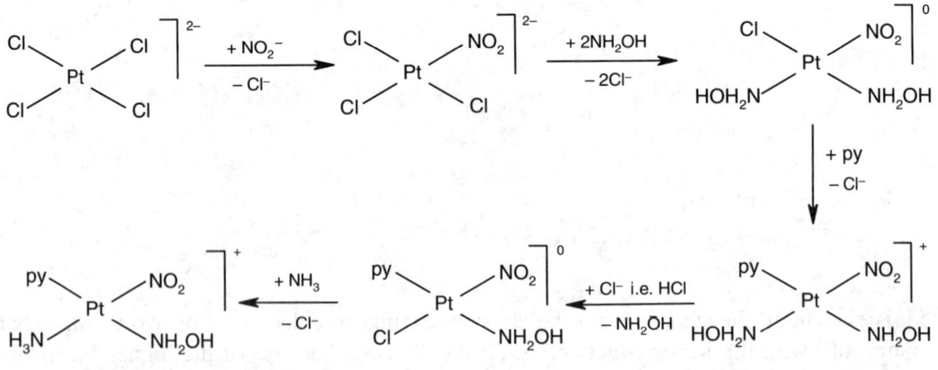

(Isomer III)

Synthesis of the isomers-I, II, III in more direct routes:

By considering the *trans*-effect order as $NO_2^- \rangle Cl^- \rangle NH_3$ and relative lability order Pt—Cl ⟩ Pt—N (controlled by bond strength), the said isomers can be synthesised in **more direct routes.**

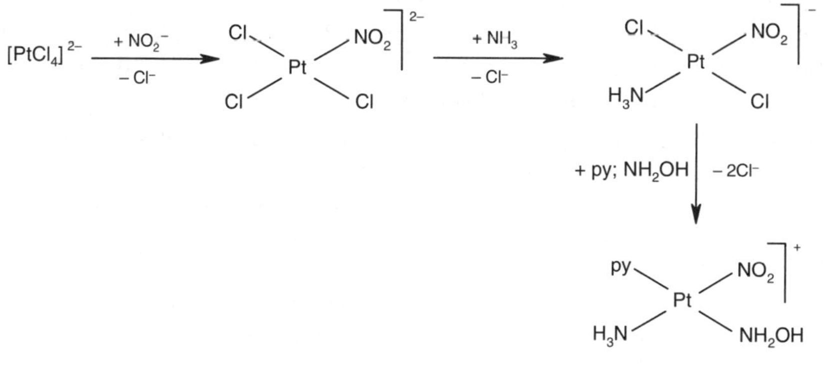

(Isomer-III)

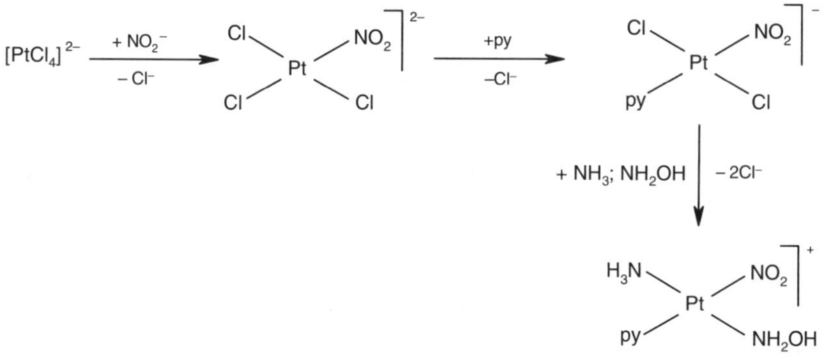

(Isomer-II)

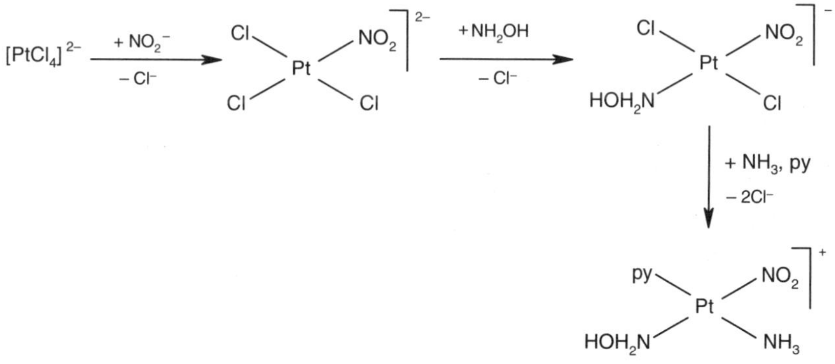

(Isomer-I)

(d) **Synthesis of the three isomers of [PtBrCl(NH_3)(py)]**: To design the routes of synthesis, we are to consider both the *trans*–effect series and bond-strength series.

***trans*–effect:** Br^- 〉 Cl^- 〉 py 〉 NH_3

bond-strength: Pt — NH_3 〉 Pt — py 〉 Pt — Br 〉 Pt — Cl

In the synthesis of **isomer I,** at the 2nd step, the *trans*–'Pt — Cl' is preferably replaced rather than the *trans*–Pt — NH_3 bond. In fact, it occurs so due to the higher Pt — NH_3 bond stength than that of the Pt — Cl bond. Other steps are in accordiance with the *trans*–effect series.

In the synthesis of the **isomer II,** at the 3rd step, the *trans*–effect operates (*cf.* though the Pt — *py* bond is stronger than the Pt — Cl bond) *i.e.* the bond *trans* to 'Cl' is replaced. But at the 2nd step, the bond strength effect operates *i.e.* the Pt — Cl bond is replaced rather than the Pt — py bond. Thus it is very difficult to rationalise the observations: ***at the 2nd step, bond strength effect wins while at the 3rd step, trans–effect wins***.

In the synthesis of **isomer III**, at the 2nd step, the *trans*–Pt — Cl bond is replaced rather than the *trans*–Pt — py bond (*cf.* both Cl and py are *trans* to Cl) because of the higher Pt — py bond strength. The 3rd step is in conformity with the *trans*–effect series.

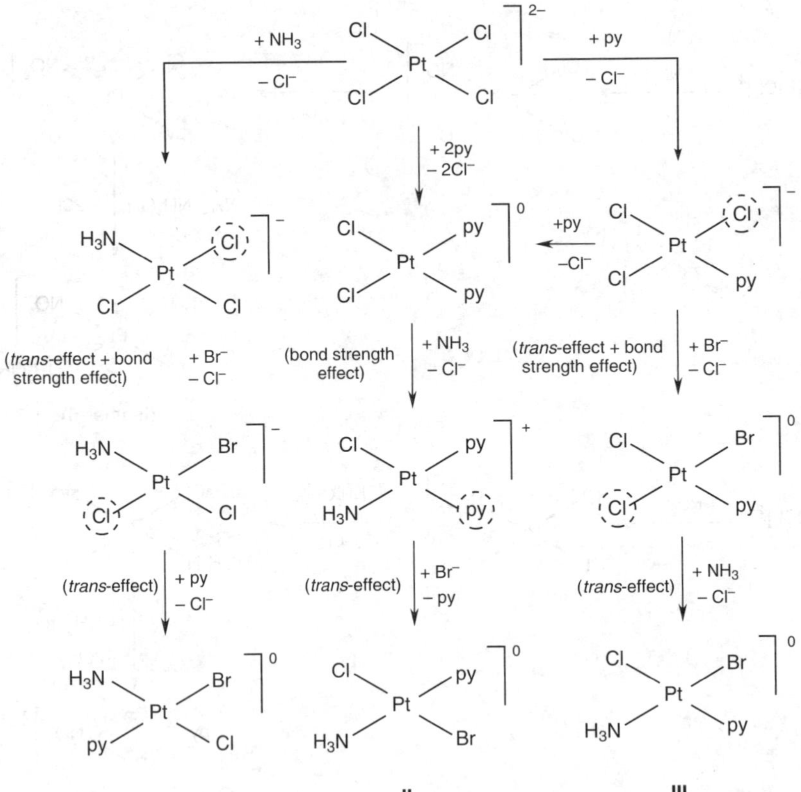

I II III

(e) **Synthesis of the three isomers of [PtCl(NH_3)(RNH_2)(NO_2)]**

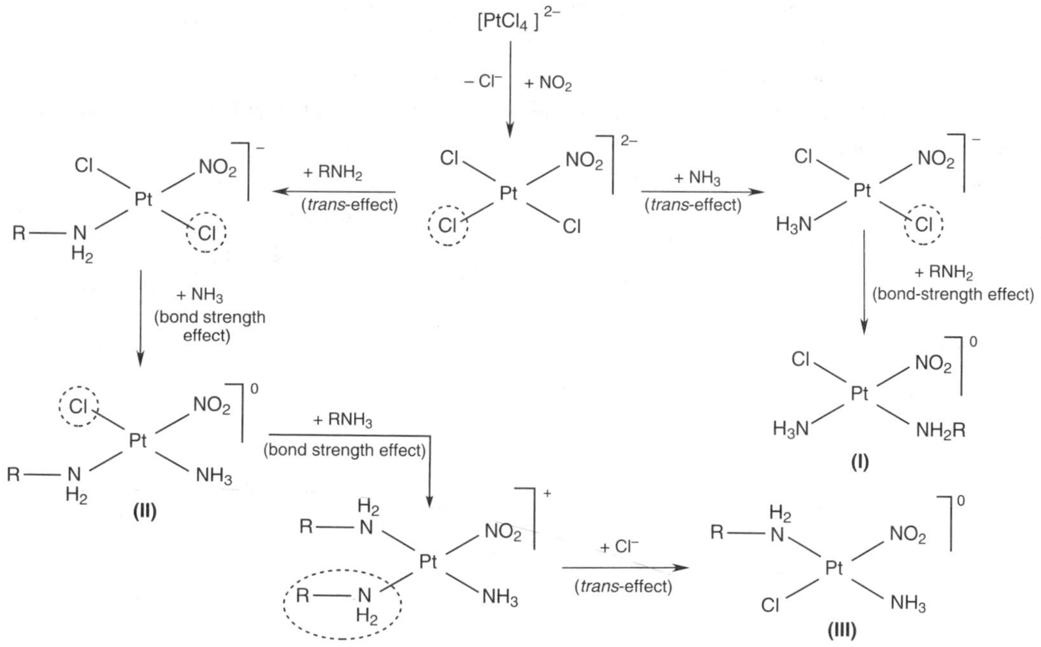

***Trans*–effect:** $NO_2^- \rangle Cl^- \rangle NH_3 \sim RNH_2$
Bond strength: Pt — N ⟩ Pt — halogen.
In the synthesis of the the above mentioned three isomers (I, II, III), both the *trans*–effect and bond-strength effect are kept in mind.

5.30.14 Application of *Trans*–effect in the Distinction of *cis–trans* Isomers of Pt(II) Complexes

This methodology was developed by the Russian Chemists by using thiourea (tu) and $S_2O_3^{2-}$ (by considering the **very high *trans*–effect of tu and $S_2O_3^{2-}$**). This principle is illustrated below.

Using tu [*i.e.* SC(NH_2)$_2$] **(N.S. Kurnakov, 1894).**

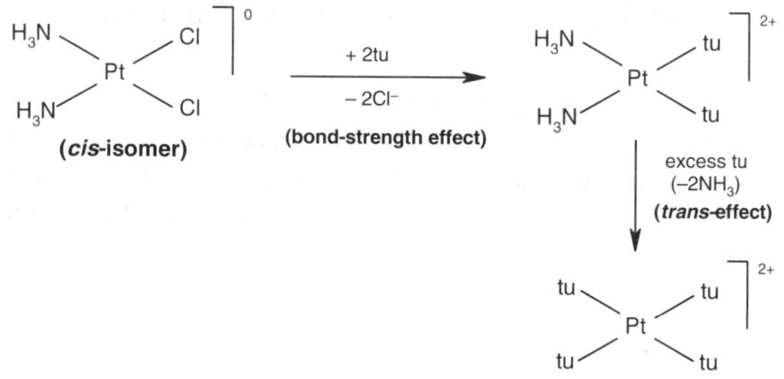

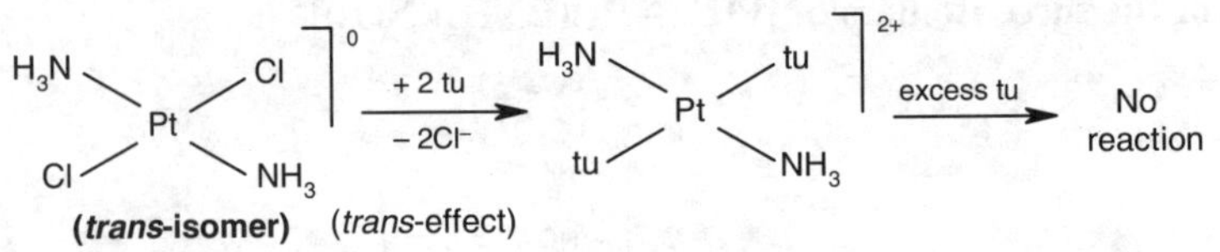

The use of this thiourea reaction to determine the *cis- trans* structures of diacidodiammineplatinum(II) is known as **Kurnakov test.**

Using $S_2O_3^{2-}$ (D.I. Ryabehikov, 1941)

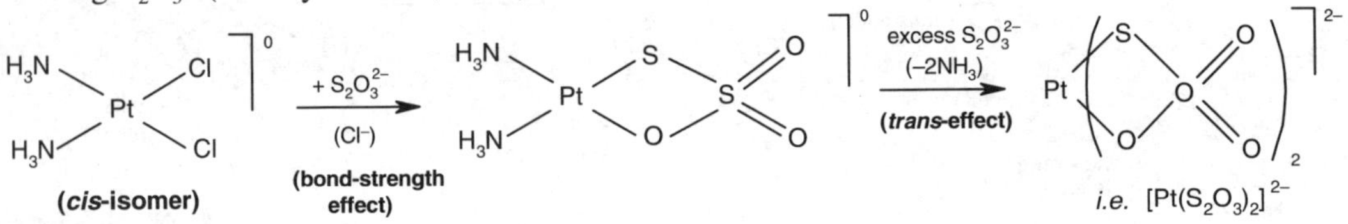

The *cis*-isomer in presence of excess thiosulfate can also partly produce $[Pt(S_2O_3)_4]^{6-}$ where $S_2O_3^{2-}$ acts as a monodentate ligand.

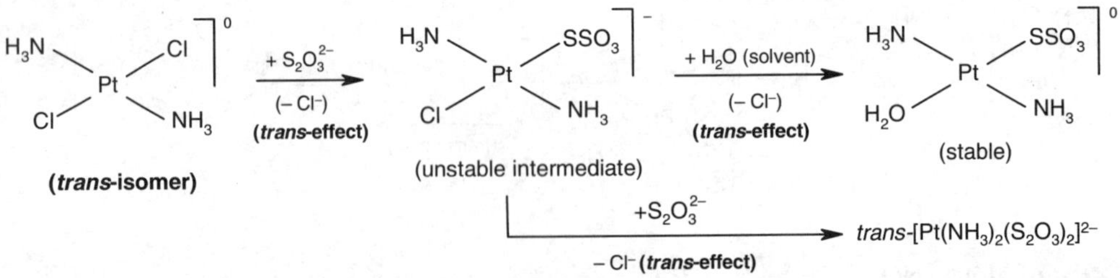

5.30.15 Theories of Trans–effect (*i.e.* Kinetic Trans–effect) and *Trans*–influence (*i.e.* Static Trans–effect)

***Trans*–influence** (*a thermodynamic phenomenon*) indicates the bond weakening phenomenon at the *trans*–position in the ground state.

***Trans*–effect** (*a kinetic phenomenon*) indicates the labilisation at the *trans*–position during the nucleophilic substitution (Werner and Chernyaev)

In most of the cases, the ligands showing the high trans–influence show also the high trans–effect. Exception: olefins show the high trans–effect but the low trans–influence.

Lowering of activation energy and *trans*–effect

The enhanced rate for a good *trans*–activating ligand is due to the reduced activation energy. This lowering of activation energy can occur **in two ways**:

- the ground state may be destabilised due to the weakening of a metal-ligand bond by a *trans*–influencing ligand; it makes the activation energy (which is the energy difference between the ground state and transition state) less; **thus the *trans*–influence may also reduce the activation energy.**
- the transition state may be stabilised to lower the activation energy.

To reduce the activation energy, either of the two factors or combination of both factors may operate.

The *trans*–effect and *trans*–influence very often run parallel in the same direction. The reaction rate (kinetic process) depends on the *activation energy*. ***Thus a good trans–directing/activating ligand must be able to lower the activation energy for the enhanced rate.*** If a good *trans*–influencing ligand shows the rate benefit then it will also contribute to lower the activation energy. Thus any theory to be proposed must take into consideration of the above fact.

A. Grinberg's polarisation theory (1935) and *trans*–influence: This polarisation theory explains the *trans*–influence (*i.e.* bond weakening at the *trans*–position). In the Pt—L bond, Pt(II) will polarise the ligand (L) to induce a dipole in the group L. This dipole in L will also polarise the metal centre *i.e.* Pt(II), to induce a dipole. **This mutual polarisation works in a synergistic fashion.** In the L—Pt—X

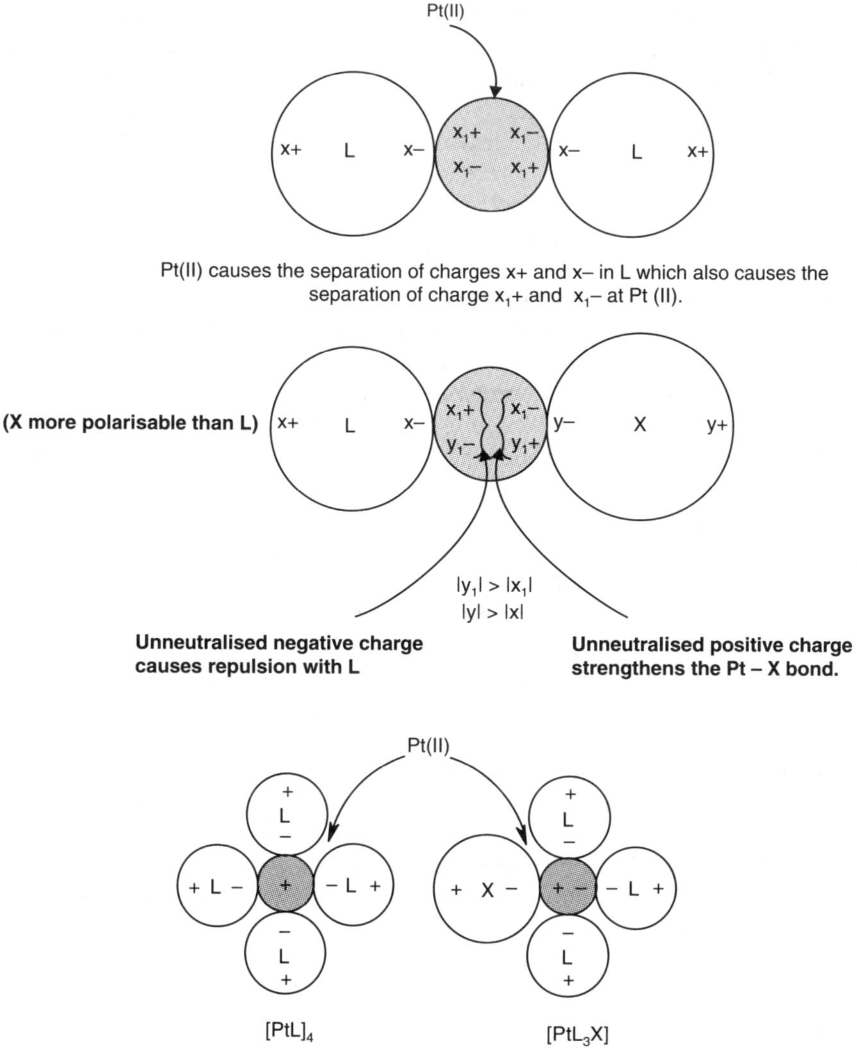

Fig. 5.30.15.1 Schematic representation of Grinberg's polarisation theory to explain the *trans–influence* (*i.e.* ground state effect) and *trans*–effect.

segment, if X is more polarisable than L, the dipole induced at Pt(II) by X opposes the dipole induced at L. Thus in a symmetrical compound $[PtL]_4$, there will be no net polarisation in any direction of the Pt — L bond. In $[PtL_3X]$ (where X is more polarisable than L), there will be a net polarisation causing a repulsion between negative charges developed in the Pt — L bond *trans–* to the Pt — X bond. This will weaken the Pt — L bond *trans* to the Pt — X bond. The degree of bond weakening in the Pl — L bond *trans* to Pt — X bond depends on the polarisability of X *i.e.* **greater the polarisability, greater the bond weakening at its *trans*–position due to the electrostatic repulsion**. Thus the *trans*–effect series of halides follow their polarisability sequence.

***trans*–effect:** $I^- \rangle Br^- \rangle Cl^-$; $NH_3 \rangle H_2O$

***polarisability*:** $I^- \rangle Br^- \rangle Cl^-$; $NH_3 \rangle H_2O$

● ***Trans*-influence *vs.* *trans*-effect in terms of polarisation theory:** *This bond weakening phenomenon (i.e. trans–influence) i.e. destabilisation in the ground state, helps in lowering the activation energy* (*i.e.* energy difference between the ground state and transition state). Thus Grinberg's polarisation theory explains the *trans*–influence and *trans*–effect (*i.e.* labilisation in the *trans* bond through the dissociative activation). In fact, if bond breaking by the leaving group contributes to attain the activated complex, then the *trans*-influence causes the *trans*-effect as in the **dissociative process.** In the **interchange (I) process,** bond breaking is only partly important to determine the activation energy. Thus the *trans*-influence can explain the *trans*–effect (kinetic factor) for the D and I_d process as in the case of octahedral Co(III) complexes (*cf.* Fig. 5.30.15.7). For the square planar Pt(II) cmplexes, the ligand substitution primarily involves the **associative activation** (mainly I_a path) where bond breaking by the leaving group is of minor importance only to determine the activation energy. Hence, to understand the origin of *trans*-effect in the square planar complexes of Pt(II), we should consider the **stabilisation of the 5-coordinate (TBP) activated complex *i.e.* transition state.** But the *trans*-influence cannot tell anything regarding the **stabilisation of transition state** which is more important than the **ground state destabilisation** in unserstanding the kinetic aspects of the reaction.

The very high polarisability of H^- is expected to show the high *trans*–influence and *trans*–effect (due to the ground state destabilisation).

● **Relative importance of the *trans*-influence and *trans*-effect among the congeners Ni(II), Pd(II) and Pt(II) in terms of the polarisation theory:** To execute the *trans*–influence according to this polarisation theory, ***the metal centre must be polarisable.*** Among Ni(II), Pd(II) and Pt(II), the most polarisable centre is Pt(II) and in fact, the *trans*–effect is mainly known to operate in the Pt(II)–complexes. In fact, *trans*–effect is almost absent in the Ni(II) and Pd(II) complexes. Because of the same ground, *trans*–effect is not known in the Pt(IV)-complexes (*cf.* Pt(IV) is much less polarisable than Pt(II)).

Limitations of the polarisation theory: ● There is a huge difference in the *trans*–labilising powers of H^- and H_2O but in terms of their bond weakening power at their respective *trans*–positions, they differ only negligibly. Thus, the bond weakening (*i.e.* ground state destabilisation due to *trans*–influence) cannot be the sole factor in realising the *trans*–effect.

● It has been already mentioned that the theory expects a high *trans*–effect for the ligands showing the high *trans*–influence (*i.e.* bond weakening or ground state destabilisation). But olefins show high the *trans*–effect with a very poor *trans*–influence.

In fact, the theory cannot explain the origin of high *trans*–effect shown by the other π–bonding ligands like C_2H_4, CO, PR_3, CN^- etc.

● Thus the concept of bond weakening (*i.e.* ground state destabilisation) is not the total argument in understanding the *trans*-effect. It must consider the stabilisation of the transition state in understanding the *trans*–effect.

Trans-influence (_i.e._ static _trans_-effect) _vs._ _Trans_-effect (_i.e._ kinetic trans-effect) in square planar Pt(II)-complexes

The *trans*-influence leads to the bond weakening for the leaving group. Thus, if the process of ligand substitution experiences a **dissociative activation,** then the *trans*-influence also causes the *trans*-effect (*i.e.* lowering of activation energy) as in the octahedral Co(III) complexes (*cf.* Fig. 5.30.15.7). But the ligand substitution at the square planar complexes of Pt(II) primarily involves the **associative activation** (I_a path) where bond breaking by the leaving group is of only minor importance. Thus, the *trans*-influence fails to explain the *trans*-effect for the Pt(II)-complexes.

B. **Pi–bonding theory and *trans*–effect in terms of the stabilisation of transition state:** This theory was developed by Orgel (1955) and Chatt (1956) independently in terms of stabilisation of the transition state to explain the high *trans*–effect shown by the π–bonding ligands like C_2H_4, CO, CN^-, etc. Such π–bonding ligands (*i.e.* **π–acid ligands**) participate in metal (*d*-orbital) → ligand π-back bonding. For this π-back bonding, the metal *dp* hybrid orbitals (*e.g.* 5*d* + 6*p* orbitals for Pt(II)) are probably more efficient than the pure *d*–orbitals. However, for the sake of simplicity, we may consider the simple *d*–orbitals of the central metal to explain the importance of such π–bonding in explaining the *trans*–effect.

The metal (*d*-orbital) → ligand π–back bonding can favour the process kinetically in two ways.

(i) **Favouring the approach of the nucleophile through electron depletion:** Because of the metal (*d*-orbital) → ligand (*i.e.* M → L) π–back bonding, the electron density in the *trans*–regions of L is reduced *to favour the approach of the incoming nucleophile* (Y). This is illustrated in Fig. 5.30.15.2. Moreover, because of this π–back bonding, the metal centre becomes more electron deficient to attract the incoming nucleophile.

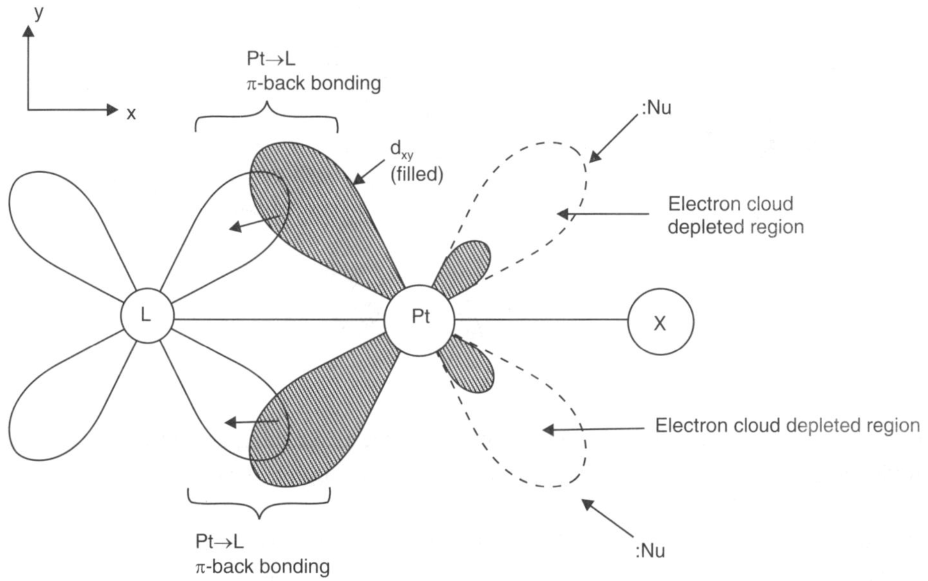

Fig. 5.30.15.2 Pi–bonding property of the *trans*–directing ligand (L) *i.e.* pi–bonding *trans*–effect theory of Orgel and Chatt (*cf.* J. Chatt, *et al., J. Chem. Soc.* (London), 4456, 1955; L.E. Orgel, *J. Inorg. Nucl. Chem.,* **1,** 137, 1956. **Note:** Pt → L π–back bonding reduces the electron density at the *trans*–regions of L to favour the approach of the nucleophile, Nu (= Y).

(ii) **Stabilisation of the 5-coordinate TBP activated complex in the associative path:** In the trigonal bipyramidal transition state, the incoming group (Nu = Y), the *trans*–directing ligand (L) and the leaving group (X) reside at the equatorial positions. The ligands at the equatorial positions can stabilise the transition state through the π–bonding. Now, if L is a good π–acceptor ligand, it can reduce the electron density along the Pt — X and Pt — Y (where Y = nucleophile Nu) bond directions (*cf.* Fig. 5.30.14.3). It will stabilise the transition state and consequently, the activation energy will be reduced. This will favour the rate process.

(iii) **Role of the π-bonding property of the entering nucleophile in the associative path:** It is evident that in the trigonal bipyramidal transition state, in the trigonal plane, L (*trans*–activating ligand), Y (incoming ligand, Nu) and X (leaving group) occupy the equatorial positions and they can stabilise the transition state. If the incoming group is a π– bonding ligand (*i.e.* π–acceptor ligand) then the transition state may be stabilised to favour the process kinetically even when L is not a good *trans*–directing ligand. ***It makes the good π–bonding ligands as the good entering ligands.***

(iv) **Role of the π-bonding property of the leaving group in the associative path:** Here it may be mentioned that if the leaving group (X) is a good π–bonding ligand then the trigonal bipyramidal geometry of the transition state will be stabilised but expulsion of the leaving group will be difficult. ***This is why, a good π–bonding ligand is a poor leaving group.*** Thus a good π-bonding leaving group (X) favours the process through the stabilisation of T.S. but disfavours the process by its poor leaving group property. This is why, π-bonding property of the entering group is more important than that of the leaving group to determine the overall rate.

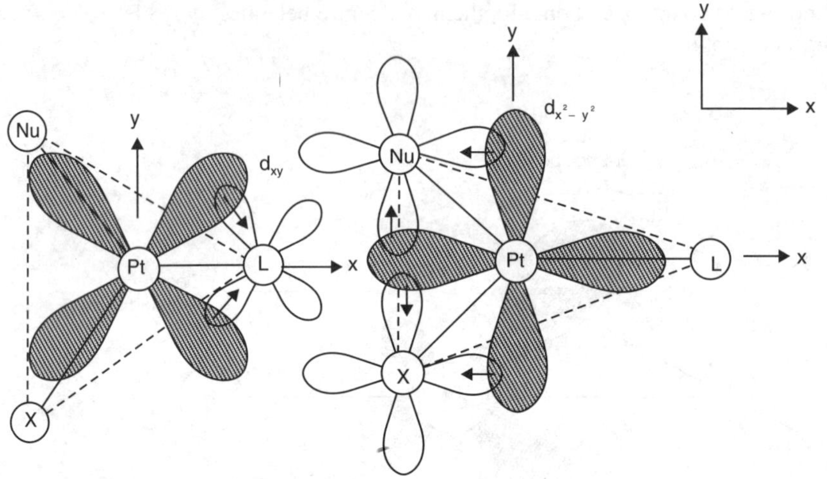

Fig. 5. 30.15.3 Positions of *the* d_{xy} and $d_{x^2-y^2}$ orbitals of Pt (II) in the trigonal bipyramidal transition state bearing the *trans*–directing group (L), leaving group (X) and incoming nucleophile (Nu *i.e.* Y) at the trigonal plane. The *trans*–axial ligands of TBP are not shown.

Note: • $Pt(d_{xy}) \xrightarrow{\pi} L$ bonding will reduce the electron density along the Pt—X and Pt—Nu bonds in the transition state

• $Pt(d_{x^2-y^2}) \xrightarrow{\pi} X$ or Nu bonding will reduce the electron density along the Pt—L bond.

(v) **Relative *trans*-directing powers of the groups in terms of their π-bonding properties:** For the square planar complex, [PtLL′X_2], there are two possible structures of the transtion state. These are: one having X, Y and L in the equatorial plane and another having X, Y and L′ in the equatorial plane (*cf.*

Fig. 5.30.15.4). The relative stabilities of these two transition states depend on the relative π–bonding properties of L and L′. Energetically, the better π–bonding ligand between L and L′ will preferably occupy the equatorial position (provided the steric factor and other factors remian the same). Thus stereochemistry of the product, [PtLL′XY] will depend on the relative π–accepting power of L and L′. If L is a better π–acceptor then L′, then Y will enter at the *trans*–position of L. If L′ is a better π–acceptor ligand than L, then Y will enter at the *trans*–position of L′.

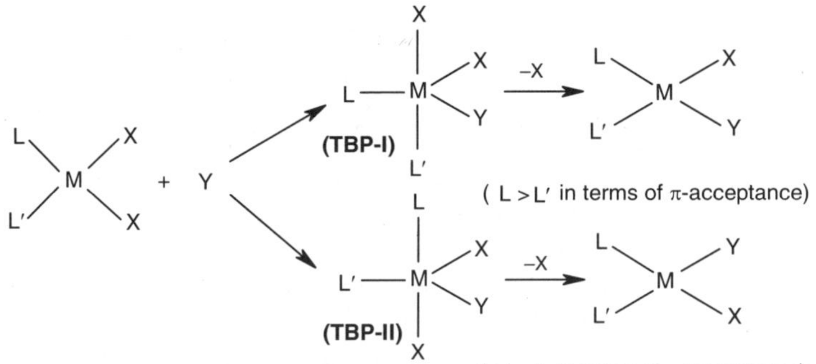

Fig. 5.30.15.4 Illustration of pi–acceptance property to determine the relative *trans*–labilising power of the nonleaving groups (L and L′) in ligand substitution process of square planar complexes of Pt(II) (*i.e.* M = Pt).

The *trans*–effect shown by the π–acidic ligands (*e.g.* C_2H_4, CO, CN^-, etc.) as the nonleaving groups (L and L′) is very often described as the ***π–trans–effect.***

(vi) **Trans-influence in terms of the π-bonding properties of the *trans*-ligands:** The π–bonding theory (called **static π–bonding theory**) can also explain the bond weakening phenomenon (*i.e. trans–influence). If two *trans*–ligands are the π–acceptor ligands (say L and L′) then they will mutually compete for the same *d*–orbital. In fact, ***in such cases, the cis–isomer (keeping L and L′– at the cis–positions) will be more stable***. If the two *trans* ligands (L and L′) are to compete for the same metal *d*–orbital, then the stronger π–acceptor ligand will predominate to weaken the bond at its *trans*–position. However, this *trans*–influence to explain the *trans*-effect suffers from the limitations as in Grinberg's polarisation theory.

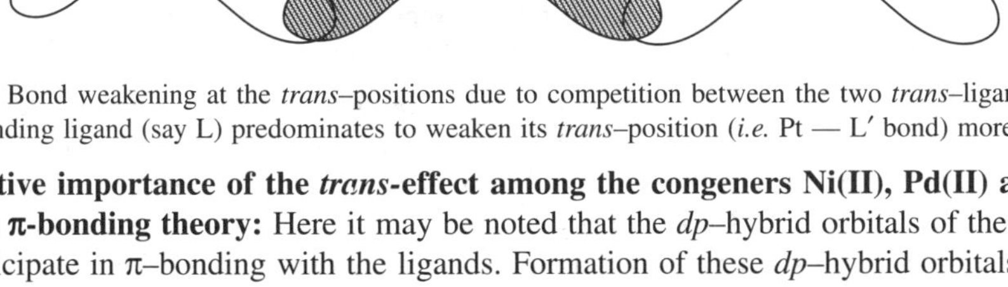

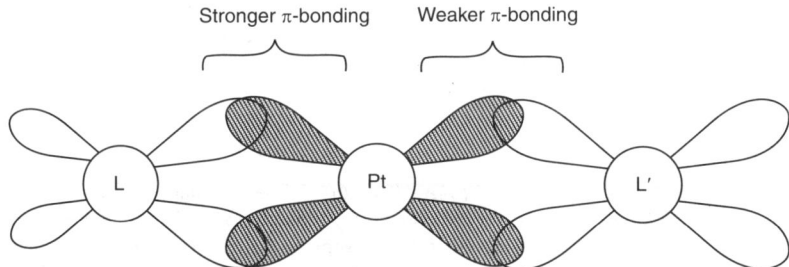

Fig. 5.30.15.5 Bond weakening at the *trans*–positions due to competition between the two *trans*–ligands (L and L′). Stronger π–bonding ligand (say L) predominates to weaken its *trans*–position (*i.e.* Pt — L′ bond) more.

(vii) **Relative importance of the *trans*-effect among the congeners Ni(II), Pd(II) and Pt(II) in terms of the π-bonding theory:** Here it may be noted that the *dp*–hybrid orbitals of the metal centre actually participate in π–bonding with the ligands. Formation of these *dp*–hybrid orbitals depends on the energy difference between the $(n-1)d$ and np orbitals. For Pt(II), energy difference between the 5*d* and 6*p* orbitals is sufficiently small to favour their mixing. For Ni(II) and Pd(II), the energy difference between the $(n-1)d$ and np orbitals is sufficiently large to disfavour their mixing. It explains why Pt(II) shows the *trans*–effect in the series Ni(II), Pd(II) and Pt(II).

(C) Sigma bonding theory and *trans*–effect in terms of transition state effect: This theory was developed by Langford and Gray (1965) to explain the high *trans*–effect shown by the σ–bonding ligands like H^-, CH_3^-, etc.

The two *trans*–ligands (L and X) are effectively competing for the same *p*–orbital for σ–bonding along the L — M — X axis (say *x*–axis). The better sigma–bonding ligand (say L) will definitely try to monopolise this *p*–orbital to weaken the bond at its *trans*–position. This will lead to ***trans*–influence (*i.e.* ground state *trans*–effect).** The electron deficiency developed in the *trans* region of the better σ–bonding ligand **will favour the approach of the entering nucleophile** in that region. However, the σ–bonding theory can also explain the **stabilisation of the trigonal bipyramidal transition state produced in the associative path.** In the L — M — X segment, if L enjoys the larger share of the same *p*–orbital for σ–bonding, there will be be a tendency in the system to change the configuration so that the same *p*–orbital can be more availed of by the better σ–bonding ligand L. This is attained by moving X out of the region of overlap in presence of the approaching ligand (Nu = Y). In the transition state, the leaving group X and entering group (Y = Nu) move to share the p_z–orbital allowing the L to have a larger share of the p_x orbital to produce the trigonal bipyramidal structure bearing L, X and Y in the equatorial plane which is the *xz*–plane (Fig. 5.28.15.6). This will stabilise the transition state, *i.e.* activation energy will be less.

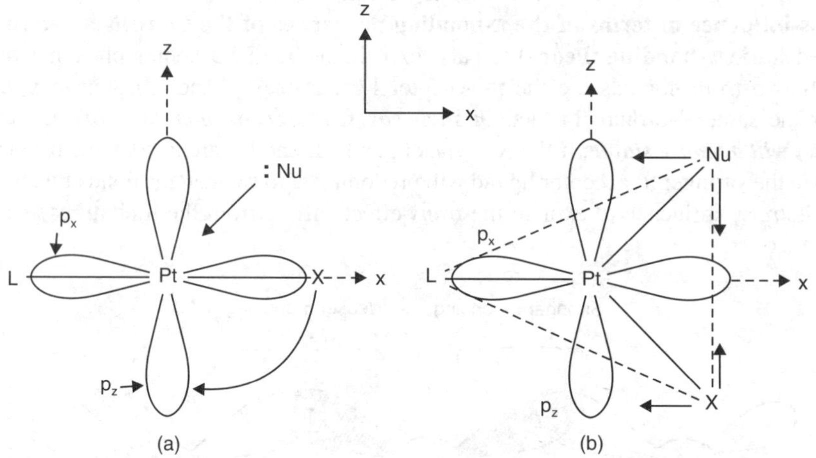

Fig. 5.30.15.6 Illustration of σ–bonding theory of *trans*–effect. (a) In the *xz*–plane (*i.e.* square plane), the ligands *L* and *X* compete for the same orbital p_x of Pt(II) for σ–bonding. (b) In the trigonal bipyramid transition state, L (*trans*–directing ligand), Nu (entering ligand) and X (leaving group) occupying the *xz*–plane (*i.e.* trigonal plane) bind with the p_x and p_z orbitals of Pt(II). The *trans*–axial ligands in the *TBP* are not shown for the sake of simplicity.

Stabilisation of the TBP activated complex by the better σ-bonding ligand: Stabilisation of a trigonal bipyramidal structure by the strong σ–bonding ligand (L) occupying an equatorial position can be rationalised in the following way. The *s*–orbital (spherical) may be considered to be equally available to the both axial and equatorial ligands. For the two axial ligands (which were initially *cis*– to the leaving group in the starting square planar complex), there is only one *p*–orbital *i.e.* one half of the metal *p*–orbital is available to each of the axial ligands. The remaining two metal *p*–orbitals are available for the three equatorial ligands *i.e.* each equatorial ligand can share two-third of a *p*–orbital. ***This is why, a better σ–bonding ligand will preferably occupy an equatorial position to stabilise the transition state to exert the trans–directing effect.*** The relative *trans*-effects of the non-π-bonding groups (*i.e.* σ-bonding groups) are: $H^- \sim PEt_3$ (*ca.* 18,000) ⟩ ⟩ CH_3^- (*ca.* 175) ⟩ Cl^- (1).

Note: ● By considering the dsp^2–hybridisation in the starting square planar complex, the *trans*–ligands actually share the $d_{x^2-y^2}$, *s* and one *p*–orbital rather than one *p*–orbital only. However, the strong sigma bonding ligand takes the larger shares of these orbitals. For the sake of simplicity, we have considered the *p*–orbital only for which the two *trans*–ligands are competing.

- In terms of the sp^3d hybridisation ($sp^2 + pd$; the sp^2-hybrid orbitals projected along the equatorial directions interact better with the ligands compared to the *pd*-hybrid orbitals projected towards the axial directions) of the TBP structure, it is expected that the better σ-donor ligands will occupy the equatorial positions (*cf. Bent's rule).*

Sigma- *vs.* pi-bonding in understranding the *trans*-effect

There is a strong controversy regarding the relative importance of the σ- and π-bonding property of the *trans*-directing ligand to cause the *trans*-effect because both the strong σ-bonding ligands (*e.g.* H^-, CH_3^- etc. having no π-bonding property) and π-bonding ligands (*e.g.* CO, CN^-, C_2H_4 etc.) are the strong *trans*-directing ligands. Some workers have argued in favour of the σ-bonding theory because the good π-bonding ligands also exercise a strong sigma effect through the synergistic interaction.

Trans–influence *vs. Trans*–effect in the Octahedral Complexes

Though these phenomena are well documented in the square planar complexes of Pt(II), these are noted also in some octahedral complexes like *trans*–$[Co(en)_2(OH_2)X]^{n+}$ (X = $S_2O_3^{2-}$, RSO_2^-, SO_3^{2-}, etc., R = alkyl group; all X–ligands bind through the S–end). The S–end of the ligands (*X*) are polarisable to induce the *trans*–influence *i.e.* bond weakening at the *trans*–position (*cf.* Grinberg's polarisation theory).

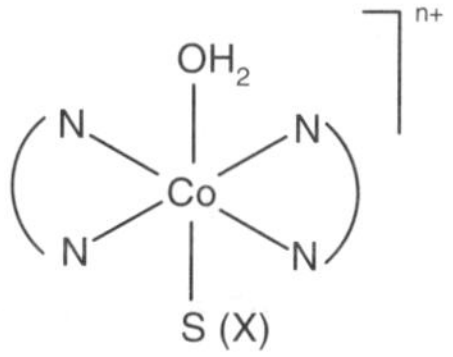

It is noted that the *trans*–bond 'Co — O' (with respect to the Co — S bond) is elongated depending on the nature of the *trans*–axial ligand (*i.e.* S–bonded X). This is called the **ground state *trans*–effect or *static trans–effect* (STE)**. It has been noted that this bond lengthening phenomenon *i.e.* STE also influence the kinetics, *i.e.* labilisation which is called the **kinetic *trans*–effect (KTE).** In fact, log (KTE) bears a linear relationship with STE *i.e.* **more the STE, more the KTE** (*cf.* J.N. Cooper *et al., Inorg. Chem.,* **19**, 2265, 1980). This relationship is quite reasonable as the **Co(III) complexes very often experience the dissociative activation in ligand substitution.**

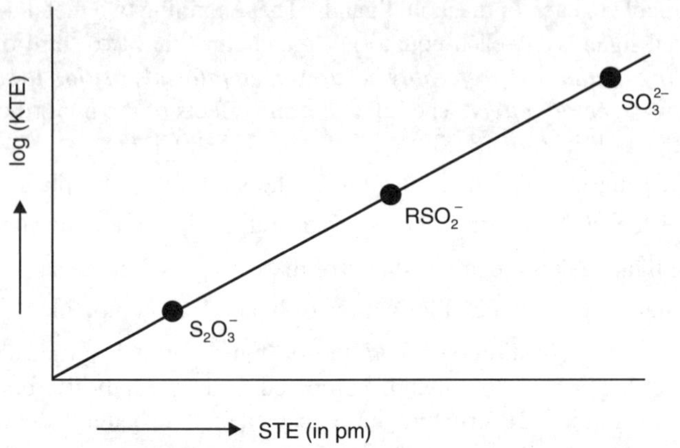

Fig. 5.30.15.7 Qualitative representation of the plot of log(KTE) *vs.* STE. KTE is measured by the water exchange rate constant; STE is measured by bond lengthening in pm in the Co—O bond; X = S–bonded ligands.

D. Cardwell's proposition in terms of electronegativity of the substituents: It is suggested that in the associative activation, square planar complex of Pt(II) adopts the **TBP activated complex** in which the **more electronegative non-*trans*-directing substituents will preferably occupy the two axial positions (Bent's rule).** The **least electronegative group (L)** acting as the *trans*-directing group, leaving group (X) and entering group (Y) will occupy the equatorial positions of the TBP (activated complex). Loss of the leaving group (X) will give back the square planar complex where L and Y remain at the *trans*-positions. Obviously, the highly electronegative groups (*e.g.* F, O, N donor sites) will preferably occupy the axial positions of the TBP activated complex **(Bent's rule)** and **such electronegative groups are the poor *trans*-directing groups**. Here it may be mentioned that it is very difficult to estimate the **effective electronegativities** of different groups present in the complex. This limitation has reduced the popularity of this approch compared to that of other theories like the π-bonding theory (Orgel and Chatt) and σ-bonding theory (Langford and Gray).

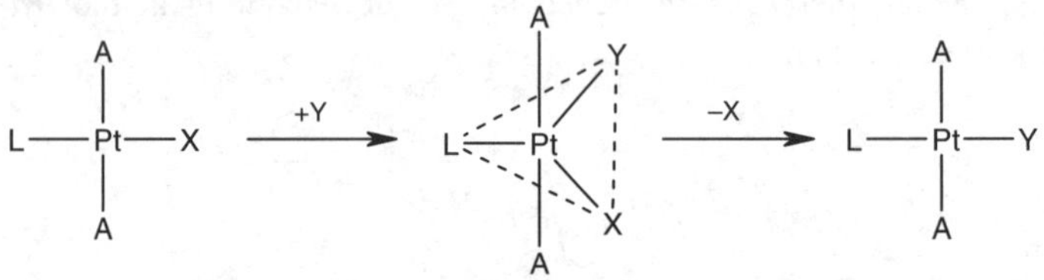

Fig. 5.30.14.8 TBP activated complex bearing two **non-*trans*-directing groups** (A,A) of high electronegativity at the axial directions; *trans*-directing group (L) of low electronegativity at the equatorial position.

Conclusion

- **Order of *trans*-influence:**

  $H^- \rangle PR_3 \rangle SCN^- \rangle I^-, CH_3^-, CO, CN^- \rangle Br^- \rangle Cl^- \rangle OH^-$

  Note: The *trans*-effect series approximately follows the same sequence with some striking exceptions for CO and CN^- showing the high *trans*-effect by dint of their strong π-bonding property.

- **Ability of π–acceptance:**

  $C_2H_4, CO \rangle CN^- \rangle NO_2^- \rangle SCN^- \rangle I^- \rangle Br^- \rangle Cl^- \rangle NH_3 \rangle OH^-$

- **Trans–effect series (average order):** It can be explained by considering both the *trans*-influence series, σ-bonding effect and π-bonding effect of the ligands.

CN^-, CO, NO, C_2H_4 〉 R_2S, AsR_3, PR_3, H^- 〉 CH_3^-, $C_6H_5^-$, $SC(NH_2)_2$ 〉 NO_2^-, I^-, N_3^-, SCN^- 〉 Br^- 〉 Cl^- 〉 py 〉 RNH_2, NH_3 〉 OH^- 〉 H_2O.

- Thus the observed *trans*–effect series can be explained by considering both the **σ–bonding and π–bonding effect** of the *trans*–activating ligand. But, their relative contribution is a matter of debate.
- *Trans*–activating ligand always occupies an equatorial position of the trigonal bipyramidal (TBP) activated complex and the non-*trans*-directing ligands occupy the axial positions of the TBP.
- The σ–bonding *trans*–ligand lowers the activation energy by—*favouring the approach of the entering nucleophile in the trans-region made electron deficient and stabilising the transition state.*
- The π–bonding *trans*–ligand lowers the activation energy by—*favouring the approach of the entering nucleophile in the trans-region due to the removal of electron density from the metal centre which becomes electron deficient and stabilisation of the transition state through π–bonding.*

5.30.16 Cis–effect in the Substitution Reactions of Square Planar Complexes (*cf.* F. Basolo, *et al., J. Chem. Soc.,* 2207, 1961)

It has been established that a good *trans*–activating group can also activate the *cis*–position. However this *cis*–effect is not so strong as the *trans*–effect. The *cis*–effect is illustrated in some examples.

In the reactions of $[PtCl_3(NH_3)]^-$, $[PtCl_3(py)]^-$, the leaving group Cl^- is *trans*– to the Pt — Cl bond, but one *cis*–ligand of the leaving group is different (*cf.* NH_3 *vs.* py) in the two reactions. The rate constant k_{py} (where *py* is the *cis*–ligand) is about 40% higher than k_{NH_3} (where NH_3 is the *cis*–ligand). ***It indicates that py shows the better cis–labilising effect than NH_3.*** Here it may be mentioned that in terms of *trans*–effect, *py* is also a better *trans*–labilising group.

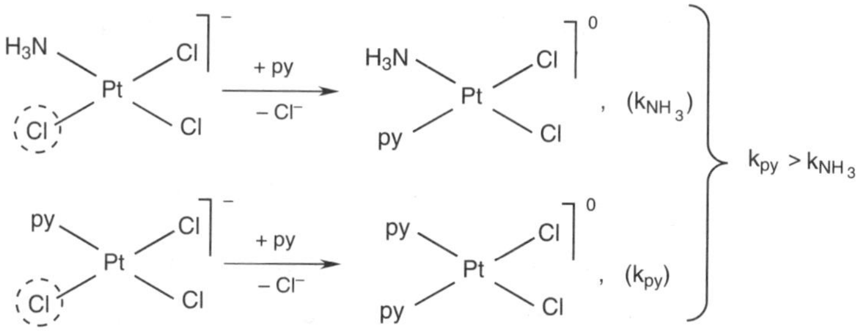

Interestingly, in the above reaction, replacement of Cl^- *cis*– to NH_3 is about 20% faster in comparison to replacement of Cl^- *cis* to NO_2 (*i.e.* $k_{NH_3} > k_{NO_2}$). It may be noted that NO_2^- is stronger *trans*–activating group. Thus we can writes *cis*–effect: py 〉 NH_3 〉 NO_2^- (*cf. trans*–effect: NO_2^- 〉 py 〉 NH_3). This finding of a limited study led to the conclusion at the early date as: ***a good trans-directing ligand is a poor cis-activating ligand.*** However, the extensive studies indicated that the said conclusion was not a general one. Rather, in most of the cases (with some exceptions) it was found: ***a good trans-activating ligand is also a good cis-activating ligand but the trans-effect is more pronounced than the cis-effect.*** It is illustrated in the following examples (Basolo *et al.,* 1961).

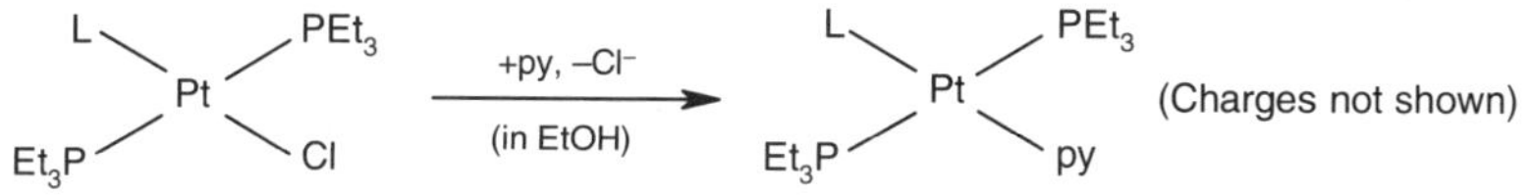

L = Cl^- $\quad C_6H_5^-$ $\quad CH_3^-$ $\quad PEt_3$ $\quad H^-$

$$\frac{k_{Y(L)}}{k_{Y(Cl)}} = 1 < 30 < 170 < 17{,}000 < 18{,}000$$ ***(i.e. trans-effect is large)***

(effect of the *trans*–labilising group L)

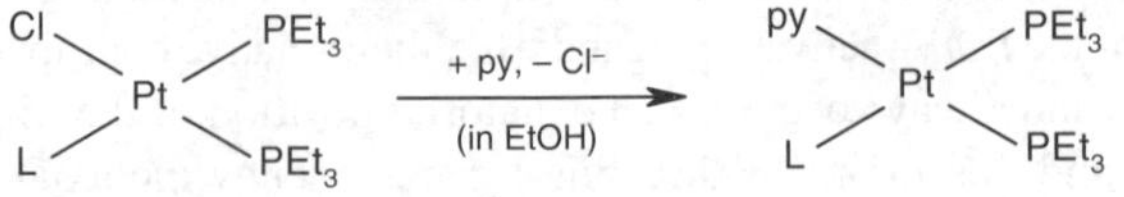

(Charges not shown)

$$L = Cl^- \quad C_6H_5^- \quad CH_3^-$$

$$\frac{k_{Y(L)}}{k_{Y(Cl)}} = 1.0 < 2.3 < 3.6$$ **(*i.e. cis*-effect is small)**

In the above reactions, the *trans*–labilising effect and *cis*–labilising effect of some groups have been measured against the standard ligand Cl^-. Both the *trans*– and *cis*–labilising effects very often run (but not always) in the same sequence, *i.e.* ***a better trans–labilising ligand is also a better cis–labilising group***. But for the ligands, Cl^-, CH_3^- and $C_6H_5^-$, the *trans*–labilising effect covers the span 1 – 170 while the *cis*–labilising effect covers the span 1 to 3.6.

Based on the kinetic data on aquation of Cl^- in different square planar complexes of Pt(II), the relative order of *cis*– and *trans*–effect are as follows:

***cis*–effect:** $C_2H_4(0.05) < Br^-(0.3) < Cl^-(0.4) < NH_3\ (1.0) \approx H_2O\ (1.0) <$ dmso *i.e.* Me_2SO (5)

***trans*–effect:** $H_2O\ (1.0) < NH_3\ (200) < Cl^-(300) < Br^-(3{,}000) < I^-(4 \times 10^4) < SCN^-(1 \times 10^5) < Me_2SO\ (2 \times 10^6) \ll C_2H_4$ (*ca.* 10^{11}).

It is evident that **the *cis*–effect is much poorer (in magnitude) than the *trans*–effect and the two series do not run always parallel** (*cf.* C_2H_4 is a very strong *trans*–activating group while it is a very poor *cis*–activating group).

● **Common origin of both the *cis*-effect and *trans*-effect:** It is believed that origin of both the *cis*–effect and *trans*–effect is the same but **the extent (*i.e.* degree) is different**. It is quite reasonable because in a square planar complex, the *trans* ligands communicate via the metal $d_{x^2-y^2}$ orbital and one metal *p*–orbital (p_x or p_y) while the *cis*–ligands communicate via the only the $d_{x^2-y^2}$ orbital. The metal *s*–orbital is equally available for both the *trans*–ligands and *cis*–ligands. This orbital involvement is understandable by considering the fact that the square planar complex is produced through the dsp^2 hybridisation $\left(d_{x^2-y^2} + s + p_x + p_y\right)$.

Thus roughly we may conclude that the two *cis*–ligands mutually interact through the metal orbitals in the same way as the two *trans*–ligands interact. Thus if a ligand can show the bond weakening and labilising effect at the *trans*–position, then the same ligand will also show the bond weakening and labilising phenomenon at the *cis*–positions also, *i.e.* ***trans*–influence, *trans*-effect** and ***cis*–influence, *cis*-effect** will appear simultaneously. They will also show the *trans*–effect and *cis*–effect in the same way.

5.31 DISSOCIATIVE PATHWAY OF SUBSTITUTION REACTIONS OF SQUARE PLANAR COMPLEXES

This is known only in some rare cases. Probably, *cis*–$[PtMe_2(SMe_2)_2]$ reacts with some didentate ligands in a dissociative path. A three coordinate (*T*–shaped) intermediate is formed and the intermediate may experience the intramolecular rearrangements giving rise to the **nonstereospecific reactions.**

5.32 SUBSTITUTION REACTIONS OF TETRAHEDRAL COMPLEXES (*cf.* Chapter 9 on Reaction Pathways of Organometallic Compounds)

The tetrahedral complexes possess relatively less cfse and very often they react very fast. In fact, substitution reactions of the tetrahedral complexes are relatively less studied.

● **Dissociative path at Ni(0) - tetrahedral complexes:** The complexes like [$Ni(CO)_4$], [$Ni(CO)_2(PR_3)_2$], etc. maintaining a formal 18e–count experience **a dissociative path** passing through the **trigonal planar activated complex.**

$$[Ni(CO)_2(PR_3)_2] + {}^*PR_3 \rightarrow [Ni(CO)_2(PR_3)({}^*PR_3)] + PR_3$$

For this ligand exchange reaction, the rate law is:

rate = k[$Ni(CO)_2(PR_3)_2$], (*i.e.* in conformity with the **dissociative path**)

Ligand replacement at [$Ni(CO)_4$] also passes through the ***dissociative path***. This is supported by the positive $\Delta S^{\neq}$ and $\Delta V^{\neq}$ values.

$$[Ni(CO)_4] + PR_3 \rightarrow [Ni(CO)_3(PR_3)] + CO$$

$$\text{rate} = k[Ni(CO)_4]$$

The reaction pathway may be outlined as follows:

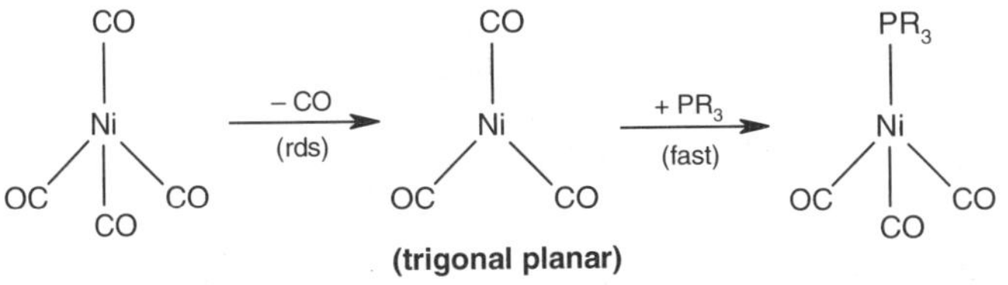

The $\Delta H^{\neq}$ value (*ca.* 105 kJ mol^{-1}) is close to the average Ni — CO bond energy. This supports the dissociative path.

● **Dissociative path for the 18e-system:** For the **18e–system**, the **dissociative (*D*) path** is favoured because of the following factors.

(i) **Very low oxidation state** does not encourage the nucleophilic attack, *i.e.* the incoming nucleophile is not attracted.

(ii) The filled *d*–orbitals [*e.g.* d^{10} configuration as for Ni(0)] cannot accommodate the nucleophilic attack. *i.e.* there is **no vacant orbital to accommodate the nucleophilic attack.**

(iii) In the dissociative path, change of the tetrahedral (sp^3) structure to the trigonal planar structure (sp^2) favours the better π– bonding *i.e.* **metal-ligand π–bonding is better in the trigonal planar structure.** This will stabilise the transition state to favour the dissociative path.

Note: Some optically active tetrahedral complexes, [M(A)(B)(C)(D)], *e.g.* [$Mn(CH_3CO_2)(Cp)(NO)(PPh_3)$] racemise through ligand dissociation giving rise to a **trigonal planar intermediate.** In the said Mn-complex, PPh_3 dissociates and here the rates of ligand (PPh_3) exchange and racemisation are the same.

● **Associative path for the systems having the formal electron count less than 18:** Complexes showing the formal electron count less than 18 experience the **associative path** in the nucleophilic substitution process. Such complexes are coordinatively unsaturated and to gain the 18e–count, they have a tendency to increase the coordination number.

[$FeBr_4$]$^-$ (13e species) undergoes the halide exchange through the **associative path.** The intermediate/activated complex is a trigonal bipyramidal one.

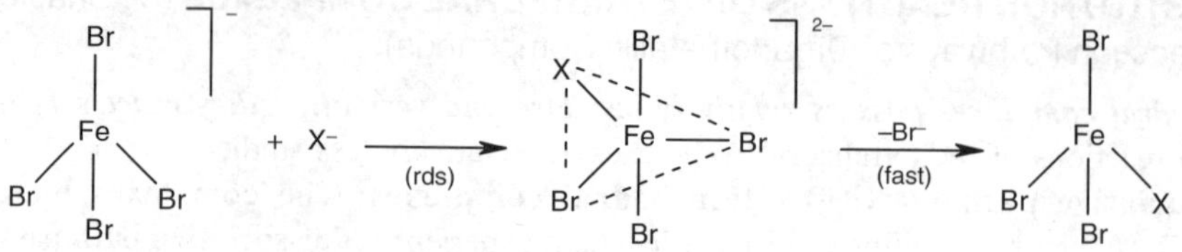

The rate law for the said ligand exchange reaction is:

rate = $k[FeBr_4^-][X^-]$, (in conformity with the associative path).

This associative path is supported by the large negaive value of $\Delta S^{\neq}$.

● **Associative path for the mixed nitrosyl-carbonyls:** The nitrosyl–carbonyls very often favour the *associative path* rather than the dissociative path. This is illustrated for the isoelectronic species (where NO has been formally considered to act as a 3e donor ligand).

$$\underbrace{[Ni(CO)_4]}_{D\text{-path}},\ \underbrace{[Co(CO)_3(NO)],\ [Fe(CO)_2(NO)_2]}_{A\text{-path}}$$

$$[Ni(CO)_4] + L \xrightarrow{-CO} [Ni(CO)_3L], \quad \text{rate} = k[Ni(CO)_4]$$

$$[Co(CO)_3(NO)] + L \xrightarrow{-CO} [Co(CO)_2(NO)L], \quad \text{rate} = k[Co(CO)_3(NO)][L]$$

$$[Fe(CO)_2(NO)_2] + L \xrightarrow{-CO} [Fe(CO)(NO)_2L], \quad \text{rate} = k[Fe(CO)_2(NO)_2][L]$$

$$(L = PR_3,\ P(OR)_3,\ AsR_3,\ \text{etc.})$$

The factors favouring the associative path for the said Co– and Fe– complexes are discussed below.

(i) The formal oxidation states for the given isoelectronic complexes are:

Fe(–II), Co(–I), Ni(0)

Oxidation state: ⟶ increasing trend

Metal to ligand π–bonding: ⟵ increasing trend

It is quite reasonable that in Ni(0), the metal → ligand π–bonding interaction is the weakest one while it is the strongest one in Fe(–II) having the lowest oxidation state. Thus, in terms of π–bonding, in the Ni–complex, the metal-ligand bond is the weakest one to favour the dissociative path. ***For the Co– and Fe–complexes, because of the higher metal-ligand bond strength, the dissociative path is relatively disfavoured and rather the associative path is favoured.***

(ii) NO can be present as NO^+ (*i.e.* effectively **3e–donor**) and it may be also present as NO^- (*i.e.* effectively **le–donor**). If it exists as NO^- (*i.e.* bent M — N — O arrangement) then the metal centre possesses the 16e–count (*i.e.* the **metal centre becomes coordinatively unsaturated**).

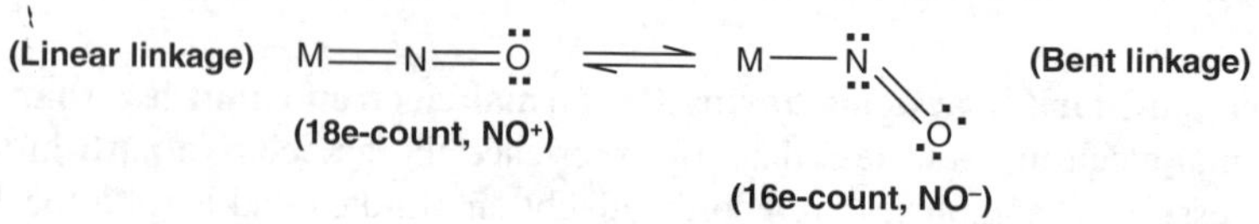

It may be argued that in the starting tetrahedral complex, NO predominantly exists as NO^+ to maintain the 18e–rule but with the approach of the incoming nucleophile, the π–electron shifting to NO makes the metal centre electron deficient to attract the nucleophile **(*cf.* electromeric effect).**

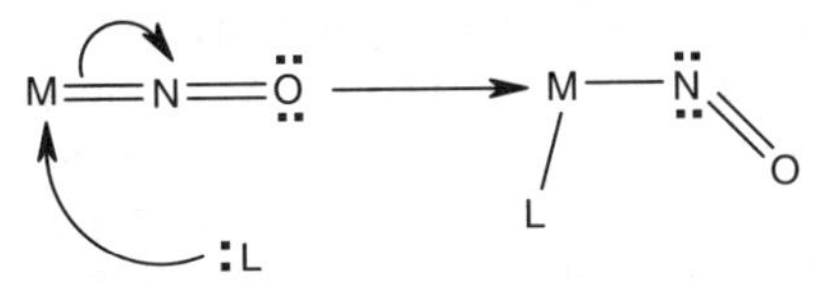

This peculiar behaviour of the 'M — NO' linkage allows the **associative character** at the metal centre. The mechanistic steps may be outlined as follows:

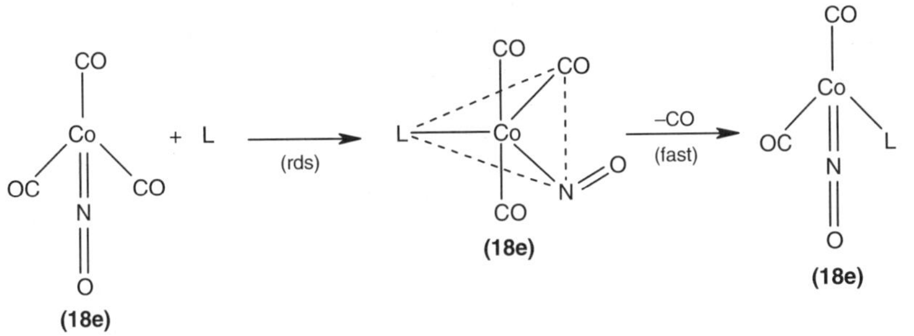

The rate law is:

$$\text{rate} = k[Co(CO)_3(NO)][L]$$

The associative path is supported by the large negative value of $\Delta S^{\neq}$ (*ca.* –92 J K^{-1} mol^{-1})

Note: Here it may be mentioned that in some metal nitrosyl-carbonyls, both the dissociative and associative path may occur simultaneously. **It will lead to the *two term rate law*.** In fact, for the exchange of CO ligand in $[Co(CO)_2(NO)(PR_3)]$, the two term rate law has been found.

$$\text{rate} = (k_d + k_a[CO])[Co(CO)_2(NO)(PR_3)]$$

k_d stands for the dissociative path while k_a stands for the associative path.

These can be explained by considering the existence of 16e and 18e species in an equilibrium.

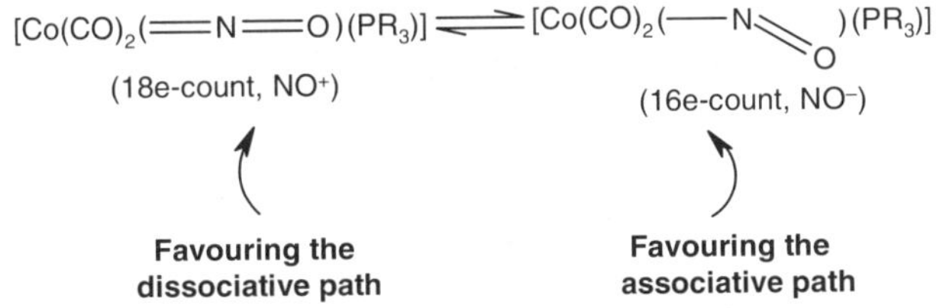

5.33 RACEMISATION OF THE TETRAHEDRAL COMPLEXES (*cf.* Sec. 5.29 for the octahedral complexes)

Very often, the optically active tetrahedral complexes rapidly undergo racemisation and they cannot be resolved. However, two types of optical isomers in the tetrahedral complexes are known (*cf.* Sec. 2.8). These are: **[M(A)(B)(C)(D)]** (having four different ligands) and **$[M(AB)_2]$** (AB = unsymmetrical didentate ligand). The different routes of racemisation for the optically active tetrahedral complexes are discussed here.

(i) **Intermolecular path — formation of a trigonal planar intermediate (*i.e.* decrease of C.N.) through dissociation of a ligand:** From the trigonal planar intermediate, both the optical isomers can be generated. $[Mn(CH_3CO_2)(Cp)(NO)(PPh_3)]$ racemises in benzence through the formation of a trigonal planar intermediate produced from the **dissociation of PPh_3**. This view is supported by the fact that the rates of racemisation and ligand (*i.e.* PPh_3) exchange are the same.

(ii) **Intermolecular path — formation of a stereomobile trigonal bipyramidal (TBP) intermediate** (*i.e.* increase of C.N.) **through a nucleophilic attack:** Nucleophilic attack can generate a **long-lived TBP intermediate** which can experience the Berry pseudorotation producing two different TBP configurations. Then expulsion of the nucleophile **(acting as a catalyst)** from the TBPs gives a racemic mixture. It happens so for the optically active **tetrahedral Si-compounds** (*cf.* Scheme 5.30.4.3).

(iii) **Intramolecular path — formation of a square planar structure (*i.e.* no change in C.N.) through an intramolecular rearrangement (*cf.* twist mechanism, Sec. 5.29):** For the $[M(AB)_2]$ type tetrahedral complexes, if the steric hindrance and rigidity of the chelate ring do not prevent the formation of the square planar structure, then the process operates.

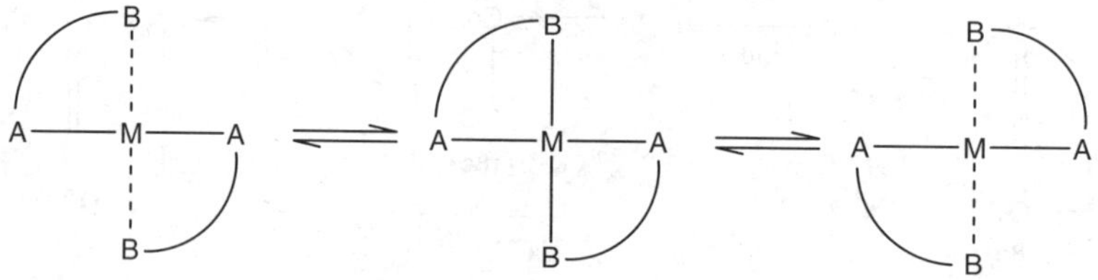

Ligand substitution Process in the Tetrahedral Compounds of C, Si, and P (*cf.* Chapter 10, Vol. 2)

Tetrahedral C-compounds can experience an **interchange path** (I_a or I_d) (*cf.* Fig. 5.30.3.2) producing a **TBP transition state (too short lived)** which cannot undergo the structural rearrangement. It leads to **Walden Inversion** for the optically active C-compounds (*cf.* Sec. 5.30.4) in the I_a-path. For the optically active tetrahedral Si-compounds, the product is a **recemic mixture.** In fact, it passes through the **A-path** where the TBP intermediate (sp^3d hybridisation) is sufficiently **long-lived to undergo the Berry pseudorotation** to give the racemic product (*cf.* Scheme 5.30.4.3).

5.34 ACID CATALYSED DISSOCIATION OF THE MACROCYCLIC AND OPEN CHAIN COMPLEXES: MACROCYCLIC EFFECT

5.34.1 Origin of the Macrocyclic Effect

Macrocyclic effect (*cf. Sec.* 4.5.4): The **macrocyclic effect** indicates the relatively enhanced stability of the metal complexes with the macrocyclic ligands compared to that of the complexes formed by the comparable open-chain ligands. This macrocyclic effect originates from both the **entropic favour** and **enthalpic favour.**

Entropic favour: The macrocyclic ligand is less flexible than the analogous open-chain ligand. During complexation by an open-chain ligand, the *cyclisation* makes ΔS more negative. The macrocyclic ligand which is **preorganised** and possesses the diminished flexibility experiences the less negative ΔS on complexation.

Enthalpic favour: The free macrocyclic ligand is more **preorganised** than the corresponding open-chain ligand. Consequently, it requires a less conformational change of the macrocyclic ligand for complexation. This **less preorganisation energy** gives an enthalpic favour. Moreover, compared to the open-chain ligand, the macrocyclic ligand in free state is less solvated because of its diminished flexibility.

Thus it requires **less desolvation energy** during the complexation. This gives also another contribution to the enthalpic favour.

Macrocyclic effect in terms of dissociation kinetics: The **disfavoured dissociation process** for the macrocyclic complex also explains the macrocyclic effect in terms of kinetics (*cf.* stability constant = formation rate constant/dissociation rate constant).

5.34.2 Mechanism of Acid Catalysed Dissociations of Some Macrocyclic and Open-chain Tetraamine Complexes as the Representative Examples

In terms of dissociation kinetics, the macrocyclic complexes are more stable compared to the open-chain counterparts. The representative examples (Fig. 5.34.2.1) of tetraamine ligands are: *tet*–a (meso), *tet*–b (*d*, *l*), *cyclam* and 3, 2, 3–*tet* (open-chain analogue).

For the macrocyclic complexes, during dechelation, the partially dislodged ligand still remains within the coordination sphere to **allow its religation** to regenerate the complex. To avoid this religation, the partially dislodged ligand is required to fold so that the dislodged ligating sites may move away from the metal centre. This is why, for the macrocyclic complexes, the dissociation process very often requires the *energetically unfavourable rearrangements* such as folding of the partially dislodged ligand within the coordination sphere before the complete dislodgement of the ligand. Moreover, because of the steric crowding, *the partially dislodged ligand and the metal centre cannot be stabilised through solvation.* Here the dissociation kinetics for the Cu(II)–complexes are compared as the representative examples.

$$\left[\text{Cu}(\text{cyclam})\right]^{2+} \underset{k_{-1},(-\text{H}^+)}{\overset{k_1,(+\text{H}^+)}{\rightleftharpoons}} \left[\text{Cu}(\text{Hcylam})\right]^{3+} \xrightarrow{k_2,(+\text{H}^+)} \text{H}_4\text{cyclam}^{4+} + \text{Cu}^{2+}_{\text{aq}}$$

$$\left[\text{Cu}(3,\,2,\,3\text{-tet})\right]^{2+} \xrightarrow{k',(+\text{H}^+)} \text{H}_4(3,\,2,\,3\text{-tet})^{4+} + \text{Cu}^{2+}_{\text{aq}}$$

The important observations are:

● **[Cu(3, 2, 3 –tet)]$^{2+}$** (*i.e. open-chain complex*) dissociates much faster than the macrocyclic complex [Cu(cyclam)]$^{2+}$.

● In contrast to **[Cu(cyclam)]$^{2+}$** (*i.e.* macrocyclic complex), dissociation of the open chain counterpart, *i.e.* [Cu(3, 2, 3–tet)]$^{2+}$ occurs *in a single stage.*

● For the **open-chain complex**, the **acid independent solvation path** makes the *major contribution* while for the corresponding **macrocyclic complex**, the **acid dependent path** is the *main contributing factor.*

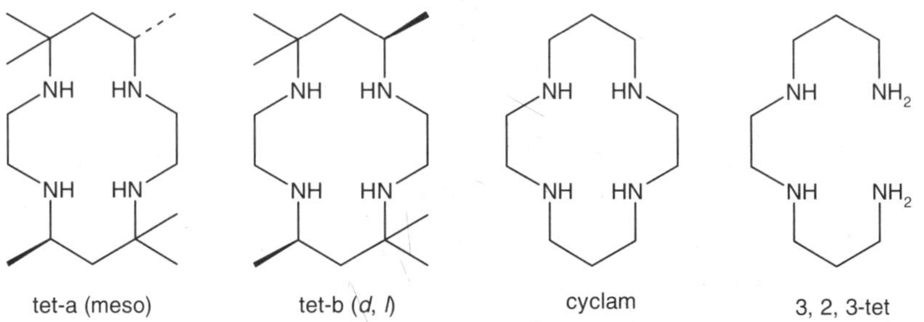

Fig. 5.34.2.1 Some representative macrocycles like 1, 4, 8, 11–tetraazacyclotetradecane (*i.e.* cyclam), C–*meso*–5, 5, 7, 12, 12, 14–hexamethyl–1, 4, 8, 11–tetraazacyclotetradecane (*i.e.* tet–*a*) (which gives both red and blue isomeric complexes); C–*rac*–5, 5, 7, 12, 12, 14–hexamethyl–1, 4, 8, 11–tetraazacyclotetradecane (*i.e.* tet–*b*) and open-chain analogue like 1, 5, 8, 12–tetraazadodecane (*i.e.* 3, 2, 3–tet).

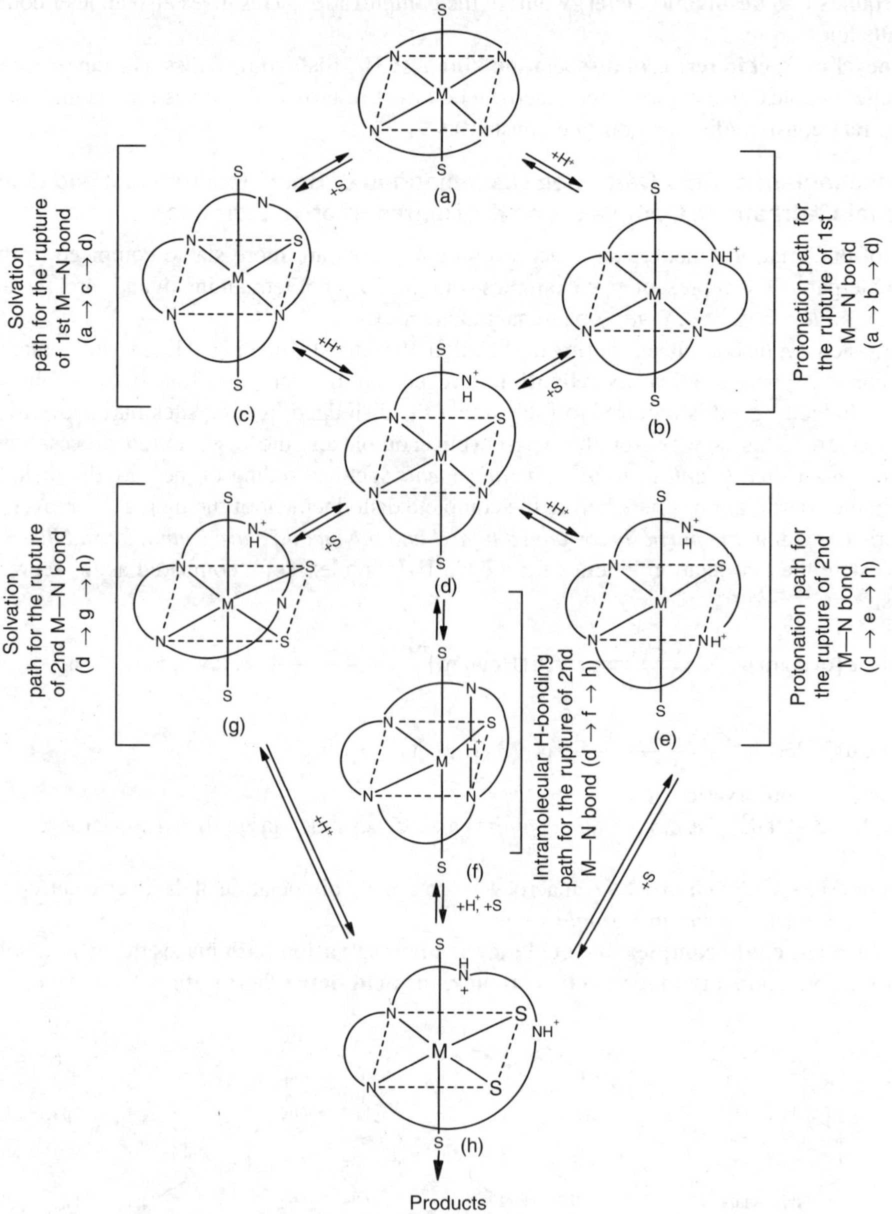

Scheme 5.34.2.1 Generalised mechanistic routes of dissociation of a macrocyclic tetraamine complex S = protic solvent like H_2O (Overall charges not shown).

Mechanistic steps of dissociation of the macrocyclic and open-chain complex: For the dissociation of tetraamine macrocyclic and open chain complexes of Cu(II), the general mechanistic steps can be outlined as in Schemes 5.33.2.1-2 (*cf.* C.S. Chung *et al.*, *Inorg. Chem.* **20,** 2152, 1981; **22,** 1017, 1983; **25,** 1940, 1986; **27,** 1880, 1988). For the macrocyclic complexes, the first step leads to cleavage of a

Cu — N bond and it can happen through the *direct protonation path* and *solvent assisted path.* The second step cleaves the second Cu — N bond that can occur through the direct protonation, solvation and intramolecular H–bonding path. Dissociation of the corresponding open-chain complex, can be also considered to occur in two steps, *i.e.* rupture of the first Cu — N bond followed by the second Cu — N bond rupture.

The *different factors disfavouring the dissociation process for the macrocyclic complex* compared to those for the open-chain complex are discussed below.

● **Relative loss of stability of the complex for the rupture of the first Cu—N bond:** Here it is important to note that rupture of the first Cu — N bond destroys only one chelate ring for the open-

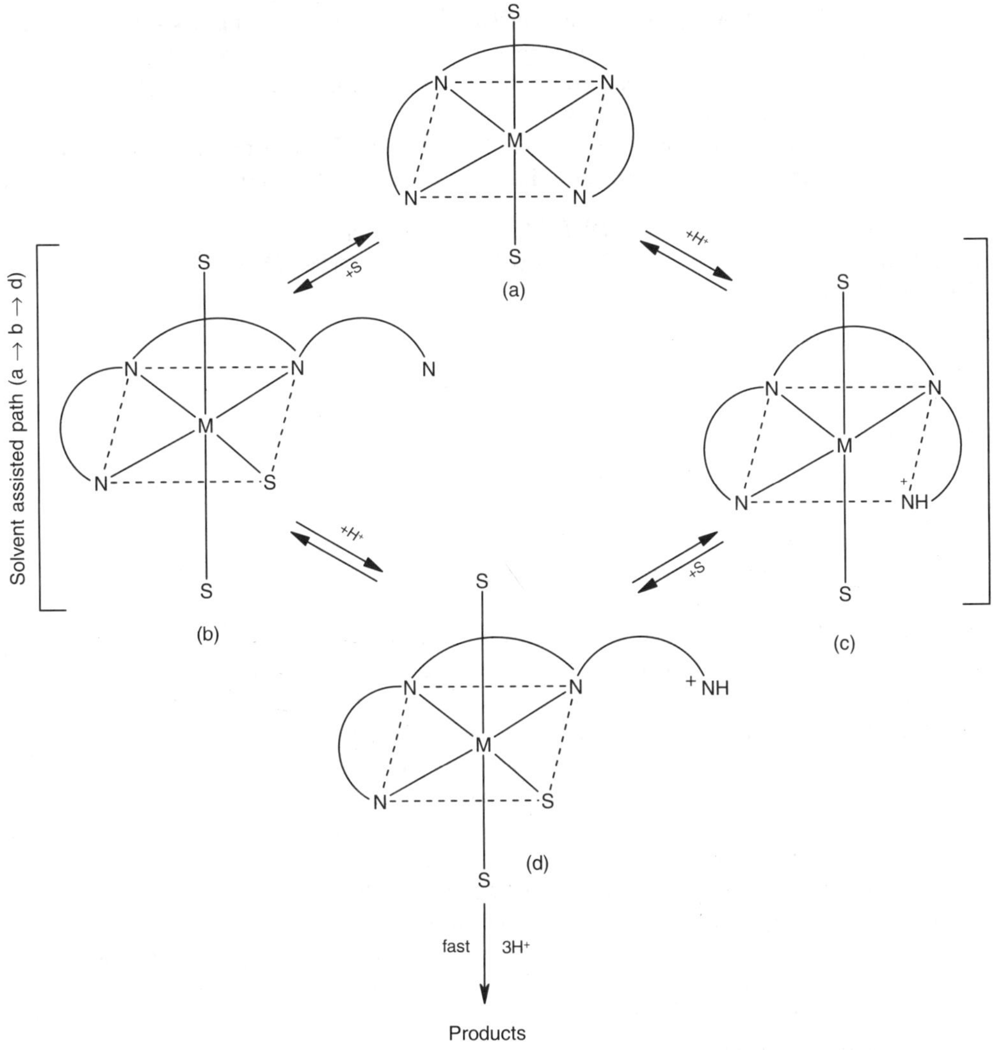

Scheme 5.34.2.2 Generalised mechanistic routes of dissociation of an open-chain tertraamine complex. S = protic solvent like H_2O (Overall charges not shown).

chain complex while it destroys two chelate rings for the corresponding macrocyclic complex. **Thus rupture of the first Cu — N bond is energetically more costly for the macrocyclic complex, and consequently this step is much slower for the macrocyclic complex.**

- **Relative solvation effect during the dissociation process:** After the cleavage of first Cu — N bond, for the open-chain complex, the dislodged N–donor site does not experience any restriction to move away from the coordination sphere and this dislodged N–site can be solvated and it also does not produce any barrier for solvation at the vacant coordination site. On the other hand, for the macrocyclic complex, the movement of the dislodged N–site is restricted and it still remains within the coordination sphere. **This disfavours the solvation of the dislodged site and vacant coordination site. It makes the rupture of Cu – N bond more costly for the macrocyclic complex.**
- **Relative ease of religation by the partially dissociated ligand:** For the macrocyclic complex, the dislodged *N–site present within the coordination sphere can be prevented from religation by protonation* or *by rearrangement* (*e.g.* ***energetically unfavourable folding***). On the other hand, the dislodged N–site for the open-chain complex can move away freely from the coordination sphere and thus the possibility of religation is less. **This is why, acid has a little effect on the dissociation kinetics for the open-chain complex and acid has a strong effect on the dissociation kinetics of the macrocyclic complex.**

5.35 REACTION MECHANISM AND REACTIVITIES OF THE COMPOUNDS OF p-BLOCK ELEMENTS: MECHANISTIC ASPECTS OF REACTIONS OF THE C–, Si–, N– AND P–COMPOUNDS

These aspects have been discussed in Chapter 10 of Vol. 2.

EXERCISE 5

A. General Type Questions

1. What are the different types of ligand substitution reactions in terms of reaction mechanism? Discuss with examples.
2. What do you mean by lability and inertness?
3. What do you mean by stoichiometric and intimate mechanism of ligand substitution reaction? How are they related?
4. Derive the rate laws for the *D*, *A* and *I* processes of ligand substitution reaction.
5. "Different reaction mechanisms may lead to the same rate law". Illustrate with examples.
6. What do you mean solvent exchange reaction? Discuss the importance of solvent exchange rate constants (k_{ex}) in the diagnosis of reaction mechanisms of other ligand substitution processes.
7. Classify the metal ions in terms of water exchange rate constants (k_{ex}). Discuss the different factors to control the water exchange rate constants.
8. What do you mean Eigen-Wilkins mechanism of ligand substitution process? How can you distinguish between the I_a and I_d path?
9. During chelation by a didentate ligand, depending on the nature of the rate determing step, it may experience the steady-state or equilibrium mechanism (*i.e.* intermediate complex bearing the didentate ligand as a mondentate ligand may remain in steady-state or equilibrium state). Explain.
10. How can you distinguish among the *D*, I_d and I_a and *A* pathways of ligand substitution reaction?
11. What do you mean *proton ambiguity* in kinetics of ligand substitution reaction? How can you solve the problem? Illustrate with an example.
12. How can you attempt to explain the lability or inertness of transition metal complexes in terms of VBT? What are the limitations in this approach?

13. What do you mean by crystal field activation energy (CFAE)? How can you explain the lability or inertness of metal complexes in terms of CFT?
14. What do you mean by $\Delta H^{\neq}$, $\Delta S^{\neq}$ and $\Delta V^{\neq}$? How can you apply the knowledge of these activation parameters to diagnose a reaction mechanism?
15. How can you experimentally determine $\Delta H^{\neq}$, $\Delta S^{\neq}$ and $\Delta V^{\neq}$? What are the relationships between E_a (activation energy in Arrhenius equation) and $\Delta H^{\neq}$; $\Delta V^{\neq}$ and $\Delta S^{\neq}$?
16. Discuss the factors controlling the $\Delta V^{\neq}$ and $\Delta S^{\neq}$ values.
17. What do you mean by LFER plot? How can you apply this plot to diagnose a reaction mechanism?
18. What do you mean by isokinetic trend? What does it signify?
19. What do you mean steric accleration and steric retardation? How can you interpret these in terms of reaction mechanism?
20. Discuss the effect of charge on a complex to diagnose a ligand substitution reaction mechanism.
21. Discuss the mechanistic aspects of aquation or acid hydrolysis of cobalt(III) complexes. Discuss the steric and electronic effects of the spectator ligands.
22. Discuss the kinetic aspects of acid catalysed and metal ion catalysed aquation of metal complexes.
23. Discuss the pseudo-substitution reaction in octahedral metal complexes.
24. Discuss the stereochemical changes in the acid hydrolysis of Co(III)–complexes.
25. Discuss the mechanism of base hydrolysis of cobalt(III)–ammine complexes. Give the experimental evidences in favour of the proposed mechanism. Critically comment on the *D* – CB, I_d – CB and *D* – IP processes in relation to such base hydrolysis reaction.
26. What are the stereochemical changes in base hydrolysis of Co(III)–ammine complexes? Compare these results with those of the acid catalysed aquation reaction.
27. Discuss the kinetic aspects of formation reactions at Fe(III).
28. Discuss the mechanism of isomerisation (both linkage and geometrical isomerisation) reaction in octahedral complexes.
29. Discuss the possible mechanistic routes of racemisation of octahedral and tetrahedral complexes.
30. What are the characteristic features of the ligand substitution reactions of square planar complexes of Pt(II)?
31. Discuss the nucleophilicity scale, *trans*–influence, *trans*–effect (both σ– and π– *trans* effect), stereospecificity, *cis–trans* isomerisation in relation to the reactivity of square planar complexes of Pt(II).
32. Discuss the application of *trans*—effect series in the preparation of isomers of Pt(II) complexes.
33. Discuss the mechanistic aspects of ligand substitution reaction in carbonyl and carbonyl-nitrosyl complexes of tetrahedral geometry.
34. Discuss the mechanism of nucleophile catalysed *cis–trans* isomerisation of square planar complexes of Pt(II). What are the other possible routes of *cis- trans-* isomerisation in square planar complexes?
35. Discuss the σ– and π–bonding theory of *trans*–effect.
36. Outline the routes to synthesise the following complexes.
    (i) isomers of $[Pt(C_2H_4)Cl_2(NH_3)]$;
    (ii) isomers of $[Pt(CO)Cl_2(py)]$.
    (iii) isomers of $[Pt(NH_3)(NH_2OH)(NO_2)(py)]^+$;
    (iv) isomers of $[PtCl(NH_3)(RNH_2)(NO_2)]$
    (v) isomers of $[PtBrCl(NH_3)py]$
37. What do you mean by macrocyclic effect? Explain its origin.
38. Compare the acid catalysed dissociation of the macrocyclic and open-chain complexes.

39. Cu(II)–complexes are, in general, labile with respect to acid catalysed dissociation. But many Cu(II)–macrocyclic complexes dissociate slowly—explain.
40. Compare the stereochemical aspects of ligand substitution process of the tetrahedral C-compounds, tetrahedral Si-compounds and square planar Pt(II)-complexes.
41. Explain the following facts in base hydrolysis of the Co(III)-ammine complexes.
(i) proton abstraction preferably occurs from the *trans*-position of the leaving group, (ii) rate enhancement by the CB, (iii) entry of H_2O not OH^- as a nucleophile at the fast step, (iv) stereochemistry of the product.

B. Justify the statements

1. Thermodynamically stable complexes are not necessarily kinetically stable.
2. The origins of thermodynamic stability and kinetic stability are different.
3. Nucleophilicity and basicity do not always run parallel.
● *Trans*-influence and *trans*-effect do not always run parallel (*cf.* octahedral Co(III)-complexes *vs.* square planar Pt(II)–complexes).
4. Associative (*A*) and dissociative (*D*) are the two-step reactions while interchange (*I*) reaction is a single-step reaction.
5. The Langford-Gray notation, *A*, I_a, I_d and *D* are the combination of notations of stoichiometric and intimate mechanisms.
● Intimate mechanism looks at the maxima while stoichiometric mechanism looks at the minima of the free energy *vs.* reaction coordinate plot.
6. Langford-Gray notation and Hughes-Ingold notation can be correlated.
7. Associative (*A*) and Dissociative (*D*) pathways involve two transition states while interchange (*I*) pathway involves a single transition state.
8. In a dissociative process, the rate law can show a complicated dependence pattern on the living group and entering group.
9. A reaction occurring in a dissociative process may show the rate dependence on the entering ligand depending on the condition.
10. An associative reaction shows a rate law that indicates a strict first-order dependence on the entering ligand concentration.
11. The interchange pathways I_a and I_d can be distinguished by considering the rate dependence on the entering ligand concentration and in terms of k_{ex} values.
12. The rate law for the *D* and *I* processes can be expressed by a general form.

$$L_5M - X + Y \rightarrow L_5M - Y + X$$

$$\text{rate} = \frac{pq[L_5M - X][Y]}{1 + q[Y]}$$

● The rate law for an associative process (*A*), can also be expressed in the above form under some restrictions.

● In the energy profile diagram (free energy *vs.* reaction coordinate), the activated complex or transition state resides at the maxima while the intermediates resides at the minima.

● The transition states and intermediates differ in their life times.

13. *D* and I_d pathways can lead to the same rate law under some conditions.
14. A formation reaction (*i.e.* replacement of the coordinated solvent) passing through a *D*–process can show a rate dependence on the nature of the entering ligand.

15. In a perfect D–process, the dependence of formation rate constant (k_f) on the nature of the entering ligand arises from the competition between the centering ligand and leaving group for the lower coordinate intermediate.
16. A D–reaction may show or may not show dependence on the nature of the entering ligand depending on the condition.
17. An I_d–reaction may show or may not show rate dependence on the nature of the entering ligand depending on the condition.
18. Very often an I_d–path may be characterised by considering the solvent exchange rate constant (k_{ex}).
19. Both the I_a– and A–paths can develop the same rate law under some conditions.
20. In the I_d– path of a formation reaction, the interchange rate constant can never exceed the solvent exchange rate constant (k_{ex}) but for the I_a–path, it may be less than or greater than the k_{ex} value depending on the condition.
21. A–path always follows a second order kinetics but I_a–path may follow both the first-order and second kinetics depending on the conditions.
22. In the octahedral complex, perfect D–process is generally found for the anation of anionic complexes like $[Fe(CN)_5(OH_2)]^{3-}$, $[Co(CN)_5(OH_2)]^{2-}$.
23. Perfect D–process is also likely to operate in the octahedral complexes bearing the strong *trans*-labilising ligands.
24. To distinguish between the A and I_a paths, it may be required to follow the effect of the concentration of the entering ligand in a *wide range*.
25. In the formation reactions of $[Pt(OH_2)_4]^{2+}$, $[Pd(OH_2)_4]^{2+}$, $[Ti(OH_2)_6]^{3+}$, A–process is argued based on the fact: no deviation from the strict first-order dependence on the entering ligand even when concentration of the ligand is very high; the said strict first-order is maintained even when the entering ligand is highly anionic to make the K_{OS} value very high.
26. Fractional order in the entering ligand concentration is an indication of the I_a–path.
27. The same rate law can be obtained for different reaction mechanisms.
28. The second order rate law is obtained for the following reaction.
    $L_5M - X + Y \rightarrow L_5M - Y + X$; rate $\propto$ $[L_5MX][Y]$
    There are several mechanisms to generate the above rate law.
29. $L_5M - X + Y \rightarrow L_5M - Y + X$
    The fractional order in $[Y]$ or saturation kinetics at a higher value of $[Y]$ is an indication of a preequilibrium step involving the both reactants followed by the rate determining step.
30. NMR peak broadening phenomenon can be applied for the determination of water exchange rate constants of metal ions.
31. $\Delta V^{\neq}$ is a very much reliable parameter to diagnose the solvent exchange reaction mechanism.
32. V^{2+} and Cr^{3+} show the very low values of k_{ex}.
33. The water exchange rate is faster for Ni^{2+} than that for Cr^{3+} though they are having the same CFAE.
34. The k_{ex} values of Cr^{2+} and Cu^{2+} are tremendously high.
35. For the bivalent metal ions of 1st transition series, the water exchange process involves the more associative character for the early members while the late members of the series adopt the more dissociative character.
36. The mechanism of solvent exchange reaction depends also on the bulkiness of solvent molecules.
37. The trivalent metal ions of 1st transition series very often adopt the associative path in water exchange process. However, their hydrolysed species, *i.e.* $[M(OH)(OH_2)_5]^{2+}$ (M = Fe, Cr, etc.) adopt the dissociative process like the bivalent metal ions in water exchange process.

38. $[Fe(OH_2)_6]^{3+}$ and $[Fe(OH)(OH_2)_5]^{2+}$ adopt different mechanistic paths in ligand substitution process including the water exchange process.
39. $[Ti(OH_2)_6]^{3+}$ shows the more associative character than $[Fe(OH_2)_6]^{3+}$ in ligand substitution process.
40. For the trivalent metal ion, $[Co(OH_2)_6]^{3+}$, the water exchange process is unusually fast.
41. $\Delta S^{\neq}$ is a reliable parameter to identify the solvent exchange path.
42. Solvent exchange rate constant (k_{ex}) is an important parameter to diagnose the reaction mechanism for different types of ligand substitution processes.
43. In **d**–activation, the ligand substitution rate constant can never exceed the k_{ex} value.
44. In **a**–activation, ligand substitution rate constant may be greater or less than the k_{ex} value depending on the condition.
45. Based on the water exchange rate constants (k_{ex}), the metal ions have been classified into four classes.
46. The parameter *ionic potential* or Z^{*2}/r plays an important role in determining the k_{ex} values.
47. Among the alkaline earth cations, Ba^{2+}, Sr^{2+} and Ca^{2+} are in Class-I, Mg^{2+} is in Class-II while Be^{2+} is in Class-III.
48. In terms of k_{ex} values, Hg^{2+} and Cd^{2+} are placed in Class-I while Zn^{2+} is placed in Class-II.
49. The pK_a–values of $[M(OH_2)_6]^{n+}$ (acting as the Brönsted acids) can be correlated with their k_{ex} values.
50. Cr^{2+} and Cu^{2+} are highly labile in terms of their k_{ex} values.
51. The lability sequence is: Ni(II) 〉 Pd(II) 〉 Pt(II)
   ● Ni^{2+} is a Class-II metal ion; Pd^{2+} is a Class-III metal ion and Pt^{2+} is a Class-IV metal ion.
52. The k_{ex} values run in the sequence:
   $V^{2+} \langle\langle Cr^{2+} \rangle Mn^{2+} \rangle Fe^{2+} \rangle Co^{2+} \rangle\rangle Ni^{2+} \langle\langle Cu^{2+} \rangle Zn^{2+}$

	● Cr^{2+} ≈	Cu^{2+} 〉〉	Mn^{2+} ≈	Zn^{2+} 〉	Fe^{2+} 〉	Co^{2+} 〉〉	Ni^{2+} 〉	V^{2+}
log k_{ex} : (25°C)	~9	~9.0	7.5	7.5	6.6	6.4	4.5	2.0

53. Generally, the Co(III) complexes are more inert than the Cr(III) complexes, but in terms of the k_{ex} values, the water exchange process in $Co(aq)^{3+}$ is much faster than that in $Cr(aq)^{3+}$.
54. Co(II) can catalyse the ligand substitution process in Co(III); Pt(II) can catalyse the ligand substitution process in Pt(IV); Cr(II) can catalyse the ligand substitution process in Cr(III).
55. Water exchange rate constants follows the sequece:
   $$Cr(aq)^{3+} \langle Rh(aq)^{3+} \langle\langle Ir(aq)^{3+}.$$
56. There is a rapid ligand exchange in $[Co(LL)_3]^{3+}$ (LL = bpy, phen, en).
57. The ionic radii of Sc^{3+}, Ti^{3+}, V^{3+} and Fe^{3+} are comparable but their k_{ex} values differ significantly.

	$[Sc(OH_2)_6]^{3+}$	$[Ti(OH_2)_6]^{3+}$	$[V(OH_2)_6]^{3+}$	$[Fe(OH_2)_6]^{3+}$
log k_{ex} ~(25°C)	7.2	5.3	2.7	2.2

58. Eigen-Wilkins mechanism actually represents the stoichiometric interchange (I) mechanism.
59. A ligand substitution reaction passing through the interchange process shows the rate law:
   $$k_{obs} \approx k_i K_{os}[Y],\ Y = \text{entering ligand}.$$
   ● On the other hand, a solvent exchange reaction passing through the interchange process gives the rate law:
   $$k_{obs} = k_i = k_{ex}$$
   (k_i, k_{ex}, K_{os} bear the usual significance).

60. $[Ni(OH_2)_6]^{2+} + L \rightarrow [Ni(L)(OH_2)_5]^{n+} + H_2O$.
In the above reaction, k_f (second order formation rate constant) varies with the nature of the entering ligand but k_f/K_{OS} remains more or less constant.
61. In the formation reaction, if the entering ligand is a didentate chelating ligand then the normal Eigen mechanism may operate or may not operate depending on the contribution of the ring closure step to the rate determining step.
62. 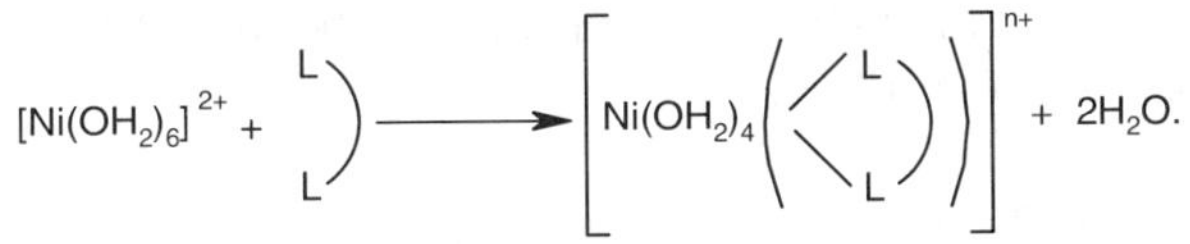

- In the above reaction, bpy reacts slower than phen.
- In the above reaction, α–ala⁻ reacts faster than β–ala⁻.

63. $[(A)Ni^{II}(OH_2)_2]^{x+} + gly^- \rightarrow [(A)Ni^{II}(gly)]^{y+} + 2H_2O$.
(**cis**–complex)
In the above ternary complex formation reaction, the normal Eigen mechanism does not operate. (A = a tetradentate and bulky ligand).
64. In the chelation reaction, sometimes the intermediate formed by the binding of the didentate ligand as a unidentate one may remain in **steady state** or **equilibrium state** depending on the conditions.
65. Both the D and I_d processes experience the **d**–activation. In the D–process, the rate constant shows some sort of dependence on the nature of the entering ligand but in the I_d process, the rate constant does not show any such dependence on the nature of the entering ligand.
66. The D and I_d processes can be distinguished by following the rate dependence on the nature of the entering ligand.
67. Both the D and I_a processes show rate the dependence on the nature of the entering ligand. They can be distinguished under some limiting conditions where the D-process becomes independent of the entering ligand but such a situation never arises for an I_a–process.
68. Proton ambiguity arises due to the presence of *isomeric reaction paths.*
69. Proton ambiguity arises generally for the trivalent metal ions prone to hydrolysis.
70. Proton ambiguity may be solved by considering the *reasonableness of the rate constant values.*
71. VBT assumes that the outer orbital complexes participate in a dissociative process for the ligand substitution reactions while the inner orbital complexes participate in an associative process.
72. The lability sequence is:
$AlF_6^{3-} > SiF_6^{2-} > PF_6^- > SF_6$.
73. The inner orbital complexes of d^3 and d^4 configurations can also participate in an associative process in ligand substitution reaction in terms of VBT.
74. The assumption, '*outer-orbital complexes are weaker than the inner orbital complexes*' is not justified always.
75. Sometimes CFAE (crystal field activation energy) becomes negative.
76. Generally, the low spin complexes are more inert than the high spin complexes.
77. d^4 (h.s.) d^7 (l.s.) and d^9 systems are highly labile.
78. The heavier congeners are more inert than the lighter congeners.
79. The associative character increases in moving down in a group while the associative character decreases in a period in moving from left to right.
80. CFT also predicts that the higher valent metal ions are more inert.

81. Generally, with the increase of oxidation state, the lability decreases but V(III) is more labile than V(II).
82. The lability orders are: Co(III) ⟩ Rh(III) ⟩ Ir(III); Ni(II) ⟩ Pd(II) ⟩ Pt(II).
83. The low spin Fe(II) (d^6) complexes dissociate slower than the Ni(II) complexes but the high spin Fe(II) complexes dissociate faster than the Ni(II) complexes.
84. Generally the low spin Fe(II)–complexes dissociate slower than the Ni(II)–complexes because of the higher CFAE (*cf.* $4Dq$ for Fe^{II} *vs.* $2Dq$ for Ni^{II}), but $[Fe(phen)_3]$ and $[Fe(bpy)_3]^{2+}$ dissociate faster than the corresponding complexes of Ni(II).

$$\left.\begin{array}{l}\left[Fe(LL)_3\right]^{2+} \longrightarrow \left[Fe(LL)_2\right]^{2+} + LL;\ k_{Fe}\\ \left[Ni(LL)_3\right]^{2+} \longrightarrow \left[Ni(LL)_2\right]^{2+} + LL;\ k_{Ni}\end{array}\right\} k_{Fe} > k_{Ni}\ (LL = \text{bpy, phen})$$

85. $$\left.\begin{array}{l}\left[Fe(LL)\right]^{2+} \xrightarrow{k_{Fe}} Fe^{2+} + LL\\ \left[Ni(LL)\right]^{2+} \xrightarrow{k_{Ni}} Ni^{2+} + LL\end{array}\right\} k_{Fe} > k_{Ni};\ (LL = \text{phen, bpy, terpy})$$

86. $$\left.\begin{array}{l}\left[Cr(bpy)_3\right]^{2+} \xrightarrow{k_{Cr}} \left[Cr(bpy)_2\right]^{2+} + bpy\\ \left[Ni(bpy)_3\right]^{2+} \xrightarrow{k_{Ni}} \left[Ni(bpy)_2\right]^{2+} + bpy\end{array}\right\} k_{Cr} > k_{Ni}$$

CFAEs for both the systems are more or less comparable but the Cr(II)–complex dissociates faster than the Ni(II)–complex.

87. $[V(phen)_3]^{2+}$ dissociates faster than $[Ni(phen)_3]^{2+}$ though their CFAEs are almost the same. Interestingly, the water exchange rate is much higher for Ni(II) than that for V(II).

$$\text{Lability order: } \underbrace{\left[V(phen)_3\right]^{2+} > \left[Ni(phen)_3\right]^{2+}}_{\text{Dissociation process}};\ \underbrace{\left[V(OH_2)_6\right]^{2+} < \left[Ni(OH_2)_6\right]^{2+}}_{\text{Water exchange process}}$$

88. The water exchange rate follows the sequence:
$[Sc(OH_2)_6]^{3+}$ ⟩ $[Ti(OH_2)_6]^{3+}$ ⟩ $[V(OH_2)_6]^{3+}$ ⟩ $[Fe(OH_2)_6]^{3+}$ ⟩⟩ $[Cr(OH_2)_6]^{3+}$
89. The systems like d^0, d^1 and d^2 generally prefer to adopt the associative path. It can be explained in terms of both VBT and CFT.
90. The d^{10} systems are expected to be labile in terms of both VBT and CFT.
91. In terms of VBT, Ni(II) (d^8) is expected to be labile but in terms of CFT, it should be inert.
92. The CFAEs of Ni(II) and Cr(III) are more or less the same, but generally the Cr(III) complexes react much slower than the Ni(II)–complexes.
93. Ni(II) centre is less labile than many other bivalent metal ions of first-transition series.
94. Generally, d^5–system reacts faster than d^4 system. But in the CN^- ligand exchange reactions of $[M(CN)_6]^{n-}$, the lability order runs as: d^4 ⟩ d^5 ⟩ d^6
Hints: Trapezoidal octahedral intermediate.
95. Effect of charge on the complex is straight-forward for the dissociative (*D*) and associative (*A*) paths but the effect is not so straight-forward for an interchange (I) process.
96. For the aquation of Cl^- in the chlorido-ammine complexes of Co(III) and Cr(III), the rate drastically increases for the Co(III) complexes with the decrease of overall positive charge on the complex, but the said change is less sensitive for the corresponding Cr(III) complexes.

97. The reactivity and choice of the mechanistic paths in the formation reactions of $[M(OH_2)_6]^{3+}$ and $[M(OH)(OH_2)_5]^{2+}$ (M = Cr^{III}, Fe^{III}) are totally different. This difference is not only due to the difference in overall charge of the complexes.
98. In the so called Eyring Equation:

$$-\ln\left(\frac{k_2 h}{k_B T}\right) = \frac{\Delta H^{\neq}}{RT} - \frac{\Delta S^{\neq}}{R},$$

the assumption of transmission coefficient (κ) to be unity is not always justified.
99. Eyring equation is not applicable always correctly for the electron transfer process, $H/H^+/H^-$–transfer processes.
100. Tunneling effect may complicate the interpretation of Eyring equation.
101. The entropy of activation ($\Delta S^{\neq}$) cannot be directly obtained from the intercept of the plot, $\ln\left(\frac{k_2 h}{k_B T}\right)$ *vs.* $\frac{1}{T}$, though the enthalpy of activation ($\Delta H^{\neq}$) can be directly obtained from the slope of the plot.
102. $\Delta S^{\neq}$ (of T.S. theory) can be correlated with the preexponential factor (*A*) of Arrhenius equation.
103. For the reactions occurring in solution, $\Delta H^{\neq}$ may be roughly taken as the E_a (activation energy obtained from Arrhenius equation).
104. In the aquation of X^- in $[Co(NH_3)_5X]^{2+}$ (X^- = Cl^-, Br^-, NO_3^-, NO_2^-, NCS^-, etc.), the enthalpy of transition (ΔH_T) is more or less the same.
105. In the aquation of X^- in $[Co(NH_3)_5X]^{2+}$ (X^- = different uninegative ligands), $\Delta H^{\neq}$ bears a linear relationship with the $\bar{\nu}$ of $[Co(NH_3)_5X]^{2+}$ ($\bar{\nu}$ = wavenumber in cm^{-1} for the first absorption band in the electronic spectra).
106. In the aquation of $[M(NH_3)_5X]^{2+}$ (M = Co, Cr; X = Cl, Br, I), the $\Delta S^{\neq}$ values follow the sequence: $\Delta S^{\neq}$: $I^- \rangle Br^- \rangle Cl^-$.
107. In the solvent exchange reactions or formation reactions involving the replacement of a coordinated solvent by a neutral ligand, $\Delta S^{\neq}$ value is a good parameter to diagnose the reaction pathway.
108. In solvent assisted dissociation (SAD) of $[M(NH_3)_5X]^{2+}$, solvation of the leaving group (X) is an important factor in attaining the transition state.
109. In a dissociative process, $\Delta S^{\neq}$ is expected to be positive while in an associative process, $\Delta S^{\neq}$ is expected to be negative but these straightforward conclusions may not operate always.
110. To interpret the value of $\Delta S^{\neq}$, it is very important to consider the possibility of *charge creation* (*i.e.* charge separation) or *charge destruction* (*i.e.*charge neutralisation) in attaining the transition state.
111. To diagnose a reaction pathway, the small values of $\Delta S^{\neq}$ are not the reliable diagnositic parameters.
112. Reactions passing through a similar pathway may be characterised by their isokinetic trends.
113. Isokinetic trend arises from the mutual compensation effect of $\Delta H^{\neq}$ and $\Delta S^{\neq}$.
114. For aquation of $[Co(NH_3)_5X]^{2+}$, LFER holds good with unit slope.

$$\left[Co(NH_3)_5(-X)\right]^{2+} + H_2O \rightleftharpoons \left[Co(NH_3)_5(OH_2)\right]^{3+} + X^-$$

115. $[M(NH_3)_5X]^{2+} \rightarrow [M(NH_3)_5(OH_2)]^{3+} + X^-$
- Slope of LEER plot: ~1.0 (for M = Co), $\langle$ 1.0 (for M = Cr, Rh).
- Aquation rate sequence: X = I $\rangle$ Br $\rangle$ Cl (for M = Co),
  X = Cl $\rangle$ Br $\rangle$ I (for M = Rh).

116. $[Cr(OH_2)_6X]^{2+} \rightarrow [Cr(OH_2)_6]^{3+} + X^-$; slope of LEER plot = ~0.55
$[Cr(NH_3)_6X]^{2+} \rightarrow [Cr(NH_3)_5(OH_2)]^{3+} + X^-$; slope of LEER plot = ~ 0.90
117. Volume of activation ($\Delta V^{\neq}$) can be obtained from the pressure dependence of rate at constant temperature.

$$k \propto \exp\left(\frac{-P\Delta V^{\neq}}{RT}\right).$$

118. There are two important factors to determine the $\Delta V^{\neq}$ values: (i) change in the number of particles in attaining the T.S.; (ii) change of electrorestriction over the solvent in attaining the T.S.
$\Delta V^{\neq} = \Delta V^{\neq}_{intrinsic} + \Delta V^{\neq}_{solvent}$
119. $\Delta V^{\neq}$ is not always a good parameter for the diagnosis of a reaction mechanism, but in some cases, it is a reliable parameter.
120. $\Delta V^{\neq}$ and $\Delta S^{\neq}$ bear a linear correlationship.
121. $[M(NH_3)_5L]^{3+} \rightarrow$ aquation of L (neutral ligand)
$\Delta V^{\neq}$ = –ve for M = Cr; $\Delta V^{\neq}$ = +ve for M = Co.
122. Aquation of X^- in $[M(NH_3)_5X]^{2+}$ (M = Co, Cr):
$\Delta V^{\neq}$ = –ve for both M = Co, Cr.
- $[M(NH_3)_5X]^{2+} + H_2O \rightarrow [M(NH_3)_5(OH_2)]^{3+} + X^-$ (hydrated), ΔV = –ve.

123. $[Co(NH_3)_5(SO_4)]^+$ and $[Co(NH_3)_5X]^{2+}$ aquate through the dissociative path but their $\Delta V^{\neq}$ values are negative (*cf.* $\Delta V^{\neq}$ of k_{ex} for $[Co(NH_3)_5(OH_2)]^{3+}$ is positive).
124. • $\left[RuCl(NH_3)_5\right]^{2+} \xrightarrow[a\text{-activation}]{\text{(aquation)}} \left[Ru(NH_3)_5(OH)_2\right]^{3+} + Cl^-,\ \Delta V^{\neq} = -30\ cm^3\ mol^{-1}$
- $\left[Ru(NH_3)_5(OH_2)\right]^{3+} + Cl^- \xrightarrow[a\text{-activation}]{\text{(anation)}} \left[RuCl(NH_3)_5\right]^{2+},\ \Delta V^{\neq} = -20\ cm^3\ mol^{-1}$
- Both the aquation and anation paths adopt the **a**–activation, but $\Delta V^{\neq}$ is less negative for the anation reaction.

125. The aquation rates of $[Co(NH_3)_5(L)]^{n+}$ differ significantly for the various types of leaving groups (L) while the formation rates of $[Co(NH_3)_5(L)]^{n+}$ from $[Co(NH_3)_5(OH_2)]^{3+}$ are more or less insensitive towards the nature of the entering group (L).
126. Anation rates of $[Ti(OH_2)_6]^{3+}$ and $[V(OH_2)_6]^{2+}$ drastically depend on the nature of the entering anion.
127. $[Cr(NH_3)_5(OH_2)]^{3+}$ and $[Cr(OH_2)_6]^{3+}$ adopt different paths for anation and their reactivities differ widely. This difference can be explained by considering relative σ–basicity of NH_3 and H_2O; relative energies of the LUMOs.
128. Associative path is more favoured for the Cr(III)–complexes than for the low spin Co(III) complexes.
- In terms of the energy of the LUMO, it is quite reasonable to expect a more associative character in the ligand substitution process of Cr(III) (t_{2g}^3) than that in the low spin complexes of Co(III) (t_{2g}^6).

129. Anation of $[Ti(OH_2)_6]^{3+}$ experiences a more associative character than that of $[V(OH_2)_6]^{2+}$.
130. $[V(OH_2)_6]^{2+} + X^- \rightarrow [V(OH_2)_5(X)]^+ + H_2O$; $X^- = Cl^-, N_3^-, NCS^-$
k_f ($dm^3\ mol^{-1}\ s^{-1}$) = 1 (Cl^-): 2 (NCS^-): 10 (N_3^-).
131. Dissociative path experiences the **steric acceleration** while associative path experiences the **steric retardation**.

132. $[Cu(N-O)_2]$ (square planar complex) shows the amino acid exchange rate sequence: glycinate $\rangle$ sarcosinate $\rangle$ N, N–dimethylglycinate.
- $[M(N-O)_3]^-$ (octahedral complex; M = Co, Ni) shows the amino acid exchange rate sequence: glycinate $\langle$ sarcosinate $\langle$ N, N–dimethylglycinate; (N – O = deprotonated form of the amino-acid)

133. Aquation rate follows the sequences:
- *trans*–$[CoCl_2(cyclam)]^+ \langle$ *trans*–$[CoCl_2(tet-b)]^+$
- $[CoCl(NH_3)_5]^{2+} \langle [CoCl(MeNH_2)_5]^{2+}$
- $[CrCl(NH_3)_5]^{2+} \rangle [CrCl(MeNH_2)_5]^{2+}$
- $[CoCl(dien)(tn)]^{2+} \rangle [CoCl(dien)(en)]^{2+}$
- $[CrCl(dien)(tn)]^{2+} \approx [CrCl(dien)(en)]^{2+}$
- *cis*–$[Co^{III}Cl(en)_2(L)]^{n+} \rangle$ *trans*–$[Co^{III}Cl(en)_2(L)]^{n+}$
  (L = spectator ligand with the π–donor properties; *e.g.* OH, Cl, NH_2 etc.)
- *cis*–$[Co^{III}Cl(en)_2(L^1)]^{n+} \rangle$ *cis*–$[Co^{III}Cl(en)_2(L^2)]^{n+}$
  L^1 = π–donor ligand (*e.g.* Cl^-, OH^-)
  L^2 = π–acceptor ligand (*e.g.* NO_2^-, CN^-)
- *trans*– $[Co^{III}Cl(en)_2(L)]^{n+} \rangle$ *cis*–$[Co^{III}Cl(en)_2(L)]^{n+}$
  L = π–acceptor ligand (*e.g.* NO_2^-, CN^-)

134. $\underset{(CA)}{\left[M(OH_2)_6\right]^{3+}} + H_2O \rightleftharpoons \underset{(CB)}{\left[M(OH)(OH_2)_5\right]^{2+}} + H_3O^+$; M = Fe, Cr

CB is more reactive than CA in ligand substitution process. CA adopts the I_a–path while CB adopts the I_d–path.

135. In an interchange process, we have the following relationships:
- $k_i = \frac{6}{8}k_{ex}$; $k_i = \frac{k_f}{K_{OS}}$, k_f = 2nd order formation rate constant
- $k_{\lim} = k_i$, $(K_{OS}[Y] \rangle\rangle 1$, Y = entering ligand).
- k_i or $k_{\lim}$ can never exceed the k_{ex} value in I_d process.
- k_i or $k_{\lim}$ may be greater or less than k_{ex} depending on the conditions in I_a process.

136. Anation of $[Co(NH_3)_5(OH_2)]^{3+}$: $k_{\lim}/k_{ex} \leq 1.0$
- Anation of $[M(NH_3)_5(OH_2)]^{3+}$: $k_{\lim}/k_{ex} \rangle\rangle 1.0$, (M = Rh, Ir).
- Formation reaction at $[Fe(OH_2)_6]^{3+}$: k_f/k_{OS} = no constancy
- Formation reaction at $[Fe(OH)(OH_2)_5]^{2+}$: k_f/K_{OS} = constant and close to k_{ex}
- Anation of $[Co(CN)_5(OH_2)]^{2-}$ or $[RhCl_5(OH_2)]^{2-}$; the double reciprocal plot, $\frac{1}{k_{obs}}$ *vs.* $\frac{1}{[Y^{n-}]}$ is

linear (Y^{n-} = entering ligand) with finite intercept and slope; $\frac{\text{slope}}{\text{intercept}} = x(\text{say})$

x depends on the nature of Y^{n-}

(**Hints:** *D*–process.)

137. The relationship expressing the effect of ionic strength (I) on rate constant can be obtained from transition state theory and Debye-Huckel theory.

$$\log k_I \approx \log k_0 + 2QZ_AZ_B\sqrt{I}$$

138. The rate dependence pattern on ionic strength depends on the nature of charge bearing properties of the reactant species in a bimolecular reaction.

139. Deviation from the linear plot of log k vs. $\sqrt{I}$ is an indication of secondary salt effect.

140. To compare the rate constants, the reactions are to be carried out at a fixed and fairly high ionic strength to be maintained by the inert electrolytes.

141. Nitrate (NO_3^-) and perchlorate (ClO_4^-) salts are generally used to maintain the ionic strength.

142. $(H_3N)_5Co^{III}$—O—C(=O)—C$_6$H$_4$—R $\quad[\ ]^{2+}$ $\xrightarrow{\text{Base hydrolysis}}$ $(H_3N)_5Co^{III}$—OH $[\ ]^{2+}$ + R—C$_6$H$_4$—CO_2^-

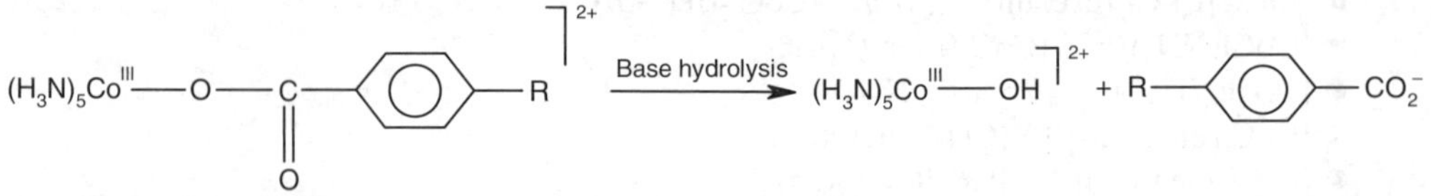

Use of Hammett equation and determination of Hammett reaction constant (ρ) for different R–groups can decide whether the base hydrolysis occurs through the Co—O or O—C bond rupture.

143. $(H_3N)_5Co^{III}$—O—C(=O)—R $\quad[\ ]^{2+}$ $\xrightarrow{\text{Base hydrolysis}}$ $(H_3N)_5Co^{III}$—OH $[\ ]^{2+}$ + R—CO_2^-

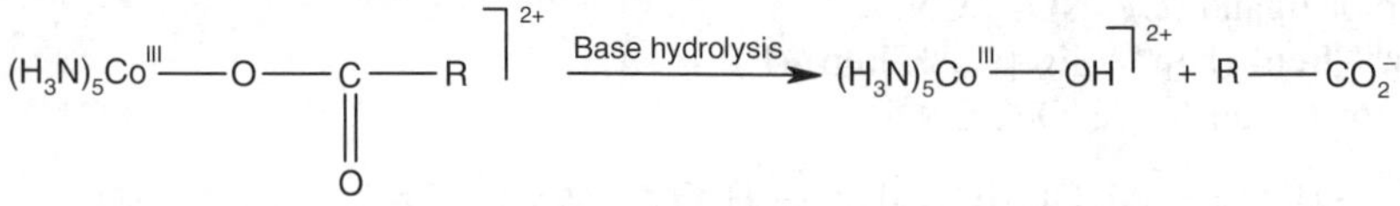

(R = alkyl groups)

Use of Taft relationship and determination of Taft reaction constant (ρ^*) can decide whether the Co—O or C—O bond rupture occurs in the above mentioned base hydrolysis reaction.

144. Solvent effect (*i.e.* dielectric constant effect) on the rate constant can tell us the charge bearing properties of the kinetically active species.

● $\log k_\varepsilon$ vs. $\dfrac{1}{\varepsilon}$ is linear (ε = dielectric constant of the medium) and sign of the slope depends on

the sign of the product of charges of the reactants in a bimolecular process.

145. $L_5M - OH_2 + Y \rightarrow L_5M - Y + H_2O$
The above formation reaction occurring through the D or I_a path will slow the rate dependence on the nature of Y. But the D– and I_a paths can be distinguished by analysing the rate dependence pattern on [Y].

146. $[Co(NH_3)_5 - X]^{2+} + H_2O \rightarrow [Co(NH_3)_5(OH_2)]^{3+} + X^-$
Rate $\propto [Co(NH_3)_5(X)^{2+}]$
Different mechanistic paths can lead to the above rate law but they can be discriminated by considering the factors controlling the rate.

147. The aquation rates follow the following sequences:
● cis–$[CoCl_2(NH_3)_4]^+ \rangle$ $trans$–$[CoCl_2(NH_3)_4]^+ \rangle\rangle$ $[CoCl(NH_3)_5]^{2+}$.
● $trans$–$[CoCl_2(N - N)_2]^+$:
N — N = $meso$ — bn $\rangle$ dl — bn $\rangle$ pn $\rangle$ en.
● $trans$–$[CoCl_2(tet - b)]^+ \rangle\rangle$ $trans$–$[CoCl_2(cyclam)]^+$

148. cis–$trans$–$[CoCl(en)_2(L)]^{n+}$ aquates (*i.e.* replacement of Cl$^-$ by H_2O)
● $k_{cis} \rangle k_{trans}$ if L is a π–donor ligand, *e.g.* Cl$^-$, OH$^-$, NCS$^-$;
● aquation of cis–isomer leads to cis–product while aquation of $trans$–isomer leads to a mixture of cis– and $trans$-isomers.
● $k_{cis} \approx k_{trans}$, if L is a weakly π–donor ligand (*e.g.* H_2O, N_3^-) and L is a non-π-bonding ligand (*e.g.* NH_3); for L = NH_3 there is a 100% retention of configuration for both isomers.

$k_{trans} \rangle k_{cis}$, if L is a strong π–acceptor ligand (*e.g.* CN^-, NO_2^-) and there is a 100% retention of configuration for both the *cis*– and *trans*– isomer.

149. $[CoCl(en)_2(R—py)]^{2+}$ aquates and the aquation rate increases with the increase of basicity of the pyridine derivative.

150. $[Co(NH_3)_5X]^{2+}$ aquates and the aquation rate follows the order:

$$k_{ex} : \xrightarrow[\text{Ligand field strength}]{X = Br^- > F^- > NCS^- > NO_2^-}$$

- The LFER plot yields slope = 1.0
- ΔH_T (= $\Delta H - \Delta H^{\neq}$) remains more or less constant for different leaving groups and it is close to $\Delta H^{\neq}$ for the water exchange process.

151. $$\left[Co(NH_3)_5\left(O—\underset{\underset{O}{\|}}{C}—R\right)\right]^{2+} \xrightarrow{H_2O} \left[Co(NH_3)_5(OH_2)\right]^{3+} + RCO_2^-$$

Basicity of RCO_2^- covers a huge range (*ca.* by a factor 10^4) while the aquation rate covers a narrower span (*ca.* by a factor 10).

- $k_{aq} : \xrightarrow[\text{Increasing basicity}]{CF_3CO_2^- > CH_2ClCO_2^- > CH_3CO_2^- > Me_3CCO_2^-}$

- It supports an interchange pathway (*i.e.* I_a) where the nucleophilic attack by H_2O plays an important role to give the transition state.
- If a better nucleophile like OH^- becomes available, then the span of rate constants will cover a larger range.

152. Aquation of *trans*–$[Co(en)_2(NH_3)(X)]^{2+}$ ($X^- = NO_3^-, Cl^-, Br^-$)
k_{aq}: $NO_3^- \rangle Br^- \rangle Cl^-$; $\Delta H^{\neq}$: $NO_3^- \rangle Br^- \rangle Cl^-$; $\Delta S^{\neq}$: $NO_3^- \rangle Br^- \rangle Cl^-$
K_{aq}: $NO_3^- \rangle Cl^- \rangle Br^-$; $\Delta H = Br^- \rangle Cl^- \rangle NO_3^-$
Hints: k_{aq} sequence is controlled by $\Delta S^{\neq}$; K_{aq} is controlled by ΔH; consider SAD mechanism.

153. Aquation of $[Cr(NH_3)_5X]^{2+}$: $X^- = Cl^-, Br^-, I^-$
k_{aq}: $I^- \rangle Br^- \rangle Cl^-$; $\Delta S^{\neq}$: $I^- \rangle Br^-, I^-$
Hint: SAD—solvent assisted dissociation mechanism.

154. $[Co(NH_3)_5X]^{n+} + H_2O \rightarrow [Co(NH_3)_5(OH_2)]^{3+} + X^{m-}$
$X^{m-} = Cl^-, Br^-, NCS^-, SO_4^{2-}$
$\Delta V^{\neq}$ (≈ ΔV) = –ve; $\Delta V^{\neq}$ (k_{ex} for $[Co(NH_3)_5(OH_2)]^{3+}$) = +ve.
In spite of the negative value of $\Delta V^{\neq}$, the process can be explained in terms of a dissociative path.
- The negative value of $\Delta V^{\neq}$ may be also explained in terms of an associative activation. However, this argument may be ruled out by comparing the $\Delta V^{\neq}$ values with the ΔV values.

155. $[Co(L)(NH_3)_5]^{3+} + H_2O \rightarrow [Co(NH_3)_5(OH_2)]^{3+} + L$
L = neutral leaving group like $O{=}SMe_2$, $O{=}CH(NH_2)$; $O{=}C(NH_2)_2$, etc.
$\Delta V^{\neq}$ (≈ΔV) = +ve
The observation is in conformity with the dissociative activation.

156. *cis–/trans*–$[Co(Cl)(en)_2(L)]^{n+} + H_2O \rightarrow [Co(en)_2(L)(OH_2)]^{n+1} + Cl^-$
L = nonleaving group
- L = π–donor ligand (*e.g.* Cl^-, NH_2^-, NCS^-), there is 100% retention of configuration for the *cis*–complex; tendency of isomerisation for the *trans*–complex: L = $OH^- \rangle NCS^- \rangle Cl^-$

● L = π–acceptor ligand (*e.g.* NO_2^-, CN^-), a non-π-bonding ligand (*e.g.* NH_3) there is a 100% retention of configuration for both the *cis*– and *trans*– complexes.

● *cis*–complex experiences 100% retention of configuration irrespective of the nature of the nonleaving group L.

157. $[CoCl(L)(N_4)]^{n+} + H_2O \rightarrow [CoL(N_4)(OH_2)]^{(n+1)+} + Cl^-$
N_4 = cyclam or tet–*b*; L = nonleaving group;
There is a 100% retention of configuration irrespective the nature of nonleaving group (L).

158. Aquation of $[Co^{III}Cl(L)(N—N)_2]^{n+}$ to $[Co^{III}(L)(N—N)_2(OH_2)]^{(n+1)+}$
● Isomerisation is accompanied with the positive $\Delta S^{\neq}$ value.
● Retention of configuration is accompanied with the relatively less and negative $\Delta S^{\neq}$ value.

159. In some cases, aquation of *cis*–$[Co^{III}Cl(L)(N - N)_2]^{n+}$ may lead to isomerisation into the *trans*–complex to some extent.

160. Aquation of *cis*–/*trans*– $[CoCl(en)_2(L)]^{n+}$:
If the dissociative path operates, then the following stereochemical results are expected.
● Λ–*cis*–complex leads to Λ–*cis*–product (58.3%), Δ – *cis* – product (25%) and *trans*–product (16.7%), *i.e.* loss of optical activity due to both racemisation and isomerisation (considering the long-lived trigonal bipyramidal intermediate).
● *trans*–complex gives 66.6% *cis*–product and 33.3% *trans*–product (considering the long lived trigonal bipyramidal intermediate).

161. Aquation of $[Cr(NH_3)_5X]^{2+}$ (X = Cl, Br, I, etc.)
● k_{aq}: $I^- \rangle Br^- \rangle Cl^-$ and $\Delta S^{\neq}$: $I^- \rangle Br^- \rangle Cl^-$
Hint: SAD path
● Group replacement factor, $GRF\left(=\dfrac{k_{R-X}}{k_{R-Cl}}\right)$: $I^- > Br^- > Cl^- >> F^-$
(**Hint:** Importance of bond breaking in the activation process)

162. Aquation of $[MCl(NH_3)_5]^{2+}$ (**I**) and $[MCl_2(NH_3)_4]^+$ (**II**)

$$\frac{k_{aq}(\mathrm{II})}{k_{aq}(\mathrm{I})} \approx 10^3\,(\mathrm{M = Co}),\ \approx 10^1\,(\mathrm{M = Cr})$$

163. Aquation of $[MCl(NH_3)_5]^{2+}$ (M = Co, Cr)
The plot of $\Delta V^{\neq}$(Co) *vs.* $\Delta V^{\neq}$(Cr) is linear, but the slope is far away from unity.

164. Hydrolysis of the Co(III)–ammine complexes show a two term rate law.
rate = k_A[complex] + k_B[complex][OH^-]
● The first term (*i.e.* first-order path) is important at pH < 8.0 while the 2nd term (*i.e.* second order path) is important at pH 〉 8.0.

165. An electrophile can also catalyse the aquation of $[Co(N_5)—X]^{2+}$.

166. The complexes $[M^{III}(NH_3)_5X]^{n+}$ can experience an acid catalysed aquation (*M* = Co, Cr; X^{m-} = F^-, NO_2^-, N_3^-, CN^-, CO_3^{2-}, SO_4^{2-}, ox^{2-}, $acac^-$, etc.)

167. $[M^{III}(NH_3)_5X]^{2+}$ (M = Co, Cr) can experience an acid catalysed aquation for $X^- = F^-$ but not for $X^- = Cl^-$, Br^-, I^-.

168. $[Co(bigH)_3]^{3+}$ is thermodynamically more stable than $[Co(en)_3]^{3+}$, but in acidic condition, $[Co(bigH)_3]^{3+}$ dissociates faster than $[Co(en)_3]^{3+}$.

169. The acid catalysed aquation rates of the biguanide complexes of Co(III) follow their thermodynamic stability orders.
● k_{aq} and thermodynamic stability order: biguanide 〉 phenyl-biguanide.

170. In acid catalysed aquation of $[M(bpy)_3]^{2+}$ (M = Ni, Fe), the rate process shows the acid dependence while the corresponding phen-complex shows no acid dependence of rate under the comparable conditions.

171. In acidic condition, $[M(phen)_3]^{2+}$ (M = Fe, Ni) experiences aquation. The rate of aquation process is independent of $[H^+]$ but the aquation can only occur in acidic condition.

172. Stepwise aquation (acid catalysed):

$$[Fe(N-N)_3]^{2+} \xrightarrow{k_1} [Fe(N-N)_2(OH_2)_2]^{2+} \xrightarrow{k_2} [Fe(N-N)(OH_2)_4]^{2+} \xrightarrow{k_3} [Fe(OH_2)_6]^{2+}$$

$k_2, k_3 \rangle\rangle k_1$, N—N = bpy, phen

- For the corresponding Ni(II)–complex, no such step-wise rate order is noticed.

173. In the bpy–complexes, though the acid catalysed one-ended dissociation of the chelate ring is possible, it is not possible for the corresponding phen-complex.
- For the dissociation of $[Fe(bpy)_3]^{2+}$ and $[Fe(phen)_3]^{2+}$, the acid dependence patterns are different.
- For $[Fe(bpy)_3]^{2+}$, the limiting values of k_{obs} at very high $[H^+]$ and low $[H^+]$ can tell us the kinetic behaviour of the 5-coordinate intermediate.

174. Different metal ions (as electrophiles) can catalyse the aquation of metal complexes bearing the chelating ligands like oxalate, malonate, acetylacetonate, aminopolycarboxylate, etc.
- The catalysing efficiency of the metal ions depends on mainly two factors: HSAB matching and complexing power.

175. $$[Co(NH_3)_5X]^{2+} \xrightarrow[(^{18}O \text{ enriched water solvent})]{Hg^{2+}} [M(NH_3)_5(OH_2)]^{3+}$$

The product bears the same $^{16}O/^{18}O$ ratio for $X^- = Cl^-, Br^-, I^-$.
(**Hint:** D–process)
- If the process is catalysed by the metal ions like Ag^+ or Tl^{3+}, the $^{16}O/^{18}O$ ratio in the product is found to depend on the nature of X^-.

(**Hint:** I_d–process)

176. $$[(H_3N)_5M-{}^*OH_2]^{3+} + HCO_3^- \rightleftharpoons [(H_3N)_5M(-{}^*O-CO_2^-)]^+ + H^+ + H_2O$$

M = Co(III), Rh(III), Ir(III).
- The above mentioned reactions are basically the pseudo-substitution reactions.
- The forward (*i.e.* formation reaction) and backward (*i.e.* aquation reaction) rate constants are more or less comparable for Co(III), Rh(III) and Ir(III).

176. $$[M(NH_3)_5X]^{(3-n)+} \xrightarrow{\text{aquation}} [M(NH_3)_5(OH)_2]^{3+} + X^{n-}$$

- $X^{2-} = CO_3^{2-}$: k_{aq} values vary within a very narrow span for M = Co(III), Rh(III) and Ir(III).
- $X^- = Br^-, NO_3^-, CH_3CO_2^-$; $X^{2-} = SO_4^{2-}$: k_{aq} values cover a wide span for M = Co(III), Rh(III) and Ir(III).

177. The formation rate of $[M(CO_3)(NH_3)_5]^{2+}$ and its aquation rate are unexpectedly high for M = Co(III), Rh(III) and Ir(III).

178. When $[Co(NH_3)_5(OH_2)]^{3+}$ is treated with $HNO_2 - NO_2^-$ buffer, it rapidly produces a red coloured solution which on standing slowly changes into the yellow coloured solution.

179. $$[M(NH_3)_5(OH_2)]^{3+} \xrightarrow[(1st\ step)]{HNO_2-NO_2^-} [M(NH_3)_5(ONO)]^{2+} \xrightarrow[(2nd\ step)]{} [M(NH_3)_5(NO_2)]^{2+}$$

M = Cr, Co, Rh, Ir.

● The first-step leading to the nitrito isomer is unusually fast (compared to the expected inertness of the metal).

● The 2nd step leading to isomerisation of the nitrito isomer to the nitro isomer is a slow reaction and the corresponding rate is compatible with the expected CFAE of the metal centre.

180. Cl^-, Br^-, etc. can catalyse the aquation of $[Co(NH_3)_5(ONO)]^{2+}$ in acidic condition.

● $$\left[(H_3N)_5Co(ONO)\right]^{2+} \xrightarrow{H_2O^*,\ H^+,\ X^-} \left[(H_3N)_5Co(OH_2)\right]^{3+} + HNO_2 + HX$$

(No incorporation of *O in the product).

181. $$\left[Co(NH_3)_5X\right]^{2+} \xrightarrow{^-OH} \left[Co(NH_3)_5(OH)\right]^{2+}$$

Rate $\propto$ [complex][OH^-]

● The said rate law can be attained by considering both the A–process and D–CB process.

182. In the base hydrolysis of $[Co(NH_3)_5X]^{2+}$, though it passes through a preequilibrium step involving OH^-, it never shows the saturation kinetics or fractional order in [OH^-].

183. Base hydrolysis of $[Co(NH_3)_5X]^{2+}$ occurs through the formation of a conjugate base (CB). The CB may be considered either to remain in a preequilibrium step or in a steady state. Both the concepts lead to the same rate law:
rate $\propto$ [complex][OH^-].

184. Base hydrolysis of $[Co(NH_3)_4(L)X]^{n+}$ leads to isomerisation for both the *cis*– and *trans*– isomers while acid hydrolysis of the complex leads to 100% retention of configuration for the *cis*–isomer and isomerisation in the *trans*–complex depending on the nature of the nonleaving group L.

185. Base hydrolysis of $[Co(NH_3)_4(L)X]^{n+}$ occurs at comparable rates for both the *cis*– and *trans*– isomers. On the other hand, in acid hydrolysis, the *cis*– and *trans*–isomer generally react at different rates and this rate difference largely depends on the nature of the nonleaving group L.

186. In acid hydrolysis of $[Co(NH_3)_5X]^{2+}$ or $[Co(L)(NH_3)_4X]^{n+}$, the 5–coordinate intermediate/activated complex may adopt the square pyramidal or trigonal bipyramidal geometry depending on the conditions but in base hydrolysis, the corresponding 5–coordinate intermediate/activated complex is always a trigonal bipyramidal.

187. If base hydrolysis of $[Co(NH_3)_5X]^{2+}$ is carried out in presence of other nucleophiles, there is a mixed product, *i.e.*$[Co(NH_3)_5(Y)]^{2+}$, though there is no rate dependence on [Y].

$$\left[Co(NH_3)_5X\right]^{2+} \xrightarrow{^-OH,\ Y^-} \left[Co(NH_3)_5(OH)\right]^{2+} + \left[Co(NH_3)_5Y\right]^{2+}$$

$$\left(Y^- = N_3^-,\ NCS^-,\ \text{etc.}\right)$$

rate $\propto$ $[Co(NH_3)_5X^{2+}][OH^-][Y^-]^0$

● Without OH^-, the entry of Y^- occurs very slowly.

● Deprotonation by OH^- preferably occurs from the NH_3 group *trans* to X (leaving group).

● The $-NH_2$ group in the CB favours the dissociative activation by its both σ- and π- donor property; but the σ-donor property is more important than the π-donor property of the $-NH_2$ group.

188. The Co(III) complexes are more inert than the Cr(III)–complexes, but base hydrolysis of $[Co(NH_3)_5X]^{2+}$ is faster than that of $[Cr(NH_3)_5X]^{2+}$.

● In terms of MOT and energy of the LUMO, an associative process is more disfavoured in the low spin Co(III) complexes than in the Cr(III) complexes.

189. Aquation of $[Co(NH_3)_5X]^{2+}$ is slower than that of $[Cr(NH_3)_5X]^{2+}$ but the reverse is true for base hydrolysis (X = Cl).
190. Base hydrolysis of $[CoCl(NH_3)_5]^{2+}$ is much faster than that for $[Co(CN)_5Cl]^{3-}$ or $[CoCl(py)_5]^{2+}$.
191. Base hydrolysis of $[CoCl(NH_3)_5]^{2+}$ occurs through the D-CB path while base hydrolysis of $[CoCl(py)_5]^{2+}$ occurs through the I_d–path.
   - Base hydrolysis of $[CoCl(NH_3)_5]^{2+}$: rate $\propto$ [complex]$[OH^-]$
   - Base hydrolysis of $[CoCl(py)_5]^{2+}$: rate $\propto$ [complex]$[OH^-]^0$
   - Base hydrolysis of $[CoBr(CN)_5]^{2+}$: rate $\propto$ [complex]$[OH^-]^0$
192. For base hydrolysis of $[Co(NH_3)_5X]^{2+}$ or $[Co(NH_3)_4X_2]^+$, there are evidences to support the I_d–CB mechanism rather than the D–CB mechanism.
193. For $[Co(NH_3)_5X]^{2+}$, the base hydrolysis rate constant is much faster than the acid hydrolysis rate constant. This rate enhancement cannot be explained by simple consideration of the better nucleophilic character of the OH^- group than that of H_2O.
194. $$\left[Co(NH_3)_5X\right]^{2+} + Y^- \xrightarrow{\text{(in DMSO)}} \left[Co(NH_3)_5(Y)\right]^{2+} + X^-$$

   The reaction is tremendously slow but it can be potentially catalysed by a trace amount of base (B) like pipyridine or OH^-. The rate law is:
   rate $\propto$ [complex][B]$[Y^-]^0$.
195. $$\left[Co(NH_3)_5X\right]^{n+} \xrightarrow{SCN^-,\ OH^-} \left[Co(NH_3)_5(OH)\right]^{2+} + \left[Co(NH_3)_5(Y)\right]^{2+}$$

   X^{m-} = $OSMe_2$, Cl^-, Br^-, OSO_3^{2-} etc., Y = N– or S–bonded SCN^-
   - It is evident that with the increase of negative charge of the leaving group, % of $[Co(NH_3)_5(Y)]^{2+}$ decreases.

X^{m-} :	$OSMe_2$,	I^-	OSO_3^{2-}
% of $[Co(NH_3)_5Y]^{2+}$:	~20%	~13%	~7%

   - This observation can be explained by the I_d–CB not by the A–mechanism.
   - The above observation indicates that the base hydrolysis occurs through the I_d–CB path not through the D–CB path.
196. Anation of $[Co(NH_3)_5X]^{2+}$:
   $$[Co(NH_3)_5X]^{2+} + Y^- \rightarrow [Co(NH_3)_5(Y)]^{2+} + X^-$$
   The said anation reaction occurs in different paths in acidic and basic conditions.
197. $[Co(N)_5X]^{2+}$ experiences *steric acceleration* in both acid hydrolysis and base hydrolysis.
   - $[M(N)_5X]^{2+}$ (M = Cr, Rh, Ir) experiences steric retardation in acid hydrolysis while steric acceleration is noted in their base hydrolysis.
   - $(N)_5$ denotes the amine ligands differing in steric crowding; X^- = $-OSO_2CF_3$.
198. Base hydrolysis of $[Co(NH_3)_5X]^{2+}$ is retarded by the addition of H_2O_2.
   - In base hydrolysis, addition H_2O_2 produces a better nucleophile HO_2^- at the cost of OH^-, but still the process is retarded.
199. There are several factors to contribute to $\Delta V^{\neq}$ and $\Delta S^{\neq}$ in base hydrolysis of $[Co(NH_3)_5X]^{2+}$ through the D–CB mechanism.
200. Base hydrolysis of $[Co(NH_3)_5X]^{2+}$ is accompanied with the positive $\Delta V^{\neq}$ values.
   - If the base hydrolysis is carried out in presence of ^{18}O-labeled water, the ($^{18}O/^{16}O$) ratio in the product matches with that of water but not with that of hydroxide.
   - The reactivities of two isomers (red and purple) of $[CoCl(NH_3)(tren)]^{2+}$ in base hydrolysis are different, *i.e.* the red isomer reacts faster than the purple isomer.

- The base hydrolysis products of the red and purple isomers of $[CoCl(NH_3)(tren)]^{2+}$ are different.

201. $$[M(N)_5X]^{2+} \xrightarrow[(k_H)]{\text{acid hydrolysis}}$$
$$[M(N)_5X]^{2+} \xrightarrow[(k_{OH})]{\text{base hydrolysis}}$$

- For M = Co, $\ln k_H$ and $\ln k_{OH}$ bear a linear relationship
- For M = Cr, Rh and Ir; there is no linear relationship between $\ln k_H$ and $\ln k_{OH}$.
- Stereochemical results for the aquation and base hydrolysis of $[CoCl_2(en)_2]^+$ are different.

(**Hints:** Electronic effect (σ- and π-donor property) of the 'NH' group of CB; long-lived 5-coordinate intermediate in base hydrolysis).

202. *cis-/trans-*$\left[Co(en)_2(L)X\right]^{m+} \xrightarrow{\text{base hydrolysis}}$ Isomerisation for both *cis-* and *trans-*isomers.

203. $\left[M(NH_3)_5X\right]^{2+} \xrightarrow{\text{Base hydrolysis}}$

- The base hydrolysis through the *D*–CB path can never show the saturation kinetics or mixed order kinetics with respect to $[OH^-]$.
- The *D*–CB and *D*–IP paths are the two limiting situations of a common intermediate. The choice of the mechanism depends on the protonic character of the N—H bond and it depends largely on the electron pushing or electron withdrawing character of X^-.

204. $\left[Cr(NH_3)_5X\right]^{2+} \xrightarrow{\text{Base hydrolysis}}$

- $X^- = NCS^-$, it occurs through the D–IP mechanism, *i.e.* mixed order kinetics or saturation kinetics in $[OH^-]$.
- $X = N_3^-$, it occurs through the D–CB mechanism, *i.e.* no mixed-order kinetics or saturation kinetics in $[OH^-]$.

205. Base hydrolysis of *trans*–$[Rh(en)_2X_2]^+$

X^-: I^- SCN^- N_3^- Br^- Cl^-

$\xrightarrow{\hspace{3em}}$ favouring D–CB path

$\xleftarrow{\hspace{3em}}$ favouring D–IP path

206. Though $[M(OH_2)_6]^{3+}$ and $[M(OH)(OH_2)_5]^{2+}$ are correlated in terms of a protonation-deprotonation equilibrium, but their reactivities in ligand substitution reactions are totally different (M = Fe, Cr).

207. Isomerisation: (in both solid and solution phase)
$$\underset{\text{(Red)}}{\left[Co(NH_3)_5(ONO)\right]^{2+}} \xrightarrow{\text{heating}} \underset{\text{(yellow)}}{\left[Co(NH_3)_5(NO_2)\right]^{2+}}$$

The process passess through an intramolecular path having the negative $\Delta V^{\neq}$ values (*cf.* Molar volumes for the *nitro–* and *nitrito–*somers are: 69.4 $cm^3\ mol^{-1}$ and 82.3 $cm^3\ mol^{-1}$ respectively)

208. Isomerisation: $[Co(NH_3)_5(SCN)]^{2+} \rightarrow [Co(NCS)(NH_3)_5]^{2+}$. It occurs through an intramolecular path ($\Delta V^{\neq}$ = –ve).

209. $M—ONO \rightleftharpoons M—NO_2$; $M—SCN \rightleftharpoons M—NCS$

The S_Ni–path (*i.e.* intramolecular path) is more probable for the nitrito-nitro isomerisation, than for the thiocynato-isothiocynato isomerisation.

210. From the solution obtained by the interaction of dil. HCl with $[Co(CO_3)(en)_2]^+$, two different isomers (*i.e.* green and violet coloured) may be isolated depending on the conditions.

211. *cis*–$[CoCl_2(en)_2]^+$ slowly isomerises to its *trans*–isomer in solution through an intermolecular path.

- The rate of isomerisation is close to the rate of Cl^{*-} exchange reaction.
- The loss of optical activity of *cis*-$[M(A—A)_2X_2]$ in solution may occur in two ways.
- The rate of loss of optical activity of *cis*-$[M(A—A)_2X_2]$ may be close to or greater than the rate of geometrical isomerisation (*i.e.* *cis* → *trans* isomerisation) depending on the condition.

212. If *cis*–$[Co(N-N)_2(X)(Y)]^{n+}$ (optically pure) undergoes isomerisation in a dissociative process through a long-lived trigonal bipyramidal intermediate, it will yield the following results:

$$cis\text{–}(ca.\ 83\%)\ \text{and}\ trans\text{–}(ca.\ 17\%)$$

$$\downarrow$$

$$\underbrace{50\%\ (\text{racemic}) + 33\%\ (\text{retention})}_{58\%\ (cis\text{-retention}) + 25\%\ (cis\text{-inversion}) + 17\%\ (trans)}$$

- If *trans*– $[Co(N-N)_2(X)(Y)]^{n+}$ undergoes isomerisation in a dissociative process through a long-lived trigonal bipyramidal intermediate, it will give the following results:

$$\underbrace{cis\text{-}(ca.\ 67\%)}_{\text{Racemic mixture}}\ \text{and}\ trans.\ (ca.\ 33\%)$$

213. Isomerisation: $trans\text{-}\left[Cr(AA)_2(OH_2)_2\right]^- \rightleftharpoons cis\text{-}\left[Cr(AA)_2(OH_2)_2\right]^-$

For A — A = malonate, $\Delta V^{\neq}$ is positive

For A — A = oxalate, $\Delta V^{\neq}$ is negative.

- The *cis-trans* isomerisation of $[Cr(C_2O_4)_2(OH_2)_2]^-$ can be catalysed by the bivalent metal ions (*e.g.* Cu^{2+}, Ni^{2+}, Co^{2+}, Ca^{2+}, etc.)
- The isomerisation and racemisation of *cis*-$[Cr(C_2O_4)_2(OH_2)_2]^-$ show $\Delta V^{\neq}$ = +ve (in neutral or slightly acidic condition), and –ve (in acidic condition).

214. Isomerisation of $[Co(CH_3COO)(en)_2(OH_2)]^{2+}$ is accompanied with positive $\Delta V^{\neq}$.

215. Racemisation of *cis*–$[M(L-L)_2X_2]$ occurs through the dissociative transformation to a trigonal bipyramidal geometry with the expulsion of one monodentate ligand X.

216. For $[Ni(N-N)_3]^{2+}$, the rate of racemisation, the rate of dissociation of *tris*– to *bis*–complex, and the rate of ligand exchange are comparable.
(N — N = bpy, phen)

217. For $[M(N-N)_3]^{2+}$, the rate of racemisation is greater than the rate of dissociation (*tris*– → *bis*–).
M = Fe, Co; N — N = bpy, phen.

218. Racemisation of $[M(C_2O_4)_3]^{3-}$ (M = Co, Cr, Rh).

- Rate of racemisation is greater than rate of oxalate exchange.
- Rate of recemisation is more or less comparable to the O–exchange in acidic condition in ^{18}O–enriched water.

- The noncoordinating 6 oxygen atoms of $[M(C_2O_4)_3]^{3-}$ (M = Co, Rh) exchange faster than the remaining 6 oxygen atoms.

219. $K_3[Cr(C_2O_4)_3] \cdot 2H_2O$ can undergo racemisation when it is heated in a sealed tube but the racemisation is not at all efficient if heated in an open tube.
220. $[Co(edta)]^-$ undergoes racemisation in alkaline solution and the process is characterised by both the high positive values of $\Delta H^{\neq}$ and $\Delta S^{\neq}$.
221. For the $[M(A — A)_3]$ type chelates, Ray-Dutt twist leads to a transition state of C_{2v} symmetry while Bailar twist leads to a transition state of D_{3h} symmetry.
   - Ray-Dutt twist is described as rhombic or tetragonal twist and Bailar twist is described as a trigonal twist.
222. From the result, it is not possible to identify the Bailar twist or Ray-Dutt twist for the $[M(A—A)_3]$ type chelates but these can be distinguished for the $[M(A — B)_3]$ type chelates.
223. In the $[M(A — A)_3]$ or $[M(A — B)_3]$ type chelates, racemisation through the Bailar twist is favoured if the bite distance of the chelating ligands is small.
   Hint: In such cases, the octahedral geometry is already distorted towards the required transition state.
   - Ray-Dutt twist is favoured for the chelating ligands of large bite distance.
224. *cis*–$[MCl_2(en)_2]Cl$ (M = Co, Cr) undergoes racemisation in solid state when heated without experiencing any geometrical isomerisation. The kinetic parameters (*i.e. k*, $\Delta H^{\neq}$, $\Delta S^{\neq}$) for the racemisation are more or less comparable for the Co(III)– and Cr(III)– complexes. These findings can be explained by the Ray-Dutt twist mechanism.
   - For *tris*-(acetylacetylacetonato)cobalt(III) (where one CH_3 group is substituted by CD_3), both the linkage isomerisation and racemisation go on simultaneously but the linkage isomerisation rate is relatively slower.
225. The racemisation of the Cr(III)–complexes like $[Cr(bpy)_3]^{3+}$, $[Cr(phen)_3]^{3+}$, $[Cr(bpy)_2(ox)]^+$, $[Cr(ox)(phen)_2]^+$ etc. experience small positive $\Delta V^{\neq}$ while their anionic complexes like $[Cr(ox)_3]^{3-}$, $[Cr(ox)_2(phen)]^-$, $[Cr(bpy)(ox)_2]^-$ experience negative $\Delta V^{\neq}$ during racemisation.
226. Racemisation of $[M(N - N)_3]^{2+}$, M = Fe, Ni; N — N = bpy or phen.
   - Rate of dissociation (*tris* → *bis*) ≈ Rate of racemisation, (M = Ni)
   - Rate of dissociation (*tris* → *bis*) ⟨ Rate of racemisation, (M = Fe)
   - $[Fe(N - N)_3]^{2+}$ can partly racemise through the twist mechanism but it is not possible for $[Ni(N - N)_3]^{2+}$
   - In strongly acidic condition, the residual rate of racemisation of $[M(N—N)_3]^{2+}$ is probably due to the twist mechanism.
   - $k_{rac} \rangle k_{diss}$ indicates $k_{rac} = k_{rac\ (inter)} + k_{rac\ (intra)}$. The intramolecular path of racemisation for the bpy-complex can occur in two ways but for the phen-complex, it can occur only in one way.
227. There are many evidences that the d^8–systems can form the stable 5-coordinate complexes.
228. $ML_3 - X + Y \rightarrow ML_3 - Y + X$
   (ligand substitution process at the square planar complexes of Pt(II))
   It shows a two term rate law:
   rate = k_s [complex] + k_Y [complex][Y]
   - k_s–path depends on the nature of the solvent not on the nature of the entering ligand Y.
   - k_Y–path varies for the different entering ligands in a particular solvent.
   - For a particular entering ligand Y, the k_Y value may differ for different solvents.
229. The rate of ligand substitution process in a square planar complex of Pt(II) depends on both the nature of entering ligand and leaving ligand.

- The rate process is strongly dependent on the nature of entering ligand while the process is mildly dependent on the nature of the leaving group.
- The ligand exchange rate of $[PtX_4]^{2-}$ follows the order: $X^- = CN^- \rangle I^- \rangle Br^- \rangle Cl^-$

230. *Trans*–effect and *trans*–influence are the two characteristic phenomena of the square planar complexes of Pt(II).
231. Ligand exchange reaction:
trans–$[PtCl_2(*NHEt_2)(PPr_3)] + NHEt_2 \rightarrow$ *trans*–$[PtCl_2(NHEt_2)(PPr_3)] + *NHEt_2$
In a noncoordinating solvent like hexane, the k_s–path makes no contribution in the overall process while in a coordinating solvent like methanol, the k_Y–path makes a negligible contribution to the overall rate process.
232. In the ligand substitution process of square planar complexes of Pt(II), the k_s–path (*i.e.* solvent dependent path) involves the formation of an intermediate having a solvent molecule as a ligand. In some cases, the proposed intermediate can be trapped.

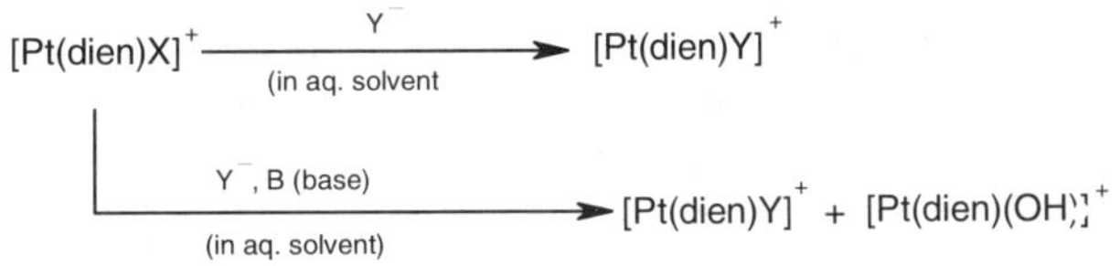

233. Compared to the Pt(II)–complexes, the Au(III)–complexes show the more associative character in the ligand substitution process.
234. A two term rate law operates for the following ligand substitution process.
$[MCl(dien)]^{n+} + Y^- \rightarrow [M(Y)(dien)]^{n+} + Cl^-$
Rate = k_s [complex] + k_Y [complex][Y^-]
$k_Y/k_s \rangle\rangle 1$ [for M = Au(III)]; $k_Y/k_s \approx 1.0$ [for M = Pt(II)].
235. Retention of stereochemical configuration is an important characteristic feature of the ligand substitution process in the square planar complexes of Pt(II).
236. Electrophilic substitution and oxidative addition followed by reductive elimination may also lead to ligand replacement in some square planar complexes of Pt(II).
237. Racemisation of $[M(N-N)_3]^{2+}$: M = Fe, Ni; N – N = bpy, phen.
At high acidity: $k_{rac} \approx 0$ for $[Ni(N-N)_3]^{2+}$; $k_{rac} \rangle 0$ for $[Fe(N-N)_3]^{2+}$.
- For $[Fe(N-N)_3]^{2+}$, racemisation can partly occur through the intramolecular twist mechanism but this remains absent for $[Ni(N-N)_3]^{2+}$.
- There are two possible intramolecular routes for racemisation of $[Fe(bpy)_3]^{2+}$ but for $[Fe(phen)_3]^{2+}$, there is only one possible intramolecular route of racemisation.
238. Nucleophilicity is a kinetic parameter while basicity is a thermodynamic parameter.
239. The nucleophilicity parameters n^0_{Pt} and $n^0_{CH_3I}$ do not run always parallel.
240. The nucleophilicity parameter n^0_{Pt} is defined as:
trans–$[PtCl_2(py)_2] + Y \rightarrow$ *trans*–$[PtCl(py)_2(Y)]^{n+} + Cl^-$

$$sn^0_{Pt} = \log\left[\frac{k_Y}{k_{CH_3OH}/[CH_3OH]}\right]$$

- Plot of log k_Y *vs.* n^0_{Pt} gives a straight line whose slope is given by s (nucleophilic discrimination constant).

● The larger value of *s* indicates that the reaction is more sensitive to the change of nucleophilicity. *s*–value of the reacting complexes changes as follows:

trans–$[PtCl_2(PEt_3)_2]$ 〉 *trans*–$[PtCl_2(py)_2]$ 〉 *trans*–$[PtCl_2(en)]$.

241. The ligands like CN^-, SCN^- are the poor leaving groups while the halides are the better leaving groups in the substitution reactions of square planar complexes of Pt(II).

242. $[PtCl(L)(PEt_3)_2]^+ + Y \rightarrow [Pt(L)(PEt_3)(Y)]^{n+} + Cl^-$
L = pyridine and its substituted derivatives.
● Both the *cis*– and *trans*–complexes show the steric retardation in the substitution process, *i.e.* with the increase of bulkiness of L, the rate decreases.
● The *cis*–complex is more sensitive to the steric crowding on L than the *trans*–complex in the said substitution reaction.

243. $[Pt(dien)X]^+ + Y^- \rightarrow [Pt(dien)(Y)]^+ + X^-$
$X^- = Cl^-, Br^-, I^-, N_3^-; Y^- = OH^-, I^-, N_3^-, NO_2^-$,
● $\Delta V^{\neq}$ is negative for both the k_s and k_Y paths; but $\Delta V^{\neq}$ is more negative for the k_s–path.
● The observations are in conformity with the I_a mechanism rather than with A–mechanism.

244. In the Pt(II)–square planar complexes, the ligand substitution reaction or ligand exchange reaction is almost insensitive to the change of overall charge of the complex.
● Cl*–exchange rate in $[PtCl(NH_3)_3]^+$, *trans*–$[PtCl_2(NH_3)_2]$ and $[PtCl_4]^{2-}$ covers the range $1.5 \times 10^{-5}\ s^{-1}$ to $3.8 \times 10^{-5}\ s^{-1}$, though the overall charge changes by 3 unit (+1 to −2). This observation is not in conformity with the pure–D or A–process.
● The ligand exchange rate for the $[PtX_4]^{2-}$ complexes follow the order:

$$[Pt(CN)_4]^{2-} \rangle [PtI_4]^{2-} \rangle [PtBr_4]^{2-} \rangle [PtCl_4]^{2-}$$

(**Hints:** Compare the relative effects of the entering group and leaving group on the rate process).

245. In terms of CFAE, formation of the square pyramidal geometry as the activated complex/intermediate in the ligand substitution process of the square planar Pt(II)–complexes is more feasible, but in reality, the process involves the trigonal bipyramidal geometry.

246. For the square planar complexes, the lability sequence runs as:
d^{10} 〉〉 d^9 〉 d^8 〉 d^7, *e.g.* Zn(II) 〉 Cu(II) 〉 Ni(II).

247. For the square planar complexes of d^8–system, the lability sequence runs as : Ni(II) 〉 Pd(II) 〉 Pt(II).

248. For the *cis–trans* isomerisation in the square planar complexes like $[PtCl_2(PR_3)_2]$:
● The isomerisation process is catalysed by a trace amount of PR_3.
● Ligand replacement process in the square planar complexes of Pt(II) is very much stereospecific. Hence, to explain the isomerisation process, we must consider a double displacement process.
● For the relatively more labile complexes $[MCl_2(PR_3)_2]$ (M = Pd, Ni), the *cis-trans* isomerisation can occur in solution even without the addition of any nucleophile.
● The corresponding Ni(II) complex may isomerise through an intramolecular path.

249. *Stereospecificity, trans*–effect and *trans*–influence — these are the three important characteristic features of the reactivity of square planar complexes of Pt(II).

250. The *trans*–effect series is constructed by comparing the relative rates of ligand substitution at the *trans*–position of the ligands.
● Thus the construction of *trans*–effect series is not absolute because the labilisation at the *trans*–position is determined by the leaving group, *trans*–directing group and incoming group jointly *in an interdependent process* in a particular complex.

● *trans*–$[PtCl(PEt_3)_2(T)] + Y^- \rightarrow$ *trans*–$[Pt(PEt_3)_2(T)Y] + Cl^-$
(i) For $Y^- = N_3^-$ or Br^-, *trans*–effect series of T: $CH_3^- \rangle C_6H_5^- \rangle Cl^-$
(ii) For $Y^- = I^-$, *trans*–effect series of T: $CH_3^- \rangle Cl^- \rangle C_6H_5^-$

251. $[PtCl_3(T)]^{n-}$; the Pt — Cl bond length changes as follows with the change of its *trans*-ligand (T). (*trans*– to T): $PEt_3 \rangle C_2H_4 \rangle Cl^-$
(**Hints:** It is due to the *trans*–influence).
252. *Trans*–influence (a thermodynamic effect) is referred to as the static *trans*–effect (STE).
253. Knowledge of both the *trans*–effect series and metal-ligand bond strength are required to design the routes of synthesis of different isomers of Pt(II)–square planar complexes.
254. The *cis/trans*–isomers of $[PtCl_2(NH_3)]$ can be distinguished by following their substitution reactions with thiourea and thiosulfate.
255. The *cis/-trans*–effect, *i.e.* lowering of activation energy can arise from two factors: ground state destabilisation, *i.e.* static kinetic effect or *trans*–influence and stabilisation of the transition state.
256. Grinberg's polarisation theory can explain the *trans*–influence better than the *trans*–effect.
● Grinberg's polarisation theory can explain the *trans*–effect series: $I^- \rangle Br^- \rangle Cl^-$; very high *trans*–directing power of H^- (assuming the *trans*-influence causes the *trans*-effect).
● Among Ni(II), Pd(II) and Pt(II), only the Pt(II) centre shows the *trans*–influence and *trans*–effect.
257. The π–bonding ligands like C_2H_4, CN^-, CO, etc., show a weak *trans*–influence but a strong *trans*–effect.
258. In terms of *trans*–influence, the ligands like H^- and H_2O differ marginally but their *trans*–labilising powers differ drastically.
259. The π–acid ligands C_2H_4, CO, CN^-, etc. are the very powerful *trans*–directing ligands.
● The π–acid ligands show both the *trans*–influence and *trans*–effect.
● The π–acid ligands are the poor leaving groups.
● The highly electronegative groups are the weak *trans*-directing ligands but the low electronegative groups are the better *trans*-directing ligands.
(**Hint:** *See* Cardwell's theory).
260. The strong σ–bonding ligands like H^-, CH_3^- etc. are the powerful *trans*–directing ligands.
● The σ–bonding ligands can also show the ground state *trans*–effect.
● Compared to the π-bonding theory, the σ-bonding theory is more general to explain the *trans*-effect of the Pt(II) complexes.
261. A *trans*–directing group can labilise (in general) its both *cis*– and *trans*–positions in the square planar complexes of Pt(II).
262. For a particular ligand, its *trans*–effect is more pronounced than its *cis*–effect in the square planar complexes.
● Static *trans*-effect (STE) bears a good correlationship with kinetic *trans*-effect (KTE) for the octahedral Co(III)-complexes but such a good correlationship is not maintained for the square planar complexes of Pt(II).
263. Origin of the *cis*–effect and *trans*–effect is the same.
264. *cis*–$[PtMe_2(SMe_2)_2]$ can experience the nonstereospecific ligand substitution reactions.
(**Hint:** Dissociative path through a trigonal planar intermediate).
265. ● $[Ni(CO)_4]$ experiences the ligand exchange:
rate ∝ $[Ni(CO)_4]$; $\Delta V^{\neq}$ and $\Delta S^{\neq}$ = +ve.
$\Delta H^{\neq}$ is close to the Ni—CO bond energy.

● $[FeBr_4]^-$ experiences the ligand exchange:

rate ∝ $[FeBr_4^-][X^-]$; $\Delta V^{\neq}$ and $\Delta S^{\neq} = -ve$

● The 18e–complexes like $[Ni(CO)_4]$, $[Ni(CO)_2(PR_3)_2]$ participate in dissociative paths.

● Very often, the nitrosyl–carbonyl mixed ligand complexes like $[Co(NO)(CO)_3]$, $[Fe(CO)_2(NO)_2]$ adopt the associative path.

● $[Ni(CO)_4]$ adopts the dissociative path while $[Co(NO)(CO)_3]$, $[Fe(CO)_2(NO)_2]$ adopt the associative path. This difference can be explained in terms of *effective electron count, metal-ligand bond strength and variable ligating behaviour of NO.*

● Some metal nitrosyl-carbonyl complexes like $[Co(CO)_2(NO)(PR_3)]$ can experience both the *dissociative* and *associative* paths simultaneously.

● Very often, it is a very difficult task to resolve the optically active pure tetrahedral complexes.

● Optically active tetrahedral complexes may undergo racemisation through both the intermolecular (both dissociative and associative) intramolecular (*i.e.* twisting) paths.

266. Macrocyclic effect (*i.e.* stability of the complexes for the macrocyclic ligand *vs.* open-chain analogous ligand) rests on both the entropic and enthalpic favour.

267. Macrocyclic complexes are more resistant (kinetically) towards dissociation compared to the corresponding open-chain complexes.

● $[Cu(cyclam)]^{2+}$ dissociates much more slowly than $[Cu(3, 2, 3\text{– tet})]^{2+}$ in acidic media.

● For the dissociation of $[Cu(cyclam)]^{2+}$, the acid dependent path is more important than in the case of $[Cu(3, 2, 3\text{– tet})]^{2+}$.

Inorganic Reaction Mechanisms and the World's Most Felicitated Coordination Chemists

Understanding of the reaction mechanism in organic chemistry is quite systematic but the inorganic chemists are yet to achieve that level. In fact, it is a very difficult task to systematize the versatile types of reactivities of more than one hundred elements. However to explore the reaction mechanisms of the metal complexes including the organometallics, a large number of world famous coordination chemists have contributed. Some of them are: M. Eigen (**Noble Prize, 1967**), F. Basolo, R.G. Pearson, R.G. Wilkins, H. Taube (**Nobel Prize, 1983),** H.B. Gray, C.H. Langford, R.A. Marcus (**Nobel Prize, 1992),** M.L. Tobe, J.H. Espenson, A.G. Sykes, R.D. Cannon, R. van Eldik, T.W. Saddle, A.W. Adamson, E.S. Gould, A.M. Sargeson, D.W. Maegerum, J.C. Bailar, P. Ray, etc.

CHAPTER 6

Reaction Mechanism: Electron Transfer Reactions and Photochemical Reactions of Metal Complexes

6.1 TYPES OF ELECTRON TRANSFER REACTIONS

- **Electron exchange reactions:** There are some electron transfer reactions in which the reactants and products are identical and there is **no net chemical change**. Such reactions are called **electron exchange reactions.** In such cases, $\Delta G^0 = 0$, *i.e.* equilibrium constant = 1. Such reactions can be studied by isotope labelling experiment. This is illustrated in the following example.

$$\left[Fe(OH_2)_6\right]^{3+} + \left[Fe^*(OH_2)_6\right]^{2+} \rightleftharpoons \left[Fe^*(OH_2)_6\right]^{3+} + \left[Fe(OH_2)_6\right]^{2+}$$

Examples of electron exchange reactions:

$[Fe(CN)_6]^{3-}$ – $[Fe(CN)_6]^{4-}$; $[Co(en)_3]^{3+}$ – $[Co(en)_3]^{2+}$; $[Co(phen)_3]^{3+}$ – $[Co(phen)_3]^{2+}$; $[Fe(bpy)_3]^{3+}$ – $[Fe(bpy)_3]^{2+}$; $[Os(bpy)_3]^{3+}$ – $[Os(bpy)_3]^{2+}$; $[Mo(CN)_8]^{3-}$ – $[Mo(CN)_8]^{4-}$

- **Net redox reactions:** There are some reactions where electron transfer leads to a **net chemical change.** Such reactions are commonly described as the **redox reactions** and for such reactions, $\Delta G^0 = -ve$, *i.e.* $K_{eq} \rangle 1$. It is illustrated in the following examples.

$$\left[Fe(phen)_3\right]^{3+} + \left[Fe(CN)_6\right]^{4-} \rightleftharpoons \left[Fe(phen)_3\right]^{2+} + \left[Fe(CN)_6\right]^{3-}$$

Examples of redox reactions

$[Mo(CN)_8]^{3-}$ – $[Os(bpy)_3]^{2+}$; $[IrCl_6]^{2-}$ – $[Fe(CN)_6]^{4-}$; $[Mo(CN)_8]^{3-}$ – $[Fe(CN)_6]^{4-}$; $[Fe(CN)_6]^{3-}$ – $[W(CN)_8]^{4-}$; $[Co(NH_3)_5Cl]^{2+}$ – $[Cr(OH_2)_6]^{2+}$; $[Ru(NH_3)_6]^{3+}$ – $[V(OH_2)_6]^{2+}$; etc.

6.2 MECHANISMS OF ELECTRON TRANSFER REACTIONS

There are two established mechanistic paths: **outer sphere mechanism** and **inner sphere mechanism**.

- **Outer sphere electron transfer (OSET) mechanism:** Electron transfer takes place from the reductant to the oxidant keeping the coordination spheres of both reactants intact. In this path, at no stage of the electron transfer process, the two reactive species remain chemically bonded through the coordination spheres.

  Such a mechanism generally operates for the substitutionally inert complexes.

- **Inner sphere electron transfer (ISET) mechanism:** During the electron transfer process, oxidants and reductants remain chemically bonded in a binuclear complex through a bridging ligand. Generally, the substitutionally inert oxidant centre provides the bridging ligand for making a bridge with the labile reductant centre. Thus in this route, the coordination spheres of both reactants are modified during the course of electron transfer.

Nobel Prizes: Electron Transfer Reactions

H. Taube: In understanding the mechanism of electron transfer process, **H. Taube** and his coworkers made the most significant contribution. This contribution was recognized by the **Nobel Prize in 1983 to H. Taube.**

R. A. Marcus: He was given the **Nobel Prize** in 1992 for his contribution to the theory of electron transfer reactions.

6.3 OUTER SPHERE ELECTRON TRANSFER (OSET) PROCESS

6.3.1 Steps of Outer Sphere Electron Transfer Reaction

For the substitutionally inert metal complexes, in the OSET process, the oxidant and reductant complexes never chemically bind through their coordination spheres. The electron transfer process may be explained in terms of a **simple collision model** in which the required steps are outlined below.

- **Step-I (Formation of a precursor cage or outer sphere complex):** The reactants *i.e.* oxidant (*Ox*) and reductant (*Red*) diffuse together to produce an **Outer Sphere** (O.S.) **complex** keeping their inner coordination spheres unchanged. This step is fast. In this O.S. complex, an electrostatic interaction as in ion-pair, ion-dipole interaction, dipole-dipole interaction, etc. can operate.

$$\text{Ox} + \text{Red} \rightleftharpoons [\text{Ox}\cdots\cdots\text{Red}],\ (\textbf{O.S. Complex})$$

- **Step II (Chemical activation of the precursor complex):** Bond distances and bond angles around each metal centre (*i.e.* in each inner coordination sphere) within the O.S. complex change to attain the geometrical parameters as in the products. *This step basically reorganizes (through the bond length and bond angle changes) the inner coordination spheres of each metal centre towards those of the products.*

At this step, the *solvent shell around the outer sphere complex is also reorganized*. This step may be represented as follows:

$$[\text{Ox}\cdots\cdots\text{Red}] \rightleftharpoons [\text{Ox}\cdots\cdots\text{Red}]^{\neq}\ (\textbf{Activated O.S. Complex})$$

The required reorganization energy needed for chemical activation of the reactants (*i.e.* oxidant and reductant) will be consumed for:

bond length changes; bond angle changes (if the stereochemical configurations of the reactants and products differ); ***spin state change*** (if the spin states of the products and reactants differ); ***solvent reorganization in the outer sphere.***

Obviously for electron exchange in the couple, $\mathbf{[Fe(CN)_6]^{3-} - [Fe(CN)_6]^{4-}}$ (where there is no spin state change; there is no change in geometrical configurations, *i.e.* both the oxidized and reduced centres are octahedral), chemical activation (*i.e.* reorganization) will **mainly involve the bond length adjustment.**

- **Step III (Electron transfer and relaxation to the successor complex):** Electron transfer occurs within the activated outer sphere complex $[\text{Ox}\cdots\text{Red}]^*$, *i.e.*

$$[\text{Ox}\cdots\cdots\text{Red}]^* \rightleftharpoons [\text{Ox}^-\cdots\cdots\text{Red}^+]\ (\textbf{Successor Complex})$$

● **Step IV (Dissociation of the successor complex to the products):** The products diffuse away from the successor complex and this step is generally fast.

$$\left[\mathrm{Ox}^{-}\cdots\cdots\mathrm{Red}^{+}\right] \rightleftharpoons \mathrm{Ox}^{-} + \mathrm{Red}^{+}$$

● **Illustration of the OSET process in the $[Fe(phen)_3]^{3+}$ – $[Fe(CN)_6]^{4-}$ system:** The above steps are illustrated in the outer-sphere oxidation of $[Fe(CN)_6]^{4-}$ by $[Fe(phen)_3]^{3+}$.

Formation of precursor complex (rapid step): The ion-pair formation is electrostatically highly favoured here.

$$\left[\mathrm{Fe(phen)_3}\right]^{3+} + \left[\mathrm{Fe(CN)_6}\right]^{4-} \rightleftharpoons \underbrace{\left[\mathrm{Fe(phen)_3}\right]^{3+}\cdot\left[\mathrm{Fe(CN)_6}\right]^{4-}}_{\textbf{O.S. precursor complex (ion-pair)}} ;(K_{OS}\,\text{very high})$$

Chemical activation followed by electron transfer (rds):

$$\left[\mathrm{Fe(phen)_3}\right]^{3+}\cdot\left[\mathrm{Fe(CN)_6}\right]^{4-} \rightleftharpoons \underbrace{\left[\mathrm{Fe(phen)_3}\right]^{2+}\cdot\left[\mathrm{Fe(CN)_6}\right]^{3-}}_{\textbf{Successor Complex}}$$

Fission of the successor complex (rapid step):

$$\left[\mathrm{Fe(phen)_3}\right]^{2+}\cdot\left[\mathrm{Fe(CN)_6}\right]^{3-} \rightleftharpoons \left[\mathrm{Fe(phen)_3}\right]^{2+} + \left[\mathrm{Fe(CN)_6}\right]^{3-}$$

● **OSET process in the $[Co(NH_3)_5(OH_2)]^{3+}$ – $[Fe(CN)_6]^{4-}$ system:** Formation of the O.S. complex (*i.e.* precursor complex) may be sometimes electrostatically favoured as in the present case. In the reduction of $[Co(NH_3)_5(OH_2)]^{3+}$ by $[Fe(CN)_6]^{4-}$, the K_{OS} value is as high as 1500 dm^3 mol^{-1}.

$$\left[\mathrm{Co(NH_3)_5(OH_2)}\right]^{3+} + \left[\mathrm{Fe(CN)_6}\right]^{4-} \rightleftharpoons \underbrace{\left[\mathrm{Co(NH_3)_5(OH_2)}\right]^{3+}\cdot\left[\mathrm{Fe(CN)_6}\right]^{4-}}_{\text{O.S. Complex}},\ K_{OS}\approx 1500\ \mathrm{dm^3mol^{-1}}$$

$$\xrightarrow{\text{e-transfer (slow, rds)}} \left[\mathrm{Co(NH_3)_5(OH_2)}\right]^{2+}\cdot\left[\mathrm{Fe(CN)_6}\right]^{3-} \xrightarrow{\text{fast}} \text{Products}$$

The rate constant of the said OSET reaction is 2×10^{-1} s^{-1}.

● **Energetics of electron exchange reaction:** In the *exchange reaction* (where $\Delta G^0 = 0$), the free energy of activation ($\Delta G^{\neq}$) may be expressed as follows:

$$\Delta G^{\neq} = \Delta G_r^{\neq} + \Delta G_i^{\neq} + \Delta G_s^{\neq}$$

$\Delta G_r^{\neq}$ = energy required to bring the oxidant and reductant at a state separated by the required distance r at which the electron transfer occurs; it will take care of **electrostatic interaction** between the reactants.

$\Delta G_i^{\neq}$ = energy required for **chemical activation,** *i.e.* to change the bond lengths and bond angles, *i.e.* energy required to reorganise the inner sphere of the reactants.

$\Delta G_s^{\neq}$ = energy required for **solvent reorganization.**

Formation of the precursor complex and specially the changes of bond length and bond angle in the precursor complex need the solvent reorganization. This solvent reorganization makes a significant contribution to $\Delta G^{\neq}$.

(**Note:** For the cross-reactions having $\Delta G^0 \neq 0$, ΔG^0 also contributes to $\Delta G^{\neq}$; *cf.* Marcus theory)

● **Rate law for the OSET path:** The process can be simply represented as follows:

$$\mathrm{Ox} + \mathrm{Red} \underset{}{\overset{K_{OS}}{\rightleftharpoons}} [\mathrm{Ox}\cdots\cdots\mathrm{Red}] \xrightarrow{k} \text{Products}$$

Here, k-stands for the rate constant of the overall process involving activation of the precursor complex, electron transfer within the activated precursor complex and dissociation to the products. In fact, the successor complex produced through electron transfer rapidly dissociates into the products. The rate law is given by:

$$\text{rate} \approx \frac{kK_{OS}[Ox][\text{Red}]}{1+K_{OS}[\text{Red}]}, \text{ ([Red]} \rangle\rangle \text{ [Ox]), } \textit{see} \text{ Sec. 5.9.3 for derivation of the rate law}$$

$$\approx kK_{OS}[\text{ox}][\text{Red}], \ (1 \rangle\rangle K_{OS}[\text{Red}], \textit{ i.e.} \text{ when } K_{OS} \text{ is small}).$$

6.3.2 Electron Transfer and Franck-Condon Principle: Chemical Activation to Overcome the Franck-Condon Barrier

(A) Activation energy to satisfy the demand of Franck-condon principle: Franck-Condon principle (*cf.* Fig. 7.4.1.3) indicates that the electronic motion is so rapid that during the electron transfer process, the nuclei including the metals and ligands do not have the time to move. Thus during the electron transfer process, there will be no change in atomic arrangements in the reactants (*cf. thermal reaction* vs. *photochemical reaction*; in photochemical reaction, Franck-Condon principle is not applicable; *see* Sec. 6.9.1, Figs. 6.9.1.1-2). To illustrate the case, let us consider the electron exchange process in the couple $[Fe(CN)_6]^{3-}/[Fe(CN)_6]^{4-}$

$$[\text{Fe(CN)}_6]^{3-} + [^*\text{Fe(CN)}_6]^{4-} \rightleftharpoons [\text{Fe(CN)}_6]^{4-} + [^*\text{Fe(CN)}_6]^{3-}; \ \Delta G^0 = 0$$

The Fe^{II}—C bond in $[Fe(CN)_6]^{4-}$ is longer than that in $[Fe(CN)_6]^{3-}$. If the electron transfer occurs at the ground state configurations (without any reorganization or chemical activation), then according to Franck-Condon principle, we shall get $[Fe(CN)_6]^{3-}$ with the longer Fe^{III}—C bonds and $[Fe(CN)_6]^{4-}$ with the shorter Fe^{II}—CN bonds. ***It happens so because during the time scale of electron transfer process, there will be no change of bond length***. Both these species thus produced are unstable and they will subsequently relax to the ground state with the release of energy. **This will contradict the 1st law of thermodynamics** because the process produces energy out of nothing. *It indicates that there must be some input of energy for the electron transfer process to occur even when $\Delta G^0 = 0$.*

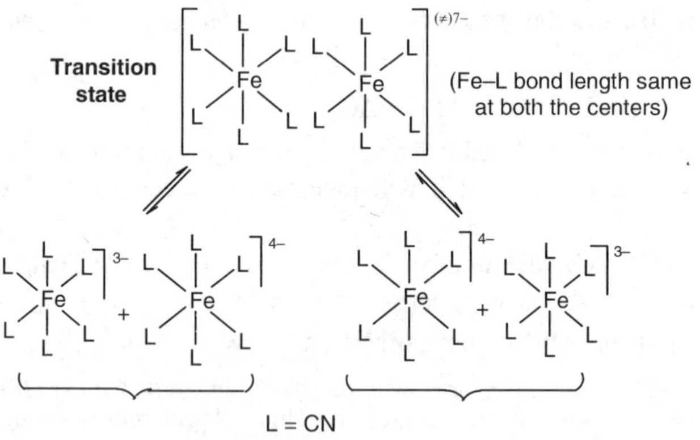

Fig. 6.3.2.1 Elongation of the Fe^{III}–CN bond in $[Fe(CN)_6]^{3-}$ and shrinkage of the Fe^{II}–CN bond in $[Fe(CN)_6]^{4-}$ to produce the transition state in which the metal-ligand bond distances are equal at the both centres.

The above discussion indicates the requirement of input of energy for the electron transfer process to take place. This will be consumed to reorganize the nuclei of the reactants before the electron transfer. This **reorganization of the nuclei** called **chemical activation** can be attained vibronically. This reorganization process will lead to shortening of the Fe^{II}—C bond in $[Fe(CN)_6]^{4-}$ and elongation of the Fe^{III}—C bond in $[Fe(CN)_6]^{3-}$. *This bond length shortening and elongation will continue until the 'Fe—C' bond length becomes equal in both the reactant species and under this condition, the participating orbitals become of the same energy to allow the electron transfer.* This requirement of reorganization of the nuclei of the reactants through the change of bond length and bond angle gives the height of the *Franck-Condon barrier* of the electron transfer process.

In the electron exchange process, in the couple $[Fe(CN)_6]^{3-}$– $[Fe(CN)_6]^{4-}$, an electron from the t_{2g} level of Fe(II) (t_{2g}^6 low-spin) is to be transferred to the t_{2g} level of Fe(III) (t_{2g}^5, low-spin). Because of the bond length differences between the species, energies of the participating orbitals differ. According to Franck-Condon principle, **energy of these participating orbitals must be the same.** The reorganization process prior to the electron transfer is illustrated in Fig. 6.3.2.1.

These are obtained through vibrational stretching and compression of the metal-ligand bonds.

Obviously, the amount of the required **reorganization energy will be less** if the geometric and structural parameters of the electron donor (*i.e.* reductant) and electron acceptor (*i.e.* oxidant) are more or less comparable. In such cases, the activation energy will be less and the rate will be faster. But if the structural parameters differ largely, then the reorganization energy measuring the activation energy will be high and consequently the electron transfer rate will be slower.

(B) Different factors to determine the magnitude of chemical activation (*i.e.* height of the Franck-Condon barrier): In the couples like $[Fe(OH_2)_6]^{3+}$–$[Fe(OH_2)_6]^{2+}$, $[Fe(CN)_6]^{3-}$–$[Fe(CN)_6]^{4-}$, etc. the **stereochemical configurations** of the Fe(III) and Fe(II) complexes are the same (*i.e.* both are octahedral), both are having the **same spin state** in each couple, but the ***bond the lengths are different.*** This is why, in such cases only the ***bond lengths (not the bond angles)*** are to be adjusted in the activation process of OSET. But when the ***stereochemical configurations*** change on addition or removal of electron, ***both the bond angles and bond lengths*** are to be adjusted in the activation process. Besides these, in OSET, the electron transfer can occur from $\pi^* \rightarrow \pi^*$ **(*i.e.* $t_{2g} \rightarrow t_{2g}$),** if the ground state electronic configuration does not allow this $\pi^* \rightarrow \pi^*$ transfer, then the reactants must be ***electronically activated*** to allow the $\pi^* \rightarrow \pi^*$ transfer. This **electronic activation** will also contribute to the chemical activation process. In some cases, removal or addition of electron may change the spin state. To accommodate this ***spin state change,*** in the activation process, electronic configurations of the reactants are to be changed accordingly. Thus the possible contributing factors in the ***chemical activation process*** of OSET are:

(i) *Bond length adjustment.*

(ii) *Bond angle adjustment (if there is a change in stereochemical configuration).*

(iii) *Electronic activation to allow the $\pi^* \rightarrow \pi^*$ transfer when the ground state configurations lead to $\sigma^* \rightarrow \sigma^*$, $\sigma^* \rightarrow \pi^*$ and $\pi^* \rightarrow \sigma^*$ electron transfer.*

(iv) *Electronic activation to accommodate the spin state change (if any).*

(C) Franck-Condon Principle and metalloproteins as the electron carriers: The metalloproteins acting as the electron carriers generally use the Fe(II)/Fe(III) couple (as in ***ferredoxins, cytochromes***) and Cu(II)/Cu(I) couple (as in ***blue proteins*** like azurin, plastocyanin, stellacyanin, etc.). The electron exchange in these couples generally occur through the *outer-sphere mechanism.*

To minimize the required ***reorganization energy***, in each couple, the following conditions should be satisfied.

- The spin state should not change during the electron transfer.
- The metal-ligand bond length in the oxidized and reduced forms of the couple should be comparable.
- The stereochemical configurations of both forms should be the same (at least comparable).

The above requirements are maintained in the electron carrier metallo-proteins. These are illustrated below.

- High-spin distorted tetrahedral Fe(III)/Fe(II) couple as in ***rubredoxin and ferredoxin***.
- Low-spin octahedral Fe(III)/Fe(II) couple as in ***cytochromes***.
- Distorted tetrahedral Cu(II)/Cu(I) couple as in ***blue proteins***.

(i) **Rubridoxin and ferredoxin:** Tetrahedral Fe(III) ($e^2t_2^3$) does not experience any Jahn-Teller distortion but Fe(II)($e^3t_2^3$) experiences the distortion. This is why, nature has placed the Fe-centre in such electron carriers ***in a distorted tetrahedral geometry to make a reasonable compromise between the preferences of Fe(II) and Fe(III)***. The bond lengths in this geometry for the oxidized and reduced centres are also comparable.

(ii) **Cytochromes:** *The octahedral geometry and spin state do not change during the electron exchange process.* The bond lengths in this geometry for the oxidized and reduced centres are also comparable. Thus, the reorganization energy is minimized to favour the electron exchange process.

(iii) **Blue proteins:** Cu(II)(d^9), the oxidized form of the couple has a strong preference for the square-planar geometry while Cu(I)(d^{10}), the reduced form of the couple has a preference to adopt the tetrahedral geometry. In a square planar geometry, Cu(II) enjoys a high *crystal field stabilization energy (cfse)* though there is a relatively more steric crowding compared to that in a tetrahedral geometry (*cf.* 90° *vs.* 109° in terms of bond angle). Cu(I) without having any cfse prefers the tetrahedral geometry to minimize the steric crowding. Here to make a reasonable compromise between the preferences of Cu(II) and Cu(I) to adopt the geometry, nature has placed the Cu-centre in a ***flattened tetrahedral geometry, an intermediate structure between the two idealized geometries, i.e. square planar and tetrahedral.*** In fact, in the said two idealized geometries, the metal-ligand bond distance also differs significantly but in the distorted geometry, *i.e.* flattened tetrahedral geometry, the bond lengths of the oxidized and reduced forms do not differ significantly. In fact, if this distortion were not done, then a large amount of reorganization energy would be required to adjust the geometry and bond length parameter.

6.3.3 Potential Energy Diagram for the Electron Transfer Process

(a) Energetics of the electron exchange reaction pathway: Now let us consider the following representative electron exchange reaction.

$$\left[Fe(OH_2)_6\right]^{3+} + \left[{}^{*}Fe(OH_2)_6\right]^{2+} \rightleftharpoons \left[Fe(OH_2)_6\right]^{2+} + \left[{}^{*}Fe(H_2O)_6\right]^{3+}$$

$$k_{ex} = 3.0\ dm^3\ mol^{-1}\ s^{-1}\ (25°C);\ \text{activation energy} = 32\ kJ\ mol^{-1}$$

In this electron exchange process, an electron from the t_{2g} level of Fe(II) ($t_{2g}^4e_g^2$) is transferred to the t_{2g} level of Fe(III) ($t_{2g}^3e_g^2$). The bond lengths are unequal, *i.e.* Fe^{III}—O (205 pm) ⟨ Fe^{II}—O (221 pm). *Thus energies of the participating orbitals are of different energy.* According to the demand of Franck-Condon principle, in the transition state, both the 'Fe—O' bond lengths should be the same. It will make the participating orbitals of the same energy. For this requirement, the Fe^{III}—O bond is to be elongated and the Fe^{II}—O bond is to be compressed in attaining the transition state (T.S.).

Free energy of activation ($\Delta G^{\neq}$) for electron exchange in the couple, $[Fe(OH_2)_6]^{3+}$ – $[Fe(OH_2)_6]^{2+}$ is the sum of three energy terms.

$$\Delta G^{\neq} = \Delta G_i^{\neq} + \Delta G_s^{\neq} + \Delta G_r^{\neq}$$

ΔG_i = Gibbs energy required for changing the bond length in the reactants, *i.e.* **for reorganising the inner coordination spheres of the reactants;** $\Delta G_s^{\neq}$ = Gibbs energy required for **solvent reorganisation**

in the outer spheres of the reactants; $\Delta G_r^{\neq}$ = Gibbs energy required to **overcome the coulombic repulsion** between the reactants when the reactants are brought together separated by the distance *r*.

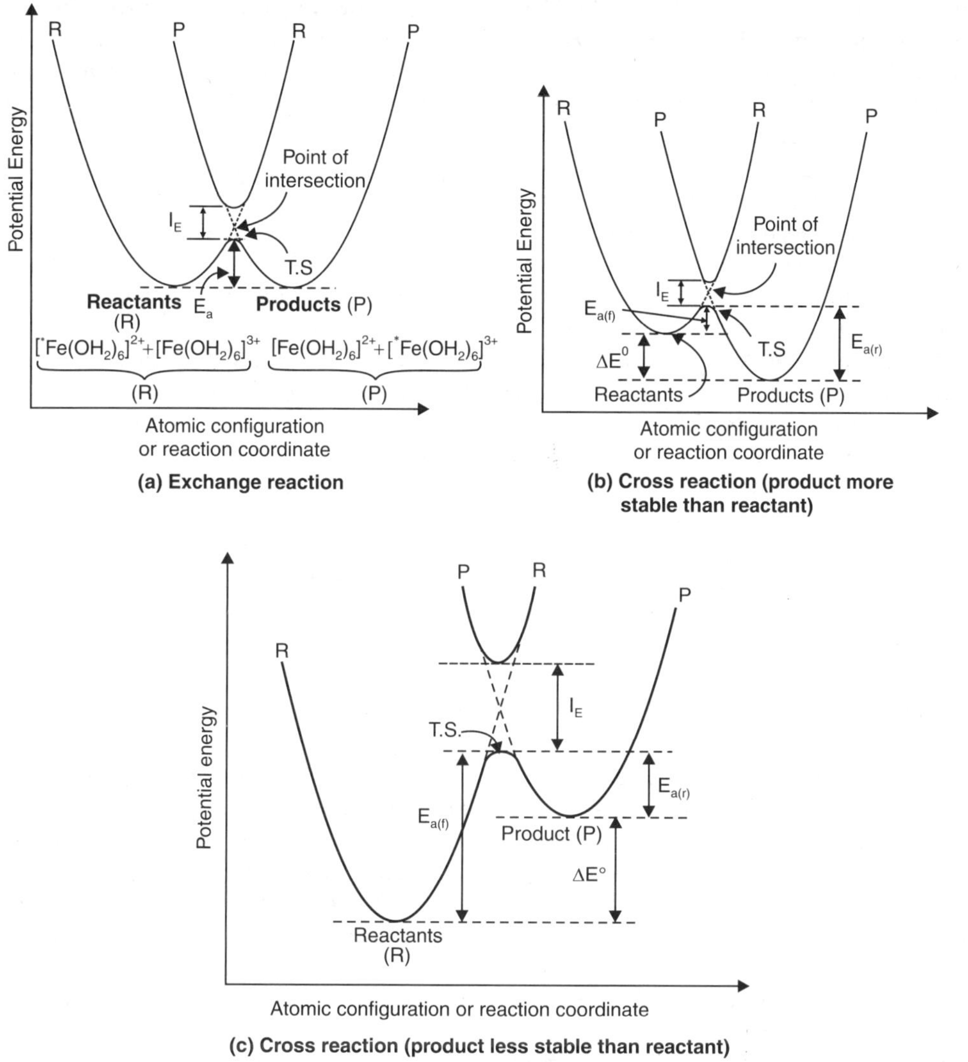

Fig. 6.3.3.1 Potential energy diagram showing the Franck-Condon barrier (denoted by E_a) in outer-sphere electron transfer (OSET) process. (T.S. for transition state; *P* for products; *R* for reactants)

(a) Electron exchange in redox couple, *i.e.* homonuclear electron transfer process as in the $[Fe(OH_2)_6]^{2+}$–$[Fe(OH_2)_6]^{3+}$ couple. (b) and (c) Cross reaction (or redox reaction), *i.e.* heteronuclear redox reaction. $E_{a(f)}$ for forward reaction and $E_{a(r)}$ for the reverse reaction.

(b) Potential energy diagrams for the outer sphere redox reaction: The metal-ligand bond deformations (required in attaining the T.S.) may be compared to the **harmonic oscillations** and the potential energy curves are of **parabolic shapes.** Thus the potential energy diagrams of the solvated reactants and products may be represented approximately by the **harmonic potential wells** of parabolic shapes. For the electron exchange reaction (*i.e.* no net chemical reaction), the reactants and products having the same energy can be represented by the same harmonic potential well but these are separated along the reaction coordinate. **At the point of intersection (shown by the dotted lines) of the potential wells** of the products and reactants, *the requirement of equal energies of the participating orbitals is attained. The point of intersection of the potential wells represents the position of* **transition state** *along the reaction coordinate.* The reaction coordinate represents the change of bond length and bond angle in the inner sphere of the involved metal centres.

In reality, the transition state (T.S.) lies below the point of intersection of the two potential energy curves representing the reactants and products. It can be explained by considering the **noncrossing rule** which states that the potential energy curves of two states of the same symmetry cannot cross and they mutually repell strongly at the hypothetical point of intersection giving rise to two energy curves—upper curve and lower curve. *The reactants follow the pathway of this lower curve (along the reaction coordinate) and transform into the products.*

The above phenomenon can also be realised in terms of mixing of the wave functions of the reactants and products. As the reaction coordinate approaches towards the T.S., *the wave-functions of the reactants and products mix and it leads to two separate energy states (separated by I_E energy) instead of two intersecting parabolas.* If there were no mutual interaction (*i.e.* no mixing), there would be two parabolas. The courses of such noninteracting parabolas are shown by the dashed lines. The extent of interaction or mixing is measured by I_E. At the T.S., the motion pushes the reactant well to slide down into the product well through the Franck-Condon barrier.

(c) I_E and position of T.S.: In terms of I_E, position of the T.S. with respect to the point of intersection of the parabolas of the reactants and products can be determined as follows:

- **Large I_E** (good coupling interaction): T.S. far below the point of intersection, *i.e.* low activation energy.
- **Small I_E** (poor coupling interaction): T.S. close to the point of intersection, *i.e.* high activation energy.

(d) Franck-Condon barrier height and I_E: If the coupling interaction measured by I_E is ***strong, then the height of Franck-Condon barrier is small*** (*i.e.* condition for small bond distortions and less inner sphere reorganisation) and the electron transfer process becomes favourable. If I_E is small (*i.e.* weak interaction, ***then the height of Franck-Condon barrier is large*** (*i.e.* condition for large bond distortion and more inner sphere reorganisation) and the electron transfer process becomes less favourable.

(e) I_E and κ_{el}: There is a relationship between ***electron transmission coefficient*** **(κ_{el})** (which indicates the probability of electron transfer at the transition state to give the product) and I_E. When I_E is large, $\kappa_{el} \approx 1.0$, *i.e.* reactants smoothly pass into the products. If I_E is small, the transition state may roll back down to the reactant side without giving the product (*i.e.* $\kappa_{el} < 1$). In fact, when I_E is small (*i.e.* poor coupling interaction), the T.S. practically lies close to the point of intersection and cosequently, even at the T.S., the reactant lies practically on the reactant parabola. In such cases, the reactant can easily roll back down along the reactant parabola without experiencing transformation into the product. ***Obviously when the T.S. lies far below the point of intersection (i.e. I_E is large for good coupling interaction), possibility of such 'roll back' from the T.S. is less.***

(f) Activation energy and cross reaction: In the same way, potential energy diagram for a cross reaction (*i.e.* net reaction, *i.e.* $\Delta G° \neq 0$) can be constructed. If the product is thermodynamically less stable than the product (*i.e. product curve runs above the reactant curve*), then the point of intersection of the potential energy curves of the reactants and products moves up and **consequently the activation energy becomes higher.** On the other hand, if the product becomes thermodynamically more stable than the reactant (*i.e. product curve lies below the reactant curve*), then the point of intersection is lowered down and consequently, the **activation energy becomes less.** It leads to the conclusion: **the thermodynamically favoured outer-sphere redox reaction is also kinetically favoured** (*cf.* Marcus theory and LFER).

Note: It may be noted that **at high exothermocity** (*i.e.* ΔE^0 is highly positive or ΔG^0 is highly negative), the crossing point rises and consequently the *electron transfer reaction rate becomes slower.* This situation is referred to as **Marcus Inverted Region.** Discussion on this aspect lies beyond the scope of this book.

6.3.4 Marcus Equation for Outer Sphere Cross-Reactions

By considering the different factors to contribute to $\Delta G^{\neq}_{12}$ (free energy of activation for the cross-reaction between Ox_1 and Red_2), Marcus and Hush theoretically developed an equation to calculate the rate constant (k_{12}) of the reaction between Ox_1 and Red_2.

$$Ox_1 + Red_2 \rightleftharpoons Red_1 + Ox_2$$

$$k_{12} = \sqrt{(k_{11}k_{22}K_{12}f_{12})} \text{ where } \log f_{12} = \frac{(\log K_{12})^2}{4\log\left(k_{11}k_{22}/Z^2\right)} \approx 0\,;\ i.e.\ f_{12} \approx 1.0$$

Where Z = number of collisions per second between the practicles (~ 10^{11} dm^3 mol^{-1} s^{-1}); k_{11} and k_{22} are electron exchange rate constants for the redox couples Ox_1–Red_1 and Ox_2–Red_2 respectively; K_{12} is the equilibrium constant which can be calculated from the E^0 values of the redox couples Ox_1–Red_1 and Ox_2–Red_2.

$$\Delta G^0_{12} = -\text{RTln}K_{12} = -nFE^0_{\text{cell}};\ E^0_{\text{cell}} = E^0_1 - E^0_2.$$

Where E^0_1 and E^0_2 are the standard reduction potentials of the couples $Ox_1 - Red_1$ and $Ox_2 - Red_2$, *i.e.*

$$Ox_1 + ne \rightleftharpoons Red_1,\ E^0_1;\ Ox_2 + ne \rightleftharpoons Red_2,\ E^0_2$$

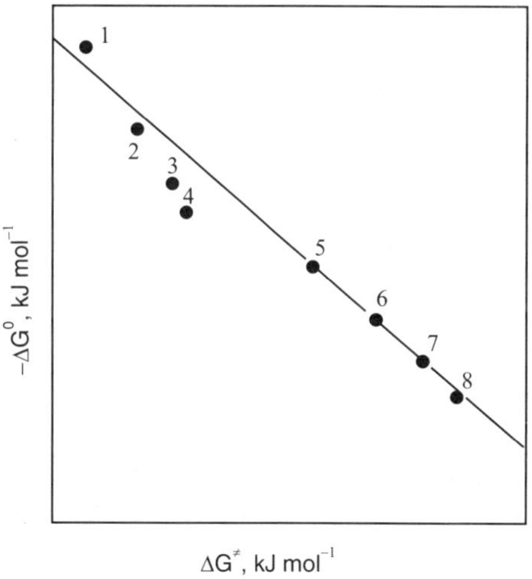

Ligand (L): 4,7–Me$_2$–phen (1); 5–Me–phen (2); phen (3); 5–Ph–phen (4); 5–Cl–phen (5); 5–SO$_3$H–phen (6); 3–SO$_3$H–phen (7); 5–NO$_2$–phen (8)

Fig. 6.3.4.1 Illustration of Marcus relationship (*i.e.* LFER plot) for the oxidation of $[Fe(L)_3]^{2+}$ (L = phen or substituted phen) complexes by Ce(IV) in aqueous sulfuric acid media (*cf.* G. Dulz *et al*, *Inorg. Chem. 2,* 917, 1963; J.D. Miller *et al*, *J. Chem. Soc. A,* 1370, 1966).

Taking $f_{12} \approx 1$, the **Marcus-Hush equation** (simply called Marcus equation) is given by:

$$k_{12} \approx \sqrt{(k_{11}k_{22}K_{12})} \text{ or, } k_{12}^2 = k_{11}k_{22}K_{12}$$

$$i.e.\ \mathbf{2\ln k_{12} = \ln k_{11} + \ln k_{22} + \ln K_{12} \Rightarrow 2\Delta G_{12}^{\neq} = \Delta G_{11}^{\neq} + \Delta G_{22}^{\neq} + \Delta G_{12}^{0}}$$

The Marcus equation indicates that the rate constant of a redox reaction depends on the *self electron exchange rate constants* (k_{11} and k_{12}) and the *equilibrium constant* (K_{12}).

The free energy of activation ($\Delta G_{12}^{\neq}$) for the electron transfer process in the said redox reaction can be given by:

$$\Delta G_{12}^{\neq} = \frac{(\Delta G_{11}^{\neq} + \Delta G_{22}^{\neq})}{2} + \frac{\Delta G_{12}^{0}}{2}, \text{ (assuming } f_{12} \approx 1)$$

The first term involving $\Delta G_{11}^{\neq}$ and $\Delta G_{22}^{\neq}$ gives the measure of **intrinsic contribution** while the second term involving ΔG_{12}^{0} gives the measure of ***thermodynamic contribution***. $\Delta G_{11}^{\neq}$ and $\Delta G_{22}^{\neq}$ are the free energies of activation for the self electron exchange process in the couples Ox_1–Red_1 and Ox_2–Red_2 respectively. Very often, *the above equation for $\Delta G_{12}^{\neq}$ is dominated by the thermodynamic part ΔG_{12}^{0}* (*i.e.* free energy change in the cross reaction) and it gives a *linear relationship* between $\Delta G_{12}^{\neq}$ and ΔG_{12}^{0}, *i.e.* linear relationship between log k_{12} and log K_{12}. Thus the Marcus equation is an example of **Linear Free Energy Relationship (LFER).**

The Marcus equation indicates that higher the equilibrium constant (K_{12}), higher the rate constant (k_{12}), *i.e.* ***thermodynamically favoured reactions are also kinetically favoured*** (*cf.* Fig. 6.3.4.2). This can also be concluded by considering the potential wells of the reactants and products (*cf.* Fig. 6.3.3.1).

The Marcus equation can theoretically calculate the rate constant of a redox reaction from the exchange rate constants and equilibrium constant that may be obtained from the E^0 values. This is of

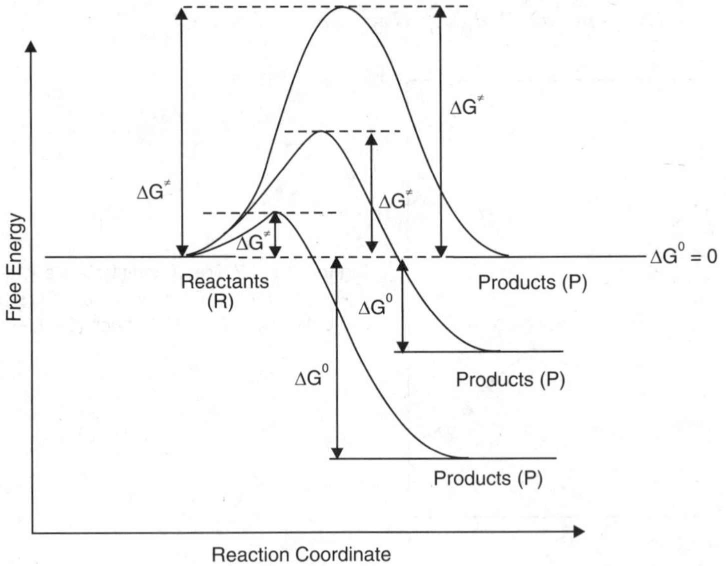

Fig. 6.3.4.2 Illustration of Marcus relationship (*i.e.* LEER plot) for the outer sphere cross reaction. ΔG^0 = free energy change in a reaction; $\Delta G^{\neq}$ = free energy of activation; higher ΔG^0 indicates lower $\Delta G^{\neq}$. **Note:** Same conclusion from Fig. 6.3.3.1.

much importance in understanding the electron transfer process and Marcus's work was recognized by **Nobel Prize in 1992.**

Let us illustrate the application of Marcus equation to calculate the rate constant of the following reaction, *i.e.* oxidation of hexacyanidoferrate(II) by octacyanidomolybdate(V).

$$\underset{(Ox_1)}{\left[Mo(CN)_8\right]^{3-}} + \underset{(Red_2)}{\left[Fe(CN)_6\right]^{4-}} \rightleftharpoons \underset{(Red_1)}{\left[Mo(CN)_8\right]^{4-}} + \underset{(Ox_2)}{\left[Fe(CN)_6\right]^{3-}} ; K_{12} = 1\times10^2$$

The corresponding electron exchange rate constants are:

$$\left[Fe(CN)_6\right]^{4-} + \left[{}^*Fe(CN)_6\right]^{3-} \longrightarrow \left[Fe(CN)_6\right]^{3-} + \left[{}^*Fe(CN)_6\right]^{4-} ; k_{11} = 7.4\times10^2\ dm^3\ mol^{-1}\ s^{-1}$$

$$\left[Mo(CN)_8\right]^{3-} + \left[{}^*Mo(CN)_8\right]^{4-} \longrightarrow \left[Mo(CN)_8\right]^{4-} + \left[{}^*Mo(CN)_8\right]^{3-} ; k_{22} = 3\times10^4\ dm^3 mol^{-1}\ s^{-1}$$

$$\log f = \frac{\left[\log\left(10^2\right)\right]^2}{4\log\left(\frac{7.4\times10^2\times3\times10^4}{10^{22}}\right)} \quad \text{leads to} f \approx 0.9$$

Table 6.3.4.1 Comparison of calculated and observed second order rate constants for the outer-sphere cross-reactions (*cf.* R. A. Marcus *et al*, *Biochim. Biophys. Acta.*, **811**, 265, 1985; T. J. Meyer *et al*; *Inorg. Chem.*, **7**, 2369, 1968; T. Przystas *et al*; *J. Am. Chem. Soc*; **95**, 5545, 1973).

Reaction	***logK₁₂***	**k_{12} ($dm^3\ mol^{-1}\ s^{-1}$)**	
		(observed)	***(calculated)***
$[Co(phen)_3]^{3+}$ + $[Ru(NH_3)_6]^{2+}$	5.42	1.5×10^4	1×10^5
$[Co(en)_3]^{3+}$ + $[V(OH_2)_6]^{2+}$	0.25	5.8×10^{-4}	7×10^{-4}
$[IrCl_6]^{2-}$ + $[Mo(CN)_8]^{4-}$	2.18	1.9×10^6	8×10^5
$[IrCl_6]^{2-}$ + $[Fe(CN)_6]^{4-}$	4.08	3.8×10^5	1×10^6
$[MnO_4]^-$ + $[Fe(CN)_6]^{4-}$	3.40	1.7×10^5	6×10^4
$[Co(bpy)_3]^{3+}$ + $[Co(terpy)_2]^{2+}$	0.55	64	40
	***E^0_{cell}* (V)**		
$[Co(phen)_3]^{3+}$ + $[Cr(OH_2)_6]^{2+}$	0.83	3×10^1	$\approx 1.1 \times 10^4$
$[Co(en)_3]^{3+}$ + $[Cr(OH_2)_6]^{2+}$	0.17	3.4×10^{-4}	$\approx 5.1 \times 10^{-4}$
$[Co(NH_3)_6]^{3+}$ + $[Cr(OH_2)_6]^{2+}$	0.51	1×10^{-3}	$\approx 1.5 \times 10^{-3}$
$[Ru(NH_3)_6]^{3+}$ + $[Cr(OH_2)_6]^{2+}$	0.51	2×10^2	$\approx 1.5 \times 10^3$
$[Ru(NH_3)_6]^{3+}$ + $[V(OH_2)_6]^{2+}$	0.355	80	$\approx 4.2 \times 10^3$
$[Co(NH_3)_6]^{3+}$ + $[V(OH_2)_6]^{2+}$	0.355	1×10^{-2}	$\approx 2.3 \times 10^{-3}$
$[Co(en)_3]^{3+}$ + $[V(OH_2)_6]^{2+}$	0.015	7.2×10^{-4}	$\approx 6 \times 10^{-4}$

$E^0_{Cr^{3+}/Cr^{2+}} = -0.41V; E^0_{V^{3+}/V^{2+}} = -0.26V$ *i.e.* $Cr(aq)^{2+}$ is a better reducing agent thermodynamically.

Note: Thermodynamically $Eu(aq)^{2+}$ ($4f^7$) is comparable to $Cr(aq)^{2+}$ as a reducing agent (*cf.* $E^0_{Eu^{3+}/Eu^{2+}} = -0.37V$). But it reduces $\left[Co(NH_3)_6\right]^{3+}$ faster ($k_{12} = 2\times10^{-2} dm^3 mol^{-1}\ s^{-1}$) than $Cr(aq)^{2+}$ (*cf.* $k_{12} = 1 \times 10^{-3}\ dm^3\ mol^{-1}\ s^{-1}$).

$$\text{Taking, } k_{12} = \sqrt{k_{11}k_{22}K_{12}} \quad (\text{assuming } f = 1)$$

$$= \sqrt{7.4\times10^2 \times3\times10^4 \times1\times10^2} \approx 4.7\times10^4 \text{ dm}^3 \text{ mol}^{-1} \text{s}^{-1}$$

The experimental value of k_{12} under the same condition is 3×10^4 dm^3 mol^{-1} s^{-1} which is very much comparable to the value obtained from Marcus equation. The Marcus equation has been tested in many cases (*cf.* Table 6.3.4.1).

Let us apply the Marcus equation for the following reactions:

$$\underset{(\text{Ox}_1)}{[\text{Co(bpy)}_3]^{3+}} + \underset{(\text{Red}_2)}{[\text{Co(terpy)}_2]^{2+}} \rightleftharpoons \underset{(\text{Red}_1)}{[\text{Co(bpy)}_3]^{2+}} + \underset{(\text{Ox}_2)}{[\text{Co(terpy)}_2]^{3+}}$$

Electron exchange process

$$[\text{Co(terpy)}_2]^{2+} - [{}^*\text{Co(terpy)}_2]^{3+} : k_{11} = 48 \text{ dm}^3 \text{ mol}^{-1} \text{s}^{-1} \ (0°\text{C})$$

$$[\text{Co(bpy)}_3]^{2+} - [{}^*\text{Co(bpy)}_3]^{3+} : k_{22} = 9.0 \text{ dm}^3 \text{ mol}^{-1} \text{s}^{-1} \ (0°\text{C})$$

$E^0[Co(bpy)_3]^{3+}/[Co(bpy)_3]^{2+} = + 0.34$ V

$E^0[Co(terpy)_2]^{3+}/[Co(terpy)_2]^{2+} = + 0.30$ V

$E^0_{cell} = (0.34 - 0.31) = 0.03$ V

$$-nFE^0_{cell} = -RT\ln K_{12} \text{ or, } \frac{nFE^0_{cell}}{RT} = \ln K_{12}$$

It leads to: $\log K_{12} = 0.553$ and $K_{12} = 3.6$

$$\log f = \frac{(0.553)^2}{4\log\frac{9\times48}{10^{22}}}, \text{ it leads to } f \approx 1.0$$

$$k_{12} = \sqrt{9\times48\times3.6\times1} \approx 40 \text{ dm}^3 \text{ mol}^{-1} \text{ s}^{-1}$$

(*cf.* experimental value, 64 dm^3 mol^{-1} s^{-1})

Conclusion: In fact, the agreement between the experimental value and calculated value is good under the conditions $K_{12} \leq 10^6$ *and* $f \geq 0.3$. **This agreement is regarded as an evidence in favour of outer sphere reaction.**

The considerable disagreement indicates a mechanism other than outer-sphere mechanism. However, **for the highly exo- and endothermic cross reactions, a modified Marcus equation is needed.**

By using the Marcus equation, it is also possible to calculate the self electron exchange rate constants. *This technique is specially important for the couples like metalloproteins for which it is difficult to directly measure the electron exchange rate constants.*

6.3.5 Nature of the Donor (*i.e.* Reductant) and Receptor (*i.e.* oxidant) Molecular Orbitals in Outer sphere Electron Transfer (OSET) Process

The donor MO (generally HOMO) of the reductant and receptor MO (generally LUMO) of the oxidant must be of the same type for the electron transfer. ***It is quite reasonable that if these involved MOs are the π^*–MOs (i.e. t_{2g} in terms of CFT), then the electron transfer process will be favourable.*** The reasons behind this preference are discussed below.

- (i) **Metal-ligand bond length change and inner-sphere reorganisation energy:** Change of electronic configuration (for the octahedral complexes) in the π^* orbital (*i.e.* t_{2g} level) does not seriously affect the metal-ligand bond length, but the metal-ligand bond length is very much sensitive to the electronic configuration in the σ^* orbital (*i.e.* e_g level in terms of CFT). It is due to the fact that the ligands are directly projected towards the lobes of e_g orbitals in the octahedral geometry.

  Thus donation of electron from the t_{2g} level ($\equiv \pi^*$) of the reductant and accommodation of the electron in the t_{2g}-level ($\equiv \pi^*$) of the oxidant do not change the bond lengths significantly at the metal centres.

  This is why, π^* (reductant)$\rightarrow \pi^*$ (oxidant) i.e. $t_{2g} \rightarrow t_{2g}$ transfer needs the less reorganization energy (i.e. low Franck-Condon barrier) and it is the most favourable path. Other possible paths like $\sigma^* \rightarrow \sigma^*$, $\sigma^* \rightarrow \pi^*$ (*i.e.* $e_g \rightarrow t_{2g}$) require the higher activation energy in terms of reorganization energy. Definitely, the ***$\sigma^* \rightarrow \sigma^*$ transfer is the most unfavourable path***.
- (ii) **Overlap integral:** In the octahedral geometry, the π^* orbitals (*i.e.* t_{2g}) are exposed everywhere except the vertices where the ligands are positioned (*cf.* Fig. 6.3.5.1). This makes the π^*–π^* overlapping interaction better. ***Consequently, in terms of orbital overlapping interaction, $\pi^* \rightarrow \pi^*$ (i.e. $t_{2g} \rightarrow t_{2g}$) electron transfer is the most favourable pathway.*** The σ^*-orbitals (*i.e.* e_g orbitals) are, in fact, projected towards the vertice, *i.e.* ligands. Thus the σ^* orbitals are more shielded compared to the π^* orbitals. In other words, direct overlap between the σ^*-donor orbital and σ^*-receptor orbital is not possible because the σ^* (*i.e.* e_g) orbitals are directly projected towards the ligands.

Different contributing factors to determine the activation energy and rate of electrons transfer in the outer sphere process

- **Electrostatic barrier in forming the precursor complex:** Sometimes, the oxidants and reductants are bearing the same charge to cause an electrostatic repulsion in the precursor complex (*i.e.* outer sphere complex) as in the electron exchange in the couple $[Fe(CN)_6]^{3-}$–$[Fe(CN)_6]^{4-}$. To overcome this repulsive barrier, some energy will be required.
- **Reorganization energy (*i.e.* Franck-Condon barrier):** To meet the demands of Franck-Condon barrier, the ***bond length and bond angles of the reactants*** are to be changed (*i.e.* inner spheres of the reactants are to be reorganised) properly before the electron transfer. Sometimes, the ground state electronic configuration may be required to change (*i.e.* **electronic activation** of the reactants) to allow the $\pi^* \rightarrow \pi^*$ transfer and to accommodate the spin state change, if any. This inner sphere reorganization energy contributes to the activation energy.
- **Nature of orbitals:** The $\pi^* \rightarrow \pi^*$ electron transfer is the most favourable path, *i.e.* it needs least activation energy. It arises so because of the ***least reorganization required*** *and* ***favoured orbital– orbital overlapping interaction.*** Obviously, the $\sigma^* \rightarrow \sigma^*$ electron transfer makes the activation energy very high.
- **Solvent reorganization energy:** Solvent reorganization in the outer sphere at the events of precursor complex formation and reorganization process makes also an important contribution to the activation energy.
- **Thermodynamic favour:** According to the Marcus Equation, more the equilibrium constant, more the rate constant.

- (iii) **Relative ease of electron transfer among the possible pathways ($\sigma^* \rightarrow \sigma^*$, $\sigma^* \rightarrow \pi^*$, $\pi^* \rightarrow \sigma^*$ and $\pi^* \rightarrow \pi^*$):** It is evident that in the outer sphere electron transfer process, the electron transfer can occur from π^* to π^* *i.e.* t_{2g} to t_{2g}. If the ground state electronic configuration of the oxidant and reductant can allow this $\pi^* \rightarrow \pi^*$ transfer (*e.g.* Fe^{II}–Fe^{III} couple, V^{II}–Ru^{III} cross reaction) *then there is no need to activate electronically the reactants, i.e.* no need of change of the electronic configuration for the electron transfer process. If the ground state electronic configuration leads to $\sigma^* \rightarrow \sigma^*$ transfer (*e.g.* Co^{II}–Co^{III} couple, Cr^{II}–Co^{III} cross reaction) or $\sigma^* \rightarrow \pi^*$ transfer (*e.g.* Cr^{II}–Ru^{III} cross reaction) or $\pi^* \rightarrow \sigma^*$ transfer (*e.g.* V^{II}–Co^{III} cross reaction), then the reactants must be properly electronically excited to allow the $\pi^* \rightarrow \pi^*$ transfer.

Obviously, for the net $\sigma^* \rightarrow \sigma^*$ transfer, both the oxidant and reductant need electronic activation; for the net $\pi^* \rightarrow \pi^*$ transfer, there is no need of electronic activation for the both oxidant and reductant.

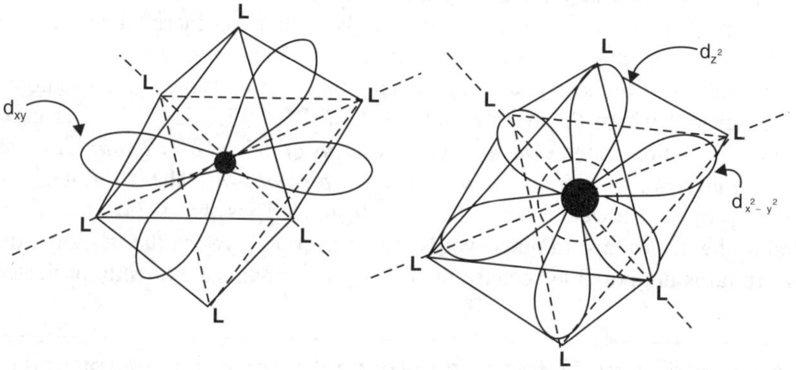

Fig. 6.3.5.1 Spatial orientation of the d_{xy} (*i.e.* t_{2g}) and $d_{x^2-y^2}$, d_{z^2} (*i.e.* e_g set) orbitals with respect to the ligands (*L*) in the octahedral geometry.

6.3.6 Illustration of the Importance of Different Factors to Control the Rate Constants of Outer Sphere Electron Transfer (OSET) Reactions

(a) **$\pi^* \rightarrow \pi^*$ *vs.* $\sigma^* \rightarrow \sigma^*$ electron transfer:** Table 6.3.6.1 indicates that in general, the $\pi^* \rightarrow \pi^*$ electron transfer process is kinetically more favoured over the $\sigma^* \rightarrow \sigma^*$ electron transfer process. The $\sigma^* \rightarrow \sigma^*$ transfer process needs more *reorganization energy* as the bond lengths differ generally significantly for the oxidant and reductant centres. Moreover, direct overlap between the donor and receptor orbitals is prevented because of the ligands which are projected towards the σ* (*i.e.* e_g) orbitals. *These aspects have been already discussed.*

(b) **Unusually slow electron exchange rate in the $[Fe(OH_2)_6]^{3+}$ – $[Fe(OH_2)_6]^{2+}$ couple in spite of the $\pi^* \rightarrow \pi^*$ transfer:** Though the process leads to the $\pi^* \rightarrow \pi^*$ electron transfer, the rate is very slow compared to the other systems involving the $\pi^* \rightarrow \pi^*$ transfer (*cf.* Table 6.3.6.1). In fact, besides the nature of the participating orbitals, many other factors are also important to determine the activation energy. In the present case, the **large bond length difference** between the two species (*cf.* Fe^{III}—OH_2 = 205 pm, Fe^{II}—OH_2 = 221 pm; Ru^{III}—NH_3 = 210 pm, Ru^{II}—NH_3 = 214 pm) requires ***a large reorganization energy*** to overcome the Franck-Condon barrier. In fact, this reorganization energy is consumed to contract the Fe^{II}—O bond length and to elongate the Fe^{III}—O bond lengths so that in the *transition state* the 'Fe—O' bond lengths become identical for the both centres (*cf.* Fig. 6.3.3.1). It may be noted that in the $[Ru(NH_3)_6]^{3+}$–$[Ru(NH_3)_6]^{2+}$ couple, the bond lengths differ by only 4 pm (*cf.* 16 pm for the Fe(III)-Fe(II) system). Consequently, it needs less reorganization energy to attain the transition state.

Table 6.3.6.1 2nd order rate constants for electron exchange process occurring through outer sphere electron transfer (OSET) process (*cf.* D.R. Stranks, *Discuss Faraday soc.*, **29**, 73, 1960., T. J. Meyer *et al.*, *Inorg. Chem.*, **7**, 2369, 1968).

	Reaction/System			*Rate Constant** ($dm^3\ mol^{-1}\ s^{-1}$)	*Net electron transfer*
(a)	$[Fe(OH_2)_6]^{2+}$	+	$[Fe(OH_2)_6]^{3+}$	2.0	
	$t_{2g}^4 e_g^2$		$t_{2g}^3 e_g^2$		
(b)	$[Fe(CN)_6]^{4-}$	+	$[Fe(CN)_6]^{3-}$	10^3	
	(t_{2g}^6)		(t_{2g}^5)		
(c)	$[Fe(phen)_3]^{2+}$	+	$[Fe(phen)_3]^{3+}$	$> 10^7$	
	(t_{2g}^6)		(t_{2g}^5)		
(d)	$[Os(bpy)_3]^{2+}$	+	$[Os(bpy)_3]^{3+}$	$> 10^5$	$t_{2g} \rightarrow t_{2g}$ *i.e.* $\pi^* \rightarrow \pi^*$
	(t_{2g}^6)		(t_{2g}^5)		
(e)	$[IrCl_6]^{3-}$	+	$[IrCl_6]^{2-}$	3×10^2	
	(t_{2g}^6)		(t_{2g}^5)		
(f)	$[Ru(NH_3)_6]^{2+}$	+	$[Ru(NH_3)_6]^{3+}$	10^2	
	(t_{2g}^6)		(t_{2g}^5)		
(g)	$[Ru(phen)_3]^{2+}$	+	$[Ru(phen)_3]^{3+}$	$> 10^7$	
	(t_{2g}^6)		(t_{2g}^5)		
(h)	$[Co(OH_2)_6]^{2+}$	+	$[Co(OH_2)_6]^{3+}$	3	
	$(t_{2g}^5 e_g^2)$		(t_{2g}^6)		
(i)	$[Co(NH_3)_6]^{2+}$	+	$[Co(NH_3)_6]^{3+}$	$< 10^{-8}$	
	$(t_{2g}^5 e_g^2)$		(t_{2g}^6)		
(j)	$[Co(en)_3]^{2+}$	+	$[Co(en)_3]^{3+}$	1.4×10^{-4}	$\sigma^* \rightarrow \sigma^*$ *i.e.* $e_g \rightarrow e_g$
	$(t_{2g}^5 e_g^2)$		(t_{2g}^6)		
(k)	$[Co(ox)_3]^{4-}$	+	$[Co(ox)_3]^{3-}$	9×10^{-7}	
	$(t_{2g}^5 e_g^2)$		(t_{2g}^6)		
(l)	$[Co(phen)_3]^{2+}$	+	$[Co(phen)_3]^{3+}$	1.1	
	$(t_{2g}^5 e_g^2)$		(t_{2g}^6)		

* Rate constants in the temperature range 0–25°C.

(c) **Unusually high electron exchange rate in the couple $[M(phen)_3]^{3+}$–$[M(phen)_3]^{2+}$ (M = Fe, Ru):** In fact, this rate benefit is common for the complexes bearing the π-*acceptor ligands* like CN^-, phen, bpy, etc. Such unsaturated π-acid ligands bear the vacant π^*–MOs that can overlap with the π^*—MOs of the complex. This delocalization of the π^*–MO of the complex onto the phenanthroline skeleton allows the mixing between the donor π^*–MO of the reductant (M_{red}) and receptor π^*—MO of the oxidant (M_{ox}), through the π^*-MOs of phenanthroline, *i.e.*

electron flow

$$(M_{red})\ \pi^*\text{–MO} \longrightarrow (phen)\ \pi^*\text{–MO} \dashrightarrow \pi^*\text{–MO}\ (phen) \longrightarrow \pi^*\text{–MO}\ (M_{ox})$$

Thus it can be assumed that the electron from the donor π^*-MO of the reductant (M_{red}) is first transferred to the π^*-MO of its ligand (phen); thence to π^* of the phen ligand attached to the oxidant and thence to the receptor π^*-MO of the oxidant (*i.e.* M_{ox}). Thus participation of the vacant π^*-MOs of the phen ligands present in the coordination spheres of the oxidant (M_{ox}) and reductant (M_{red}) centres favour the electron transfer process. It happens so for the other complexes bearing the π-acid ligands having the vacant π^*–MOs.

(d) **Bond length reorganisation energy in the complexes bearing the π-acid ligands:** Besides participation of the vacant π^*-MOs of the π-acid ligands in the electron transfer process as stated above, the π-acid ligands (*i.e.* **strong field ligands**) produce **another favourable effect,** *i.e.* geometrical parameters including the bond length for the both oxidized and reduced centres are very much comparable. For example, in the couples $[M(CN)_6]^{3-}$–$[M(CN)_6]^{4-}$ (M = Fe, Mn), $[Fe(phen)_3]^{3+}$–$[Fe(phen)_3]^{2+}$, the oxidized and reduced centres differ by one electron **in the nonbonding t_{2g} level** (*e.g.* t_{2g}^6 *vs.* t_{2g}^5 for the Fe(II)–Fe(III) couple, t_{2g}^5 *vs.* t_{2g}^4 for the Mn(II)–Mn(III) couple). *Consequently, bond lengths do not differ significantly in each couple. Moreover, the* ***metal → ligand π-back bonding*** *is more important for the lower valent complex. It will contract the bond length and the bond length becomes comparable to that in the higher valent complex, i.e. oxidized centre.* This explains the higher electron exchange rate in the $[Fe(CN)_6]^{4-}$–$[Fe(CN)_6]^{3-}$ couple than in the $[Fe(OH_2)_6]^{3+}$–$[Fe(OH_2)_6]^{2+}$ couple.

Favourable factors (summary) in the electron exchange process in the couples $[Fe(CN)_6]^{4-}$–$[Fe(CN)_6]^{3-}$, $[FeL_3]^{2+}$–$[FeL_3]^{3+}$ (L = bpy, phen).

(i) ***No electronic activation*** is required for the favourable $\pi^* \rightarrow \pi^*$ (*i.e.* $t_{2g} \rightarrow t_{2g}$) electron transfer.

(ii) Bond length difference in Fe^{II}–L and Fe^{III}–L (L = π-acid ligand like CN^-, bpy, phen, etc.) is small. Generally, the Fe^{II}–L bond is lognger than the Fe^{III}–L bond, but the more favoured ligand → metal π-back bonding in the Fe(II)-complexes reduces the bond length. ***It makes the bond lengths in both the Fe(II) and Fe(III) complexes comparable.***

(iii) Electronic configurations of the low spin Fe(II) ($t_{2g}^6 e_g^0$) and Fe(III) (t_{2g}^5) differ only in the t_{2g}-level. It makes also the ***bond lengths in the Fe(II) and Fe(III) complexes comparable.***

(iv) In the electron transfer process, there is ***no change in the spin state.***
All these above factors make the ***inner-sphere reorganization energy*** **less,** *i.e.* height of the Franck-Condon barrier becomes low.

(v) Besides the above factors, participation of the vacant π^*–MOs of the ligands (*i.e.* CN^-, phen, bpy) in mixing of the π^*–MO of the donor site and π^*–MO of the receptor site facilitates the electron transfer process.

Note: It is worth mentioning that in the Co(III)/Co(II) couple, in the presence of strong field ligands like π-acid ligands, ***electronic activation of Co(II) becomes easier.*** This gives an additional rate benefit (*cf.* (e) of this section).

(e) **Slow electron exchange rate in the Co^{III}–Co^{II} couple:** It experiences the net $\sigma^* \rightarrow \sigma^*$ electron transfer. In the octahedral geometries, the σ^* (*i.e.* e_g) orbitals are directly projected towards the ligands. Consequently, the $\sigma^* - \sigma^*$ overlap between the receptor and donor orbital is not possible. *To execute the electron transfer process, the electronic configurations are to be changed so that the $\pi^* \rightarrow \pi^*$ transition can occur.* It makes the activation energy high. Besides this, the oxidized centre Co^{III} is generally in low spin state (*i.e.* t_{2g}^6) while the reduced centre Co^{II} is in high spin state (*i.e.* $t_{2g}^5 e_g^2$). **Thus it also needs the spin state change.**

In fact, change of the electronic configuration to allow the $\pi^* \rightarrow \pi^*$ electron transfer and requirement of the spin state change are the **two major barriers** to make the activation energy high.

Let us compare the following two redox couples:

$[Ru(NH_3)_6]^{3+}$–$[Ru(NH_3)_6]^{2+}$	$[Co(NH_3)_6]^{3+}$–$[Co(NH_3)_6]^{2+}$
($t_{2g}^5 e_g^0$) (t_{2g}^6)	($t_{2g}^6 e_g^0$) ($t_{2g}^5 e_g^2$)
$t_{2g} \rightarrow t_{2g}$ **transfer (Net)**	$e_g \rightarrow e_g$ **transfer (Net)**

$\overbrace{\text{Ru}^{\text{III}}\text{–NH}_3 = 210 \text{ pm} \quad \text{Ru}^{\text{II}}\text{–NH}_3 = 214 \text{ pm}}$ *i.e.* bond length difference by 4 pm

$\overbrace{\text{Co}^{\text{III}}\text{–NH}_3 = 194 \text{ pm} \quad \text{Co}^{\text{II}}\text{–NH}_3 = 211 \text{ pm}}$ *i.e.* bond length difference by 17 pm

Bond length effect: It is evident that in terms of bond length difference, the required *inner sphere reorganization energy* to overcome the Franck-Condon barrier is much more higher for the $[Co(NH_3)_6]^{3+}$–$[Co(NH_3)_6]^{2+}$ couple.

Spin state effect: Co(III) is in low spin state while Co(II) is in high spin state ($t_{2g}^5e_g^2$). Simple addition of one electron to Co(III) gives $t_{2g}^6 e_g^1$ which is not the ground state electronic configuration of Co(II). To attain the ground state electronic configuration of Co^{II}, it needs a rearrangement of electronic configuration.

$$\underset{\text{(Ground state, low spin)}}{\text{Co(III)}\left(t_{2g}^6e_g^0\right)} \xrightarrow{+e} \underset{\text{(Excited state, low spin)}}{\left[\text{Co(II)}\left(t_{2g}^6e_g^1\right)\right]} \longrightarrow \underset{\text{(Ground state, high spin)}}{\text{Co(II)}\left(t_{2g}^5e_g^2\right)}$$

Similarly, simple removal of an electron from Co(II) does not generate the oxidized centre Co(III).

$$\underset{\text{(Ground state, high spin)}}{\text{Co(II)}\left(t_{2g}^5e_g^2\right)} \xrightarrow{-e} \underset{\text{(Excited state)}}{\left[\text{Co(III)}\left(t_{2g}^5e_g^1\right)\right]} \longrightarrow \underset{\text{(Ground state, low spin state)}}{\text{Co(II)}\left(t_{2g}^5e_g^0\right)}$$

Thus reaction between Co(III) and Co(II) at their ground states produce both the products in the excited states. ***This causes a barrier.***

Possible paths of electron transfer to the execute the $\pi^* \rightarrow \pi^*$ *i.e.* $t_{2g} \rightarrow t_{2g}$ transfer:

$$\textbf{Path I:}\ \left.\begin{matrix}\text{Co(II)}\\+\\\text{Co(III)}\end{matrix}\right\}: \begin{matrix}t_{2g}^5e_g^2\\+\\t_{2g}^6e_g^0\end{matrix} \rightleftharpoons \begin{matrix}t_{2g}^6e_g^1\\+\\t_{2g}^5e_g^1\end{matrix} \xrightleftharpoons{\pi^*\rightarrow\pi^*} \begin{matrix}t_{2g}^5e_g^1\\+\\t_{2g}^6e_g^1\end{matrix} \rightleftharpoons \begin{matrix}t_{2g}^6e_g^0\\+\\t_{2g}^5e_g^2\end{matrix} : \left\{\begin{matrix}\text{Co(III)}\\+\\\text{Co(II)}\end{matrix}\right.$$

$$\textbf{Path II:}\ \left.\begin{matrix}\text{Co(II)}\\+\\\text{Co(III)}\end{matrix}\right\}: \begin{matrix}t_{2g}^5e_g^2\\+\\t_{2g}^6e_g^0\end{matrix} \rightleftharpoons \begin{matrix}t_{2g}^5e_g^2\\+\\t_{2g}^4e_g^2\end{matrix} \xrightleftharpoons{\pi^*\rightarrow\pi^*} \begin{matrix}t_{2g}^4e_g^2\\+\\t_{2g}^5e_g^2\end{matrix} \rightleftharpoons \begin{matrix}t_{2g}^6e_g^0\\+\\t_{2g}^5e_g^2\end{matrix} : \left\{\begin{matrix}\text{Co(III)}\\+\\\text{Co(II)}\end{matrix}\right.$$

$$\textbf{Path III:}\ \left.\begin{matrix}\text{Co(II)}\\+\\\text{Co(III)}\end{matrix}\right\}: \begin{matrix}t_{2g}^5e_g^2\\+\\t_{2g}^6e_g^0\end{matrix} \rightleftharpoons \begin{matrix}t_{2g}^6e_g^1\\+\\t_{2g}^4e_g^2\end{matrix} \xrightleftharpoons{\pi^*\rightarrow\pi^*} \begin{matrix}t_{2g}^5e_g^1\\+\\t_{2g}^5e_g^2\end{matrix} \rightleftharpoons \begin{matrix}t_{2g}^6e_g^0\\+\\t_{2g}^5e_g^2\end{matrix} : \left\{\begin{matrix}\text{Co(III)}\\+\\\text{Co(II)}\end{matrix}\right.$$

$$\textbf{Path IV:}\ \left.\begin{matrix}\text{Co(II)}\\+\\\text{Co(III)}\end{matrix}\right\}: \begin{matrix}t_{2g}^5e_g^2\\+\\t_{2g}^6e_g^0\end{matrix} \rightleftharpoons \begin{matrix}t_{2g}^5e_g^2\\+\\t_{2g}^5e_g^1\end{matrix} \xrightleftharpoons{\pi^*\rightarrow\pi^*} \begin{matrix}t_{2g}^4e_g^2\\+\\t_{2g}^6e_g^1\end{matrix} \rightleftharpoons \begin{matrix}t_{2g}^6e_g^0\\+\\t_{2g}^5e_g^2\end{matrix} : \left\{\begin{matrix}\text{Co(III)}\\+\\\text{Co(II)}\end{matrix}\right.$$

Path-I needs activation of both centres and both the products are in the excited state.

Path-II needs activation of one centres, *i.e.* Co(III) and one product remains in the excited state.

Path-III needs activation of both centres and one product remains in the excited state.

Path-IV needs activation of one reactant, *i.e.* Co(III) but both the products remain in the excited state.

Co(III):	$t_{2g}^6e_g^0$	$\longrightarrow$	$t_{2g}^5e_g^1$	$\longrightarrow$	$t_{2g}^4e_g^2$	
cfse:	$24Dq_o - 2P$ (I)		$14Dq_o - P$ (II)		$4Dq_o$; (III)	$10Dq_o$ and P for Co(III)
Co(II):	$t_{2g}^5e_g^2$	$\longrightarrow$	$t_{2g}^6e_g^1$			
cfse:	$8Dq_o$ IV		$18Dq_o - P$; V			$10Dq_o$ and P for Co(II)

(To calculate the cfse, only the additional pairing (P) compared to that in the free state is considered.)

Path-I: $IV + I \rightleftharpoons V + II \xrightleftharpoons{\pi^* \to \pi^*} II + V \rightleftharpoons I + IV$

Path-II: $IV + I \rightleftharpoons IV + III \xrightleftharpoons{\pi^* \to \pi^*} III + IV \rightleftharpoons I + IV$

Path-III: $IV + I \rightleftharpoons V + III \xrightleftharpoons{\pi^* \to \pi^*} II + IV \rightleftharpoons I + IV$

Path-IV: $IV + I \rightleftharpoons IV + II \xrightleftharpoons{\pi^* \to \pi^*} III + V \rightleftharpoons I + IV$

Co(III) activation: It leads to ***low spin to high spin,*** *i.e.* $t_{2g}^6e_g^0 \to t_{2g}^5 e_g^1 \to t_{2g}^4 e_g^2$. The increase in electron density in the e_g-orbital will elongate the bond length and there is a loss of cfse.

Co(II) activation: It will lead to ***high spin to low spin state,*** *i.e.* $t_{2g}^5e_g^2 \to t_{2g}^6e_g^1$. The decrease in electron density in the e_g-level will contract the bond length and there is a loss of cfse (because P > $10Dq_o$).

Thus the bond length change in the activation of Co(III) and Co(II) centre will be reflected in the reorganization energy to increase the activation energy.

It is evident that activation energy for Co(III) (*i.e.* low spin to high spin, I→ II→ III) *increases with the increase of crystal field strength while activation energy for* Co(II) (*i.e.* high spin to low spin) *decreases with the increase of crystal field strength.* It indicates ***an opposing effect of the crystal field strength.*** In fact, it is very difficult to correlate the electron exchange rate in the Co(III)-Co(II) couple with the crystal field strength. The crystal field strength of the ligands runs in the sequence:

$$H_2O \langle \text{ oxalate } \langle \text{ edta } \langle NH_3 \langle \text{ en } \langle \text{ phen.}$$

The electron exchange rate in the Co(III)-Co(II) couple has been found **very fast at both ends of the series** (*i.e.* for both weak and strong field ligand) and the rate is **very slow at the middle of the series.** In fact, the rate becomes minimum at NH_3 and it is not surprising. At this intermediate ligand field strength, activation is not favoured energetically both for Co(II) and Co(III). For the strong field ligands, activation of Co(II) (h.s. → l.s.) is favoured and for the weak field ligands, activation of Co(III) (l.s. → h.s.) is favoured.

Co(II)-Activation *vs.* Co(III)-Activation

If cfse of the low-spin Co(III) the complexes does not differ drastically from that of the high spin Co(III) complexes (*i.e.* under the condition, $\Delta_o \geq P$), Co(II)-activation plays a crucial role in the electron exchange process in the Co(II)-Co(III) couple. For most of the $[CoL_6]^{2+}$ complexes, $\Delta_o \approx 10$ kK and $P \approx 20$ kK. Thus the Co(II) activation energy is given by $P - \Delta_o \approx 10$ kK ≈ 125 kJ mol^{-1} for the conversion of high spin ($t_{2g}^5 e_g^2$) to low spin ($t_{2g}^6e_g^1$).

(f) Unusually high electron exchange rate in the couple $[Co(OH_2)_6]^{3+}$ – $[Co(OH_2)_6]^{2+}$: For $[Co(OH_2)_6]^{3+}$, the $10Dq_o$ value and P (pairing energy) are pretty close (*i.e.* $10Dq_o \approx$ P, critical $10Dq_o$ value) and it allows the spin isomerism.

$$\left[Co(OH_2)_6\right]^{3+}\left(l.s.,\ t_{2g}^6 e_g^0\right) \rightleftharpoons \left[Co(OH_2)_6\right]^{3+}\left(h.s.,\ t_{2g}^4 e_g^2\right)$$

It has been estimated that the low spin $[Co(OH_2)_6]^{3+}$ is more stable than the high spin $[Co(OH_2)_6]^{3+}$ by only 25 kJ mol^{-1}. ***Thus it is possible to activate $[Co(OH_2)_3]^{3+}$ for the electron transfer at the cost of a small amount of energy.*** The high spin Co(III) can react **(Path-II)** with the high spin Co(II) easily to give the observed rate benefit.

The reaction path may be outlined as follows:

$$\underset{(t_{2g}^6 e_g^0)}{\left[Co(OH_2)_6\right]^{3+}} + \underset{(t_{2g}^5 e_g^2)}{\left[Co(OH_2)_6\right]^{2+}} \xrightleftharpoons{\text{Activation of Co(III)}} \underset{(t_{2g}^4 e_g^2)}{\left[Co(OH_2)_6\right]^{3+}} + \underset{(t_{2g}^5 e_g^2)}{\left[Co(OH_2)_6\right]^{2+}}$$

$$\xrightleftharpoons{\pi^* \rightarrow \pi^*} \underset{(t_{2g}^5 e_g^2)}{\left[Co(OH_2)_6\right]^{2+}} + \underset{(t_{2g}^4 e_g^2)}{\left[Co(OH_2)_6\right]^{3+}} \qquad \textbf{Path - II}$$

$$\xrightleftharpoons{\text{Relaxation of Co(III) to ground state}} \underset{(t_{2g}^5 e_g^2)}{\left[Co(OH_2)_6\right]^{2+}} + \underset{(t_{2g}^6 e_g^0)}{\left[Co(OH_2)_6\right]^{3+}}$$

Here the required activation energy for Co(III) (*i.e.* conversion of low spin state to high spin state) is very small. Relaxation of Co(III) (*i.e.* h.s. → l.s.) is energetically favourable. **It makes the rate so high.**

Here it is worth mentioning that the electron exchange in the couple $[Co(OH_2)_6]^{3+}/[Co(OH_2)_6]^{2+}$ can also partly go through the **inner sphere electron transfer path** (ISET path) by using the weak bridging property of OH_2 (*cf.* Secs. 6.4.4, 6.6 for detailed discussion).

(g) Unusually high electron exchange rate in the $[Co(phen)_3]^{3+}$–$[Co(phen)_3]^{2+}$ couple: It has been already mentioned that for the very strong field ligands, the electron exchange process in the Co(III)/Co(II) couple is kinetically favoured. ***It is mainly due to the low cost of energy for activation of Co(II).*** In the presence of strong field ligand, high spin Co(II) can be easily converted into low spin Co(II) ($t_{2g}^6 e_g^1$). For the π-acid ligands, the better Co(II) → L π-bonding makes the Co(II)–L bond relatively shorter, *i.e.* the ***Co(III)–L and Co(II)–L bond lengths do not differ significantly.*** It makes the inner sphere reorganisation energy less. In addition to this, the vacant π^*–MO of phen ligand can participate in the electron transfer process. The role of π-acid ligands in the rate acceleration of electron transfer in the OSET path has been discussed in detail in (c) of this section. Electron-flow occurs as follows through the vacant π^*–MOs of phen.

$$\text{Co(II)}(\pi^*\text{–MO}) \longrightarrow \text{phen}(\pi^*\text{-MO}) \dashrightarrow (\text{phen})(\pi^*\text{–MO}) \longrightarrow \text{Co(III)}(\pi^*\text{–MO})$$

(h) **$[Cr(OH_2)_6]^{2+}$ *vs.* $[V(OH_2)_6]^{2+}$ as the reducing agents:** The important kinetic and thermodynamic features of the reducing agents $[Cr(OH_2)_6]^{2+}$ and $[V(OH_2)_6]^{2+}$ are compared below.

(i) E^0 (Cr^{3+}/Cr^{2+}) = –0.41 V, $E^0(V^{3+}/V^{2+})$ = –0.26 V, *i.e.* **Cr^{2+} is a better reducing agent than V^{2+} in terms of thermodynamics.** V^{3+}, V^{2+}, Cr^{3+} and Cr^{2+} simply denote their aqua complexes.

(ii) k_{ex} (Cr^{3+}–Cr^{2+}) ≈ 2×10^{-5} dm^3 mol^{-1} s^{-1} ⟨ k_{ex} (V^{3+}–V^{2+}) ≈ 10^{-2} dm^3 mol^{-1} s^{-1}. It indicates that **kinetically V^{2+} is a better reducing agent** in a cross reaction because the rate constant (k_{12})

for the cross-reaction depends on the exchange rate constants k_{11} and k_{22} in terms of Marcus Equation: $k_{12} \approx \sqrt{k_{11}k_{22}K_{12}}$.

(iii) $[Cr(OH_2)_6]^{3+}$ $(t_{2g}^3 e_g^0)$–$[Cr(OH_2)_6]^{2+}$ $(t_{2g}^3 e_g^1)$: σ^* (*i.e.* e_g) $\rightarrow$ σ^* (*i.e.* e_g) transfer.
$[V(OH_2)_6]^{3+}$ $(t_{2g}^2 e_g^0)$–$[V(OH_2)_6]^{2+}$ $(t_{2g}^3 e_g^0)$: π^* (*i.e.* t_{2g}) $\rightarrow$ π^*(*i.e.* t_{2g}) transfer.

This is why, k_{ex} for the couple V^{3+}–V^{2+} is higher than that for the Cr^{3+}–Cr^{2+} couple.

Table 6.3.6.2 Kinetic parameters of some cross reactions occurring through outer sphere electron transfer (OSET) process (*cf.* T. J. Meyer *et al., Inorg. Chem.,* **7,** 2369, 1968; R. A. Marcus *et al, Biochim. Biophys. Acta.,* **811,** 265, 1985).

Oxidant	*Reductant*	*Net electron transfer*	E^0_{12} **(volt)**	k_{12} **(dm³ mol⁻¹ s⁻¹)**
$[Co(NH_3)_6]^{3+}$	$[Cr(OH_2)_6]^{2+}$	$\sigma^* \rightarrow \sigma^*$	0.51	1×10^{-3}
$[Co(en)_3]^{3+}$	$[Cr(OH_2)_6]^{2+}$	$\sigma^* \rightarrow \sigma^*$	0.17	3.5×10^{-4}
$[Co(phen)_3]^{3+}$	$[Cr(OH_2)_6]^{2+}$	$\sigma^* \rightarrow \sigma^*$	0.83	3×10^{1}
$[Co(NH_3)_6]^{3+}$	$[V(OH_2)_6]^{2+}$	$\pi^* \rightarrow \sigma^*$	0.37	1×10^{-2}
$[Co(en)_3]^{3+}$	$[V(OH_2)_6]^{2+}$	$\pi^* \rightarrow \sigma^*$	0.015	7.2×10^{-4}
$[Co(phen)_3]^{3+}$	$[V(OH_2)_6]^{2+}$	$\pi^* \rightarrow \sigma^*$	0.675	4×10^{3}
$[Fe(OH_2)_6]^{3+}$ +	$[Cr(OH_2)_6]^{2+}$	$\sigma^* \rightarrow \pi^*$	1.18	2.3×10^{3}
$[Fe(OH_2)_6]^{3+}$ +	$[V(OH_2)_6]^{2+}$	$\pi^* \rightarrow \pi^*$	1.03	2×10^{4}
$[Ru(NH_3)_6]^{3+}$	$[V(OH_2)_6]^{2+}$	$\pi^* \rightarrow \pi^*$	0.355	8×10^{1}
$[Ru(NH_3)_6]^{3+}$	$[Cr(OH_2)_6]^{2+}$	$\sigma^* \rightarrow \pi^*$	0.51	2×10^{2}
$[Co(NH_3)_6]^{3+}$	$[Cr(bpy)_3]^{2+}$	$\pi^* \rightarrow \sigma^*$	–	7×10^{2}
$[Co(NH_3)_6]^{3+}$	$[Ru(NH_3)_6]^{2+}$	$\pi^* \rightarrow \sigma^*$	≈0.0	1×10^{-2}
$[Co(NH_3)_6]^{3+}$	$[Eu(OH_2)_6]^{2+}$		0.48	2×10^{-2}

(i) **Electronic activation for different paths in OSET:** OSET occurs through the $\pi^* \rightarrow \pi^*$ transfer. By keeping this view in mind, we can conclude: for the $\sigma^* \rightarrow \sigma^*$ net transfer, both the oxidant and reductant are to be electronically activated; for the $\pi^* \rightarrow \sigma^*$ net transfer, only the oxidant is to be electronically activated; for the $\sigma^* \rightarrow \pi^*$ net transfer, the reductant is to be electronically excited.

(j) **$[Ru(NH_3)_6]^{3+}$–$[Cr(OH_2)_6]^{2+}$ *vs.* $[Co(NH_3)_6]^{3+}$–$[Cr(OH_2)_6]^{2+}$:** Both the reactions are thermodynamically equally favourable ($E^0_{12} \approx 0.51$ V) but the rate of reduction of $[Ru(NH_3)_6]^{3+}$ is about 10^5 faster than the rate of reduction of $[Co(NH_3)_6]^{3+}$. In terms of Marcus equation, this rate benefit is mainly due to the higher electron exchange rate constant of the couple $[Ru(NH_3)_6]^{3+}$–$[Ru(NH_3)_6]^{2+}$.

k_{ex} in $[Ru(NH_3)_6]^{3+}$–$[Ru(NH_3)_6]^{2+} \approx 2 \times 10^2 \gg k_{ex}$ in $[Co(NH_3)_6]^{3+}$–$[Co(NH_3)_6]^{2+} \approx 10^{-3}$

$\pi^* \rightarrow \pi^*$ transfer	$\sigma^* \rightarrow \sigma^*$ transfer.
Ru^{III}–N = 210 pm,	Co^{III}–N = 194 pm
Ru^{II}–N = 214 pm,	Co^{II}–N = 211 pm.

The difference in the k_{ex} values (in dm³ mol⁻¹ s⁻¹) is quite reasonable in terms of the above parameters (discussed earlier).

In other words, for the first reaction (*i.e.* reduction of Ru^{III}), the $\sigma^* \rightarrow \pi^*$ transfer process needs only the electronic activation of the reductant while for the second reaction (*i.e.* reduction of Co^{III}) the $\sigma^* \rightarrow \sigma^*$ process needs the electronic activation of both the reductant and oxidant. It makes the rate of the second reaction slower.

(k) **$[Ru(NH_3)_6]^{3+}$–$[V(OH_2)_6]^{2+}$ *vs.* $[Co(NH_3)_6]^{3+}$–$[V(OH_2)_6]^{2+}$:** The first reaction is about 10^4 times faster than the second reaction though both the reactions are equally favoured thermodynamically. This can be explained by considering the higher exchange rate in the $[Ru(NH_3)_6]^{3+}$–$[Ru(NH_3)_6]^{2+}$ couple (*cf.* Marcus equation) and the favourable $\pi^* \rightarrow \pi^*$ transfer in the first reaction. The second reaction experiencing the $\pi^* \rightarrow \sigma^*$ transfer, needs the electronic activation of the oxidant.

(l) **Co(III)–Ru(II) *vs.* Ru(III)-Ru(II):** The examples are:

$$\left[Ru(NH_3)_5(py)\right]^{3+} + \left[Ru(NH_3)_6\right]^{2+} \left(k_{12} = 1.4 \times 10^6\, dm^3\, mol^{-1}\, s^{-1},\ \log K_{12} \approx 4.4\right);$$

$$\left[Ru(bpy)(NH_3)_4\right]^{3+} + \left[Ru(NH_3)_5(py)\right]^{2+} \left(k_{12} = 1.1 \times 10^8\, dm^3\, mol^{-1}\, s^{-1},\ \log K_{12} = 3.4\right);$$

$$\left[Co(phen)_3\right]^{3+} + \left[Ru(NH_3)_6\right]^{2+} \left(k_{12} = 1.5 \times 10^4\, dm^3\, mol^{-1}\, s^{-1},\ \log K_{12} = 5.4\right), \text{etc.}$$

The electronic configurations are:

$$Co(III)\ (t_{2g}^6 e_g^0),\ Co(II)\ (t_{2g}^5 e_g^2);\ Ru(III)\ (t_{2g}^5 e_g^0),\ Ru(II)\ (t_{2g}^6 e_g^0)$$

Thus the Co(III)–Ru(II) system involves the net transfer $\pi^* \rightarrow \sigma^*$ (*i.e.* $t_{2g} \rightarrow e_g$) while the Ru(III)–Ru(II) reaction involves the net transfer $\pi^* \rightarrow \pi^*$ transfer. This is why, the Ru(III)-Ru(II) reaction is faster than the Co(III)–Ru(II) reaction, in general, in spite of the more thermodynamic favour in the Co(III)-Ru(II) reaction. *In other words,* $[Ru(aq)]^{2+}$ or $[Ru(NH_3)_6]^{2+}$ *is a good reducing agent in the OSET path as it requires no electronic activation. It may be noted that Ru(II) being inert cannot generally act as a reducing agent in the ISET process.*

(m) **Comparison of reducing activity of $[Cr(OH_2)_6]^{2+}$ and $[V(OH_2)_6]^{2+}$:** Thermodynamically Cr^{2+} is a better reducing agent but kinetically V^{2+} is a better reducing agent in the OSET process. In fact, V^{2+} reduces Co(III) or Ru(III) complexes much faster than Cr^{2+}. For the reduction of Co(III) complex by Cr^{2+}, it experiences the $\sigma^* \rightarrow \sigma^*$ net transfer and it requires the electronic activation of both the oxidant and reductant. While in the reduction of Co(III) complexes by V^{2+} through the $\pi^* \rightarrow \sigma^*$ net transfer, it needs the electronic activation of the oxidant only. Similarly, for the reduction of Ru(III) complexes by V^{2+} through the $\pi^* \rightarrow \pi^*$ net transfer, it needs no electronic activation of the both oxidant and reductant. On the other hand, for the reduction of Ru(III) complexes by Cr^{2+}, through the $\sigma^* \rightarrow \pi^*$ net transfer, it needs an electronic activation of the reductant. In terms of Marcus equation, the rate benefit for the reduction by V^{2+} over the reduction by Cr^{2+} arises from the higher electron exchange rate constant in the V^{3+}–V^{2+} couple ($\pi^* \rightarrow \pi^*$ transfer) compared to that in the Cr^{3+}–Cr^{2+} couple ($\sigma^* \rightarrow \sigma^*$ transfer). In other words, for the required $\pi^* \rightarrow \pi^*$ electron transfer, in cross reactions, the reductant V^{2+} needs no electronic activation while the reductant Cr^{2+} needs an electronic activation. It makes the activation energy less for the reduction by V^{2+}.

For a common oxidant, if the rate of reduction by V^{2+} is found faster than the rate of reduction by Cr^{2+}, then it can be concluded that the redox process involves the outer sphere electron transfer mechanism.

6.3.7 Effect of External Ions on Electron Transfer Rate in the OSET Process

Sometimes, there is an **electrostatic barrier in forming the precursor outer sphere complex** as in the case of electron exchange in the couples $[Co(NH_3)_6]^{3+}$– $[Co(NH_3)_6]^{2+}$, $[Fe(CN)_6]^{3-}$– $[Fe(CN)_6]^{4-}$, $[MnO_4]^-$– $[MnO_4]^{2-}$, etc. Oppositely charged ions can bridge the oxidant and reductant bearing same type of charge to reduce the electrostatic repulsive barrier.

- **Alkali metal ions (M^+)** can catalyze the electron exchange rate in the couple MnO_4^-– MnO_4^{2-}. M^+ probably bridges them in the outer-sphere complex as follows:
$MnO_4^{2-}\cdots M^+\cdots MnO_4^-$, *i.e.* $[MnO_4\cdots M\cdots MnO_4]^{2-}$. The observed catalytic efficiency runs as:
$$Cs^+ \rangle K^+, Na^+ \rangle Li^+.$$
- **Alkali metal ions (M^+) and alkaline earth metal ions (M^{2+})** are also found to catalyze the electron exchange process in the $[Fe(CN)_6]^{4-}$–$[Fe(CN)_6]^{3-}$ couple. The transition state may involve the following bridges (with M^+).
$$[Fe(CN)_6\cdots M\cdots Fe(CN)_6]^{6-},\ [M\cdots Fe(CN)_6\cdots M\cdots Fe(CN)_6]^{5-},$$
$$[M\cdots Fe(CN)_6\cdots M\cdots Fe(CN)_6\cdots M]^{4-}.$$
The catalyzing efficiency runs as:
$$Cs^+ \rangle Rb^+ \rangle K^+, NH_4^+ \rangle Na^+ \rangle Li^+ \rangle H^+;\ Sr^{2+} \rangle Ca^{2+} \rangle Mg^{2+}.$$
It is suggested that to minimize the electrostatic repulsion in the outer-sphere complex of transition state, the **partially desolvated cation** bridges between the exchanging anions. The bridging cation should not be very large. *In fact, the large cation like Ph_4As^+ is not effective in catalyzing process.* ***At the same time, the energy required to desolvate the catalyzing ion should not be large. This is why, Li^+ which requires a high desolvation energy acts as a poor catalyst.***
- For the exchanging cations, a suitable anion can bridge to catalyze the process. In fact, it has been found that OH^- can catalyze the electron exchange process in the $[Co(NH_3)_6]^{3+}$–$[Co(NH_3)_6]^{2+}$ couple. This can occur in **two ways to stabilize the outer sphere complex in the transition state.**

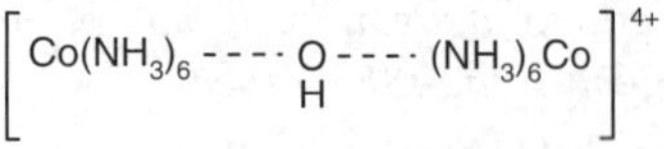

I

Simple bridging by OH−

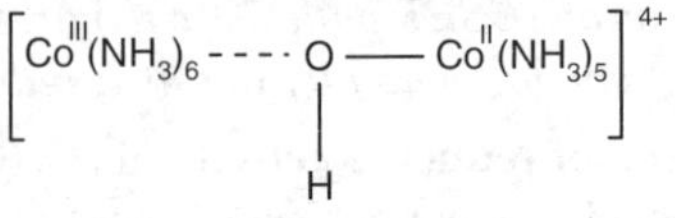

II

Bridging by the coordinated OH group present in the Co^{II}-centre.

Co(II) centre is labile and it can produce $[Co(NH_3)_5(OH)]^+$ and this coordinated OH group can bridge with the coordination sphere of Co(III) through H-bonding. Participation of the substituted Co(II)-complex, *i.e.* $[Co(NH_3)_5(OH)]^+$ in the electron exchange process is supported by the fact that the product contains the $[Co(NH_3)_5(OH)]^{2+}$ species. Formation of the Co(III)-complex as the $[Co(NH_3)_5(OH)]^{2+}$ can be explained by considering the transition state (**II**). Cl^- is also found to catalyze the electron exchange process in the $[Co(NH_3)_6]^{2+}$–$[Co(NH_3)_6]^{3+}$ couple and the catalysis may originate in the same way.

In fact, in the outer sphere complex, aqua and hydroxo-aqua complexes can participate in H-bonding interaction.

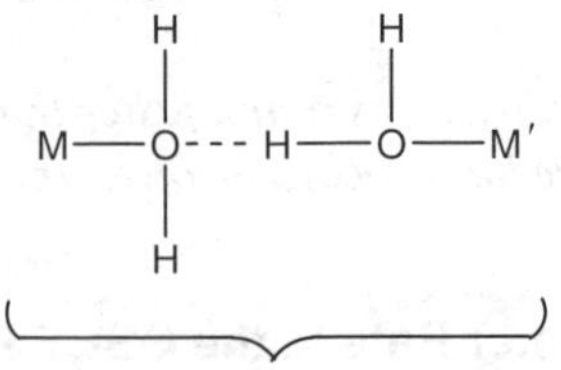

aqua-aqua

III

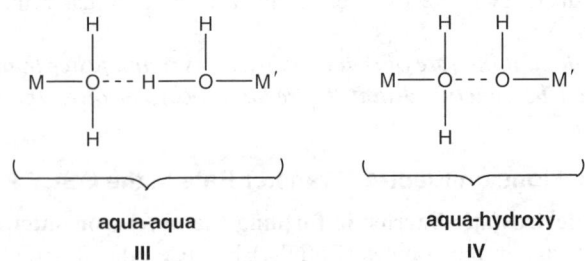

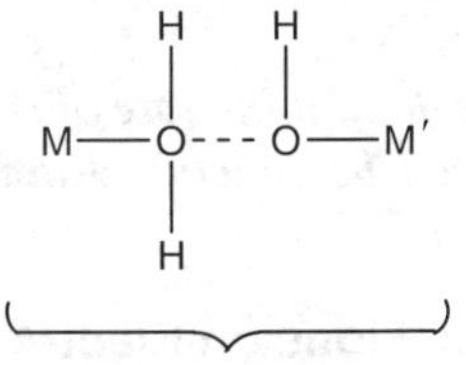

aqua-hydroxy

IV

Obviously the 'OH' group is more effective than H_2O in carrying out the said H-bonding interaction as in **IV**.

6.4 INNER SPHERE ELECTRON TRANSFER (ISET) PROCESS (*cf.* H. Taube, *J. Chem. Edu.*, **45**, 452, 1968)

The reduction of $[Co(NH_3)_6]^{3+}$ by $[Cr(OH_2)_6]^{2+}$ is slow ($k \approx 10^{-4}$ dm^3 mol^{-1} s^{-1}) while the reduction of $[CoCl(NH_3)_5]^{2+}$ by the same reducing agent is reasonably much faster ($k \approx 6 \times 10^5$ dm^3 mol^{-1} s^{-1}) under the comparable conditions. It may be noted that the thermodynamic driving force for both the reactions is more or less comparable.

This tremendous rate enhancement by a factor about 10^9 indicates that the reduction pathways are different for $[Co(NH_3)_6]^{3+}$ and $[CoCl(NH_3)_5]^{2+}$ (*cf.* Sec. 6.4.4 for the classic work of **H. Taube** to establish the ISET path)

$$\left[Co(NH_3)_6\right]^{3+} + \left[Cr(OH_2)_6\right]^{2+} \xrightarrow{H^+} \left[Co(OH_2)_6\right]^{2+} + \left[Cr(OH_2)_6\right]^{3+} + 6NH_4^+$$

$$(t_{2g}^6\, e_g^0) \qquad (t_{2g}^3 e_g^1) \qquad (t_{2g}^5 e_g^2) \qquad (t_{2g}^3 e_g^0)$$

i.e. $e_g \rightarrow e_g$ transfer, *i.e.* $\boldsymbol{\sigma^* \rightarrow \sigma^*}$ **transfer.**

$$\left[CoCl(NH_3)_5\right]^{2+} + \left[Cr(OH_2)_6\right]^{2+} \xrightarrow{H^+} \left[Co(OH_2)_6\right]^{2+} + \left[CrCl(OH_2)_5\right]^{2+} + 5NH_3$$

$$(t_{2g}^6 e_g^0) \qquad (t_{2g}^3 e_g^1) \qquad (t_{2g}^5 e_g^2) \qquad (t_{2g}^3 e_g^0)$$

i.e. $e_g \rightarrow e_g$ transfer $\equiv$ $\boldsymbol{\sigma^* \rightarrow \sigma^*}$ **transfer.**

Besides the rate enhancement, in the second reaction, another important observation is that it produces the chlorido-complex of Cr(III) which is a very inert centre. It indicates that during the electron transfer process, chloride is transferred to the reducing agent. From the knowledge of anation rate of $[Cr(OH_2)_6]^{3+}$ (*i.e.* $[Cr(OH_2)_6]^{3+} + Cl^- \rightarrow [CrCl(OH_2)_5]^{2+} + H_2O$, $k \approx 10^{-8}$ dm^3 mol^{-1} s^{-1} which is much slower than the rate constant for the said redox reaction coupled with the Cl^- transfer) and isotope labelling experiment, it was proved that Cl^- does not enter into the coordination sphere of Cr(III) after its formation. To explain the experimental findings, *i.e.* ***rate enhancement and electron transfer accompanied with chloride transfer to the reducing agent,*** it was required to consider that the second reaction does not go on through the outer-sphere mechanism and it occurs through a different mechanism which involves the electron transfer and chloride transfer in a concommitant step. This mechanistic path is described as the **inner sphere electron transfer (ISET) process.**

6.4.1 Steps of the Inner Sphere Electron Transfer Process

The inner sphere mechanism involves the formation of a *binuclear complex* (called **precursor complex**) from the reactants followed by the electron transfer across the bridging ligand within the binuclear complex. This electron transfer produces the *successor binuclear complex* that undergoes fission to give the products.

To allow the formation of the *binuclear precursor complex,* generally the oxidant (that is substitutionally inert) possesses a bridging ligand (say *X*) that leads to a nucleophilic substitution at the reducing centre (which is substitutionally labile). Thus, the **prerequisite conditions** for the formation of the binuclear precursor complex are: *presence of a ligand having the bridging capacity (i.e. capable of making a bridge) in the coordination sphere of the inert oxidant and sufficient lability of the reducing centre to experience a nucleophilic substitution by the bridging ligand provided by the oxidant.* In some rare cases, (*e.g.* reduction of $HCrO_4^-$ by $[Fe(CN)_6]^{4-}$) the inert reducing centre provides the bridging ligand towards the labile oxidant.

Thus between the reactants, one (generally the oxidant) is inert and the other (generally the reductant) is labile. The inert centre will provide the bridging ligand. This is the condition in general.

The *elementary steps* of an ISET process can be represented as follows in general.

Note: Presence of a bridging ligand with the inert oxidant does not always ensure its reduction by a the reductant through the ISET path, *e.g.* **$[Ru(NH_3)_5X]^{2+}$ + $[V(aq)]^{2+}$: OSET path** (*cf.* V^{2+} : d^3 inert)

Step-I (Formation of the binuclear precursor complex):

$$\text{Ox—X} + H_2O\text{—Red} \rightleftharpoons [\text{Ox—X—Red}]\ (\text{precursor complex}) + H_2O$$

Considering the reaction in aqueous solution, the reducing centre (Red) is to be sufficiently labile to experience a nucleophilic substitution by the ligand *X* (that can act as a bridging ligand) present in the coordination sphere of the starting oxidant (*i.e.* Ox—X) which is itself inert.

(**Note:** Generally, the inert oxidant centre provides the bridging ligand. In some rare cases, *e.g.* reduction of $[HCrO_4]^-$ by $[Fe(CN)_6]^{4-}$, the inert reductant provides the bridging ligand to the labile oxidant.)

Step-II (Chemical activation of the precursor complex):

$$[\text{Ox—X—Red}] \rightleftharpoons [\text{Ox—X—Red}]^{\neq}\ (\textbf{Activated Complex})$$

This step needs the necessary inner sphere reorganizations required for the electron transfer.

This step may require the change of electronic configuration (*i.e. electronic activation*), bond length adjustment and adjustment of other geometrical parameters. ***All these changes are required to overcome the Franck-Condon barrier.***

Step-III (Electron transfer within the activated precursor complex): This leads to the formation of a *binuclear successor.*

$$\overset{\curvearrowleft\, e}{[\text{Ox—X—Red}]^{\neq}} \xrightleftharpoons{e\text{-transfer}} [\text{Ox}^- \text{—X—}^+\text{Red}]\ (\textbf{Successor complex})$$

Step-IV (Fission of the binuclear successor complex to the products): This step will lead to the mononuclear products. The position of bond cleavage depends on the relative ease (both thermodynamically and kinetically) of cleavage of the bonds Ox^-–X and X–$^+$Red.

$$\text{Ox}^- \underset{\uparrow}{\vdots} \text{X—}^+\text{Red} + OH_2 \rightleftharpoons \text{Ox}(OH_2)^- + \text{X—}^+\text{Red}$$

Point of cleavage (**leading to transfer of '*X*' to the reducing agent**).

OR

$$\text{Ox}^- \text{—X} \underset{\uparrow}{\vdots} {}^+\text{Red} + H_2O \rightleftharpoons \text{Ox}^- \text{—X} + H_2O\text{—}^+\text{Red}$$

Point of cleavage (**leading to retention of '*X*' with the oxidizing agent**).

Thus possibility of transfer or retention of the bridging ligand (X) depends on the relative lability of the Ox^- and Red^+ centres and the relative bond strengths of Ox^-–X and Red^+–X bonds. **Thus the ligand transfer is not an essential condition for the ISET process.**

It may be noted that in the OSET process, during the course of electron transfer, the coordination spheres of the reactants remain unchanged and *no chemical bonding interaction occurs between the coordination spheres.* On the other hand, in the ISET process, coordination spheres of the reactants are linked through a bonding interaction. *Thus besides the electron transfer, in the ISET process, bond formation and bond breaking are also required in the overall process.*

6.4.2 Rate Law for the ISET Process

By considering the overall process to be irreversible, the process can be simply represented as follows:

$$\text{Ox—X} + \text{Red} \underset{k_{-1}}{\overset{k_1}{\rightleftharpoons}} \underset{\mathbf{I}}{[\text{Ox—X — Red}]} \xrightarrow{k} \text{Products}$$

Here k denotes the overall rate constant for the 2nd, 3rd and 4th steps shown in Sec. 6.4.1. Thus k stands for the overall process involving activation of the binuclear precursor complex, electron transfer within the activated complex and dissociation of the successor into the final products. Under the **steady state condition** of the intermediate binuclear complex (**I**), we can write:

$$\frac{d[\text{I}]}{dt} = 0 = k_1[\text{Ox –X}][\text{Red}] - k_{-1}[\text{I}] - k[\text{I}]$$

i.e.

$$[\text{I}] = \frac{k_1[\text{Ox—X}][\text{Red}]}{k_{-1} + k}$$

Thus the rate is given by:

$$\text{rate} = k[\text{I}] = \frac{kk_1[\text{OX—X}][\text{Red}]}{k_{-1} + k}$$

Under different conditions, the above rate law reduces as follows:

Case-I: When $k \rangle\rangle k_{-1}$, the rate is given by:

$$\text{rate} \approx k_1[\text{Ox—X}][\text{Red}]$$

In such cases, the rate determining step is the formation of the precursor binuclear complex. *Thus the rate determining step is shifted to the step of nucleophilic substitution at the reducing centre.* If the reducing centre experiences the substitution through a **dissociative path**, then the rate becomes insensitive to the nature of the bridging ligand X (that acts as the entering ligand in the substitution process).

Obviously, shifting of the rate determining step to the nucleophilic substitution step at the reducing centre occurs when the reducing agent is relatively inert (*e.g.* $[V(aq)]^{2+}$, d^3 system).

Case-II: When $k_{-1} \rangle\rangle k$, the rate is given by:

$$\text{rate} \approx \frac{kk_1}{k_{-1}}[\text{Ox—X}][\text{Red}] = kK[\text{Ox—X}][\text{Red}],$$

$$\left(K = \frac{k_1}{k_{-1}}\right).$$

In such cases, the rate determining step involves the activation of the binuclear precursor complex and electron transfer within the binuclear precursor complex leading to the binuclear successor complex *(considering fission of the successor complex into the products to be fast).*

Note: In some cases, the step leading to the fission of the successor complex is very much kinetically disfavoured (compared to the other steps – formation of the binuclear complex, and electron transfer within the binuclear complex) and the rate determining step is shifted to this step. In such cases, the overall reaction can be represented as follows:

$$\text{Ox — X} + \text{Red} \overset{K_{eq}}{\rightleftharpoons} \left[\text{Ox}^- \text{— X — }^+\text{Red}\right], \text{(cumulative expression of Steps I, II and III)}$$

$$\left[\text{Ox}^- \text{— X — }^+\text{Red}\right] \xrightarrow[(rds)]{k} \text{product (mononuclear complexes)}.$$

The rate of the overall process can be given as follows.

$$\text{rate} = \frac{kK_{eq}[\text{Ox}—\text{X}][\text{Red}]}{1+K_{eq}[\text{Red}]}, \left(\text{when } [\text{Red}]_T >> [\text{Ox}—\text{X}]_T\right), (\textit{see } \text{Sec. 5.9.3 for derivation})$$

$$\approx kK_{eq}[\text{Ox}—\text{X}][\text{Red}], \text{ when } 1 >> K_{eq}[\text{Red}]$$

where K_{eq} is the overall equilibrium constant of the steps I, II and III leading to the formation of the binuclear successor complex, and *k actually stands for the rate constant of the step leading to the product from the successor complex through the bond fission.* Such a situation arises when both the metal centres (*i.e.* Ox^- and ^+Red) present within the successor complex are very much kinetically inert in the substitution process.

6.4.3 Effect of the Nature of Donor Orbital (*i.e.* HOMO) of the Reductant and Receptor Orbital (*i.e.* LUMO) of the Oxidant in the ISET Process

It has been already pointed out that in the OSET process, the electron transfer occurs from the π^*-*MO* of the reductant to the π^*-*MO* of the oxidant in the activated outer-sphere complex. Thus if $\pi^* \rightarrow \pi^*$ (*i.e.* $t_{2g} \rightarrow t_{2g}$) transfer is possible from the ground state electronic configurations of the reactants then the process can go on without any electronic activation in the OSET process. *Such OSET reactions are very fast.* In the OSET process, the $\sigma^* \rightarrow \sigma^*$ net transfer (*i.e.* $e_g \rightarrow e_g$ transfer) needs electronic activation of both the oxidant and reductant. *The process is energetically costly and such OSET reactions are tremendously slow.*

Astonishingly, *in the ISET process, for the* $\sigma^* \rightarrow \sigma^*$ *net transfer (i.e.* $e_g \rightarrow e_g$*) process, the process is highly favoured and for the* $\pi^* \rightarrow \pi^*$ *net transfer (i.e.* $t_{2g} \rightarrow t_{2g}$*) the process is tremendously disfavoured.* In fact, no ISET reaction involving the net $\pi^* \rightarrow \pi^*$ transfer is known and **all known reactions experiencing the $\pi^* \rightarrow \pi^*$ net transfer undergo the OSET process.** Table 6.4.3.1 illustrates the effect of the nature of HOMO and LUMO of the reactants on the reaction rates passing through the OSET and ISET processes.

Table 6.4.3.1 Effect of the nature of the donor and receptor orbitals on the rate constants of the reactions passing through the OSET and ISET process.

Donor orbital of the reductant (i.e. HOMO of the reactant)	*Receptor orbital of the oxidant (i.e. LUMO of the oxidant)*	*Net transfer*	*Examples*	$\frac{k_{ISET}}{k_{OSET}}$ *ratio*
π^*	π^*	$\pi^* \rightarrow \pi^*$	V^{2+}–Ru^{3+}	Data not available
π^*	σ^*	$\pi^* \rightarrow \sigma^*$	V^{2+}–Co^{3+}	~10^4
σ^*	π^*	$\sigma^* \rightarrow \pi^*$	Cr^{2+}–Ru^{3+}	~10^2
σ^*	σ^*	$\sigma^* \rightarrow \sigma^*$	Cr^{2+}–Co^{3+}	~10^{10}

$V^{2+}(t_{2g}^3e_g^0)$; $Cr^{2+}(t_{2g}^3e_g^1)$; $Co^{3+}(t_{2g}^6e_g^0$, l.s.); $Co^{2+}(t_{2g}^5e_g^2$, h.s.); $Ru^{3+}(t_{2g}^5e_g^0$, l.s.); $Ru^{2+}(t_{2g}^6e_g^0$, l.s.).

The preference behind the $\sigma^* \rightarrow \sigma^*$ transfer (*i.e.* $e_g \rightarrow e_g$) in the ISET process can be explained by considering the **orbital requirement for the formation of the bridging segment M_{ox}–X–M_{red}** in the binuclear precursor complex. Considering the bridge along the *z*-axis, *the* d_{z^2} *orbitals* (*i.e.* e_g orbitals) *of* M_{ox} *and* M_{red} *can combine with the filled* p_z *orbital of the bridging ligand* X^- (say Cl^-). This will produce **three three-centre *MOs*** (one bonding, one nonbonding and one antibonding). This **bonding scheme** is very much comparable to that in XeF_2 compound.

If it is assumed that the d_{z^2} orbital of reductant (M_{red}) bears an unpaired electron; the d_{z^2} orbital of the oxidant (M_{ox}) is vacant and the p_z orbital of X^- (say Cl^-) is filled in, then these three electrons are distributed as σ_b^2, $\sigma_{nb}^1$and σ_{ab}^0. (*i.e.* **3c–3e bonding**). *The placement of the unpaired electron from the d_{z^2} orbital of the reductant into the nonbonding MO (i.e. σ_{nb}) shifts partially the electron to the M_{ox} because the three centre nonbonding MO (i.e. σ_{nb})* is delocalized over the three centres. The bond order in each M_{ox}–X and X–M_{red} bond in the precursor complex is half. Thus the precursor complex formation yields the following results:

partial transfer of the electron from M_{red} to M_{ox}; bond weakening in the original 2c–2e M_{ox}–X bond (i.e. bond order decreases from 1 to 1/2) and a new bond formation (of half bond order) by the bridging atom with the M_{red}.

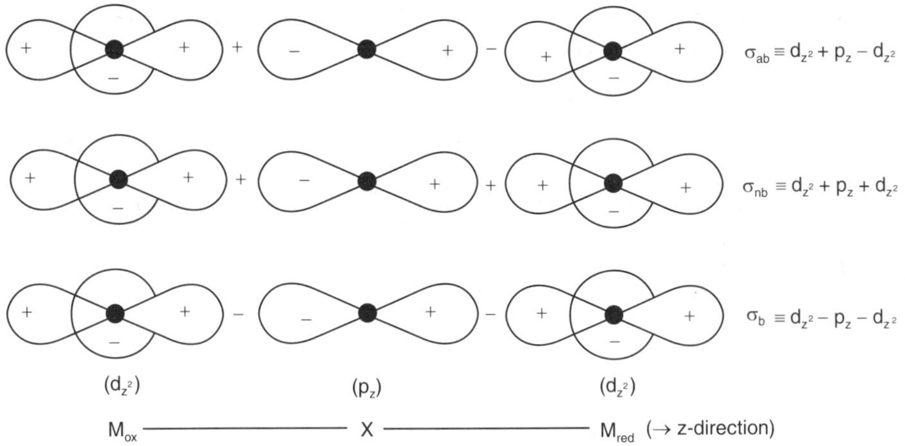

Fig. 6.4.3.1 Orbital picture for the formation of the bridging segment M_{ox}—X—M_{red} in the precursor complex (σ_b, σ_{nb} and σ^* denote the three three-centre MOs–bonding, nonbonding and antibonding respectively).

Now, if the precursor complex approaches smoothly towards the products along the reaction coordinate, the electron in the nonbonding *MO* (*i.e.* σ_{nb}^1) moves out of the M_{red}–X region into the region of M_{ox}–X. This leads to the electron transfer from M_{red} to M_{ox}.

Thus for the electron transfer from the reductant to the oxidant in the ISET process, the donor orbital (HOMO) of the reductant and the receptor orbital (LUMO) of the oxidant must be of σ^* type (*i.e.* e_g in terms of CFT). Formation of the three centre delocalized σ–*MO* allows **the electron transfer through this delocalized MO.** It explains that in the ISET process, electron transfer occurs from $\sigma^* \rightarrow \sigma^*$ (*i.e.* $e_g \rightarrow e_g$).

If the ground state electronic configurations of the reactants do not allow this $\sigma^* \rightarrow \sigma^*$ transfer path then the reactants will be required to be electronically activated.

Obviously, for the net $\pi^* \rightarrow \pi^*$ transfer process, it needs an electronic activation of both the oxidant and reductant and consequently the activation energy in the ISET process will be very high. In fact, the systems having the $\pi^* \rightarrow \pi^*$ net transition adopt the OSET pathway where there is no requirement of electronic activation. In the same line of argument, for the $\sigma^* \rightarrow \pi^*$ net transfer process, the oxidant is to be electronically activated for the required $\sigma^* \rightarrow \sigma^*$ transfer path in the ISET process. Similarly, for the net $\pi^* \rightarrow \sigma^*$ net transfer, the reductant centre needs an **electronic activation** in the ISET path.

6.4.4 Comparison of Reaction Rates for the Reduction of Co(III) Complexes by $[Cr(OH_2)_6]^{2+}$ through the OSET and ISET Processes

The most authentic evidence in favour of the ISET process was provided by **H. Taube** and coworkers (1953–54) who observed the tremendous rate enhancement for the reduction of $[CoCl(NH_3)_5]^{2+}$ by $[Cr(OH_2)_6]^{2+}$ compared to the reduction of $[Co(NH_3)_6]^{3+}$ by the same reducing agent.

$$[Co(NH_3)_6]^{3+} + [Cr(OH_2)_6]^{2+} \xrightarrow{H^+} [Co(OH_2)_6]^{2+} + [Cr(OH_2)_6]^{3+} + 6NH_4^+ \quad (\mathbf{k \approx 10^{-4}\, dm^3\, mol^{-1}\, s^{-1}})$$

$$[CoCl(NH_3)_5]^{2+} + [Cr(OH_2)_6]^{2+} \xrightarrow{H^+} [Co(OH_2)_6]^{2+} + [CrCl(OH_2)_5]^{2+} + 5NH_4^+ \quad (\mathbf{k \approx 6 \times 10^5\, dm^3\, mol^{-1}\, s^{-1}})$$

They argued that the difference in the rate and in the nature of the product, *i.e.* Cr(III), can be explained by considering the reduction of $[CoCl(NH_3)_5]^{2+}$ by $[Cr(aq)]^{2+}$ through the ISET path.

To explain the products in the reduction of $[CoCl(NH_3)_5]^{2+}$ by $[Cr(aq)]^{2+}$, in terms of the OSET and ISET paths, the following sets of reactions are to be proposed.

OSET path:

(a) $$[Co^{III}Cl(NH_3)_5]^{2+} + [Cr^{II}(OH_2)_6]^{2+} \longrightarrow [Co^{II}Cl(NH_3)_5]^{2+} + [Cr(OH_2)_6]^{3+}$$

(b) $$[Co^{II}Cl(NH_3)_5]^{+} \xrightarrow[\text{(fast)}]{H^+,\, H_2O} [Co(OH_2)_6]^{2+} + Cl^- + 5NH_4^+$$

(c) $$[Cr(OH_2)_6]^{3+} + Cl^- \xrightarrow[(\mathbf{k \approx 3\times10^{-8}\, dm^3\, mol^{-1}\, s^{-1}})]{\text{anation}} [CrCl(OH_2)_5]^{2+} + H_2O$$

ISET path:

(d) $$[(H_3N)Co^{III} - Cl]^{2+} + [Cr(OH_2)_6]^{2+} \rightleftharpoons [(H_3N)_5Co^{III} - Cl - Cr^{III}(OH_2)_5]^{4+} + H_2O \quad \text{(Precursor complex)}$$

(e) $$[(H_3N)_5Co^{III} - Cl - Cr^{II}(OH_2)_5]^{4+} \xrightarrow{rds} [(H_3N)_5Co^{II} - Cl - Cr^{III}(OH_2)_5]^{4+} \quad \text{(Successor complex)}$$

(f) $$\xrightarrow[\text{(fast)}]{H^+,\, H_2O} [Co(OH_2)_6]^{2+} + 5NH_4^+ + [CrCl(OH_2)_5]^{2+}$$

To explain the products in the OSET path, it is to be considered that after the generation of Co(II) and Cr(III) centres through the electron transfer, they are to experience ligand substitution process. Co(II) (high spin) being labile can rapidly respond to the reaction (b), but $[Cr(OH_2)_6]^{3+}$ (d^3 system) being kinetically inert undergoes anation (c) very slowly. In fact, the 2nd order rate constant for Cl^- anation of $[Cr(OH_2)_6]^{3+}$ is of the order of 10^{-8} $dm^3\, mol^{-1}\, s^{-1}$ which is about 10^{13} slower than the overall reaction rate. It indicates that the independent anation of $[Cr(OH_2)_6]^{3+}$ after its generation through the electron transfer cannot explain the result. **If the OSET path were true, then the overall reaction rate could not generate the 2nd order rate constant greater than 10^{-8} dm^3 mol^{-1} s^{-1} which is the 2nd order rate constant for Cl^- anation of $[Cr(OH_2)_6]^{3+}$.** Moreover, the isotope labeling experimental result is not tenable with the OSET path. If the reaction is carried out in presence of excess labeled $^*Cl^-$ (radioactive Cl^-), then the final product, *i.e.* $[CrCl(OH_2)_5]^{2+}$ should also contain the labeled chloride, but no $[Cr^*Cl(OH_2)_5]^{2+}$ was detected.

The above findings can be explained by considering the reaction to pass through the ISET path where the ligand transfer and electron transfer occur simultaneously. Actually in the ISET path, electron transfer within the activated precursor complex produces a binuclear successor complex of Cr^{III} and Co^{II} having the bridge Cr^{III}—Cl—Co^{II}. In this binuclear complex, Cr^{III} (d^3) is inert, but Co^{II} ($t_{2g}^5e_g^2$) is labile. In fact, as soon as the successor complex is produced, it undergoes fission through the cleavage of Co^{II}–Cl bond and it leads to the transfer of Cl^- to the reducing centre giving rise to $[Cr^*Cl(OH_2)_5]^{2+}$ as the final product. ***In fact, this step reaction (i.e. reaction f) is so fast that it is reasonable to conclude that the electron transfer (i.e. reaction e) and ligand transfer occur simultaneously.*** This pathway does not require the separate anation at $[Cr(OH_2)_6]^{3+}$ through a different reaction. *Consequently, if the reaction is carried out in presence of $^*Cl^-$, this labeled chloride is not found in the coordination sphere of the product Cr(III).*

$[Co(NH_3)_6]^{3+}$ fails to react with $[Cr(OH_2)_6]^{2+}$ through the ISET path because the NH_3 ligands present with the Co(III) (*i.e.* oxidant) *fail to act as the bridging ligands.* This is why, the reaction goes through the OSET path for which the activation energy is very high because of the $\sigma^* \rightarrow \sigma^*$ net transfer in the reaction.

Characteristic features of reduction of $[Co^{III}(NH_3)_5X]^{n+}$ by $[Cr(OH_2)_6]^{2+}$ in the ISET path: In fact, the reduction rate of $[Co^{III}(NH_3)_5X]^{n+}$ by $[Cr(OH_2)_6]^{2+}$ depends largely on the nature of X.

(i) **For X = NH_3:** ISET is not possible because it cannot act as a bridging ligand. The reaction occurs through the OSET path (with a very high activation energy).

(ii) **For X = H_2O, OH^-:** ^{18}O-labeling experiment indicates the transfer of X, during the electron transfer process.

$$\left[(H_3N)_5Co-{}^*OH_2\right]^{3+} + \left[Cr(OH_2)_6\right]^{2+} \xrightarrow{H^+} \left[Co(OH_2)_6\right]^{2+} + 5NH_4^+ + \left[(H_2O)_5Cr({}^*OH_2)\right]^{3+}$$

'$^-$OH' is a much better bridging ligand than H_2O and the rate is much increased for X = OH^-. This is why, **$[Co^{III}(NH_3)_5(OH_2)]$ reacts much faster at the relatively higher pH** where the aqua-complex produces the hydroxido-complex through deprotonation.

Table 6.4.4.1 The second order rate constant for reduction of $[Co^{III}(NH_3)_5X]^{n+}$ by $[Cr(OH_2)_6]^{2+}$

X =	*NH_3*	*OH_2*	*^-OH*	*Cl^-*	*N_3^-*	*NCS^-*	*SCN^-*
k (dm^3 mol^{-1} s^{-1}) = (25°C)	9×10^{-5}	0.5	1.5×10^6	6×10^5	3×10^5	20	1.8×10^5

Here it is worth mentioning that though there are evidences to support the fact that H_2O as a bridging ligand can mediate the ISET process in some cases as in the reduction of $[Co(NH_3)_5(OH_2)]^{3+}$ by $[Cr(OH_2)_6]^{2+}$, very often its poor bridging capacity leads to the OSET process. This is why, the electron exchange in the couple, $[Fe(OH_2)_6]^{3+}$ – $[Fe(OH_2)_6]^{2+}$ occurs through the OSET process, but at the relatively higher pH, the couple $[Fe(OH)(OH_2)_5]^{2+}$ – $[Fe(OH_2)_6]^{2+}$ experiences the ISET process for the electron exchange because the OH group attached to Fe(III) can act as a good bridging ligand (*cf.* Sec. 6.4.10).

OSET *vs.* ISET for the reaction [Co(NH$_3$)$_5$(OH$_2$)]$^{3+}$ + [Cr(aq)]$^{2+}$

Though for X = OH$_2$, the rate is increased by a factor about 10^4 compared to the rate for X = NH$_3$, there is a doubt whether the reaction switches over from outer-sphere (for X = NH$_3$) to inner-sphere (for X = OH$_2$). In fact, H$_2$O is not expected to act as a good bridging ligand. It may be argued that for both X = NH$_3$, OH$_2$, the reaction, *i.e.* [Co^{III}(NH$_3$)$_5$X]$^{2+}$ + [Cr(aq)]$^{2+}$ (X = NH$_3$, H$_2$O) goes through the OSET path. *The rate benefit for X = H$_2$O is probably due to the better H-bonding interaction between the coordination spheres of the two reactants in the outer-sphere precursor complex.* In fact, this H-bonding interaction gives the very high K_{OS} value to favour the overall rate process in the OSET process.

However, there are some evidences to support the fact that the reaction [Co(NH$_3$)$_5$(OH$_2$)]$^{3+}$ + [Cr(aq)]$^{2+}$ passes through the ISET path involving the following precursor inner sphere complex.

$$\left[(H_3N)_5Co \; \text{----} \; \overset{\displaystyle H}{\underset{\displaystyle H}{\overset{\diagdown}{\underset{\diagup}{O^*}}}} \text{----} Cr(OH_2)_5 \right]^{5+}$$

The transfer of labeled H$_2$O* to the Cr-centre, *i.e.* formation of [Cr(OH$_2$)$_5$(*OH$_2$)]$^{3+}$ when [(H$_3$N)$_5$Co(*OH$_2$)]$^{3+}$ reacting with [Cr(aq)]$^{2+}$, supports the ISET path of the reaction. There are also other evidences to support the proposed ISET path. The rate increases with the increases of pH. At higher pH, [Co(NH$_3$)$_5$(OH$_2$)]$^{3+}$ hydrolyses to generate [Co(NH$_3$)$_5$(OH)]$^{2+}$ that can more favourably participate in the ISET path. **[Co(NH$_3$)$_5$(*OH$_2$)]$^{3+}$ (labeled with ^{18}O) reacts slower than the unlabeled complex [Co(NH$_3$)$_5$(OH$_2$)]$^{3+}$.** Heavier H$_2$^{18}O moves slower than H$_2$^{16}O. It also supports the transfer of the bridging H$_2$O* along with the electron transfer (*i.e.* supporting the ISET path). Interestingly, both the labeled and unlabeled [Ru(NH$_3$)$_6$]$^{3+}$ reacts at comparable rates with [Cr(aq)]$^{2+}$. It is reasonable because [Ru(NH$_3$)$_6$]$^{3+}$ reacts with [Cr(aq)]$^{2+}$ in the OSET path where there is no question of ligand transfer.

The aspect of **H$_2$O *vs.* OH$^-$ as a bridging ligand** in the ISET process has been discussed in Sec. 6.4.10.

(iii) **For X = Cl$^-$, N$_3$$^-$:** The rate is significantly high because they can act as the good bridging ligands.

(iv) **For X = NCS$^-$ and SCN$^-$:** [Co(NCS)(NH$_3$)$_5$]$^{2+}$ reacts much slower than [Co(NH$_3$)$_5$(SCN)]$^{2+}$ by a factor about 10^4. The Cr(III) products are also different.

$$\left[Co(NCS)(NH_3)_5 \right]^{2+} \xrightarrow{[Cr(aq)]^{2+}, H^+} \left[Co(OH_2)_6 \right]^{2+} + 5NH_4^+ + \left[Cr(OH_2)_5(SCN) \right]^{2+}$$
$$\text{(main product)}$$

$$\left[Co(NH_3)_5(SCN) \right]^{2+} \xrightarrow{[Cr(aq)]^{2+}, H^+} \left[Co(OH_2)_6 \right]^{2+} + 5NH_4^+ + \left[Cr(NCS)(OH_2)_5 \right]^{2+}$$
$$\text{(main product)}$$

The product analysis indicates that for the first reaction, the free *S*-end of NCS$^-$ binds with the Cr(II) centre in the precursor binuclear complex while in the second reaction, the free *N*-end binds with the Cr(II) centre. These are:

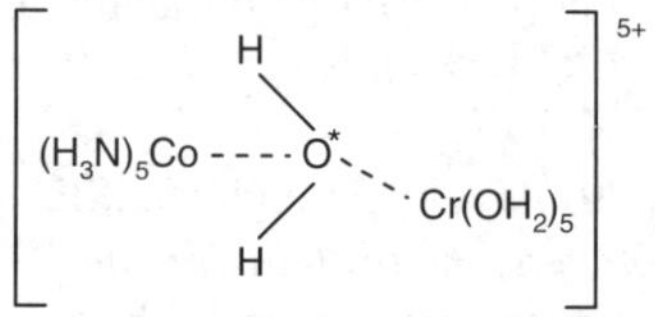

$$\left[(H_3N)_5Co^{III}—NCS\right]^{2+} + \left[Cr(OH_2)_6\right]^{2+} \rightleftharpoons \left[(H_3N)_5Co^{III}—NCS—Cr^{II}(OH_2)_5\right]^{4+} + H_2O$$

$$\Big\downarrow \text{rds}$$

$$\left[Co(OH_2)_6\right]^{2+} + 5NH_4^+ + \left[(H_2O)_5Cr(SCN)\right]^{2+} \xleftarrow[\text{fast}]{H^+,\ H_2O} \left[(H_3N)_5Co^{II}\cdots NCS—Cr^{III}(OH_2)_5\right]^{4+}$$

i.e. $\left[Cr(OH_2)_5(SCN)\right]^{2+}$
(green)

$$\left[(H_3N)_5Co^{III}—SCN\right]^{2+} + \left[Cr(OH_2)_6\right]^{2+} \rightleftharpoons \left[(H_3N)_5Co^{III}—SCN—Cr^{II}(OH_2)_5\right]^{4+} + H_2O$$

$$\Big\downarrow \text{rds}$$

$$\left[Co(OH_2)_6\right]^{2+} + 5NH_4^+ + \left[(H_2O)_5Cr(NCS)\right]^{2+} \xleftarrow{H^+,\ H_2O} \left[(H_3N)_5Co^{II}\cdots SCN—Cr^{III}(OH_2)_5\right]^{4+}$$

i.e. $\left[Cr(NCS)(OH_2)_5\right]$
(purple)

$[Cr(OH_2)_6]^{2+}$ being a relatively harder centre prefers the N-binding site and the *S*-site of NCS^- is not preferred by $[Cr(OH_2)_6]^{2+}$. Thus, coordinated —*S*CN can act as a better bridging ligand then *coordinated —N*CS. **This explains the rate enhancement for Co^{III}—SCN.**

(**Note:** For Co^{III}–NCS, only ***remote attack*** on S-end is possible but for Co^{III}–SCN, both ***adjacent attack*** on S site and ***remote attack*** on N-site are possible. However, for Co^{III}–SCN, the remote attack is the predominant path. The aspect has been discussed in Sec. 6.4.9).

The **purple isomer**, $[Cr(NCS)(OH_2)_5]^{2+}$ **thermodynamically is more stable** than the **green isomer**, $[Cr(OH_2)_5(SCN)]^{2+}$ (*cf.* **symbiotic theory** of HSAB principle).

6.4.5 Types of Inner Sphere Electron Transfer (ISET) Reactions Depending on the Position of the Rate Determining Step

(A) Type-I (electron transfer step is the rate determining step): The overall process consists of different steps (*cf.* Secs. 6.4.1, 6.4.2). If the electron transfer within the precursor complex to generate the successor complex (*i.e.* formation of the precursor complex in the pre-equilibrium step is fast; fission of the successor complex into the products is also fast), is the slowest step then it gives the following rate law (*cf.* Sec. 6.4.2 for the rate law derivation):

$$\text{rate} \approx k\left(\frac{k_1}{k_{-1}}\right)[Ox][Red], \left(\text{under the condition } k_{-1} >> k\right)$$

$$= kK[Ox][Red]$$

Most of the ISET reactions belong to this type of reaction.

For the electron transfer within the precursor complex in the inner-sphere process, it may require the electronic activation of the reactants; lowering of energy of the receptor orbital of the oxidant (*cf.* Sec. 6.4.14). *It also needs the reorganization* in the inner-sphere of the reactants *to overcome the Franck-Condon barrier.* To overcome this **Franck-Condon barrier**, it needs the elongation of bond length in the oxidant and shortening of bond length in the reductant in the activated complex. This bond length rearrangement energy depends on the ligand field strength and electronic configuration (*cf.* Sec. 6.4.14). In fact, for this electron transfer step, a Marcus type relationship also exists.

The dinuclear complex, $\left[Cl(bpy)_2Ru^{II} - N \bigcirc N - Ru^{III}(bpy)_2Cl \right]^{3+}$ involving the Ru^{II} and Ru^{III}

centres has been treated as a **model complex** of the precursor complex.

(B) Type-II (formation of the binuclear precursor complex is the rate determining step): If the *reductant* (*e.g.* $[V(aq)]^{2+}$) is relatively inert, then the substitution at the reductant leading to the dinuclear precursor complex is the rate determining step (*i.e.* electron transfer leading to the successor complex and fission of the successor complex into the product are the relatively faster steps), then the rate for this **substitution controlled ISET process** is given by:

$$\text{rate} \approx k_1[\text{Ox}][\text{Red}], \text{ (under the condition, } k \gg k_{-1}; \text{ } see \text{ Sec. 6.4.2)}$$

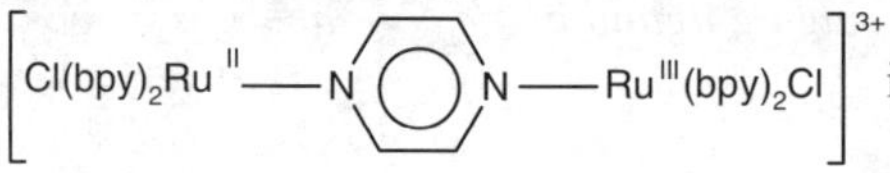

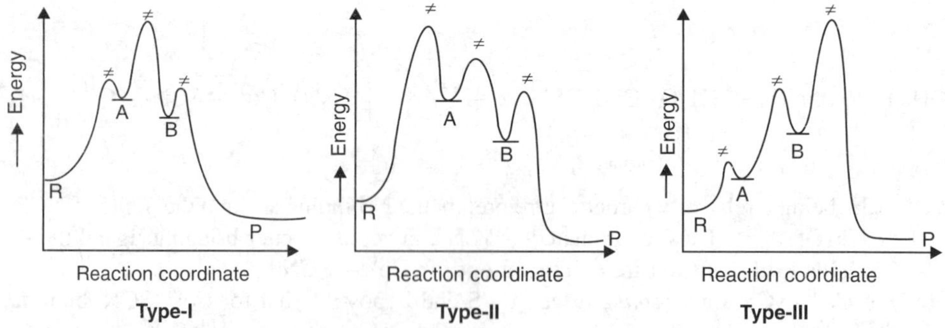

R = Reactants; A = Binuclear precursor intermediate complex, B = Binuclear successor intermediate complex, P = product. ≠ denotes the corresponding activated state or transition state.

        Type-I : Electron transfer step (*i.e.* A→ B) rate determining step.
        Type-II : Formation of the binuclear precursor complex (*i.e.* R→ A) rate determining step.
        Type-III: Formation of products from the binuclear successor complex (*i.e.* B→ P)

Fig. 6.4.5.1 Reaction profile of inner sphere electron transfer reactions.

   (a) $[V(aq)]^{2+}$ as a reducing agent in the substitution controlled ISET process: *Substitution controlled inner sphere reduction* by $[V(aq)]^{2+}$ (d^3) is well documented. Let us consider the lability of some common reducing agents often employed in the ISET process.

Water exchange: $\left[M(OH_2)_6 \right]^{2+} + H_2O^* \longrightarrow \left[M(OH_2)_5 (^*OH_2) \right]^{2+} + H_2O$; M = V, Cr, and Fe

                          $[V(OH_2)_6]^{2+}$ $[Cr(OH_2)_6]^{2+}$ $[Fe(OH_2)_6]^{2+}$

k_{ex} (s^{-1}): 10^2 10^9 10^6

   The reaction rates and their activation parameters for the reduction of different Co(III)–complexes by $[V(aq)]^{2+}$ are very much comparable with those of water exchange and nucleophilic substitution process of $[V(aq)]^{2+}$ (*cf.* Tables 6.4.5.1–2). These features for reduction by $[V(aq)]^{2+}$ are in direct contrast with those found for the reductions of Co(III)-complexes by $[Fe(aq)]^{2+}$ and $[Cr(aq)]^{2+}$. In fact, for reductions by $[Fe(aq)]^{2+}$ and $[Cr(aq)]^{2+}$, the rate determining step is the electron transfer step leading to the successor complex (*i.e.* **Type I ISET**).

Reduction of $[Ru^{III}(NH_3)_5X]^{2+}$ by $[V(aq)]^{2+}$: OSET or ISET

The presence of a suitable bridging ligand in the coordination sphere of an inert oxidant does not ensure its reduction by a reductant through the ISET path. The situation is illustrated in the following redox reaction.

$$\left[Ru^{III}(NH_3)_5X\right]^{2+} + \left[V(OH_2)_6\right]^{2+} \longrightarrow \left[Ru^{II}(NH_3)_5X\right]^{+} + \left[V(OH_2)_6\right]^{3+},$$

$$k \approx 5 \times 10^3 \text{ dm}^3 \text{ mol}^{-1} \text{ s}^{-1} (\pi^* \rightarrow \pi^* \text{ net transfer})$$

The reaction rate is much faster than the rate of nucleophilic substitution process at $[V(aq)]^{2+}$. *It clearly indicates that the above redox reaction does not pass through the ISET path.*

In spite of the presence of a suitable bridging ligand (X) with the oxidant Ru^{III}-complex, its reduction by $[V(aq)]^{2+}$ does not occur through the ISET process. Rather, the reaction occurs through the **OSET process**. In fact, the given reaction experiences the net $\pi^* \rightarrow \pi^*$ transfer which is the **most favourable path in the OSET process** (*cf.* reduction of the low-spin Co(III)-complexes by V(II) experiences the $\pi^* \rightarrow \sigma^*$ net transfer which requires only the electronic activation at the oxidant in the OSET process). The **$\pi^* \rightarrow \pi^*$ transfer is the most unfavourable path for the ISET process** because it needs the electronic activation of both the oxidant and reductant. Moreover, **inertness of $[V(aq)]^{2+}$ puts a mild barrier for the ISET process.**

Thus, if the reduction of Ru(III) by V(II) is supposed to pass through the ISET path, it will require the electronic activation of both the oxidant and reductant for the required $\sigma^* \rightarrow \sigma^*$ transfer at the kinetic step. On the other hand, if the reaction occurs through the OSET path, it needs no electronic activation. Consequently, it passes through the OSET process in which the activation energy is relatively less.

Table 6.4.5.1 Rate constants and activation parameters for reduction of Co(III)–complexes by $[V(aq)]^{2+}$ (25°C) **NET ELECTRON TRANSFER:** $\pi^* \rightarrow \sigma^*$ *i.e.* $t_{2g} \rightarrow e_g$. (*cf.* J. Espenson, *Inorg Chem.*, **4**, 121, 1965; J. A. Stritar *et al. ibid*, **8**, 2284, 1969).

Oxidant Co(III)-complex	k ($dm^3\ mol^{-1}\ s^{-1}$)	$\Delta H^{\neq}$ ($kJ\ mol^{-1}$)	$\Delta S^{\neq}$ ($J\ K^{-1}\ mol^{-1}$)
$[CoCl(NH_3)_5]^{2+}$	8.0		
$[CoBr(NH_3)_5]^{2+}$	25.0		
$[CoI(NH_3)_5]^{2+}$	13.0		
$[Co(NH_3)_5N_3]^{2+}$	13.0		
$[Co(NCS)(NH_3)_5]^{2+}$	0.3		
$[Co(C_2O_4H)(NH_3)_5]^{2+}$	12.5	51.0	–54.0
cis–$[Co(en)_2(NH_3)(N_3)]^{2+}$	10.3	53.0	–50.0
cis–$[Co(en)_2(N_3)(OH_2)]^{2+}$	16.5	51.0	–50.0
trans–$[Co(en)_2(N_3)_2]^{+}$	26.5	51.0	–46.0

$[V(aq)]^{2+}$ ($t_{2g}^3 e_g^0$); $[V(aq)]^{3+}$ ($t_{2g}^2 e_g^0$); Co^{3+} (t_{2g}^6, l.s.); Co^{2+} ($t_{2g}^5 e_g^2$, h.s.).

Table 6.4.5.2 Kinetic parameters (25°C) of substitution at $[V(OH_2)_6]^{2+}$

Process	*Rate Constant*	$\Delta H^{\neq}$ (kJ mol^{-1})	$\Delta S^{\neq}$ (J K^{-1} mol^{-1})
Water exchange	1×10^2 s^{-1}	68.5	−23
Anation by NCS^-	28 dm^3 mol^{-1} s^{-1}	56.5	−29

Tables 6.4.5.1-2 indicate that the kinetic parameters for the reduction of Co(III)-complexes by $[V(aq)]^{2+}$ are comparable with those of the ligand substitution reactions of $[V(aq)]^{2+}$.

It is interesting to note that for the reductions of $[Co^{III}(NH_3)_5X]^{2+}$ by $[V(aq)]^{2+}$, *the rates are more or less insensitive to the nature of the bridging ligands.* It is quite reasonable. The rate determining step is the nucleophilic substitution at $[V(aq)]^{2+}$ and the nucleophilic substitution at $[V(aq)]^{2+}$ occurs through the **dissociative path.** *Consequently, the substitution process is more or less insensitive to the nature of the nucleophile (i.e. X in the present case).*

(b) Magnitude of the rate constant, a diagnostic parameter for the mechanistic paths of reductions by [V(aq)]$^{2+}$: If the rate constants for the reduction of metal complexes by $[V(aq)^{2+}]$ are close to those of nucleophilic substitution process at $[V(aq)]^{2+}$, then the said redox reactions are reasonably considered to pass through the ISET path where the ***rate determining step is the dissociative nucleophilic substitution at [V(aq)]$^{2+}$ to produce the binuclear precursor complex.*** On the other hand, if the rate constants of the redox reactions are substantially higher than the rate constants of substitution reactions of $[V(aq)]^{2+}$, then the redox reactions are supposed to pass through the OSET path rather than the ISET path.

$[Co^{III}(NH_3)_5X]^{2+} + [V(aq)]^{2+}$ ($\pi^* \rightarrow \sigma^*$ net transfer): $k_{el} \approx k_{subs}$; **substitution controlled ISET process**

$[Ru(NH_3)_5X]^{2+} + [V(aq)]^{2+}$ ($\pi^* \rightarrow \pi^*$ net transfer): $k_{el} \gg k_{subs}$; **OSET process**

$[Fe(aq)]^{3+} + [V(aq)]^{2+}$ ($\pi^* \rightarrow \pi^*$ net transfer): $k_{el} \approx 2 \times 10^4$ dm^3 mol^{-1} s^{-1} $\gg k_{subs}$; **OSET process**

$k_{el} \Rightarrow$ rate constant for the given redox reaction

$k_{subs} \Rightarrow$ rate constant for the substitution process of $[V(aq)]^{2+}$.

(c) [Cr(aq)]$^{2+}$ and [Fe(aq)]$^{2+}$as the reducing agents in the ISET process: The kinetic parameters for the reduction of Co(III)-complexes by these reducing agents in the ISET process are given in Table 6.4.5.3.

Table 6.4.5.3 Rate constants (at 25° C) for reduction of some Co(III)-complexes by $[Cr(aq)]^{2+}$ ($t_{2g}^3 e_g^1$) and $[Fe(aq)]^{2+}$ ($t_{2g}^4 e_g^2$) (*cf.* J. A. Stritar *et al, Inorg Chem.*, **8**, 2284, 1969; J. Espenson *et al, ibid,* **4**, 121, 1965).

Oxidant	***k* (dm^3 mol^{-1} s^{-1}) (*for* Cr^{II})**	Transfer	***k* (dm^3 mol^{-1} s^{-1}) (*for* Fe^{II})**	Transfer
$[CoCl(NH_3)_5]^{2+}$	6×10^5	**Net $\sigma^* \rightarrow \sigma^*$ transfer**	1.5×10^{-3}	**Net $\pi^* \rightarrow \sigma^*$ transfer**
$[CoBr(NH_3)_5]^{2+}$	1.5×10^6		7.3×10^{-4}	
$[CoN_3(NH_3)_5]^{2+}$	3×10^5		8.5×10^{-3}	
$[Co(NCS)(NH_3)_5]^{2+}$	20		$\langle 3 \times 10^{-6}$	
$[Co(C_2O_4H)(NH_3)_5]^{2+}$	4×10^2		4.5×10^{-1}	

k_{ex} for water exchange process: k_{ex} for $[Fe(OH_2)_6]^{2+} = 10^6$ s^{-1}; k_{ex} for $[Cr(OH_2)_6]^{2+} = 10^9$ s^{-1}.

Note: In the reduction of Co(III)-complexes by Fe(II), to understand the net transfer of electron, let us consider their ground state electronic configurations.

$$\mathrm{Co(III)}\left(t_{2g}^6 e_g^0\right) \xrightarrow{+e} \mathrm{Co(II)}\left(t_{2g}^5 e_g^2\right);\ \mathrm{Fe(II)}\left(t_{2g}^4 e_g^2\right) \xrightarrow{-e} \mathrm{Fe(III)}\left(t_{2g}^3 e_g^2\right)$$

The net transfer involves $t_{2g} \rightarrow e_g$, *i.e.* $\pi^* \rightarrow \sigma^*$. Thus it needs the electronic activation of Fe(II) for the ISET path. But, if we consider the transfer of an electron from Fe(II) in the following way:

$$(\text{Fe}^{\text{II}})\,t_{2g}^4 e_g^2 \xrightarrow{-e} (\text{Fe}^{\text{III}})\,t_{2g}^4 e_g^1 \ \textbf{(excited state)} \xrightarrow{\text{relaxation}} \text{Fe(III)}\left(t_{2g}^3 e_g^2\right)$$

then it will not need the electronic activation of Fe(II) to execute the $\sigma^* \rightarrow \sigma^*$ transfer in ISET path, but it will generate Fe(III) in an excited state.

Reduction of the Co(III) complexes by [Cr(aq)]$^{2+}$ is much faster than the reduction by [Fe(aq)]$^{2+}$. Both [Cr(*aq*)]$^{2+}$ and [Fe(*aq*)]$^{2+}$ are the labile centres (*cf.* in contrast to [V(*aq*)]$^{2+}$) and consequently the step leading to the formation of the binuclear precursor complex cannot be the rate determining step. In fact, for the reduction by [Cr(*aq*)]$^{2+}$ and [Fe(*aq*)]$^{2+}$, the electron transfer step leading to the successor complex is the rate determining step (**Type-I**, *cf.* Fig. 6.4.5.1).

Thermodynamically [Cr(*aq*)]$^{2+}$ is a better reducing agent than [Fe(*aq*)]$^{2+}$. It is evident from the reduction potentials of the couples.

$$E^0_{\text{Cr}^{3+}/\text{Cr}^{2+}} = -0.41 \text{ V}; \quad E^0_{\text{Fe}^{3+}/\text{Fe}^{2+}} = +0.76 \text{ V}$$

Reduction of the Co^{III}-complexes by [Cr(*aq*)]$^{2+}$ involves the net $\sigma^* \rightarrow \sigma^*$ transfer while reduction of the Co^{III}-complexes by [Fe(*aq*)]$^{2+}$ involves the net $\pi^* \rightarrow \sigma^*$ transfer. Thus for the reduction by Cr^{2+} through the ISET path, there is no need of electronic activation while for the reduction by [Fe(*aq*)]$^{2+}$ through the ISET path, it needs the electronic activation of the reductant. ***Thus the thermodynamic disfavour and unfavourable nature of the HOMO of the reductant are jointly responsible for the lower rates of reduction of the Co(III) complexes by Fe(II) compared to those for the reduction by Cr(II).***

(d) Reduction of [Co^{III}(NH$_3$)$_5$X]$^{2+}$ by Eu^{2+} in the ISET process (*See* Question No. 92): As in the case of reduction of Co^{III}-complexes by [V(*aq*)]$^{2+}$ through the ISET path, the reduction rate by Eu^{2+} ($4f^7$) is also insensitive to the nature of the bridging ligand (X). *It indicates that the reduction of [Co^{III}(NH$_3$)$_5$X]$^{2+}$ complexes by Eu^{2+} is also an example of* **substitution controlled inner sphere electron transfer reaction.** However, some workers have argued the reduction by [Eu(aq)$^{2+}$] as the **outer sphere electron transfer reaction** (*cf.* reduction by inert [Cr(bpy)$_3$]$^{2+}$ through the OSET path is insensitive to the nature of X). It may be noted that both the **products Co(II) and Eu(III) are highly labile to release the bridging ligand to the solvent.**

Table 6.4.5.4 Second order rate constants (25°C) for the reduction of Co(III)-complexes by [V(*aq*)]$^{2+}$ and [Eu(*aq*)]$^{2+}$. (**Substitution controlled ISET process**)

Co(III)-complex (oxidant)	***k* (dm^3 mol^{-1} s^{-1}) by [V(aq)]$^{2+}$**	***k* (dm^3 mol^{-1} s^{-1}) by [Eu(aq)]$^{2+}$**	***k* (dm^3 mol^{-1} s^{-1}) by [Cr(bpy)$_3$]$^{2+}$**
[CoCl(NH$_3$)$_5$]$^{2+}$	10	4×10^2	8×10^5
[Co(NH$_3$)$_5$(N$_3$)]$^{2+}$	13	2×10^2	4×10^4
[Co(NH$_3$)$_5$(NO$_3$)]$^{2+}$	–	1×10^2	–

$$E^0_{V^{3+}/V^{2+}} = -0.26 \text{ V}; \quad E^0_{Eu^{3+}/Eu^{2+}} = -0.37 \text{ V}$$

(e) Electronic activation of [V(aq)]$^{2+}$ in the ISET path: Here it may be mentioned that for the reduction of the Co(III) complexes by [V(*aq*)]$^{2+}$, the rate determining step is the substitution at [V(*aq*)]$^{2+}$ centre by the bridging ligand present with the Co(III) centre. The net transfer $\pi^* \rightarrow \sigma^*$ (*i.e.* $t_{2g} \rightarrow e_g$) also requires the activation of [V(*aq*)]$^{2+}$ from $t_{2g}^3 e_g^0$ to $t_{2g}^2 e_g^1$ to execute the $\sigma^* \rightarrow \sigma^*$ transfer (*i.e.* $e_g \rightarrow e_g$) at the kinetic step of electron transfer in the inner sphere path. *This electronic activation also puts a barrier* *compared to the reduction by Cr^{2+} where the net transfer is $\sigma^* \rightarrow \sigma^*$.* Here it may be mentioned that $[V(aq)]^{2+}$ is a poorer reducing agent than $[Cr(aq)]^{2+}$ in a thermodynamic sense. (*cf.* Secs. 6.3.6 and Table 6.3.6.2).

(C) **Type-III (fission of the successor complex to the final products is the rate determining step):** It happens so when the electron transfer generates *both the metals centres inert in the successor binuclear complex.* It is classically illustrated in the reduction of $[Ru^{III}(NH_3)_5X]^{2+}$ by $[Cr(aq)]^{2+}$.

$$\left[(H_3N)_5Ru^{III}-X\right]^{2+} + \left[Cr(OH_2)_6\right]^{2+} \rightleftharpoons \left[(H_3N)_5Ru^{III}-X-Cr^{II}(OH_2)_5\right]^{4+} + H_2O$$

$$\left[(H_3N)_5Ru^{III}-X-Cr^{II}(OH_2)_5\right]^{4+} \xrightarrow[e\text{-transfer}]{\text{fast}} \left[(H_3N)_5Ru^{II}-X-Cr^{III}(OH_2)\right]^{4+} \xrightarrow{k\ (\text{rds})} \text{Product}$$

The above scheme leads to the following rate law (*cf.* Secs. 5.9.3; 6.4.2):

$$\text{Rate} = \frac{kK_{eq}}{1+K_{eq}[Cr^{II}]}\left[Ru(NH_3)_5X^{2+}\right]_T[Cr^{II}];\ \text{in presence of excess reductant}\ \left(i.e.\ [Cr^{II}]_T \gg [Ru^{III}]_T\right)$$

i.e. $$k_{obs} = \frac{kK_{eq}[Cr^{II}]}{1+K_{eq}[Cr^{II}]},\ K_{eq} = \text{overall equilibrium constant of the process leading to the formation of the binuclear precursor complex}$$

In such cases, there is a bright possibility to *characterize the successor binuclear complex.*

Note: The lone pair on the 'N' site of the $-CONH_2$ group cannot be used for the nucleophilic attack on the reductant to make a bridge because the **lone pair is engaged in resonance with the carbonyl group.** Moreover, attack by the N-site of the $-CONH_2$ group will produce a **nonconjugated system** that will not allow the ISET process. These aspects have been discussed in Sec. 6.4.9.

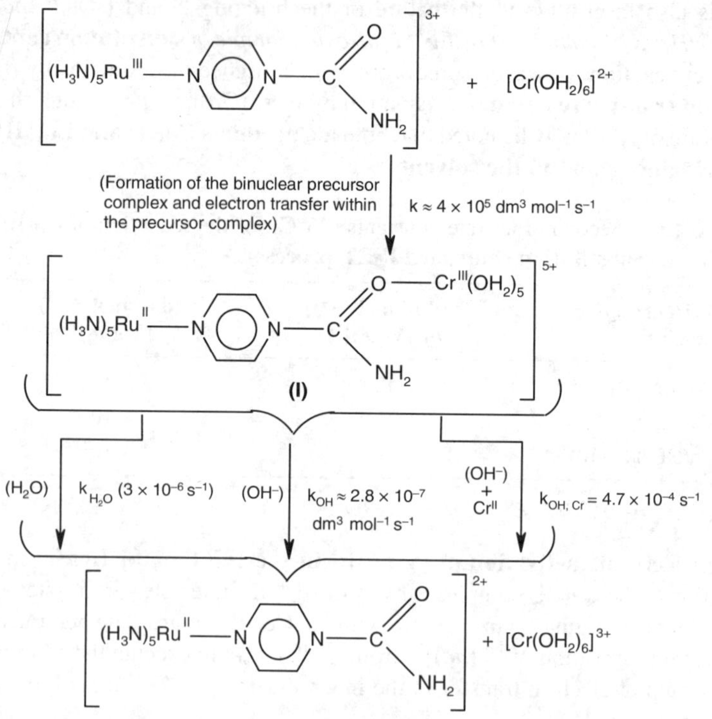

(a) **Reduction of $[Ru^{III}(NH_3)_5X]$ by $[Cr(aq)]^{2+}$ through the Type-III ISET process (X = isonicotinamide)** (*cf.* R. G. Gaunder *et al, Inorg. Chem.*, **9**, 2627, 1970; **14**, 1283, 1975)**:** In the reduction of pentaammineisonicotinamideruthenium(III) by $[Cr(aq)]^{2+}$, formation of the precursor complex and electron transfer within the binuclear complex leading to the binuclear successor complex bearing Ru(II) and Cr(III) become complete within the time scale of mixing. *It gives the orange-yellow binuclear successor complex which is sufficiently stable to be characterized.* In fact, in this successor complex, both the metal centres are inert. These inert centres are: Ru(II) ($t_{2g}^6e_g^0$) and Cr(III) ($t_{2g}^3e_g^0$). This successor complex slowly decomposes to $[Ru(NH_3)_5(\text{isonicotinaimde})]^{2+}$ and $[Cr(aq)]^{3+}$. Fission of this successor is catalyzed by OH^- alone and jointly by Cr(II) + OH^-.

Between the Ru^{II} ($t_{2g}^6e_g^0$) and Cr(III) ($t_{2g}^3e_g^0$) centres, ***the Ru^{II}-centre is more inert*** and the fission (*i.e.* nucleophilic substitution by water) occurs at the Cr^{III}-centre.

(i) In the k_{H_2O} path of decomposition of the binuclear successor complex, the Cr^{III}–O bond undergoes cleavage by the nucleophile H_2O.

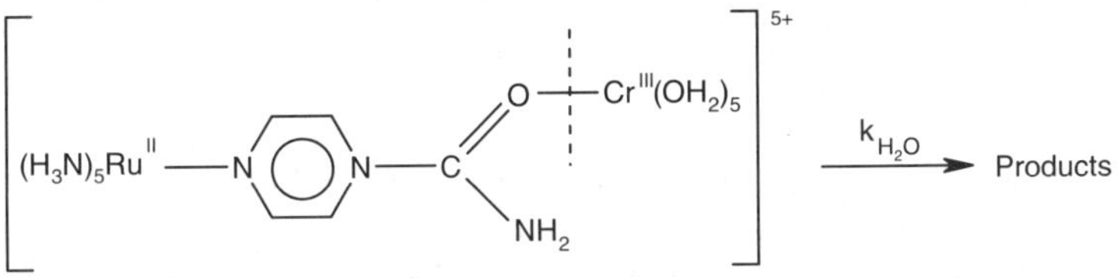

I

(ii) In the k_{OH} path of fission of the successor complex, at higher pH, deprotonation of the coordinated H_2O present with the Cr^{III}-centre generates the OH group that can show the ***cis*– labilizing effect** to favour the nucleophilic replacement of the Cr^{III}–O bond.

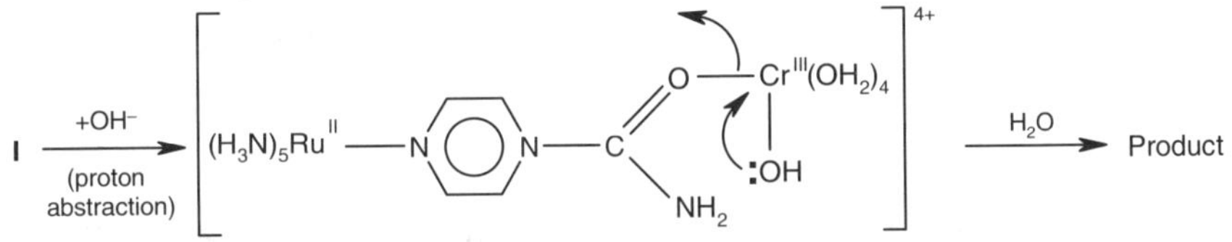

(iii) In the $k_{OH, Cr}$ path, the OH group generated in the coordination sphere of Cr(III) can act as a bridging ligand to make a bridge with the catalyst $[Cr(aq)]^{2+}$ which is labile.

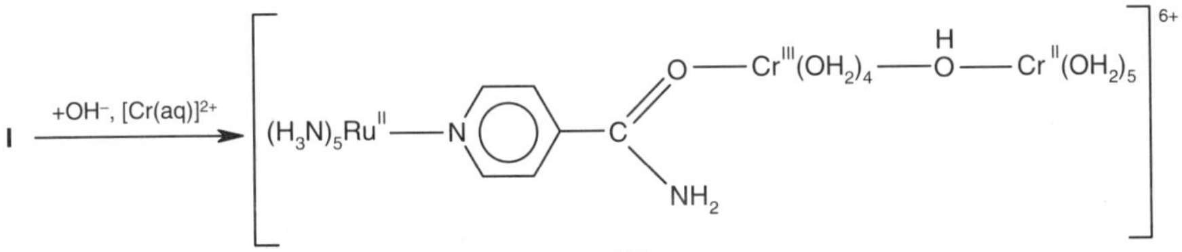

(II)

In the trinuclear complex (**II**), an inner sphere electron transfer occurs in the segment,

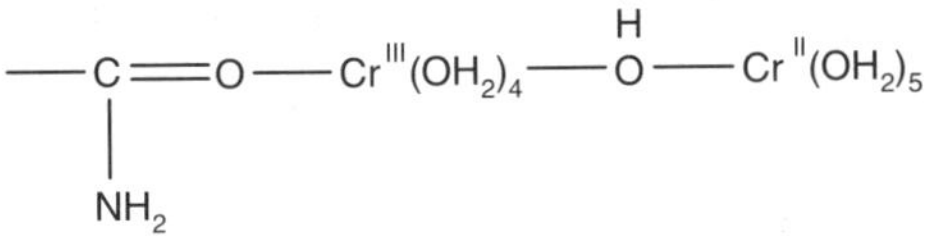

to generate the Cr^{II}-centre that remains linked with Ru^{II} through the bridging ligand. Cr^{II} being labile experiences a rapid bond fission. This is illustrated below.

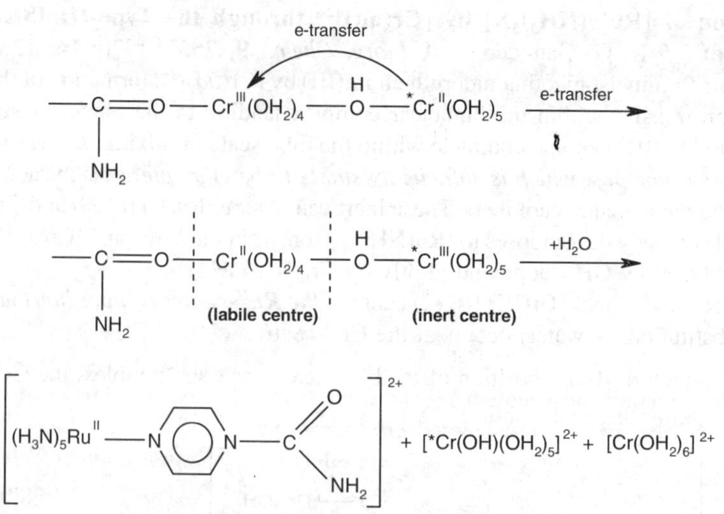

It is evident that the above step regenerates the catalyst $[Cr(aq)]^{2+}$.

For the fission of the binuclear successor complex of Ru^{II} and Cr^{III} bridged by isonicotinamide ligand, if labeled $[^{*}Cr(aq)]^{2+}$ is added as a catalyst, then the radioactivity gets confined in the Cr(III)-product. It supports the above mentioned step.

Characterization of dinuclear successor complexes in the ISET process

Reactants	**Species characterized***
$[RuCl_6]^{3-} + [Cr(OH_2)_6]^{2+}$	$[Cl_5Ru^{II}—Cl—Cr^{III}(OH_2)_5]^-$
$[VO(OH_2)_5]^{2+} + [Cr(OH_2)_6]^{2+}$	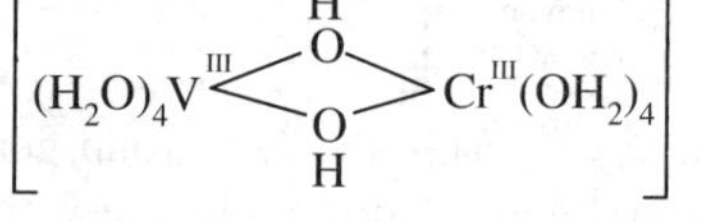
$[VO(OH_2)_5]^{2+} + [V(OH_2)_6]^{2+}$	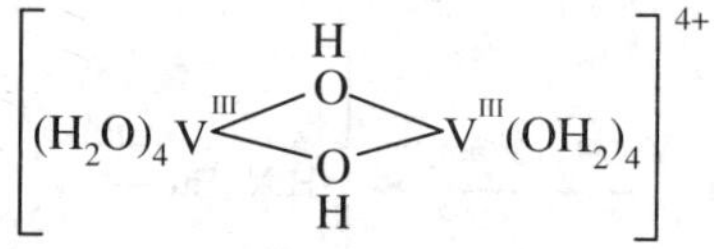
$[Fe(CN)_6]^{3-} + [Co(edta)]^{2-}$	$[(NC)_5Fe^{II}—CN—Co^{III}(edta)]^{5-}$
$[IrCl_6]^{2-} + [Cr(aq)]^{2+}$	$[Cl_5Ir^{III}—Cl—Cr^{III}(OH_2)_5]$
$[IrCl_6]^{2-} + [Co(CN)_5]^{3-}$	$[Cl_5Ir^{III}—Cl—Co^{III}(CN)_5]^{5-}$
$[Fe(CN)_6]^{3-} + [Co(CN)_5]^{3-}$	$[(NC)_5Fe^{II}—CN—Co^{III}(CN)_5]^{6-}$

* Both the **metal centres are substitutionally inert.** $edta^{4-}$ can show the **flexidentate character** to accommodate the bridging ligand.

(b) **Reduction of $[VO(OH_2)_5]^{2+}$ by $[V(aq)]^{2+}$ through the Type-III ISET process:** Another example of **Type-III** ISET process (*i.e. formation of a stable successor binuclear complex that decomposes slowly*) is the reduction of $[VO(OH_2)_5]^{2+}$ by $[V(aq)]^{2+}$.

$$[VO(OH_2)_5]^{2+} + [V(OH_2)_6]^{2+} \longrightarrow \left[(H_2O)_4 V^{III} \underset{\underset{H}{O}}{\overset{\overset{H}{O}}{<>}} V^{III}(OH_2)_4 \right]^{4+} \xrightarrow[(slow)]{fission} Product$$

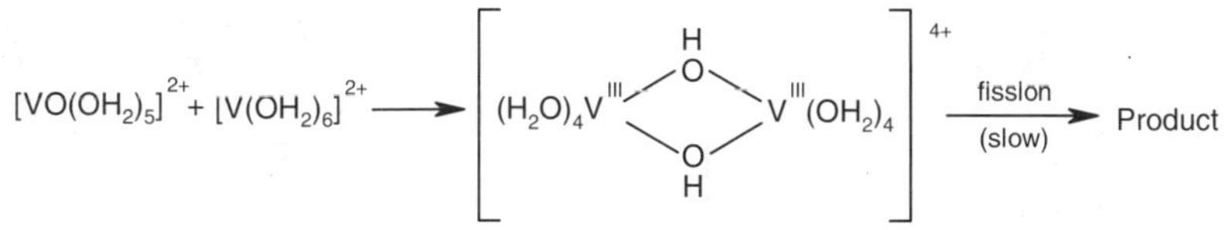

(Stable successor complex)

The said successor complex can also be generated from the hydrolysis of V(III)-solution.

(c) **Reduction of cis-$[Ru^{III}Cl_2(NH_3)_4]^+$ by $[Cr(aq)]^{2+}$ through the Type-III ISET process:** The successor complex has been spectroscopically characterized in the following reaction.

$$cis\text{-}\left[Ru^{III}Cl_2(NH_3)_4 \right]^+ + \left[Cr(OH_2)_6 \right]^{2+} \xrightarrow{fast} \left[(H_3N)_4 ClRu^{II} - Cl - Cr^{III}(OH_2)_5 \right]^{3+}_{\textbf{Stable}}$$

$$Products \xleftarrow{slow}$$

In the successor complex both the metal centres, *i.e.* Ru^{II} ($t_{2g}^6 e_g^0$) and Cr^{III} ($t_{2g}^3 e_g^0$) are inert.

(d) **Reduction of $[Fe(CN)_6]^{3-}$ by $[Co(CN)_5]^{3+}$ through the Type-III ISET process:** In the reduction of $[Fe(CN)_6]^{3-}$ by $[Co(CN)_5]^{3-}$, the produced successor binuclear complex is so stable that it can be isolated as a Ba^{2+}-salt.

$$\left[Fe(CN)_6 \right]^{3-} + \left[Co(CN)_5 \right]^{3-} \xrightarrow{fast} \left[(NC)_5 Fe^{II} - CN - Co^{III}(CN)_5 \right]^{6-} \textbf{(stable)}$$

$$H_2O \downarrow slow$$

$$\left[Fe(CN)_6 \right]^{4-} + \left[Co(CN)_5(OH_2) \right]^{2-}$$

In the successor complex, both the low-spin metal centres, *i.e.* Fe^{II} ($t_{2g}^6 e_g^0$) and Co^{III} ($t_{2g}^6 e_g^0$) are kinetically inert in the ligand substitution process. Here the **relatively weaker bond 'Co^{III}—N'** **undergoes cleavage** by the nucleophilic attack of H_2O (*cf.* 'N'-end of CN^- is a weak field ligand).

6.4.6 Possibility of Characterization of the Precursor Binuclear Complex

There are very few cases where the precursor dinuclear complex is sufficiently stable to be characterized. The favourable situations are: *thermodynamically stable precursor complex and the electron transfer is not very fast*. If formation of the complex is thermodynamically favoured then it will lead to an accumulation of the complex to a significant extent and if the electron transfer is not very fast then the life time of the complex increases.

● Reduction of the nitrilotriacetic acid complex of pentaaminecobalt(III) by $[Fe(aq)]^{2+}$ produces a binuclear precursor complex (life time $\approx$ 10 s) that can be characterized.

$$\left[(H_3N)_5 Co^{III}(ntaH_2) \right]^{2+} + \left[Fe(aq) \right]^{2+} \longrightarrow \left[Co(aq) \right]^{2+} + \left[Fe^{III}(nta)(OH_2)_2 \right] + 2H^+$$

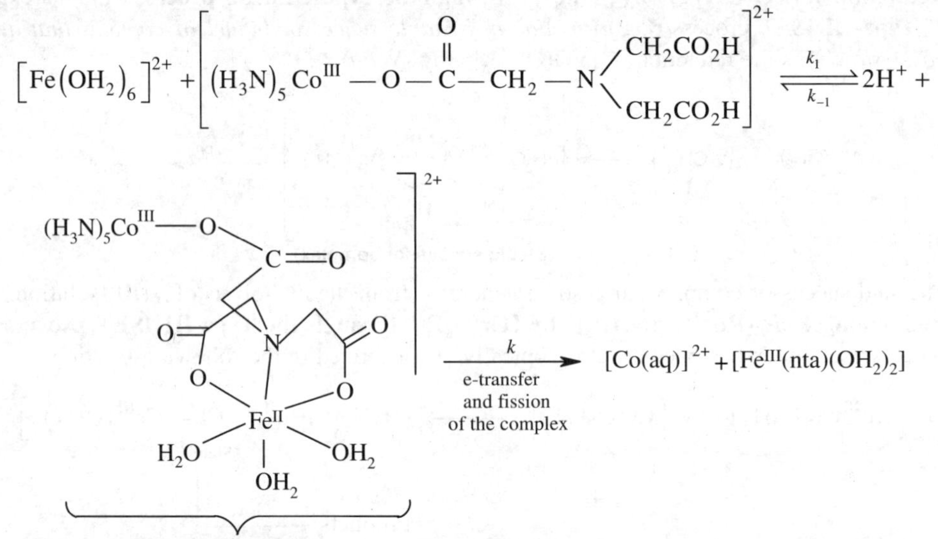

$$\left[Fe(OH_2)_6\right]^{2+} + \left[(H_3N)_5\,Co^{III}-O-\overset{\overset{\displaystyle O}{\parallel}}{C}-CH_2-N\begin{array}{l}CH_2CO_2H\\ \\CH_2CO_2H\end{array}\right]^{2+} \underset{k_{-1}}{\overset{k_1}{\rightleftharpoons}} 2H^+ +$$

$$\xrightarrow[\substack{\text{e-transfer}\\ \text{and fission}\\ \text{of the complex}}]{k} \quad [Co(aq)]^{2+} + [Fe^{III}(nta)(OH_2)_2]$$

Precursor Complex

The overall rate constant is: $k_f = K_{eq}k = 1 \times 10^5$ dm^3 mol^{-1} s^{-1} where $K_{eq} = \frac{k_1}{k_{-1}}$ (25° C) and this second order rate constant is very much higher compared to those of other reactions of reductions of Co(III)-complexes by $[Fe(aq)]^{2+}$ (*cf.* Table 6.4.5.3).

	$[CoCl(NH_3)_5]^{2+}$–$[Fe(aq)]^{2+}$	$[Co(SCN)(NH_3)_5]^{2+}$–$[Fe(aq)]^{2+}$	$[Co(C_2O_4H)(NH_3)_5]^{2+}$–$[Fe(aq)]^{2+}$
k_f: (dm^3 mol^{-1} s^{-1}) 25°C	1.4×10^{-3}	1.2×10^{-1}	4.3×10^{-1}

The apparently very high value of k_f for the reduction of $[Co(NH_3)_5(ntaH_2)]^{2+}$ by $[Fe(aq)]^{2+}$ is due to the very high value of K_{eq} (which was estimated to be 1.1×10^6 dm^3 mol^{-1}). This yields $k = 9.4 \times 10^{-2}$ s^{-1}. This value for the electron transfer step is quite reasonable compared to the rate constants for the reductions of other Co(III) complexes by $[Fe(aq)]^{2+}$ through the ISET path (*cf.* Table 6.4.5.3).

- Fairly stable precursor complex has been reported in the following reaction.

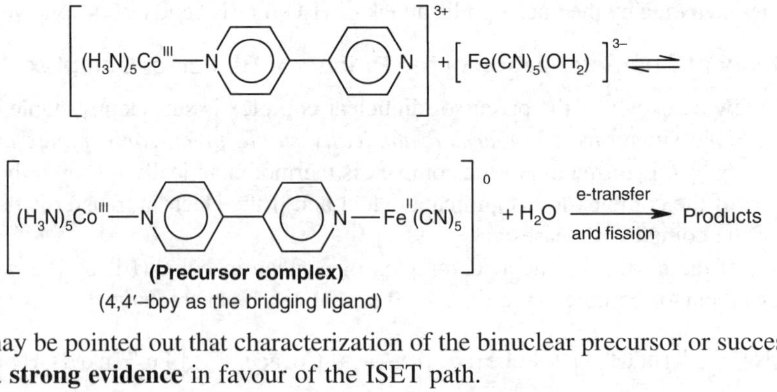

(Precursor complex)
(4,4′–bpy as the bridging ligand)

Note: Here it may be pointed out that characterization of the binuclear precursor or successor complex is regarded as a **strong evidence** in favour of the ISET path.

6.4.7 Effect of the Nature of HOMO of the Reductant and LUMO of the Oxidant on the Rate Process of Inner-Sphere Reaction

It has already been pointed out that in the ISET path, the kinetic step of electron transfer occurs from the σ^*-HOMO ($\equiv e_g$) of the reductant to the σ^*-LUMO ($\equiv e_g$) of the oxidant.

Thus if the ground state electronic configurations allow the $\sigma^* \rightarrow \sigma^*$ transfer then there will be no requirement of electronic activation of the reactants. In other cases, proper electronic activation will be needed for the required $\sigma^* \rightarrow \sigma^*$ transfer.

The above requirement of orbital matching explains the following rate order (*cf.* Table 6.4.5.1, 3).

	Co(III)—Cr(II)	〉	Co(III)—V(II)	〉	Co(III)—Fe(II)
Net transfer:	$(\sigma^* \rightarrow \sigma^*)$		$(\pi^* \rightarrow \sigma^*)$		$(\pi^* \rightarrow \sigma^*)$

Here it may be pointed out that for the said reduction by V(II) in the ISET path, the overall process is controlled by the substitution rate at V(II). In terms of the E^0 values of the M^{3+}/M^{2+} couples, the reducing power (thermodynamic sense) runs as: $[Cr(aq)]^{2+} \rangle [V(aq)]^{2+} \rangle\rangle [Fe(aq)]^{2+}$. **Thus to explain the order of rate constants, both the nature of the HOMO-LUMO and E⁰ values are to be considered simultaneously.**

6.4.8 Possibility of Ligand Transfer in the ISET Process

Possibility of the ligand transfer depends on the relative lability and bond strength of the metal centres present in the successor binuclear complex formed after the electron transfer. There may be **three possibilities** (A, B and C).

(A) Examples where the ligand transfer occurs to the reducing agent: It happens so when the reduced metal centre is more labile than the oxidized metal centre in the binuclear successor complex.

● **Reduction of the Co(III)-complexes by $[Cr(aq)]^{2+}$:** It leads to the ligand transfer to the Cr-centre.

$$\left[(H_3N)_5Co^{III}-X\right]^{n+} + \left[Cr(OH_2)_6\right]^{2+} \rightleftharpoons \left[(H_3N)_5Co^{III}-X-Cr^{II}(OH_2)_5\right]^{(n+2)+} + H_2O$$

$$\downarrow \text{e-transfer}$$

$$\left[Co(OH_2)_6\right]^{2+} + 5NH_4^+ + \left[(H_2O)_5Cr^{III}-X\right]^{n+} \xleftarrow[(+H_3O^+)]{\text{fast, fission}} \left[\underset{(t_{2g}^5e_g^2,\ \text{labile})}{(H_3N)_5Co^{II}} \vdots X - \underset{(t_{2g}^3e_g^0,\ \text{inert})}{Cr^{III}(OH_2)_5}\right]^{(n+2)+}$$

Point of nucleophilic substitution

X = OH^-, halide, pseudohalide, $SO_4^{2-}, PO_4^{3-}, H_2O$, etc.

That the entry of X^- in the coordination sphere of Cr(III) does not occur through an independent nucleophilic substitution at Cr(III) was proved by an isotope labeling experiment and comparing the substitution rate of Cr(III) with the rate of the overall redox reaction leading to the ligand transfer. These aspects have been discussed in Sec. 6.4.4.

● **Reduction of the Co(III)-complexes by $[V(aq)]^{2+}$:** Reduction of the Co^{III}-complex by $[V(aq)]^{2+}$ also leads to ligand transfer through the ISET path.

$$\left[L_5Co^{III}-X\right]+\left[V(OH_2)_6\right]^{2+} \rightleftharpoons \left[L_5Co^{III}-X-V^{II}(OH_2)_5\right]+H_2O$$

$$\downarrow \text{e-transfer}$$

$$\left[L_5Co^{II}-OH_2\right]+\left[X-V^{III}(OH_2)_5\right] \xleftarrow[\text{fission}]{\text{fast}} \left[\underset{\substack{\textbf{More labile}\\(t_{2g}^5 e_g^2)}}{L_5Co^{II}} \vdots X-\underset{\substack{\textbf{More inert}\\(t_{2g}^2)}}{V^{III}}(OH_2)_5\right]+H_2O$$

$$\downarrow \text{fast (further aquation)}$$

$$[Co(OH_2)_6]^{2+} + 5L$$

(*cf.* k_{ex} of the water exchange process is higher for $[Co(aq)]^{2+}$ than that for $[V(aq)]^{3+}$).

- **Reduction of the Co(III)-complexes by $[Fe(aq)]^{2+}$:** Similarly, reduction of the Co(III)-complexes by $[Fe(aq)]^{2+}$ is expected to give the ligand transfer product through the ISET path. In the successor complex, there are Co(II) and Fe(III) centres but Co(II) is more labile than Fe(III). It will lead to the ligand transfer.

$$\left[L_5Co^{III}-X\right]+\left[Fe(OH_2)_6\right]^{2+} \xrightarrow{ISET} \left[Co(aq)\right]^{2+} + 5L + \left[Fe^{III}(OH_2)(X)\right]$$

(*cf.* k_{ex} of water exchange : $[Co(OH_2)_6]^{2+} \rangle [Fe(OH_2)_6]^{3+}$).

$$\left[(H_3N)_5Co^{III}(ntaH_2)\right]^{2+}+\left[Fe(OH_2)_6\right]^{2+} \longrightarrow \left[Co(aq)\right]^{2+}+\left[Fe^{III}(nta)(OH_2)_2\right]+5NH_3.$$

Note: Because of the **lability of V(III) (t_{2g}^2) and Fe(III) ($t_{2g}^3 e_g^2$)**, the products V^{III}—X, Fe^{III}—X will hydrolyse to give the free ligand (X) in solution.

- **Reduction of the Cr(III)-complexes by $[Cr(aq)]^{2+}$:** Oxidation of $[Cr(aq)]^{2+}$ by Cr(III) leads to ligand transfer in the ISET path. This can be understood by an isotope labeling experiment.

$$\left[L_5Cr^{III}-X\right]+\left[{}^*Cr(aq)\right]^{2+} \rightleftharpoons \left[L_5Cr^{III}-X-{}^*Cr^{II}(aq)\right]$$

$$\downarrow \text{e-transfer}$$

$$\left[Cr(aq)\right]^{2+}+5L+\left[{}^*Cr^{III}(aq)(X)\right] \xleftarrow[\text{(fast)}]{\text{fission}} \left[L_5Cr^{II} \vdots X-{}^*Cr^{III}(aq)\right]$$

In the binuclear successor complex, Cr(II)($t_{2g}^3 e_g^1$) is more labile than Cr^{III}(t_{2g}^3).

(B) Examples where the bridging ligand is retained with the oxidizing centre: It happens so when the reduced centre is more inert than the oxidized centre.

- **Reduction of $[IrCl_6]^{2-}$ by $[Cr(aq)]^{2+}$:** Reduction of hexachloridoiridate(IV) by $[Cr(aq)]^{2+}$ leads to retention of the bridging ligand (*i.e.* ***no bridging ligand transfer***) with the oxidant.

$$[Cl_5IrCl]^{2-} + [Cr(OH_2)_6]^{2+} \rightleftharpoons [Cl_5Ir^{IV} - Cl - Cr^{II}(OH_2)_5]^0 + H_2O$$

$$\downarrow \text{e-transfer}$$

$$[IrCl_6]^{3-} + [Cr(OH_2)_6]^{3+} \xleftarrow{\text{fission}} [Cl_5Ir^{III} - Cl \cdots Cr^{III}(OH_2)_5]^0$$

Cl_5Ir^{III}: More inert (t_{2g}^6); $Cr^{III}(OH_2)_5$: Relatively less inert (t_{2g}^3)

Point of nucleophilic substitution

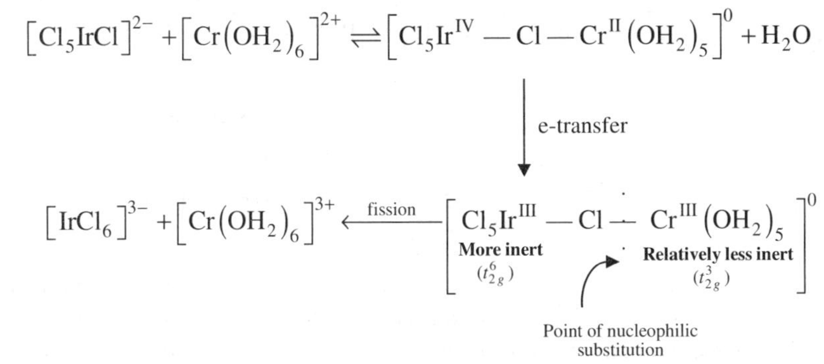

The Ir(III)–Cl bond is more inert and stronger than the Cr(III)–Cl bond. This is why, fission of the successor complex occurs through the nucleophilic attack on the Cr(III)–Cl bond.

- **Oxidation of $[Co(CN)_5]^{3-}$ by $[Fe(CN)_6]^{3-}$:** Oxidation of pentacyanidocobaltate(II) by hexacyanidoferrate(III) through the ISET path does not lead to transfer of the bridging ligand.

$$[(NC)_5Fe^{III} - CN]^{3-} + [Co(CN)_5]^{3-} \rightleftharpoons [(NC)_5Fe^{III} - C{\equiv}N - Co^{II}(CN)_5]^{6-}$$

$$\downarrow \text{e-transfer}$$

$$[Fe(CN)_6]^{4-} + [Co(CN)_5(OH_2)]^{2-} \xleftarrow{\text{fission}} [(NC)_5Fe^{II} - C{\equiv}N \cdots Co^{III}(CN)_5]^{6-}$$

$(NC)_5Fe^{II}$: $(t_{2g}^6e_g^0)$; $Co^{III}(CN)_5$: $(t_{2g}^6e_g^0)$

Strong field site; Weak field site

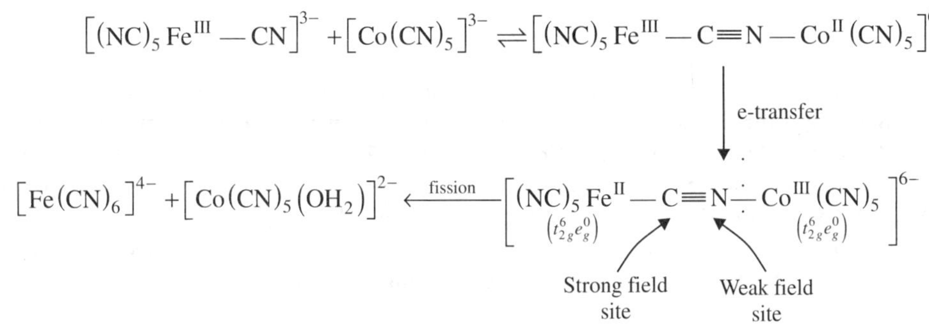

***C*-bound CN^- is a strong field ligand while *N*-bound CN^- is a relatively weaker field ligand.** This is why, the C-bound CN^- ligand stabilizes Fe(II) (t_{2g}^6) in $[Fe(CN)_6]^{4-}$ more than the N-bound CN^- ligand stabilizing Co(III) (t_{2g}^6) in $[Co(CN)_5(NC)]^{3-}$. This is also expected from the **principle of symbiosis and principle of HSAB theory**. This is why, the bond cleavage preferably occurs at the 'Co^{III}–N' bond.

- **Oxidation of $[Co(edta)]^{2-}$ by $[Fe(CN)_6]^{3-}$:** Oxidation of $[Co(edta)]^{2-}$ by $[Fe(CN)_6]^{3-}$ through the ISET path does not lead to transfer of the bridging ligand.

$$[Co(edta)]^{2-} + [Fe(CN)_6]^{3-} \longrightarrow [Co(edta)]^{-} + [Fe(CN)_6]^{4-}$$

Here it should be pointed out that both $[Fe(CN)_6]^{3-}$ and $[Fe(CN)_6]^{4-}$ are inert. The oxidant $[Fe(CN)_6]^{3-}$ provides the bridging ligand to the labile reductant $[Co(edta)]^{2-}$. *It may require opening of one acetate arm in $[Co(edta)]^{2-}$ to accommodate the bridging ligand. After the electron transfer, the low spin Co(III)-centre bearing the relatively weaker field ligands (3N + 3O) is more labile than the low spin Fe(II)-centre in $[Fe(CN)_6]^{4-}$ (bearing the very strong field ligands).* It leads to retention of the bridging ligand.

- **Reduction of the Ru(III)-complexes by $[Cr(aq)]^{2+}$:** Reduction of $[Ru(NH_3)_5X]^{n+}$ by $[Cr(aq)]^{2+}$ through the ISET path does not experience the ligand transfer.

$$\left[\left(H_3N\right)_5 Ru^{III} - X\right] + \left[Cr(aq)\right]^{2+} \rightleftharpoons \left[\left(H_3N\right)_5 Ru^{III} - X - Cr^{II}\left(OH_2\right)_5\right]$$

$$\downarrow \text{e-transfer}$$

$$\left[\underset{(t_{2g}^6,\,\text{inert})}{\left(H_3N\right)_5 Ru^{II}} - X - \underset{(t_{2g}^3,\,\text{inert})}{Cr^{III}\left(OH_2\right)_5}\right]$$

In the successor complex, both the metal centres (*i.e.* low spin Ru^{II} of t_{2g}^6 configuration and Cr^{III} of t_{2g}^3 configuration) are inert to nucleophilic substitution. Thus the successor complex is sufficiently stable to be characterized. *Nucleophilic substitution at both the centres are known to cause fission of the successor complex.*

For $X = Cl^-$, formation of $[CrCl(OH_2)_5]^{2+}$ in the product indicates rupture of the Ru^{II}–Cl bond, *i.e.* ISET reaction with the ligand transfer.

For X = isonicotinamide, formation of the product $[Ru(NH_3)_5(\text{isonicotinamide})]^{2+}$

i.e.
$$\left[(H_3N)_5 Ru^{II} - N \bigcirc - CONH_2\right]^{2+}$$

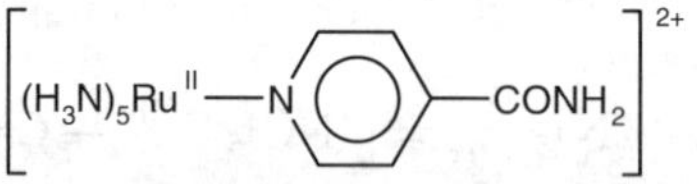

indicates the rupture of the Cr^{III}–Cl bond, *i.e.* ISET reaction without the ligand transfer.

(C) Examples where the bridging ligand is released to the solution when both the oxidized metal centre and the reduced metal centre (*e.g.* Co^{II}, $t_{2g}^5 e_g^2$) **are labile:** Reduction of $[Co(NH_3)_5X]^{2+}$ by Fe^{2+}, V^{2+}, Eu^{2+} and Cu^+ through the ISET path (with the transfer of X^-) will produce the **labile products FeX^{2+}, VX^{2+}, EuX^{2+} and CuX^+** which will hydrolyse to release the bridging ligand in solution. Among these, EuX^{2+} and CuX^+ are **extremely labile** to hydrolyse and in some cases FeX^{2+} and VX^{2+} have been characterised. It may be noted that the product Co(II) is highly labile.

6.4.9 Electrophilic Remote and Adjacent Attack on the Bridging Ligand by the Reducing Agent

(A) Remote Attack

- For the **monoatomic bridging ligands** like Cl^-, Br^-, OH^- (assuming H not to bridge), etc, the same atom bridges the metal centres. There is no other possibility. This is described as the **adjacent attack** on the bridging ligand/site (generally present with the oxidising centre) by the **reducing agent as an electrophile.**

- For the N-bonded NCS^- (*i.e.* $M_{ox}\!-\!\overset{..}{N}\!=\!C\!=\!\overset{..}{S}$) and CN^- ligand (*i.e.* M_{ox}–C ≡ N), the free ligating site (*i.e.* S for –NCS; N for –CN) present at the remote area acts as the nucleophile. This is described as the **remote attack.**

(Almost linear)

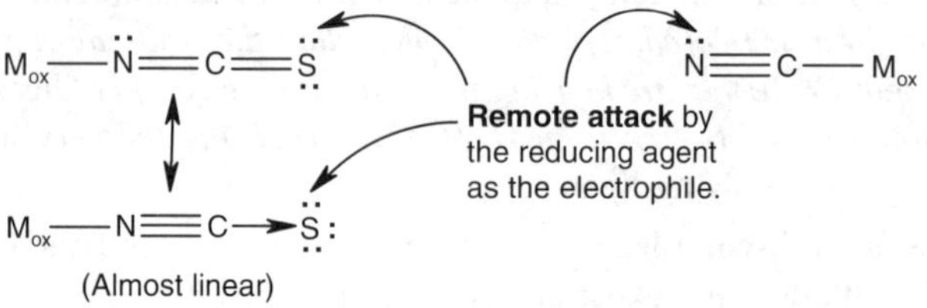

(**Note:** The lone pair on N of the –NCS linkage cannot be used for the nucleophilic attack because of the steric reason and this lone pair is engaged in resonance as shown).

In such cases, depending on the possibility of remote and adjacent attack, the ligand transfer in the ISET process can lead to the generation of linkage isomers which may not be thermodynamically stable. It is illustrated in the following examples.

(i) **$[Co^{III}(NCS)(NH_3)_5]^{2+}$ + $[Cr(OH_2)_6]^{2+}$:** Reduction of $[Co(NCS)(NH_3)_5]^{2+}$ by $[Cr(aq)]^{2+}$ produces the linkage isomer $[Cr(OH_2)_5(SCN)]^{2+}$ which on standing slowly isomerizes into the thermodynamically more stable linkage isomer, $[Cr(OH_2)_5(NCS)]^{2+}$.

$$\left[(H_3N)_5Co^{III}-NCS\right]^{2+}+\left[Cr(OH_2)_6\right]^{2+}\rightleftharpoons\underset{\text{(Remote attack on —NCS)}}{\left[(H_3N)_5Co^{III}-NCS-Cr^{II}(OH_2)_5\right]^{4+}}$$

$$\downarrow \text{e-transfer}$$

$$\begin{array}{c}5NH_4^+ +\left[Co(aq)\right]^{2+}+\\ \left[Cr^{III}(OH_2)_5(SCN)\right]^{2+}\\ \textbf{(green isomer,}\\ \textbf{kinetically controlled product)}\end{array}\xleftarrow[\text{(fission)}]{+H_3O^+,\ \text{fast}}\left[(H_3N)_5Co^{II}\vdots NCS-Cr^{III}(OH_2)_5\right]^{4+}$$

$$\downarrow \text{slow (isomerisation)}$$

$$\underset{\textbf{(purple isomer)}}{\left[(H_2O)_5Cr^{III}(NCS)\right]^{2+}}\ \textit{i.e.}\left[Cr(NCS)(OH_2)_5\right]^{2+}\ (\textbf{thermodynamically controlled product})$$

Cr^{III}-being hard prefers the harder donor site *N* compared to the softer site *S*. This is why, $[Cr(OH_2)_5–SCN]^{2+}$ slowly isomerizes into the more stable linkage isomer bearing the *N*-bonded SCN^-.

(ii) $$\left[(H_3N)_5Co^{III}-NCS\right]^{2+}+\left[Co^{II}(CN)_5\right]^{3-}\rightleftharpoons\left[(H_3N)_5Co^{III}-NCS-Co^{II}(CN)_5\right]^{-}$$

$$\downarrow \text{e-transfer}$$

$$\begin{array}{c}5NH_4^+ +\left[Co(aq)\right]^{2+}+\\ \left[Co(CN)_5(SCN)\right]^{3-}\\ \textbf{(kinetically controlled product)}\end{array}\xleftarrow[\text{(fast)}]{H_3O^+}\left[(H_3N)_5Co^{II}\vdots NCS-Co^{III}(CN)_5\right]^{-}$$

(iii) $$\left[(H_3N)_5Co^{III}-CN\right]^{2+} + \left[Co(CN)_5\right]^{3-} \rightleftharpoons \left[(H_3N)_5Co^{III}-CN-Co^{II}(CN)_5\right]^{-}$$

e-transfer

$$\left[(H_3N)_5Co^{II} \vdots CN-Co^{III}(CN)_5\right]^{-} \xrightarrow{H_3O^+,\ fast} 5NH_4^+ + \left[Co(aq)\right]^{2+} + \left[(CN)Co^{III}(CN)_5\right]^{3-}$$

(*i.e.* $[Co(CN)_5(NC)]^{3-}$ **Kinetically controlled linkage isomer**)

slow (isomerisation) ↓

$$\left[Co(CN)_6\right]^{3-}$$

(Thermodynamically controlled product)

(iv) **Remote attack on the –NO_2 (*N*-nitro) linkage** is illustrated below.

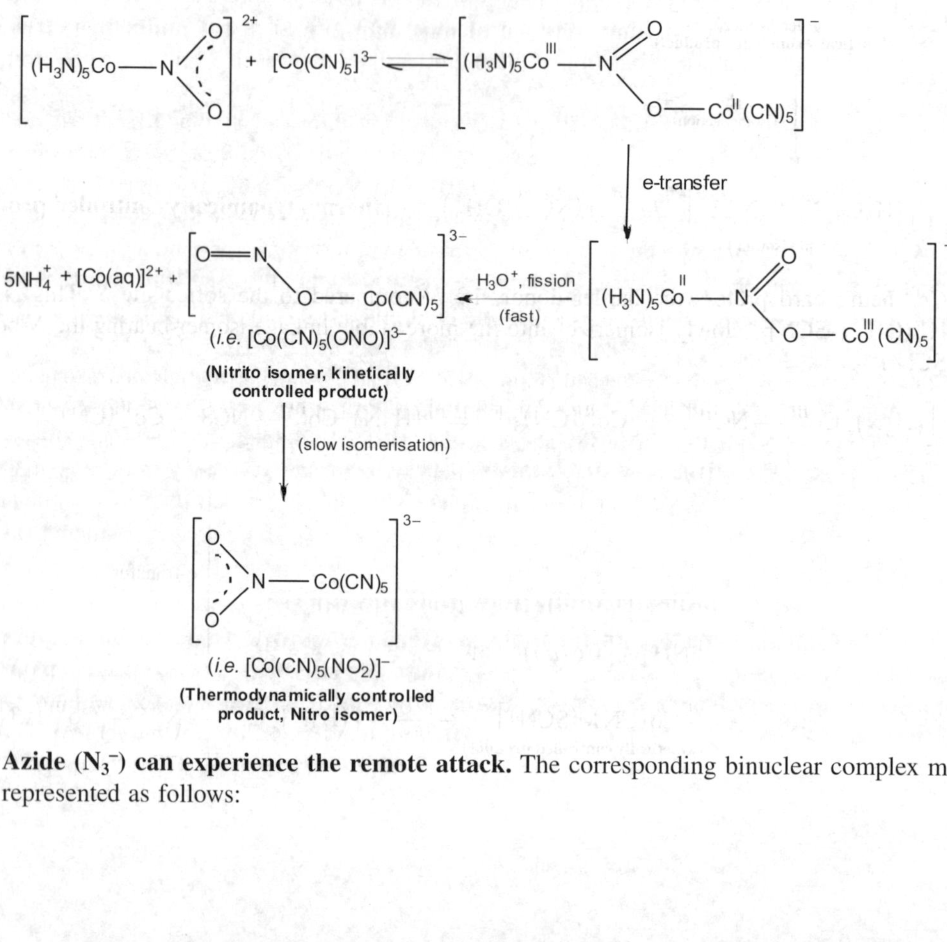

(v) **Azide (N_3^-) can experience the remote attack.** The corresponding binuclear complex may be represented as follows:

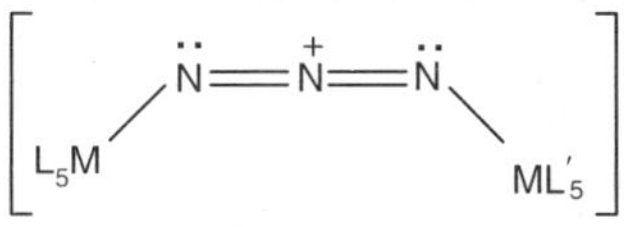

- **Failure of the product analysis technique to distinguish between the remote and adjacent attack:** *For the symmetrical bridges as in the case of azide and carboxylate, it is not possible to distinguish between the remote attack and adjacent attack from the product analysis.* Because, both the possibilities will lead to the same product.

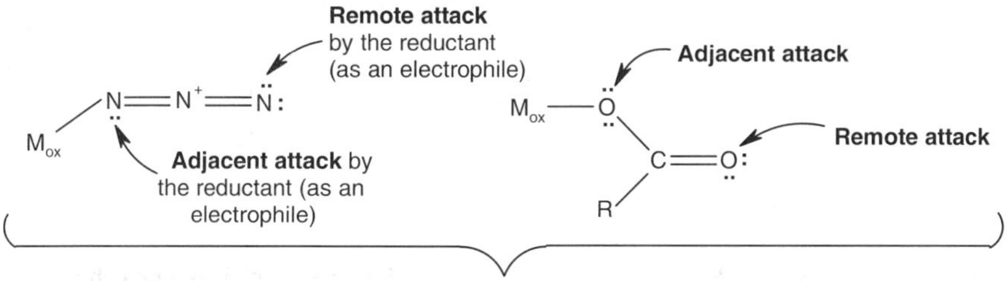

In such cases, other arguments are needed to distinguish between the possibilities– adjacent and remote attack. **Isotope labeling experiment** may be helpful to solve the problem.

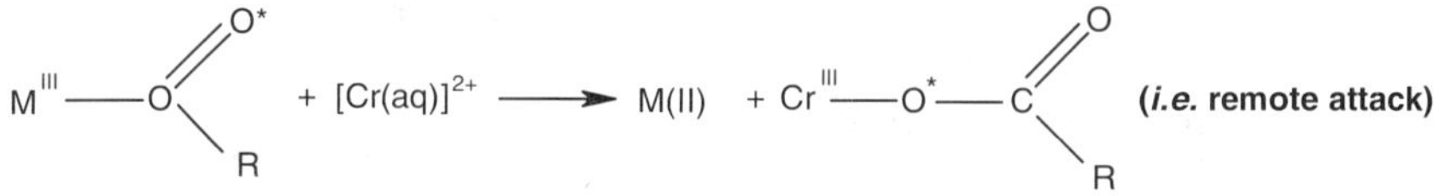

(vi) **Isonicotinamide (*i.e.* pyridine carboxamide) as a bridging ligand experiences a remote attack on the carbonyl–O.**

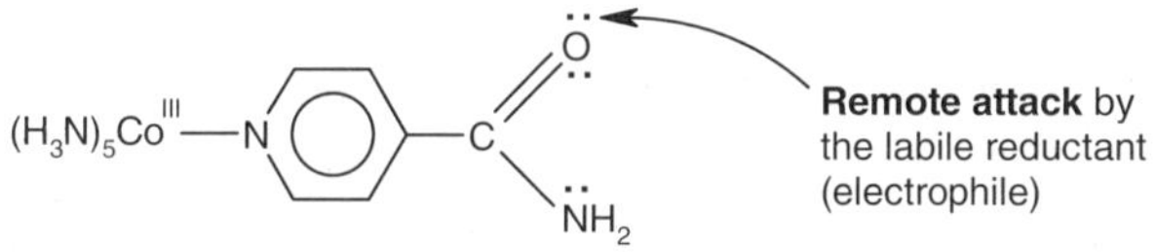

(**Note:** In such cases, remote attack on 'N' fails for the ISET process because of **two reasons: poor basicity** (*i.e.* nucleophilicity) of 'N' and **failure to resonance mechanism**. The lone pair on the $—\ddot{N}H_2$ group is engaged in resonance with the 'CO' group. Thus, the $—\ddot{N}H_2$ group is not sufficiently basic to act as the nucleophile 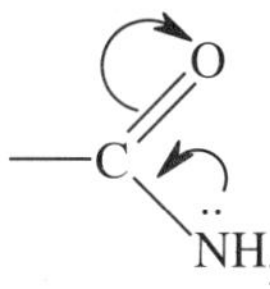 . These aspects are illustrated in the following example where the bridging property of the **formamide group** is discussed).

(vii) **Remote attack on the coordinated formamide group:**

$$\left[(H_3N)_5Co^{III}{-}\underset{H_2}{N}{-}C \begin{smallmatrix} \overset{\cdot\cdot}{O} \\ \\ H \end{smallmatrix} \right]^{3+} + [Cr(aq)]^{2+} \rightleftharpoons \left[(H_3N)_5Co^{III}{-}\underset{H_2}{N}{-}C \begin{smallmatrix} \overset{\cdot\cdot}{O}{-}Cr^{II}(OH_2)_5 \\ \\ H \end{smallmatrix} \right]^{5+}$$

i.e. **ISET involving the remote attack.**

(H_3O^+) | e-transfer and ligand transfer

$$5NH_4^+ + [Co(aq)]^{2+} + \left[(H_2O)_5Cr^{III}{-}O \begin{smallmatrix} \\ =C \\ \\ NH_2 \end{smallmatrix} \overset{H}{} \right]^{3+}$$

$$\left[(H_3N)Co^{III}{-}O{=}C \begin{smallmatrix} \overset{\cdot\cdot}{N}H_2 \\ \\ H \end{smallmatrix} \right]^{3+} + [Cr(aq)]^{2+} \Longrightarrow$$

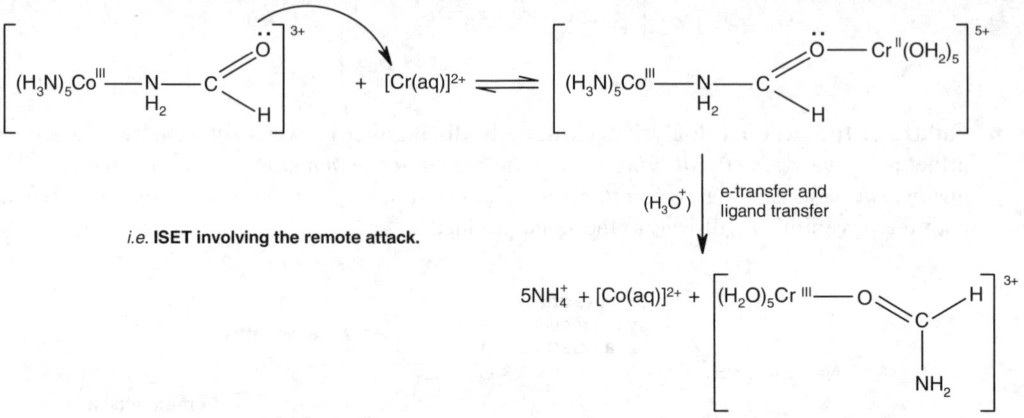

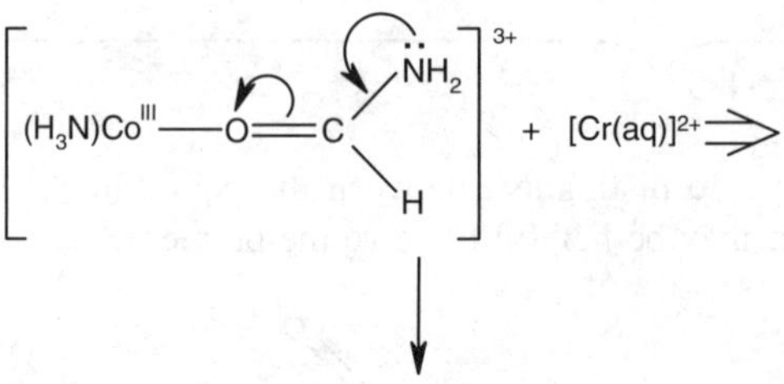

Remote attack is not possible to allow the ISET process because the NH$_2$ group is not sufficiently basic and the remote attack on N fails to give a conjugated system required for the resonance transfer of electron.

Reaction passing through the OSET path

(viii) **Formation of a conjugated system through the remote attack—a prerequisite condition for ISET (*i.e.* resonance mechanism of ISET):** It has been suggested that the remote attack in the ISET process must produce a **conjugated system** involving the oxidant, bridging ligand and reductant. This is called the **resonance mechanism** of ISET reaction. In fact, the remote attack by

the S-end of –NCS, N-end of –CN, N-end of –N$_3$, O– end of –NO$_2$, O-end of $-C\begin{smallmatrix} O \\ \\ NH_2 \end{smallmatrix}$ can

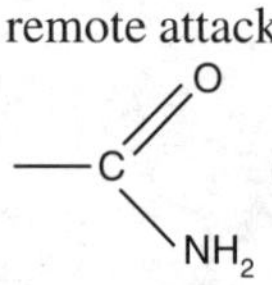

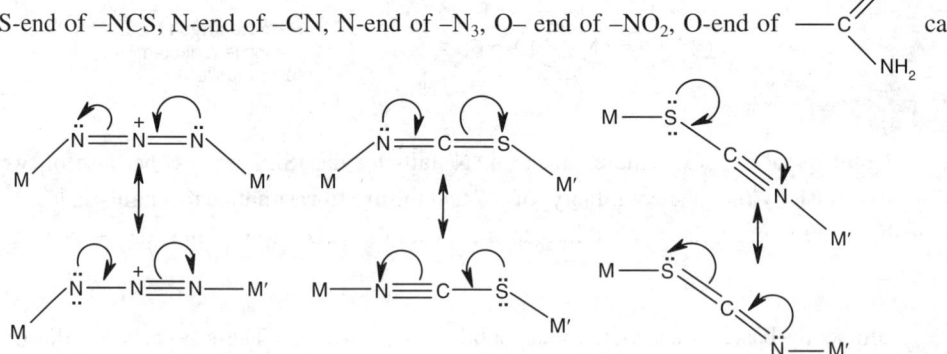

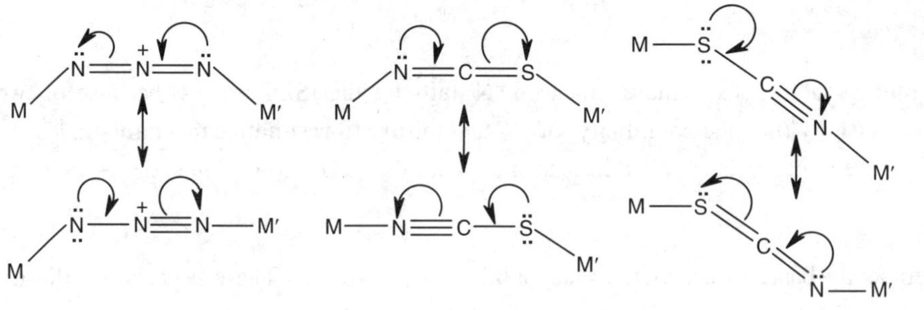

(The bridging atoms are having the **intermediate hybridisation state** because of the resonance. The bond angle M — S — C, 107° close to the tetrahedral angle, 109° indicates that S predominantly exists in *sp*^{3}-state).

constitute the conjugated system required for the resonance mechanism. In such cases, participation of the π-electron cloud and π-orbitals of the bridging ligand mediate the electron transfer.

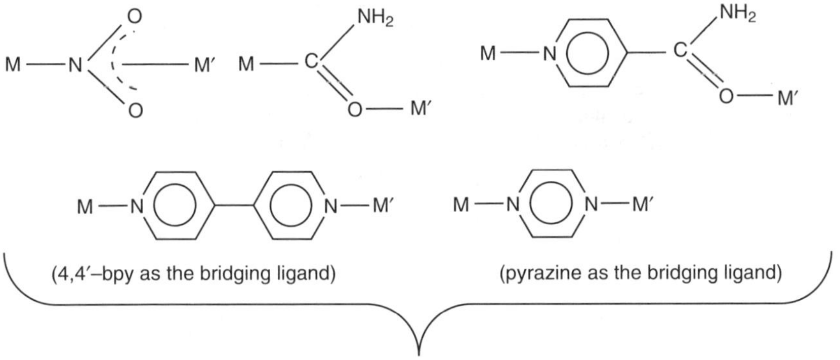

Conjugated system and participation of π-electron system through the remote attack
(Resonance mechanism of ISET)

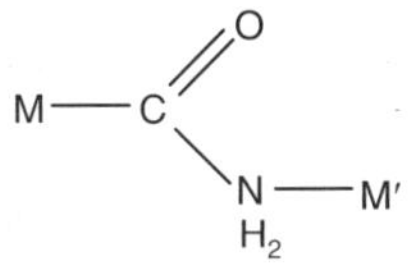

(Nonconjugated system and no π-electron system in the remote attack, failure of Resonance mechanism)

(B) Adjacent Attack

S-bonded thiocyanate (*i.e.* –SCN) can lead to both the possibilities– adjacent attack and remote attack by the reducing agent.

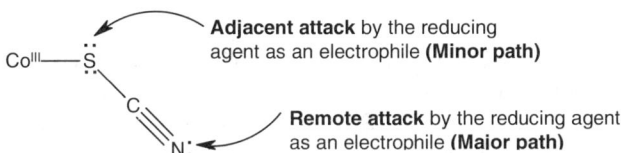

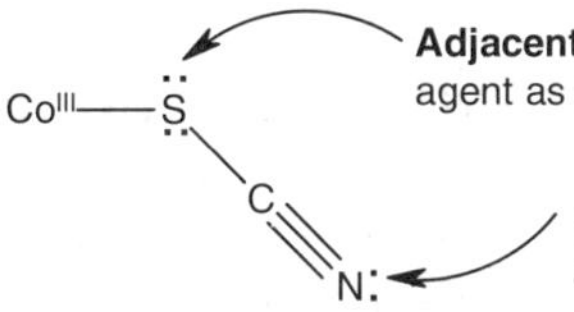

● **$[Co(NH_3)_5(SCN)]^{2+}$ + $[Cr(OH_2)_6]^{2+}$:** Here the ISET process can proceed through the both remote and adjacent attack. **Attack on the S-end** is not thermodynamically favoured by the **HSAB principle** for the hard reducing agent like $[Cr(aq)]^{2+}$. Attack on the S-end (*i.e.* adjacent attack) is also **sterically disfavoured.** But the S-site is kinetically good to mediate the electron transfer.

$$\left[(H_3N)_5Co(SCN)\right]^{2+} + \left[Cr(aq)\right]^{2+} \xrightarrow{\text{Both remote and adjacent attack}} \underset{ca.\ 70\%}{\left[(H_2O)_5Cr^{III}(NCS)\right]^{2+}} + \underset{ca.\ 30\%}{\left[(H_2O)_5Cr^{III}(SCN)\right]^{2+}}$$

Both Cr^{III} and Cr^{II} being hard prefer the *N*-end of SCN^-. Thus $[(H_2O)_5Cr^{III}(NCS)]^{2+}$, *i.e.* $[Cr(NCS)(OH_2)_5]^{2+}$ is the **major product.**

$$[(H_3N)_5Co^{III}-\!\!-SCN]^{2+} + [Cr(OH_2)_6]^{2+}$$

Remote attack
(Major route)

Adjacent attack
(Minor route)

$$\left[(H_3N)_5Co^{III}-S\!\!\diagdown_{C}\diagdown_{N-Cr^{II}(OH_2)_5}\right]^{4+}$$

$$\left[(H_3N)_5Co^{III}-S-Cr^{II}(OH_2)_5\diagdown_{C}\diagdown_{N}\right]^{4+}$$

e-transfer followed
by bond fission

e-transfer followed
by bond fission

$[(H_3N)_5Co^{II}(OH_2)]^{2+} + [(H_2O)_5Cr^{III}(NCS)]^{2+}$
i.e. $[Co(NH_3)_5(OH_2)]^{2+}$ *i.e.* $[Cr(NCS)(OH_2)_5]^{2+}$

**Kinetically controlled product is
also thermodynamically stable
(purple isomer)**

$[(H_3N)_5Co^{II}(OH_2)]^{2+} + [(H_2O)_5Cr^{III}(SCN)]^{2+}$
i.e. $[Co(NH_3)_5(OH_2)]^{2+}$ *i.e.* $[Cr(OH_2)_5(SCN)]^{2+}$

**(Kinetically controlled product,
green isomer)**

slow
isomerisation

$[(H_2O)_5Cr^{III}-\!\!-(NCS)]^{2+}$
i.e. $[Cr(NCS)(OH_2)_5]^{2+}$

**(Thermodynamically controlled
product, purple isomer)**

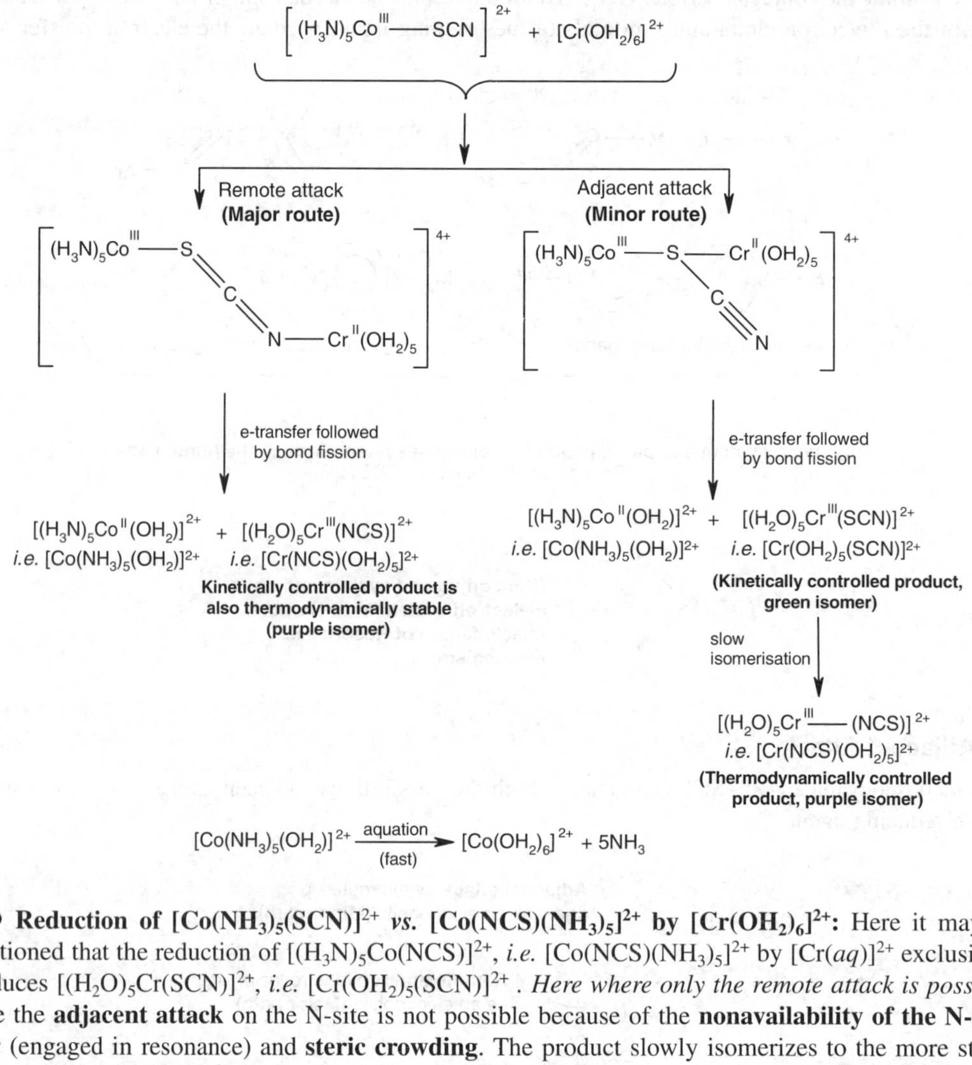

$$[Co(NH_3)_5(OH_2)]^{2+} \xrightarrow[\text{(fast)}]{\text{aquation}} [Co(OH_2)_6]^{2+} + 5NH_3$$

● **Reduction of $[Co(NH_3)_5(SCN)]^{2+}$ *vs.* $[Co(NCS)(NH_3)_5]^{2+}$ by $[Cr(OH_2)_6]^{2+}$:** Here it may be mentioned that the reduction of $[(H_3N)_5Co(NCS)]^{2+}$, *i.e.* $[Co(NCS)(NH_3)_5]^{2+}$ by $[Cr(aq)]^{2+}$ exclusively produces $[(H_2O)_5Cr(SCN)]^{2+}$, *i.e.* $[Cr(OH_2)_5(SCN)]^{2+}$. *Here where only the remote attack is possible.* Here the **adjacent attack** on the N-site is not possible because of the **nonavailability of the N-lone pair** (engaged in resonance) and **steric crowding**. The product slowly isomerizes to the more stable isomer $[(H_2O)_5Cr(NCS)]^{2+}$. Reduction of $[(H_3N)_5Co(NCS)]^{2+}$, *i.e.* $[Co(NCS)(NH_3)_5]^{2+}$ by $[Cr(aq)]^{2+}$ producing the both linkage isomers of Cr(III) occurs very slowly compared to the reduction of $[(H_3N)_5Co(SCN)]^{2+}$ *i.e.* $[Cr(OH_2)_5(SCN)]^{2+}$.

$(cf.$ Table 6.4.10.2$)$ $\begin{cases} [(H_3N)_5Co^{III}(NCS)]^{2+} + [Cr(aq)]^{2+}: k = 20 \text{ dm}^3 \text{ mol}^{-1} \text{ s}^{-1} \text{ at } 25°C. \\ [(H_3N)_5Co^{III}(SCN)]^{2+} + [Cr(aq)]^{2+}: k = 2 \times 10^5 \text{ dm}^3 \text{ mol}^{-1} \text{ s}^{-1} \text{ at } 25°C. \end{cases}$

The higher rate for the second reaction is due to the favoured remote attack involving the harder N-site. The slowness of the first reaction is due to the remote attack on the softer S-site. The first reaction can only lead to this remote attack while the second reaction can lead to both the remote and adjacent attack. The rate benefit in the second reaction is due to the **better HSAB matching and less steric crowding in the remote attack.**

6.4.10 Role of the Bridging Ligand in the ISET Reaction (*cf.* H. Taube *et al., Acc. Chem. Res.*, **2**, 321, 1969; A. Haim, *ibid*, **8**, 265, 1965)

The important functions of the bridging ligands in the ISET process are:

(i) **Kinetic Contribution–** favouring the **electron tunneling** from the reducing agent to the oxidant; this occurs for the monoatomic bridges and conjugated systems (**resonance transfer**) involving the oxidant, bridging ligand and oxidant.

(ii) **Kinetic Contribution–** reducible bridging ligand can participate in the redox process to favour the ISET process through the **chemical mechanism.**

(iii) **Kinetic contribution**– reducing the electrostatic repulsion between the positively charged oxidant (*e.g.* M^{III}) and reductant (*e.g.* M^{II}) metal centres.

(iv) **Thermodynamic Contribution–** through chelation, stability of the binuclear precursor complex may be enhanced to favour the overall rate process (*cf.* $k_f = kK_{eq}$, rate law).

Role of the bridging ligands in **chemical and tunneling mechanism** of the electron transfer will be discussed separately in Secs. 6.4.11.12. Here we shall mainly discuss the effect of stabilization of the binuclear precursor complex on the overall rate process (*cf.* Sec. 6.4.6).

- **(a) Effect of the bridging ligand depending on the position of rds in the ISET process:** All the bridging ligands are not equally efficient in stabilizing the binuclear complexes and mediating the electron transfer step. ***However, effect of the bridging ligand is noticeable only when formation of the precursor complex is rapid*** (*i.e.* reducing centre is substitutionally labile) and the overall rate is given by (under the condition, $k_{-1} >> k$):

$$\text{rate} \approx \frac{k_1}{k_{-1}} k[\text{Ox}—\text{X}][\text{Red}] = kK[\text{Ox}—\text{X}][\text{Red}] \quad (cf.\ \text{Sec. 6.4.2})$$

If the reducing centre (*e.g.* $[V(aq)]^{2+}$, $[Eu(aq)]^{2+}$, etc.) is substitutionally inert then the overall rate process is controlled by the nucleophilic substitution rate at the reducing centre and the overall rate is given by:

$$\text{rate} \approx k_1[Ox—X][\text{Red}] \quad (cf.\ \text{Sec. 6.4.2})$$

If the substitution experiences the **dissociative activation** as in the case of $[V(aq)]^{2+}$, $[Eu(aq)]^{2+}$, the nucleophilic substitution rate is insensitive to the nature of the bridging ligand that acts as the entering ligand. ***In such substitution controlled ISET process, effect of the bridging ligand nature is insignificant.***

- **(b) H_2O *vs.* ^-OH as the bridging ligand:** ^-OH is a much better bridging ligand than water.

This is why reduction of $[Co(NH_3)_5(OH)]^{2+}$ by $[Cr(aq)]^{2+}$ is much faster than the reduction of $[Co(NH_3)_5(OH_2)]^{3+}$ by $[Cr(aq)]^{2+}$ (*See* Sec. 6.4.4 for discussion of the ISET path in the reduction of $[Co(NH_3)_5(OH_2)]^{3+}$ by $[Cr(aq)]^{2+}$).

The electron exchange rate in the couple $[Fe(OH_2)_6]^{3+}$–$[Fe(OH_2)_6]^{2+}$ is highly pH dependent. The rate increases with the pH. With the increase of pH, $[Fe(aq)]^{3+}$ hydrolyses to $[Fe(OH)(OH_2)_5]^{2+}$ which can provide the 'OH' group for the bridge formation with the labile centre $[Fe(aq)]^{2+}$. *In fact, the electron exchange in the aqua couple, i.e.* $[Fe(H_2O)_6]^{3+}$*–*$[Fe(H_2O)_6]^{2+}$ *occurs through the OSET path* *while the electron transfer in the couple, [Fe(OH)(OH$_2$)$_5$]$^{2+}$–[Fe(OH$_2$)$_6$]$^{2+}$ occurs more rapidly through the ISET path.*

[Fe(OH$_2$)$_6$]$^{3+}$–[Fe(OH$_2$)$_6$]$^{2+}$: OSET path (H_2O cannot acts as a bridging ligand)

[Fe(OH)(OH$_2$)$_5$]$^{2+}$–[Fe(OH$_2$)$_6$]$^{2+}$: ISET path (OH group can act as a much better bridging group; k_{ex} (ISET) >> k_{ex} (OSET)).

Effect of the externally added anions on the ISET path

Sometimes, the rate is altered in presence of the externally added ions and the product is also altered. Reduction of [CoCl(NH$_3$)$_5$]$^{2+}$ by [Cr(*aq*)]$^{2+}$ in presence of the foreign anions like $P_2O_7^{4-}$, gives the Cr(III) product bearing the $P_2O_7^{4-}$ ligand. This happens so due to the substitution at the labile Cr(II)-centre before the electron transfer step. Thus the precursor complex may be proposed as follows:

$$\left[(H_3N)_5Co^{III}—Cl—Cr^{II}(OH_2)_4(P_2O_7^{4-})\right]$$

$$\Big\downarrow \text{ e-transfer and fission of the successor complex}$$

$$\left[Co(aq)\right]^{2+} + 5NH_3 + \left[ClCr^{III}(OH_2)_4(P_2O_7)\right]^{2-} \quad i.e. \left[CrCl(OH_2)_4(P_2O_7)\right]^{2-}$$

- **(c) N_3^- *vs.* NCS^- as the bridging ligands:** Remote attack occurs for both the azide and N-bonded thiocyanate but azide is a better bridging ligand than the N-bonded thiocyanate.

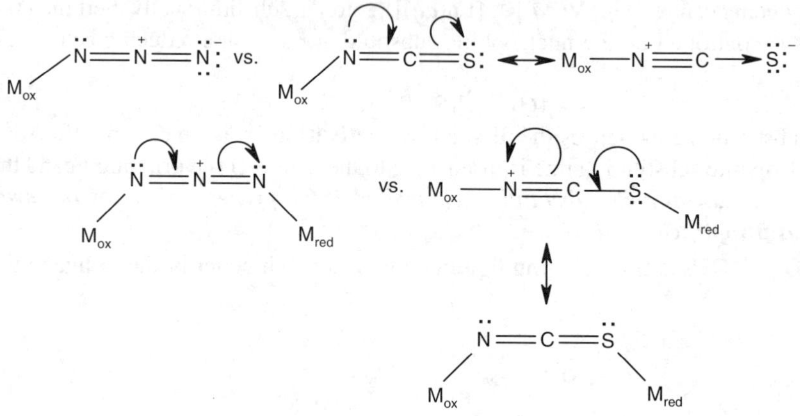

Attack on the S-end by the reducing agents like [Cr(aq)]$^{2+}$, [V(aq)]$^{2+}$, [Fe(aq)]$^{2+}$, etc. does not give a good **hard-soft matching** in terms of the HSAB principle. The S-end is soft while the said reducing agents and their oxidized centres are hard. ***Thus azide gives a better hard-soft matching.***

Besides the **HSAB matching**, azide makes a **symmetrical bridge** but NCS^- cannot make a such symmetrical bridge. Moreover, in terms of **resonance mechanism of ISET** (involving a conjugated system), the azide bridged complex can act better than the thiocyanate bridged complex. ***This is why, for X = N_3^-, the rate is higher than that for X = NCS^-.*** The conclusions are:

$$k_{N_3}/k_{NCS} >> 1 \Rightarrow \textbf{ISET process}$$

$$k_{N_3}/k_{NCS} \approx 1 \Rightarrow \textbf{OSET process or substitution controlled ISET process}$$

Note: Compared to the N-bonded thiocyanate, the S-bonded thiocyanate can give the better bridge for the remote attack on N. This gives also a better hard-soft matching. In fact, Co^{III}–SCN is reduced faster than Co^{III}–NCS in the ISET process. This aspect has been already discussed.

Table 6.4.10.1 Second order rate constants (25° C) for the reduction of azido and N-bonded thiocyanato complexes by different reducing agents. (*cf.* J. Espenson, *Inorg. Chem.*, **4**, 121, 1965).

Oxidant	*Reductant*	*k* (dm³ mol⁻¹ s⁻¹)		k_{N_3}/k_{NCS}	*Mechanism*
		k_{N_3}	k_{NCS}		
$[Co(NH_3)_5X]^{2+}$	$[Cr(aq)]^{2+}$	3×10^5	20.0	1.5×10^4	ISET
$[Co(NH_3)_5X]^{2+}$	$[V(aq)]^{2+}$	13.0	0.30	43.5	substitution controlled
$[Co(NH_3)_5X]^{2+}$	$[Eu(aq)]^{2+}$	2×10^2	0.70	2.8×10^2	ISET* or OSET*
$[Co(NH_3)_5X]^{2+}$	$[Fe(aq)]^{2+}$	8.5×10^{-3}	$< 3 \times 10^{-6}$	$>10^3$	ISET
$[Co(NH_3)_5X]^{2+}$	$[Cr(bpy)_3]^{2+}$	4×10^4	1×10^4	4.0	OSET**
$[Cr(OH_2)_5X]^{2+}$	$[Cr(aq)]^{2+}$	6.1	1.4×10^{-4}	4.3×10^4	ISET
$[Co(NH_3)_5X]^{2+}$	$[Co(CN)_5]^{3-}$	1.5×10^6	1×10^6	1.5	Substitution controlled ISET*
$[Co(NH_3)_5X]^{2+}$	$[Ru(NH_3)_6]^{2+}$	1.2	–	–	OSET

* Reducing centre V^{2+} (t_{2g}^3) is relatively inert; ISET process is concluded from the evidences like comparison with the kinetic parameters of the **substitution reaction at this reducing centre** and product analysis (transfer of the bridging ligand to the reducing centre). For Eu^{2+}, it may be simple OSET or substitution controlled ISET.

** Low spin (t_{2g}^4) **$[Cr(bpy)_3]^{2+}$ being** substitutionally inert participates in OSET.

Table 6.4.10.2 Second order rate constants (25°C) for the reduction of *N*-bonded and *S*-bonded thiocynato complexes (–X = –*N*CS or –*S*CN).

Oxidant	*Reductant*	*k* (dm³ mol⁻¹ s⁻¹)		k_{SCN}/k_{NCS}	*Mechanism*
		k_{SCN}	k_{NCS}		
$[Co(NH_3)_5X]^{2+}$	$[V(aq)]^{2+}$	30	0.3	10^2	ISET*
$[Co(NH_3)_5X]^{2+}$	$[Cr(aq)]^{2+}$	2×10^5	20	10^4	ISET
$[Co(NH_3)_5X]^{2+}$	$[Fe(aq)]^{2+}$	0.12	$< 3 \times 10^{-6}$	$> 4 \times 10^4$	ISET

* Substitution controlled ISET and this is why the ratio k_{SCN}/k_{NCS} is less.

- **(d) –*NCS*⁻ *vs.* –*SCN*⁻ as the bridging ligands:** For M_{ox}–NCS, an adjacent attack on 'N' is not possible because of the reasons: the lone pair on 'N' is engaged in **resonance** and **steric hindrance.** Thus, for M_{ox}–NCS, ***only the remote attack on S*** by the reducing agent is possible while for M_{ox}–SCN ***both the remote attack on 'N' and adjacent attack on 'S'*** are possible giving rise to a mixture of isomeric products (if ligand transfer in the ISET path is possible). If the reducing agents are **hard** (*e.g.* $[V(aq)]^{2+}$, $[Cr(aq)]^{2+}$, $[Fe(aq)]^{2+}$, etc.) then they prefer the *N*-site over the *S*-site. Consequently the rate of reduction of M_{ox}–SCN is much faster than the rate of reduction of M_{ox}–NCS which can only provide the *S*-site for bridging (*cf.* Sec. 6.4.9 for reduction of $[Co(NH_3)_5(SCN)]^{2+}$ *vs.* $[Co(NCS)(NH_3)_5]^{2+}$ by $[Cr(OH_2)_6]^{2+}$). The following factors explain the higher rate of reduction of M_{ox}–SCN in the ISET path.

(i) **M_{ox}–SCN:** Both S (adjacent attack) and N (remote attack) sites are available for coordination with the reducing agents (*e.g.* V^{2+}, Cr^{2+}, Fe^{2+}, etc.) but the remote attack on N gives the more favourable path.

(ii) **M_{ox}–NCS:** Only S– (*i.e.* remote attack) site is available for coordination with the reducing agent.

(iii) **M_{ox}—SCN—M_{red}** $\rangle$ **M_{ox}—NCS—M_{red}** (in terms of stability, *cf.* HSAB principle).

- **(e) Chelation and coordinating power of the bridging ligand to determine the efficiency of the ISET path:** Stability of the binuclear precursor complex largely depends on the efficiency of coordination by the bridging ligand with the reducing centre. The steric effect and possibility of chelation are the important factors to determine the stability of the binuclear complex. It is illustrated in the following examples:

(i) **Effect of steric crowding on the bridging ligand in the reduction of**

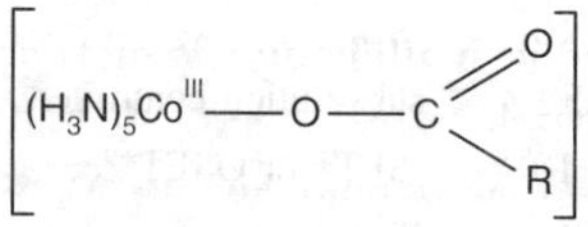

$$\left[(H_3N)_5Co^{III}\!\!-\!\!O\!\!-\!\!C \overset{\displaystyle O}{\underset{\displaystyle R}{<}} \right]^{2+} \quad \textbf{by } [Cr(aq)]^{2+}:$$

R =	H	CH$_3$	CMe$_3$
k (dm^3 mol^{-1} s^{-1}) =	7.2 $\rangle\rangle$	0.35 $\rangle\rangle$	9.7×10^{-3}

Obviously with the increase of bulkiness of R, the increased steric crowding disfavours the coordination by the carbonyl oxygen (*i.e.* remote attack) with the reducing metal centre.

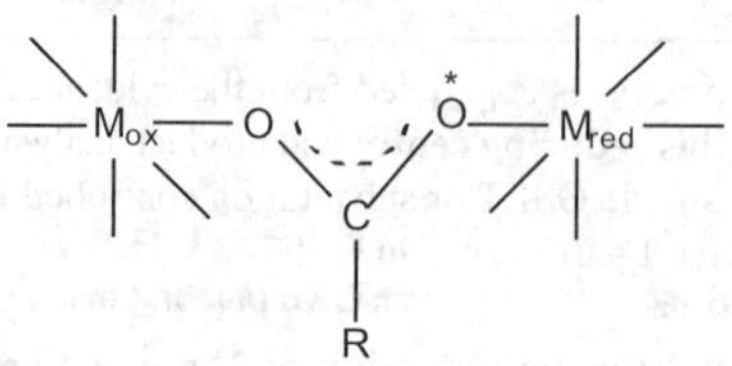

(Remote attack)
established by isotope
labeling experiment

Note: For R = H (*i.e.* formate), the bridging ligand is itself reducible and it can lead to the **chemical mechanism of ISET**. This also partially contributes for the observed rate enhancement.

(ii) **Effect of chelation by the bridging ligand in the reduction of** $\left[(H_3N)_5Co^{III}\!\!-\!\!O\!\!-\!\!C\overset{\displaystyle O}{\underset{\displaystyle R}{<}} \right]^{2+}$

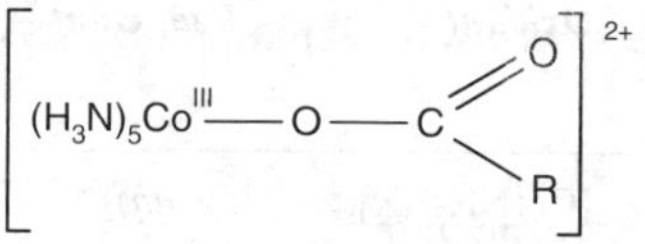

by $[Cr(aq)]^{2+}$:

R =	Ph	CH$_2$OH
k (dm^3 mol^{-1} s^{-1}):	0.15 $\langle$	3.0

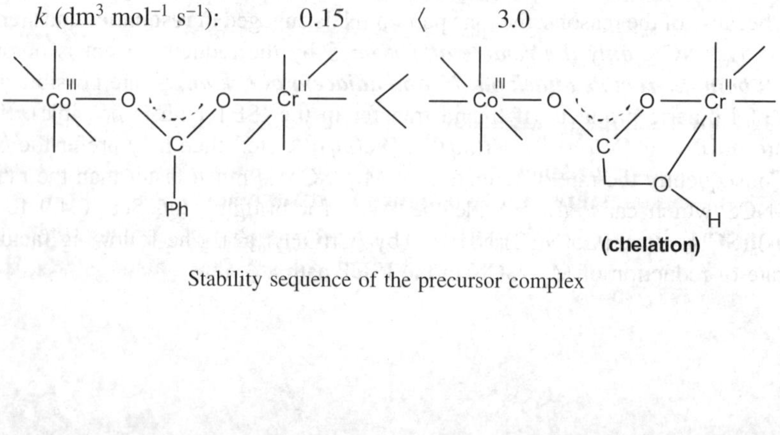

(chelation)

Stability sequence of the precursor complex

(**Note:** For the bridging ligand $^{-}OCOCH_2OH$ (*i.e.* glycolate), another factor comes to play. This reducible ligand can lead to the **chemical mechanism of electron transfer** to enhance the rate.)

(iii) For the **more powerful chelating bridging ligands,** the rates are dramatically increased. It is illustrated below for the reduction of $[(H_3N)_5\ Co^{III}—X]$ by $[Cr(aq)]^{2+}$.

X =	**glycolate**	**malonate**	**pyridine-2-carboxylate**	**salicylate**
k ($dm^3\ mol^{-1}\ s^{-1}$) =	3.0	2.5×10^3	2×10^5	2×10^8

(iv) **Effect of chelation to stabilize the binuclear precursor complex** is illustrated in the efficiency of reduction of $[Co^{III}(NH_3)_5X]^{2+}$ by $[Fe(aq)]^{2+}$.

	$[CoCl(NH_3)_5]^{2+}$ + $[Fe(aq)]^{2+}$	$[Co(C_2O_4H)(NH_3)_5]^{2+}$ + $[Fe(aq)]^{2+}$	$[Co(NH_3)_5(ntaH_2)]^{2+}$ + $[Fe(aq)]^{2+}$
k ($dm^3\ mol^{-1}\ s^{-1}$):	1.3×10^{-3}	0.43	1×10^5

Chloride ligand cannot cause any chelation; coordinated bioxalate ($C_2O_4H^-$) can provide two bridging sites to make a chelation; unidentatedly coordinated $ntaH_2^-$ can provide three binding sites (1N + 2-acetate – O) to stabilize the 'binuclear complex'. In fact, the binuclear precursor complex for $X^- = ntaH_2^-$ has been characterized and the formation constant (= $K = \frac{k_1}{k_{-1}} \approx 10^6$ $mol^{-1}\ dm^3$) is very high.

(v) **Effect of the stability of the binuclear complex** is illustrated in the reduction of $[Co(NH_3)_5X]^{2+}$ by $[Eu(aq)]^{2+}$.

Stability: $EuF^+ > EuCl^+ > EuBr^+ > EuI^+$; **Rate for X:** F > Cl > Br > I

- **(f) Identification of the nature of chelation by the bridging ligand from the product analysis:** Chelation by the bridging ligand is also known in many other cases. *The nature of chelation is very often identified from the product analysis.* Reduction of $[Co(C_2O_4)(NH_3)_4]^+$ by $[Cr(aq)]^{2+}$ produces $[Cr(C_2O_4)(OH_2)_4]^+$. It suggests the following precursor complex. It explains the presence of bidentate oxalate in the product Cr(III)-complex, *i.e.* $[Cr(C_2O_4)(OH_2)_4]^+$.

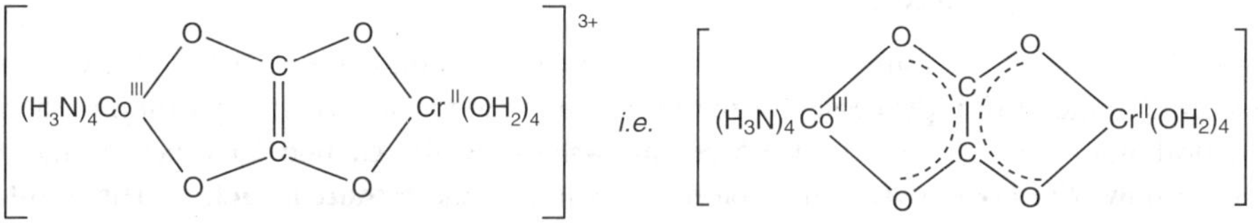

(**Note:** Reducible character of the bridging oxalate ligand may lead to the **chemical mechanism of ISET** to cause the rate enhancement.)

Reduction of $[Co(NH_3)_5(OCOCH_2CO_2H)]^{2+}$ by $[Cr(aq)]^{2+}$ gives the product $[Cr(malonate)(OH_2)_4]^+$ indicating formation of the following precursor complex.

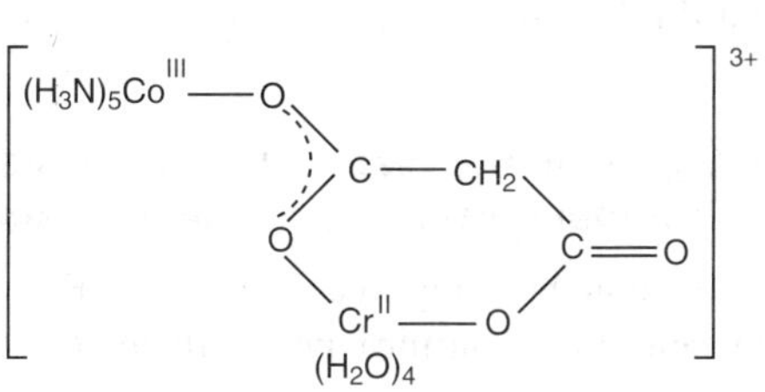

- **(g) Switch-over from the OSET to ISET path depending on the nature of the bridging ligand in the reduction of** 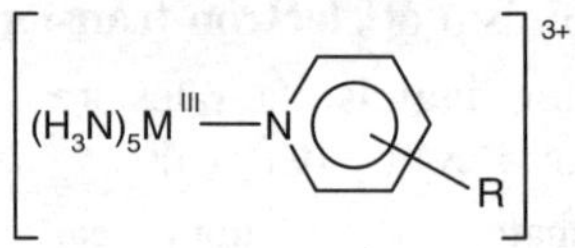 **by $[Cr(aq)]^{2+}$ (M = Co, Ru):** Rate sequence is:

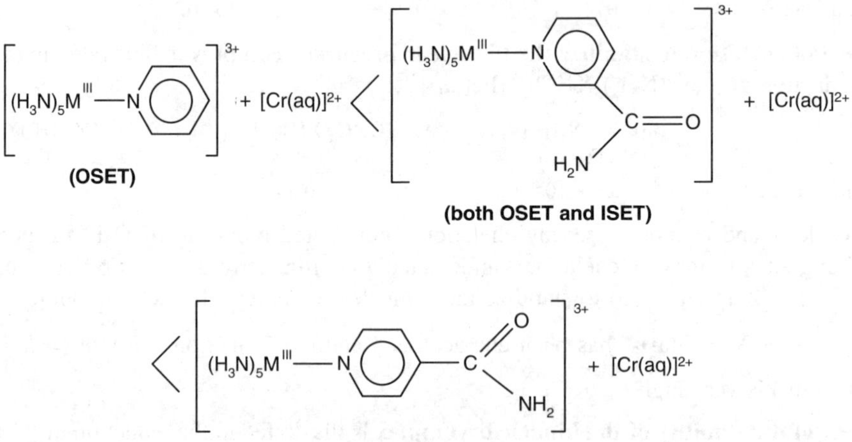

(ISET)

		$[(H_3N)_5Co^{III}\text{–}L]^{3+} + [Cr(aq)]^{2+}$: (Net $\sigma^* \rightarrow \sigma^*$ transfer)	
L =	**py**	**nicotinamide**	**isonicotinamide**
k ($dm^3\ mol^{-1}\ s^{-1}$):	4×10^{-3} (OSET)	~10^{-2} (OSET) ~10^{-2} (ISET)	18.0 (ISET)
		$[(H_3N)_5Ru^{III}\text{–}L]^{3+} + [Cr(aq)]^{2+}$: (Net $\sigma^* \rightarrow \pi^*$ transfer)	
k ($dm^3\ mol^{-1}\ s^{-1}$):	3.5×10^3 (OSET)	~10^4 (OSET) ~10^4 (ISET)	4×10^5 (ISET)

Pyridine cannot act as a bridging ligand. This is why, the complex $[(H_3N)_5M^{III}(py)]$ experiences the reduction through the **OSET process. Isonicotinamide** (amide group at the *p*-position in the pyridine ring) as a bridging ligand can experience a **remote attack on the carbonyl oxygen** by the reducing agent and it allows the *ISET path.* It may be noted that here the **remote attack on the amide-N site cannot occur** because the N-lone pair is engaged in resonance with the carbonyl group and it leads to the failure of the ISET **resonance mechanism** when bridged through the amide-N site (*cf.* Secs. 6.4.9, 6.4.11). Nicotinamide (*i.e.* amide group at the *m*-position in the pyridine ring) as a bridging ligand allows **both the OSET and ISET paths simultaneously.** In fact, the bridging group (*i.e.* $CONH_2$) at the *meta* position experiences two obstacles in the ISET path– **steric crowding and hindered conjugation.** This is why, both the OSET and ISET paths operate and the rate constants in both the paths are more or less comparable.

6.4.11 Role of the Bridging Ligand in the Electron Transfer Step – Tunneling Transfer, Resonance Transfer and Chemical Mechanism

- **Tunneling transfer:** Montoatomic bridging ligands like Cl^-, F^-, etc. simply helps the **tunneling of electron** from the reductant to oxidant through the three-centred σ-type delocalized MO.

- **Resonance transfer:** When there is a **conjugation** in the segment, oxidant–bridging ligand–reductant **through the remote attack** on the bridging ligand and the bridging ligand is **not reducible, tunneling of electron** is enhanced. This tunneling mechanism is described as the **resonance transfer.**

  It happens so when the **unsaturated nonreducible bridging ligands** (*e.g.* N_3^-, NCS^-, RCO_2^-, $RCONH_2$, etc.) experience the **remote attack** by the electrophilic reducing agent.

Outer Sphere Electron Transfer (OSET) in the Bridged Precursor Complex

This happening illustrates the fact that formation of the dinuclear precursor complex through the remote attack *does not always ensure the operation of an ISET mechanism.* An ISET path involving the remote attack needs the proper conjugation to allow the **resonance transfer mechanism** to operate. If this requirement is not fulfilled, then within the binuclear complex, the OSET path will operate instead of the ISET path. The following example of long-lived precursor complex illustrates the phenomenon.

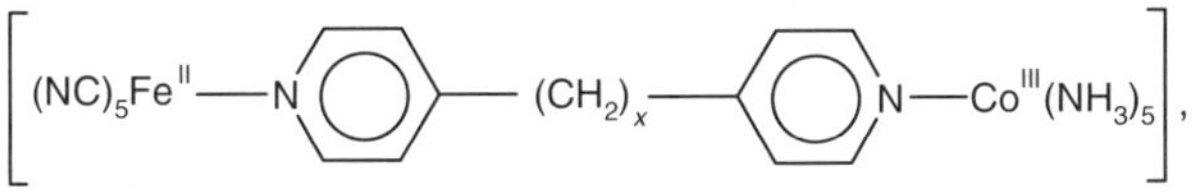

$$\left[(NC)_5Fe^{II}-N\!\!\bigcirc\!\!-(CH_2)_x-\!\!\bigcirc\!\!N-Co^{III}(NH_3)_5\right], \ x = 1-3, \textbf{(OSET path)}$$

The $(CH_2)_x$ moiety is saturated and the *presence of it prevents the electron tunneling through resonance transfer.* Thus the ISET path is prevented. However, **flexibility of the $(CH_2)_x$ moiety** can allow the oxidizing and reducing centres to come in proximity and this favours the simple OSET path.

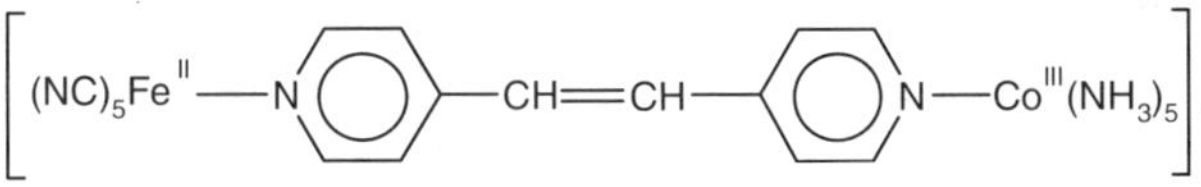

$$\left[(NC)_5Fe^{II}-N\!\!\bigcirc\!\!-CH\!=\!CH-\!\!\bigcirc\!\!N-Co^{III}(NH_3)_5\right], \textbf{(ISET path)}$$

$$\left[(NC)_5Fe^{II}-N\!\!\bigcirc\!\!-\overset{\displaystyle C}{\underset{\displaystyle O}{\|}}-\!\!\bigcirc\!\!N-Co^{III}(NH_3)_5\right], \textbf{(OSET path)}$$

Here the presence of the $\diagdown C=O$ group cannot extend the conjugation in the complete segment, **reductant-bridged ligand–oxidant.** Because of this incomplete conjugation in the segment, electron transfer through the resonance mechanism cannot occur. Interestingly, in contrast to the above example, the examples having the bridging ligand $X = N_3^-$, CN^-, NCS^-, RCO_2^-, etc. can allow the ISET path through resonance.

- **Chemical mechanism:** When the bridging organic ligands (*e.g.* nicotinamide, isonicotinamide, formate, etc.) are reducible, then at the first step (*probably the rds*), the electron is transferred to the bridging ligand which is temporarily reduced to a *radical ion.* At the next step (faster step), the electron from the reduced bridging ligand (*i.e.* radical ion) is transferred to the oxidant. This stepwise **pathway** is described as the **chemical or radical ion mechanism.** It is illustrated below for the reduction of $M^{III}-X$ by Cr^{II} (M = Co, Cr, Ru).

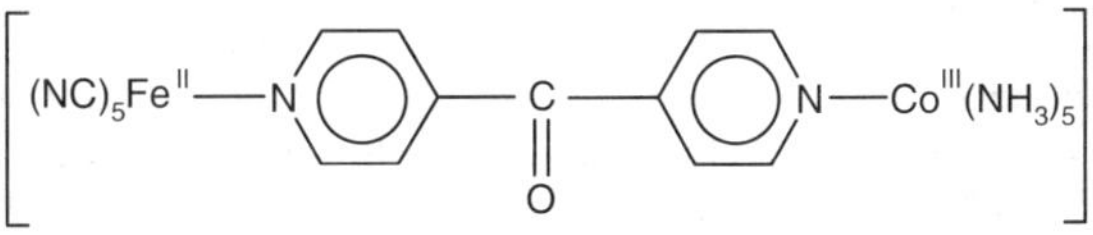

$$M^{III}—X + Cr^{II} \xrightarrow[\text{step-I}]{} M^{III}—X—Cr^{II} \xrightarrow[\text{step-II}]{(rds)} M^{III}—X^{\dot{-}}—Cr^{III}$$

$$\downarrow \text{step-III}$$

$$M^{II} + Cr^{III}X \xleftarrow[\text{step-IV}]{} M^{II}—X—Cr^{III}$$

Step-I: Formation of the binuclear complexes.
Step-II (rate determining step): *e*-transfer to the bridging ligand from the reducing agent.
Step-III: *e*-transfer from the reduced ligand to the oxidant.
Step-IV: Fission of the binuclear successor complex.

(**Note:** If the life period of the **temporarily reduced bridging ligand** (*i.e.* $X^{\dot{-}}$, **radical ion**) is too short, then the **radical ion mechanism** becomes equivalent to the **resonance tunneling** (*i.e.* resonance transfer). It happens so (*i.e.* step-III, IV very fast) when the electron carrier orbital of the bridging ligand does not experience any **symmetry mismatch barrier** to transfer the electron to the receptor orbital of the oxidant. Such a barrier does not exist in the case of reduction of Ru^{III}—X by Cr(II) where the bridging ligand X is reducible. This aspect will be illustrated in detail later.)

Pathways of electron transfer via the bridging ligand (X) in the ISET process

- **Tunneling mechanism** (Nonreducible bridging ligand)
  - **Path-I** — **Simple tunneling through the 3-centred delocalised σ-MO** (X^- = F^-, Cl^-, OH^-, etc.)
  - **Path-II** — **Resonance transfer** through the conjugated system involving the **delocalised π-MOs of the bridging ligand (Resonance tunneling)** (X^- = unsaturated non-reducible ligands like N_3^-, RCO_2^-, etc. experiencing the remote attack)
- **Chemical mechanism** (Bridging ligand reducible)
  - **Path-III** — **Ion radical mechanism** (Temporarily reduced ligand, *i.e.* $X^{\dot{-}}$ is relatively long-lived; **hindered full length conjugation** (*e.g.* X = maleate) or **symmetry mismatch barrier** between the electron carrier ligand orbital and the receptor orbital of the oxidant, *e.g.* M^{III}—X + Cr^{II}, M = Co, Cr, X = isonicotinamide, —OCOCHO, etc.)
  - **Path-IV ≡ Path-II** — **Resonance transfer** (Temporarily reduced ligand *i.e.* $X^{\dot{-}}$ is too short-lived; **no hindered conjugation and no symmetry mismatch barrier** between the electron carrier ligand orbital and the receptor orbital of the oxidant; *e.g.* Ru^{III}—X + Cr^{II}; X = isonicotinamide, —OCOCHO, etc.)

Scheme 6.4.11.1 Classification of ISET reaction depending on the nature of mechanistic pathway of electron transfer through the bridging ligand.

(**Note:** Sometimes, the bridging ligand is itself reduced and stabilised. The reduced ligand fails in turn to reduce the oxidant of the binuclear complex. It may happen for the organic bridging ligand having a nitro group that receives the remote attack by the reducing agent. Subsequently, the nitro-compound is reduced to the stable amine compound (*i.e.* $-NO_2$ to $-NH_2$).

(a) **Reduction mechanism of the Co(III), Cr(III) and Ru(III) complexes by $[Cr(aq)]^{2+}$ through the ISET path controlled by the nature of the bridging ligand:**

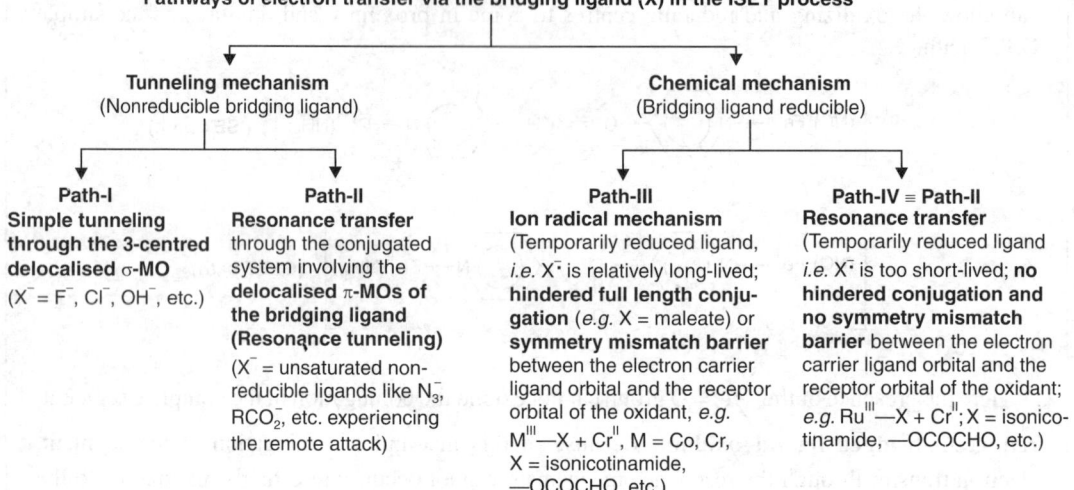

Conclusion: $\begin{cases} k_{\mathrm{Co^{III}}}/k_{\mathrm{Cr^{III}}} \sim 10^6;\ k_{\mathrm{Co^{III}}}/k_{\mathrm{Ru^{III}}} \sim 10^1 \Rightarrow \textbf{tunneling mechanism} \\ k_{\mathrm{Co^{III}}}/k_{\mathrm{Cr^{III}}} \sim 10^1;\ k_{\mathrm{Co^{III}}}/k_{\mathrm{Ru^{III}}} \sim 10^{-5} \Rightarrow \textbf{chemical mechanism} \end{cases}$

$$[M^{III}(NH_3)_5X]\text{—}[Cr(aq)]^{2+}:\ k_{\mathrm{Co^{III}}}/k_{\mathrm{Cr^{III}}} \sim 10^6\ \ (X = Cl, F, OH, OCOCH_3, \text{etc.})$$

$$k_{\mathrm{Co^{III}}}/k_{\mathrm{Cr^{III}}} \sim 10^1\ \ (X = \text{isonicotinamide})$$

Thermodynamically, the Co(III) and Ru(III) complexes are more or less comparable as the oxidizing agents but the Cr(III) complexes are relatively the poorer oxidizing agents (thermodynamic sense). It is evident from their E^0 (standard reduction potentials) values of aqua-complexes.

$$E^0 = 1.8\text{ V for } [Co(aq)]^{3+}/[Co(aq)]^{2+};\ E^0 = -0.41\text{V for } [Cr(aq)]^{3+}/[Cr(aq)]^{2+}.$$

Table 6.4.11.1 Comparison of rate constants for the reduction of Co(III), Cr(III) and Ru(III) complexes by $[Cr(aq)^{2+}]$ to illustrate the effect of the bridging ligand.(*cf.* Work of H. Taube, *Inorg. Chem.*, **8**, 2281, 1969; *ibid,* **9**, 2627, 1970; *J. Am. Chem. Soc.*, **90**, 1162, 1968; E. S. Gould, *J. Am. Chem. Soc.*, **96**, 2373, 1974).

Oxidant	**k (dm³ mol⁻¹ s⁻¹)**	*Remark*
$[CoF(NH_3)_5]^{2+}$	2.5×10^5	**Simple tunneling** through the delocalised 3-centred σ-MO (*i.e.* bridging ligand acts as a mediator of electron flow)
$[CrF(OH_2)_5]^{2+}$	2.6×10^{-2}	
$[CoCl(NH_3)_5]^{2+}$	6×10^5	
$[CrCl(OH_2)_5]^{2+}$	90.0	
$[RuCl(NH_3)_5]^{2+}$	3.5×10^4	
$[Co(NH_3)_5(OCOCH_3)]^{2+}$	0.30	**Simple resonance tunneling ISET path** through the conjugated system involving the delocalised π-MOs of the bridging ligand (a remote attack leads to the resonance transfer).
$[Cr(OCOCH_3)(OH_2)_5]^{2+}$	$\sim 10^{-4}$	
$[Co(NH_3)_5(OCOCHO)]^{2+}$	$\sim 10^4$	
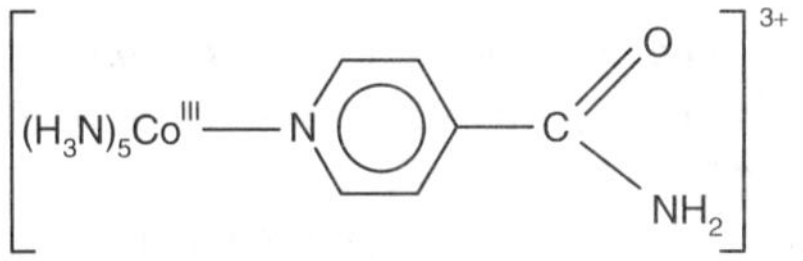	17.4	**Radical-ion mechanism (multistep process)** (remote attack)
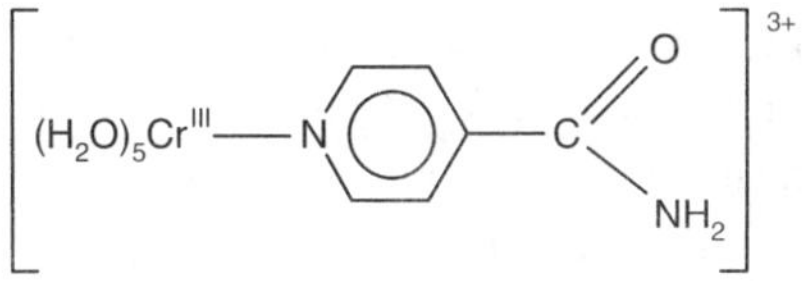	1.8	**Radical-ion mechanism (multistep process)**
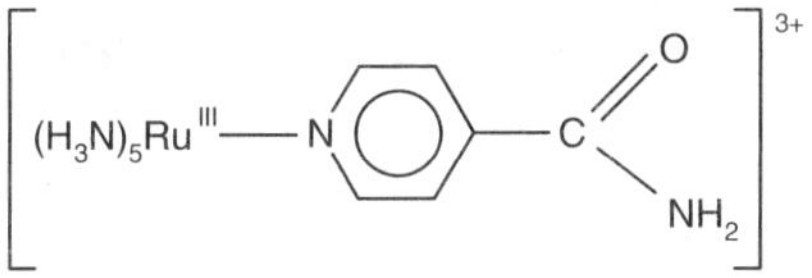	4×10^5	**Resonance transfer** (*i.e.* **direct tunneling process**)

The net transfer in the following redox reactions are:

	Co(III)–Cr(II)	**Cr(III)–Cr(II)**	**Ru(III)–Cr(II)**
Net transfer	$\sigma^* \rightarrow \sigma^*$	$\sigma^* \rightarrow \sigma^*$	$\sigma^* \rightarrow \pi^*$
	(*i.e.* $e_g \rightarrow e_g$)	(*i.e.* $e_g \rightarrow e_g$)	(*i.e.* $e_g \rightarrow t_{2g}$)

In the ISET process, the higher rate of reduction of the Co(III) complexes compared to the rate of reduction of the Cr(III)-complexes by Cr(II) is mainly due to the thermodynamic favour. The **higher electron affinity of Co(III) compared to that of Cr(III)** is evidenced by the fact that under the comparable conditions, it is relatively easier to produce Co(II) from Co(III) than to produce Cr(II) from Cr(III). This gives the better thermodynamic favour for the reduction of Co(III) to Co(II). It is also evident from their E^0-values (*cf.* 1.8 V vs. −0.41 V for their aqua complexes).

Thermodynamic favour for both the reactions, *i.e.* Co(III)–Cr(II) and Ru(III)–Cr(II) is more or less comparable but the higher rate (*cf.* X = Cl, $k_{Co}/k_{Ru} \approx 20$) in the ISET process for the reaction Co(III)–Cr(II) is due to the **orbital favour** (*i.e.* **no requirement of electronic activation**). The reaction, Ru(III)–Cr(II) in the ISET path needs an electronic activation of the oxidant for the required $\sigma^* \rightarrow \sigma^*$ transfer in the kinetic step. Thus by considering the both **thermodynamic factor and HOMO-LUMO nature**, we expect the following rate sequence:

rate (ISET path): Co(III)–Cr(II) 〉 Ru(III)–Cr(II) 〉〉 Cr(III)–Cr(II)

(b) **Tunneling mechanism for the nonreducible bridging ligand (*e.g.* OH^-, F^-, Cl^-, $CH_3CO_2^-$):**

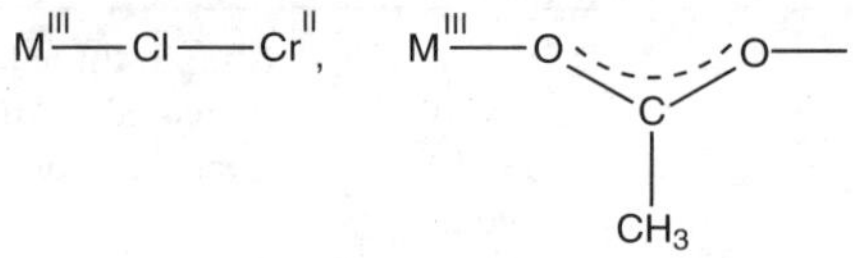

For the chlorido or fluorido bridged binuclear complexes, the simple electron tunneling occurs through the bridge (*i.e.* **via the 3-centred delocalized σ-MO**). In such cases, the chemical mechanism, leading to F^{2-} or Cl^{2-} is not possible.

For $CH_3CO_2^-$ as a bridging ligand, the remote attack can establish a conjugated system through which electron tunneling can occur by the **resonance transfer**. In such nonreducible bridging ligands, electron tunneling leads to an electron transfer. The observed sequence of the rate of reaction in such ISET processes is in conformity with the expected sequence from the consideration of thermodynamic favour and HOMO-LUMO nature. ***The Co(III)-complexes are reduced faster than the Cr(III) complexes by Cr(II) mainly due to the thermodynamic favour. Faster reduction of Co^{III}–X compared to that of Ru^{III}–X is mainly due to the kinetic favour ($\sigma^* \rightarrow \sigma^*$ vs. $\sigma^* \rightarrow \pi^*$).***

(c) **Chemical mechanism for the reducible bridging ligands** (*e.g.* $^-$OCOCHO, nicotinamide, etc.): In such cases the chemical or ion radical mechanism may operate and the *rate determining step* is the transfer of an electron from Cr(II) to the bridging ligand and this involves the $\pi^* \rightarrow \pi^*$ transfer. The reducing agent **$[Cr(aq)]^{2+}$ ($t_{2g}^3 e_g^1$) needs an electron activation** for the transfer of an electron into the vacant π^*-MO of the ligand like isonicotinamide. At the next faster step, the *e*-transfer occurs from the π^*–MO of the reduced bridging ligand ($L^{\bar{\cdot}}$) to the σ^*–MO of the oxidant (*i.e.* Co^{III}, Cr^{III}). This step needs an electronic activation of M(III) (*cf.* $\pi^* \rightarrow \sigma^*$ transfer forbidden) so that electron transfer from $L^{\bar{\cdot}}$ to M(III) can occur in the $\pi^*(L^{\bar{\cdot}}) \rightarrow \pi^*(M^{III})$ path. **This activation is not required for Ru(III)**. The whole process may be simply represented as follows:

$$\underset{i.e.\ (\pi^*)^3(\sigma^*)^1}{Cr^{II}\left(t_{2g}^3 e_g^1\right)} \xrightarrow{\text{activation}} \underset{i.e.\ (\pi^*)^4(\sigma^*)^0}{Cr^{II}\left(t_{2g}^4 e_g^0\right)} \xrightarrow[(\pi^* \rightarrow \pi^*)]{+L} Cr^{III}\left(t_{2g}^3\right) + L^{\bar{\cdot}}$$

$$M(III) \xrightarrow{\text{activation}} M(III)^*;\ L^{-} + M(III)^* \xrightarrow[(\pi^* \to \pi^*)]{\text{fast}} L + M(II),\ M = Co, Cr$$

Thus the reduction of both the Co^{III}– and Cr^{III}– complexes experience the common rate determining step, i.e. electronic activation of Cr(II) and then e-transfer from the π–MO of Cr(II) to the vacant π^*–MO of the bridging ligand.* In such cases, assuming the activation energy for Co(III) and Cr(III) (to provide the vacant π* orbital) more or less comparable, it is expected that both the Co(III) and Cr(III) complexes bearing the same bridging ligand will be reduced at the comparable rates in spite of the higher thermodynamic favour for the Co(III)-complexes. Experimental findings also support this view (*cf.* Table 6.4.11.1)

The above chemical mechanism of electron transfer is experimentally verified for the reduction of Co(III) and Cr(III) complexes bearing the same reducible ligands like isonicotinamide. *If the same mechanism is expected to operate for the reduction of the corresponding Ru(III)-complex then its rate constant should be comparable with those for the reductions of the Co(III) and Cr(III)-complexes.*

In spite of the similar thermodynamic favour for the Co^{III}– and Ru^{III}– complexes, the Ru(III) complexes are reduced remarkably faster than the Co(III) and Cr(III) complexes through the chemical mechanism of ISET path. Interestingly, for the nonreducible inorganic bridging ligands like Cl, both the Ru(III) and Co(III) complexes are reduced by Cr(II) at comparable rates (within a factor of 10–20) where the **simple tunneling mechanism operates.** In the simple tunneling mechanism, the Co^{III}-complexes are reduced faster by a factor of 10– 20 compared to the Ru^{III}-complexes but the rates differ dramatically for the chemical mechanism. It demands an explanation.

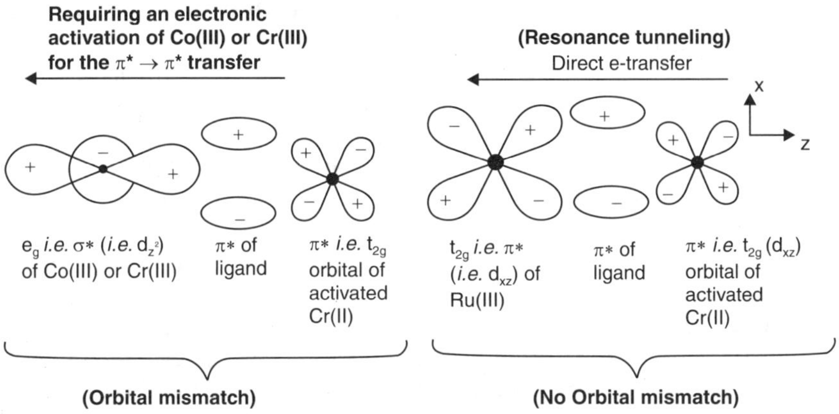

Fig. 6.4.11.1 Symmetries of the receptor orbitals of the oxidants (*i.e.* Co^{III}, Cr^{III}, Ru^{III}) and electron carrier orbital of the bridging ligand.

For the reduction of the Co(III) or Cr(III) complex, the receptor orbital is σ^* (*i.e.* e_g) while for the reduction of the Ru(III) complex, the receptor orbital is π^* (*i.e.* t_{2g})

$$\underset{(t_{2g}^6 e_g^0,\ l.s.)}{Co(III)} + \underset{(t_{2g}^3 e_g^1)}{Cr(II)} \longrightarrow \underset{(t_{2g}^5 e_g^2,\ h.s.)}{Co(II)} + \underset{(t_{2g}^3 e_g^0)}{Cr(III)};\ (\sigma^* \to \sigma^* \text{ transfer})$$

$$\underset{(t_{2g}^3 e_g^0)}{Cr(III)} + \underset{(t_{2g}^3 e_g^1)}{Cr(II)} \longrightarrow \underset{(t_{2g}^3 e_g^1)}{Cr(II)} + \underset{(t_{2g}^3)}{Cr(III)};\ (\sigma^* \to \sigma^* \text{ transfer})$$

$$\underset{(t_{2g}^5 e_g^0,\ l.s.)}{Ru(III)} + \underset{(t_{2g}^3 e_g^1)}{Cr(II)} \longrightarrow \underset{(t_{2g}^6 e_g^0,\ h.s.)}{Ru(II)} + \underset{(t_{2g}^3 e_g^0)}{Cr(III)};\ (\sigma^* \to \pi^* \text{ transfer})$$

In the chemical mechanism, *activated Cr(II)* transfers an electron from its π*–MO to the vacant π*–MO of the reducible ligand and thence to the receptor orbital of the oxidant. In the case of the Co(III) and Cr(III) oxidants, the receptor orbital is of σ*-symmetry while in the case of the Ru(III) oxidant, the receptor orbital is of π*-symmetry. Thus to transfer the electron from the ligand π*-orbital to the receptor orbital of oxidant, ***it needs some sort of activation in the case of the Co(III) or Cr(III) complexes.*** But no such activation is needed for the Ru(III) complex that provides the receptor orbital of π*-symmetry. ***This is why, the reduction of some Ru(III) complex is faster than the reduction of the Co(III) or Cr(III) complexes through the chemical mechanism.***

Chemical mechanism of ISET: Radical-ion mechanism *vs.* resonance tunneling mechanism (*cf.* Fig. 6.4.11.1)

$$\left.\begin{array}{l}\left[Co^{III}(NH_3)_5X\right]^{3+} \xrightarrow{[Cr(aq)]^{2+}} \\ \left[Cr^{III}(OH_2)_5X\right]^{3+} \xrightarrow{[Cr(aq)]^{2+}}\end{array}\right\}$$

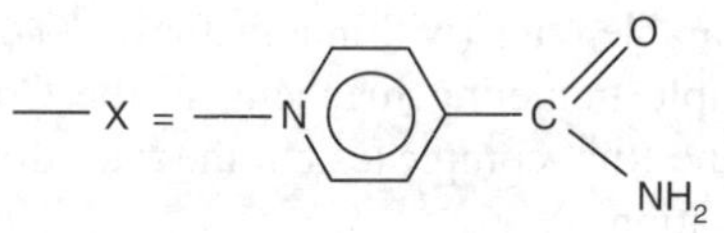

(pyridine carboxamide)

$$\left[Ru(NH_3)_5X\right]^{3+} \xrightarrow{[Cr(aq)]^{2+}} \}$$

- **Radical ion mechanism** (*e*-transfer from the t_{2g}, *i.e.* π* orbital of activated Cr^{2+} to the e_g, *i.e.* σ* orbital of the oxidant (*i.e.* Co^{III}, Cr^{III}) through the π* orbital of the bridging ligand. It leads to: **stepwise process** due to the **mismatch in symmetry** between the receptor orbital and electron carrier orbital of the ligand. Here the oxidant is to be activated to provide the π*-orbital as the receptor orbital.
- **Direct tunneling through the resonance mechanism** (*e*-transfer from the π* orbital of activated Cr^{2+} to the π* orbital of Ru^{III} through the π* orbital of the ligand: direct *transfer* experiencing ***no barrier from the symmetry mismatch of the orbitals.***)

- In the **radical ion mechanism,** electron transfer occurs in a **multistep process** involving the participation of the bridging ligand **chemically** in the redox process.
- The chemical mechanism becomes equivalent to the **resonance tunneling mechanism** when the temporarily reduced bridging ligand is too short-lived. In this **resonance mechanism,** electron transfer occurs **directly** where the bridge simply acts as a mediator of electron flow. When the life time of the temporarily reduced bridging ligand is neither too short-lived nor long-lived (*i.e.* intermediate character as in the partially hindered conjugation), it becomes difficult to distinguish between the two types of chemical mechanism.

- **(d) Chemical mechanism of the reduction of Co(III), Cr(III) *vs.* Ru(III) by $[Cr(aq)]^{2+}$ – nature of the receptor orbital:** By changing the bridging ligand Cl^- (a representative nonreducible ligand) to isonicotinamide (a representative reducible ligand), the rate constants are dramatically affected for the reduction of the Co(III) and Cr(III) complexes (*cf.* Table 6.4.11.1). But there is no such change for the reduction of the Ru(III)-complexes. In the cases of the Co(III) and Cr(III) complexes, there is a change of mechanistic path of electron transfer, *i.e.* **simple tunneling** (for $X = Cl^-$) **to chemical mechanism** (for X = isonicotinamide). *But, for the chemical mechanism to operate for the reduction of Co(III) and Cr(III) complexes, there is a mismatch in symmetry between the receptor orbital of the oxidant and electron carrier orbital of the reduced ligand.* ***Thus this chemical mechanism needs the stepwise passage of the electron for the reduction of***

the Co(III) and Cr(III) complexes. For the reduction of the Ru(III) complexes, no such barrier from the symmetry mismatch appears and the electron transfer can directly occur from the electron carrier ligand orbital to the receptor orbital of Ru(III). ***This favoured direct electron passage is very often described as the resonance transfer or resonance tunneling mechanism.***

- (e) **Distinction between the two possible routes of chemical mechanism of ISET process – electron transfer through resonance mechanism *vs.* radical ion mechanism:** From, the above discussion, it appears that the difference between the radical-ion mechanism (*i.e.* chemical mechanism) and resonance transfer mechanism lies in the ***life-time of the reduced radical ion,*** *i.e.* **temporarily reduced ligand**. In the case of **direct** transfer, *i.e.* **resonance transfer** (experiencing no barrier from the orbital symmetry mismatch), definitely the life time of the radical ion is too short to be detected.

For the reduction of the Co(III) and Cr(III) complexes by $[Cr(aq)]^{2+}$, the radical ion (*i.e.* temporarily reduced isonicotinamide, $H_2NCOC_5H_4N$) is **relatively long-lived** and there are experimental evidences to support the existence of the radical ion, *i.e. radical ion chemical mechanism.* In the reduction of the Co(III) complexes by Eu(II), there are also evidences in favour of the proposed radical intermediate. But, for the reduction of analogous Ru(III) compounds by $[Cr(aq)]^{2+}$, the temporarily reduced bridging ligand is **too short lived** to be detected. It occurs through the resonance transfer (*i.e.* **direct transfer**) where there is **no symmetry mismatch** between the electron carrier ligand orbital and the receptor orbital of Ru(III).

- (f) **Role of the 'R' group in dictating the mechanism of reduction of $[M(NH_3)_5(OCOR)]^{2+}$ by $[Cr(aq)]^{2+}$:**

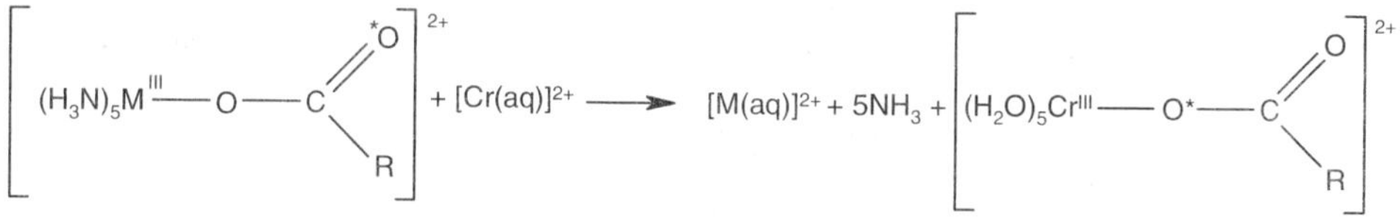

The important observations are:

(i) If the R-group is **saturated** and 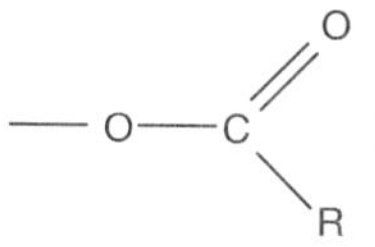 moiety is **not reducible**, the rate constants are comparable for different R-groups. In such cases, $[Cr(aq)]^{2+}$ makes a remote attack on the carbonyl oxygen and then the electron transfer occurs through the simple resonance tunneling. Examples of such bridging groups (OCOR) are: *acetate, butyrate, succinate, etc.*

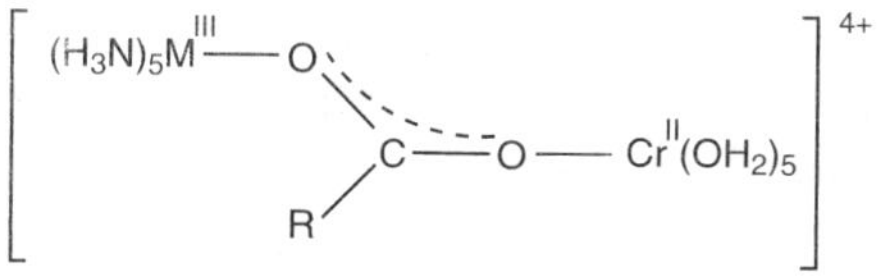

R = CH_3 (acetate)
= C_3H_7 (butyrate)
= $(CH_2)_2CO_2H$ (acid succinate)

(Binuclear precursor complex)

(ii) If the R-group is **unsaturated (specially to maintain the conjugation)** and the OCOR group is **reducible** then, the rate constants are remarkably affected depending on the nature of M(III) (*cf.* Table 6.4.11.1). It happens so for oxalate, maleate, fumarate, etc. In such cases, the free carboxylate group (which is a better coordinating group than the carbonyl group) coordinates with Cr(II) and the established conjugation (due to unsaturation) facilitates the *resonance transfer* if there is ***no symmetry mismatch barrier*** between the electron carrier bridging ligand orbital and the receptor orbital of M(III). If this symmetry mismatch barrier exists, then the ***radical ion mechanism*** — a multistep process is perferred (*cf.* reduction of Co^{III}, Cr^{III} *vs.* Ru^{III} by Cr^{II}, earlier discussion). However, the positively charged metal centres are kept well separated in such binuclear complexes and it reduces both the steric and electrostatic repulsion between the M(III) and Cr(II) centre. The bridge for the fumaric acid is shown below.

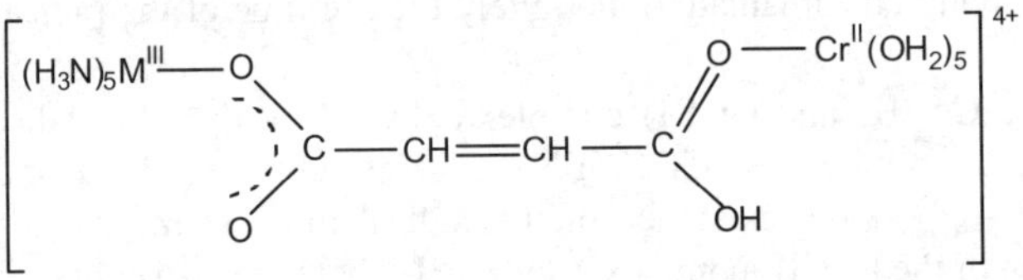

$(HO_2C)CH{=}CH(CO_2H)$
trans-form: fumaric acid
cis-form: maleic acid

Thus the increased rate (compared to those of the cases having the bridging ligand like acetate) arises due to the factors: **better bridging** by using the free carboxylate group, **unsaturation in the bridging ligand** allows the better electron flow (*i.e. resonance transfer through the conjugated system*) and avoiding the electrostatic and steric repulsion between the M(III) and Cr(II) centres which are kept separated by the bridging ligand.

It has been suggested by some workers that for oxalic acid, formic acid maleic acid, in the bridge, M^{III}—ligand—Cr^{II}, ***full-length conjugation is not attained.*** For maleic acid, it is due to the steric factor. In such cases, the enhanced rate is believed to be due to their reducible character and the reaction is believed to pass through the ***chemical mechanism*** (for M^{III} = Co^{III}, Cr^{III}, Ru^{III}, *i.e.* irrespective of the symmetry mismatch barrier) where the bridging ligand is temporarily reduced as an intermediate one.

ISET: Chemical mechanism *vs.* Tunneling mechanism (*cf.* Fig. 6.4.11.1)

- Sometimes, in both the cases, the vacant low lying orbital of the bridging ligand participates and very often, it is very difficult to distinguish between these two mechanisms of electron transfer.

$$\left[M^{III}(NH_3)_5X\right] \xrightarrow{[Cr(aq)]^{2+}} M(II) + Cr^{III}X + 5NH_3$$

Rate of the above redox reaction depends on the reducibility of X. For X = $-OCOCH_3$ **(nonreducible bridging ligand)** the rate is slow; for X = –OCOCHO (glyoxylate) *i.e.* **(reducible bridging ligand)** the rate is fast. For X = –OCOCHO, the presence of a vacant low lying orbital in this bridging ligand facilities the *e*-transfer process through the chemical mechanism. The chemical mechanism may be of two types: radical ion mechanism and direct resonance tunneling depending on the nature of X and the receptor orbital of M^{III}.

6.4.12 Multiple Bridging in the Activated Complex in the ISET Process

There are examples in which multiple bridges remain present in the activated complex in the ISET process. Some examples are given below. *The nature of bridging is very often identified from the product analysis.*

(i) **Reduction of *cis*-$[Cr(OH_2)_4X_2]^+$ by $[Cr(aq)]^{2+}$ ($X = N_3^-$, RCO_2^-)**

$$cis\text{-}\left[Cr(OH_2)_4X_2\right]^+ + \left[{}^*Cr(aq)\right]^{2+} \longrightarrow \left[Cr(aq)\right]^{2+} + cis\text{-}\left[{}^*Cr(OH_2)_4(X)_2\right]^+$$

Transfer of two *cis*-ligands (X) indicates the nature of bridging in the **activated complex** as follows:

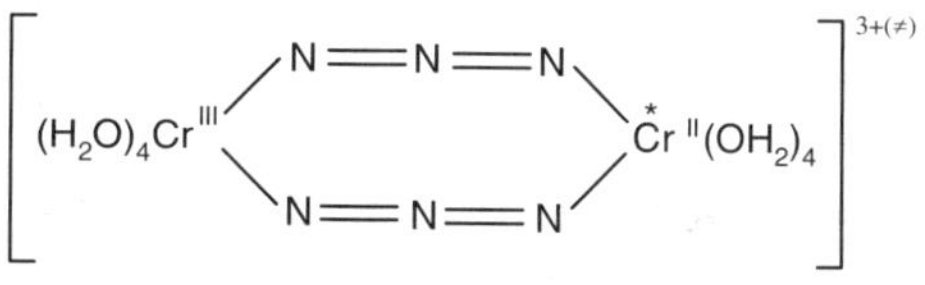

Though there is a double bridge, but one electron transfer occurs. Probably, the double bridging stabilizes the binuclear precursor complex more. It has been found that for $X = N_3^-$, the rate constant increases by 30 times compared to the single bridge.

$$\frac{k_{(\text{double bridge})}}{k_{(\text{single bridge})}} = 30.$$

(ii) **Reduction of *cis*-$[Co(NH_3)_4(N_3)_2]^+$ or *cis*-$[Co(en)_2(N_3)_2]^+$ by $[Cr(aq)]^{2+}$** also passes through the double bridge made by the azides in *cis*-position.

Note: Reduction of ***cis*-$[CrF_2(OH_2)_4]^+$ by $[Cr(aq)]^{2+}$** occurs through the single bridge though the two *cis*-fluorides are in a suitable position to bridge.

$$cis\text{-}\left[CrF_2(OH_2)_4\right]^+ + \left[{}^*Cr(aq)\right]^{2+} \longrightarrow \left[Cr(aq)\right]^{2+} + F^- + \left[{}^*CrF(OH_2)_5\right]^{2+}$$

i.e. only one F^- is transferred.

Probably, for the monoatomic bridging ligands like F^-, the double bridge will produce a strain to destabilize the binuclear system.

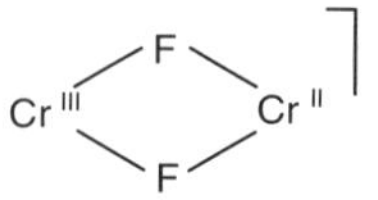

(strained)

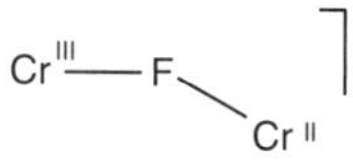

(less strain)

(iii) **Reduction of $[Co(edta)]^-$ by $[Cr(aq)]^{2+}$** gives the product $[Cr(edta)(OH_2)_3]^-$ (where edta acts as a tridentate ligand). It indicates that in the binuclear precursor complex, edta provides three binding sites to coordinate with the reducing agent.

(iv) The **reduction of $[Co(NH_3)_5(ntaH)]^+$ by $[Fe(aq)]^{2+}$** indicates that nta provides three binding sites to coordinate with Fe(II) in the precursor complex. This aspect has been discussed earlier. The structure of the binuclear precursor complex is as follows:

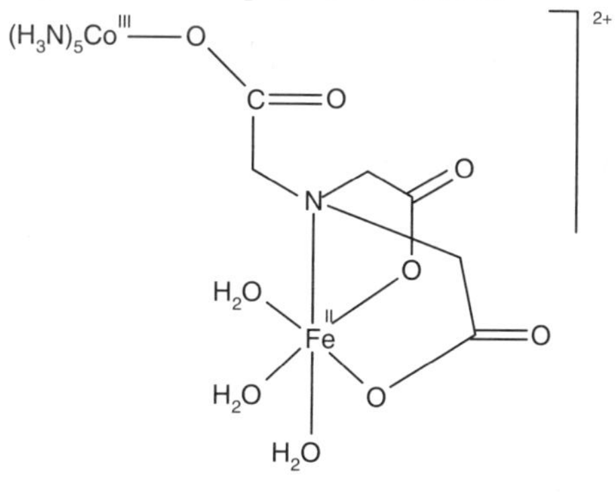

(v) **Reduction of $[Co(C_2O_4)(NH_3)_4]^+$ by $[Cr(aq)]^{2+}$** produces $[Cr(C_2O_4)(OH_2)_4]^+$ suggesting the following binuclear precursor complex.

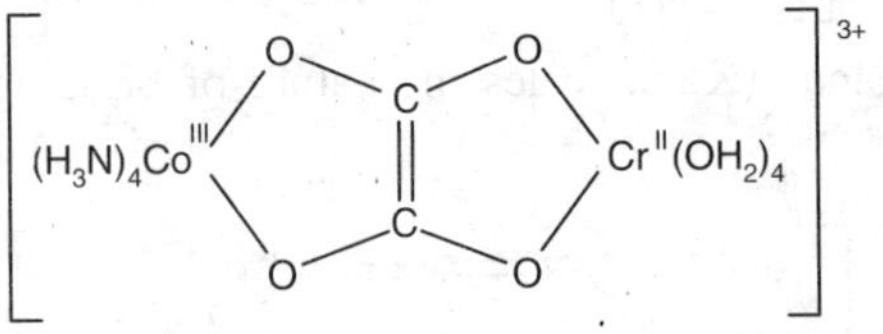

(vi) **Reduction of $[Co(NH_3)_5(OCOCH_2CO_2H)]^{2+}$ by $[Cr(aq)]^{2+}$** gives $[Cr(malonate)(OH_2)_4]^+$. It suggests the following binuclear precursor.

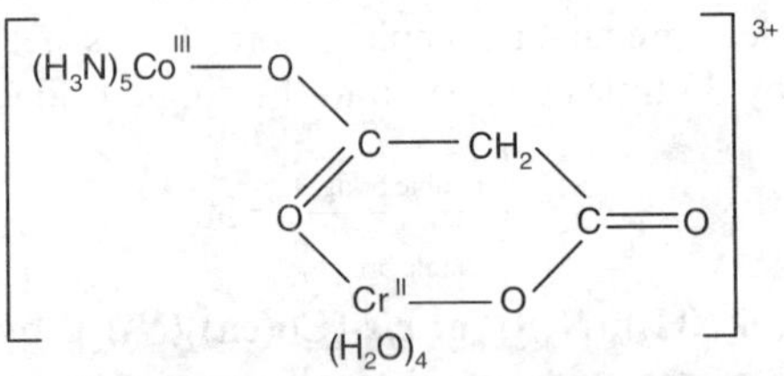

(vii) In the **reduction of $[Co^{III}(acac)_2(en)]^+$ by $[Cr(aq)]^{2+}$**, different products of Cr(III)-complexes are found. These are: $[Cr(aq)]^{3+}$, $[Cr(acac)(aq)]^{2+}$, $[Cr(acac)_2(aq)]^+$. It can be reasonably concluded that $[Cr(aq)]^{3+}$ is produced through the OSET path; $[Cr(acac)(aq)]^{2+}$ is produced through the single acetylacetonate bridged ISET path and $[Cr(acac)_2(aq)]^+$ is produced through the double acetylacetonate bridged ISET path. The probable structures of the binuclear bridged successor complexes are given below.

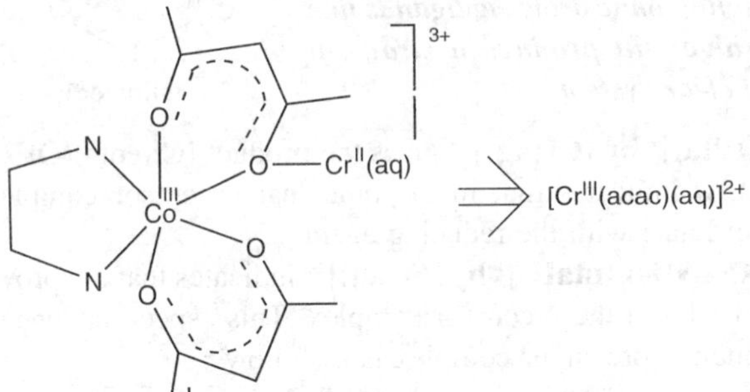

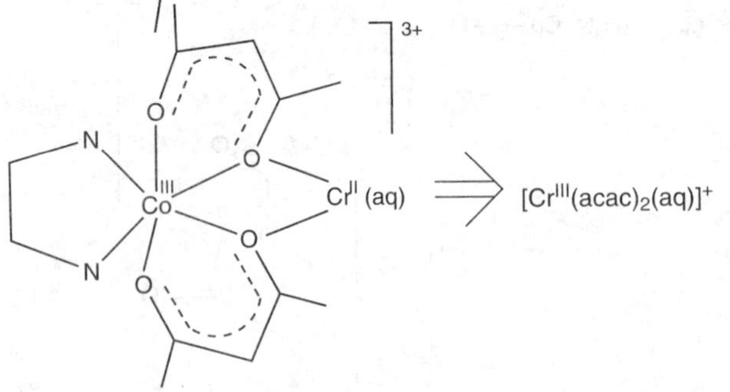

Note: In all the above given examples, though multibridges are formed in the binulcear precursor complexes, but there is only one electron transfer. It may be pointed out that in Pt(II)–Pt(IV) exchange path, two electron transfer can occur through a single bridge like Pt^{II}–Cl–Pt^{IV} (*cf.* Sec. 6.4.15).

6.4.13 Bridging Ligand Provided by the Reducing Agent in the ISET Process

In the examples discussed in all earlier sections, it was noted that the bridging ligand is provided by the oxidant (which is also substitutionally inert). However, in the reduction of the chromic acid ($HCrO_4^-$) by $[Fe(CN)_6]^{4-}$, it was noted that the *inert reductant provides the bridging ligand.* It is supported by the characterization of the following product species.

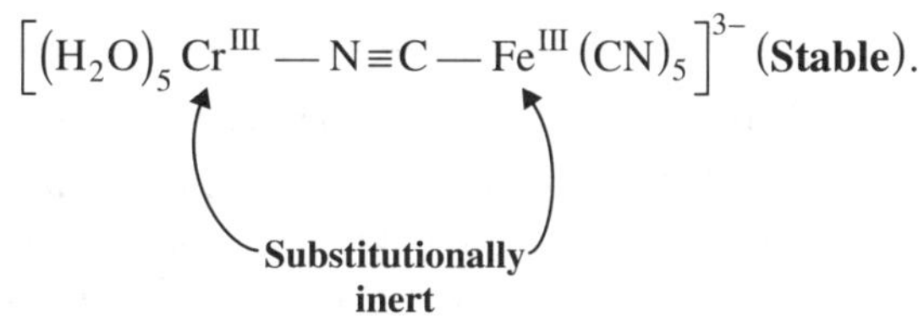
$\left[(H_2O)_5\,Cr^{III} - N\equiv C - Fe^{III}\,(CN)_5\right]^{3-}$ (**Stable**).
Substitutionally inert

Cr(VI) experiences 3e-reduction to Cr(III) at the successive 1e-transfer step giving rise to Cr(V) and Cr(IV) intermediates. All these higher valent species, *i.e.* Cr(VI), Cr(V) and Cr(IV) are labile while Cr(III) is inert. The above mentioned product (binuclear) is produced through the reduction of Cr(IV) by $[Fe(CN)_6]^{4-}$ in the ISET path.

6.4.14 Effect of the Non-bridging Ligand on the Rate Process in the ISET Path: Ligand Field Strength Controlling Energy of the Receptor Orbital of the Oxidant and Franck-Condon Barrier

Energy of the receptor orbital (σ*) of the oxidant: It has been noted that the rate of reduction of *trans*-$[M^{III}(NH_3)_4(L)X]$ by $[Cr(aq)]^{2+}$ or $[Fe(aq)]^{2+}$ in the ISET process drastically depends on the ligand field strength of L (*trans* to the bridging ligand X). It is quite reasonable by considering the nature of the receptor orbital of the oxidant. For the oxidants Co(III) (t_{2g}^6) and Cr(III) (t_{2g}^3), the receptor orbital is σ* (*i.e.* e_g orbital which is d_{z^2} by considering the bridge along the *z*-direction). Energy of this receptor orbital is very much sensitive to the nature of the *trans*-ligand *L* (when X is kept the same for comparison) compared to the *cis*-ligands. *With the increase of the ligand field strength of L, energy of the e_g-orbital (i.e. d_{z^2}) will increase, i.e.* ***the receptor orbital will be destabilized more by the stronger field ligand L and it will be stabilized more for the weaker field ligand L.***

To receive the electron in the vacant e_g (*i.e.* σ*) orbital of the oxidant, energy of this receptor orbital must be lowered. This can be done by moving the *trans*-ligands *L* and *X* outwards (*i.e. L* and *X* move in opposite directions). If the ligand field strength of *L* is varied (for a particular bridging ligand *X*), then for the stronger field ligand, more distance is to be traveled outwards by *L* to lower the energy of the receptor orbital. **Consequently, more activation energy will be needed for the stronger field ligand L.** The ligand field strength is: en~NH_3 ⟩⟩ H_2O. Thus, *trans*-$[CoCl(en)_2(NH_3)]^{2+}$ should react slower than *trans*-$[CoCl(en)_2(OH_2)]^{2+}$. This expectation is experimentally verified.

Table 6.4.14.1 Effect of the nonbridging ligands on rate constants for the ISET reactions (*cf.* A. Haim, *J. Am. Chem. Soc.*, **86**, 2352, 1964; R. D. Cannon *et al. J. Am. Chem. Soc.*, **87**, 5264, 1965)

Oxidant	*Reducing agent*	*k* ($dm^3\ mol^{-1}\ s^{-1}$) *25°C*
trans– $[Co(en)_2(OH)(OH_2)]^{2+}$	$[Cr(aq)]^{2+}$	2.6×10^6
trans– $[Co(en)_2(NH_3)(OH)]^{2+}$	$[Cr(aq)]^{2+}$	2.0×10^5
trans– $[CoCl(en)_2(OH_2)]^{2+}$	$[Fe(aq)]^{2+}$	0.25
trans– $[CoCl(en)_2(NH_3)]^{2+}$	$[Fe(aq)]^{2+}$	6.5×10^{-5}
$[CrCl(NH_3)_5]^{2+}$	$[Cr(aq)]^{2+}$	8.8×10^{-2}
trans– $[CrCl(NH_3)_4(OH_2)]^{2+}$	$[Cr(aq)]^{2+}$	1.3

$$trans\text{–}\left[CrCl(NH_3)_4(^*OH_2)\right]^{2+} + \left[Cr(aq)\right]^{2+} \longrightarrow \text{products (ISET path).}$$

It has been noted that *trans*– $[CrCl(NH_3)_4(^{18}OH_2)]^{2+}$ *reacts slower than trans*– $[CrCl(NH_3)_4(^{16}OH_2)]^{2+}$. This observation supports the requirement of bond length elongation in the *trans*-axial directions to lower the energy of the receptor e_g-orbital. *Obviously the movement of heavier isotope needs more energy.*

Ligand field strength effect and Franck-Condon barrier: To overcome the Franck-Condon barrier, bond length is to be elongated in the oxidant and bond length is to be shortened in the reductant. Thus in the reductant, presence of the stronger field ligand will require less shortening of the bond length. This will reduce the activation energy. In fact, it has been experimentally verified that the Co(III)-complexes are reduced by the Cr(II)-complexes in the following order of rate constants.

$$\underset{\xleftarrow[\text{Reducing power (kinetically)}]{}}{\xleftarrow[\text{Increasing ligand field strength}]{\left[Cr(en)_2(OH_2)_2\right]^{2+} > \left[Cr(en)(OH_2)_4\right]^{2+} > \left[Cr(OH_2)_6\right]^{2+}}}$$

6.4.15 Two Electron Transfer in the ISET Process

Here we shall illustrate the case of Pt(II)–Pt(IV) system where there is only one bridging ligand in the binuclear precursor complex but two electrons are transferred through the bridge followed by the transfer of the bridging ligand.

Pt(II) which is relatively more labile than Pt(IV) is known to **catalyze the exchange of free Cl^-** with the Cl^- bound to Pt(IV) in the following reaction.

$$trans\text{-}\left[PtCl_2(en)_2\right]^{2+} + {}^*Cl^- \xrightarrow[\text{(catalyst)}]{[Pt(en)_2]^{2+}} trans\text{-}\left[Pt^*ClCl(en)_2\right]^{2+} + Cl^-$$

The proposed mechanism involves the rapid coordination of free $^*Cl^-$ to $[Pt(en)_2]^{2+}$ to give a five coordinate intermediate followed by the formation of the precursor bridged complex utilizing the vacant coordination site of $[PtCl^*(en)_2]^+$. Then electron transfer occurs within the activated precursor complex followed by the fission of the successor complex. This fission will transfer the bridging ligand.

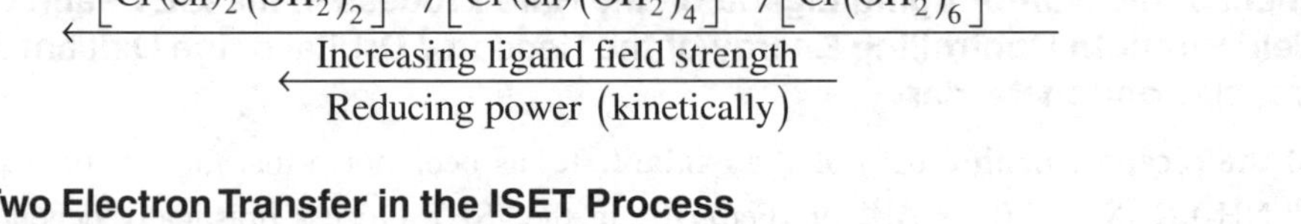

$trans\text{-}[PtCl_2(en)_2]^{2+}$ + $[*PtCl*(en)_2]^+$ $\rightleftharpoons$

2e-transfer

$trans\text{-}[*Pt*ClCl(en)_2]^{2+}$ + $[Pt(en)_2]^{2+}$ + Cl^- $\longleftarrow$

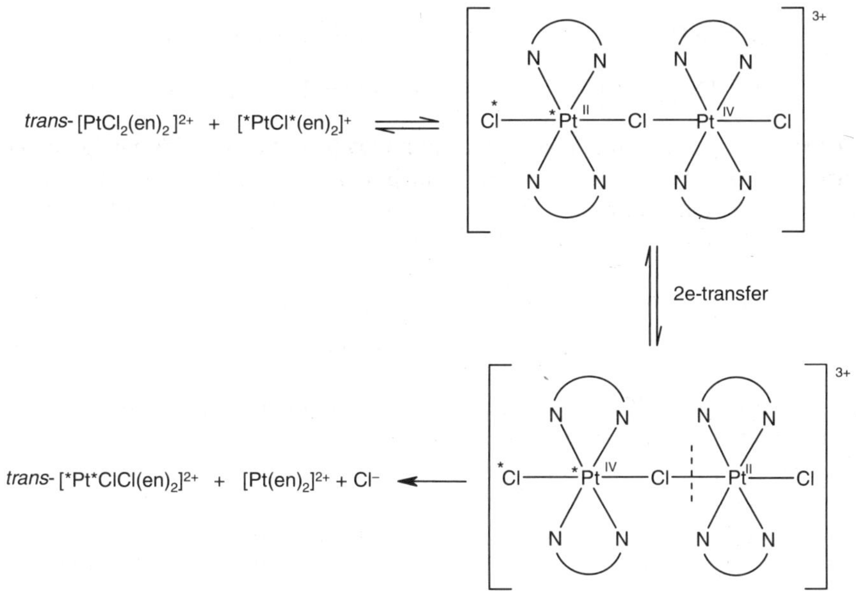

By using the isotopically labelled Pt-centre, the above mechanism has been proved. The rate law is given by:

$$\text{rate} \propto [\text{Pt(II)}][\text{Pt(IV)}][\text{Cl}^-]$$

The following ligand substitution reaction actually represents a **Pt(II)-catalysed process**. Pt(II) may be generated in the reaction through the reduction of Pt(IV) by NO_2^- or it may be added externally.

$trans\text{-}[PtCl_2(en)_2]^{2+} + NO_2^- \rightarrow [PtCl(en)_2(NO_2)]^{2+} + Cl^-$

rate $\propto[\text{Pt(II)}][\text{Pt(IV)}][NO_2^-]$.

Here it may be mentioned that *between the Pt(II) and Pt(IV) centres, Pt(II)-centre is relatively more labile.* In fact, it is the common observation that in a particular system, if the lower valent centre is more labile and the higher valent centre is inert then the lower valent centre catalyzes the substitution process at the higher valent centre (*cf.* Sec. 6.6).

6.5 MECHANISTIC CRITERIA AND THE FACTORS FAVOURING THE OUTER AND INNER SPHERE ELECTRON TRANSFER PROCESSES

(A) Factors favouring the OSET process:

- formation of the binuclear bridged complex is denied;
- $\pi^* \rightarrow \pi^*$ net transfer;
- less inner sphere reorganization energy to overcome the Franck-Condon barrier;
- higher equilibrium constant for the overall reaction.

Bridge formation leading to the dinuclear complex is disfavoured. It may happen in many ways:

(i) The reductants like **$[Cr(bpy)_3]^{2+}$** (t_{2g}^4), **$[Ru(NH_3)_6]^{2+}$** (t_{2g}^6), etc. are tremendously **substitutionally inert** and their associated ligands, *i.e.* bpy, NH_3, etc. cannot function as the bridging ligands. In such cases, the inert oxidants like $[Co^{III}(NH_3)_5X]$ can provide the bridging ligand (X) but *formation of the binuclear precursor complex is prevented because of the inertness of the reducing centre.*

$$\left.\begin{matrix}[M^{III}(NH_3)_5X]+[Cr(bpy)_3]^{2+} \longrightarrow OSET \\ [M^{III}(NH_3)_5X]+[Cr(bpy)_3]^{2+} \longrightarrow OSET\end{matrix}\right\} \begin{matrix}M^{III}=Co^{III},\ Ru^{III} \\ X^-=Cl^-,\ Br^-,\ N_3^-\ \text{etc.}\end{matrix}$$

(ii) The reductant may be substitutionally inert but **it may provide the bridging ligand.** For example, the inert reductant $[Fe(CN)_6]^{4-}$ can provide the bridging ligand, *i.e.* C-bound cyanido ligand but if the **oxidant is not labile** then formation of the dinuclear species will be prevented. Consequently, the reactions occur through the OSET process.

$$\left.\begin{matrix}[IrCl_6]^{2-}+[Fe(CN)_6]^{4-} \longrightarrow \\ [Fe(phen)_3]^{3+}+[Fe(CN)_6]^{4-} \longrightarrow\end{matrix}\right\} \begin{matrix}\text{OSET process.} \\ \text{(both the oxidant and reductant are inert)}\end{matrix}$$

(**Note:** The reaction of $HCrO_4^-$ with $[Fe(CN)_6]^{4-}$ *goes* through *the ISET process* where the —CN group bridges with the labile oxidant).

(iii) The oxidant is inert and it fails to provide any suitable bridging ligand. In such cases, formation of the dinuclear precursor complex is prevented even when the reductant is labile. Reduction of $[Co(phen)_3]^{2+}$, $[Fe(bpy)_3]^{3+}$, $[M(NH_3)_6]^{3+}$, $[M(NH_3)_5(py)]^{3+}$ (M = Co, Ru), by the even labile reductants like $[Cr(aq)]^{2+}$, $[Fe(aq)]^{2+}$ occurs through the OSET process.

$$\left.\begin{matrix}[M(LL)_3]^{3+}+[M'(aq)]^{2+} \longrightarrow \\ [M(NH_3)_6]^{3+}+[M'(aq)]^{2+} \longrightarrow \\ [M(NH_3)_5(py)]^{3+}+[M'(aq)]^{2+} \longrightarrow\end{matrix}\right\} \begin{matrix}\text{OSET process} \\ M=Co,\ Ru;\ LL=bpy,\ phen \\ M'=Cr,\ Fe\end{matrix}$$

cf. $[M(NH_3)_5X]^{2+} + [M'(aq)]^{2+} \rightarrow$ product; ISET; $X^- = Cl^-, Br^-, OH^-, SCN^-, N_3^-, RCO_2^-$, etc.

(B) Factors favouring the ISET process

- formation of binuclear bridged complex is possible;
- bridging ligand is suitable to mediate the electron transfer process;
- $\sigma^* \rightarrow \sigma^*$ net transfer;
- less reorganization energy to overcome the Franck-Condon barrier;
- higher equilibrium constant for the overall reaction;

Formation of the binuclear bridged precursor complex is favoured under the following conditions.

(i) One of the reactant is labile while the other reactant is inert but capable of providing a bridging ligand. *Generally, the inert oxidant provides the bridging ligand and the reductant is labile.*

Formation of this dinuclear bridged complex (where the bridging ligand is suitable to mediate the electron transfer process) is the primary requirement for the operation of the ISET process. But fulfillment of this condition cannot ensure the ISET process and it happens so in the case of reduction of $[Ru(NH_3)_5X]^{2+}$ by $[V(aq)]^{2+}$ ($\pi^* \rightarrow \pi^*$ net transfer). This reaction occurs through the OSET path (*see* Sec. 6.4.5). The reaction experiences the net $\pi^* \rightarrow \pi^*$ transfer which is a favourable situation for the OSET path but **it is an unfavourable situation for the ISET path** because it needs the electronic activation of both the oxidant and reductant for the ISET path. Moreover, the inertness of the $[V(aq)]^{2+}$ centre produces a barrier for the ISET path. These are the reasons for the OSET path.

If the bridging ligand is a long chain saturated moiety, then even within the binuclear precursor complex, the OSET path instead of the ISET path operates (*see* Sec. 6.4.11).

(C) Experimental Evidences in Favour of the ISET Process

(i) **Detection of the binuclear complex:** If it is possible to **detect experimentally the binuclear complex** (*i.e.* successor complex and / or precursor complex) under the experimental condition, then it can be concluded without any hesitation that the reaction is passing through the ISET process.

(ii) **Ligand transfer** (*see* Sec. 6.4.8)**:** If the electron transfer is accompanied with the **ligand transfer** and overall reaction rate is much faster than the substitution rate of the metal centre to which the ligand is transferred, *then it can be concluded without any hesitation that the reaction is passing through the ISET process.*

Here it may be mentioned that the ligand transfer is not a necessary condition for the ISET process. Thus, in absence of the ligand transfer or release of the ligand into the solution, other arguments are needed to substantiate the occurrence of the ISET process.

(iii) **Rate dependence on the nature of the bridging ligand:** If nature of the bridging ligand affects the overall rate process, then it supports the occurrence of the ISET process. But, for the substitutionally controlled ISET process, the rate does not depend on the nature of the bridging ligand. Here it is assumed that the substitution experiences the *d*-activation.

- It has been already mentioned that azide bridge is more efficient than the N-bonded thiocyanate bridge in the overall electron transfer process. **Thus the criterion, $k_{N_3} >> k_{NCS}$ is an argument in favour of the ISET process.** Similarly, for the reducing agents like $[Cr(aq)]^{2+}$, $[Fe(aq)]^{2+}$, the ***S*-bonded NCS^- can act as a better bridging ligand than the *N*-bonded** *NCS^-* and the observation, $k_{SCN} \gg k_{NCS}$ is an evidence in favour of the ISET process. The $k_{OH} > k_{H_2O}$ observation is also in favour of the ISET process. These aspects have been discussed in detail in Sec. 6.4.10.

$$\left[M^{III}(NH_3)_5X\right]+\left[M'(aq)\right]^{2+} \longrightarrow \text{Product};\ M = Co^{III}, Ru^{III}, \textit{etc.},\ M' = Fe^{II}, Cr^{II}.$$

$$X = -N_3^-, -NCS^-, -SCN^-, OH^-, H_2O.$$

$$k_{N_3} \gg k_{NCS};\ k_{SCN} \gg k_{NCS};\ k_{OH} > k_{H_2O} \Rightarrow \textbf{ISET process.}$$

- Sometimes, the overall reaction rate increases with the increase of ***coordinating power*** of the ***bridging ligand.*** *It supports the ISET process* (*cf.* Sec. 6.4.10).
- If the **remote electrophilic attack** on the bridging ligand by the reducing agent can establish a conjugated system (*i.e. if the bridging ligand bears the proper unsaturation*), then it favours *the* ***electron tunneling through the resonance mechanism.*** This observation supports the ISET process (*cf.* Secs. 6.4.9, 11).

If the bridging ligand (*e.g.* oxalate, formate, glyoxylate, maleate, glycolate, isonicotinamide, etc.) is **reducible**, then the electron transfer process is favoured due to the ***chemical* or *ion-radical mechanism***. Such an observation is conformity with the ISET process. (*cf.* Sec. 6.4.11).

(**Note:** If the ISET process is ***substitution controlled,*** *i.e. rate determining step is the formation of the binuclear precursor complex,* and if the substitution process occurs through the ***dissociative path,*** then even in the ISET process, the overall rate is insensitive to the nature of the bridging ligands. It happens so for the reduction by $[V(aq)]^{2+}$. Then the kinetic parameters of the overall reaction are found to be comparable with those of the substitution process at $[V(aq)]^{2+}$. This aspect has been discussed in detail in Sec. 6.4.5.)

(iv) ***Volume of activation (ΔV≠)***: It may be used to diagnose the reaction pathway. In the OSET path where formation of the precursor complex enhances the charge (*e.g.* electron exchange in the couple $[Fe(aq)]^{2+}$–$[Fe(aq)]^{3+}$), **ΔV≠ is expected to be negative** because of the more electrorestriction imposed on the surrounding solvents by the activated complex. However, the charge destruction (*i.e.* ion pair formation) in the precursor complex formation may lead to the +ve value of ΔV≠. In the ISET path, formation of the binuclear precursor complex releases solvent from the labile centre. ***It generally makes ΔV≠ value positive.*** These are illustrated in the following examples.

$[Fe(OH_2)_6]^{3+}$–$[Fe(OH_2)_6]^{2+}$ (electron exchange): $\Delta V^{\neq} = -12 \pm 2\ cm^3\ mol^{-1}$ (⇒ **OSET path**)

$[Fe(OH)(OH_2)_5]^{2+}$–$[Fe(OH_2)_6]^{2+}$ (electron exchange): ΔV≠ is slightly positive. (⇒ **ISET path**)

$[Co(NH_3)_5X]^{2+}$–$[Fe(OH_2)_6]^{2+}$ (X = F, Cl, Br, N_3): $\Delta V^{\neq} = 8 - 14\ cm^3\ mol^{-1}$ (⇒ **ISET path**)

6.6 LIGAND REPLACEMENT REACTIONS THROUGH ELECTRON TRANSFER

If for a particular metal ion, the higher valent centre is relatively inert but the lower valent state is relatively labile, then the lower valent state can catalyze the substitution process at the higher valent inert centre. Such examples are found for:

Pt(II)–Pt(IV); Cr(III)–Cr(II); Co(III)–Co(II).

This aspect is illustrated in different examples

(i) **Unusually high water exchange rate for $[Co(aq)]^{3+}$** (*see* Sec. 5.13): The water exchange rate at $[Co(OH_2)_6]^{3+}$ is unexpectedly high. The high reduction potential of the $[Co(aq)]^{3+}/[Co(aq)]^{2+}$ couple (E^0 = 1.8 V) allows the oxidation of H_2O (*cf.* E^0 for O_2/H_2O = 1.23 V) by $[Co(aq)]^{3+}$ but it occurs very slowly (kinetic reason, *i.e.* over-potential for gas evolution). Thus, $[Co(aq)]^{3+}$ solution is always contaminated by a trace amount of $[Co(aq)]^{2+}$ which is labile and its water exchange rate is very high.

Rapid water exchange: $\left[Co(OH_2)_6\right]^{2+} + 6H_2O^* \rightleftharpoons \left[Co(^*OH_2)_6\right]^{2+} + 6H_2O$

Fairly rapid electron exchange (OSET): $[Co(OH_2)_6]^{3+} + [^*Co(^*OH_2)_6]^{2+} \rightleftharpoons [^*Co(^*OH_2)_6]^{3+} + [Co(OH_2)_6]^{2+}$

The fairly rapid electron exchange between $[Co(OH_2)_6]^{3+}$ and $[^*Co(OH_2)_6]^{2+}$ leads to the apparently enhanced overall water exchange rate in $[Co(OH_2)_6]^{3+}$. The electron exchange rate in the $[Co(OH_2)_6]^{3+}/[Co(OH_2)_6]^{2+}$ couple is fairly fast even through the OSET process (*see* Sec. 6.3.6). The electron exchange through the **ISET path cannot be completely ruled out** as H_2O can act also as a poor bridging ligand (*cf.* Sec. 6.4.4, **ISET path**).

$$\left[Co(NH_3)_5(^*OH_2)\right]^{3+} + \left[Cr(aq)\right]^{2+} \longrightarrow \left[Co(aq)\right]^{2+} + 5NH_3 + \left[Cr(^*OH_2)(OH_2)_5\right]^{3+}$$ **(ISET path)**

$$\left[Co(OH_2)_6\right]^{3+} + \left[^*Co(^*OH_2)_6\right]^{2+} \rightleftharpoons \left[Co(OH_2)_6\right]^{2+} + \left[^*Co(^*OH_2)_6\right]^{3+}$$ **(OSETpath)**

$$\left[(H_2O)_5Co^{III}-OH_2\right]^{3+}+\left[{}^*Co({}^*OH_2)_6\right]^{2+} \rightleftharpoons \left[(H_2O)_5Co^{III}-\underset{H}{\overset{H}{O}}-{}^*Co^{II}({}^*OH_2)_5\right]^{5+}$$

$$\xrightarrow[H_2O]{\text{e-transfer, ISET path}} \left[Co(OH_2)_6\right]^{2+}+\left[{}^*Co(OH_2)({}^*OH_2)_5\right]^{3+}$$

Thus the **rapid water exchange** at $[Co(aq)]^{2+}$ (present in a catalytic amount) followed by the favourable electron exchange **(mainly OSET path, and partly ISET path)** in the couple $[Co(aq)]^{3+}$–$[Co(aq)]^{2+}$ explains the unusually high water exchange rate at $[Co(aq)]^{3+}$ (*cf.* Sec. 5.13). **Because of the critical 10 *Dq* value, $[Co(aq)]^{3+}$ remains in a low spin ⇌ high spin equilibrium and this is the another reason for the overall enhanced electron exchange rate. This aspect has been discussed in Sec. 6.3.6.** Here it is worth mentioning that the said critical 10 Dq value is also another reason for the enhanced water exchange rate at $[Co(aq)]^{3+}$ (*see* Sec. 5.13).

(ii) **Catalysis by $[Cr(aq)]^{2+}$ in the water exchange process of $[Cr(aq)]^{3+}$:** The water exchange rate at $[Cr(aq)]^{3+}$ is exceedingly slow but in presence of a trace amount of $[Cr(aq)]^{2+}$, the water exchange process at $[Cr(aq)]^{3+}$ is tremendously accelerated. This route leading to this observed catalysis is the same as in the case of $[Co(aq)]^{3+}$ discussed above.

Note: In the case of $[Co(aq)]^{3+}$, the catalyst $[Co(aq)]^{2+}$ is spontaneously generated from $[Co(aq)]^{3+}$ but for $[Cr(aq)]^{3+}$, the catalyst $[Cr(aq)]^{2+}$ is to be externally added. This difference is reasonable in terms of the E^0-values of the M^{III}/M^{II} couples.

E^0 for $[Co(aq)]^{3+}/[Co(aq)]^{2+} = 1.8$ V; E^0 for $[Cr(aq)]^{3+}/[Cr(aq)]^{2+} = -0.41$ V; E^0 for $O_2/H_2O = 1.23$ V, *i.e.* $[Co(aq)]^{3+}$ is a much stronger oxidizing agent and it can slowly oxidize water and it is itself reduced to $[Co(aq)]^{2+}$.

(iii) **Solubilization of anhydrous $CrCl_3$:** Anhydrous $CrCl_3$ is insoluble in water but it can be solubilized in presence of a catalytic amount of $[Cr(aq)]^{2+}$. This can be explained in terms of the ISET path. $[Cr(aq)]^{2+}$ can be generated from $CrCl_3$ by using some suitable reducing agent.

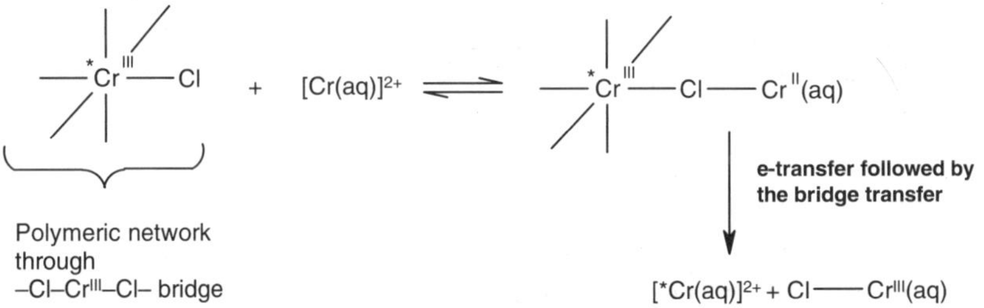

The polymeric network involving the –Cl–Cr^{III}–Cl– segments (present in anhydrous $CrCl_3$) is insoluble because of the **inertness of Cr^{III}-centre**. However, the Cr^{III}-centre (shown as *Cr) of the polymer is converted into the labile Cr^{II}-centre through an ISET path in presence of $[Cr(aq)]^{2+}$. *Rapid substitution at this labile centre breaks down the polymer.*

(iv) **Catalysis by Pt(II) in the ligand exchange or ligand replacement at Pt(IV):** This has been illustrated in Sec 6.4.15.

(v) **Catalysis by $[Cr(aq)]^{2+}$ in the substitution process of Cr(III):** This has been illustrated already on the fission of the inert successor complex, $Ru^{II}–X–Cr^{III}$ (*cf.* Sec. 6.4.5c).

● The **aquation of NH_3 ligands in $[Cr(NH_3)_5X]^{2+}$** is very slow but the process is potentially catalyzed by $[Cr(aq)]^{2+}$.

$$\left[Cr^{III}(NH_3)_5X\right]^{2+} + 5H_3O^+ \longrightarrow \left[Cr(OH_2)_5X\right]^{2+} + 5NH_4^+;\ (X = Cl,\ Br,\ I)$$

The process is catalyzed by $[Cr(aq)]^{2+}$ as shown below.

$$\left[(H_3N)_5{}^*Cr^{III}-X\right]^{2+} + \left[Cr(aq)\right]^{2+} \rightleftharpoons \left[(H_3N)_5{}^*Cr^{III}-X-Cr^{II}(OH_2)_5\right]^{4+}$$

$$\downarrow \text{e-transfer (ISET path)}$$

$$5NH_4^+ + \left[{}^*Cr(aq)\right]^{2+} + \left[Cr(OH_2)_5X\right]^{2+} \xleftarrow[(H_3O^+)]{\text{fast}} \left[(H_3N)_5{}^*Cr^{II}-X-Cr^{III}(OH_2)_5\right]^{4+}$$

● $[Cr(aq)]^{2+}$ can catalyze the **ligation or substitution at $[Cr(aq)]^{3+}$** (inert centre) as follows:

$$\left[Cr(aq)\right]^{2+} + L \rightleftharpoons \left[Cr^{II}(aq)(L)\right],\ (\text{rapid})$$

Electron exchange:

$$\left[Cr^{II}(aq)(L)\right] + \left[{}^*Cr^{III}(aq)\right] \longrightarrow \left[Cr^{III}(aq)(L)\right] + \left[{}^*Cr^{II}(aq)\right];\ (\textbf{OSET path, Cr}^{\textbf{III}}\textbf{ - inert centre})$$

The rate is given by: rate = $k[Cr^{III}][Cr^{II}][L]$

Note: The electron transfer reaction:

$$\left[Cr^{II}(aq)L\right] + \left[Cr^{III}(aq)\right] \longrightarrow \left[Cr^{III}(aq)L\right] + \left[Cr^{II}(aq)\right],$$

can occur through the **OSET path or through the ISET path by using the weak bridging property of H_2O**. The process experiences a net $\sigma^* \to \sigma^*$ transfer, a favourable situation for the ISET path. In this ISET path, H_2O bound to Cr(III) makes a bridge with the labile Cr(II) centre.

● Cr(II) also catalyzes the reverse reaction, *i.e.* **aquation of L in $[Cr^{III}(L)(aq)]$ where L is a nonbridging ligand through the OSET path.**

Electron exchange:

$$\left[(aq)Cr^{III}-L\right] + \left[Cr^{II}(aq)\right] \longrightarrow \left[(aq)Cr^{II}-L\right] + \left[Cr^{III}(aq)\right],\ (\textbf{OSET path, L nonbridging ligand})$$

$$\downarrow \text{fast}$$

$$\left[Cr^{II}(aq)\right] + L$$

The rate is given by:

$$\text{rate} = k\left[Cr^{III}(L)\right]\left[Cr^{II}\right]$$

Here 'L' is supposed to have no bridging property. **If L can bridge**, then the above reaction will go through the ISET path and L will be present with the Cr^{III}-centre (*i.e.* **no effective aquation of L**). This is illustrated below:

$$\left[(aq)Cr^{III}-L\right]+\left[Cr^{II}(aq)\right]\rightleftharpoons\left[(aq)Cr^{III}-L-Cr^{II}(aq)\right]$$

$$\downarrow \text{ e-transfer and fission} \quad \textbf{ISET path}$$

$$\left[(aq)Cr\right]^{II}+\left[L-Cr^{III}(aq)\right]$$

(vi) **Reaction of $[Co^{III}(NH_3)_5X]$ with CN^-:** The nature of the product depends on the nature of X.

$\left[Co^{III}(NH_3)_5X\right]\xrightarrow{+CN^-}$

- $[Co^{III}(CN)_5X]$; when X (having good bridging property) = Cl^-, Br^-, I^-, N_3^-, OH^-, $S_2O_3^{2-}$, etc.
- $[Co^{III}(CN)_6]^{3-}$; when X (no bridging or poor bridging property) = NH_3, CO_3^{2-}, SO_4^{2-}, RCO_2^-, etc.
- $[Co^{III}(CN)_5X]$ + $[Co(CN)_6]^{3-}$; when X (intermediate in bridging property) = F^-, NO_3^-, etc.

The observations can be explained by *considering the coexistence of a trace amount of Co(II) as an impurity.* In presence of CN^-, it gives an equilibrium mixture of $[Co(CN)_5]^{3-}$ and $[Co(CN)_6]^{4-}$, *i.e.*:

$$Co(II)+5CN^-\rightleftharpoons\left[Co(CN)_5\right]^{3-};\text{(fast and high }K_{eq})$$

$$\left[Co(CN)_5\right]^{3-}+CN^-\rightleftharpoons\left[Co(CN)_6\right]^{4-}\text{ (very low }K_{eq})$$

$[Co(CN)_5]^{3-}$ as a reducing agent can experience the ISET path while $[Co(CN)_6]^{4-}$ as a reducing agent can experience the OSET path.

- **ISET path:** $\left[(H_3N)_5Co^{III}-X\right]+\left[Co(CN)_5\right]^{3-}\rightleftharpoons\left[(H_3N)_5Co^{III}-X-Co^{II}(CN)_5\right]$

$$\downarrow \text{ e-transfer}$$

$$5NH_3+\left[Co(aq)\right]^{2+}+\left[X-Co^{III}(CN)_5\right]\xleftarrow{\text{fast}}\left[(H_3N)_5Co^{II}-X-Co^{III}(CN)_5\right]$$

This ISET path is favoured if X is a good bridging ligand. In fact, in this route $[Co^{III}(CN)_5(S_2O_3)]^{4-}$ was synthesized by **Prof. Ray (1927)**. For this process, the rate is given by:

$$\text{rate} = k_{IS}[(H_3N)_5Co^{III}X][Co^{II}].$$

- **OSET path:** $\left[(H_3N)_5Co^{III}-X\right]+\left[Co^{II}(CN)_6\right]^{4-}\rightleftharpoons\underbrace{\left[(H_3N)_5Co^{III}-X\right]\cdot\left[Co^{II}(CN)_6\right]^{4-}}_{\text{O.S. precursor complex}}$

$$\downarrow \text{ e-transfer}$$

$$\left[Co^{III}(CN)_6\right]^{3-}+5NH_3+X+\left[Co(aq)\right]^{2+}\xleftarrow{\text{(fast)}}\underbrace{\left[(H_3N)_5Co^{II}-X\right]\cdot\left[Co^{III}(CN)_6\right]^{3-}}_{\text{O.S. successor complex}}$$

The OSET path is favoured when X is having no bridging or very poor bridging property. For this process, the rate is given by:

$$\text{rate} = k_{OS}[(H_3N)_5Co^{III}X][Co^{II}][CN^-].$$

When the bridging property of X is of an intermediate nature, both the OSET and ISET paths can operate.

6.7 NON-COMPLEMENTARY ELECTRON TRANSFER REACTIONS

The reactions (*e.g.* Fe^{III}–Fe^{II}, Co^{III}–Fe^{II}, Co^{III}–Cr^{II}, Co^{III}–V^{II}, Ru^{III}–Cr^{II}, Pt^{IV}–Pt^{II}, etc.) already discussed involve the oxidants and reductants which undergo the same formal oxidation state change. These are called the **complementary electron transfer reactions**. These involve the 1e-oxidant–1e-reductant, 2e-oxidant–2e-reductant, etc. and stoichiometries are 1:1. But there are reactions in which the changes of oxidation states by the oxidant and reductant are not the same. These involve the 1e-oxidant–2e-reductant, 2e-oxidant–1e-reductant, 3e-oxidant–1e-reductant, etc. Such reactions are referred to as the **noncomplementary electron transfer reactions.** Consequently, in these reactions, the stoichiometries are not 1:1. Some examples are:

$$2Fe(III) + Sn(II) \longrightarrow 2Fe(II) + Sn(IV)$$
$$Tl(III) + 2Fe(II) \longrightarrow Tl(I) + 2Fe(III)$$
$$Pt(IV) + 2Cr(II) \longrightarrow Pt(II) + 2Cr(III)$$
$$Cr(VI) + 3Fe(II) \longrightarrow Cr(III) + 3Fe(III)$$
$$Mn(VII) + 5Fe(II) \longrightarrow Mn(II) + 5Fe(III)$$

In such cases, the electron transfer *generally occurs* ***in consecutive 1e-transfer steps***. In fact, 2e or multi-electron transfer cannot occur in a single step. Stepwise 1e-transfer steps are illustrated below.

$$Tl(III) + Fe(II) \underset{k_{-1}}{\overset{k_1}{\rightleftharpoons}} Tl(II) + Fe(III)$$

$$Tl(II) + Fe(II) \xrightarrow{k_2} Tl(I) + Fe(III)$$

By considering the steady state condition to Tl(II), we get the following rate law:

$$\frac{d[Tl(II)]}{dt} = 0 = k_1[Tl(III)][Fe(II)] - k_{-1}[Tl(II)][Fe(III)] - k_2[Tl(II)][Fe(II)]$$

i.e.
$$[Tl(II)] = \frac{k_1[Tl(III)][Fe(II)]}{k_{-1}[Fe(III)] + k_2[Fe(II)]}$$

$$\text{Rate} = -\frac{d[Tl(III)]}{dt} = k_2[Tl(II)][Fe(II)] = \frac{k_1k_2[Tl(III)][Fe(II)]^2}{k_{-1}[Fe(III)] + k_2[Fe(II)]}$$

Reduction of Cr(VI) (3e-oxidant) by the 1e-reductant Fe(II) occurs as follows:

$$Cr(VI) + Fe(II) \underset{k_{-1}}{\overset{k_1}{\rightleftharpoons}} Cr(V) + Fe(III)$$

$$Cr(V) + Fe(II) \xrightarrow{k_2} Cr(IV) + Fe(III)$$

$$Cr(IV) + Fe(II) \xrightarrow{\text{fast}} Cr(III) + Fe(III)$$

$$-\frac{d[\text{Cr(VI)}]}{dt} = \frac{k_1k_2[\text{Cr(VI)}][\text{Fe(II)}]^2}{k_{-1}[\text{Fe(III)}] + k_2[\text{Fe(II)}]};$$ applying the steady condition to Cr(V).

6.8 REDUCING ACTIVITY OF THE SOLVATED ELECTRON

Hydrated electrons can be generated in many ways:

(a) **Radiolysis of water** by the ionizing radiations like X-rays and γ-rays.

$$H_2O \xrightarrow{h\nu} H^+_{aq} + OH + e^-_{aq}$$

(b) **Irradiation by UV light** on solution containing the reducing agents (Red) like I^-, $[Fe(CN)_6]^{4-}$, M^{2+} (bivalent metal ions of the 1st transition series), etc.

$$\text{Red} \xrightarrow{h\nu} \text{Red}^+ + e^-_{aq}$$

The hydrated electron has the following **properties**:

(i) absorbtion in the red region (λ_{max} = 720 nm; ε_M = 15,800 mol^{-1} cm^{-1}); (ii) high reducing power indicated by the estimated reduction potential –2.7 V; (iii) half life < 10^{-3} s; (iv) reducing activity – almost diffusion controlled rate.

Because of the very short life-time, its characterization is a relatively difficult task. The following reactions are responsible for the rapid destruction of the hydrated electron.

$$2H_2O + 2e^-_{aq} \longrightarrow H_2 + 2OH^-_{aq} \quad (k \approx 10^{10}\ \text{dm}^3\ \text{mol}^{-1}\ \text{s}^{-1} \text{ at } 25°\text{C, pH}\sim 11)$$

$$H_3O^+_{aq} + e^-_{aq} \longrightarrow H + H_2O \quad (k \approx 10^{10}\ \text{dm}^3\ \text{mol}^{-1}\ \text{s}^{-1} \text{ at } 25°\text{C, pH} = 2-4)$$

Thus in acidic condition, both the reducing species H and e^-_{aq} coexist.

$$H_2O + e^-_{aq} \longrightarrow H + OH^- \quad (k \approx 16\ \text{dm}^3\ \text{mol}^{-1}\ \text{s}^{-1} \text{ at } 25°\text{C, pH} = 8.4)$$

H_2 can produce H as follows:

$$H_2 + OH \longrightarrow H_2O + H$$

The reactivities of H and e^-_{aq} are different. N_2O reacts with both e^-_{aq} and H to yield N_2 but their rates differ by a factor 10^4.

$$N_2O + e^-_{aq} \longrightarrow N_2 + O^-,\ O^- + H_2O \longrightarrow OH + OH^-$$

Hydrated electrons generally react very fast close to diffusion controlled rate.

Because of very powerful reducing activity (both thermodynamically and kinetically) of the hydrated electrons, these are used to generate the unstable lower valent cations. Thus the generated lower valent cations can be studied.

6.9 PHOTOCHEMICAL REACTIONS: BOTH SUBSTITUTION AND REDOX PROCESSES IN COORDINATION COMPOUNDS

6.9.1 Photochemical Excitation

(A) Amount of energy absorption: Absorption of photon from the UV-visible regions (*i.e.* 200 to 600 nm) is equivalent to *ca.* 600–200 kJ mol^{-1}. Thus absorption of a photon by a metal complex can raise the energy by a few hundred kilojoules per mole. These energies are quite larger than the typical activation energies required for most of the reactions. In fact, absorption of photon may lead to both nonredox (*e.g.* substitution, isomerization, racemization) and redox reactions. But very often, the backward reaction of the primary photoexcitation process may prevent the net progress of the reaction. This is why, it is required to carry out the process in such a condition that can avoid the backward reaction.

Cage-effect

The species generated in the primary photochemical process remain confined in a cage at the moment of their formation. The cage is surrounded by the solvent. If they cannot escape from the cage to the bulk then they will recombine to generate the starting species to reduce the quantum yield. This phenomenon is called the cage effect.

(B) Fate of the photochemically excited species: The photochemically excited state may participate in several kinds of energy transfer processes as stated below.

(i) **Luminescence:** The excess energy is lost in a *radiative process* (*i.e.* light energy is remitted) and the excited state returns to the ground state. This may lead to phosphorescence and fluorescence.

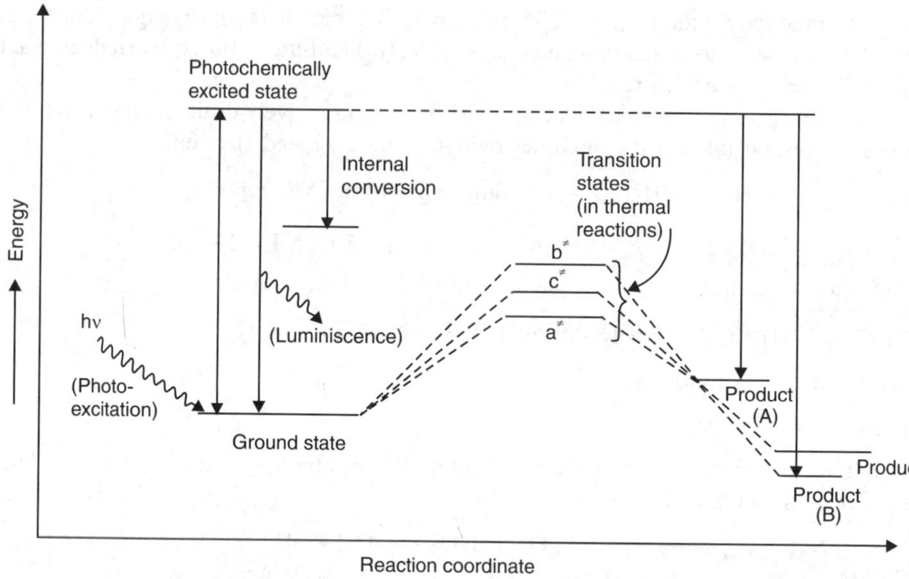

Fig. 6.9.1.1 Fate of a photochemically excited state and comparison (*cf.* Fig. 6.9.1.2) of the photochemical and thermal (with the activation energies; **activated states $a^{\neq}$, $b^{\neq}$, $c^{\neq}$ correspond to the products A, B and C respectively**) reactions. **Note:** Photochemically all the products (A, B, C) are accessible but in the thermal reaction (*i.e.* dark reaction), the *product (A) is thermodynamically inaccessible and the product (B) is kinetically inaccessible* (*i.e.* high activation energy). But all these products may obtained through the photochemical excitation.

If the luminescence persists even after removal of the exciting source, it is called *phosphorescence* and otherwise it is called *fluorescence* which stops as soon as the exciting source is removed.

(ii) **Internal conversion process***:* The excess energy may be converted into vibrational energy and dissipated thermally.

(Note: In the context of photochemical reactions, luminescence and molecular vibrations are of no interest).

(iii) **Long-lived excited state:** Such excited states are important to participate in different types of reactions.

Photochemical *vs.* Thermal reactions (*i.e.* Dark reactions)

● **Product:** Figs. 6.9.1.1, 2 illustrate that the photochemically excited state can participate in the reactions to give the products which may be thermodynamically less stable compared to the starting complex. Sometimes, it may give the products for which activation energy (in thermal reactions) may be very high. *In fact, many thermodynamically and kinetically unfavourable thermal reactions can occur through the photochemical assistance.* In Fig. 6.9.1.1, in thermal path, the product (A) is thermodynamically inaccessible and the product (B) is kinetically inaccessible (*i.e.* high activation energy). **But these products are photochemically accessible.** Here it must be mentioned that the paths of photochemical and thermal reactions are different. It is illustrated in the following example:

● **$[Co(NH_3)_6]^{3+}$:** When an aqueous solution of $[Co(NH_3)_6]^{3+}$ is irradiated by the light of lower wavelength ($\lambda < 330$ nm), it readily aquates to $[Co(NH_3)_5(OH_2)]^{3+}$. But, an acidic solution of $[Co(NH_3)_6]^{3+}$ can be boiled for hours without any change because the process is kinetically disfavoured in spite of the thermodynamic favour.

● **$[CoBr(NH_3)_5]^{2+}$:** Thermal reaction of $[CrBr(NH_3)_5]^{2+}$ leads to slow aquation of Br^- but photochemical reaction of the same complex leads to aquation of NH_3 (*cf.* **Adamson's rule**).

● **Frank-Condon principle:** In the photochemical electron transfer process, there is no requirement bond length adjustment before the electron transfer, *i.e.* **Franck-Condon principle does not operate.** But in the thermal electron transfer process, Franck-Cordon principle is to be obeyed.

(C) Modes of photochemical excitation in the transition metal complexes: The excited state can be attained in different possible routes.

(i) *Ligand field (LF) transition (i.e. d–d transition):* The ligand field excited state arises from the promotion of an electron from the nonbonding t_{2g} orbitals to the antibonding e_g^* orbital, *i.e.* $t_{2g} \rightarrow e_g$ in terms of CFT (considering the octahedral symmetry). The increased electron density in the antibonding orbital causes the M–L bond weakening. This bond weakening is more severe for the stronger field ligands for which the e_g orbital possesses the more antibonding character. *This bond weakening process can lead to various types of reactions, e.g.* ligand replacement, isomerization and racemization.

The *d–d* transitions generally occur in the visible range.

(ii) *Charge transfer (CT) transition*: Sometimes, absorption of radition can lead to the charge transfer transition (*i.e.* metal → ligand, *i.e.* MLCT or ligand → metal, *i.e.* LMCT) to generate the *charge transfer excited state.* Such excited states very often lead to the photo-redox reactions.

(iii) *Intra-ligand transfer:* This transition is mainly localized on the ligands to generate the *intra-ligand excited state* that may initiate the ligand rearrangement or reaction of the ligand with the other substrates.

(iv) *Charge transfer to solvent (MSCT)* (Sec. 6.9.8): Charge transfer to solvent leading to an oxidized metal centre and a solvated electron (e_{solv}) which can lead to different types redox reactions.

$$[Fe(CN)_6]^{4-} \xrightarrow{h\nu} [Fe(CN)_6]^{3-} + e_{aq}^-;\ [Fe(OH_2)_6]^{2+} \xrightarrow{h\nu} [Fe(OH_2)_6]^{3+} + e_{aq}^-$$

(v) **Intervalence transfer (IT):** Absorption of irradiation may lead to electron transfer from one metal centre to another metal centre in the dinuclear and polynuclear complexes. It can initiate the redox reaction. It is illustrated in the following examples.

$$[(H_3N)_5Co^{III}—NC—Ru^{II}(CN)_5]^- \xrightarrow{h\nu} Co^{2+} + 5NH_3 + [Ru(CN)_6]^{3-}$$

CT excitation *vs.* LF excitation

- **Ligand field (LF) excitation** ($d \rightarrow d$ excitation) changes the **angular distribution** of the electron but not the **radial distribution.** In the **charge-transfer excitation** (M $\rightarrow$ L or L $\rightarrow$ M), the radial distribution of the electron density is changed.
- When the ligand field band and charge transfer band overlap, the LF excitation may also cause the **photoredox reaction** as in the Co(III)-complexes.
- **Ligand field (LF) excitation** generally leads to photo-substitution, photo-isomerization and photo-racemization reactions.
- **Charge-transfer (CT) excitation** generally initiates the photo-redox reactions. However, a CT excitation in $[\text{CoCl(NH}_3)_5]^{2+}$ leads to aquation by an *indirect path.*

$$\left[(\text{NH}_3)_5 \text{Co}^{\text{III}} - \text{Cl}\right]^{2+} + \text{H}_2\text{O} \xrightarrow[(\lambda < 350 \text{ nm})]{h\nu} \left[(\text{NH}_3)_5 \text{Co}^{\text{III}} - \text{OH}_2\right]^{3+} + \text{Cl}^-$$

In fact, the primary step of photo-excitation leads to the homolytic fission of the Co^{III}–Cl bond causing the reduction of Co^{III} to Co^{II} which is labile.

The mixed valence compounds like $\left[\text{Cl(bpy)}_2\text{Ru}^{\text{II}} - \text{N} \bigcirc \text{N} - \text{Ru}^{\text{II}}\text{(bpy)}_2\text{Cl} \right]^{3+}$ experience the intervalence transfer through the **bridging ligand pyrazine.**

$$\left[(\text{H}_3\text{N})_5\text{Ru}^{\text{II}} - \text{N} \bigcirc - \bigcirc \text{N} - \text{Ru}^{\text{III}}\text{(NH}_3)_5 \right]$$

$$\xrightarrow[\lambda = 1050 \text{ nm}]{h\nu} \left[(\text{H}_3\text{N})_5\text{Ru}^{\text{III}} - \text{N} \bigcirc - \bigcirc \text{N} - \text{Ru}^{\text{II}}\text{(NH}_3)_5 \right]$$

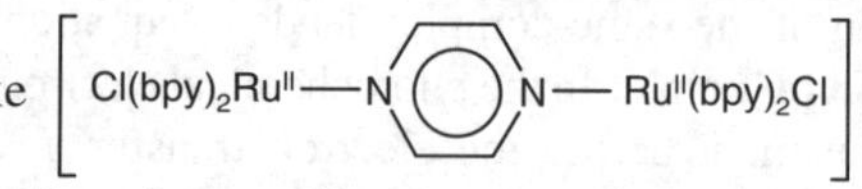

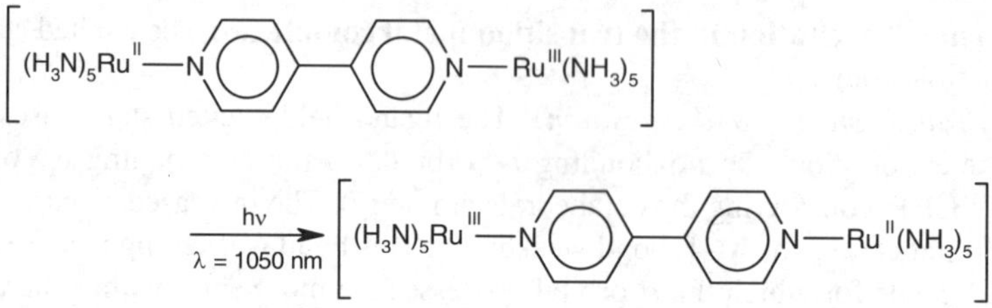

This IT occurs in the near infra-red region through the **bridging ligand 4, 4′-bipyridine.**

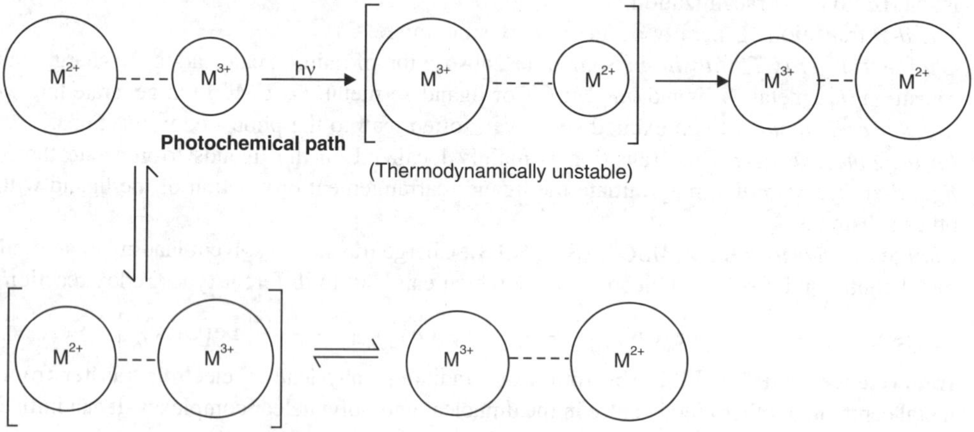

Fig. 6.9.1.2 Comparison of the photochemical and thermal homonuclear electron transfer in a homonuclear mixed valence systems, *e.g.* $\text{M}^{2+} \cdots \text{M}^{3+}$.

Photochemical and thermal IT in a homonuclear mixed valence system, say $M^{2+} \cdots M^{3+}$ can be compared. In the photochemical process, electron transfer from M^{2+} to M^{3+} can occur **without any preorganization (*i.e.* bond length adjustment).** But in the thermal process, before the electron transfer, the bond length must be adjusted (as demanded by **Franck-Condon principle**). These aspects are illustrated in Fig. 6.9.1.2.

(vi) **Photodissociation of the metal-metal bond in the dinuclear complexes having the M—M multiple bond:** Irradiation by light may lead to $\delta \rightarrow \delta^*$ transition to cause photodissociation of the metal-metal bond. Irradiation by light of 500 nm on $\left[(PhO)_2 PO_2\right]Mo \equiv Mo\left[O_2P(OPh)_2\right]$ having the $Mo \equiv Mo$ quadruple bonding can cause fission of the metal-metal bond under the suitable conditions to favour the oxidation of the Mo centre.

(D) Controlled Reactions — Application of the Photochemical Reactions: Selectivity of photochemical reactions may be used to control the reactions. It is illustrated in the following examples.

(i) LF excitation of $[Co(NH_3)_5N_3]^{2+}$ causes the aquation of NH_3 while CT excitation of the same complex causes the redox decomposition of the complex.

$$\left[Co(NH_3)_5(N_3)\right]^{2+} \xrightarrow[\text{(green light)}]{\text{LF excitation}} \left[Co(NH_3)_4(N_3)(OH_2)\right]^{2+} + NH_3, \text{ (\textit{cf.} \textbf{Adamson's rule})}.$$

$$\left[Co(NH_3)_5(N_3)\right]^{2+} \xrightarrow[\text{(UV light)}]{\text{CT excitation}} Co(II) + NH_3 + N_2$$

Thus selection of wavelength of the irradiation can give the different products.

(ii) In thermal reaction (*i.e.* dark reaction), $[CrBr(NH_3)_5]^{2+}$ aquates Br^- while in photochemical reaction it aquates NH_3 (*see* **Adamson's rule**).

$$\left[CrBr(NH_3)_5\right]^{2+} \xrightarrow[\text{(thermal reaction)}]{k_BT\text{ (slow)}} \left[Cr(NH_3)_5(OH_2)\right]^{3+} + Br^-$$

$$\left[CrBr(NH_3)_5\right]^{2+} \xrightarrow[\text{(rapid)}]{h\nu} \left[CrBr(NH_3)_4(OH_2)\right]^{2+} + NH_3$$

The details of the mechanistic aspects of the above reactions will be discussed later.

6.9.2 Prompt and Delayed Reactions

In some cases, the photochemically excited states dissociate promptly and in some cases, the excited states have the relatively long lifetimes.

- (a) **Prompt reactions** (*occurring very fast and in less than 10 ps*): Some representative examples are discussed below.

(i) **Substitution reactions in metal carbonyls:**

$$\underset{\textbf{(18e)}}{\left[Cr(CO)_6\right]} \xrightarrow{h\nu\text{ (UV)}} \underset{\textbf{(16e)}}{\left[Cr(CO)_5\right]} + CO,$$

$$\left[Cr(CO)_5\right] + L \longrightarrow \left[Cr(CO)_5L\right]$$

Absorption of UV radiation by $[Cr(CO)_6]$ leads to $\pi \rightarrow \sigma^*$ electronic transition and this excitation populates the electron density in the antibonding orbital to weaken the Cr—CO bond. This leads to dissociation of the Cr—CO bond to produce a **16e species** $[Cr(CO)_5]$ (square pyramidal) which is kinetically very much active.

$$\left[Mn_2(CO)_{10}\right] \xrightarrow[\text{(fast)}]{h\nu} 2\left[\bullet Mn(CO)_5\right]$$

$$[\bullet Mn(CO)_5] + CCl_4 \xrightarrow{\text{slow}} [Mn(CO)_5Cl] + \bullet CCl_3$$

In this binuclear complex, the $\sigma \rightarrow \sigma^*$ transition leads to metal-metal bond rupture to produce the free radicals that can participate in free radical mechanism with CCl_4, $CHCl_3$, etc. $[Fe_2(CO)_9]$, $[Co_2(CO)_8]$ can also participate in the similar reactions.

(ii) **Photo-aquation of $[CoCl(NH_3)_5]^{2+}$:** $[Co^{III}Cl(NH_3)_5]^{2+} \xrightarrow[(\lambda < 350\text{ nm})]{h\nu} [Co^{II}(NH_3)_5]^{2+} + \bullet Cl$

$$\downarrow H_2O$$

$$[Co(NH_3)_5(OH_2)]^{3+} + Cl^-$$

Scission of the Co—Cl bond leads to reduction of Co^{III} to Co^{II} which is a labile centre and experiences a rapid aquation followed by reoxidation by Cl. This type of ***photoredox process*** leading to substitution is common with the Co(III) and Ru(II) complexes. This actually involves the *charge transfer* (CT) leading to M^-—L^+ or M^+—L^-.

The ***quantum yield*** (measuring the amount of reaction per mole of photons absorbed) of the above reaction increases with the decrease of wavelength. *In fact, the excess energy (= hν–bond energy) remaining after the bond fission is available to the dissociated fragments and it increases the probability of escaping away of the dissociated fragments from each other in solution.* Thus the possibility of backward reaction (*i.e.* recombination or **cage effect**) decreases.

(iii) ***Photosubstitution in $[Cr^{III}(NH_3)_xX_y]^{n+}$ ($x + y = 6$):*** Very often, photosubstitution in these cases leads to different products compared to those obtained in the thermal substitutions. In the thermal substitution reactions, halides (X^-) are substituted while in photosubstitutions, NH_3 may also be replaced. The photoexcitation may lead to the following excited states:

$^4T_{1g}$ (*i.e.* $t_{2g}^2e_g^1$) and ^2E_g (**low spin** t_{2g}^3, *i.e.* (↑↓)(↑)).

$$Cr^{3+}(t_{2g}^3,\ ^4A_{2g}) \xrightarrow{h\nu} Cr^{3+}(t_{2g}^2e_g^1,\ ^4T_{1g})$$

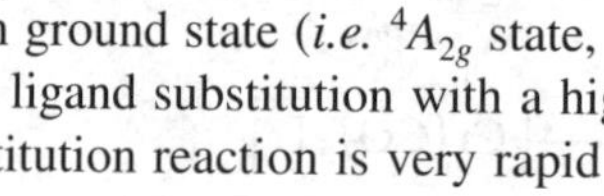

$(^4A_{2g}$ *i.e.* high spin $t_{2g}^3)$ $\qquad$ $(^2E_g$ *i.e.* low spin $t_{2g}^3)$

In the $^4T_{2g}$ or $^4T_{1g}$ state, accumulation of electron density in the antibonding orbital weakens the metal-ligand bond to favour the **dissociative process of substitution** while in the ^2E_g state, a vacancy created in the t_{2g} level can favour the **associative process of substitution**. Here it is worth mentioning that the Cr(III) complexes in ground state (*i.e.* $^4A_{2g}$ state, t_{2g}^3) experience the **associative activation** in thermal or dark reaction for ligand substitution with a high activation energy.

The following photosubstitution reaction is very rapid and it occurs within less than 5 ps.

$$[Cr(NH_3)_6]^{3+} + H_2O \xrightarrow{h\nu} [Cr(NH_3)_5(OH_2)]^{3+} + NH_3$$

- (b) **Delayed Reactions:** Sometimes, the photochemically excited state is long-lived and it can participate in delayed reactions.

$$[Ru(bpy)_3]^{2+} \xrightarrow[\lambda = 590\text{ nm}]{h\nu} [Ru(bpy)_3^*]^{2+}$$

$[Ru(bpy)_3^*]^{2+}$ is relatively long-lived to participate in different types of redox reactions to be discussed later. Photochemically, the excited $[Cr(bpy)_3^*]^{2+}$ state possesses a long-lived ^2E state and it participates in several delayed redox reactions. In fact, $[Cr(bpy)_3^*]^{2+}$ is a good oxidizing agent ($E^0 = +1.3$ V). **These are discussed in detail in Secs. 16.16.1–2, Vol. 3.**

6.9.3 Examples of Photochemical Reactions of Metal Carbonyls

Some representative examples are given below.

(i) $$\underset{(\mathbf{18e})}{[M(CO)_6]} \xrightarrow[(\pi\rightarrow\sigma^*)]{h\nu} [M(CO)_6^*] \longrightarrow \underset{(\mathbf{16e})}{[M(CO)_5]} + CO$$

$$\underset{(M = Cr, Mo, W)}{[M(CO)_5]} + L \longrightarrow [M(CO)_5L]$$

The photoexcitation ($\pi \rightarrow \sigma^*$) increases the electron density in the antibonding orbital to weaken the metal-ligand bond to favour the above mentioned **dissociative process.**

(ii) $$[Mn_2(CO)_{10}] \xrightarrow[(\sigma\rightarrow\sigma^*)]{h\nu} [Mn_2(CO)_{10}^*] \longrightarrow 2[\bullet Mn(CO)_5],$$

$$[\bullet Mn(CO)_5] + CCl_3—Cl \longrightarrow [MnCl(CO)_5] + \bullet CCl_3$$

$[\bullet Mn(CO)_5]$ participates in **free radical reaction mechanism**. Similarly, $[Fe_2(CO)_9]$, $[Co_2(CO)_8]$ can also participates in such photosubstitution reactions.

In fact, the photoexcitation ($\sigma \rightarrow \sigma^*$) enhances the electron density in the antibonding orbital to favour the metal-metal bond rupture.

(iii) Photochemical dimerization of $[Fe(CO)_5]$ occurs as follows.

$$[Fe(CO)_5] \xrightarrow{h\nu} [Fe(CO)_5^*] \longrightarrow [Fe(CO)_4] + CO$$

$$[Fe(CO)_5] + [Fe(CO)_4] \longrightarrow [Fe_2(CO)_9].$$

(iv) $$[Ni(CO)_4] \xrightarrow{h\nu} \text{solid product.}$$

The reaction probably occurs as follows:

$$[Ni(CO)_4] \xrightarrow{h\nu} [Ni(CO)_3] + CO;\ [Ni(CO)_3] \xrightarrow{h\nu} Ni(CO)_2 + CO$$

$$Ni(CO)_2 \longrightarrow \text{solid product.}$$

6.9.4 Photochemical Reactions in the Co(III) Complexes

In the Co(III)-complexes, light of about 550 nm causes the ***ligand field excitation*** while light of about 370 nm causes the ***charge transfer excitation,*** *i.e.*

$[Co(NH_3)_6]^{3+}$, $[Co(NH_3)_5X]^{2+}$, $[Co(NH_3)_5(SO_4)]^+$, (X = Cl, Br, I, N_3, NO_2, etc.)

hν ($\lambda \approx 370$ nm) → **Charge transfer (CT) excitation**

hν ($\lambda \approx 550$ nm) → **Ligand field (LF) excitation** (increasing electron density in e_g^*)

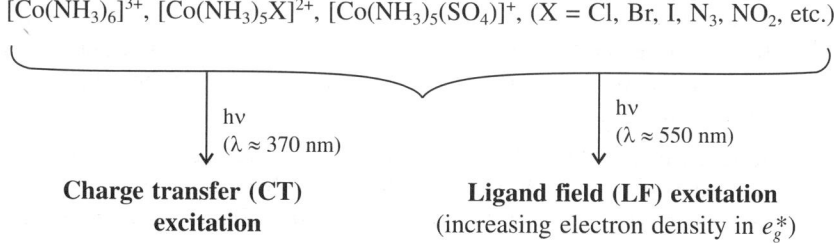

(**Note:** For the Co(III)-complexes, the CT band very often overlaps with the LF band. Hence, even for the LF excitation, it is very difficult to prevent the photoredox reaction that generates the reactive Co(II)-species. Though, in general, CT excitation leads to the photoredox reaction.)
Photochemical LF excitation ($t_{2g} \rightarrow e_g$ transition) leads to promotion of an electron from the nonbonding t_{2g} orbital to the antibonding e_g^* orbital. It weakens the metal-ligand bond. **Thus the LF excitation generally favours the substitution and isomerisation reaction through the dissociative process.** For the stronger field ligands, the e_g-level possesses the more antibonding character and the LF excitation weakens the metal-ligand bond more. In fact, in general, **quantum yield increases with the ligand field strength.**

For $[Co(CN)_6]^{3-}$ and $[Co(CN)_5(X)]^{3-}$, the ligand field strength is sufficiently high and the light of 370 nm causes the *ligand field excitation rather than the charge transfer excitation.* Ligand field excitation weakens the metal-ligand bond in the excited state because of the enhanced electron density in the antibonding orbital.

Generally, CT excitation favours the redox process. However, it depends on the nature of X (*i.e.* leaving group). For example, CT excitation in $[Co(NH_3)_5X]^{2+}$ causes aquation of X^- for X = Cl and redox decomposition of the complex for X = I.

The photochemical reactions of Co(III)-complexes are believed to proceed through the following steps. It is illustrated for the aquation reaction.

Step-I: $$\underset{\left(\begin{smallmatrix}X^- = \text{halide or}\\ \text{pseudohalide}\end{smallmatrix}\right)}{\left[L_5Co^{III}-X\right]} \underset{}{\overset{h\nu}{\rightleftharpoons}} \underset{\left(\begin{smallmatrix}\text{outer sphere}\\ \text{complex}\end{smallmatrix}\right)}{\left[L_5Co^{II}\right]\bullet X},$$ ***(i.e. CT excitation** (λ ≤ 350 nm) **causing the homolytic bond cleavage)***

The first-step (*i.e.* homolytic bond cleavage) is basically a ***photoredox process.*** Where Co(III) is reduced to Co(II) and X^- is oxidized to X. The free radicals (•X) can be detected by Flash-photolysis.

Similar reactions are also known for the Rh(III) and Ir(IV)-complexes.

$$\left[Rh(NCS)(NH_3)_5\right]^{2+} \xrightarrow{h\nu} \left[Rh(NH_3)_4\right]^{2+} + NH_3 + \bullet NCS;$$

$$\left[IrCl_6\right]^{2-} + H_2O \xrightarrow{h\nu} \left[IrCl_5(OH_2)\right]^{2-} + \bullet Cl$$

In all these cases, one electron transfer occurs to give the changes:

$$[Co^{III} \rightarrow Co^{II},\ Rh^{III} \rightarrow Rh^{II},\ Ir^{IV} \rightarrow Ir^{III}]$$

Step-II: $$\left[L_5Co^{II}\right]\bullet X + H_2O \longrightarrow \underset{(\text{outer sphere complex})}{\left[L_5Co^{II}(OH_2)\right]\bullet X}, \ (\text{high-spin } Co^{II} \text{ is labile})$$

Step-III:

$$\underset{\text{Outer sphere complex}}{\left[L_5Co^{II}(OH_2)\right]\bullet X}$$

(left arrow) $[L_5Co^{II}(OH_2)] + X$ (if X is not a good electron acceptor, *i.e.* oxidation of Co^{II} to Co^{III} is not possible).

(right arrow) $[L_5Co^{III}(OH_2)] + X^-$ (if X is a good electron acceptor)

If X cannot oxidize Co(II) to Co(III), then the free radical X (=1, N_3, etc.) comes out from the outer-sphere complex to participate in different reactions. X can combine to generate X_2. It happens so for X = I as in the case of $[Co(NH_3)_5I]^{2+}$. For X = Br, released Br_2 will oxidise the released NH_3.

$$\left[Co\left(NH_3\right)_5 I\right]^{2+} \xrightarrow[\lambda=370,550\ nm]{h\nu} Co(II) + I_2 + 5NH_3$$

For $[Co(NH_3)_5I]^{2+}$, at both $\lambda = 370, 550$ nm, the redox reaction is predominant while for $[CoCl(NH_3)_5]^{2+}$ the aquation is the predominant reaction, *i.e.*

$$\left[CoCl\left(NH_3\right)_5\right]^{2+} \xrightarrow[\lambda=370,550\ nm]{h\nu} \left[Co\left(NH_3\right)_5\left(OH_2\right)\right]^{3+} + Cl^-$$

It is reasonable as Cl is a much better electron acceptor than I. It is worth mentioning that aquation of $[Co(CN)_5X]^{3-}$ gives the following results.

$$\left[Co(CN)_5 X\right]^{3-} \xrightarrow[\lambda=370\ nm]{h\nu} \left[Co(CN)_5\left(OH_2\right)\right]^{2-} + X^-$$

Quantum yield (ϕ): $I^- \rangle Br^- \rangle Cl^-$; *$Co^{III}$–X bond strength:* $Cl^- \rangle Br^- \rangle I^-$

It indicates that the first step leading to the homolytic cleavage is the important step.

$$\left[(NC)_5 Co^{III} — X\right] \xrightarrow{h\nu} \left[(NC)_5 Co^{II}\right] \bullet X, \left(i.e.\ \text{reduction of } Co^{III}\right)$$

Here the electron transfer from X^- to Co^{III} is most favoured for the least electronegative X. It follows the sequence: $I^- \rangle Br^- \rangle Cl^-$. It is the **sequence of quantum yield.**

Table 6.9.4.1 Photochemical reactions of Co(III) complexes (*cf.* A. W. Adamoson, *Disc. Faraday Soc.*, **29**, 163, 1960).

Complex	*Irradiating radiation* (nm)	*Product*	*Quantum Yield* ϕ (mol einstein^{-1})
$[Co(NH_3)_6]^{3+}$	370	No reaction	–
$[Co(NH_3)_5X]^{2+}$			
X = Cl	370	$[Co(NH_3)_5(OH_2)]^{3+}$	0.011
	550	–do–	0.001
X = I	370	Co(II), NH_3, I_2	0.66
	550	–do–	0.10
X = N_3^*	370	Co(II)	0.44
	550	–do–	0.011
$[Co(CN)_5X]^{3-}$			
X = CN	370	$[Co(CN)_5(OH_2)]^{2-}$	0.9
X = Cl	370	–do–	0.3
X = Br	370	–do–	0.7
X = I	370	–do–	0.95
$[Co(C_2O_4)_3]^{3-}$	370	Co(II), CO_2	1.0

*Azide may be oxidised to N_2.

● **LF excitation vs CT excitation:** Generally, CT excitation causes the photoredox reaction while LF excitation causes the photosubstitution reaction. In the Co(III)-complexes, very often, the LF band and CT band overlap to complicate the observation. However, the following examples illustrate the effect of wavelength of irradiation.

$$\left[Co\left(NH_3\right)_5\left(N_3\right)\right]^{2+} \xrightarrow[\text{(green light)}]{\text{(LF excitation)}} \left[Co\left(NH_3\right)_4\left(N_3\right)\left(OH_2\right)\right]^{2+} + NH_3,$$

$\left(\textbf{aquation of } NH_3\right)$ (*cf.* **Adamson's rule**)

$$\left[Co(NH_3)_5(N_3)\right]^{2+} \xrightarrow[\text{UV-light}]{\text{(CT excitation)}} \left[Co(NH_3)_5\right]^{2+} + \bullet N_3 \longrightarrow Co(II) + NH_3 + N_2$$

(Redox decomposition).

In the following example, CT(LMCT) excitation causes the **photoisomerization.**

$$\left[Co(NH_3)_5(NO_2)\right]^{2+} \xrightarrow{h\nu(CT)} \left\{\left[Co(NH_3)_5\right]^{2+} \bullet NO_2\right\} \xrightarrow{\text{Isomerization}} \left[Co(NH_3)_5(OND)\right]^{2+};$$

(Linkage isomer)

6.9.5 Photochemical Reactions of the Cr(III) Complexes

For the Cr(III) complexes, analogous to those of the Co(III) complexes, **there is no redox reaction under the comparable conditions.** It is reasonable. For the Co(III)-complexes, Co(II) can be generated as an intermediate, *i.e.* reduction of Co(III) to Co(II) is feasible. On the other hand, for the Cr(III) complexes, it is unlikely to generate the Cr(II) intermediate state. This can be rationalised in terms of their E^0-values, *e.g.* 1.8 V for $[Co(aq)]^{3+}/[Co(aq)]^{2+}$ and –0.42 V for $[Cr(aq)]^{3+}/[Cr(aq)]^{2+}$ couple, *i.e.* reduction of Co(III) to Co(II) is thermodynamically easier. Other characteristic features of the photochemical reactions of the Cr(III) complexes are discussed below.

The quantum yield increases with the increase of temperature. It indicates the ***requirement of a thermal activation energy***. It is suggested that after the *primary photochemical process,* it requires the thermal activation energy for the subsequent reactions. Consequently, the overall process is slow. **Here it may be mentioned that for the analogous Co(III)-complexes, there is no need of such thermal activation energy**. Because, for the Co(III)-complexes, after the primary photochemical process, a very

Differences in photochemical reactions of the analogous Co(III) and Cr(III) complexes

- For the Co(III)-complexes, there may be both substitution and redox reactions depending on the conditions, but for the Cr(III) complexes, redox reaction does not occur.
- For the Co(III) complexes, photoexcitation can generate **a reactive intermediate Co(II)** (which is labile and reducing) while for the Cr(III)-complexes no such reactive intermediate like Cr(II) is generated. In fact, photoexcitation of the Cr(III) complexes, probably gives a **long-lived ^2E_g** state that subsequently participates in the reactions ***requiring*** *a the thermal activation energy.*
- For the Co(III) complexes, after the primary photoexcitation, *there is no need of thermal activation energy for the subsequent reactions.* ***The Cr(III) complexes need a thermal activation energy for the subsequent reactions.***
- Reactions of the Cr(III) complexes are slower.
- For the Cr(III) complexes, **quantum yield is more or less insensitive to the wavelength in the range 300-700 nm.** For the Co(III) complexes, light of 370 nm favours the CT excitation for the redox process while light of 550 nm leads to ligand field excitation to favour the substitution reaction.
- For the Cr(III)-complexes, two possible **reactive intermediates,** *i.e.* ^2E_g and $^4T_{2g}$ are suggested. The ^2E_g state (low-spin t_{2g}^3) having a vacant orbital in the t_{2g}-level experiences an **associative activation** while the **$^4T_{2g}$ state** ($t_{2g}^2\, e_g^1$) having an enhanced electron density in the antibonding orbital experiences the **dissociative activation** in the subtitution process. LF excitation in the Co(III) complexes favours the **dissociative activation** due to the weakening of metal ligand bond.

reactive centre which is Co(II) is generated. In the case of Cr(III) complexes, there is no such possibility of generation of Cr(II) species. In fact, even after the photoexcitation, Cr(III)-state (an inert centre) is still maintained. It is suggested that photoexcitation of the Cr(III) complexes produces **a relatively long-lived doublet state ^2E_g which subsequently participates in the substitution reactions.** This long-lived ^2E_g state requires the thermal activation energy for its subsequent substitution reactions.

(A) **High activation energy (*ca.* 60 kJ mol^{-1}) and effect of temperature:** The quantum yield (ϕ) is *almost insensitive to the wavelength of irradiation in the range 300-700 nm* but *the quantum yield increases with the rise of temperature.* In fact, this high activation energy (*ca.* 60 kJ mol^{-1}) is unusual for a photochemical reaction. This can be explained by considering the requirement of a thermal activation after the photoexcitation. This step of thermal activation is only possible, if the photochemically produced intermediate is sufficiently long-lived (life-time $\geq$ 1 ms). Thus to explain the effect of temperature on ϕ, we must consider an intermediate (produced photochemically) that will survive at least 1 ms. The probable intermediate ^2E_g and/or $^4T_{2g}$ (to be discussed later) is still Cr(III) (d^3) which is inert and it needs a thermal activation for substitution or racemisation.

Table 6.9.5.1 Photochemical reactions of the Cr(III) complexes (*cf.* A. W. Adamson *et al*, *J. Am. Chem. Soc.*, **80**, 3865, 1958; R. A. Plane *et al*, *ibid*, **79**, 3343, 1957; *Inorg. Chem.*, **3**, 231, 1964).

Complex/System	*Wavelength of irradiation* λ **(nm)**	*Reaction product*	*Quantum Yield* ϕ **(mol einstein^{-1})**
$[Cr(NH_3)_6]^{3+}$	320–600	$[Cr(NH_3)_5(OH_2)]^{3+}$	0.32
$[Cr(NH_3)_5(OH_2)]^{3+}$	320–600	$[Cr(NH_3)_4(OH_2)_2]^{3+}$	0.25
$[Cr(NH_3)_4(OH_2)_2]^{3+}$	320–700	$[Cr(NH_3)_3(OH_2)_3]^{3+}$	0.16
$[Cr(NH_3)_3(OH_2)_3]^{3+}$	320–700	$[Cr(NH_3)_2(OH_2)_4]^{3+}$	0.014
$[Cr(NH_3)_2(OH_2)_4]^{3+}$	320–700	$[Cr(NH_3)(OH_2)_5]^{3+}$	0.0018
$[Cr(NH_3)_5(NCS)]^{2+}$	560	$[Cr(NH_3)_5(OH_2)]^{5+}$	0.013
$[Cr(OH_2)_6]^{3+} + Cl^-$	370	$[CrCl(OH_2)_5]^{2+}$	0.006
$[Cr(OH_2)_6]^{3+} + NCS^-$	400	$[CrNCS(OH_2)_5]^{2+}$	0.0024
$(+)[Cr(C_2O_4)_3]^{3-}$	370	Racemization (no decomposition)	0.045

(B) Nature of the intermediate state produced through the primary photoexcitation process in the case of Cr(III) complexes: To understand this aspect, it is required to consider the energy level diagrams of the Cr(III) complex (Fig. 6.9.5.1).

The *spin allowed transitions* in the Cr(III) complexes are:

$$^4A_{2g}(F) \xrightarrow{\nu_1} {}^4T_{2g}(F),\ {}^4A_{2g}(F) \xrightarrow{\nu_2} {}^4T_{1g}(F),\ {}^4A_{2g}(F) \xrightarrow{\nu_3} {}^4T_{1g}(P).$$

The third transition (*i.e.* υ_3) very often merges with the strong charge transfer band. The spin forbidden transition is:

$$^4A_{2g}(F) \longrightarrow {}^2E_g(F)$$

^2E_g state arises from the low spin configuration of t_{2g}^3, *i.e.* (↑↓)(↑)( ) where one level is remaining vacant.

(a) **Doublet state, *i.e.* ^2E_g as the reactive intermediate:** It is suggested that the excited quartet states [*i.e.* $^4T_{1g}(P)$, $^4T_{1g}(F)$ and $^4T_{2g}(F)$] will rapidly revert to the ground state but in this deexcitation process,

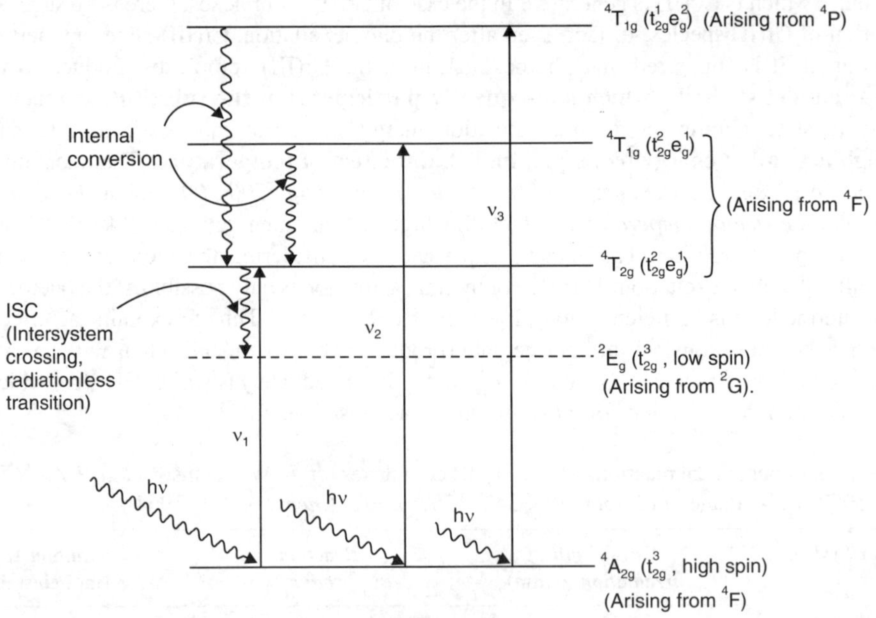

Fig. 6.9.5.1 Energy level diagram of the octahedral complexes of Cr(III) and transition in terms of **Jablonski diagram.** ^2E_g state actually represents two states ^2E_g and $^2T_{1g}$ arising from ^2G, *i.e.* $^2T_{1g}(G)$, $^2E_g(G) = {}^2E_g(G)$ (*cf.* Fig. 7.14.5.6).

population density of the doublet state (*i.e.* ^2E_g) will increase to some extent and *this doublet state is relatively long-lived because the transition from the* ^2E_g *state to the ground* $^4A_{2g}$ *state is spin forbidden.* ***This is why, some workers believe that*** ^2E_g ***is the intermediate generated from the photoexcitation process.***

The ^2E_g excited state (low spin t_{2g}^3) generates a vacant orbital at the t_{2g} level. Availability of this vacancy at the t_{2g} level favours the **associative activation in the photosubstitution process.** It may be noted that in the thermal substitution reactions, Cr(III) also adopts the associative path with a high activation energy.

The quantum yield (ϕ) is not sensitive to the wavelength of irradiaion in the region 300-700 nm. It indicates that the reactive intermediate is the **lowest excited state** which is not produced directly through the photochemical excitation. **It supports the ^2E_g state as the reactive intermediate.** Thus, if the ^2E_g state is considered to be responsible as the reactive intermediate then the photochemical yield should depend on the life period of this doublet state. **It has been noted that the *quantum yield for the photoaquation of*** $[Cr(NH_3)_x(OH_2)_{6-x}]^{3+}$ ***decreases gradually from*** $[Cr(NH_3)_6]^{3+}$ ***to*** $[Cr(OH_2)_6]^{3+}$. This observation can be explained by considering the energy difference between the states ^2E_g and $^4T_{2g}$. This energy difference is maximum for $[Cr(NH_3)_6]^{3+}$ while it is minimum (*ca.* 30 kJ mol^{-1}) for $[Cr(OH_2)_6]^{3+}$ (*cf.* NH_3 is stronger field ligand than H_2O). **In the case of $[Cr(OH_2)_6]^{3+}$, by using the thermal energy, population density of the ^2E_g state may decrease due to the promotion to the $^4T_{2g}$ state and thence to the $^4A_{2g}$ state** (ground state).

$$^2E_g \xrightarrow{\text{thermally}} {}^4T_{2g} \xrightarrow{\text{(allowed)}} {}^4A_{2g} \text{ (ground state)}$$

This ***indirect path*** (described as the **leakage path**) reduces the life-time or population density of the ^2E_g state which is supposed to be the reactive intermediate. This indirect path to reduce the life time of

^2E_g state is most efficient for the $[Cr(OH_2)_6]^{3+}$ complex in the series $[Cr(NH_3)_x(OH_2)_{6-x}]^{3+}$. *The efficiency of this leakage mechanism increases with the decrease of x in the series.* It explains the increasing trend of quantum yield for the photoaquation of $[Cr(NH_3)_x(OH_2)_{6-x}]^{3+}$ with the increase of *x*.

(b) **Quartet state as the reactive intermediate:** ***Some workers believe that $^4T_{2g}$ state, i.e. quartet state is the reactive intermediate*** having the electronic configuration $t_{2g}^2\, e_g^1$. This state experiences the **Jahn-Teller distortion.** Because of this J.T. distortion, the geometry of the excited quartet state is different from that of the nondistorted ground state. *Because of this difference in symmetry between the distorted excited state and nondistorted ground state, the life-time of the excited quarlet state increases.* Because of the J.T. distortion, the excited quartet state is also labile.

In the $^4T_{2g}$ state ($t_{2g}^2\, e_g^1$), accumulation of electron density in the antibonding orbital weakens the metal-ligand bond. This favours the **dissociative activation** in photosubstitution. The J.T. distortion also favours the **dissociative activation.** Here it is worth mentioning that the ^2E_g state and $^4A_{2g}$ ground state (*i.e.* thermal reaction) favour the associative activation.

Nature of the reactive intermediate: ^2E_g *vs.* $^4T_{2g}$ (F)

- Enhanced life-time of the ^2E_g-state arises due to the **spin-forbidden transition** to the $^4A_{2g}$ ground state. However, its life time may decrease through the ***indirect leakage path.***
- Enhanced life time of the $^4T_{2g}$-state arises due to the **symmetry forbidden transition** to the ground state which is nondistorted while the excited $^4T_{2g}$ state is distorted (J.T. distortion).
- In the photosubstitution process, the ^2E_g state (low spin t_{2g}^3) favours the **associative activation** while the $^4T_{2g}$ state ($t_{2g}^2 e_g^1$) favours the **dissociative activation.** In the thermal reaction, the ground state adopts the associative path with a high activation energy.
- In fact, in the photochemical reactions of the Cr(III) complexes, there are evidences for both the ^2E_g (doublet state) and $^4T_{2g}$ (quartet state) states to represent the reactive intermediate.

(C) **Adamson's rules:** Adamson had formulated some **empirical rules** to explain the results of photochemical reactions of $[Cr(NH_3)_5X]^{2+}$ (X^- = halide).

(i) In $[Cr(NH_3)_5X]^{2+}$, the ligands are at the ends of three mutually perpendicular axes. On irradiation, the ligands will be labilized most along the axis having the lowest average ligand field strength.

(ii) If there are two different ligands along the labilized axis, the ligand having the higher ligand field strength will be preferably labilized.

Thus for $[Cr(NH_3)_5X]^{2+}$, the *trans-* ligands NH_3 and X will be labilized most and between these two ligands, NH_3 (having higher ligand field strength) will be preferably labilized.

$$\left[CrBr(NH_3)_5\right]^{2+} + H_2O \xrightarrow{h\nu} \left[CrBr(NH_3)_4(OH_2)\right]^{2+} + NH_3$$

However, there are exceptions to these rule.

The $\Delta V^{\neq}$ values for photoaquation of $[Cr(NH_3)_5X]^{2+}$ and $[Cr(NH_3)_6]^{3+}$ are negative for aquation of both NH_3 and X^-. **These data support the I_d-mechanism where the dissociated fragments in the T.S. put an enhanced electrorestriction over the solvent.** This is reasonable as the excited state ($^4T_{2g}$) accumulates more electron density in the antibonding e_g^* orbital from the nonbonding t_{2g}-orbital and this will weaken the metal-ligand bond to favour the dissociative activation. It may be noted that in the thermal reactionsm, the Cr(III)-complexes adopt the associative activation.

(D) **Interpretation of Adamson's Rule in photosubstitution reaction in terms of crystal field theory** (Ref. A.W. Adamson, *Coord. Chem. Rev.,* **3,** 169-188, 1968): The rules are applicable for both the Co(III) (d^6, low-spin, *i.e.* t_{2g}^6) and Cr(III) (d^3, *i.e.* t_{2g}^3) complexes for which the ligand field transitions are quite similar (*cf.* Figs. 7.14.5.5, 7.15.2.4). But, for the Co(III) complexes, the LF band and CT band very often overlap and the Co(III)-complexes very often experience the *photo-redox reactions* because of the ease of reduction of Co(III) to Co(II). Here we shall illustrate the rule taking the Cr(III)-complexes as the representative ones.

It is suggested that the *quartet excited state* ($^4T_{2g}$ for $t_{2g}^2e_g^1$) of the Cr(III)-complexes is the **reactive intermediate** for the photosubstitution process. The excited quartet state ($t_{2g}^2e_g^1$) experiences the *Jahn-Teller distortion* generating the elongated two *trans*-L—M—L bonds (*c.f.* *z*-out distortion). Because of this significant difference in the geometries of the undistorted ground state ($^4A_{2g}$, t_{2g}^3) and excited state ($^4T_{2g}$, t_{2g}^2 e_g^1), *life time of the excited quartet state is increased as the transition,* $^4T_{2g} \rightarrow {}^4A_{2g}$ *experiences the* ***Franck-Condon barrier.*** For the difference in the structures of $^4A_{2g}$ and $^4T_{2g}$ states, the Franck-Condon vertical transition ($^4A_{2g} \rightarrow {}^4T_{2g}$) produces the excited state (*i.e.* $^4T_{2g}$ state) at a higher vibrational level (*cf.* Fig. 7.4.1.3).

In the *unsymmetrical octahedral complexes,* the average ligand field strength along the *three mutually perpendicular axes* will not be the same and the $^4T_{1g}$ and $^4T_{2g}$ states will split. This splitting pattern for the *cis-/trans*-[$Cr^{III}A_4X_2$] complex is shown in Fig. 6.9.5.2. For the [$Cr^{III}A_4X_2$] complex, orientations of the ligands along the mutually perpendicular axes are as follows:

(i) ***cis*-complex** (two types of axes): A—A, A—X, A—X; **Average ligand field strength:** A—A axis ⟩ A—X axis.

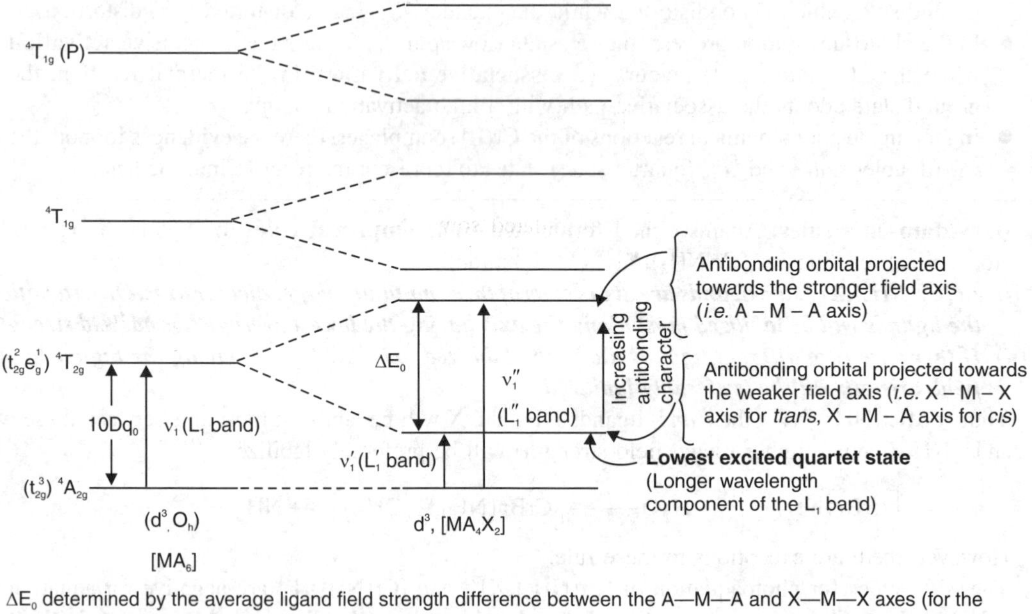

ΔE$_0$ determined by the average ligand field strength difference between the A—M—A and X—M—X axes (for the *trans*-complex) and between the A—M—A and A—M—X axes (for the *cis*-complex). **ΔE$_0$ (trans): ΔE$_0$ (*cis*) = 2:1.**

Fig. 6.9.5.2 Term diagram for the d^3 (*e.g.* Cr^{III}) octabedral (O_h) and less symmetric (D_{4h}, C_{2v}) complexes (*cf.* Figs. 7.14.5.5, 7.15.2.1).

Note: Even for the perfectly octahedral complexes like [$Cr^{III}L_6$], the $^4T_{2g}$ (*i.e.* $t_{2g}^2e_g^1$) state experiences a distortion (J.T. distortion) and in fact, this excited state is structurally different from the ground state ($^4A_{2g}$, t_{2g}^3). Thus, the vertical Franck-condon transition, $^4A_{2g} \rightarrow {}^4T_{2g}$ produces the excited at a higher vibrational level. The excited state may thermally equilibrate before emission to have the lower vibrational level at the excited state. Then emission (due to the return to the $^4A_{2g}$ state) will occur to a higher vibrational level of the ground state (*cf.* Fig. 7.4.1.3).

(ii) ***trans*-complex** (two types of axes): A—A, A—A, X—X; **Average ligand field strength:** A—A axis ⟩ X—X axis.

(*cf.* positions of A (=NH_3, say) and X^- (halide, pseudohalide, say) in the spectrochemical series; NH_3 is a stronger field ligand than X^-).

The magnitude of splitting (ΔE_0) of the quartet state ($^4T_{2g}$) depends on the difference in the average ligand field strength between the two types of axes. The average ligand field strength difference between the A—A and X—X axes in the *trans*-complex is twice the difference between the A—A and A—X axes in the *cis*-complex. This is why, in the *trans*-complex, splitting of the quartet state is more. Because of this splitting, the lowest ligand field band (say L_1, $\nu_1 = {}^4A_{2g} \rightarrow {}^4T_{2g}$) will undergo splitting into two components, say $L_1'(\nu_1')$ and $L_1''(\nu_1'')$ where $\nu_1'' \rangle \nu_1'$. The ν_1' transition (longer wavelength component of the L_1 band) gives the **lowest quartet excited state** (in terms of the spin allowed ligand field excitation) which accumulates the electron density in the antibonding orbital to weaken the metal-ligand bond. The **L_1' band (*i.e.* ν_1' transition) leads to the accumulation of electron density in the antibonding orbital projected towards the weaker ligand field axes X—M—X (for the *trans*-complex) and A—M—X (for the *cis*-complex).** Less splitting of the $^4T_{2g}$ state corresponds to the higher value of ν_1'. Obviously, the higher value of ν_1' indicates the higher antibonding electron density in the lowest excited quartet state, *i.e.* higher metal-ligand bond weakening in the lowest excited quartet state.

- **Rule 1:** Thus for the unsymmetrical octahedral complexes, we may say that, in general, *the longer wavelength component of the L_1 band corresponds to the accumulation of electron density in the antibonding orbital directed along the weaker field axis. Thus the axis having the lower average ligand field strength will be labilised more in terms of the dissociative activation.* **It leads to the Rule 1,** if we assume that **all the photochemically excited states will rapidly convert into the lowest quartet excited state** through the different possible ways as shown in Jablonski diagram (*cf.* Fig. 6.9.5.1). *Thus it is reasonable to expect that regardless of the wavelength of irradiation, the complex will attain the same lowest excited quartet state (i.e.* ***same quantum yield****).*
- **Rule 2:** Higher ligand field strength (*i.e.* more Δ_o) makes, in general, the e_g orbitals of more antibonding character. Thus the ligand field excitation (*i.e.* $t_{2g} \rightarrow e_g$) weakens the metal-ligand bond more where Δ_o is more. This is why, in general, the quantum yield for the photochemical reactions passing through the dissociative activation increases with the ligand field strength. This explains the Rule 2.

Illustration: (i) Anion of **Reinecke's salt,** *i.e.* ***trans*-$[Cr(NCS)_4(NH_3)_2]^-$** possess two different types of axes, *i.e.* H_3N—Cr—NH_3 (**stronger field axis**) and SCN—Cr—NCS (**weaker field axis** which experiences the better labilisation). According to Rule 1, it should have the **photoaquation of thiocyanate** not NH_3. In fact, it happens so.

$$trans\text{-}\left[Cr(NCS)_4(NH_3)_2\right]^- \xrightarrow[(+H_2O)]{h\nu} \left[Cr(NCS)_3(NH_3)_2(OH_2)\right] + NCS^-$$

(ii) In **$[Cr(NCS)(NH_3)_5]^{2+}$**, the two axes are: H_3N—Cr—NH_3 (**stronger field axis**) and H_3N—Cr—NCS (**weaker field axis** which experiences the better labilisation). In fact, it experiences the **photo-aquation of NH_3** which is a stronger field ligand than NCS^- (Rule 1, 2).

$$\left[Cr(NCS)(NH_3)_5\right]^{2+} \xrightarrow[(+H_2O)]{h\nu} \left[Cr(NCS)(NH_3)_4(OH_2)\right]^{2+} + NH_3$$

(*cf.* Dark or thermal reaction leads to aquation of NCS^-).

(iii) In **$[Cr(NH_3)_5(N_3)]^{2+}$**, the two axes are: H_3N—Co—NH_3 (**stronger field axis**) and H_3N—Co—N_3 (**weaker field axis** which experiences the better labilisation), and NH_3 is a stronger field ligand than N_3^-. It leads to the **photoaquation of NH_3** (Rule 1 and 2).

$$\left[Co(NH_3)_5(N_3)\right]^{2+} \xrightarrow[\text{(green light irradiation)}]{LF \text{ excitation}} \left[Co(NH_3)_4(N_3)(OH_2)\right]^{2+} + NH_3$$

(iv) Photochemical behaviour of the *cis-trans* isomers of **$[Cr(en)_2(OH)_2]^+$:**

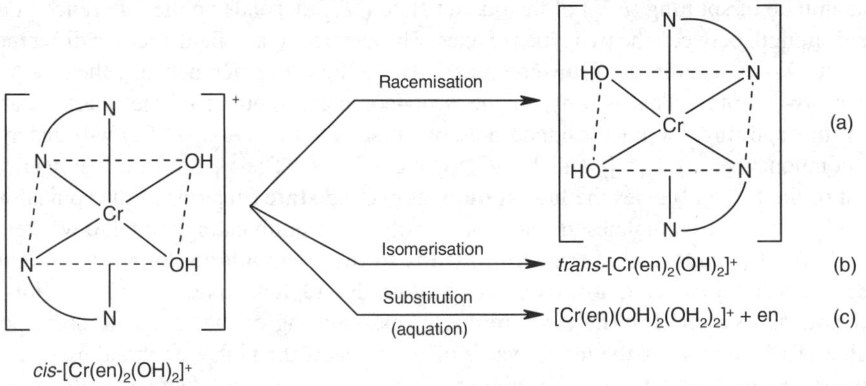

Racemisation (a) and isomerisation (b) of *cis*-$[Cr(en)_2(OH)_2]^+$ occur through the **one-ended dissociation** of ethylenediamine (en) chelate (producing an intermediate of lower coordination number) followed by the reattachment by the dissociated end of en to the metal centre. *Photochemically the en-aquation and isomerisation reactions (b,c) are important while the predominant thermal reaction is only isomerisation.* **Thermal reaction cannot aquate en.**

In *cis*-$[Cr(en)_2(OH)_2]^+$ the two ligand field axes are:

(en)N—Cr—N(en) (**stronger field axis**) and (en)N—Cr—OH (**weaker field axis**). According to Rule 1, the more labilised axis is (en)N—Cr—OH and according to Rule 2, the (en)N—Cr bond is labilised more. Thus **photoaquation of en** (reaction c) and **photo-isomerisation** (reaction b) can be explained through the rupture of (en)N—Cr bond. In the corresponding *trans*-complex, the weaker field axis is HO—Cr—OH. It predicts (Rule 1) the **O-exchange** and isomerisation through the formation of a stereomobile intermediate of lower coordination number. For the *trans*-compound, **en-aquation is not expected** and it has been experimentally verified.

- *cis*-$[Cr(en)_2(OH)_2]^+$ (**weaker field axis:** (en)N—Cr—OH): **en-aquation** and **isomerisation but no O-exchange.**
- *trans*-$[Cr(en)_2(OH)_2]^+$ (**weaker field axis**: HO—Cr—OH): O-exchange and isomerisation but no en-aquation.

These predictions have been qualitatively verified.

Limitations of the Adamson's empirical rules: There are many experimental findings to contradict the Adamson's rules. For example, in photoaquation of $[Cr(NH_3)_5X]^{2+}$ (X^-= halide or pseudo-halide), photoaquation occurs along the *trans*-H_3N—Cr—X axis (Rule 1) and it should give the *trans*-$[Cr(NH_3)_4(OH_2)(X)]^{2+}$ but this expectation is not always satisfied. Formation of the *cis*-product has been explained by considering the *stereomobile character* of the photochemically excited states of Cr(III) of lower coordination number.

6.9.6 Photolysis of $[M(CN)_8]^{3-}$ and $[M(CN)_8]^{4-}$ (M = Mo, W) in Aqueous Solution

The quantum yield (ϕ) for photolysis of $[M(CN)_8]^{3-}$ is greater than 1. To explain this higher value of ϕ, it has been argued for the generation of OH radical (in the primary photochemical process) which subsequently participates in the reaction. The suggested reaction paths are:

$$[M(CN)_8]^{3-} + H_2O \xrightarrow{h\nu} [M(CN)_8]^{4-} + H^+ + OH$$

$$[M(CN)_8]^{3-} + OH \longrightarrow [M(CN)_8]^{4-} + H^+ + \frac{1}{2}O_2$$

$$[M(CN)_8]^{3-} + H_2O + OH \longrightarrow [M(CN)_8]^{4-} + H^+ + 2OH$$

6.9.7 Photolysis of the Oxalato-Complexes, $[M(ox)_3]^{3-}$ (M = Fe, Mn, Co)

The said *tris*-oxalato-complexes undergo the redox decomposition during the photolysis. The proposed mechanism involving the intermediates M(IV) and M(II) is outlined below.

$$[M^{III}(C_2O_4)_3]^{3-} \xrightarrow{h\nu} [M^{III}(C_2O_4)_3^*]^{3-} \longrightarrow [M^{II}(C_2O_4)_2]^{2-} + C_2O_4^{-}$$

$$[M^{III}(C_2O_4)_3]^{3-} + C_2O_4^{-} \longrightarrow [M^{IV}(C_2O_4)_3]^{2-} + C_2O_4^{2-}$$

$$\downarrow$$

$$[M^{II}(C_2O_4)_2]^{2-} + 2CO_2$$

The overall process of photolysis of $[M(C_2O_4)_3]^{3-}$ may be shown as follows:

$$2[M(C_2O_4)_3]^{3-} \xrightarrow{h\nu} 2[M(C_2O_4)_2]^{2-} + 2CO_2 + C_2O_4^{2-}$$

The quantum yield (ϕ) for such photochemical redox decomposition processes depends on the wavelength of irradiation but it remains more or less constant in the charge-transfer region and it decreases in the d-d band region. For example, for $[Fe(C_2O_4)_3]^{3-}$, the ϕ value (0.4–0.6) is fairly high in the CT region while the ϕ value is very small in the region of spin-forbidden d-d band region (*ca.* $\rangle$ 650 nm). **The higher quantum yield (ϕ) for the light of shorter wavelength arises due to the fact:** allowed transition in the CT band region, excess energy allows the degraded species $[M(C_2O_4)_2]^{2-}$ and $C_2O_4^{-}$ to escape away from each other and preventing their recombination. At higher wavelength (*i.e.* low-energy light), recombination of the oxalate radical with $[M(C_2O_4)_2]^{2-}$ is more probable (**cage effect**) to reduce the ϕ value.

Note: ● Some of the oxalato-complexes find uses in ***actiometer***. $[Fe(ox)_3]^{3-}$ can be used in this purpose. UO_2^{2+} – oxalate system can also be used in acitometer. It may be noted that UO_2^{2+} can cause photochemical oxidation of different types of organic and inorganic substrates.

● The complexes of Co(III), Mn(III) and Fe(III), *i.e.* **$[M(C_2O_4)_3]^{3-}$ (M = Co, Mn, Fe) experience the redox decomposition photochemically** but the corresponding Cr(III)-complex, *i.e.* $[Cr(C_2O_4)_3]^{3-}$ does not experience any such photochemical redox decomposition. The proposed mechanistic paths involve the participation of intermediates M(IV) and M(II). Thus formation of these oxidation states should be energetically feasible. In fact, **photochemical stability of $[Cr(C_2O_4)_3]^{3-}$** arises due to the fact that the generation of Cr(II) is not likely. However, the Cr(III)-complex experience the *photorecimization.*

6.9.8 Photochemical Reactions of the Fe(II) and Fe(III) Complexes

- Photolysis of an acidic aqueous solution of Fe(II) can lead to oxidation of Fe(II) to Fe(III).

$$[Fe(OH_2)_6]^{2+} \xrightleftharpoons{h\nu} [Fe(OH_2)_6]^{3+} + e^-_{aq}$$

After the photochemical primary process, different sets of reactions depending on the pH and concentration of the solution are set up.

- Photochemical excitation of $[Fe(OH_2)_6]^{3+}$ or $[Fe(OH_2)_5X]^{2+}$ (X = OH, Cl, Br, N_3, etc.) in the CT region (ligand to metal) will lead to oxidation of the coordinated ligands including the solvent H_2O.

$$[Fe(OH_2)_5X]^{2+} \xrightarrow[(CT\ \text{band})]{h\nu} [Fe(OH_2)_5]^{2+} + X$$

X can subsequently oxidize other reducing agent including water.

- Photolysis of $[Fe(C_2O_4)_3]^{3-}$ has been discussed in Sec. 6.9.7.
- Photolysis of $[Fe(CN)_6]^{4-}$ in aqueous solution involves the following processes.

$$[Fe(CN)_6]^{4-} \xrightleftharpoons{h\nu} [Fe(CN)_6]^{3-} + e^-_{aq},$$ **charge transfer to solvent** (MSCT), (mainly CT excitation)

$$[Fe(CN)_6]^{4-} \xrightarrow[H_2O]{h\nu} [Fe(CN)_5(OH_2)]^{3-} + CN^-,$$ (mainly LF excitation)

The released electron, *i.e.* e^-_{aq} can be trapped by using **N_2O as a scavanger** to generate the OH radical which can subsequently oxidize $[Fe(CN)_6]^{4-}$.

$$N_2O + e^-_{aq} \longrightarrow N_2 + O^-,\ O^- + H^+ \longrightarrow OH;\ [Fe(CN)_6]^{4-} + OH \longrightarrow [Fe(CN)_6]^{3-} + OH^-$$

- Photochemical reaction of $[Fe(CN)_6]^{3-}$ in aqueous solution is more complicated. The events are shown below.

$$[Fe(CN)_5(OH_2)]^{3-} + HCN + OH \xleftarrow[(H_2O)]{h\nu} [Fe(CN)_6]^{3-} \xrightarrow[(H_2O)]{h\nu} [Fe(CN)_5(OH_2)]^{2-}$$

(Redox aquation) (Aquation)

$$[Fe(CN)_5(OH_2)]^{3-} \downarrow \cdots \downarrow [Fe(OH_2)_6]^{2+} \xrightarrow{[Fe(CN)_6]^{3-}} \textbf{Prussian blue}$$

$$[Fe(CN)_5(OH_2)]^{2-} \downarrow \cdots \downarrow [Fe(OH_2)_6]^{3+} \xrightarrow{OH^-} Fe(OH)_3\downarrow$$

$$H_2O + OH + [Fe(CN)_6]^{3-} \longrightarrow [Fe(CN)_6]^{4-} + H^+ + 2OH$$

Thus the produced $[Fe(CN)_6]^{4-}$ may also experience photochemical reactions as stated above.

The above photochemical reactions explain why the **$[Fe(CN)_6]^{4-}$ and $[Fe(CN)_6]^{3-}$ solutions on standing give the blue precipitate of prussian blue.**

6.9.9 Photochemical Redox Process in $[M(aq)]^{n+}$

It can lead to oxidation or reduction of metal centre depending on the conditions.

Reducing metal centre : $[M(OH_2)_x]^{n+} \xrightarrow[(CT:\ M\rightarrow L)]{h\nu} [M(OH)(OH_2)_{x-1}]^{n+} + H$

It happens so for $[Fe(OH_2)_6]^{2+}$. Subsequently, H-radical can show the reducing activity.

Oxidising metal centre : $$\left[M(OH_2)_x\right]^{n+} \xrightarrow[(CT:\ L\to M)]{h\nu} \left[M(OH_2)_5\right]^{(n-1)+} + H^+ + OH$$

It happens so for $Tl(aq)^{3+}$, $Ce(aq)^{4+}$. Then the OH radical can oxidize the available reducing agents.

$$OH + e \longrightarrow OH^-$$

6.9.10 Photochemical Substitution Process in Pt(IV)

The ligand exchange and substitution reactions are photochemically carried out at the Pt(IV) centre which is expected to be inert in thermal activation process.

$$\left[PtCl_6\right]^{2-} + {}^*Cl^- \xrightarrow{h\nu} \left[PtCl_5^*Cl\right]^{2-} + Cl^- \ ; \ \left[PtCl_6\right]^{2-} + X^- \xrightarrow{-h\nu} \left[PtCl_5X\right]^{2-} + Cl^-$$

$$\left[PtCl_6\right]^{2-} + H_2O \xrightarrow{h\nu} \left[PtCl_5(OH)\right]^{2-} + H^+ + Cl^-$$

It is believed that in the primary photochemical step, homolytic bond cleavage generates the Pt(III) centre (d^7) which is labile.

$$\left[PtCl_6\right]^{2-} \xrightarrow{h\nu} \left[Pt^{III}Cl_5\right]^{2-} + Cl, \ \left[Pt^{III}Cl_5\right]^{2-} + H_2O \rightarrow \left[PtCl_4(OH)\right]^{2-} + H^+ + Cl^-,$$

$$\left[Pt^{III}Cl_5\right]^{2-} + X^- \longrightarrow \left[PtCl_4X\right]^{2-} + Cl^-$$

$$\left[PtCl_4X\right]^{2-} + \left[PtCl_6\right]^{2-} \xrightarrow[\text{(inner sphere)}]{e\text{-transfer}} \left[PtCl_5X\right]^{2-} + \left[PtCl_5\right]^{2-}$$

The process proceeds as follows (ISET with ligand transfer):

$$\left[PtCl_4X\right]^{2-} + \left[PtCl_6\right]^{2-} \longrightarrow \left[\begin{matrix} Cl_5Pt^{IV} & — Cl — & Pt^{III}Cl_4X \\ (d^6) & & (d^7) \\ \text{(inert)} & & \text{(labile)} \end{matrix}\right]^{4-}$$

$$\xrightarrow{1e\text{-transfer}} \left[Cl_5Pt^{III} \,\vdots\, Cl — Pt^{IV}Cl_4X\right]^{4-} \longrightarrow \left[PtCl_5\right]^{2-} + \left[PtCl_5X\right]^{4-}$$

($X^- = {}^*Cl^-$ or other nucleophile)

Thus the generated $[PtCl_5]^{2-}$ species can participate in chain reaction and it leads to the high quantum yields.

The ligand exchange reaction in $[PtBr_6]^{2-}$ passes through the similar path generating Pt(II) at the primary photochemical step. The Pt(II) centre (distorted structure due to the removal of a *cis*– pair of Br-atoms from $[PtBr_6]^{2-}$) is labile to experience the rapid ligand exchange.

$$\left[PtBr_6\right]^{2-} \xrightarrow{h\nu} \left[PtBr_4\right]^{2-} + Br_2 \ ; \ \left[PtBr_4\right]^{2-} \xrightarrow[\text{(ligand exchange)}]{{}^*Br^-} \left[Pt^*Br_4\right]^{2-}$$

$${}^*Br^- + \left[Pt^*Br_4\right]^{2-} + \left[PtBr_6\right]^{2-} \xrightarrow[\text{(inner sphere)}]{e\text{-transfer}} \left[Pt^*Br_5Br\right]^{2-} + Br^- + \left[PtBr_4\right]^{2-}$$

The mechanistic aspects of the above reaction leading to bromide exchange have been discussed in Sec. 6.4.15.

Thus, $[PtBr_4]^{2-}$ is generated to carry out the chain reaction. It explains the high quantum yield.

6.9.11 Photochemistry of $[Ru(bpy)_3]^{2+}$ and Possibility of Photochemical Splitting of Water

These aspects have been discussed in detail in Secs. 16.16.1-2 of Vol. 3.

6.9.12 TiO_2 as a Green Photocatalyst in Removing Water Pollutants and Air Pollutants

These aspects have been discussed in detail in Secs. 16.16.3 of Vol. 3.

6.9.13 Possibility of Direct Photochemical Reduction of Nitrogen

These aspects have been discussed in detail in Secs. 16.16.4 of Vol. 3.

6.9.14 Chemistry of Photography

These aspects have been discussed in Sec. 16.16.5 of Vol. 3.

EXERCISE 6

A. General Type Questions

1. Illustrate the electron exchange and redox reactions.
2. Compare the mechanistic steps of OSET and ISET process.
3. What do you mean by Franck-Condon barrier in the electron transfer process? How does it contribute to the activation energy of the electron transfer process. Illustrate with examples.
4. Discuss the factors favouring the OSET process. Give the rate law.
5. What are the types of receptor and donor orbitals in the OSET process?
6. Discuss the Marcus equation in relation to Cross-reactions passing through the OSET process.
7. What do you mean by ISET process? Discuss the steps of ISET process. Deduce the rate law.
8. Classify the ISET reaction depending on the position of rate determining step.
9. Discuss the factors favouring the ISET process.
10. Discuss the possibility of characterization of binuclear complexes involved in the ISET process.
11. Discuss the ISET reaction experiencing the concomitant ligand transfer.
12. Give the examples where the ISET reaction experiences no ligand transfer.
13. What do you mean by the substitution controlled ISET reaction?
14. Discuss the role of a bridging ligand in the ISET process.
15. How does the bridging ligand help in the tunneling of electron from the reductant to oxidant?
16. How can you distinguish between the chemical and tunneling mechanism of electron transfer?
17. What are the conditions of electron transfer through the resonance mechanism in the ISET process?
18. Discuss the effect of nonbridging ligands in the ISET process.
19. Discuss the mechanistic criteria of the OSET and ISET process.
20. How can you distinguish between the OSET and ISET process?
21. Discuss the ligand replacement reactions through the electron transfer process.
22. Illustrate the complementary and noncomplementary electron transfer reactions.
23. Discuss the properties of solvated electron.
24. What do you mean by the photochemical reaction? How does it differ from the thermal reaction?
25. What are the different types of photochemical reaction? What do you mean by the delayed and prompt photochemical reaction?
26. What are the ways of photochemical excitation in the coordination compounds?
27. Discuss the photochemical substitution reaction in the metal carbonyls.
28. Compare the photochemical reactions of the Co(III) and Cr(III) complexes.

29. Discuss the mechanism of photolysis of $[M(ox)_3]^{3-}$ (M = Co, Fe).
30. Discuss the photochemical reactions of the Pt(II), Pt(IV), Fe(II) and Fe(III) complexes.
31. Discuss the photochemistry of $[Ru(bpy)_3]^{2+}$. Discuss the photolysis of water by using the photocatalyst $[Ru(bpy)_3]^{2+}$.
32. Discuss the use of TiO_2 as a green photocatalyst. Why is it considered as a green catalyst in removing the air pollution?
33. State the Adamson's rules in connection with the photosubstitution processes in the Cr(III)- and Co(III)- complexes. Explain the Rules in terms of CFT and illustrate with the suitable examples.

B. Explain the following statements

1. Electron exchange and redox reactions are basically of the same type.
2. The rate law of an OSET reaction is : rate = $kK_{os}[Ox][Red]$.
3. In the electron exchange process, energies of the participating donor and receptor orbitals should be the same.
4. Potential energy diagram of the electron transfer process can give us the measure of the activation energy.
5. Marcus equation is a representation of linear free energy relationship (LFER).
6. In the electron exchange process, Franck-Condon barrier gives the measure of activation energy.
7. In the electron carrier metalloproteins like ferredoxins, blue proteins, nature has placed the metal centres in the distorted geometries rather than in the regular geometries.
8. Electronic activation may be an important factor to determine the activation energy of an electron transfer process.
9. In the OSET process, $\pi^* \rightarrow \pi^*$ (*i.e.* $t_{2g} \rightarrow t_{2g}$) transfer is the most favourable pathway for the octahedral complexes.
10. The electron exchange rate in the $[Fe(aq)]^{3+}$—$[Fe(aq)]^{2+}$ couple is very slow though it gives the $\pi^* \rightarrow \pi^*$ transfer, a favourable situation for an OSET path.
11. The electron exchange rates in the couples $[Fe(CN)_6]^{3-}$—$[Fe(CN)_6]^{4-}$, $[Fe(phen)_3]^{3+}$—$[Fe(phen)_3]^{2+}$ are fairly high.
12. The electron exchange rate in the couple $[Co(aq)]^{3+}$—$[Co(aq)]^{2+}$ is fairly fast though the electron exchange rate in the couple $[Co(NH_3)_6]^{3+}$—$[Co(NH_3)_6]^{2+}$ is very slow.
13. Electron exchange rate in $[Ru(NH_3)_6]^{3+}$— $[Ru(NH_3)_6]^{2+}$ is much higher than that in $[Co(NH_3)_6]^{3+}$—$[Co(NH_3)_6]^{2+}$.
14. The electron exchange rate in $[Mn(phen)_3]^{3+}$—$[Mn(phen)_3]^{2+}$ is fairly high.
15. The electron exchange rate in the Co(III)—Co(II) couple depends on the ligand field strength of the involved ligand. *Crystal field strength*: $H_2O \langle$ oxalate $\langle$ edta $\langle$ $NH_3 \sim$ en $\langle$ phen. The electron exchange rate is fast for the both weak (*e.g.* H_2O) and strong (*e.g.* phen) field ligands.
16. The electron exchange rates in the couples, $[Co(NH_3)_6]^{3+}$–$[Co(NH_3)_6]^{2+}$ and $[Co(phen)_3]^{3+}$ – $[Co(phen)_3]^{2+}$ differ widely.
17. The electron exchange rate in the $[Cr(aq)]^{3+}$–$[Cr(aq)]^{2+}$ couple is much slower than that in the $[V(aq)]^{3+}$–$[V(aq)]^{2+}$ couple.
18. In the OSET process, $[V(aq)]^{2+}$ is a better reducing agent than $[Cr(aq)]^{2+}$ kinetically but the reverse is true thermodynamically.
19. The rate of reduction of $[Ru(NH_3)_6]^{3+}$ by $[V(aq)]^{2+}$ is much faster than the rate of reduction of $[Co(NH_3)_6]^{3+}$ by the same reducing agent though the thermodynamic favour in both the reactions is more or less the same.

20. $[Ru(aq)]^{2+}$ and $[Ru(NH_3)_6]^{2+}$ are the good reducing agents in the OSET path.
21. The Ru(III)–Ru(II) reaction is faster than the Co(III)–Ru(II) reaction.
22. The electron exchange reactions in the couples like $[Fe(CN)_6]^{3-}$– $[Fe(CN)_6]^{4-}$, $[MnO_4]^-$– $[MnO_4]^{2-}$ are catalysed by the alkali metal ions. The catalyzing efficiency of the alkali metal ions runs as:
$$Cs^+ \rangle Rb^+ \rangle K^+, NH_4^+ \rangle Na^+ \rangle Li^+$$
23. The electron exchange process in $[Co(NH_3)_6]^{3+}$–$[Co(NH_3)_6]^{2+}$ can be catalyzed by the anions.
24. The electron exchange rate in $[Co(NH_3)_6]^{3+}$– $[Co(NH_3)_6]^{2+}$ is remarkably increased in an alkaline solution.
25. Compared to $[Co(NH_3)_6]^{3+}$, the reduction of $[CoCl(NH_3)_5]^{2+}$ by $[Cr(aq)]^{2+}$ is much faster.
26. $\left[CoCl(OH_2)_5\right]^{2+} + \left[Cr(aq)\right]^{2+} \longrightarrow \left[Co(aq)\right]^{2+} + 5NH_3 + \left[Cr(Cl)(OH_2)_5\right]^{2+}$

 The experimental findings of the above reaction cannot be explained by an OSET path but by an ISET path.
27. Depending on the position of rate determining step, the ISET reaction can be classified into three groups.
28. Rate law for the ISET reactions depends on the position of rate determining step.
29. In an ISET path, $\sigma^* \rightarrow \sigma^*$ (*i.e.* $e_g \rightarrow e_g$) transfer is the most favourable path for the M—X type oxidant (X = monoatomic bridging ligand).
30. There is no example of ISET reaction experiencing the net $\pi^* \rightarrow \pi^*$ transfer.
31. Reduction of $[Co(NH_3)_5(OH_2)]^{3+}$ by $[Cr(aq)]^{2+}$ occurs through the ISET path not through the OSET path. The rate of the reaction depends on pH.
32. $\left[Co(NH_3)_5X\right]^{2+} + \left[Cr(aq)\right]^{2+} \longrightarrow$ Product
    - Rate for X = $NH_3 \langle OH_2 \langle {}^-OH$; X = $N_3^- \approx$ *S*CN$^-$ $\rangle\rangle$ *N*CS$^-$
    - For X = –*N*CS, only one linkage isomer of Cr(III) is obtained; for X = –*S*CN, a mixture of linkage isomers of Cr(III) is obtained.
33. $\left[Co(NH_3)_5X\right]^{2+} + \left[M(aq)\right]^{2+} \longrightarrow$ Product; M = V, Cr, Fe
    - For reduction by $[V(aq)^{2+}]$, the rate is more or less insensitive to the nature of X while for reduction by $[Cr(aq)]^{2+}$ or $[Fe(aq)]^{2+}$, the rate is sensitive to the nature of X (bridging ligand).
    - The reactivity order is: $[Cr(aq)]^{2+} \rangle [V(aq)]^{2+} \rangle [Fe(aq)]^{2+}$.
    - For M = V, Fe and Eu, the bridging ligand X^- sometimes gets released in the solution.
34. $\left[M(NH_3)_5X\right]^{2+} + \left[V(aq)\right]^{2+} \longrightarrow$ Products, M = Co, Ru.
    - If the rate is much faster than the ligand substitution rate of $[V(aq)]^{2+}$, the reaction may be argued to pass through the OSET path.
    - If the rate is comparable to the rate of nucleophilic substitution rate of $[V(aq)]^{2+}$, then the reaction may be argued to pass through the ISET path.
    - For M = Co(III), it is an ISET reaction while for M = Ru(III), it is an OSET reaction.
35. $\left[Co(NH_3)_5X\right]^{2+} + \left[Eu(aq)\right]^{2+} \longrightarrow$ Products

 For X = Cl^-, N_3^- and NO_3^-, the rate is more or less the same and X^- is released to the solution.
36. $[RuCl_6]^{3-}$– $[Cr(aq)]^{2+}$; $[V(aq)]^{4+}$ + $[Cr(aq)]^{2+}$; $[IrCl_6]^{2-}$–$[Cr(aq)]^{2+}$.
 The above reactions (ISET path) can generate the stable dinuclear complexes that can be decomposed easily in presence of $[Cr(aq)]^{2+}$ and ^-OH.

37. $[IrCl_6]^{2-} + [Co(CN)_5]^{3-}$; $[Fe(CN)_6]^{3-} + [Co(CN)_5]^{3-}$; $HCrO_4^- + [Fe(CN)_6]^{4-}$
    The above reactions can produce the stable binuclear complexes.

38. There are many ISET reactions where there is no ligand transfer.

39. $[Co(NH_3)_5(ntaH_2)]^{2+}$ is reduced much faster than $[CoCl(NH_3)_5]^{2+}$ by $[Fe(aq)]^{2+}$.

40. There are some favourable situations for characterizing the binuclear precursor complexes in the ISET path.

41. Electron transfer leading to the simultaneous ligand transfer is a convincing evidence in favour of an ISET path.

42. Ligand transfer is not an essential condition for an ISET path.
    ● Sometimes, the bridging ligand of the ISET path is released to the solution.

43. Product analysis is helpful to predict the nature of the precursor complex in an ISET path.

44. Reduction of $[CoCl(NH_3)_5]^{2+}$, and $[Co(NCS)(NH_3)_5]^{2+}$ by $[Cr(aq)]^{2+}$ can occur through the adjacent attack and remote attack on the bridging ligand by the reducing agent respectively.

45. Reduction of $[Co(NH_3)_5(SCN)]^{2+}$ by $[Cr(aq)]^{2+}$ can occur through the both remote and adjacent attack.

46. Reduction of $\left[(H_3N)_5Co{-}N\bigcirc{-}C\!\!\underset{NH_2}{\overset{O}{{}}} \right]^{3+}$ by $[Cr(aq)]^{2+}$ experiences the remote attack on

    the carbonyl–O not on the NH_2 group.

47. $[(H_3N)_5Co^{III}{-}OCOR]^{2+}$ can experience the remote attack but $\left[(H_3N)_5 Co^{III}{-}O{=}C\!\!\underset{H}{\overset{NH_2}{{}}} \right]^{3+}$

    cannot experience an remote attack in reduction through the ISET path.

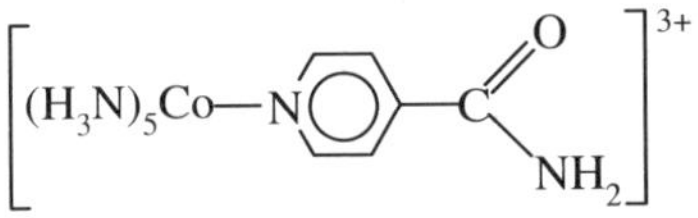

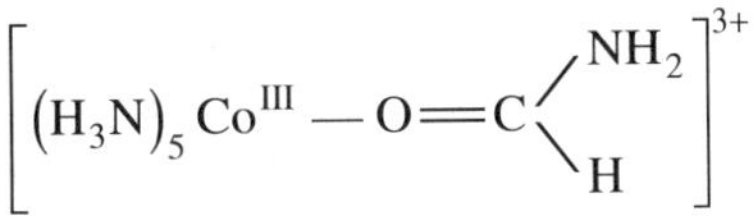

48. Reduction of $[(H_3N)_5CoCN]^{2+}$ by $[Co(CN)_5]^{3-}$ in presence of excess CN^- gives a mixture of products $[Co(CN)_5(NC)]^{3-}$ and $[Co(CN)_6]^{3-}$ in the final solution. On standing, it gives only $[Co(CN)_6]^{3-}$.

49. $\left[Co(NH_3)_5(NO_2) \right]^{2+} + \left[Co(CN)_5 \right]^{3-} \longrightarrow \left[Co(CN)_5(NO_2) \right]^{3-} + \left[Co(aq) \right]^{2+} + 5NH_3$

    The product can be explained in terms of remote attack not in terms of adjacent attack but apparently it indicates an adjacent attack.

50. $\left[Co(NH_3)_5(OCOR) \right]^{2+} + \left[Cr(aq) \right]^{2+} \longrightarrow$

    The simple product analysis cannot distinguish between the remote and adjacent attack.

51. The adjacent attack in ISET generally leads to electron transfer by a tunneling mechanism but the remote attack for the nonreducible bridging ligands leads to electron transfer through a resonance transfer mechanism.

52. The electron exchange rate of the $[Fe(aq)]^{3+}$–$[Fe(aq)]^{2+}$ couple increases with the increase of pH.

53. Reduction of $[CoCl(NH_3)_5]^{2+}$ by $[Cr(aq)]^{2+}$ in presence of N_3^- gives the product $[CrCl(N_3)(OH_2)_4]^+$.

54. ISET reduction of $[Co(NH_3)_5X]^{2+}$ by $[M(aq)]^{2+}$:

    For $X = N_3^-$ and NCS^-: $k_{N_3}/k_{NCS} >> 1 \Rightarrow$ ISET process.

    $(M = V, Cr, Eu, Fe)$ $k_{N_3}/k_{NCS} \approx 1 \Rightarrow$ OSET and/or ISET path.

55. $\left[Co(NH_3)_5X\right]^{2+} + \left[Cr(aq)\right]^{2+} \longrightarrow$ Product

- $X = -NCS^-$ or $-SCN^-$: $k_{N_3} >> k_{NCS} \Rightarrow$ ISET process.

  $k_{N_3} \approx k_{NCS} \Rightarrow$ OSET and/or ISET process.

- The rate constant varies as:

X =	$-OCOCH_3$ (acetate)	$-OCOCH_2OH$ (glycolate)	$-OCOH$ (formate)	$-OCOCHO$ (glyoxylate)	$-OCOCO_2H$ (acid oxalate)	$-OCOCO_2^-$ (oxalate)
k ($dm^3\ mol^{-1}\ s^{-1}$):	0.35	3.1	7.2	10^4	1×10^2	4.5×10^4

56. Reduction of $[Ru(NH_3)_5X]^{2+}$ by $[Ru(aq)]^{2+}$ occurs through the OSET path not by the ISET path.

57. $\left[(H_3N)_5M^{III}-X\right]+\left[Cr(aq)\right]^{2+} \longrightarrow$ Product, M = Co, Ru. Different mechanisms operate for:

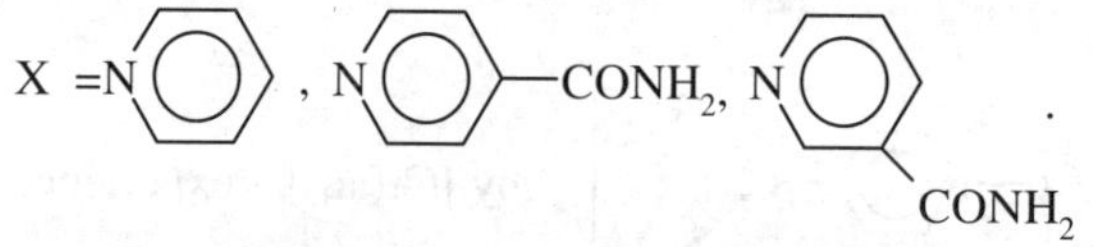

58. The following binuclear complexes experience the OSET path.

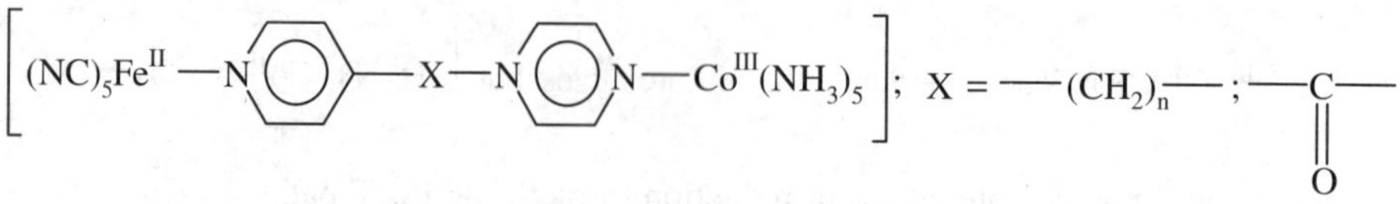

59. The chemical mechanism of electron transfer in the ISET path is a multistep process.
60. Sometimes, it may be difficult to distinguish between the resonance mechanism and chemical mechanism for electron transfer in the ISET process.
61. $\left[(H_3N)_5Co^{III}-X\right]+\left[Cr(aq)\right]^{2+} \longrightarrow$ Product

- The rate constants varies as:

X =	succinate	glycolate	malonate	pyridine-2-carboxylate	salicylate
k ($dm^3\ mol^{-1}\ s^{-1}$) =	1.0	3.0	2.5×10^3	2×10^5	2×10^8

- For X = –O–COCHO (glyoxylate) the rate is faster than that for $X = -OCOCH_3$.
- For X = benzoate, acetate and propionate, the rate is relatively slower than that for X = formate, oxalate, maleate, fumarate. ($HO_2CCH{=}CHCO_2H$, *cis*-form = fumaric acid; *trans*-form = maleic acid)
- For the reduction of $\left[(H_3N)_5Co^{III}-O-C(=O)R\right]$ by $\left[Cr(aq)\right]^{2+}$, the rate varies as:

  R = CH_3 (acetate), C_3H_7 (butyrate), C_6H_5 (benzoate), $(CH_2)_2CO_2H$ (acid succinate) $\langle$ CH_2OH (glycolate) $\langle$ CHO (glyoxylate).

62. $\left[M^{III}(NH_3)_5X\right]+\left[Cr(aq)\right]^{2+} \longrightarrow$ Products; M = Co, Cr, Ru.

- For X = F, Cl, $OCOCH_3$ etc; $\frac{k_{Co(III)}}{k_{Cr(III)}} \geq 10^6$; $\frac{k_{Ru(III)}}{k_{Cr(III)}} \geq 10^5$; $\frac{k_{Co(III)}}{k_{Ru(III)}} \approx 10$
- $\frac{k_{Co(III)}}{k_{Ru(III)}} \approx 10^1$ (for X = F, Cl) but $\frac{k_{Ru(III)}}{k_{Co(III)}} > 10^5$ (for X = isonicotinamide)
- For X = 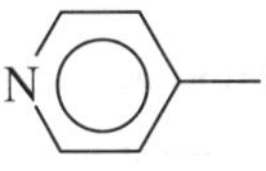—$CONH_2$, $\frac{k_{Co(III)}}{k_{Cr(III)}} \approx 10^1$; $\frac{k_{Ru(III)}}{k_{Cr(III)}} \geq 10^5$, $\frac{k_{Ru(III)}}{k_{Co(III)}} > 10^5$
- For X = 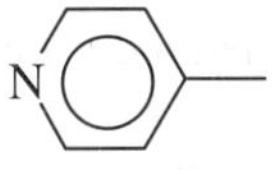—$CONH_2$, Co(III) and Cr(III) experience the ion radical chemical mechanism of electron transfer while Ru(III) experiences the resonance tunneling mechanism for electron transfer.

63. *cis* – $\left[Cr(H_2O)_4X_2\right]^+ + \left[{}^*Cr(aq)\right]^{2+} \longrightarrow$ Products
- For X = N_3^-, the product is *cis*– $[{}^*Cr(OH_2)_4X_2]^+$
- For X = F^-, the product is *cis*– $[{}^*Cr(OH_2)_5X]^{2+}$

64. Reduction of $[HCrO_4]^-$ by $[Fe(CN)_6]^{4-}$ generates the product, $[(H_2O)_5Cr-N{\equiv}C-Fe(CN)_5]^{3-}$

65. $\left[Co^{III}(acac)_2(en)\right]^+ + \left[Cr(aq)\right]^{2+} \longrightarrow \left[Cr(aq)\right]^{3+} + \left[Cr(acac)(aq)\right]^{2+} + \left[Cr(acac)_2(aq)\right]^+$
The product analysis indicates the different parallel paths of reduction of the Co(III) complex.
- $\left[Co(edta)\right]^- + \left[Cr(aq)\right]^{2+} \longrightarrow \left[Co(aq)\right]^{2+} + \left[Cr(edta)(OH_2)_3\right]^-$

The product can predict the structure of the activated complex.

66. Reduction of *trans*– $[Co^{III}(en)_2(L)(OH)]$ by $[Cr(aq)]^{2+}$.
- Rate is faster for L = H_2O than for L = NH_3
- $[Cr(aq)]^{2+}$ reduces *trans*– $[CrCl(NH_3)_4(OH_2)]^{2+}$ faster than $[CrCl(NH_3)_5]^{2+}$.

67. Reduction rate of the Co(III)-complexes by Cr(II) follows the sequence:
$[Cr(en)_2(OH_2)_2]^{2+} \rangle [Cr(en)(OH_2)_4]^{2+} \rangle [Cr(OH_2)_6]^{2+}$.
68. $[Pt(en)_2]^{2+}$ can catalyze the chloride exchange process in *trans*-$[PtCl_2(en)_2]^{2+}$.
69. The reducing agents like $[Ru(aq)]^{2+}$, $[Ru(NH_3)_6]^{2+}$, $[Cr(bpy)_3]^{2+}$, $[Fe(CN)_6]^{4-}$, $[Co(CN)_6]^{4-}$ are selective for the OSET process.
70. Reduction of $[M(phen)_3]^{3+}$, $[M(NH_3)_6]^{3+}$, $[M(NH_3)_5(py)]^{3+}$ (M = Co, Ru) cannot occur through the ISET path.
71. $\Delta V^{\neq}$ value may be helpful to distinguish between the OSET and ISET process.
72. The water exchange rate at $[Co(aq)]^{3+}$ is unusually high.
73. The water exchange process at $[Cr(aq)]^{3+}$ can be catalyzed by $[Cr(aq)]^{2+}$.
74. Anhydrous $CrCl_3$ is insoluble in water but it goes into solution easily in the presence of a catalytic amount of $[Cr(aq)]^{2+}$.

75. Generally, the Co(III) complexes are more inert than the Cr(III) complexes but for the water exchange process, the reverse is true.
76. Aquation of NH_3 in $[Cr(NH_3)_5X]^{2+}$ (X with bridging property) can be catalyzed by $[Cr(aq)]^{2+}$.
   - Aquation of L in $[Cr^{III}(aq)(L)]$ can be catalyzed by $[Cr(aq)]^{2+}$, if L does not have any bridging property.
77. Reduction of $[Co^{III}(NH_3)_5X]$ by $[Co(aq)]^{2+}$ in presence of excess CN^- leads to different products depending on the bridging property of X.
   - $[Co^{III}(NH_3)_5X] \xrightarrow[(+\text{excess } CN^-)]{+\text{trace amount of Co(II)}}$ different types of products depending on the nature of X.
78. Generally the *ligand field* (LF) excitation in the coordination compounds leads to the nonredox processes, *e.g.* substitution, isomerization, racemization while the *charge transfer* (CT) excitation leads to the redox reactions.
   - Photochemically it is possible to have the products which are thermodynamically and kinetically inaccessible in thermal reactions.
   - In the photochemical reactions, the Co(III) complexes require no thermal activation but the analogous Cr(III) complexes require a thermal activation.
   - In the photochemical reactions of the Cr(III)- complexes, the quantum yield increases with the increase of temperature.
79. Charge transfer excitation in $[CoCl(NH_3)_5]^{2+}$ leads to aquation while charge transfer excitation in $[Co(C_2O_4)_3]^{3-}$ leads to redox decomposition.
80. $\left[Co^{III}(NH_3)_5X\right]^{2+} \xrightarrow[(H_2O)]{h\nu}$ Product.
   The nature of product depends largely on the nature of X.
   - For X = Cl, it leads to aquation while for X = I, it leads to redox decomposition.
   - The quantum yield for the photochemical aquation of $[CoCl(NH_3)_5]^{2+}$ increases with the energy of the irradiating radiation.
81. The nature of photosubstitution in $[Cr^{III}(NH_3)_x(Y)_y]^{n+}$ $(x + y = 6)$ is different from the thermal substitution process.
   - Adamson's rule can predict the nature of product in the photosubstitution of the Cr(III) and Co(III) complexes.
82. In the photochemical reaction of $[Cr^{III}(NH_3)_xX_y]^{n+}$ $(x + y = 6)$, the reactive intermediate state may be characterized by ${}^4T_{2g}$ and/or ${}^2E_g$ state.
83. Photochemical aquation of $[Co(CN)_5X]^{3-}$:

$$\left[Co(CN)_5(X)\right]^{3-} \xrightarrow[(\lambda=370\text{ nm})]{h\nu} \left[Co(CN)_5(OH_2)\right]^{2-} + X^-;\ \text{Quantum Yield: } X = I \rangle Br \rangle Cl.$$

- $\left[Co(NH_3)_5(N_3)\right]^{2+} \xrightarrow{h\nu(LF)} \left[Co(NH_3)_4(N_3)(OH_2)\right]^{2+} + NH_3$; **(Adamson's rule)**
- $\left[Co(NH_3)_5(N_3)\right]^{2+} \xrightarrow{h\nu(CT)} Co(II) + NH_3 + N_2$
- $\left[CrBr(NH_3)_5\right]^{2+} \xrightarrow[(\text{dark reaction})]{k_BT} \left[Cr(NH_3)_5(OH_2)\right]^{2+} + Br^-$
- $\left[CrBr(NH_3)_5\right]^{2+} \xrightarrow[(\text{photochemical reaction})]{h\nu} \left[CrBr(NH_3)_4(OH_2)\right]^{2+} + NH_3$; **(Adamson's rule)**.

84. Photochemical reactions of the Co(III) complexes are generally faster than those of the Cr(III) complexes though the reverse is true for the thermal reactions.
    - In the photochemical reactions of the Cr(III)-complexes, the quantum yield depends on the temperature, but such a situation does not prevail for the Co(III)-complexes.
    - Primary photochemical process in the Cr(III)-complexes generates a relatively long-lived reactive intermediate which requires a thermal activation.
85. Photochemical reaction of $[Cr(NH_3)_x(OH_2)_{6-x}]^{3+}$.

$$\left[Cr(NH_3)_x(OH_2)_{6-x}\right]^{3+} \xrightarrow[(H_2O)]{h\nu} \left[Cr(NH_3)_{x-1}(OH_2)_{6-x+1}\right]^{3+}$$

The quantum yield increases with the increase of *x*
    - The quantum yield is maximum for $[Cr(NH_3)_6]^{3+}$ (*i.e.* $x = 6$) and it is minimum for $[Cr(H_2O)_6]^{3+}$ (*i.e.* $x = 0$).
    - For the Cr(III)-complexes, the excited states ^2E_g and $^4T_{2g}$ favour different pathways of photosubstitution.
86. Photolysis of $[M(C_2O_4)_3]^{3-}$ (M = Co, Mn, Fe):
    - The quantum yield depends on the λ of the irradiation.
    - The quantum yield is maximum for the CT excitation and it is minimum for the LF excitation.
    - The said complexes experience the photo-redox decomposition while the corresponding Cr(III) complex experiences the photoracemization.
87. On standing, $[Fe(CN)_6]^{3-}$ produces a blue stain.
88. The photochemical redox process in $[M(OH_2)_x]^{n+}$ depends on the nature of the charge transfer excitation, *i.e.* M→ L or L → M.
89. Photolysis of $[M(CN)_8]^{3-}$ (M = Mo, W) in aqueous solution shows the quantum yield > 1.
90. $[PtCl_6]^{2-}$ and $[PtBr_6]^{2-}$ can experience photochemical ligand exchange and substitution reactions where quantum yield is very often greater than unity.
91. Reduction of $[CoCl(edta)]^{2-}$ and $[Co(edta)(OH_2)]^{-}$ by different reducing agents gives the different rate ratios for the chlorido and aqua complexes (*Ref.* H. Ogino *et al, J. Chem. Soc. Dalton Trans.* 894, 1981)

Reducing agent	k_{Cl}/k_{aqua}
$[Cr(aq)]^{2+}$	2×10^3
$[Fe(aq)]^{2+}$	$\rangle\ 3 \times 10^2$
$[Fe(CN)_6]^{4-}$	33
$[Ti(aq)]^{3+}$	31

(**Hints:** ● ISET: The reductants $[Cr(aq)]^2$ ($\sigma^* \rightarrow \sigma^*$) and $[Fe(aq)]^{2+}$ ($\pi^* \rightarrow \sigma^*$) for the chlorido complex. ● For the aqua-complex, both ISET (Cl, a better bridging ligand than H_2O) and OSET for the reductants $[Cr(aq)]^{2+}$ and $[Fe(aq)]^{2+}$ ● OSET: both the chlorido- and aqua- complexes for $[Fe(CN)_6]^{4-}$ (substitutionally inert) and $[Ti(aq)]^{3+}$ ($\pi^* \rightarrow \sigma^*$ net transfer, requiring an electronic activation of Ti(III) in the ISET path).
92. Reduction of $[Co(NH_3)_5X]^{2+}$ (X = F, Cl, Br, I) by $[Eu(aq)]^{2+}$ gives the rate constant order as follows:
k ($dm^3\ mol^{-1}\ s^{-1}$) = 2.6×10^4 (F) 〉 4×10^2 (Cl) 〉 2.5×10^2 (Br) 〉 1.1×10^2 (I)
(**Hints:** Thermodynamic stabilities: EuF^+ 〉 $EuCl^+$ 〉 $EuBr^+$ 〉 EuI^+; ISET path; X^- is transferred to the solution; extreme lability of EuX^{2+} leads to its hydrolysis).
93. Reduction of $[Co^{III}(NH_3)_5X]$ (X = NH_3, H_2O, halide, nitrate, etc.) by $[Cr(bpy)_3]^{2+}$ and $[Ru(NH_3)_6]^{2+}$ occurs exclusively through the OSET path.

CHAPTER 7

Electronic Spectra of Metal Complexes

7.1 INTRODUCTION TO ELECTROMAGNETIC RADIATION AND ABSORPTION SPECTRUM

When an electromagnetic radiation is allowed to pass through a matter, the radiation may be partly absorbed and the remaining portion of the radiation will be transmitted. The portion absorbed can be identified through the analysis of the radiation (obtained after interaction with the matter) by a prism. This analysis shows a spectrum with gaps at certain wavelengths where the interacting matter has absorbed the incident electromagnetic radiation. Such an analysed spectrum is called **absorption spectrum** where absorption versus wavelength or frequency is plotted (Fig. 7.1.1).

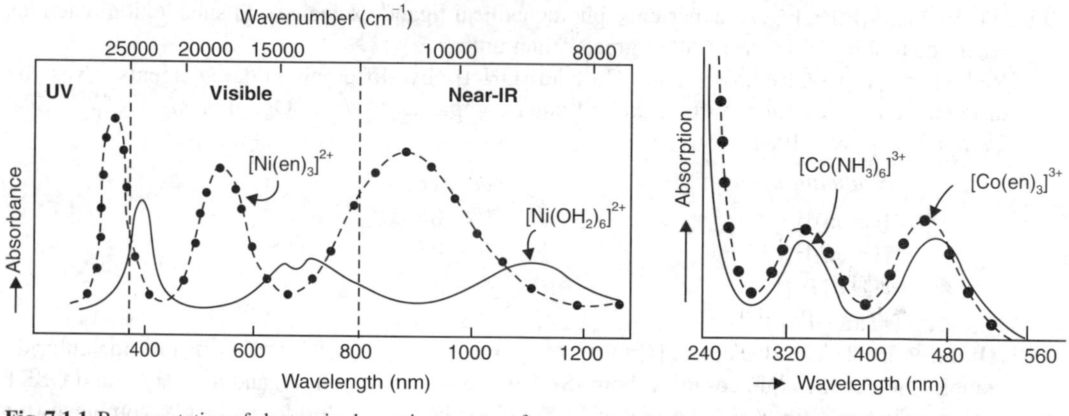

Fig. 7.1.1 Representation of electronic absorption spectra of some representative complexes like $[Ni(OH_2)_6]^{2+}$, $[Ni(en)_3]^{2+}$, $[Co(NH_3)_6]^{3+}$ and $[Co(en)_3]^{3+}$ in aqueous solution. (*See* Fig. 7.14.4.2 for the Ni(II)-complexes; Fig. 7.15.1.3 for the Co(III) complexes).

Depending on the nature of interaction of the electromagnetic radiation with matter, the absorption peaks may be sharp or broad.

During the process of absorption, the molecules or atoms of the matter are excited from a lower energy state to a higher energy state. If the molecules or atoms return to the lower energy state by radiating the excess energy, then an emission spectrum is generated.

The different regions of electromagnetic radiation and the process of interaction with matter are shown in Table 7.1.1.

Table 7.1.1 Range of electromagnetic radiation and process of interaction with matter.

Spectral region	*Wavelength* (nm)	*Wavenumber* (cm^{-1})	*Frequency* ($\times 10^{-12}$ Hz)	*Process of interaction*
(i) γ–radiation	⟨ 0.1	⟩ 10^8	⟩ 3×10^6	Nuclear reactions
(ii) X–rays	0.1 – 10	$10^6 - 10^8$	$3\times10^4 - 3\times10^6$	Transition of inner electrons in atoms or molecules
(iii) Ultraviolet	10 – 360	$2.5\times10^4 - 1\times10^6$	$8\times10^2 - 3\times10^4$	Transition of outer electrons
(iv) Visible	360 – 800	$12.5\times10^3 - 2.5\times10^4$	$4\times10^2 - 8\times10^2$	Transition of outer electrons
(v) Near infrared	800 – 2500	$3\times10^3 - 12.5\times10^3$	100 – 400	Bond vibration
(vi) Infrared	$3\times10^3 - 3\times10^4$	330 – 3300	10 – 100	Bond vibration
(vii) Far-infrared	$3\times10^4 - 3\times10^5$	10 – 330	1 – 10	Molecular rotation
(viii) Microwave	$3\times10^5 - 1\times10^9$	0.01 – 10	$3\times10^{-4} - 1.0$	Molecular rotation
(ix) Radio-frequency	$1\times10^9 - 1\times10^{13}$	$1\times10^{-6} - 1\times10^{-2}$	$1\times10^{-8} - 3\times10^{-4}$	Change of spin

$c = \nu\lambda = 3 \times 10^{10}$ cm s^{-1}; $\bar{\nu}$ (wavenumber) = $1/\lambda$

A. Energy associated with the radiation

- $E = h\nu = hc/\lambda = hc\bar{\nu}$ *i.e.* $\bar{\nu} \Rightarrow E = hc\bar{\nu}$
- $\bar{\nu} = \frac{1}{\lambda}$, 1 kilokaiser(kK) = 1000 cm^{-1}
- 1 kK per molecule $= 10^3$ cm^{-1} molecule^{-1} $\equiv hc \times 1000$ cm^{-1} molecule^{-1}

$$= 6.626 \times 10^{-34} \text{ J s} \times 3 \times 10^8 \text{ m s}^{-1} \times 10^3 \times 10^2 \text{ m}^{-1} \text{ molecule}^{-1}$$
$$= 1.987 \times 10^{-20} \text{ J molecule}^{-1}$$
$$= 1.987 \times 10^{-20} \times N_0 \text{ J mol}^{-1} \approx 11.96 \text{ kJ mol}^{-1}$$

$$\boxed{i.e.\ 1 \text{ kK} = 1000 \text{ cm}^{-1} = 11.96 \text{ kJ mol}^{-1}}$$

- 1 eV = 1.602×10^{-19} J; 1 eV molecule^{-1} = 1.602×10^{-19} J $\times N_0$ mol^{-1}

$$= 96.484 \text{ kJ mol}^{-1}$$

Similarly, 1 eV molecule^{-1} = 23.06 kcal mol^{-1}

- Say, λ = 200 nm = 200×10^{-9} m

$$\nu = \frac{c}{\lambda} = \frac{3\times10^8 \text{ m s}^{-1}}{200\times10^{-9} \text{ m}} = 1.5\times10^{15} \text{ Hz}$$

$$E = hc\bar{\nu} = h\nu = 6.626\times10^{-34} \text{ J s}\times1.5\times10^{15} \text{ Hz}$$
$$= 9.94 \times 10^{-19} \text{ J} = 9.94 \times 10^{-19} \times N_0 \text{ J mol}^{-1}$$

$$= 9.94 \times 10^{-19} \times 6.023 \times 10^{23} \text{ J mol}^{-1}$$

$$= \frac{60.2 \times 10^4}{10^3} \text{kJ mol}^{-1} = 602 \text{ kJ mol}^{-1}$$

i.e. electromagnetic radiation of 200 nm corresponds to 602 kJ mol^{-1}

B. Colour observed

When visible light (380 – 750 nm) is absorbed by a matter, the observed colour of the matter is given by the transmitted light which is not absorbed. ***Thus we see the complementary colour.*** This is illustrated below.

λ (nm)	*$\bar{\nu}$ (kK)*	***Colour of the absorbed light***	***Colour of the transmitted light (i.e. complementary colour)***
380 – 450	26.3 – 22.2	Violet	Yellow-Green
450 – 495	22.2 – 20.0	Blue	Yellow
495 – 570	20.2 – 17.5	Green	Purple
570 – 590	17.5 – 16.9	Yellow	Blue
590 – 620	16.9 – 16.1	Orange	Green-blue
620 – 750	16.1 – 12.5	Red	Blue-green

Thus the yellow colour of K_2CrO_4 solution is due to the absorption of blue light and the blue colour of $CuSO_4$ solution arises due to the absorption of yellow colour.

The relationship between the absorbed colour and the observed colour (*i.e.* complimentary colour) is represented in **artist's wheel** (Fig. 7.1.2).

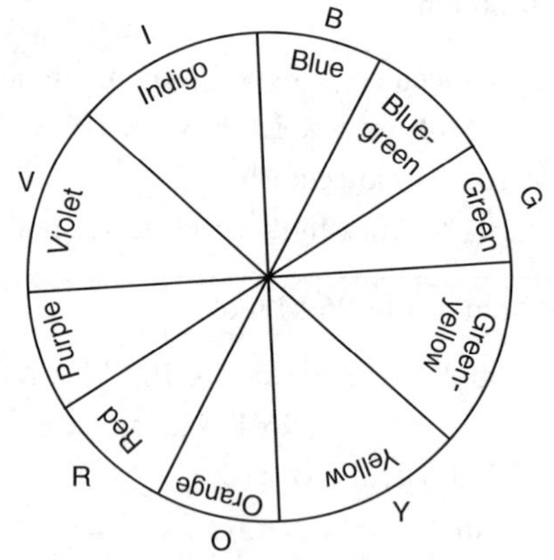

Fig. 7.1.2 Representation of the **artist's wheel** indicating the colours and their corresponding complementary colour.

By considering the absorption spectrum and the **artist's wheel,** we can predict the colour of a substance. The absorption spectra of $CuSO_4$ in presence of excess NH_3 solution and aqueous solution of $NiSO_4$ are given in Fig. 7.1.3.

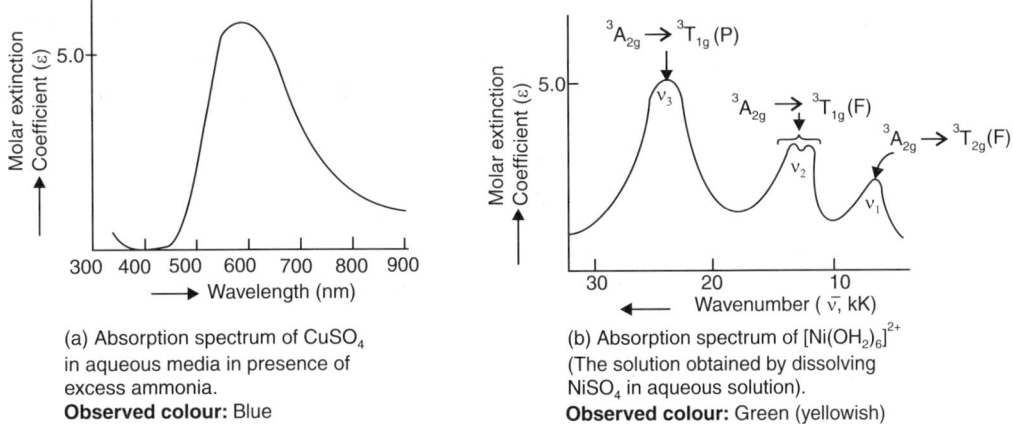

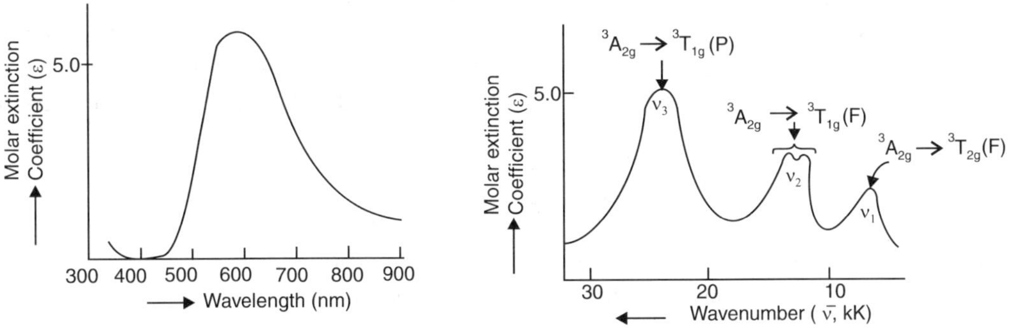

(a) Absorption spectrum of $CuSO_4$ in aqueous media in presence of excess ammonia.
Observed colour: Blue

(b) Absorption spectrum of $[Ni(OH_2)_6]^{2+}$ (The solution obtained by dissolving $NiSO_4$ in aqueous solution).
Observed colour: Green (yellowish)

Fig. 7.1.3 Understanding of the observed colour of (a) aqueous slution of $CuSO_4$ in presence of excess NH_3; (b) aqueous solution of $NiSO_4$ in terms of their absorption spectra (*cf.* Fig. 7.1.1).

- **Cu^{2+} and excess NH_3 solution:** The strong absorption occurs above 450 nm and the absorption maximum lies in the range of 600 nm (*i.e.* yellow light). Thus it looks blue. In fact, the solution absorbs minimum in the range 380 – 500 nm (*i.e.* blue light).

- **Aqueous solution of $NiSO_4$:** Spectrum of $NiSO_4$ in aqueous solution represents the electronic transitions in $[Ni(OH_2)_6]^{3+}$. The transitions (*cf.* Figs. 7.1.1, 3) are:

strongest absorption band at 26,000 cm^{-1}; the band at 14,000 cm^{-1} is actually a doublet one (15,200 cm^{-1}; 13,800 cm^{-1}) and the third band at 8,500 cm^{-1} (near infrared region).

Thus the absorption occurs in three regions: (i) strongest band at around 26,000 cm^{-1} (*i.e.* violet region) giving the **complementary colour: yellow-green;** (ii) second absorption band at around 14,000 cm^{-1} (*i.e.* red region) giving the **complementary colour: blue-green;** (iii) third absorption band at around near infrared region (at about 8,000 cm^{-1}) contributes nothing to the observed colour.

Thus, the resultant colour is determined by the absorption bands at 26,000 cm^{-1} and 14,000 cm^{-1} and the **solution looks green (yellowish).** The solution absorbs minimum in the region of blue-green (*i.e.* around 20,000 cm^{-1}).

Spectra in solids — polarisation studies, diffuse reflectance spectrum: Generally, molecules of lower symmetry (compared with that of cubic symmetry) possess the **dichroic properties,** *i.e.* **anisotropic absorption spectra.** It arises so when the electronic transition is allowed in certain directions. A beam of polarised light is used to record the absorption spectra in different crystal orientations. By knowing the orientation of the molecules with respect to the crystal axes, polarisation properties of the electronic spectra are determined. **Thus the single crystal may be studied with the polarised light in different directions to identify the allowed electronic transitions in different directions.** *Obviously, the question of polarisation does not arise if the studies are carried out in a solution phase because in the solution phase, the molecules are randomly oriented with respect to the incident light beam.*

For recording the **diffuse reflectance spectrum,** the sample is ground finely and is generally mixed with a suitable white material like $MgCO_3$ powder and the light reflected from the mixture is examined to identify the absorption bands. It is not possible to obtain the molar extinction coefficient in this method of study. This method is generally practised when the sample is insoluble or unstable in solution.

7.2 LAWS OF LIGHT ABSORPTION: LAMBERT-BEER LAW

When a beam of light is incident on a homogeneous solution, a part of the light is absorbed by the solution and the rest is transmitted (ignoring the reflection by the solution). The absorption of the incident light of a particular wavelength is governed by Lambert-Beer law which may be expressed in the following form.

$$\log\left(\frac{I}{I_0}\right) = \log T = -\varepsilon cl, \text{ or } A = \log\left(\frac{1}{T}\right) = \varepsilon cl$$

where I_0 = intensity of the incident radiation of a particular wavelength.

I = intensity of the transmitted light of the same wavelenth.

$T = \frac{I}{I_0}$ = transmittance (dimensionless)

$A = \log(1/T) = \log\left(\frac{I_0}{I}\right)$ = absorbance or optical density (OD) (**dimensionless**)

l = path-length through which the incident radiation passes through the absorbing solution.

ε = molar absorptivity or molar absorption coefficient or molar extinction coefficient.

dimension: $(\text{concentration})^{-1}\,(\text{length})^{-1} = (\text{mol dm}^{-3})^{-1}(\text{cm})^{-1} = \text{dm}^3\ \text{mol}^{-1}\ \text{cm}^{-1}$.

$$= 10^3\ \text{cm}^3\ \text{mol}^{-1}\ \text{cm}^{-1} = \text{cm}^2 \times \frac{1}{10^{-3}\,\text{mol}}$$

$$= \text{cm}^2\ \text{mmol}^{-1}$$

The higher value of ε indicates the higher absorbance.

7.3 TYPES OF ELECTRONIC SPECTRA OF TRANSITION METAL COMPLEXES

(i) **Ligand Field Spectra,** ***i.e.*** ***d–d*** **spectra for the transition metal complexes:** For the transition metal ions, in presence of the ligands, the degeneracy of the *d*–orbitals is lifted and these split into different sets of energy levels. In the light of CFT, these different orbital sets are the pure *d*-orbitals though LFT or ACFT considers the incorporation of ligand orbital character into the metal orbital sets. However, these orbitals are predominantly enriched in the character of the *d*–orbitals of metals. Thus both the CFT and LFT describe roughly the same nature of these orbitals. The electron transition occurs among these energetically nondegenerate orbital sets. In terms of the crystal field terminology, these transitions at the metal centre are generally described as the *d–d* transitions.

These *d–d* transitions may cover the range infrared, visible and ultraviolet regions but the bands in the infrared region are not experimentally accessible and the transitions in the ultraviolet region are very often overshadowed by the charge transfer (CT) bands. In the octahedral geometry, the transition may be simply represented as follows:

$$d^n\text{–system: } t_{2g}^{n-m}\, e_g^{\,m} \rightarrow t_{2g}^{n-(m+1)}\, e_g^{m+1}.$$

(ii) **Charge Transfer (CT) Spectra:** In such transitions, an electron transition occurs between the two different moieties, *i.e.* metal and ligand in the complex. These are of two types:

metal to ligand ($M \rightarrow L$), *i.e.* **MLCT**; ligand to metal ($L \rightarrow M$), *i.e.* **LMCT**

In the $M \rightarrow L$ transitions, the metal electrons suffer a transition into the vacant ligand orbital while in the $L \rightarrow M$ transitions, the ligand electrons move into the vacant metal orbitals. This type of electron movement is equivalent to the redox process (*i.e.* in the $M \rightarrow L$ transitions, the metal centre is oxidised and the ligand is reduced while in the $L \rightarrow M$ transitions, the ligand is oxidised and the metal centre is reduced). This is why, the charge transfer (CT) bands are also described as the **redox bands.** Sometimes, the CT band arises in ion-pairs and such bands are called the **IPCT bands.** Metal to metal charge transfer **(MMCT)** bands are important in mixed-valence compounds (*e.g.* **prussian blue**).

In terms of the CFT model, origin of a CT band in a coordination compound cannot be explained but the MOT can explain this. In the terms of MOT, in the $M \rightarrow L$ transitions, the electrons move from the MOs having predominantly the metal orbital character into the vacant MOs primarily constructed by the ligand orbitals. Similarly, in the $L \rightarrow M$CT band, the electrons move from the MOs primarily centred on the ligands to the MOs primarily centred on the metal.

(iii) **Ligand Spectra:** Such spectra arise due to the characteristic absorptions by the ligands themselves, *e.g.* $n \rightarrow \sigma^*$ transition as for water as a coordinated ligand; $\pi \rightarrow \pi^*$ and $n \rightarrow \pi^*$ transition as for the unsaturated ligands. However, these transitions may be partly modified in the complex.

(iv) **Counter-ion spectra:** Sometimes, the characteristic absorption by the counter-ions residing outside the coordination sphere give the counter-ion spectra (*cf.* **IPCT bands**).

7.4 SELECTION RULES FOR THE ELECTRONIC TRANSITIONS

7.4.1 Selection Rules and Illustrations

● **Laporte or orbital selection rule:** The electronic transition having $\Delta l = \pm 1$ are only allowed. The $s \rightarrow s, p \rightarrow p, d \rightarrow d, f \rightarrow f$ transitions having $\Delta l = 0$ are not allowed. In terms of symmetry, the orbitals can be classified as *g* (*gerade* for the presence of centre of inversion) and *u* (*ungerade* for the absence of centre of inversion). The *d* and *s* orbitals are of the *g*–symmetry while the *p*–orbitals are of the *u*–symmetry. In a centrosymmetric environment, electron transition can only occur between the orbitals having the different symmetry properties, *i.e.* ***g → u and u → g transitions are allowed but the g → g and u → u transitions are not allowed.***

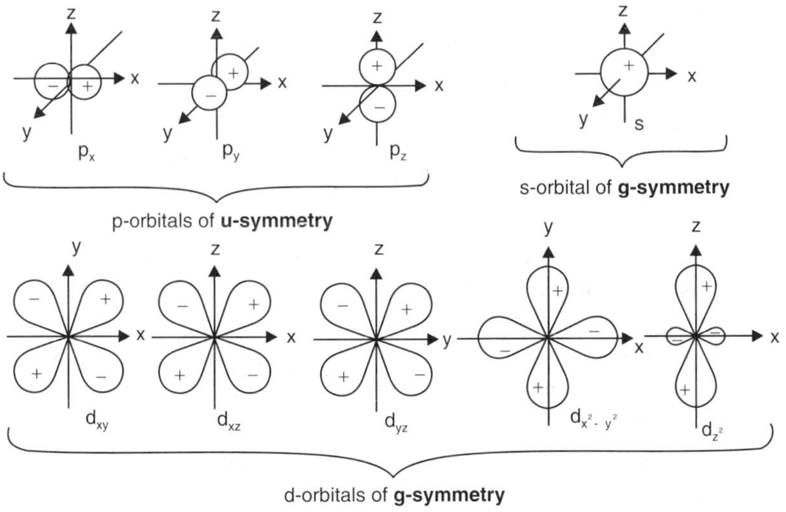

Fig. 7.4.1.1 Symmetry properties of the *s, p–* and *d*–orbitals.

The orbital selection rule is also referred to as the **parity selection rule** which **states that electron transition can only occur between the states of opposite parity.**

Parity is defined in terms of the behaviour of the orbital with respect to inversion of the space coordinates through the origin, *i.e.* x, y, z are replaced by $-x$, $-y$, $-z$. If $\psi(x, y, z) = \psi(-x, -y, -z)$ as in $\cos(x) = \cos(-x)$, then the wave function is described to have the **even parity.** On the other hand, if $\psi(x, y, z) = -\psi(-x, -y, -z)$, then the orbital is described to have the **odd parity**. Obviously, the even parity indicates the presence of centre of inversion (i) and the odd parity indicates the absence of centre of inversion (i). Thus the even parity means the g symmetry and the odd parity means u the symmetry.

If in a complex, there is no centre of inversion, the Laporte selection is not applicable. *Thus the rule is applicable for an octahedral complex but not applicable for a tetrahedral complex.*

- **Spin selection rule:** The electron transition can only occur between the states having the same spin multiplicity, *i.e.* $\Delta S = 0$.

- **Number of electron participating in transition:** One electron transfer at a single step is allowed but more than one electron transfer at a single step is forbidden.

$t_{2g}^{n}\, e_g^{m} \rightarrow t_{2g}^{(n-1)}\, e_g^{(m+1)}$, *i.e.* $1e$ transition **(allowed)**

$t_{2g}^{n}\, e_g^{m} \rightarrow t_{2g}^{(n-2)}\, e_g^{(m+2)}$, *i.e.* $2e$ transition **(forbidden).**

- **Examples of allowed and forbidden transitions:**

The following transition is forbidden.

$t_{2g}^{3}\, e_g^{2}$ (*e.g.* $[Mn(OH_2)_6]^{2+}$) $\rightarrow t_{2g}^{2}\, e_g^{3}$ **(spin forbidden and Laporte forbidden)**

($S = 5/2$, *i.e.* $2S + 1 = 6$) ($S = 3/2$, *i.e.* $2S + 1 = 4$)

The following transition is spin allowed.

$t_{2g}^{1}\, e_g^{0}$ (*e.g.* $[Ti(OH_2)_6]^{3+}$) $\rightarrow t_{2g}^{0}\, e_g^{1}$ **(spin allowed but Laporte forbidden)**

However, both the ***d-d transitions are the Laporte forbidden transitions.***

In the case of $[V(OH_2)_6]^{3+}$, between the following transitions, the transition leading to the simultaneous transition of two electrons is forbidden.

$t_{2g}^{2}\, e_g^{0}$ (*e.g.* $[V(OH_2)_6]^{3+}$) $\rightarrow t_{2g}^{0}\, e_g^{2}$ **(spin forbidden but Laporte forbidden)**

$t_{2g}^{2}\, e_g^{0} \rightarrow t_{2g}^{1}\, e_g^{1}$ **(spin allowed but Laporte forbidden).**

- **Reason behind the orbital selection rule:** When an electron interacts with the electric field of a light wave, it will move in the direction of the field. ***This movement of an electronic charge will cause a change in the dipole moment.*** Thus the interaction between the incident light and electron is only possible if it can cause a change in the dipole moment.

If an electron moves from a s–orbital to another s–orbital (say $1s \rightarrow 2s$), then spherical charge distribution pattern remains unchanged and this migration cannot create any dipole moment. Consequently, the ***s → s transition is not allowed.***

Similarly, in the ***p → p transition*** (say, $2p \rightarrow 3p$), the charge distribution pattern remains unchanged and consequently this transition cannot make any dipole moment. This is why, this transition is also forbidden. In general, the transitions possessing $\Delta l = 0$ are forbidden.

Now let us consider the ***s → p transition*** (say, $1s \rightarrow 2p_z$ when the electric field of the incident radition lies along the z–direction). In this transition, the charge distribution pattern changes (*i.e.* spherical to dumbell-shaped) to cause a change in the dipole moment. Thus the ***s → p transition (i.e. $\Delta l = -1$) is allowed.***

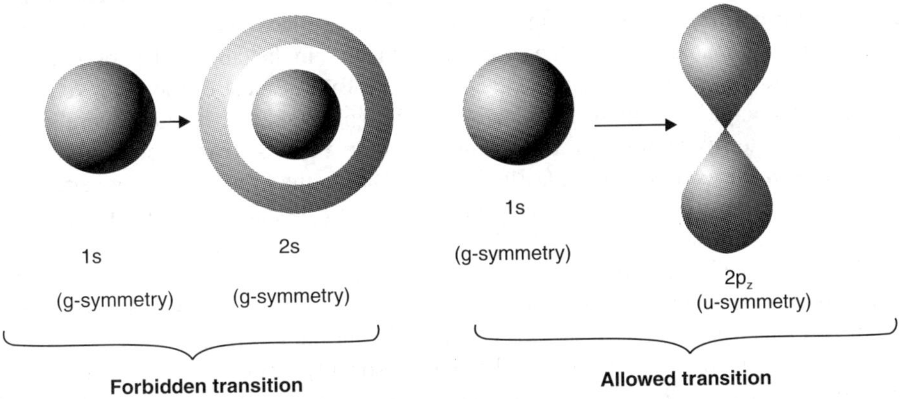

Fig. 7.4.1.2 Illustration of Laporte-selection rule.

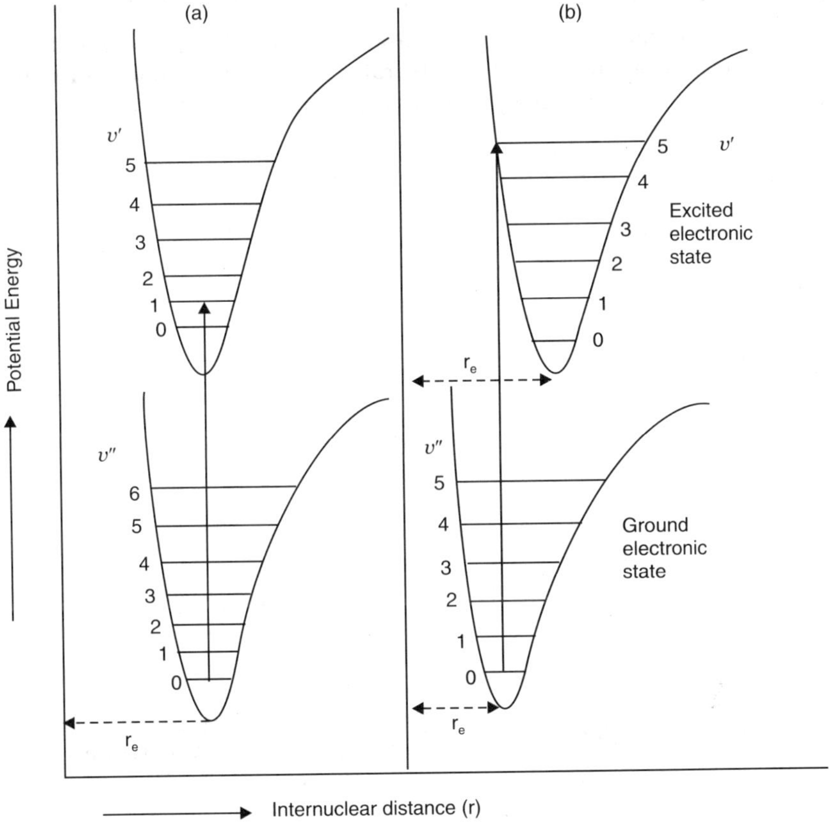

Fig. 7.4.1.3 Illustration of Franck-Condon **vertical transition** by considering the **Morse potential energy curve** (*i.e.* plot of potential energy *versus* internuclear separation). (a) r_e (ground state) = r_e (excited state), (b) r_e (ground state) $\langle$ r_e (excited state) where r_e denotes the equilibrium internuclear separation.

● **Franck-Condon transition and internuclear separation:** If the **internuclear separation** at equilibrium (r_e) remains the same for both the ground and excited electronic state then ($\upsilon'' = 0 \rightarrow \upsilon' = 0$) transition is the most probable one as in the case of *f–f* transition in the lanthanides for which the deeply seated 4*f* levels hardly participate in metal-ligand bonding. But, in most of the cases (*e.g.* $t_{2g} \rightarrow e_g$ in transition metal O_h complexes), the internuclear separation in the excited state is generally longer than that in the ground state. **In such cases, the Franck-Condon vertical transition leads to the higher vibrational levels in the excited state (*cf.* Fig. 7.4.1.3).** During the Franck-Condon vertical transition, the quantum mechanical restriction (if any) for $\upsilon'' \rightarrow \upsilon'$ must be taken into consideration, *i.e.* for a particular value of *r*, all the possible vertical transitions between the vibrational levels of the two electronic states may not be quantum mechanically allowed (*cf.* condition of vibronic coupling).

The time taken for an electronic transition is much smaller than the time taken to execute a vibration to change the internuclear separation. In other words, during the electronic transition (*i.e.* transition from one electronic state to another), the nuclei can be reasonably assumed to remain fixed because the ***electronic transition is much faster than the molecular vibration.*** This is the basis for **Franck-Condon principle** and **it demands the vertical transition** as shown in Fig. 7.4.1.3.

7.4.2 Quantum Mechanical Explanation of the Orbital Selection Rule

Any electronic transition must be associated with a change in the dipole moment in the system. Mathematically, the **transition dipole moment integral** or in short, **transition moment integral** (I) for one-electron transition from the orbital ψ_m to the orbital ψ_n must be nonzero, *i.e.*

$$I = \int_{-\infty}^{+\infty} \psi_m \mu \psi_n d\tau \neq 0$$

where μ = **dipole moment operator which is an odd function.**

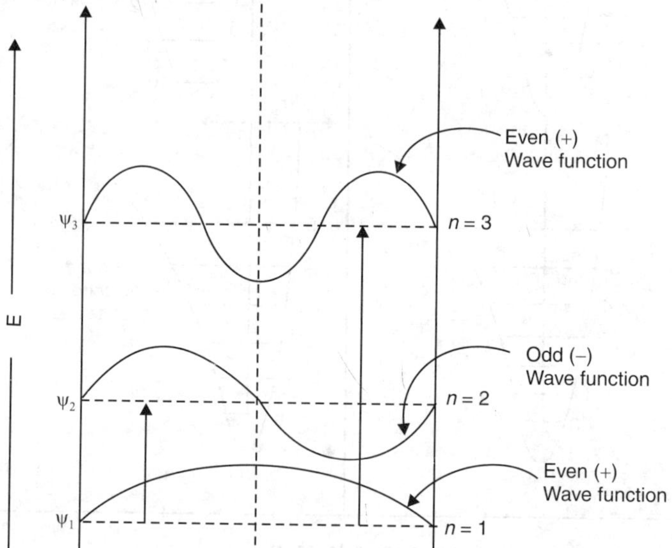

Fig. 7.4.2.1 Even and odd wave functions and allowed electronic transitions.

Odd and even (parity) function in terms of centre of inversion

Even (parity) function: $\psi(x, y, z) = \psi(-x, -y, -z)$; *e.g.* $\cos x = \cos(-x)$

Odd (parity) function: $\psi(x, y, z) = -\psi(-x, -y, -z)$; *e.g.* $\sin x = -\sin(-x)$

Now let us consider the electron (to experience the electronic transition) to reside as a particle in a one-dimensional box where the nucleus is placed at the centre of the box (Fig. 7.4.2.1). Here the dipole moment given by 'ex' where e = charge of the electron and x = distance between the electron and nucleus ***changes its sign at the mid point of the box.*** **Thus, the dipole moment operator (μ) is an odd function.**

The result of integration over an odd (*i.e.* anti-symmetric) function is zero while the integration over an even (*i.e.* symmetric) function is nonzero. ***Thus to make the transition moment integral (I) nonzero, the integrand must be overall an even function.***

Properties of the odd (*i.e. u*) and even (*i.e. g*) functions

- The total integration of an **odd function** is zero because the result of integration over the one-half of the function will produce the same result with an opposite sign obtained from the integration over the other half of the function.
- The total integration of an even function is nonzero.

Physical significance of the transition moment integral: When an electron interacts with the electric field of the incident light wave, the electron will move in the direction of the field. This movement of an electronic charge will cause a change in the dipole moment. Thus the *transition dipole moment integral (I)* gives a measure of the impulse created by an electron transition to the electromagnetic field of light. A large impulse indicates an intense transition and a zero impulse indicates a forbidden transition.

As the dipole moment operator (μ) is an odd function, the product of ψ_m and ψ_n must be an odd one so that (odd) × (odd) becomes even. This is why, if ψ_m is a symmetric wave function (*i.e.* g) then ψ_n must be an antisymmetric wave function (*i.e.* u) because (even) × (odd) becomes odd. ***Thus in an electronic transition, between the orbitals involved, if one orbital is a symmetric one then the other orbital must be an antisymmetric one.*** This is why, the electronic transition between the wave functions having the same l values are fobidden, *i.e.* $s — s$, $p — p$, $d — d$, $f — f$ transitions are forbidden but the $s — p$, $p — d$ and $d — f$ transitions are allowed. Here it may be noted that the p–orbitals are odd wave functions (*i.e.* u–symmetry) while the d–orbitals are even wave functions (*i.e.* g–symmetry).

7.4.3 Quantum Mechanical Explanation for the Forbidden Many-electron Transition

For many-electron transition, the transition moment integral $\int \psi_m \mu \psi_n \partial\tau$ becomes zero when ψ_m and ψ_n are expressed as the products of one-electron wave functions. **It happens so because the one-electron wave functions are mutually orthogonal.** This is why, in general, many-electron transitions are forbidden.

Note: Here it may be mentioned that the many-electron wave functions cannot be strictly expressed as the products of one-electron wave functions. In fact, they are the linear combinations of the one-electron wave functions. Thus, the integral I is nonvanishing but vanishingly small. This is why, the two-electron transitions are noted with weak intensities in some cases as in $[V(aq)]^{2+}$.

7.4.4 Quantum Mechanical Explanation of the Spin Selection Rule

To understand this selection rule, the total wave function is to be expressed in terms of orbital wave function (ψ^o) and spin wave function (ψ^s).

i.e. $\psi_{total} = \psi_{orbital} \times \psi_{spin} = \psi^o \times \psi^s$.

It leads to:

$$I = \int \psi_m \mu \psi_n \partial\tau$$
$$= \int \psi_m^o \psi_m^s \mu \psi_n^o \psi_n^s \partial\tau$$

The **dipole moment operator (μ)** has nothing to do with the spin wave function and we can express the above integral as follows:

$$I = \int \psi_m^o \mu \psi_n^o \partial\tau \int \psi_m^s \psi_n^s \partial\tau$$

Here the second integral involving the spin wave functions (*i.e.* $\int \psi_m^s \psi_n^s \partial\tau$) is zero when the spin states differ. Thus for $\Delta S = 0$, I becomes only nonzero, *i.e.* electron transition can occur only between the states having the same spin multiplicity.

7.5 RELAXATION OF THE SELECTION RULES

7.5.1 Relaxation of the Laporte Selection Rule: Vibronic Coupling, d–p Mixing and Symmetry Lowering

In the transition metal complexes, the electronic transitions like $t_{2g} \rightarrow e_g$ (in the O_h complexes), $e \rightarrow t_2$ (in the T_d complexes) basically involve the $d \rightarrow d$ transitions. These are the Laporte forbidden transitions and they should be colourless. But in reality, many octahedral and tetrahedral complexes like $[Ti(OH_2)_6]^{2+}$, $[Ni(NH_3)_6]^{2+}$, $[CoCl_4]^{2-}$, etc. are sufficiently coloured to be characterised even by our naked eyes. Actually, the Laporte selection rule may be relaxed in different ways. These are discussed below.

Electronic Energy State and Vibronic Transition

In an electronic state, there are many vibrational levels (*see* Fig. 7.4.1.3). Distribution of the molecules in the different vibrational energy levels is governed by the Maxwell-Boltzmann distribution law:

$$N_{ex} = N_0 \exp(-E/k_B T)$$

N_{ex} and N_0 denote the number of molecules in the excited state and total number of molecules respectively, T is the temperature in Kelvin scale, k_B is the Boltzmann constant. At room temperature, the thermal energy measured by $k_B T$ is about 200 cm^{-1}. ***Thus at room temperature, upper or excited electronic states are seldom populated thermally and almost all the molecules are in the lowest vibrational state of the ground electronic state.***

The vibrational transitions require much lower energy than that required for an electronic transition. ***Thus the vibrational transitions are expected to be mixed up with the electronic transition. This gives the vibronic transition.***

(A) Vibronic coupling (Van-Vleck) in centrosymmetric systems: In a centrosymmetric system, *e.g.* $[Ni(NH_3)_6]^{3+}$, $[Co(NH_3)_6]^{3+}$, etc. all the *d*–orbitals are considered to have the *g*–symmetry (*i.e.* gerade symmetry) and the $d \rightarrow d$ (*i.e.* $g \rightarrow g$) transitions are forbidden. However, this restriction can be relaxed through the coupling of electronic and vibrational wave functions. This coupling is described as **vibronic coupling** ('*vib*' from the term *vibrational* and '*ronic*' from the term *electronic*). The metal-ligand bonds are always vibrating. These vibrations may be both *symmetric* (*i.e.* *g*–symmetry or even function) and *antisymmetric*–(*i.e.* *u*–symmetry or odd function) with respect to the centre of inversion (*i*). Each electronic energy level is associated with the several vibrational energy levels (Fig. 7.4.1.3). During **the vibronic transition, the transition probability** between the lower state (say ground state) and upper state (say, excited state) can be determined by considering the vibronic wave functions ($\psi_{vibronic}$) of the two states.

$$\psi_{vibronic} = \psi_{el} \times \psi_{vib}$$

[**Note:** The total wave function (ψ) is the product of an electronic wave function (ψ_{el}), a vibrational wave function (ψ_{vib}) and a rotational wave function (ψ_{rot}) (*cf.* **Born-Oppenheimer approximation** which states that none of these three components of the complete wave function are interdependent).

$$\psi = \psi_{el}\, \psi_{vib}\, \psi_{rot}$$

The wave functions ψ_{el} and ψ_{vib} are not totally independent but ψ_{rot} can be treated as independent of the other two wave functions, *i.e.* ψ_{el} and ψ_{vib}. Thus, reasonably, we can write the vibronic wave function as follows:

$$\psi_{vibronic} = \psi_{el}\, \psi_{vib}]$$

Thus the vibronic coupling (*i.e.* an electronic transition coupled with the vibrational transition) indicates the **breakdown of the Born-Oppenheimer approximation because the vibronic coupling leads to the conclusion that an electronic transition is not quite independent of a vibrational transition.**

Assuming the vibronic transition to be spin-allowed (*i.e.* $\int \psi_s^{gd}\, \psi_s^{ex}\, d\tau$ is nonzero) and omitting the integral involving the spin wave functions, the following **transition moment integral** (I) can be considered to understand the **transition probability** in a vibronic transition.

$$I = \int \psi_{el}^{gd}\, \psi_{vib}^{gd}\, \mu\, \psi_{el}^{ex}\, \psi_{vib}^{ex}\, d\tau, \text{ ('gd' for ground state, 'ex' for excited state).}$$

To make the transition allowed, the above integral must be nonzero, *i.e.* the product $\psi_{el}^{gd}\, \psi_{vib}^{gd}\, \mu\, \psi_{el}^{ex}\, \psi_{vib}^{ex}$ must be a symmetric function (*g*-symmetry).

In a centrosymmetric system (*e.g.* O_h-complex), ψ_{el} (*i.e.* electronic wave function) is symmetric, *i.e.* both ψ_{el}^{gd} and ψ_{el}^{ex} are of *g*-symmetry.

Thus the product of electronic wave functions of the ground state and excited state is of *g*-symmetry, *i.e.* $\psi_{el}^{gd} \times \psi_{el}^{ex} \Rightarrow g \times g \Rightarrow g$–symmetry (*i.e.* even function).

The dipole moment operator (μ) is itself of *u*–symmetry (*i.e.* odd function). It leads to:

$$\psi_{el}^{gd} \times \psi_{el}^{ex} \times \mu \Rightarrow g \times g \times u \Rightarrow u\text{–symmetry (i.e. odd function)}$$

Thus to make the integral nonzero, the product of vibrational wave functions of the two electronic states must be an odd function, *i.e.*

$$\psi_{vib}^{gd} \times \psi_{vib}^{ex} \Rightarrow u\text{–symmetry (i.e. odd function)}$$

Thus if one vibrational wave function is of g–type then the vibrational wave function of the other electronic state must be of u–type.

Then we can have (assuming ψ_{vib}^{gd} of *g*-symmetry):

$$\underbrace{\psi_{el}^{gd} \times \psi_{vib}^{gd} \times \mu}\times \underbrace{\psi_{el}^{ex} \times \psi_{vib}^{ex}} \Rightarrow (g \times g \times u) \times (g \times u) \Rightarrow u \times u \Rightarrow g\ (\text{i.e. even function})$$

Thus, the transition is possible only when the g-type vibration of the ground electronic state is coupled with the u–type vibration of the excited electronic state and vice-versa.

At the lowest vibrational state (*i.e.* $v = 0$) *of an electronic state,* ψ_{vib} *is totally symmetric* (*i.e. g-type symmetry*). Thus, if we consider the transition from the lowest vibrational state (*i.e.* ψ_{vib}^{gd} is of *g*-type symmetry) in an octahedral complex (a centrosymmetric system), then for the nonzero value of the transition moment integral, ψ_{vib}^{ex} must be of *u*-type symmetry because ψ_{vib}^{gd} is of *g*-type symmetry. Thus the following product leads to a symmetric or even function.

$$\underset{\text{(Allowed vibronic transitions)}}{\psi_{el}^{gd} \times \psi_{vib}^{gd} \times \mu \times \psi_{el}^{ex} \times \psi_{vib}^{ex}} \Rightarrow (g \times g \times u) \times (g \times u) \Rightarrow g \text{ (i.e. even function)}$$

In an octahedral complex, the T_{1u} and T_{2u} normal modes of vibration can destroy the centre of symmetry (*i*), *i.e.* these vibrations are of *u*-symmetry. This is why, in the octahedral (O_h) complexes, the electronic transitions are allowed through the coupling of a T_{1u} or T_{2u} vibration (*i.e.* simultaneous excitation to a T_{1u} or T_{2u} vibration).

ψ_{vib} is always totally symmetric (*i.e.* *g*-symmetry) for the lowest vibration level (*i.e.* $v = 0$) in an electronic state (*e.g.* ψ_{vib}^{gd} is of A_{1g} symmetry for O_h). Thus we can conclude that the electronic transition coupled with the **$v^{gd} = 0 \rightarrow v^{ex} = 0$ vibrational transition is not allowed** because both ψ_{vib}^{gd} and ψ_{vib}^{ex} are of *g*-symmetry, *i.e.* the product $\psi_{vib}^{gd} \times \psi_{vib}^{ex}$ is of *g*-symmetry. **Thus the 0 → 0 vibrational transition makes the transition moment integral zero** when the electronic transition is itself forbidden.

$\psi_{vib}^{gd} \times \psi_{vib}^{ex} \Rightarrow g \times g \Rightarrow g$-symmetry; $\psi_{el}^{gd}\ \mu\ \psi_{el}^{ex} \Rightarrow g \times u \times g \Rightarrow u$-symmetry;

$\psi_{el}^{gd}\ \mu \psi_{el}^{ex} \times \psi_{vib}^{gd}\ \psi_{vib}^{ex} \Rightarrow g \times u \times g \times g \times g \Rightarrow u$-symmetry (for the $0 \rightarrow 0$ transition)

It may be noted that for the **tetrahedral complexes,** the $0 \rightarrow 0$ transition is allowed (**provided** $r_e^{ex} \approx r_e^{gd}$, *cf.* Fig. 7.7.2.1) as the electronic transition is not always forbidden.

Note: The vibrational wave functions ψ_{vib}^{gd} and ψ_{vib}^{ex} of different electronic states are not orthogonal. The vibrational overlap integral, $\int \psi_{vib}^{gd}\ \psi_{vib}^{ex}\, d\tau$ for the vibrational wave functions of the two electronic states is called **Franck-Condon overlap** and square of the vibrational overlap integral is described as the **Franck-Condon factor (FCF) to be discussed later.**

(B) Physical picture of vibronic coupling: Even in a centrosymmetric system, some vibrations can destroy the centre of symmetry (*i*) temporarily. In an octahedral complex, the T_{1u} and T_{2u} vibrations can distort the octahedral complex in such a way that the centre of symmetry is removed temporarily. *When the vibration removes the centre of symmetry, g-character of the electronic states is lost and at that condition, the orbital selection rule, i.e. parity selection rule is not applicable and the electronic transition becomes allowed.*

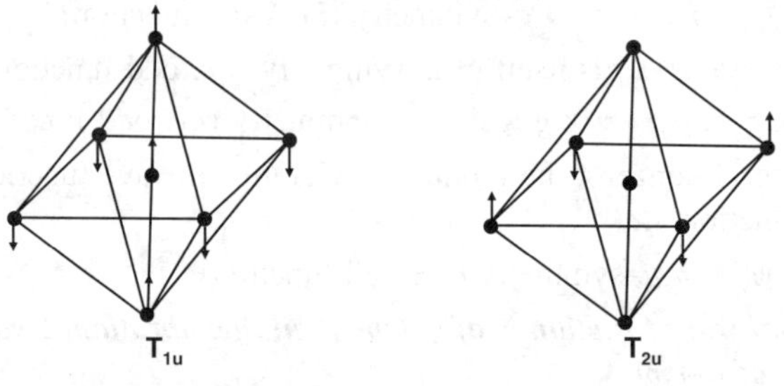

Fig. 7.5.1.1 T_{1u} and T_{2u} normal modes of vibration (destroying the centre of symmetry through the displacement of atoms) in an octahedral complex ML_6 belonging to the O_h point group.

Note: There is another type of T_{1u} vibration (in O_h) removing the centre of symmetry (*see* Appendix 12B).

Here it should be mentioned that the loss of centre of symmetry **(temporarily by an odd vibration and permanently by a ligand environment)** *does not ensure all the electronic transitions allowed* (*e.g.* in a tetrahedral complex, the $A_2 \rightarrow T_1$ transition is itself allowed but the $A_2 \rightarrow T_2$ transition is forbidden; *see* Solved Problem 15). However, loss of centre of symmetry and lowering of symmetry by any mechanism makes the electronic transition more allowed **(in terms of probability),** in general. But, if incorporation of the odd vibrational representation in the transition moment integral produces a **totally symmetric represention A_1 or A_{1g}**, then the transition is definitely allowed. This aspect will be illustrated later.

For a nonlinear n-atomic molecule, there will be $3n - 6$ **normal modes of vibration** and these can be characterised by means of their symmetry properties. For an octahedral complex, the 15 normal modes of vibration are distributed as (*cf.* Table 7.23.3 and Appendix 12B).

$$A_{1g} + E_g + 2T_{1u} + T_{2u} + T_{2g}$$

The odd vibrational modes destroying the centre of sysmmetry of the octahedral complex are T_{1u} and T_{2u} (Fig. 7.5.1.1).

(C) Understanding of vibronic coupling in terms of group theory: The condition for an allowed vibronic transition can be realised in terms of group theory. However, it is easier to understand this in terms of the symmetry arguments. For an allowed vibronic transition, the following transition moment integral must be nonzero.

$$I = \int(\psi_{\text{el}}^{\text{gd}}\,\psi_{\text{vib}}^{\text{gd}})\mu(\psi_{\text{el}}^{\text{ex}}\,\psi_{\text{vib}}^{\text{ex}})\,d\tau$$

For the nonzero value of I, the product $(\psi_{\text{el}}^{\text{gd}}\,\psi_{\text{vib}}^{\text{gd}})\ \mu\ (\psi_{\text{el}}^{\text{ex}}\,\psi_{\text{vib}}^{\text{ex}})$ must be totally symmetric or the product will contain at least one **totally symmetric representation term of the point group** (*e.g.* A_{1g} term for the O_h system). Assuming the vibrational transition from the lowest vibrational state (*i.e.* $v = 0$, at the ground electronic state), **we can conclude $\psi_{\text{vib}}^{\text{gd}}$ to be totally symmetric** (*e.g.* $\psi_{\text{vib}}^{\text{gd}}$ is A_{1g} for O_h). The direct product of $(\psi_{\text{el}}^{\text{gd}}\,\mu\psi_{\text{el}}^{\text{ex}}(\equiv g \times u \times g \Rightarrow u)$ does not contain any term of totally symmetric representation but to make the integral I nonzero, the direct product of $\psi_{\text{el}}^{\text{gd}}\mu\psi_{\text{el}}^{\text{ex}}\psi_{\text{vib}}^{\text{ex}}$ must contain at least one term of totally symmetric representation of the same point group (*e.g.* A_{1g} for O_h). *It is possible, if there is any normal mode of vibration of ψ_{vib}^{ex} whose representation belongs to at least one of the representations resulted from the direct product* $\psi_{\text{el}}^{\text{gd}}\mu\psi_{\text{el}}^{\text{ex}}$. In other words, as the direct product of $\psi_{\text{el}}^{\text{gd}}\mu\psi_{\text{el}}^{\text{ex}}$ is of u-symmetry (*cf.* $g \times u \times g \Rightarrow u$), then $\psi_{\text{vib}}^{\text{ex}}$ must be of **identical u-symmetry** to make I non-zero.

● **Vibronic coupling in the centrosymmetric O_h-system:** Let us illustrate for the transition, **$^3T_{1g}(F) \rightarrow {}^3T_{2g}$(F) in the d^2–case.** To get the result of direct product, we can consider the Table 7.5.1.1. The derivation of the Table is beyond the scope of the book. For this purpose, the readers may consult any book on Group Theory.

Table 7.5.1.1 Direct products applicable for the tetrahedral (T_d) and octahedral (O_h) complexes.

	A_1	A_2	E	T_1	T_2
A_1	A_1	A_2	E	T_1	T_2
A_2	A_2	A_1	E	T_2	T_1
E	E	E	$A_1 + A_2 + E$	$T_1 + T_2$	$T_1 + T_2$
T_1	T_1	T_2	$T_1 + T_2$	$A_1 + E + T_1 + T_2$	$A_2 + E + T_1 + T_2$
T_2	T_2	T_1	$T_1 + T_2$	$A_2 + E + T_1 + T_2$	$A_1 + E + T_1 + T_2$

By using the above Table, we get the results of direct product of T_{1g} and T_{2g}.

$$T_{1g} \times T_{2g} = A_{2g} + E_g + T_{1g} + T_{2g}$$

In the O_h system, the Cartesian coordinate axes have the same symmetry property resembling the symmetry property of the p–orbitals. It is T_{1u} which is the symmetry property of the dipole moment component, *i.e.* $\mu_x = \mu_y = \mu_z = \mu$ transforms into the T_{1u} species (*see* Character Table).

$$T_{1g} \times \mu \times T_{2g} = T_{1g} \times T_{1u} \times T_{2g} = T_{1g} \times T_{2g} \times T_{1u} = (A_{2g} + E_g + T_{1g} + T_{2g}) \times T_{1u}$$
$$= A_{2g} \times T_{1u} + E_g \times T_{1u} + T_{1g} \times T_{1u} + T_{2g} \times T_{1u}$$
$$= (T_{2u}) + (T_{1u} + T_{2u}) + (A_{1u} + E_u + T_{1u} + T_{2u}) + (A_{2u} + E_u + T_{1u} + T_{2u})$$
$$= A_{1u} + A_{2u} + 2E_u + 3T_{1u} + 4T_{2u}$$

Thus for the transition, $^3T_{1g} \rightarrow {}^3T_{2g}$ in the octahedral d^2–case, all the terms of the direct product $T_{1g} \times T_{1u} \times T_{2g}$ ($\equiv \psi_{el}^{gd} \times \mu \times \psi_{el}^{ex}$) are of *u*-symmetry, as expected. If any one of the terms present in the direct product $T_{1g} \times T_{1u} \times T_{2g}$ matches in terms of symmetry with a normal mode of vibration of an O_h-system, then the direct product of $\psi_{el}^{gd}\mu\psi_{el}^{ex}\psi_{vib}^{ex}$ (assuming ψ_{vib}^{gd} to be totally symmetric, *i.e.* $^1A_{1g}$ in O_h) will contain the term which is totally symmetric (*i.e.* A_{ig} for O_h) and I becomes nonzero. The normal modes of vibration of an octahedral (O_h) complex are of the following symmetries (*cf.* Table 7.23.3).

$$A_{1g}, E_g, 2T_{1u}, T_{2g}, T_{2u}$$

Thus the T_{1u} and T_{2u} vibrations match with the terms obtained from the direct product $T_{1g} \times T_{1u} \times T_{2g}$ ($\equiv \psi_{el}^{gd}\mu\psi_{el}^{ex}$) for the $T_{1g} \rightarrow T_{2g}$ transition. ***This is why, the electronic transition is vibronically allowed through the possible coupling of the T_{1u} and T_{2u} normal mode of vibration*** (*i.e.* ψ_{vib}^{ex} must be of T_{1u} and/or T_{2u} symmetry) and then the direct $\psi_{el}^{gd}\mu\psi_{el}^{ex}\psi_{vib}^{ex}$ will contain the totally symmetric A_{1g} term (*irreducible representation*) (*cf.* both the direct products $T_{1u} \times T_{1u}$ and $T_{2u} \times T_{2u}$ contain the A_{1g} term; *see* Table 7.5.1.1).

In other words, we may conclude that the direct product $\psi_{el}^{gd}\mu\psi_{el}^{ex}$ for an electronic transition in an octahedral system will produce the terms of *u*-symmetry (*cf.* $g \times u \times g \Rightarrow u$) and at least there must be one term of T_{1u} or T_{2u} symmetry in the direct product and then coupling of the T_{1u} or T_{2u} symmetry vibration (ψ_{vib}^{ex}) will make the electronic transition vibronically allowed. **If there is no T_{1u} or T_{2u} term in the direct $\psi_{el}^{gd}\mu\psi_{el}^{ex}$ (electronic transition in an O_h system), then the transition is not vibronically allowed.** Here it is worth mentioning that in making the above conclusion, it has been assumed that the electronic transition occurs from $\upsilon = 0$ (*i.e.* lowest vibrational level at the ground electronic state) for which ψ_{vib}^{gd} is totally symmetric (*i.e.* $^1A_{1g}$ for O_h). Here we shall illustrate this principle by taking some examples.

Table 7.5.1.2 Illustration of the allowdness of electronic transition assuming the transition from $\upsilon = 0$ (lowest vibrational level of the ground electronic state, *i.e.* ψ_{vib}^{gd} is totally symmetric).

	Transition ($\psi_{el}^{gd} \rightarrow \psi_{el}^{ex}$)	***Direct product*** $\Gamma(\psi_{el}^{gd})\ \Gamma\ \mu\ \Gamma\ (\psi_{el}^{ex})$	***Conclusion***
Electronic transitions in the low-spin octahedral Co(III)-complex, *e.g.* $[Co(NH_3)_6]^{3+}$	$^1A_{1g} \rightarrow {}^1T_{1g}$	$A_{1g} \times T_{1u} \times T_{1g} = A_{1u} + E_u + T_{1u} + T_{2u}$	The direct of $\psi_{el}^{gd}\mu\psi_{el}^{ex}$ contains the term T_{1u} and/or T_{2u}; the electronic transitions are vibroni-cally allowed through the coupling of T_{1u} and/or T_{2u} vibration of ψ_{vib}^{ex}. ***($T_{1u} \times T_{1u}$ or $T_{2u} \times T_{2u}$ always contains the A_{1g} term in the direct product result).***
	$^1A_{1g} \rightarrow {}^1T_{2g}$	$A_{1g} \times T_{1u} \times T_{2g} = A_{2u} + E_u + T_{1u} + T_{2u}$	
Electronic transition in the octahedral d^1-system, *e.g.* $[Ti(OH_2)_6]^{3+}$	$^2T_{2g} \rightarrow {}^2E_g$	$T_{2g} \times T_{1u} \times E_g = A_{1u} + A_{2u} + 2E_u + 2T_{1u} + 2T_{2u}$	
Electronic transition in $[Cr(NH_3)_6]^{3+}(d^3, O_h)$	$^4A_{2g} \rightarrow {}^4T_{2g}$	$A_{2g} \times T_{1u} \times T_{2g} = A_{1u} + E_u + T_{1u} + T_{2u}$	
	$A_{2g} \rightarrow A_{2g}$	$A_{2g} \times T_{1u} \times A_{2g} = T_{1u}$	

● **Vibronic coupling for the $A_2 \rightarrow T_2$ Transition in the noncentrosymmetric T_d system (d^7)**

We have already mentioned that even in a **noncentrosymmetic system** all the possible electronic transitions may not be allowed (*e.g.* $A_2 \rightarrow T_2$ transition in the noncentrosymmetric d^7 T_d-system is forbidden) and the forbidden transitions may be made allowed through the vibronic coupling. The condition is the same, *i.e.*

$\int \psi_{el}^{gd} \mu \psi_{el}^{ex} \psi_{vib}^{ex} d\tau \neq 0$, assuming ψ_{vib}^{gd} to be totally symmetric (*e.g.* A_1 for T_d)

i.e. the above **matrix element** must contain the totally symmetric representation of the group.

In a T_d system, the dipole moment operator (μ) transforms into the T_2 species. The following direct product result is to be analysed for the $A_2 \rightarrow T_2$ **transition.**

$A_2 \times \mu \times T_2 = A_2 \times T_2 \times T_2 = T_1 \times T_2 = A_2 + E + T_1 + T_2$, (by using the Table 7.5.1.1).

The direct product result doest not bear the **totally symmetric irreducible representation A_1** and consequently the transition is forbidden. But, coupling of any vibration of A_2, E, T_1 and T_2 symmetry will make the transition allowed. For the T_d point group, normal modes of vibration (*cf.* Table 7.23.3) belong to $A_1 + E + 2T_2$ *i.e.* E or T_2 vibration can be coupled to make the transition vibronically allowed.

It may be noted that in a T_d-system, the **$A_2 \rightarrow T_1$ transition is allowed by its own right** and it is evident from the following triple product result.

$A_2 \times \mu \times T_1 = A_2 \times T_2 \times T_1 = A_2 \times (A_2 + E + T_1 + T_2) = A_1 + E + T_2 + T_1$ (*cf.* Table 7.5.1.1).

The result contains the **totally symmetric representation term A_1** of the group and it makes the transition moment integral nonzero.

Condition for Vibronic Transition in a Centrosymmetric System

- A suitable combination of the metal-ligand vibrations of the lower state (*i.e.* ground state) and upper state (*i.e.* excited state) can allow the vibronic transition partially allowed even in a centrosymmetric system. Thus in a centrosymmetric octahedral complex, the $d \rightarrow d$ transition (*i.e.* $g \rightarrow g$ transition) becomes partially allowed, if the ground state electronic wave function is combined with a g–type (*i.e.* symmetric) vibrational wave function and the excited state (*i.e.* upper state) electronic wave function is combined with a u–type vibrational wave function and vice-versa. In such cases, the so called $d \rightarrow d$ transition (*i.e.* $g \rightarrow g$) will have the components like $g \rightarrow u$ or $u \rightarrow g$ which make the transition partially allowed. It is reasonable to assume that the electronic transition occurs from the lowest vibrational state ($\upsilon = 0$ at the ground electronic state) that makes ψ_{vib}^{gd}totally symmetric (*i.e.* A_{1g}-symmetry in O_h). ***Under this condition, the electronic transition becomes vibronically allowed through the coupling of an u-type vibration (i.e. ψ_{vib}^{ex} is of u-type symmetry) — so that the matrix element $\langle \psi_{el}^{gd} \times \mu \times \psi_{el}^{ex} \times_{vib}^{ex} \rangle$ contains the totally symmetric representation term of the point group.***
- Vibronic couping is applicable to both the centrosymmetric and noncentrosymmetric systems.

(D) Understanding of peak broadening through vibronic coupling: Here we shall illustrate this aspect by considering the transition, $^3T_{1g}(F) \rightarrow {}^3T_{2g}(F)$ (in a d^2-octahedral complex). The direct product $T_{1g} \times T_{1u} \times T_{2g} (\equiv \psi_{el}^{gd} \times \mu \times \psi_{el}^{ex})$ $(= A_{1u} + A_{2u} + 2E_u + 3T_{1u} + 4T_{2u})$ contains the T_{1u} and T_{2u} terms.

Thus the **vibrations of T_{1u} symmetry and T_{2u} symmetry** when coupled with the electronic transition, $^3T_{1g} \rightarrow {}^3T_{2g}$, the transition becomes vibronically allowed. The final product contains three T_{1u} terms and four T_{2u} terms. Thus, coupling of a set of T_{2u} metal-ligand vibrations splits the peak into 4 components (*i.e.* subpeaks). Similarly, coupling of a set of T_{1u} metal-ligand vibration splits the peak into three components. *In an octahedral complex, there are two distinct sets of T_{1u} vibration and one set of T_{2u} vibration.* Thus coupling of 2 sets of T_{1u} vibration and one set of T_{2u} vibration produces **10 components**

or sub-peaks (= 3 + 3 + 4) of the main peak. These subpeaks are separated by only *ca.* 200 cm^{-1} which represents the energy of metal-ligand bond vibrations. ***Thus the sub-peaks simply broaden the peak. This is the mechanism of peak broadening by vibronic coupling.***

(E) Allowedness of electronic transition in the tetrahedral complexes (noncentrosymmetric systems) in terms of group theory and *d-p* mixing: We have seen that in the octahedral complexes (centrosymmetric systems), the pure electronic transitions are not allowed because the direct product $\psi_{el}^{gd}\mu\psi_{el}^{ex}$ does not contain the totally symmetric term A_{1g}. However, the vibronic coupling (*i.e.* electronic transition coupled with the T_{1u} and /or T_{2u} vibration) can make the transitions allowed. Let us consider, the noncentro-symmetric tetrahedral complexes, *e.g.* $[CoX_4]^{2-}$, X^- = Cl^-, Br^-, SCN^- where the spin-allowed electronic transitions are:

$$^4A_2 \rightarrow {}^4T_1 \text{ and } {}^4A_2 \rightarrow {}^4T_2$$

In a T_d complex, the coordinates *x*, *y* an *z* generate a basis for the T_2 representation (*see* Character Table). **Thus μ transforms as T_2.** To understand the allowedness of the above electronic transitions, we are to consider the results of direct products.

$A_2 \times T_2 \times T_1$ (for $A_2 \rightarrow T_1$ transition) and $A_2 \times T_2 \times T_2$ (for $A_2 \rightarrow T_2$ transition).

$^4A_2 \rightarrow {}^4T_1$ transition: $\psi_{el}^{gd} \times \mu \times \psi_{el}^{ex} \equiv A_2 \times T_2 \times T_1 = A_1 + E + T_1 + T_2$ (*cf.* Table 7.5.1.1).

The **totally symmetric** A_1 irreducible representation is present in the direct product and consequently the **electronic transition is orbitally allowed.**

Now let us consider the **$^4A_2 \rightarrow {}^4T_2$ transition** for which the direct product $A_2 \times T_2 \times T_2$ reduces as follows:

$\psi_{el}^{gd} \times \mu \times \psi_{el}^{ex} = A_2 \times T_2 \times T_2 = A_2 + E + T_1 + T_2$, (*cf.* Table 7.5.1.1.)

Since there is no A_1 **(totally symmetric irreducible representation)** term in the direct product, **the transition is forbidden.** However, it gains some intensity through the vibronic coupling (*i.e.* coupling of *E* and/or T_2 normal mode of vibration of the T_d group). ***This is why, intensity of the $A_2 \rightarrow T_1$ transition is about 10 - 100 times higher than that of the $A_2 \rightarrow T_2$ transition.***

In fact, in a T_d complex, the *p*-orbitals, d_{xy}, d_{yz} and d_{xz} orbitals belong to the T_2 representation and it allows the ***d-p* mixing**. This is why, the electronic transition consists of different allowed components: $d \rightarrow p$ or $p \rightarrow d$. These aspects will be discussed later in detail (*cf.* Sec. 7.5.1G-K).

In the **spectrum of $[CoCl_4]^{2-}$** (in $Cs_2[CoCl_4]$), the 0 → 0 line is the most intense one and it indicates that ***r_e is more or less the same for both the ground ($^4A_2$) state and excited ($^4T_1$) state*** (*cf.* $^4A_2 \rightarrow {}^4T_1$ transition is orbitally allowed) (*cf.* Fig. 7.7.2.1).

Let us illustrate and compare the electronic transitions, $T_{2g} \rightarrow E_g$ **(d^1, d^6; O_h complexes)** and $E \rightarrow T_2$ **(d^1, d^6; T_d complexes).**

- **$T_{2g} \rightarrow E_g(O_h)$:** $\psi_{el}^{gd} \times \mu \times \psi_{el}^{ex} = T_{2g} \times T_{1u} \times E_g = A_{1u} + A_{2u} + 2T_{1u} + 2T_{2u} + 2E_u$

Absence of the totally symmetric representation (A_{1g}) in the direct product makes the transition orbitally forbidden but vibronic coupling (coupling of T_{1u} and/or T_{2u} vibration, *i.e.* ψ_{vib}^{ex} is of T_{1u} or T_{2u} symmetry) makes the transition allowed.

- **$E \rightarrow T_2 (T_d)$:** $\psi_{el}^{gd} \times \mu \times \psi_{el}^{ex} = E \times T_2 \times T_2 = A_1 + A_2 + 2E + 2T_1 + 2T_2$ (*cf.* Table 7.5.1.1). Presence of the totally symmetric representation (A_1) in the direct product makes the transition orbitally allowed.

Note: In an electronic transition, there is ***no restriction on the change of vibrational quantum number*** (*i.e.* $\Delta v = 0, \pm1, \pm2, ...$) for the ***vertical Franck-Condon transition provided the electronic transition (e.g. $A_2 \rightarrow T_1$ in a tetrahedral complex) is itself allowed by its own right.*** Thus, the most probable vibrational transition $v = 0 \rightarrow v = 0$ is possible under the conditions:

(i) Internuclear distances in the ground and excited electronic states are more or less the same, *i.e.* r_e(ground) ≡ r_e(excited).

If r_e(excited) > r_e(ground), then the $0 \rightarrow 0$ transition is no longer the most probable one in terms of **Franck-Condon vertical transition** (*cf.* Fig. 7.4.1.3). In such cases, vibrational transition occurs to the higher vibrational states of the excited electronic state.

(ii) The electronic transition is itself allowed (*i.e.* vibronic coupling is not required to make the transition allowed).

(F) Noncentrosymmetric condition imposed in the octahedral complexes by crystal packing and intrinsic asymmetry in the structure of polyatomic ligands: It has been already mentioned that the asymmetrical vibrations (*i.e.* T_{1u} and T_{2u}) can destroy the centre of symmetry in the O_h-systems and then the Laporte selection rule is not applicable. Sometimes, crystal packing in solid state and intrinsic asymmetry in the structure of polyatomic ligands may lead the O_h complexes to depart from the perfect centrosymmetric condition. It allows the electronic transition.

(G) *d–p* mixing (*i.e.* $g - u$ mixing) in the noncentrosymmetric systems like the T_d complexes: Let us illustrate the case with the tetrahedral system which lacks in centre of symmetry. This is why, the orbital groups t_2 (= d_{xy}, d_{yz}, d_{zx}) and $e\left(= d_{x^2-y^2}, d_{z^2}\right)$ do not possess the g or u properties (more correctly, the MOs of a tetrahedral complex do not possess the g or u property). However, the atomic orbitals from which the t_2 and e stets are derived are possessing the *parity properties, i.e.* the d–atomic orbitals are of g–symmetry.

In a tetrahedral system, in terms of MOT, the e–set is basically nonbonding and these are purely metal d–orbitals $\left(i.e., d_{x^2-y^2} \text{ and } d_{z^2}\right)$. ***On the other hand, the t_2 – MOs are constituted by the overlap of metal d–orbitals (i.e. d_{xy}, d_{yz} and d_{zx}) and metal p-orbitals with the ligand p–orbitals (cf. Fig. 3.17.7.1).*** Thus, the transition, $e \rightarrow t_2$ bears the following components:

$$d(g) \rightarrow d(g) \ (i.e.\ \Delta l = 0),\ d(g) \rightarrow p(u) \ (i.e.\ \Delta l \neq 0),$$

Thus, though the d → d transition component is forbidden but the d → p transition component is allowed. This is why, the intensity of spectral transitions in the tetrahedral complexes is about 10^2 times higher than that of the centrosymmetric octahedral complexes.

Note: In a tetrahedral complex, in terms of MOT, the transition is: $e \rightarrow t_2^*$ and it is actually MLCT (metal → ligand CT), *i.e.* d(metal) → p(ligand).

(H) Characteristics of *d–p* mixing: In a noncentrosymmetric systems, the d–p mixing is possible because in such systems, it is not realistic to consider the d–orbitals as g–type and p–orbitals as u–type. In fact, in a T_d-complex, the metal d_{xy}, d_{yz} and d_{xz} orbitals and p-orbitals of both the metal and ligand belong to the T_2-representation. But, in a centrosymmetric system, the d–p mixing is not possible because of their different symmetry properties, *i.e.* d as g–type and p–as u–type (*cf.* O_h *vs.* T_d).

(I) Possibility of *d–p* mixing in the octahedral complexes: An asymmetric metal-ligand bond vibration can destroy the centrosymmetric property of the complex at that particular moment. At that moment, the metal $(n - 1)d$ orbital can mix with the metal np orbital. This **instantaneous** ***d–p*** **mixing** can relax the Laporte selection rule. But efficiency of this d–p mixing can be reviewed in terms of their energy difference.

d–p Mixing: Energetics of mixing of the metal and ligand p-orbitals with the metal d-orbitals

In the noncentrosymmetric systems, the d-p mixing is possible. In the transition metal complexes, the np orbitals are of much higher energy than the $(n - 1)d$ orbitals. **Thus the intermixing of the metal *np* and (*n* – 1)d orbitals is energetically inefficient.** In fact, for the $3d$- and $4p$ orbitals, it is about only 10%. However, **in terms of MOT, the ligand *p*-orbitals can mix efficiently with the metal *d*-orbitals.**

The $(n-1)d$ orbitals and np orbitals of the metal centre differ significantly in energy and their mixing is not very much efficient. **In fact, for the first transition series, intermixing of the 3d and 4p orbitals is not very much efficient.**

(J) Possibility of *d–p* mixing in terms of MOT in the tetrahedral complexes: The *d–p* mixing, *i.e.* **mixing of the metal *d*–orbitals with the ligand *p*–orbitals** leading to the MOs is quite efficient in the noncentrosymmetric systems like tetrahedral complexes. In a tetrahedral complex, the metal t_2–orbitals (*i.e.* d_{xy}, d_{yz} and d_{zx}) can mix with the ligand *p*–orbitals to produce the t_2-MOs.

In fact, the t_2-MOs of a tetrahedral complex are constructed by combining both the metal $(n-1)d$ and *np* orbitals with the ligand *p*-orbitals (*cf.* Fig. 3.17.7.1). But, due to the energy mismatch (*i.e.* metal *np* orbitals are of much higher energy than the $(n-1)d$ orbitals), participation of the metal *p*–orbitals is relatively less important in constructing the t_2-MOs. This is why, for the sake of simplicity, here we are considering formation of the t_2-MOs through the combination of only metal *d*-orbitals (*i.e.* d_{xy}, d_{yz} and d_{zx}) with the ligand *p*-orbitals.

t_2–MO: $\psi_{MO} = x\psi_d^{t_2} + \sqrt{(1-x^2)}\sum \psi_{LGO}$, x gives the measure of mixing, *i.e.* covalence.

The *e* level is constituted by the pure metal *d*–orbitals (*i.e.* $d_{x^2-y^2}, d_{z^2}$). Thus the $e \rightarrow t_2$ transition leads to the following **transition moment integral** (I).

$$I = \int \psi_d^e \, \mu \, \psi_{MO} \, d\tau, \ \mu = \text{dipole moment operator}$$

$$= x\int \psi_d^e \mu \psi_d^{t_2} d\tau + \sqrt{(1-x^2)} \int \psi_d^e \mu \sum \psi_{LGO} d\tau$$

The first integral consisting of pure $d \rightarrow d$ transition is zero but the second integral is nonzero because of the *d* (metal) $\rightarrow$ *p* (ligand) transition which basically represents a MLCT band.

(K) Relaxation of the Laporte selection rule through the *d–p* mixing in the octahedral *vs* tetrahedral complexes:

- ***The ligand field transition ($t_{2g} \rightarrow e_g$) is forbidden in the octahedral complexes but the ligand field transition ($e \rightarrow t_2$) in the tetrahedral complexes is partly allowed because of the efficient d–p mixing (mixing of metal d–orbital and ligand p–orbital) in the noncentrosymmetric tetrahedral complexes.*** For the tetrahedral complexes, the electronic transition should be interpreted in terms of MOT not in terms of CFT. In the noncentrosymmetric T_d-complexes, the *e*-set ($d_{x^2-y^2}$, d_{z^2}) is basically representing the **nonbonding metal d-orbitals** while the t_2-set is constituted by the extensive overlap between the metal *d*-orbitals (*i.e.* d_{xy}, d_{yz}, d_{xz}) and ligand orbitals (*i.e. p* or hybrid orbitals). Thus the ***$e \rightarrow t_2$ transition should not be considered as the pure d–d transition.***
- Here it may be mentioned that in both the octahedral (*i.e.* centrosymmetric) and tetrahedral (*i.e.* noncentrosymmetric) systems, vibronic coupling to relax the Laporte selection rule is applicable. Mixing of the $(n-1)d$ orbital with the *np* orbitals of metals is not efficient because of an energetic ground but the metal *d*–orbital can mix with the ligand *p* orbitals efficiently as in the tetrahedral complexes. ***This d–p mixing mechanism is more efficient than the vibronic coupling to relax the Laporte selection rule. This is why, the ligand field transitions in the tetrahedral systems are more intense compared to those of the octahedral systems.***
- **$[Co(NH_3)_6]^{3+}$** is centrosymmetric while **$[Co(NH_3)_5X]^{2+}$** is noncentrosymmetric (C_{4v} symmetry). Thus the *d–p* mixing is possible in $[Co(NH_3)_5X]^{2+}$ while it is not possible in $[Co(NH_3)_6]^{3+}$. This is why, the ligand field transitions in $[Co(NH_3)_5X]^{2+}$ are more intense. However, the mixing of 3*d* and 4*p* orbitals is not efficient because of an energetic ground. Thus this only can weakly relax the selection rule.

- In the same way, we can explain the relatively higher intensity of the bands in **$[M(en)_3]^{3+}$** (M = Cr, Co) having the D_3 symmetry compared to those of $[M(NH_3)_6]^{3+}$ of O_h symmetry. **In fact, lowering of symmetry can allow the transition partially allowed.**

In the D_3 point group, the orbitals p_x, p_y, d_{xz}, d_{yz}, d_{z^2} and $d_{x^2-y^2}$ belong to the E species. This allows the d–p mixing and the electronic transition becomes partly allowed.

(L) Effect of temperature on vibronic coupling: According to Franck-Condon principle, only **vertical transitions** (Fig. 7.4.1.3) between the two electronic states can occur. If the temperature is lowered (*i.e.* low thermal energy), only fewer vibrational levels of the electronic states are populated. ***To allow the vertical Franck-Condon transition between the two electronic states, inhabitated portions of the two potential energy curves must lie vertically above or below each other.*** When fewer vibrational levels are inhabited at the ground electronic state (*i.e.* ψ_{el}^{gd}), *the number of ways of vibrational transitions from the ground electronic state to the upper electronic state is decreased, i.e.* the number of ways of combination of ψ_{vib}^{gd} and ψ_{vib}^{ex} decreases.

For an ***allowed vibronic transition,*** the product of $\psi_{el}^{gd}\psi_{vib}^{gd}\mu\psi_{el}^{ex}\psi_{vib}^{ex}$ must be of g-symmetry. For an octahedral complex, ψ_{el}^{gd} and ψ_{el}^{ex} are of g-symmetry and μ is of u-symmetry. Hence for an allowed vibronic transition, the product of $\psi_{vib}^{ex} \times \psi_{vib}^{gd}$ must be of u-symmetry. It is possible if ψ_{vib}^{gd} is of g-symmetry and ψ_{vib}^{ex} is of u-symmetry or ψ_{vib}^{gd} is of u-symmetry and ψ_{vib}^{ex} is of g-symmetry (*cf.* $g \times u \Rightarrow u$). *If different vibrational levels in the ground electronic state are occupied then the number of possible ways to satisfy the required condition of vibronic coupling through the combination of* ψ_{vib}^{gd} *and* ψ_{vib}^{ex} *increases, i.e.* chance of the allowed vibronic transition through the coupling of suitable vibrations of the ground and excited electronic states increases. At a lower temperature, fewer vibrational levels at the ground electronic state are occupied and it reduces the possibilities of the appropriate combinations of ψ_{vib}^{gd} and ψ_{vib}^{ex} to make the product $\psi_{vib}^{gd} \times \psi_{vib}^{ex}$ of u-symmetry (*cf.* $\upsilon = 0$, ψ_{vib} is of g-symmetry; with the lowering of temperature population density at $\upsilon = 0$ increases) **but $\upsilon = 0 \rightarrow \upsilon' = 0$ is a forbidden transition when the electronic transition is not allowed by its own right.** Thus, relaxation of the Laporte selection rule through the vibronic coupling becomes less efficient at a lower temperature (*cf.* Sec. 7.5.1R). Intensity (measured by **oscillator strength *f***) of a vibronic transition depends on temperature (*i.e.* intensity increases with the increase of temperature) as follows:

$$f \text{ (measuring intensity)} \propto \coth (h\Delta\nu/2k_BT), \quad \coth x = \frac{e^x + e^{-x}}{e^x - e^{-x}}$$

where $h\Delta\nu$ = difference in energy between the levels corresponding to the two wave functions.

Thus the relaxation of Laporte selection rule through the vibronic coupling becomes less efficient at lower temperature. In other words, the metal-ligand bond vibration is the key player for relaxation in vibronic transition. Lowering of temperature restricts the metal-ligand bond vibration to disfavour the relaxation through vibronic coupling. This is why, in such cases, the intensity of the band decreases with the lowering of temperature. **Thus, intensity of such bands depends on temperature** and this is why, such bands are called **hot bands.**

Here it may be noted that the d–p mixing (mixing of metal d–orbital with the p–orbitals of ligands through the molecular orbital formation) to relax the Laporte selection rule does not depend on temperature.

Note: Electronic transition is much faster than the time required to execute a vibration. Thus during electronic transition, the nuclei may be considered to remain fixed. Thus it is considered that the electronic transition occurs so rapidly that during the electronic transition, the metal-ligand bond distance remains unchanged. This is the basis of **Franck-Condon vertical transition (*cf.* Fig. 7.4.1.3).**

(M) Intensity stealing — a special case of vibronic coupling: If there are two excited electronic states (*i.e.* upper states) and to one of which a transtion from a particular electronic level is allowed while the transition to the other excited state is forbidden, then the forbidden transition becomes partially allowed. When the forbidden and allowed excited states lie close, ***some suitable mode of vibration can allow the mixing of the electronic wave functions of the forbidden excited state and the allowed excited state.*** In this vibronic coupling process, transition to the forbidden excited state becomes partially allowed and intensity of the band is increased. This phenomenon is called **intensity stealing.** This mixing between the forbidden and allowed excited states becomes less efficient if the energy difference between the states is high.

Generally when the ligand field band (d–d) lies close to a charge-transfer band (which is allowed), the intensity of the ligand field band is increased through the intensity stealing phenomenon. For this reason, the ligand field bands moving towards the blue region of wavelength gain more intensity because in this region, the high energy charge transfer bands generally prevail.

Intensity of ligand field bands: O_h *vs.* T_d *Complexes*

Vibronic coupling (O_h) *vs. d–p* mixing (T_d): In the centrosymmetric octahedral complex, intensity of the ligand field band is earned through the **vibronic coupling** by relaxing the Laporte Selection Rule. In the noncentrosymmetric tetrahedral complex, this vibronic coupling is also applicable. ***But, in the tetrahedral system, the more efficient mechanism is the d–p mixing (mixing of the metal t_2 d-orbitals with the ligand p-orbitals) to relax the Laporte Selection Rule.*** This *d–p* mixing is not efficient in the centrosymmetric octahedral system.

Thus, an octahedral system enjoys only vibronic coupling to relax the Laporte selection Rule while a tetrahedral system enjoys both the vibronic coupling and *d–p* mixing. But the *d–p* mixing route is more efficient than the vibronic coupling to relax the Laporte Selection Rule. ***This is why, the tetrahedral complexes are more intense in colour.*** Here it is worth mentioning that the an asymmetric metal-ligand bond vibration can instantaneously destroy the centrosymmetric character and at that moment metal $(n-1)d$ the orbital can mix with the metal *np*–orbital. This instantaneous *d-p* mixing can partly relax the Laporte selection rule. But this mixing is not efficient because of the energetic ground.

Intensities of the spectral transitions in the T_d complexes **are not affected much by temperature** because efficiency of the *d–p* mixing which relaxes the Laporte selection rule does not depend on temperature. But intensities of the spectral transitions in the O_h complexes decreases with temperature as the efficiency of the vibronic coupling to relax the selection rule decreases with the decrease of temperature.

Thus mixing of the metal d-orbitals with the ligand p-orbitals is favoured both mechanistically and energetically in the noncentrosymmetric T_d complexes while the instantaneous mixing of the metal $(n-1)d$ orbitals and *np* orbitals is inefficient in the centrosymmetric O_h complexes both mechanistically and energetically.

(N) Lowering of symmetry to relax the Laporte selection rule in terms of group theory: It has been already mentioned that in the noncentrosymmetric complexes like $[M(NH_3)_5X]^{2+}$ (M = Co, Cr) (C_{4v} symmetry) and $[Co(en)_3]^{3+}$ (D_3 symmetry), the ligand field bands are more intense compared to those in $[M(NH_3)_6]^{3+}$ of O_h symmetry. This has been explained by considering the possibility of *d-p* mixing. This can also be understood in the following manner.

● $O_h \rightarrow D_3$**:** In $[Cr(NH_3)_6]^{3+}$, the spin-allowed transition, ${}^4A_{2g} \rightarrow {}^4T_{2g}$ may be considered. In this centrosymmetric O_h complex, both the excited and ground electronic states are of *g*–symmetry and the dipole moment operator is of u–symmetry (*i.e.* T_{1u}). It indicates: $g \times u \times g = u$ and the corresponding transition moment integral is zero. Thus it represents a **forbidden electronic transition.** In moving from the O_h to D_3 point group *e.g.* $[Cr(en)_3]^{3+}$, the T_{2g} term of O_h system splits into the A_1 and E states and the A_{2g} term transforms into A_2 in D_3 (*cf.* Fig. 7.5.1.2). Thus, the relevant transitions in the D_3-point group are:

$$^4A_2 \rightarrow {}^4A_1;\ {}^4A_2 \rightarrow {}^4E.$$

Now let us consider the transition moment integral for the transition, $^4A_2 \rightarrow {}^4A_1$ where $\psi^g_{el} \times \psi^{ex}_{el} = A_2 \times A_1 = A_2$. The dipole moment operator components μ_x and μ_y are of E symmetry while the μ_z component is of A_2 symmetry (*see* **Character Table, Appendix 12A**). Thus we can write:

$\psi^{gd}_{el}\ \psi^{ex}_{el}\ \mu = A_2 \times A_1 \times E = E$ (for the μ_x and μ_y component); (A_1 term absent in the product)

$= A_2 \times A_1 \times A_2 = A_1$ (for the μ_z component)

It indicates that the transition is allowed for the μ_z component of the dipole moment because in the direct product result, the totally symmetric irreducible representation (A_1) is present. But the transition for the μ_x and μ_y components is forbidden.

In moving from **$[Cr(NH_3)_6]^{3+}$ (O_h) to $[Cr(en)_3]^{3+}$ (D_3)**, the symmetry is reduced and we can compare and analyse the spectral transitions in these two point groups (*cf.* Fig. 7.5.1.2).

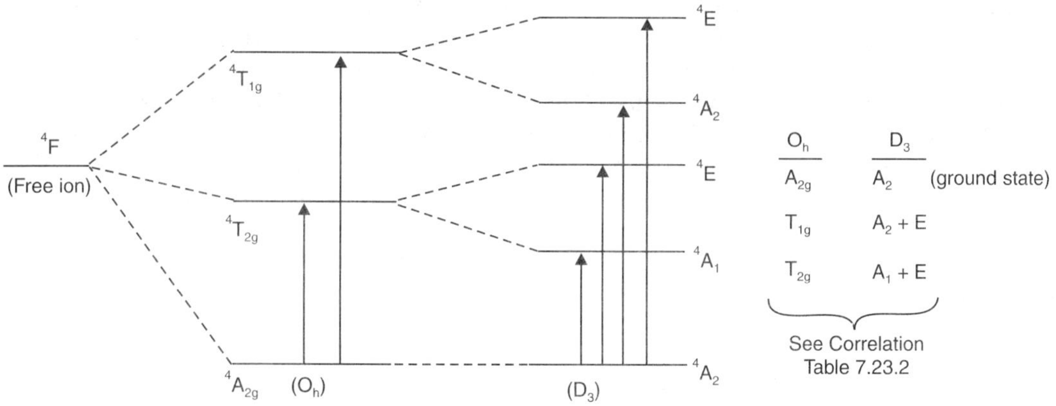

Fig. 7.5.1.2 Spectral transitions in the O_h and D_3 point groups for Cr^{3+} (d^3). The 3rd ligand field transition, $^4A_{2g} \rightarrow {}^4T_{1g}$ (P) not shown here and this transition gets very often obscured due to the CT band. In the O_h-system, all the electronic transitions are orbitally forbidden.

The results of **all the electronic transitions in D_3 point group** are given below (*cf.* Table 7.23.1 for the direct product result) to understand the electronic polarisation.

Transitions	$\Gamma\psi^{gd}_{el}\ \Gamma\mu_{x,y}\ \Gamma\psi^{ex}_{el}$	$\Gamma\psi^{gd}_{el}\ \Gamma\mu_z\ \Gamma\psi^{ex}_{el}$
$A_2 \rightarrow A_1$	$A_2 \times E \times A_1 = E$	$A_2 \times A_2 \times A_1 = A_1$
$A_2 \rightarrow A_2$	$A_2 \times E \times A_2 = E$	$A_2 \times A_2 \times A_2 = A_2$
$A_2 \rightarrow E$	$A_2 \times E \times E = A_1 + A_2 + E$	$A_2 \times A_2 \times E = E$

The direct product contains the totally symmetric irreducible representation A_1 in two transitions and these are allowed and remainings are forbidden. These are:

	$A_2 \rightarrow A_1$	$A_2 \rightarrow A_2$	$A_2 \rightarrow E$
For μ_z (II to C_3 axis):	Allowed	Forbidden	Forbidden
For $\mu_{x,y}$ ($\perp$ to C_3 axis):	Forbidden	Forbidden	Allowed

It indicates **electronic polarisation** of the incident radiation. These predictions have been experimentally verified.
Note: The details of direct product (*see* Table 7.23.1) and character table (*see* Appendix 12A) are not discussed here.

Possibility of *d–p* mixing in D_3-symmetry: Here it may be pointed out that in the D_3 point group, the orbitals (p_x, p_y), (d_{xz}, d_{yz}) and $\left(d_{x^2-y^2}, d_{z^2}\right)$ transform as the *E*-species. This allows the $(n-1)d - np$ mixing to some extent depending on the energy difference between the metal $(n-1)d$ and *np* orbitals. This route can partially relax the orbital selection rule but it is absent in the O_h-complex.

$[Cr(NH_3)_6]^{3+}$ *vs.* $[Cr(en)_3]^{3+}$, *i.e.* O_h *vs.* D_3: Now let us compare the **spectral characteristics of $[Cr(NH_3)_6]^{3+}$ (O_h) and $[Cr(en)_3]^{3+}$ (D_3).**

	${}^4A_{2g} \rightarrow {}^4T_{2g}$	${}^4A_{2g} \rightarrow {}^4T_{1g}$
$[Cr(NH_3)_6]^{3+}$	21,550 cm^{-1}; (30)	28,500 cm^{-1}, (27)
$[Cr(en_3)]^{3+}$	21,850 cm^{-1}; (76)	28,450 cm^{-1}, (60)

It is evident that in moving from the O_h to D_3 symmetry, the ***spectral bands are not split***, *i.e.* both $[Cr(NH_3)_6]^{3+}$ and $[Cr(en)_3]^{3+}$ are showing the same spectral pattern (*see* Fig. 7.14.5.3). Thus, ***approximately, both the complexes are of octahedral symmetry,*** *i.e.* $[M(N)_6]^{3+}(O_h)$. The intensities (measured by ε, given in parenthesis, *cf.* Table 7.14.5.1) of the en-complex of D_3 symmetry are approximately twice as large as those of the ammine complex of O_h symmetry. *This slightly enhanced intensity for the en-complex arises from the possibility of d–p mixing in D_3-symmetry.* Thus, basically in both the ammine and en complexes, the electronic transitions are the **vibronic transitions.**

● ***$O_h \rightarrow D_3$ vs. $O_h \rightarrow C_{2v}, D_{4h}$*:** Though in moving from $[M(NH_3)]^{3+}$ (O_h, M = Co, Cr) to $[M(en)_3]^{3+}$ (D_3), the spectral characteristics are not significantly changed (*i.e.* ***chelation by en does not effectively change the O_h symmetry***), in moving from the O_h to D_{4h} symmetry (as in *trans*-$[MCl_2(NH_3)_4]^+$, *trans*-$[MCl_2(en)_2]^+$) and C_{2v} symmetry (as in *cis*-$[MCl_2(NH_3)_4]^+$, *cis*-$[MCl_2(en)_2]^+$), the spectral features are remarkably changed (*cf.* Secs. 7.14.4, 7.14.5, 7.15.2, 7.15.3). The *trans*-complexes of $[MCl_2(NH_3)_4]^+$ or $[MCl_2(en)_2]^+$ are still centrosymmetric and these are more symmetric than the corresponding *cis*-complexes of C_{2v} symmetry. ***In fact, the cis-complexes show the more intense bands.*** In general, we can conclude:

the d–d transitions are more allowed for the complexes of lower symmetry.

Vibronic Coupling in the Transitions in $[Co(NH_3)_6]^{3+}$ (*cf.* Table 7.5.1.2)

${}^1A_{1g} \rightarrow {}^1T_{1g}$ and ${}^1A_{1g} \rightarrow {}^1T_{2g}$ transitions; normal modes of vibration = $3n - 6 = 15$ for MN_6 moiety and these are: A_{1g}, $2E_g$, $2T_{1u}$, T_{2u}, T_{2g} where the odd $M — N$ vibrations are T_{1u} and T_{2u}; $A_{1g} \times T_{1g} = T_{1g}$; $A_{1g} \times T_{2g} = T_{2g}$; dipole moment operator = T_{1u}. It leads to:

$$A_{1g} \times T_{1g} \times \mu = T_{1g} \times T_{1u} = A_{1u} + E_u + T_{1u} + T_{2u};$$

$$A_{1g} \times T_{2g} \times \mu = T_{2g} \times T_{1u} = A_{2u} + E_u + T_{1u} + T_{2u}$$

Combination of the odd vibration leads to: $T_{1u} \times T_{1u} \Rightarrow A_{1g}$ (one term) and $T_{2u} \times T_{2u} \Rightarrow A_{1g}$ (one term). Thus the transitions are **vibronically allowed** (*see* Table 7.5.1.1 for the direct product).

● ***$O_h \rightarrow D_{4h}$*:** Now let us consider the lowering of symmetry leading to O_h to D_{4h}. The *z*–out or *z*–in tetragonal distortion converts the O_h point group into the D_{4h} point group. Let us consider the transition, ${}^3T_{1g} \rightarrow {}^3A_{2g}$ in a d^2–octahedral complex. The dipole moment component μ (= $\mu_x = \mu_y = \mu_z$) is a T_{1u} species. The direct product is:

$$T_{1g} \times \mu \times A_{2g} = T_{1g} \times T_{1u} \times A_{2g} = T_{1u} + T_{2u} + E_u + A_{2u} \text{ (cf. Table 7.5.1.1)}$$

The direct product does not contain the totally symmetric A_{1g} term and consequently the **electronic transition is forbidden.** However, coupling of the odd T_{1u} and T_{2u} vibrations with the electronic transition will allow the transition.

Now let us consider the fate of the transition in the D_{4h} point group which still maintains the centre of symmetry (*i.e.* centrosymmetry).

$T_{1g}(O_h)$ is converted into $E_g + A_{2g}$ (in D_{4h}); $A_{2g}(O_h)$ is converted into B_{1g} (in D_{4h}); the dipole moment components in D_{4h} symmetry are: μ_x, μ_y of E_u symmetry and μ_z of A_{2u} symmetry.

Thus the ${}^3T_{1g} \rightarrow {}^3A_{2g}$ transition in O_h represents the transitions, E_g, $A_{2g} \rightarrow B_{1g}$ in D_{4h}. The direct product for the transitions in the D_{4h} point group becomes:

$$\left.\begin{aligned}&(E_g + A_{2g})E_u \times B_{1g} + (E_g + A_{2g}) \times A_{2u} \times B_{1g}\\ &= \underbrace{(A_{1u} + A_{2u} + B_{1u} + B_{2u} + E_u)}_{(\text{For } \mu_x, \text{ and } \mu_y)} + \underbrace{(E_u + B_{1u})}_{(\text{For } \mu_z)}\end{aligned}\right\} (cf. \text{ Table 7.23.1})$$

For moving from the O_h to D_{4h} symmetry, though the symmetry is lowered, the **above transitions are still forbidden** because the direct products of $\psi_{el}^{gd} \times \mu \times \psi_{el}^{ex}$ do not contain the totally symmetric irreducible representation A_{1g} (*cf.* D_{4h} still maintains the centre of inversion). However, the **vibronic coupling can allow the transitions.** The normal modes of vibration of D_{4h} symmetry are.

$2A_{1g}$, B_{1g}, B_{2g}, E_g, $2A_{2u}$, B_{1u}, $3E_u$ (*cf.* Table 7.23.3 and Appendix 12B).

Coupling of the odd vibrations, *i.e.* A_{2u}, B_{1u} and E_u can remove the forbiddeness of the transitions.

(O) Vibronic polarisation and dichroism: Let us illustrate it for the electronic transitions in *trans*-$[CoCl_2(en)_2]^+$ (D_{4h} symmetry) where the spin allowed transitions are (*cf.* Fig. 7.15.2.4c):

$$\left.\begin{aligned}&{}^1A_{1g} \rightarrow {}^1E_g \text{ and } {}^1A_{1g} \rightarrow {}^1A_{2g} \text{ (Corresponding transition in } O_h\text{: } {}^1A_{1g} \rightarrow {}^1T_{1g})\\ &{}^1A_{1g} \rightarrow {}^1B_{2g} \text{ and } {}^1A_{1g} \rightarrow {}^1E_g \text{ (Corresponding transition in } O_h\text{: } {}^1A_{1g} \rightarrow {}^1T_{2g})\end{aligned}\right\} (cf. \text{ Fig. 7.15.2.1})$$

In D_{4h}, μ_z (along the C_4 axis) transforms as A_{2u} while μ_x and μ_y perpendicular to z (C_4) axis transform as E_u (*see* Character Table, Appendix 12A). The normal modes of vibration are:

$2A_{1g}$, B_{1g}, B_{2g}, E_g, $2A_{2u}$, B_{1u}, $3E_u$ (*cf.* Table 7.23.3).

Transitions	$\Gamma\psi_{el}^{gd}\ \Gamma\mu_{x,y}\ \Gamma\psi_{el}^{ex}$	$\Gamma\psi_{el}^{gd}\ \Gamma\mu_z\ \Gamma\psi_{el}^{ex}$	
${}^1A_{1g} \rightarrow {}^1A_{2g}$	$A_{1g} \times E_u \times A_{2g} = E_u$	$A_{1g} \times A_{2u} \times A_{2g} = A_{1u}$	(*cf.* Table 7.23.1)
${}^1A_{1g} \rightarrow {}^1B_{2g}$	$A_{1g} \times E_u \times B_{2g} = E_u$	$A_{1g} \times A_{2u} \times B_{2g} = B_{1u}$	
${}^1A_{1g} \rightarrow {}^1E_g$	$A_{1g} \times E_u \times E_g = A_{1u} + A_{2u} + B_{1u} + B_{2u}$	$A_{1g} \times A_{2u} \times E_g = E_u$	

The **odd vibrations** are: A_{2u}, B_{1u} and E_u. Hence, if in the direct product $\psi_{el}^{gd}\mu\psi_{el}^{ex}$, there is any term matching with the symmetries of these odd vibrations, then the transition is vibronically allowed. *Thus all the above transitions are vibronically allowed but the transition $A_{1g} \rightarrow A_{2g}$ is vibronically forbidden for z-polarisation.*

The band at ~16000 cm^{-1} is due to ${}^1A_{1g} \rightarrow {}^1E_g$ and the band at ~22,000 cm^{-1} is due to ${}^1A_{1g} \rightarrow {}^1A_{2g}$ (*cf.* Fig. 7.15.2.4c). These E_g and A_{2g} components are derived from the T_{1g} state of O_h symmetry. The ${}^1A_{1g} \rightarrow {}^1E_g$ transition is vibronically allowed in all directions while the ${}^1A_{1g} \rightarrow {}^1A_{2g}$ transition is **vibronically forbidden for *z*-polarisation.** *In fact, the band at ~22,000 cm^{-1} is strongly polarised and this band remains absent for light polarised parallel to z (C_4) axis but it exists for light perpendicular to z-axis. The band at ~16,000 cm^{-1} shows no such polarisation effect.* The broad band at ~27,000 cm^{-1} is due to the ${}^1A_{1g} \rightarrow E_g$, B_{2g} (states arising from the T_{2g} state of O_h symmetry). This dichroism of *trans*-$[CoCl_2(en)_2]^+$ has been experimentally verified (*cf.* Fig. 7.15.2.4c).

Note: For the results of direct product in D_{4h} point group, the readers are to consult any standard book on group theory (*see* Table 7.23.1).

(P) Different factors (summary) responsible for the origin of colour in the octahedral complexes: These are summarised here.

(i) **Loss of centre of symmetry:** Electronic transitions become more allowed when the centre of symmetry is lost. It may be mentioned that in absence of the centre of symmetry, *probability of the allowedness of the electronic transition is increased* though all the possible electronic transitions may not be allowed (*cf.* in the noncentrosymmetric T_d system the $A_2 \rightarrow T_2$ transition is not itself allowed by its own right, *see* solved problem 15).

Loss of the centrosymmetric character allows the ***d–p* mixing** which in turn makes the electronic transitions more allowed due to the $d \rightarrow p$ and/or $p \rightarrow d$ component in the electronic transition.

Loss of the centre of symmetry can occur in different ways:

- **asymmetric metal-ligand bond vibration** (temporary loss of centre of symmetry); • **crystal packing** in solid state may lead to distortion to remove the centre of symmetry; • **intrinsic asymmetry in the polyatomic ligands** (generally organic ligands); • symmetry reduction or **symmetry lowering** due to the ligand environment (*e.g.* $O_h \rightarrow C_{4v}, C_{2v}, D_3$, etc.)

(ii) **Vibronic coupling:** Coupling of a suitable odd vibration may lead to the electronic transition allowed under the condition: transition moment with the incorporation of the odd vibrational representation leads to the totally symmetric representation of the group, *i.e.* the matrix element $\langle \psi_{el}^{gd} \times \mu \times \psi_{el}^{ex} \times \psi_{vib}^{ex} \rangle$ must contain the totally symmetric representation of the group (*e.g.* $^1A_{1g}$ in O_h) assuming ψ_{vib}^{gd} to be totally symmetric.

(iii) ***cis- vs. trans*-complex:** The *trans*-complexes being more symmetric give the less intense bands than the *cis*-complexes, *e.g.* for $[CoCl_2(en)_2]^+$, the *trans*-isomer (D_{4h}, centrosymmetric) gives the less intense bands than the *cis*-isomer (C_{2v}).

(iv) **Greater covalency:** It will establish the overlap between the ligand orbital and metal orbital. It will generate the more intense bands. For example, in a **tetrahedral complex**, the $e \rightarrow t_2$ transition does not really describe the $d \rightarrow d$ transition. The e-level is basically constituted by the nonbonding $d_{x^2-y^2}$ and d_{z^2} metal orbitals but the t_2-level is constituted by the overlap of metal d-orbitals (d_{xy}, d_{yz}, d_{xz}) and ligand p-orbitals or s–p hybrid orbitals. If the covalency can be introduced into the octahedral complexes, then the intensity of the bands will increase. In fact, the **higher oxidation state** will favour the better covalency (*cf.* Fajan's rule) to give the more intense bands.

(v) **Intensity stealing effect:** If the strong CT band appears close to the ligand field bands, then the so called weak ligand field bands become more intense due to the intensity stealing effect.

(Q) Vibronic spectra and selection rule for the vibrational transition: Vibronic spectra represent the electronic spectra associated with the vibrational transitions from the ground electronic state to the excited electronic state (*cf.* Figs. 7.4.1.3, 7.7.2.1). Vibronic spectra appear for all the allowed electronic transitions. At normal temperature, the lowest vibrational level (*i.e.* $\upsilon = 0$) is most populated and the vibrational transition will predominantly occur from $\upsilon = 0$. For the vibrational transition, $\upsilon = 0 \rightarrow \upsilon'$ (*cf.* Fig. 7.7.2.1) the **condition of vertical Franck-Condon transition** (*i.e.* r_e^{gd} for $\upsilon_0 \approx r^{ex}$ for υ') **must be maintained.** If the electronic transition is itself allowed, then there is no quantum restriction for the vibrational transition but if the electronic transition is to be allowed through vibronic coupling then there will be a ***quantum restriction for the vibrational transition.***

Case I (Electronic transition allowed itself): Some of the electronic transitions become allowed by their own right in the noncentrosymmetric systems. In such cases, there is no selection rule for $\Delta\upsilon$ ($= \upsilon \sim \upsilon'$, υ denotes the vibrational level at the ground electronic state and υ' denotes the vibrational level at the excited electronic state) besides the condition of Franck-Condon vertical transition. In such cases, $\Delta\upsilon$ may be 0, ±1, ±2,.... . However, the **Franck-Condon factor** will determine the relative intensities (*cf.* Sec. 7.7.2) of the transitions.

Case II (Electronic transition allowed through the vibronic coupling): In such cases, besides the condition of Franck-Condon vertical transition, the matrix element $\langle \psi_{el}^{gd}\ \psi_{vib}^{gd}\ \mu\ \psi_{el}^{ex}\ \psi_{vib}^{ex}\rangle$ must contain the totally symmetric representation of the group. In an octahedral system (which is centrosymmetric), for the vibronically allowed electron transition, the product $\psi_{vib}^{gd} \times \psi_{vib}^{ex}$ must be of u-symmetry (*i.e.* T_{1u} or T_{2u}). In such cases, the transitions with $\Delta v = 0$ (*e.g.* $v = 0 \rightarrow v' = 0$, $v = 1 \rightarrow v' = 1$, etc.) are the Laporte forbidden transitions but the vibrational transitions like $v = 0 \rightarrow v' = 1$, $v = 1 \rightarrow v' = 2$, $v = 1 \rightarrow v' = 0$, etc. may relax the Laporte selection rule if $\psi_{vib}^{gd} \times \psi_{vib}^{ex}$ is of u-symmetry.

(R) Peak broadening due to vibrational transitions and effect of temperature (*cf.* Secs. 7.7.2, 3): If the internuclear distance (*i.e.* metal-ligand bond distance) does not change during the electronic transition (*i.e.* $r_e^{gd} = r_e^{ex}$), then **the peak becomes sharp** while the **broader peaks** will appear under the condition, $r_e^{gd} \langle\ r_e^{ex}$, because of the more possible vibrational transitions.

For the **allowed electronic transition and $\mathbf{r_e^{gd} = r_e^{ex}}$,** at normal temperature, $v = 0 \rightarrow v' = 0$ (*i.e.* $0 \rightarrow 0$) transition is the most intense one (giving a sharp peak) but at the higher temperature $v = 1 \rightarrow v' = 1$, $v = 2 \rightarrow v' = 2$, etc. transitions become important when the higher vibrational levels at the ground electronic state become populated. It should be remembered that for the **vibronically allowed electronic transitions, even for $\mathbf{r_e^{gd} = r_e^{ex}}$, the $\mathbf{0 \rightarrow 0}$ transition is forbidden,** and the $0 \rightarrow 1$ transition may become important. At higher temperature, $1 \rightarrow 0$, $1 \rightarrow 2$ transitions may become important.

At the low temperature, the lowest vibrational level (*i.e.* $v = 0$) is the most populated one and the vibrational transition will occur from the $v = 0$ level. But at higher temperatures, the population density at the higher vibrational level (*i.e.* $v = 1, 2,...$) increases according to the Boltzmann distribution law and the vibrational transitions from these higher levels (*i.e.* $v = 1, 2,...$) will **broaden the peak towards the longer wavelength. Thus at the lower temperature, the peak becomes relatively sharper.**

Case I (electronically allowed transition and $\mathbf{r_e^{gd} \approx r_e^{ex}}$): At the low temperature, the $v = 0$ level (at the ground electronic state) is the most populated one and the $v = 0 \rightarrow v' = 0$ transition (having the highest franck-condon factor, *cf.* Sec. 7.7.2) is the most intense one. At the higher temperatures, population density at the higher vibrational levels (*i.e.* $v = 1, 2, ...$) will increase according to Boltzmann's distribution law and the transitions like, $1 \rightarrow 0$ (*i.e.* $v = 1$, $v' = 0$), $2 \rightarrow 1$, $3 \rightarrow 2$, $1 \rightarrow 1$, $2 \rightarrow 2$, etc. will occur at the lower frequencies compared to that of the $0 \rightarrow 0$ transition (*cf.* ΔE_{ex} = energy difference between the successive vibrational levels at the excited electronic state $\langle\ \Delta E_{gd}$ = energy difference between the successive vibrational levels at the ground state). Thus at higher temperature, then intensity of the $0 \rightarrow 0$ transition will decrease because the population density at $v = 0$ of the ground electronic state will decrease and the band will be broadened towards the longer wavelength but *the total oscillator strength (f) will remain unchanged.* Thus in such cases, we can conclude:

with the rise of temperature, the peak will be broadened towards the longer wavelength, and the molar extinction coefficient will decrease for the 0 → 0 transition but the total oscillator strength f (determined by the area of the peak) will remain unchanged.

Case II (vibronically allowed but electronically forbidden transition with $\mathbf{r_e^{gd} \approx r_e^{ex}}$): Here the $0 \rightarrow 0$ transition (most probable one in terms of Franck-Condon Factor) is forbidden but the $0 \rightarrow 1$ transition is allowed. With the lowering of temperature, population density at $v = 0$ will increase and it will make the $0 \rightarrow 1$ transition more intense. At the higher temperature, population density at the higher vibrational levels (*i.e.* $v = 1, 2,...$) of the ground electronic state will increase and it will allow the transitions like $1 \rightarrow 0$, $1 \rightarrow 2$, $2 \rightarrow 1$, $2 \rightarrow 3'$, etc. that will occur at the longer wavelength compared to that of the $0 \rightarrow 1$ transition. With the lowering of temperature, the population density at $v = 0$ increases but the $v = 0 \rightarrow v' = 0$ transition having the highest Franck-Condon factor is forbidden. *It will definitely enhance the intensity of the $v = 0 \rightarrow v' = 1$ transition but it will not be able to make up the* *loss of intensity earned from the transitions occurring from the higher levels, i.e.* $\upsilon = 1, 2, 3,...$ *This is why, with the lowering of temperature the peak becomes sharper but the oscillator strength f decreases for the vibronically allowed electronic transitions.* (*cf. f* **is independent of temperature for the allowed electronic transitions**). In such cases, we can conclude:

with the rise of temperature, the peak broadens towards the longer wavelength and intensity measured by the total oscillator strength will also increase *(i.e.* ***vibronic coupling becomes more efficient at the high temperature.***

Note: If $\Delta E_{gd} = \Delta E_{ex}$, *i.e.* energy separation between the adjacent vibrational levels in the ground and excited electronic states is the same, then all the transition $0 \rightarrow 0$, $1 \rightarrow 1$, $2 \rightarrow 2$, ... (*i.e.* $\Delta\upsilon = 0$) will coincide, *i.e.* no peak broadening effect. If $\Delta E_{gd} \rangle \Delta E_{ex}$, then $1 \rightarrow 1$, $2 \rightarrow 2$, $1 \rightarrow 0$, etc. transitions will broaden the peak towards the longer wavelength **compared to that of the** $\mathbf{0 \rightarrow 0}$ **transition.** Similarly, for the vibronically allowed transition, under the condition, $\Delta E_{ex} \langle \Delta E_{gd}$, the transitions $1 \rightarrow 2$, $2 \rightarrow 3$, $1 \rightarrow 0$, $2 \rightarrow 1$, $3 \rightarrow 2$ will occur at the longer wavelength compared to that of the $\mathbf{0 \rightarrow 1}$ **transition.**

7.5.2 Relaxation of the Spin Selection Rule

According to this rule, the singlet $\rightarrow$ triplet or triplet $\rightarrow$ singlet transition is not possible. In other words, the electronic transition can occur only between the states having the same spin multiplicity. When the ground state and excited state differ in spin multiplicity, the **spin-orbit coupling phenomenon** can allow the mixing of two states. In such cases, the transiton becomes partially allowed. It may be noted that the ***spin-orbit coupling is efficient when the energy difference between the ground state and excited state is small.***

If we consider that the ground state is singlet and the excited state is triplet, then spin-orbit coupling can mix some triplet character of the excited state into the ground state and similarly some character of the singlet ground state can be introduced into the excited state. Consequently, the so called, singlet $\rightarrow$ triplet transition will have the components singlet $\rightarrow$ singlet and tripet $\rightarrow$ triplet which are allowed. Thus, the so called spin forbidden transition becomes partially allowed. For example, in $[Mn(OH_2)_6]^{2+}$ the ground state bears $S = {}^5\!/_2$ while the excited state bear $S = {}^3\!/_2$. The mixing through spin-orbit coupling can introduce some character of $S = {}^3\!/_2$ into the ground state and some character of $S = {}^5\!/_2$ into the excited state. Thus the transition becomes partially allowed. Thus, the spin forbidden transition becomes partially allowed.

7.6 BAND INTENSITY AND THE PARAMETERS MEASURING THE BAND INTENSITY

Intensities of the different types of transitions: These are compared in Table 7.6.1.

Parameters measuring the band intensity: Intensities of different bands are very often compared by their ε_{max} (molar absorptivity or molar extinction coefficient at λ_{max} of absorption) values. This procedure of comparison of band intensities is true if the bands under comparison are *sharp, symmetric* and of *the same bandwidth.* But very often, the bands are *unsymmetric* and they are of *different bandwidths.* Bandwidth may change with temperature. Thus, to measure the band intensity, ε_{max} is not an accurate parameter and for this purpose, **we should measure the area under the band** and this can be obtained by integrating as follows:

$$I = \int_{\bar{\nu}_1}^{\bar{\nu}_2} \varepsilon_{\bar{\nu}} d\bar{\nu},$$

I = Integrated band intensity; $\bar{\nu}_1$ and $\bar{\nu}_2$ are the starting and ending wave numbers of the band; $\varepsilon_{\bar{\nu}}$ (in litre mol^{-1} cm^{-1}) = molar absorptivity at $\bar{\nu}$. *I represents the area under the absorption curve.* If $\bar{\nu}$ is

Table 7.6.1 Intensities in different types of electronic transitions

	Type of transition	*Example*	**ε (mol^{-1} dm^3 cm^{-1}) (or mol^{-1} litre cm^{-1})**
(i)	Both Laporte forbidden* and spin forbidden	$[Mn(OH_2)_6]^{2+}$, $[Fe(OH_2)_6]^{3+}$	10^{-1} to 10^{-2}
(ii)	Spin allowed but Laporte forbidden*	$[Ti(OH_2)_6]^{3+}$ $[Ni(OH_2)_6]^{2+}$	~10^1
(iii)	Spin allowed and Laporte partly allowed (through *d–p* mixing in the noncentrosymmetric systems as in T_d complexes)	$[CoCl_4]^{2-}$	10^2 to 10^3
(iv)	Spin allowed, Laporte forbidden (but **intensity stealing** phenomenon)	Octahedral complexes with the ligands like acetylacetonate, P or As the donor ligands	~10^3
(v)	Charge transfer band (allowed transition)	$[TiCl_6]^{2-}$, $[MnO_4]^-$ $[Fe(OH_2)_5(SCN)]^{2+}$, etc.	10^3 to 10^5

* Partly allowed due to vibronic coupling.

measured in cm^{-1}, I has the unit litre mol^{-1} cm^{-2} or 10^3 cm mol^{-1}. For a symmetrical band, we can reasonably write:

$$I = \int_{\bar{\nu}_1}^{\bar{\nu}_2} \varepsilon_{\bar{\nu}} d\bar{\nu} = \varepsilon_{max} \Delta\bar{\nu}_{1/2}, (\Delta\bar{\nu}_{1/2} \text{ called } \textbf{half bandwidth} \text{ measured in cm}^{-1}, \textit{cf.} \text{ Sec. 7.7}).$$

I (**integrated band intensity**) can be expressed in terms of **oscillator strength** (*f*) as follows:

$$f \propto I \text{ or, } f = 4.315 \times 10^{-9} \int_{\bar{\nu}_1}^{\bar{\nu}_2} \varepsilon_{\bar{\nu}} d\bar{\nu}$$

$$= 4.315 \times 10^{-9} \varepsilon_{max} \Delta\bar{\nu}_{1/2}, \left(\text{for a symmetric band}\right).$$

7.7 BANDWIDTH: SYMMETRIC AND ASYMMETRIC BANDS

Half bandwidth $\left(\Delta\bar{\nu}_{1/2}\right)$ or simply called bandwidth of a symmetric absorption peak is defined as the width of the absorption peak in wave number at $\varepsilon = \frac{1}{2}\varepsilon_{max}$. This definition works well for the symmetric bands (Fig. 7.7.1).

Very often, bandwidth of the electronic transitions in the complexes is of the order of 1000 cm^{-1}. But the **Franck-Condon vertical transition** between the two states of energy E_2 and E_1 should lead to a sharp peak of a definite frequency $(\nu)\left(= \frac{E_2 \sim E_1}{h}\right)$. Many factors are responsible for widening the bands. These are discussed here.

Note: For the free ion (*i.e.* gaseous ion), the width of the spectral lines is less than 1 cm^{-1} while for a spin allowed ligand field transition, the band is quite broad and the bandwidth is in the order of 1000 cm^{-1}.

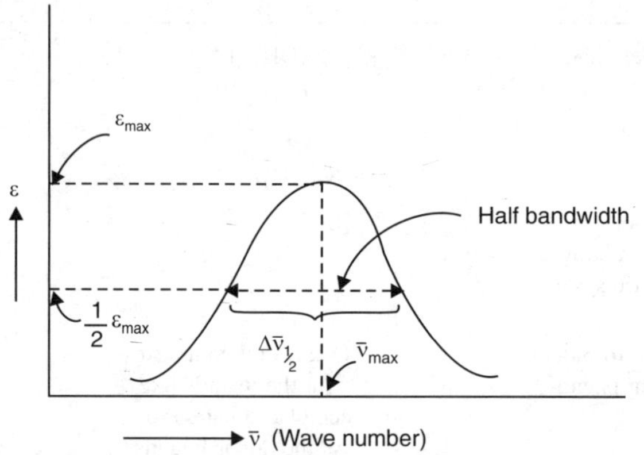

Fig. 7.7.1 Representative illustration of bandwidth for a symmetric peak.

7.7.1 Bandwidth and Difference (Δx) between the Internuclear Distances of the Ground State and Excited States

The bandwidth is a function of $\Delta x (= r_e^{gd} \sim r_e^{ex})$, difference between the internuclear distances (*i.e.* metal-ligand bond distances) of the ground and excited states. Δx depends largely on the electronic configurations in the e_g levels of the ground state and excited state. In the transition, $t_{2g} \rightarrow e_g$, the electron density in the e_g level of the excited state inccreases and it elongates the metal-ligand bond distance. Consequently, in such transitions, the metal-ligand bond distance in the excited state is longer than that in the ground state. Here it is worth mentioning that during the electronic transition, according to Franck-Condon principle, there will be no change in the internuclear distance. ***This is why, the electronically excited molecules are in the vibrationally excited levels having the same bond distance of the ground state electronic configuration to execute the Franck-Condon vertical transition*** (*cf.* Fig. 7.4.1.3).

For $[Cr(OH_2)_6]^{3+}$, the transition, $^4A_{2g}\,(t_{2g}^3 e_g^0) \rightarrow {}^4T_{2g}\,(t_{2g}^2\ e_g^1)$ at 17,400 cm^{-1} shows the **bandwidth of about 1600 cm^{-1}.** On the other hand, the spin forbidden transition, $^4A_{2g}\,(t_{2g}^3 e_g^0) \rightarrow {}^2E_g$ (low-spin t_{2g}^3) shows a **sharp peak.** For the spin-allowed transition, $t_{2g}^3\, e_g^0 \rightarrow t_{2g}^2\, e_g^1$, the bond length in the excited state is longer, *i.e.* Δx is high while for the spin-forbidden transition, t_{2g}^3 (high-spin, *i.e.* three unpaired electron) $\rightarrow t_{2g}^3$ (low-spin, *i.e.* two paired electrons and one unpaired electron), there is no change in the electron density in the e_g level, *i.e.* bond length in both the excited state and ground state is more or less the same and Δx is almost zero.

In fact, the higher value of Δx indicates the higher bandwidth. **In gereral, the spin-allowed transitions (causing a change in the electron density in the e_g–level) produce the broad peaks while the spin-forbidden transitions (generally causing no change in the electron density in the e_g-level) produce the sharp peaks.** For example, the spin-forbidden transitions in $[Mn(OH_2)_6]^{2+}$ produce the weak but sharp peaks. These aspects of band broadening have been discussed in Sec. 7.7.2.

Very often, structural parameters of the excited state in the $t_{2g} \rightarrow e_g$ transition differ significantly from those of the ground state and energy states of the excited state may undergo splitting (*cf.* for d^3, the excited state $t_{2g}^2e_g^1$ experiences the J.T. distortion to split the energy levels). ***Splitting of the energy levels causes the peak broadening and sometimes peak splitting.***

7.7.2 Peak Broadening Vibrational Transitions in Terms of Franck-Condon Vertical Transitions and Franck-Condon Factor

At normal temperature, the ground vibrational level (*i.e.* $\upsilon = 0$) will be most populated at any electronic state and for $\upsilon = 0$, the highest **vibrational probability density** (determined $\psi_{vib}\psi^*_{vib}$ or ψ^2_{vib}, Fig. 7.7.2.1) is **localised in a relatively smaller region.** On the other hand, for the higher values of υ, the probability density is distributed in different regions with some nodal points. ***For the higher vibrational levels, higher probability density regions prevail towards the walls of the potential energy well. The intense transition is possible when the high vibrational probability regions for the ground and excited electronic states occur at the same internuclear distance.*** This is why, when the minima of the Morse potential curves lie for the same internuclear distance (*i.e.* $r_e^{gd} = r_e^{ex}$), the $0 \rightarrow 0$ transition (*i.e.* $\upsilon = 0 \rightarrow \upsilon'_{min} = 0$) gives the most intense transition for the allowed electronic transition. In such cases, valu of the **Franck-Condon overlap integral** ($= \int\psi^{gd}_{vib}\psi_{vib}{}^{ex}d\tau$) becomes high and consequently the **Franck-Condon Factor (FCF)** which is the square of the vibrational overlap integral becomes high. Here it is worth mentioning that even for an allowed electronic transition, if the overlap between the initial and final vibrational wave functions is negligibly small (*i.e.* FCF is small), the transition becomes practically forbidden. This is understandable in terms of the following **transition moment integral** (R) for an allowed electronic transition.

$$R = \int \psi^{gd}_{vib}\psi^{ex}_{vib}\left[\int \psi^{gd}_{el}\mu_{el}\psi^{ex}_{el}d\tau\right]d\tau = R_e\int \psi^{gd}_{vib}\psi^{ex}_{vib}d\tau$$

$$R_e = \int \psi^{gd}_{el}\mu_{el}\psi^{ex}_{el}d\tau$$

- **Variation of FCF for the most probable vibrational transition:** The most probable (*i.e.* most intense) vibrational transition is: $v = 0 \rightarrow \upsilon_{min}$, where υ'_{min} gives the minimum value of vibrational level (at the excited electronic state) for which r^{ex} becomes equal to r_e^{gd} *i.e.* $r_e^{gd} \approx r^{ex}$. The FCF varies as:

FCF: $\upsilon = 0 \rightarrow \upsilon'_{min} = 0 >> \upsilon = 0 \rightarrow \upsilon'_{min} = 1 >> \upsilon = 0 \rightarrow \upsilon'_{min} = 2 >> ...$

(FCF becomes maximum for the $\upsilon \rightarrow \upsilon'$ transition when υ and υ' become minimum)

With the increase of the vibrational number, the number of probability regions increases. This is why, FCF decreases with the increase of υ' and then the peak broadening transitions with the less intensity become important to maintain the constancy of total oscillator strength (*f*) for an **allowed electronic transition.**

- **Peak broadening vibrational transitions:** Besides the most probable vibrational transition, *i.e.* $\upsilon = 0 \rightarrow \upsilon'_{min}$ (the minimum value of υ'_{min} for which $r_e^{ex} \approx r_e^{gd}$), other vibrational transitions of the lower intensity will cause the peak broadening. If $0 \rightarrow 2$ transition is the most probable (*i.e.* most intense) transition, then the less intense transitions like $0 \rightarrow 3$, $0 \rightarrow 4$, $0 \rightarrow 5$, ... will broaden the peak towards the lower wavelength (*i.e.* higher frequency) and the less intense transitions like $0 \rightarrow 1$, will broaden the peak towards the longer wavelength (*i.e.* lower frequency) region.

Case I ($r_e^{gd} \approx r_e^{ex}$): In such cases (*cf.* Fig. 7.7.2.1a) of the allowed electronic transition, FCF for the most probable transition, *i.e.* $0 \rightarrow 0$ transition (at the ordinary condition when the higher vibrational levels, *i.e.* $\upsilon = 1, 2,...$ at the ground electronic state are not appreciably populated) is almost unity and consequently, the other possible peak broadening transitions (towards the higher frequency) are not important. It **makes the peak sharp** (*i.e.* high ε-value) to maintain the constancy in the total oscillator strength *f* for the **allowed electronic transition.** It happens so for the ***f* → *f* transitions** in the lanthanides,

spin forbidden $d \rightarrow d$ transitions in the d^3 system (O_h complexes) and d^5 system where the condition $r_e^{gd} \approx r_e^{ex}$ is maintained.

Here it may be mentioned that for an electronic transition allowed through the vibronic coupling, even for the $r_e^{gd} = r_e^{ex}$ condition, the $\upsilon = 0 \rightarrow \upsilon' = 0$ transition is forbidden and in such cases the most probable (*i.e.* most intense) transition is $\upsilon = 0 \rightarrow \upsilon' = 1$.

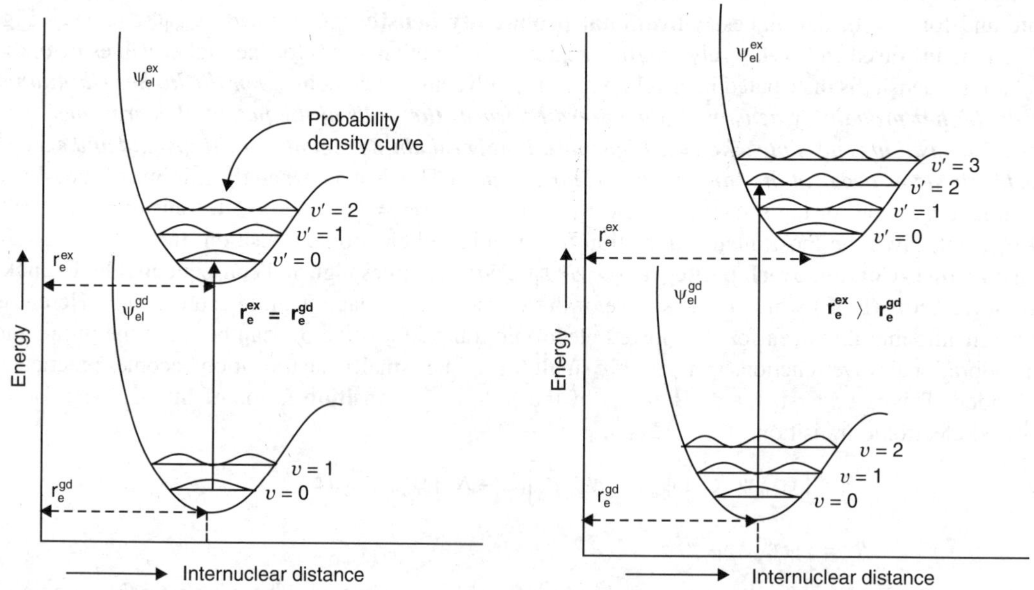

(a) r_e^{ex} ($\upsilon' = 0$) = r_e^{gd} ($\upsilon = 0$)
Most probable transition:
$\upsilon = 0 \rightarrow \upsilon' = 0$
(Sharp peak)

(b) r_e^{ex} ($\upsilon' = 2$) $\approx$ r_e^{gd} ($\upsilon = 0$)
Most probable transition:
$\upsilon = 0 \rightarrow \upsilon' = 2$
(Broad peak)

Fig. 7.7.2.1 Potential energy curves (*i.e.* Morse potential energy curres) for the ground and excited electronic states illustrating the Franck-Condon vertical transition. (a) No change in internuclear distance (*i.e.* metal-ligand bond distance) in the electronic transition. (b) Intenuclear distance (r_e) is longer for the excited electronic state.

Note: There is no quantum restriction for $\Delta\upsilon$, if, the electronic transitions are assumed to be allowed by their own right. If vibronic coupling allows the electronic transition, then besides the condition of $r_e^{gd} = r^{ex}$ (for υ'), there will be a quantum restriction for υ'. In an octahedral system, $\psi_{vib}^{gd} \times \psi_{vib}^{ex}$ must be of *u*-symmetry. For $\upsilon = 0$, in an octahedral system, ψ_{vib}^{gd} is of *g*-symmetry (more correctly A_{1g}) and υ' must make ψ_{vib}^{ex} of *u*-symmetry (*i.e.* T_{1u} or T_{2u} symmetry).

Thus in an octahedral system, $\upsilon = 0 \rightarrow \upsilon' = 0$, $\upsilon = 1 \rightarrow \upsilon' = 1$, etc. transitions are Laporte forbidden even when the vertical transition is possible. However, $\upsilon = 0 \rightarrow \upsilon' = 1$, $\upsilon = 1 \rightarrow \upsilon' = 2$, $\upsilon = 1 \rightarrow \upsilon' = 0$ etc. the transitions may not be Laporte forbidden.

Case II ($r_e^{ex} \rangle r_e^{gd}$): In such cases (*cf.* Fig. 7.7.2.1b), FCF for the most probable transition, $\upsilon = 0 \rightarrow \upsilon'_{min} (> 0)$, is relatively smaller (*i.e.* away from unity). Consequently, the peak broadening vibrational transitions (*i.e.* $0 \rightarrow \upsilon' > \upsilon'_{min}$ causing peak broadening towards the higher frequency; $0 \rightarrow \upsilon' < \upsilon_{min}$ causing peak broadening towards the lower frequency) become important to maintain the constancy in the total oscillator strength f for an allowed electronic transition. **It makes the peak broad.**

For the **octahedral complexes**, t_{2g}(≡ NBMO) → e_g (≡ ABMO) transitions make the metal-ligand bond distance longer in the excited state, *i.e.* $r_e^{ex} \rangle r_e^{gd}$ ***and the peaks become broader.***

Effect of temperature on the bandwidth (*cf.* Sec. 7.5.1R): It has been already explained that due to the increase of population density at the higher vibrational levels (*i.e.* υ = 1, 2, ...) at the higher temperature, the peak is broadened towards the longer wavelength. At the lower temperature, the peak becomes relatively sharper. The total oscillator strength (*f*) remains constant for the allowed electronic transitions but *f*-decreases with the lowering of temperatures for the electronic transitions allowed through the vibronic coupling.

7.7.3 Molecular Vibration to Broaden the Peak

Due to the metal-ligand bond vibration, the metal-ligand distance (*r*) varies. The 10*Dq* value is highly sensitive to this change (*cf.* $10Dq \propto \frac{1}{r^6}$; approximate relation). If energy of the ground state and excited state between which the transition occurs, do not vary in the same way (quantitatively) with the change of 10*Dq* value, then the energy difference between the states is different for the different values of 10*Dq* (*i.e.* the energy difference is different for the different values of the metal–ligand distance). In such cases, the transition between the states will cover a wide range of wavenumber. But if the concerned states run parallel with the 10*Dq* value (*i.e.* ligand field strength), then the energy difference between the states remains fixed during the change of metal-ligand distance due to the metal-ligand bond vibration. In such cases, the metal-ligand bond vibration cannot widen the peak. These are qualtitatively illustrated in Fig. 7.7.3.1.

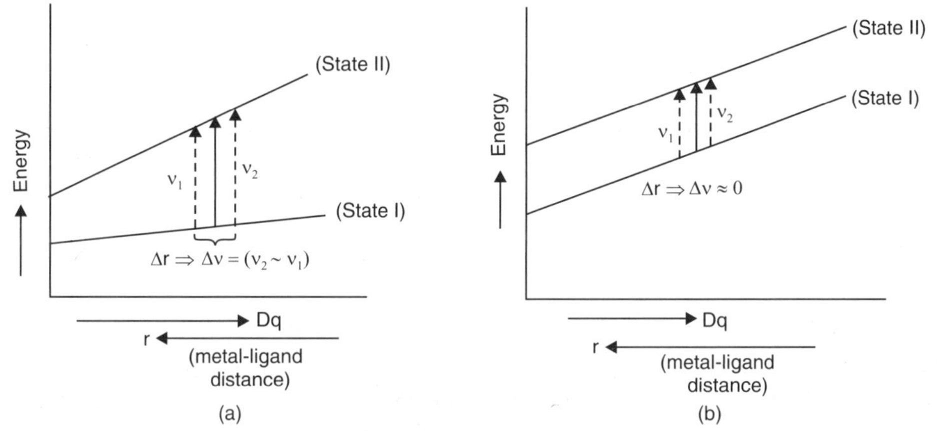

Fig. 7.7.3.1 Qualitative representation of the variation of energy of different states with *Dq* which is a function of metal-ligand distance (*r*). (a) If the metal-ligand symmetric vibration causes the change by Δr in *r* value, it would lead to $\Delta\nu$ band-width. (b) The change of Δr causes no change in the ν value of transition.

The effect of metal-ligand bond vibration on the bandwidth can be better understood by considering the **Tanabe-Sugano diagram**. This diagram will be discussed later.

It has been already mentioned that the **vibronic coupling** is responsible for the relaxation of Laporte selection rule in the centrosymmetric octahedral complexes. Thus intensity of the band/peak depends on the success of vibronic coupling in such cases but the vibronic coupling will broaden the peak at the same time. *With the lowering of temperature, the vibronic coupling becomes less efficient* (*cf.* Sec. 7.5.1) *and the intensity of the peak decreases but the peak becomes sharper.*

● **Interpretation of band broadening due to molecular vibration in terms of Tanabe-Sugano diagram:** The T.S. diagram records the relative change of energy of the different states with the change of crystal field strength (10*Dq*). Here we shall illustrate the effect in some representative cases.

(a) d^2–case: The **spin-allowed transitions** are:

$$^3T_{1g}(F) \rightarrow {}^3T_{1g}(P);\ {}^3T_{1g}(F) \rightarrow {}^3T_{2g}(F);\ {}^3T_{1g}(F) \rightarrow {}^3A_{2g}(F)$$

From the T.S. diagram (Fig. 7.7.3.2), it is evident that the energy of the states $^3T_{1g}(P)$ and $^3T_{2g}(F)$ vary almost in the same way with the change of 10*Dq* (*i.e.* the $^3T_{1g}(P)$ and $^3T_{2g}(F)$ states run parallel). On the other hand, the $^3A_{2g}(F)$ state is steeper. **This is why, the peak due to the transition to $^3A_{2g}(F)$ is expected to be broader than the peaks due to the transitions to $^3T_{1g}(P)$ and $^3T_{2g}(F)$.** In fact, the peaks due to the transitions to $^3T_{1g}(P)$ and $^3T_{2g}(F)$ are of almost comparable bandwidth—as expected from the T.S. diagram.

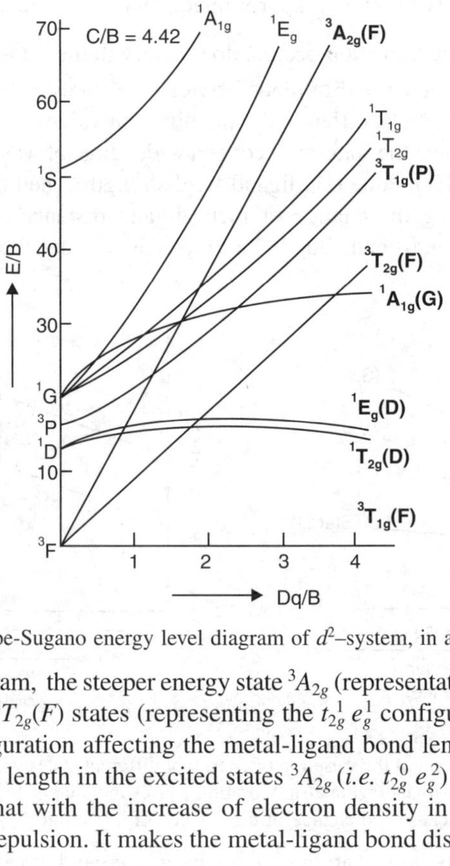

Fig. 7.7.3.2 Tanabe-Sugano energy level diagram of d^2–system, in an octahedral field.

In the Tanabe-Sugano diagram, the steeper energy state $^3A_{2g}$ (representation of the $t_{2g}^0\, e_g^2$ configuration) compared to the $^3T_{1g}(P)$ and $^3T_{2g}(F)$ states (representing the $t_{2g}^1\, e_g^1$ configuration) can be rationalised in terms of the electronic configuration affecting the metal-ligand bond length. Compared to the ground state $^3T_{1g}$(*i.e.* $t_{2g}^2\, e_g^0$), the bond length in the excited states $^3A_{2g}$ (*i.e.* $t_{2g}^0\, e_g^2$) and $^3T_{2g}$, $^3T_{1g}(P)$ (*i.e.* $t_{2g}^1 e_g^1$) is longer. It is due to the fact that with the increase of electron density in the e_g level, the approaching ligands experience the more repulsion. It makes the metal-ligand bond distance longer. The bond length sequence is:

$$t_{2g}^0\, e_g^2 \rangle\, t_{2g}^1 e_g^1 \,\rangle\, t_{2g}^2\, e_g^0$$

Thus the bond length difference (Δx) between the ground state and excited is maximum for the $t_{2g}^2 \rightarrow t_{2g}^0\, e_g^2$ transition while it is relatively less for the $t_{2g}^2 \rightarrow t_{2g}^1\, e_g^1$ transition. It explains why the bandwidth in the $^3T_{1g}(F) \rightarrow {}^3A_{2g}$ transition is much higher than that in the $^3T_{1g} \rightarrow {}^3T_{2g}$ transition.

The **spin-forbidden transitions to the states** $^1A_{1g}$(G) and $^1E_g(D)$, $^1T_{2g}(D)$ produce the very **sharp peaks.**

$$^3T_{1g}(F) \rightarrow {}^1A_{1g}(G);\ {}^3T_{1g}(F) \rightarrow {}^1E_g(D);\ {}^3T_{1g}(F) \rightarrow {}^1T_{2g}(D).$$

In fact, in the T.S. diagram, the states **$^1A_{1g}(G)$, $^1E_g(D)$ and $^1T_{2g}(D)$ run almost parallel to the ground $^3T_{1g}(F)$ state.** This is why, the metal-ligand bond vibration cannot broaden the peaks of these spin-forbidden transitions and **the peaks are sharp.**

These four states, *i.e.* $^3T_{1g}(F)$, $^1A_{1g}(G)$, $^1E_g(D)$ and $^1T_{2g}(D)$ arise from the **rearrangement of the two electrons in the t_{2g} level.** The ground state $^3T_{1g}(F)$ represents the high-spin electronic configuration t_{2g}^2, *i.e.* the two unpaired electrons are distributed in the three orbitals of t_{2g} level, while the said singlet states arise from the **low-spin electronic configuration of t_{2g}^2**, *i.e.* the paired electrons are distributed in the t_{2g}–orbitals. *It is obvious that the electronic transition from the $^3T_{1g}(F)$ ground state to the said singlet states merely involves the rearrangement of the electrons within the t_{2g} orbitals and it does not depend on Δ which gives the measure of energy separation between the t_{2g} and e_g levels.*

(b) d^3–system: The following **spin-forbidden transitions are very sharp.**

$$^4A_{2g}(F) \rightarrow {}^2E_g(G);\ {}^4A_{2g}(F) \rightarrow {}^2T_{1g}(G);\ {}^4A_{2g}(F) \rightarrow {}^2T_{2g}(G).$$

The sharpness of the above spin-forbidden transitions arises because the excited states $^2E_g(G)$, $^2T_{1g}(G)$ and $^2T_{2g}(G)$ run parallel to the ground state $^4A_{2g}(F)$ in the T.S. diagram. The reason behind this observation can be explained by considering the electronic configurations generating the $^4A_{2g}(F)$ and $^2E_g(G)$ terms. The high spin t_{2g}^3 configuration gives the $^4A_{2g}$ state while the low spin t_{2g}^3 configuration (*i.e.* one level in t_{2g} remains vacant) gives the ^2E_g state. Thus the $^4A_{2g} \rightarrow {}^2E_g$ transition **(intra-configurational transition)** does not change the electron density in the e_g-level and consequently the bond length does not change, *i.e.* bond length remains the same in both the ground ($^4A_{2g}$) and excited (^2E_g) states.

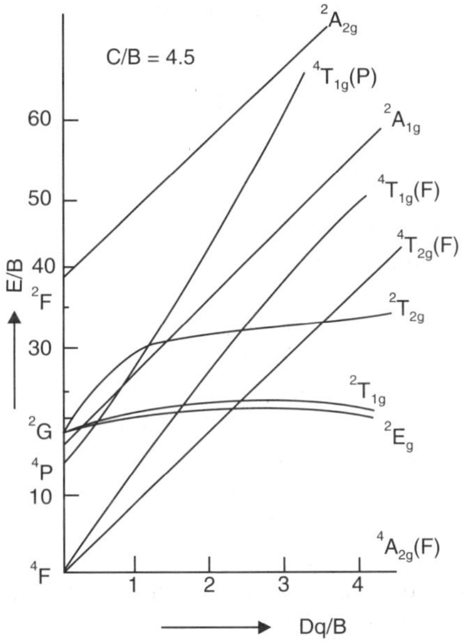

Fig. 7.7.3.3 Tanabe-Sugano energy level diagram of the d^3–system in an octahedral field.

Here it is important to mention that the role of Cr(III) (a d^3–system) doped in the Al_2O_3 lattice in **ruby laser** can be rationalised from the T.S. diagram of d^3–system. The emission of light from the excited $^2E_g(G)$ state to the ground state $^4A_{2g}(F)$ gives almost **a monochromatic laser beam** because these two states run parallel in the T.S. diagram.

7.7.4 Spin-Orbit Coupling (*cf.* Sec. 8.13) to Broaden the Peak

Spin-orbit coupling is more important for the heavier transition metals (*i.e.* 4*d* and 5*d* metals) and λ (spin-orbit coupling constant) is only in the range of 100 - 1000 cm^{-1} for the 3*d* metal ions. Spin-orbit coupling can broaden the peak in two ways:

(i) **Allowing the spin-forbidden transitions:** The spin selection rule is strictly applicable for the systems obeying the Russell-Saunders coupling. **The spin-orbit coupling can allow the mixing of different spin states and in such cases S is not a valid quantum number.** In fact, spin-orbit coupling makes the spin-forbidden transitions more intense and the intensity increases as the spin-orbit coupling increases. Very often, the spin forbiddent and spin allowed allowed transitions give the overlapping bands (which may not be resloved) to broaden the peak.

(ii) **Splitting of T-term:** Spin-orbit coupling can split the T-term to give the fine structure band or broadened band. This aspect is discussed in this section at a greater length.

In an atom, the terms like ^3P, ^3F can split through the spin-orbit coupling.

$$\underset{(L=1,\ S=1)}{^3P} \longrightarrow {}^3P_2,\ {}^3P_1,\ {}^3P_0; \quad \underset{(L=3,\ S=1)}{^3F} \longrightarrow {}^3F_4,\ {}^3F_3,\ {}^3F_2$$

The energy difference between these successive energy levels depends on the values of J and it is governed by **Lande-interval** rule, $\Delta E = E_J - E_{J-1} \propto J$ or, $\Delta E = \lambda J$.

Spin-orbit coupling constant (*cf.* Sec. 8.13)

It is generally denoted by ζ (zeta) which is the proportionality constant to express the spin-orbit coupling interaction energy, *i.e.*

$$E_{\text{spin-orbit}} \propto \vec{l}\,\vec{s}\cos(\vec{l},\vec{s})$$
$$= \zeta\vec{l}\,\vec{s}\cos(\vec{l},\vec{s})$$

The $\vec{l}$ and $\vec{s}$ vectors couple to generate the $\vec{j}$ vector. This proportionality constant ζ is used to express the spin-orbit interaction in a single electron. For a multielectronic system, it is replaced by λ which is related with ζ as follows:

$$\lambda = \pm\,\zeta/(2S) = \pm\,\zeta/n, \quad S = \frac{n}{2}$$

The + sign is applicable for the half-filled and less than half-filled shell (*e.g.* d^{1-4}) and – sign is applicable for more than half-filled shell (*e.g.* d^{6-9}).

J ranges from $|L - S|$ to $|L + S|$ and energy separation between the two successive levels specified by J and $J + 1$ is given by **Lande Interval Rule.**

$$\Delta E_{J,\,J+1} = \lambda(J + 1).$$

Energy (with respect to that of the unsplit Russel-saunders term of a particular level) of a level characterised by the J value is given by:

$$E_J = \frac{1}{2}\lambda[J(J+1) - L(L+1) - S(S+1)]$$

Thus for the $^3F_2$ state (ground level, d^2-system), its energy is given by:

$$\frac{1}{2}\lambda[2(2+1) - 3(3+1) - 1(1+1)] = -4\lambda.$$

Energy of an state generated through the spin-orbit coupling depends on the value of spin-orbit coupling constant λ. The energy is given by: $E_J = \frac{1}{2}\lambda[J(J+1) - L(L+1) - S(S+1)]$, with respect to the energy of the unsplit Russet-Saunders term.

Spin-orbit coupling can also split the crystal field **T-terms into different energy levels.** When a *T*–term splits, energy difference between the lowest and highest state may lie in the range 100 – 1000 cm^{-1}. The splitting of *T*–terms through spin-orbit coupling is shown in Figs. 7.7.4.1 and 2.

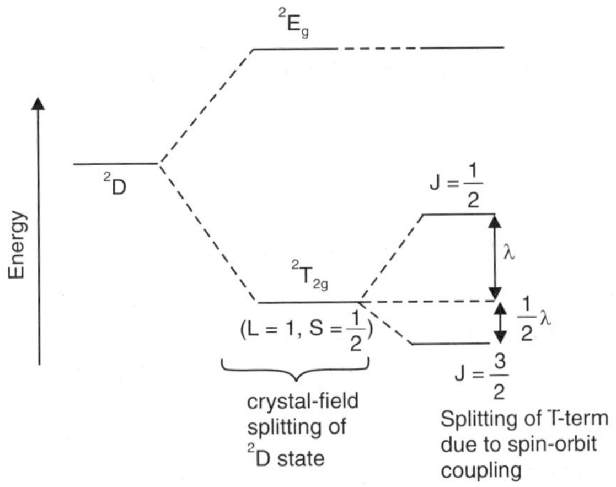

Fig. 7.7.4.1 Spin-orbit splitting of the $^2T_{2g}$ term of ^2D state of d^1–configuration in the octahedral field (*cf.* Fig. 8.20.3.1 for details).

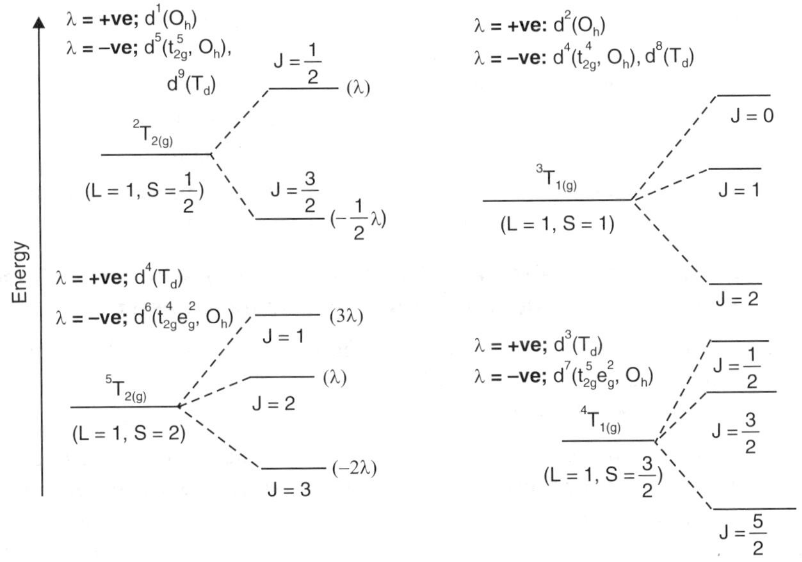

Fig. 7.7.4.2 Splitting of different the T-terms due to spin-orbit coupling. ($J = |L + S|$ to $|L - S|$) (*cf.* **Sec. 8.20.3** for determination of *L* and *S* values of a T-term). **Note:** Energy level order is shown for λ = +ve. For λ = –ve, the energy level order will be inverted.

If in the electronic transition, one term is of *T*–type, then a particular transition will cover the range of some hundred cm^{-1}. Thus it should create a **fine structure** components of the band. But for the first transition series (specially in solution), it is overshadowed by the **vibrational effect which produces the continuous bands** (*cf.* Fig. 7.7.3.1). This is why, the fine structure components of the bands due to spin-orbit coupling cannot be seen.

For the spin-forbidden transitions where peak broadening by vibration is not important, the fine structure of the peak can be detected. For the heavier transition metal ions of the 2nd and 3rd transition series, where the spin-orbit coupling constant is very high, peak broadening due to spin-orbit coupling is of much importance.

Here it may be mentioned that **the splitting of *A* and *E* terms due to spin-orbit coupling does not occur.**

Thus we can conclude as follows:

$$\textbf{Crystal field state:} \quad \underbrace{A_{1g}, A_{2g}, E_g}$$
$$\textbf{Spin-orbit coupling components:} \quad 1\,(i.e.\text{ no splitting})$$

Crystal field state:	$^2T_{2g}$	$^3T_{1g}$	$^4T_{1g}$	$^5T_{2g}$	$^3T_{2g}$	$^4T_{2g}$	$^5T_{2g}$
Spin-orbit coupling components:	2	3	3	3	3	3	3

Splitting of *T*–terms in tetrahedral fields also follows the same pattern.

(**Note:** ● It may be noted that in presence of a magnetic field, each level of a particular value of *J* can further split into the levels characterised by M_J value which can range from +*J* to –*J*.

● The spin-orbit coupling constant λ is positive for the half-filled and less than half-filled level (*i.e.* d^1 to d^5) and it is negative for more than half-filled (*i.e.* d^6 to d^9) level.

● For the states having the different *J* values, their energy order is determined by Hund's 3rd rule which states that for the half-filled or less than half-filled level, the state having the lowest *J* value is the most stable one and for the more than half-filled level, the state having the highest *J* value is the most stable one.)

Now let us consider the following transitions in d^2 and d^8–octahedral systems.

$$^3A_{2g} \rightarrow {}^3T_{2g},\ {}^3T_{1g}(\text{F}),\ {}^3T_{1g}(\text{P}) \qquad (d^8\text{–system})$$
$$^3T_{1g} \rightarrow {}^3T_{2g},\ {}^3T_{1g}(\text{P}),\ {}^3A_{2g}(\text{P}) \qquad (d^2\text{–system}).$$

Because of the splitting of *T*–terms due to spin-orbit coupling, the peak in each case of the above mentioned system should contain the fine structure components. This will broaden the peak.

7.7.5 Peak Broadening due to Jahn-Teller Distortion (*i.e.* Departure from Cubic Symmetry) and due to Reduction in Symmetry by the Ligands

(A) Jahn-Teller distortion effect: It has been mentioned that in an octahedral system, if the t_{2g} or e_g level is unsymmetrically filled in then degeneracy of that level is lifted. If the transition,

$$t_{2g}^m\, e_g^n \rightarrow t_{2g}^{m-1}\, e_g^{n+1}$$

experiences the J.T. distortion in the ground state and/or excited state, then splitting of the energy levels will allow more than one electronic transition. This multiple transition will broaden the peak. In some cases, the peak may split into different components. These aspects have been illustrated in Sec. 3.11.3. The J.T. distortion is also possible in the tetrahedral systems and the J.T. distortion can also broaden the peak in a tetrahedral system. However, this effect is relatively less important in the tetrahedral systems.

Here it should be mentioned that in the case of **dynamic J.T. distortion,** the crystallographic data indicate no distortion but the spectral property can reveal the distortion. **Thus effects of both the**

dynamic and static J.T. distortion are reflected on the spectra. This aspect has been discussed in Secs. 7.11, 7.14, and 3.11.3, 4.

(B) Reduction in Symmetry: When the symmetry of an octahedral ($O_h \rightarrow C_{4v}$, C_{2v}, D_{4h}, D_3, etc.) or a tetrahedral ($T_d \rightarrow D_{2d}$, etc.) complex is lowered by the ligand environment, additional bands (which may not be always resolved) appear in their spectra and very often they broaden the peak. It occurs due to the splitting of energy levels. For example, the bands of $[CoCl(NH_3)_5]^{2+}$ (C_{4V}) (purple coloured) are broader than those of $[Co(NH_3)_6]^{3+}$(O_h) (orange yellow coloured) though their band positions are almost similar indicating the effective O_h symmetry.

7.7.6 Band Shapes: Factors Controlling the Band Shape

Very often the bands are symmetrical of the **Gaussian shape** and in such cases, the distribution of ε for the different values of frequency (ν) is given by:

$$\varepsilon = \varepsilon_0 \exp[-\alpha(\nu - \nu_0)^2]$$

where α is a constant, ν_0 is the frequency at the centre of the band having the extinction coefficient ε_0. The area of a such symmetric peak is given by $\varepsilon_0 \Delta\nu_{1/2}$ (which is related with the oscillator strength) where $\Delta\nu_{1/2}$ is the half bandwidth.

But sometimes, the peaks are not symmetric and the factors which are responsible for band broadening are also responsible for the asymmetric peaks. The factors are discussed below.

(i) **Vibrational interaction (*cf.* Secs. 7.5.1 Q, R; 7.7.1–3 for details):** In vibronic coupling, vibrational quantun levels couple with the electronic quantum levels (*cf.* Fig. 7.4.1.3). The transitions like (0, 0), (1, 1), (2, 2), etc. are Laporte forbidden while the transitions like (0, 1), (1, 0), (2, 1), etc. may be allowed. In the bracket ( ), the first term indicates the vibrational quantum number of the ground state while the second term indicates the vibrational quantum number of the excited state.

Vibronic spectra (*i.e.* electronic spectra associated with a change in vibrational energy) are generally **asymmetric** due to the vibrational transitions between the ground and excited electronic states. The **most probable vibrational transition** (*i.e.* highest Franck-Condon Factor, FCF) in vibronic spectra is: $\upsilon^{gd} \rightarrow \upsilon^{ex}$ where υ^{gd} and υ^{ex} (which denote the vibrational quantum numbers at the ground and excited electronic states respectively) are having the minimum values for the allowed Franck-Condon transition. This will give the highest FCF. *Thus in the vibronic spectra, the band maximum does not represent exactly the electronic energy difference between the ground and excited electronic states.* Probabilities of other vibrational transitions of higher frequency and lower frequency are not always symmetrically distributed around the most probable vibrational transition which is determined by different factors like temperature, the relative values of r_e^{gd} and r_e^{ex}, the possible values of υ^{gd} and υ^{ex} for the vibronic coupling to relax the Laporte selection rule (*cf.* Fig. 7.7.2.1). These aspects are discussed in detail in Secs. 7.5.1, Q, R; 7.7.1-3.

With the increase of temperature, excited vibrational levels of the ground electronic state are populated according to Boltzmann's distribution law. **Transition from the higher vibrational levels of the ground electronic state will occur at a lower frequency.** It will produce a tail towards the lower frequency to introduce the asymmetric shape of the peak. This aspect has been illustrated in detail in Sec. 7.5.1Q, R.

(ii) **Spin-orbit coupling:** It has been already mentioned that due to spin-orbit coupling, the *T*–term is split and the splitting very often occurs asymmetrically. Thus the peak becomes asymmetric.

(iii) **Low-symmetry introduced by the ligands and Jahn-Teller effect:** It splits the *E* and *T*–terms. The *T*–term splits into two components: one component maintains the two fold orbital degeneracy of a particular energy; another component is nondegenerate of a different energy. **Thus the T–term splits asymmetrically. The *E*–term splits itself symmetrically.** But the resultant transitions due to splitting of *E* and *T* terms give the asymmetric shape of the peak. In fact, very often the resultant peak is the sum of different peaks (which are themselves symmteric individually and differing slightly in energy). When the individual peaks are of large difference in energy, separate symmetrical peaks will arise. Effect of J.T. distortion on spectra has been separately discussed in Sec. 3.11.

Conclusion: The above discussion indicates that there are many reasons for an asymmetric shape of the peaks. But, spectral bands of transition metal complexes in solution are very often symmetrical. It is believed that the resultant effects of the above mentioned causes make the peaks not very much asymmetric.

7.8 CHARACTERISTIC FEATURES OF THE ISOSBESTIC POINT

If two species remain in a dynamic equilibrium and their absorption spectra overlap at a point, the point is called **isosbestic point** that does not change with the position of the equilibrium. **Concentrations of the species remaining in an equilibrium may change depending on the condition but absorption at the isosbestic point will not change.** ***In fact, at the isosbestic point, the molar extinction coefficients of the species are the same.***

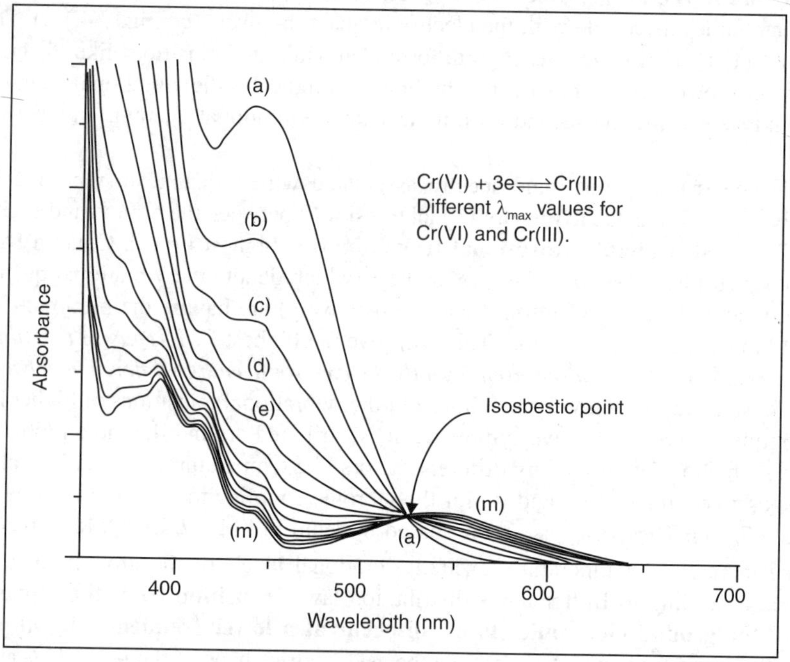

Fig. 7.8.1 A representative example of scanned spectrum in the oxidation of hexitols by chromic acid in presence of 2, 2′–bipyridine as a catalyst. (a) at the beginning; (m) at about after 50 min. Representing the gradual conversion of Cr(VI) into Cr(III)–species (Recorded in the author's laboratory).

If in a particular reaction, the reactant and product show an isosbestic point, the dynamic spectra nicely explain the characteristic property of the isosbestic point. As the reaction proceeds, spectra of either side of the isosbestic point will change, *i.e.* absorbance in one side (*i.e.* product) will increase while absorbance in the other side (*i.e.* reactant) will decrease but the absorbance at the isosbestic point will not change.

In general, the acid-base indicators show the different colours (*i.e.* spectra) for the acid (InH^+) and base (In) forms.

$$In + H^+ \rightleftharpoons InH^+$$

In such cases, if the spectra of an indicator is followed at different pH values, then the isosbestic point will be illustrated.

7.9 CRYSTAL FIELD TERMINOLOGY IN ELECTRONIC SPECTRA: ORIGIN OF THE ELECTRONIC SPECTRA AND LUMINESCENCE

7.9.1 Russel-Saunders Coupling (L–S Coupling) and Term Symbols

The details of Russel-Saunders coupling scheme have been discussed in "Fundamental Concepts of Inorganic Chemistry"– Volume 1. This scheme is applicable for the light elements ($Z = 30$). Thus it works well for the 1st transition series and for the heavier elements, *j–j* coupling scheme is required.

In terms of *L–S* coupling, the term symbol is expressed as:

$^{2s+1}L_J$ where $2S+1$ gives the spin multiplicity,

L =	0	1	2	3	4	5	6 ....
Symbol =	S	P	D	F	G	H	I ....

7.9.2 Ground State Term Symbols for the 3d^n Configurations and Hund's Rules

These are discussed in detail in "Fundamental Concepts of Inorganic Chemistry"– Volume 1. Here we give the results for the $3d^n$ configurations.

Table 7.9.2.1 Ground state terms for the d^n configurations.

$3d^n$	m_l					M_L	M_S	Term symbol
	+2	+1	0	−1	−2			
$3d^1$(*e.g.* Ti^{3+})	↑	–	–	–	–	2	1/2	$^2D_{3/2}$
$3d^2$(*e.g.* V^{3+})	↑	↑				3	1	$^3F_2$
$3d^3$(*e.g.* Cr^{3+})	↑	↑	↑			3	3/2	$^4F_{3/2}$
$3d^4$(*e.g.* Cr^{2+})	↑	↑	↑	↑		2	2	$^5D_0$
$3d^5$(*e.g.* Mn^{2+})	↑	↑	↑	↑	↑	0	5/2	$^6S_{5/2}$
$3d^6$(*e.g.* Fe^{2+})	↑↓	↑	↑	↑	↑	2	2	$^5D_4$
$3d^7$(*e.g.* Co^{2+})	↑↓	↑↓	↑	↑	↑	3	3/2	$^4F_{9/2}$
$3d^8$(*e.g.* Ni^{2+})	↑↓	↑↓	↑↓	↑	↑	3	1	$^3F_4$
$3d^9$(*e.g.* Cu^{2+})	↑↓	↑↓	↑↓	↑↓	↑	2	1/2	$^2D_{5/2}$

7.9.3 Terms Arising from the d^n–Configuration

The term symbol of the ground state can be easily obtained with the help of Hund's rules. Besides the ground state, there are many possible excited states. To obtain the term symbols for such excited states, we are to consider the M_L and M_S values for the corresponding excited states. A particular electronic arrangement with the specific m_l and m_s values represents a **microstate.** Thus a microstate represents the allowed combination of the m_l and m_s values of a particular electronic configuration.

The procedure for determination of the microstates for different d^n configurations has been discussed in "Fundamental Concepts of Inorganic Chemistry"—Volume 1. Here it may be mentioned that according to **hole formalism rule,** d^n and d^{10-n} configurations give the same terms. The possible terms arising from the different d^n–configurations are given in Table 7.9.3.1.

(a) Hole formalism or equivalency: Russel-Saunders terms for the d^n and d^{10-n} configurations are the same. Let us illustrate for the d^2 and d^8 systems. The d^8 system may result by adding two positrons in the filled d^{10} electronic configuration. A filled shell does not contribute anything towards the resultant angular momenta, *i.e.* $L = 0$, $S = 0$. Thus, the addition of two positrons to the d^{10} electronic configuration will give the values of L and S determined by the two positrons, *i.e. the Russell-Saunders terms due to the d^{10}-electron ($L = 0$, $S = 0$) + d^2-positron will arise solely from the d^2-positronic configuration.* In terms of the interelectronic repulsion, the positrons are equivalent to the electrons. Thus the Russel-Saunders terms for the d^2 and d^8 systems will be the same. In other words, in terms of L–S coupling, **the d^8–electronic configuration is equivalent to the d^2 positronic system which is equivalent to the d^2 electronic configuration.** Because of the same ground, the p^n and p^{6-n} configurations will have the same Russel-Saunders terms. But an electron is most stable where a positron is least stable.

Table 7.9.3.1 Terms for different d^n–configurations

Configuration	*Terms (after considering Pauli exclusion principle)*
d^1, d^9	$^2\mathbf{D}$
d^2, d^8	$^3\mathbf{F}$, ^3P, ^1G, ^1D, ^1S
d^3, d^7	$^4\mathbf{F}$, ^4P, ^2H, ^2G, ^2F, $^2D(2)$, ^2P
d^4, d^6	$^5\mathbf{D}$, ^3H, ^3G, $^3F(2)$, ^3D, $^3P(2)$, ^1I, $^1G(2)$, ^1F, $^1D(2)$, $^1S(2)$
d^5	$^6\mathbf{S}$, ^4G, ^4F, ^4D, ^4P, ^2I, ^2H, $^2G(2)$, $^2F(2)$, $^2D(3)$, ^2P, ^2S

The ground state terms are shown in bold letters. The number 2, 3, etc. in parentheses denotes the number of times the term appears.

(b) Relative energies of the different terms arising from the d^2–configuration (an illustrative example): After considering the Pauli exclusion principle, we get the following terms for a d^2 configuration.

$$d^2: {}^3F, {}^3P, {}^1G, {}^1D, {}^1S$$

Hund's spin multiplicity rule (1st rule) states that the most stable state possesses the highest spin multiplicity value $(2S + 1)$. ***Hund's orbital multiplicity rule (i.e. 2nd Rule)*** states that for the different terms having the same spin multiplicity, the stability increases with the increase of orbital multiplicity $(2L + 1)$. Thus the energy order runs as:

$$^3F < {}^3P < {}^1G < {}^1D < {}^1S$$

The energy order may be $^3F < {}^3P < {}^1D < {}^1G < {}^1S$ under some conditions (to be discussed later in this section). Under some conditions, $^3P > {}^1D$ energy order **(contradicting the Hund's rule)** may also appear.

Each term can split depending upon the J-values. The energy order for the states differing only in the J values (*i.e.* same L and S values) can be determined by using the ***Hund's 3rd rule*** (*i.e.* rule of J) which states that for a half-filled or less than half-filled level (*e.g.* p^1 to p^3, d^1 to d^5), the lowest J value will indicate the most stable state while for a more than half-filled level (*e.g.* p^4 to p^6, d^6 to d^{10}), the highest J value indicates the most stable state. Thus the term ^3F generates the states $^3F_4$, $^3F_3$ and $^3F_2$ through the spin-orbit coupling interaction and their energy order is:

$$^3\boldsymbol{F}: {}^3F_2 \langle {}^3F_3 \langle {}^3F_4 \ (d^2 \text{ — less than half-filled})$$

Similarly, we can write for ^3P:

$$^3\boldsymbol{P}: {}^3P_0 \langle {}^3P_1 \langle {}^3P_2$$

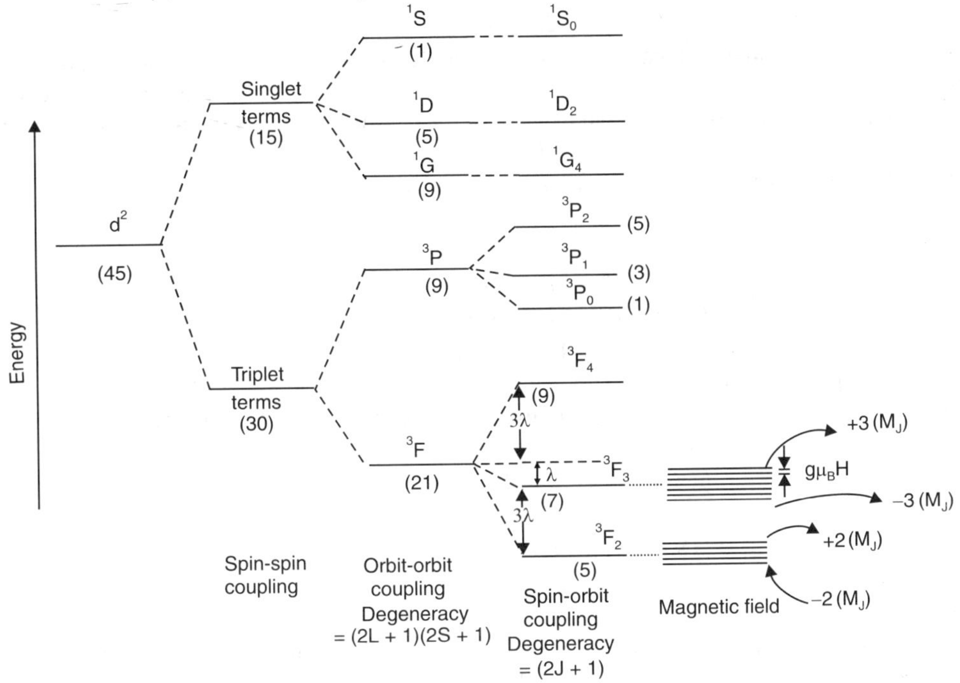

Fig. 7.9.3.1 Schematic representation of the splitting of energy states for d^2-system (Total degeneracy of each level is indicated in the parenthesis). Energy order according to Hund's rule but the actual energy order of the different states depend on the B/C ratio to be discussed later.

Considering the **further splitting of the *J*–values** in presence of a magnetic field into the M_J values which can range from $+J$ to $-J$, (*i.e.* splitting into $2J + 1$ levels, *i.e.* each state is $(2J + 1)$ fold degenerate), the energy states arising from the d^2–configuration are shown in Fig. 7.9.3.1.

The energy separation between the states differing by only J values is governed by **Lande Interval Rule** (*cf.* Sec. 8.13). It states that the energy separation between the successive levels having the J values as J and $J + 1$ is proportional to the higher value of J, *i.e.* $J + 1$, *i.e.*

$$\Delta E_{J,\ J+1} \propto (J + 1), \text{ or } \Delta E_{J,\ J+1} = \lambda\ (J+1)$$

The proportionality constant λ is called the **spin-orbit coupling constant.** The spin-orbit coupling leads to the energy levels characterised by the J values and energy of a such level is determined by the following relation.

$$E_J = \frac{1}{2}\lambda[J(J + 1) - L(L + 1) - S(S + 1)], \text{ (with respect to that of the unsplit term)}$$

Thus for $^3F_2$ (ground level, d^2 configuration), its energy is given by:

$$\frac{1}{2}\lambda[2(2 + 1) - 3(3 + 1) - 1(1 + 1)] = -4\lambda.$$

Splitting of a level having a particular J value leads to $2J + 1$ sublevels which are equally spaced and separation energy is $g\mu_B H$ where g is called the **Lande splitting factor,** μ_B stands for the **Bohr magneton** and H gives the magnetic field strength. The Lande splitting factor g is given by the **Lande formula.**

$$g = 1 + \frac{J(J+1) + S(S+1) - L(L+1)}{2J(J+1)}$$

(c) Energy of the states in terms of Condon-Shortley parameters and Racah parameters: It is illustrated for the states arising from a d^2 configuration. The energies of the states arising due to the **interelectron repulsion** can be expressed in terms of the **Condon-Shortley parameters** (F_0, F_2 and F_4) or the empirical **Racah parameters** (A, B and C). In terms of the Condon-Shortley parameters, energies of the states arising from a d^2–configuration are given below:

$$E(^1G) = F_0 + 4F_2 + F_4$$
$$E(^3F) = F_0 - 8F_2 - 9F_4$$
$$E(^1D) = F_0 - 3F_2 + 36F_4$$
$$E(^3P) = F_0 + 7F_2 - 84F_4$$
$$E(^1S) = F_0 + 14F_2 + 126F_4$$

F_0 is common to all terms.

In terms of the Racah parameters **(measuring the interelectron repulsion energies),** energies of the states of a d^2–configuration are given as follows:

$$E(^1G) = A + 4B + 2C$$
$$E(^3F) = A - 8B$$
$$E(^1D) = A - 3B + 2C$$
$$E(^3P) = A + 7B$$
$$E(^1S) = A + 14B + 7C$$

Here A is common to all terms. To consider the relative energies (which are important in understanding the spectral properties), it is not required to know the value of A. To know the energy difference (= $15B$) between the ^3P and ^3F terms, we need only the knowledge of the B term. Similarly, we need the knowledge of only the B term for the energy difference between the ^1D and ^1G terms. To know the energy difference between the ^3F and ^1D states (differing in spin multiplicities), we require the knowledge of both B and C.

Under the condition, $C \rangle 5B$, energy order of the terms is :

$$^3F \langle\, ^3P \langle\, ^1D \langle\, ^1G \langle\, ^1S \text{ (cf. Fig. 7.9.3.1 and energy order of } ^1D \text{ and } ^1G)$$

This order is **approximately** the same as obtained by considering the Hund's rules. However, under the condition, $C \langle 5B$, ^3P lies above ^1D (*cf.* **contradicting the Hund's rules**) and it happens so for Ti^{2+}. Thus for Ti^{2+}, the observed order is: $^3F \langle\, ^1D \langle\, ^3P \langle\, ^1G \langle\, ^1S$. ***In fact, the orbital multiplicity rule (i.e. Hund's second rule) is not reliable for predicting the order of excited states but it can predict the ground state (i.e. highest L).*** It has been noted that for most of the metal ions of 1st transition series, the condition $C \approx 4B$ is maintained.

The Condon-Shortley parameters are related with the Racah parameters as follows:

$$A = F_0 - 49F_4,\ B = F_2 - 5F_4,\ C = 35F_4,\ F_2 = B + \frac{C}{7}$$

The Racah parameters can be obtained from a linear combination of the exchange and Coulomb integrals. For the first transition metal ions, the ratio C/B is essentially constant at a value of 4 and B is in the order of 1000 cm^{-1}. The interelectron repulsion parameters B and C are decreased in the complexes compared to their values in the free ions. The B is decreased to about 70% during the complexation.

The energy difference between the two states for a d^2–configuration can be given as follows:

$$E(^3P) - E(^3F) = 15(F_2 - 5F_4) = 15B$$

$$E(^1D) - E(^3F) = 5(F_2 + 9F_4) = 5B + 2C.$$

Similarly, it can be shown that for a d^3–configuration (*e.g.* V^{2+}), the energy difference between the ^4F and ^4P states is $15B$.

Table 7.9.3.1 Energy differences for some of the terms for different d^n configurations

		Energy difference	
Configuration	***Term interval***	***F parameter***	***B and C Parameters***
d^2, d^8	$^1S - {}^3F$	$22F_2 + 135F_4$	$22B + 7C$
	$^1G - {}^3F$	$12F_2 + 10F_4$	$12B + 2C$
	$^3P - {}^3F$	$15F_2 - 75F_4$	$15B$
	$^1D - {}^3F$	$5F_2 + 45F_4$	$5B + 2C$
	$^1S - {}^1D$	$17F_2 + 90F_4$	$17B + 5C$
	$^3P - {}^1D$	$10F_2 - 120F_4$	$10B - 2C$
d^3, d^7	$^2P - {}^4F$	$9F_2 + 60F_4$	$9B + 3C$
	$^2G - {}^4F$	$4F_2 + 85F_4$	$4B + 3C$
	$^2H - {}^4F$	$9F_2 + 60F_4$	$9B + 3C$
	$^4P - {}^4F$	$15F_2 - 75F_4$	$15B$
	$^2G - {}^2H$	$-5F_2 + 25F_4$	$-5B$
d^4, d^6	$^3D - {}^5D$	$16F_2 + 60F_4$	$16B + 4C$
	$^3G - {}^5D$	$9F_2 + 95F_4$	$9B + 4C$
	$^3H - {}^5D$	$4F_2 + 120F_4$	$4B + 4C$
	$^3D - {}^3H$	$12F_2 - 60F_4$	$12B$
	$^3G - {}^3H$	$5F_2 - 25F_4$	$5B$
d^5	$^4G - {}^6S$	$10F_2 + 125F_4$	$10B + 5C$
	$^4F - {}^6S$	$22F_2 + 135F_4$	$22B + 7C$
	$^4D - {}^6S$	$17F_2 + 90F_4$	$17B + 5C$
	$^4P - {}^6S$	$7F_2 + 210F_4$	$7B + 7C$
	$^2I - {}^6S$	$11F_2 + 225F_4$	$11B + 8C$
	$^2H - {}^6S$	$13F_2 + 285F_4$	$13B + 10C$

Thus it is evident that the energy difference between the states having the same spin multiplicity is a function of the single Racah parameter B while the energy difference between the states of different spin multiplicities is a function of two Racah parameters B and C (cf. Table 7.9.3.1).

In Tanabe-Sugano diagram, energy (E) and field strength ($10Dq$) are to be expressed in terms of the parameter B and the plot is E/B *vs.* $10Dq/B$.

7.9.4 Terms Arising in Presence of Ligand Fields and Electronic Transitions Developing the Ligand Field Bands: Crystal Field Spitting of Free Ion Terms

The terms of free ions undergo splitting in the presence of ligands. Splitting of terms of free ions in **cubic fields**—cubic, octahedral and tetrahedral is shown in Table 7.9.4.1.

Table 7.9.4.1 Free ion terms and their splitting in cubic crystal fields.

Free ion term	*Orbital multiplicity (2L + 1)*	*Terms arising in crystal field Spectroscopic States*
S	1	A_1
P	3	T_1
D	5	$E + T_2$
F	7	$A_2 + T_1 + T_2$
G	9	$A_1 + E + T_1 + T_2$
H	11	$E + T_1 + T_2 + T_1$
I	13	$A_1 + A_2 + E + T_1 + T_2 + T_2$

Note:

(i) Splitting of the terms in the presence of a crystal field, may be compared with the splitting pattern of the orbitals in a **cubic crystal field**.
Single s–orbital (spherically symmetrical): no splitting
Three p–orbital: no splitting
Five d–orbital: Splitting into two groups t_2 and e (one triply degenerate and another one doubly degenerate)
Seven f–orbitals: Splitting into three groups: t_1 (triply degenerate), t_2 (triply degenerate) and a_2 (nondegenerate)

(ii) The capital letter T, E, A, B denote the **orbital degeneracy.** T stands for 3–fold degeneracy, E stands for 2–fold degeneracy; A or B stands for no degeneracy.

(iii) The *numerical supercripts* (*e.g.* 2, 3, 4) represent the spin multiplicity.

(iv) The numerical subscripts in T_1, T_2, A_2, etc. are originated from group theory. These are beyond the scope of this book.

(v) The subscripts g and u denote *gerade* (*i.e.* even) or *ungerade* (u). The label g is applicable for the systems having an inversion centre, *e.g.* regular octahedron and the label u appears for the systems lacking in centre of symmetry as in a tetrahedron.

(vi) During the splitting of free ion term, **orbital multiplicity (2L + 1) is retained in the sum of the terms.** It is illustrated below:
$F(L = 3, \textit{i.e. } 2L + 1 = 7) \rightarrow A_2\,(1) + T_1\,(3) + T_2\,(3)$
i.e. total orbital multiplicity = 1 + 3 + 3 = 7.

(b) Splitting pattern of the free ion terms (D, F) in a cubic crystal field: Orbital splitting of the D and F terms are shown in Fig. 7.9.4.1. (*See* Solved Problem 14 for **energy calculation**).

Let us illustrate for the d^1 and d^2 configurations which give the ground state terms D and F respectively. For the d^1 system, t_{2g}^1 corresponds to $^2T_{2g}$ (crystal field energy = $-4Dq$) and $t_{2g}^0\ e_g^1$ (excited state) corresponds to ^2E_g (crystal field energy = + $6Dq$). These energy values are given with respect to that of the unsplit ^2D state.

For the d^2 system, the ^3F term splits into the $^3T_{1g}$(F), $^3T_{2g}$(F) and $^3A_{2g}$(F) states and there is a $^3T_{1g}$(P) term arising from the excited ^{3}P state. **Thus there are two $^3T_{1g}$ states and they tend to mix.** This mixing is insignificant in a strong crystal field where the states are widely apart but the mixing is possible to some extent in a weak crystal field. No such problem exists for the $^3T_{2g}$(F) ($\equiv t_{2g}^1 e_g^1$, *i.e.* first excited state) and $^3A_{2g}$(F) ($\equiv t_{2g}^0\ e_g^2$, *i.e.* second excited state) states. Thus for the strong field ligand, the ground state $^3T_{1g}$(F) corresponds to $t_{2g}^2\ e_g^0$ but in a weak field, considering the mixing between the $^3T_{1g}$(F) and $^3T_{1g}$(P) states, the ground state may be written as $t_{2g}^x\ e_g^y$ where $x + y = 2$ and $x < 2$. From Fig. 7.9.4.1, we can write:

Ground state $^3T_{1g}(F) \equiv t_{2g}^x\ e_g^y$, crystal field energy = $(-4x + 6y)Dq = pDq$ (say).

1st excited state $^3T_{2g}(F) \equiv t_{2g}^1\ e_g^1$, crystal field energy = $+2Dq$.

2nd excited state $^3A_{2g}(F) \equiv t_{2g}^0\ e_g^2$, crystal field energy = $+12Dq$.

i.e. $3p + 3(2Dq) + 1(12Dq) = 0$, or $3p = -18Dq$; or, $p = -6Dq$. (**Considering the orbital degeneracy** and energy conservation principle).

It leads to: $(-4x + 6y)Dq = p = -6Dq$; and $x + y = 2$

Solving these two equations we get: $x = 1.8$, $y = 0.2$. **Thus, in a weak crystal field, the ground state $^3T_{1g}$(F) represents $t_{2g}^{1.8}\ e_g^{0.2}$ not t_{2g}^2** (*cf.* **Solved Problem 14**). In fact, in a weak crystal field, the interelectronic repulsion shifts some electron density from the t_{2g} level to the e_g level.

In a strong crystal field, the ground state (*i.e.* t_{2g}^2) is so stabilised that no electron cloud moves from the t_{2g} level to the e_g level.

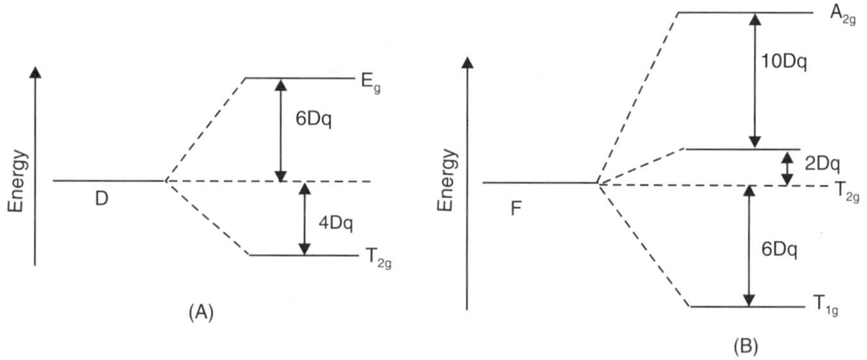

Fig. 7.9.4.1 Splitting of $D(d^1)$ and $F(d^2)$ terms in a weak octahedral field (*cf. see* Solved Problem 14 for the energy values).

(**Note:** Energy values are given in Fig. 7.9.4.1 without considering the symmetry interaction energy in the case of splitting of the F–term).

(c) d^n and d^{10-n} systems; d^n and d^{n+5} systems: According to the rule of **hole formalism, the d^n and d^{10-n} systems** should have the same Russel-Saunders terms and it arises due to the equivalence of the **d^n (electronic) configuration** with the **d^n (positronic configuration)**. It is illustrated for the d^2 and d^8 *i.e.* d^{10-2} electronic configurations. A d^8 electronic configuration is obtained by adding 2 positrons in a filled d^{10} electron subshell for which L = 0, S = 0. Thus, the Russel-saunders terms for d^{10} electron + d^2 positron will be determined by the d^2 positronic configuration, *i.e.* we can write: Russel-Saunders terms of the d^8 electronic configurations and d^{10-8} *i.e.* d^2 positronic configurations are the same. **Russel-Saunders terms for the d^n electronic and d^n positronic systems are the same.** *But an electron is most stable where a positron is least stable.*

Thus both d^8 and d^2 electronic configurations will give the same Russel-Saunders terms. It leads to: d^n electronic configuration ≡ d^n positronic configuration ≡ d^{10-n} positronic configuration ≡ d^{10-n} electronic configuration. It leads to: in terms of Russel-Saunders terms, d^n electronic configuration ≡ $(10 - n)$ holes in the d^{10} electronic configuration, *i.e.* **d^n configuration is hole equivalent of d^{10-n} configuration. But an electron is most stable where a positron is least stable.** This is why, the pattern of splitting of the spectroscopic states arising from the d^n and d^{10-n} electronic configurations is of opposite nature, *i.e.*

$$d^n\text{–diagram} = \text{inverse of } d^{10-n} \text{ diagram.}$$

Russel-Saunders terms (*i.e.* L value) for **d^n and d^{n+5}** are the same in an weak field condition because the half-filled d^5 system generates a zero orbital angular momentum ($L = 0$). However, they differ in spin multiplicity only. It is illustrated below:

d^2 and d^7: ${}^xF(x = 3 \text{ for } d^2, x = 4 \text{ for } d^7)$;

d^3 and d^8: ${}^xF(x = 4 \text{ for } d^3, x = 3 \text{ for } d^8)$;

d^4 and d^9: ${}^xD(x = 5 \text{ for } d^4, x = 2 \text{ for } d^9)$,

This is why, **splitting pattern of the spectroscopic terms is the same for d^n and d^{n+5} configurations.** The splitting patterns of the d^n–systems in a weak octahedral field are shown in Table 7.9.4.2.

Table 7.9.4.2 Spectroscopic terms (in weak octahedral field) for different d^n–ions (Splitting pattern with reference to the diagrams *A* and *B* given in Fig. 7.9.4.1).

d^n	*Ground State (free ion)*	*Energy level diagrams (Spetroscopic terms)*	*Important excited state (free ion)*
d^1	^2D	A	–
d^2	^3F	B	^3P
d^3	^4F	Inverted B	^4P
d^4	^5D	Inverted A	–
d^5	^6S	No splitting (*i.e.* $^6A_{1g}$)	–
d^6	^5D	A	–
d^7	^4F	B	^4P
d^8	^3F	Inverted B	^3P
d^9	^2D	Inverted A	–
d^{10}	^1S	No splitting	–

(d) Electronic configurations of the spectroscopic states developed in a cubic crystal field for the d^2– and d^3– configurations: Now let us illustrate the possible spectral transitions for a **d^2–configuration** for which the important terms in free ion are ^3F and ^3P. The possible electronic configurations are:

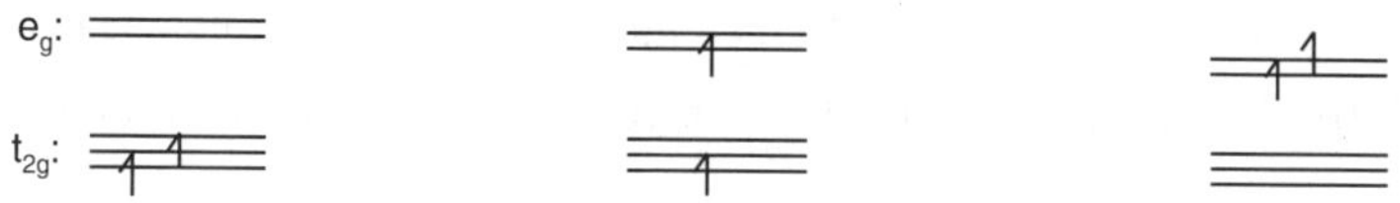

($t_{2g}^2\,e_g^0$, ground state, $^3T_{1g}(F)$, *i.e.* orbital degeneracy = 3)

($t_{2g}^1 e_g^1$, excited state; $^3T_{1g}$(P) + $^3T_{2g}$(F) *i.e.* total orbital degeneracy = 6)

($t_{2g}^0\,e_g^2$, excited state; $^3A_{2g}$ *i.e.* orbital degeneracy = 1)

Now let us consider the mode of distribution of electrons in different states.

● In the **ground state t_{2g}^2** the two electrons will be singly occupying any two of the three t_{2g} orbitals keeping their spins parallel (*i.e.* spin multiplicity = 3). These two electrons can be distributed in three ways (*i.e.* orbital degeneracy = 3).

$$(d_{xy})^1(d_{xz})^1;\ (d_{xz})^1(d_{yz})^1;\ (d_{xy})^1(d_{yz})^1$$

This state (orbital degeneracy = 3 denoted by T, spin multiplicity = 3) is represented by $^3T_{1g}(F)$

● For the ***d*³-system** (S = 3/2), the important free ion terms are ^{4}F and ^{4}P. The ground state and two excited states are as follows:

Ground state: $t_{2g}^3 e_g^0$, $^4A_{2g}(F)$ (orbitally singlet state); **First excited state:** $t_{2g}^2 e_g^1$, $^4T_{1g}(F)+{}^4T_{2g}(F)$, *i.e.* orbitally 6-fold degenerate; **Second excited state:** $t_{2g}^1 e_g^2$, $^4T_{1g}(P)$ (*i.e.* orbitally 3-fold degenerate).

● For the ***d*²-system, in the excited state $t_{2g}^1\,e_g^1$,** two different energy states develop differing in the degree of interelectron repulsion. **If the electron in the t_{2g} level resides in the d_{xy} orbital, then the electron in the $d_{x^2-y^2}$ orbital of the e_g-set will experience more inter-electron repulsion compared to the electron placed in the d_{z^2} orbital of the e_g-set. Thus the electronic configuration, $(d_{xy})^1\left(d_{x^2-y^2}\right)^1$, is less stable than the electronic configuration $(d_{xy})^1\left(d_{z^2}\right)^1$.**

Similarly, $(d_{xz})^1\left(d_{z^2}\right)^1$ is less stable than $(d_{xz})^1\left(d_{x^2-y^2}\right)^1$ and $(d_{yz})^1\left(d_{z^2}\right)^1$ is less stable than $(d_{yz})^1\left(d_{x^2-y^2}\right)^1$. Thus, we can have two different sets of electron distribution for the excited state $t_{2g}^1 e_g^1$.

$$\left.\begin{array}{l}\textbf{Lower energy set:}\ \underbrace{\left(d_{xy}\right)^1\left(d_{z^2}\right)^1;\ \left(d_{xz}\right)^1\left(d_{x^2-y^2}\right)^1;\ \left(d_{yz}\right)^1\left(d_{x^2-y^2}\right)^1}_{\text{Orbital degeneracy=3. Spin multiplicity=3; symbol= }{}^3\mathbf{T_{2g}(F)}.}\end{array}\right\}\ t_{2g}^1\,e_g^1$$

$$\left.\begin{array}{l}\textbf{Higher energy set:}\ \underbrace{\left(d_{xy}\right)^1\left(d_{x^2-y^2}\right)^1;\ \left(d_{xz}\right)^1\left(d_{z^2}\right)^1;\ \left(d_{yz}\right)^1\left(d_{z^2}\right)^1}_{\text{Orbital degeneracy=3. Spin multiplicity=3; symbol=}{}^3\mathbf{T_{1g}(P)}.}\end{array}\right\}\ t_{2g}^1 e_g^1$$

The higher interelectron repulsion in the second set represented by $^3T_{1g}(P)$ makes the energy state less stable compared to the $^3T_{2g}(F)$ state.

● **d², d³ system: $t_{2g} \rightarrow e_g$ electron transition gives two peaks—why?** The energy difference between the t_{2g} level and e_g level is Δ_o. Thus for the $t_{2g}^x\,e_g^0 \rightarrow e_g^1 t_{2g}^{x-1}$ (where x = 2, 3), electron promotion (*e.g.* initial state: $t_{2g}^2\,e_g^0$; final state; $t_{2g}^1\,e_g^1$), we can expect one peak ($h\nu = \Delta_o$). But the electron-electron repulsion in the excited state $t_{2g}^1 e_g^1$ gives two different energy states, *i.e.* $^3T_{1g}$(P) and $^3T_{2g}$(F). It explains the origin of two peaks for the $t_{2g} \rightarrow e_g$ electron promotion. Similarly, the $t_{2g}^3\,e_g^0 \rightarrow t_{2g}^2 e_g^1$ transition gives two peaks because the $t_{2g}^2\,e_g^1$ excited state gives two energy states $^4T_{1g}(F)$ and $^4T_{2g}(F)$ which differ in electron-electron repulsion.

● For the d^2-system, in the **excited state e_g^2,** the unpaired two electrons can be distributed in a single way, *i.e.* $\left(d_{x^2-y^2}\right)^1\left(d_{z^2}\right)^1$. It leads to orbital degeneracy = 1 and spin multiplicity = 3. The state is respresented by $^3A_{2g}$.

The total orbital degeneracy of the $^3F(L = 3)$ and $^3P(L = 1)$ terms is: $(2 \times 3 + 1) + (2 \times 1 + 1) = 10$.

Table 7.9.4.3 Spectroscopic transitions for the d^n–configuration in a weak octahedral field

Free ion		***Complex (O_h)***				***Transitions (Spin allowed)***
		Ground state		***Excited state***		
Elect. config.	***Term***	***Elect. config.***	***Term***	***Elect. config.***	***Term***	
d^1	^2D	$t_{2g}^1\, e_g^0$	$^2T_{2g}$	$t_{2g}^0\, e_g^1$	^2E_g	$^2T_{2g}(F) \rightarrow {}^2E_g$
d^2	^3F, ^3P*	$t_{2g}^2\, e_g^0$	$^3T_{1g}(F)$	$t_{2g}^1\, e_g^1$	$^3T_{2g}(F) + {}^3T_{1g}(P)$	$^3T_{1g}(F) \rightarrow {}^3T_{2g}(F)$
				$t_{2g}^0\, e_g^2$	$^3A_{2g}$ (F)	$^3T_{1g}(F) \rightarrow {}^3T_{1g}(P)$
						$^3T_{1g}(F) \rightarrow {}^3A_{2g}$
						(Spin forbidden)
d^3	^4F, ^4P*	$t_{2g}^3\, e_g^0$	$^4A_{2g}(F)$	$t_{2g}^1\, e_g^2$	$^4T_{1g}(P)$	$^4A_{2g} \rightarrow {}^4T_{2g}(F)$
				$t_{2g}^2\, e_g^1$	$^4T_{1g}(F) + {}^4T_{2g}(F)$	$^4A_{2g} \rightarrow {}^4T_{1g}(F)$
						$^4A_{2g} \rightarrow {}^4T_{1g}(P)$
d^4	^5D	$t_{2g}^3\, e_g^1$	^5E_g	$t_{2g}^2\, e_g^2$	$^5T_{2g}$	$^5E_g \rightarrow {}^5T_{2g}$
d^5	^6S	$t_{2g}^3\, e_g^2$	$^6A_{1g}$	$t_{2g}^2\, e_g^3$	$^4T_{1g} + {}^4T_{2g}$	**Spin-forbidden transitions**
d^6	^5D	$t_{2g}^4\, e_g^2$	$^5T_{2g}$	$t_{2g}^3\, e_g^3$	^5E_g	$^5T_{2g} \rightarrow {}^5E_g$
d^7	^4F, ^4P*	$t_{2g}^5\, e_g^2$	$^4T_{1g}(F)$	$t_{2g}^3\, e_g^4$	$^4A_{2g}$	$^4T_{1g}(F) \rightarrow {}^4T_{2g}(F)$
				$t_{2g}^4\, e_g^3$	$^4T_{1g}(P) + {}^4T_{2g}(F)$	$^4T_{1g}(F) \rightarrow {}^4A_{2g}(F)$
						$^4T_{1g}(F) \rightarrow {}^4T_{1g}(P)$
d^8	^3F, ^3P*	$t_{2g}^6\, e_g^2$	$^3A_{2g}(F)$	$t_{2g}^4\, e_g^4$	$^3T_{1g}(P)$	$^3A_{2g} \rightarrow {}^3T_{2g}(F)$
	$^3T_{1g}(F)$			$t_{2g}^5\, e_g^3$	$^3T_{1g}(F) + {}^3T_{2g}(F)$	$^3A_{2g} \rightarrow {}^3T_{1g}(F)$
						$^3A_{2g} \rightarrow {}^3T_{1g}(P)$
d^9	^2D	$t_{2g}^6\, e_g^3$	^2E_g	$t_{2g}^5\, e_g^4$	$^2T_{2g}$	$^2E_g \rightarrow {}^2T_{2g}$.

* Excited state with the same spin multiplicity of the ground state.

Note: In representing the transition, the **standard convention** is: excited state ← ground state, *i.e.* excited state is to be written first. ***Thus the transition in the d^9–system should be written as: $^2T_{2g} \leftarrow {}^2E_g$.*** However, in this book, this rigid convention is not followed.

The total orbital degeneracy arising from the $^3T_{1g}(F)$, $^3T_{1g}(P)$, $^3T_{2g}(F)$ and $^3A_{2g}(F)$ is also 10. This is illustrated below.

The development of energy states for the ^3F and ^3P terms of the free ion is as follows:

$^3\boldsymbol{F}$ (orbital degeneracy = 2 × 3 + 1 = 7)
- (i) $(d_{xy})^1 (d_{xz})^1; (d_{xz})^1 (d_{yz})^1; (d_{xy})^1 (d_{yz})^1; \Rightarrow {}^3\boldsymbol{T_{1g}(F)}$
- (ii) $(d_{xy})^1 (d_{z^2})^1; (d_{xz})^1 (d_{x^2-y^2})^1; (d_{yz})^1 (d_{x^2-y^2})^1 \Rightarrow {}^3\boldsymbol{T_{2g}(F)}$
- (iii) $(d_{x^2-y^2})^1 (d_{z^2})^1 \Rightarrow {}^3\boldsymbol{A_{2g}(F)}$

(Total orbital degeneracy = 3 + 3 + 1)

$^3\boldsymbol{P}$ (orbital degeneracy = 2 × 1 + 1 = 3)
- (i) $(d_{xy})^1 (d_{x^2-y^2})^1; (d_{xz})^1 (d_{z^2})^1; (d_{yz})^1 (d_{z^2})^1 \Rightarrow {}^3\boldsymbol{T_{1g}(P)}$

(Total orbital degeneracy = 3).

(e) Group theoretical treatment to determine the terms from the electronic configuration: Terms arising from the electronic configurations can be determined theoretically from the direct product rule (*cf.* Table 7.23.1). It is illustrated below. For details, the readers are advised to consult the book on group theory.

$t_{2g}^2 \rightarrow t_{2g}^1 \times t_{2g}^1 \rightarrow {}^3T_{1g} + {}^1T_{2g} + {}^1E_g + {}^1A_{1g}$; Total degeneracy = 3 × 3 + 1 × 3 + 1 × 2 + 1 × 1 = 15

$t_{2g}^1 e_g^1 \rightarrow t_{2g}^1 \times e_g^1 \rightarrow {}^1T_{1g} + {}^1T_{2g} + {}^3T_{1g} + {}^3T_{2g}$; Total degeneracy = 1 × 3 + 1 × 3 + 3 × 3 + 3 × 3 = 24

$e_g^2 \rightarrow e_g^1 \times e_g^1 \rightarrow {}^1E_g + {}^3A_{2g} + {}^1A_{1g}$; Total degeneracy = 1 × 2 + 3 × 1 + 1 × 1 = 6

$t_{2g}^3 \rightarrow t_{2g}^1 \times t_{2g}^2 \rightarrow T_{2g} \times \left({}^3T_{1g} + {}^1T_{2g} + {}^1E_g + {}^1A_{1g}\right) \rightarrow {}^4A_{2g} + {}^2T_{1g} + {}^2T_{2g} + {}^2E_g;$

$t_{2g}^2 e_g^1 \rightarrow E_g \times \left({}^3T_{1g} + {}^1T_{2g} + {}^1E_g + {}^1A_{1g}\right) \rightarrow {}^4T_{1g} + {}^4T_{2g} + 2{}^2T_{1g} + 2{}^2T_{2g} + 2{}^2E_g + {}^2A_{1g} + {}^2A_{2g}$

$t_{2g}^1 e_g^2 \rightarrow T_{2g} \times \left({}^3A_{2g} + {}^1E_g + {}^1A_{1g}\right) \rightarrow {}^4T_{1g} + 2{}^2T_{1g} + 2{}^2T_{2g}.$

The results are summarised as follows:

$t_{2g}^6 \Rightarrow {}^1A_{1g}$; t_{2g}^5 and $t_{2g}^1 \Rightarrow {}^2T_{2g}$; t_{2g}^2 and $t_{2g}^4 \Rightarrow {}^3T_{1g} + {}^1T_{2g} + {}^1E_g + {}^1A_{1g}$;

$t_{2g}^3 \Rightarrow {}^4A_{2g} + {}^2E_g + {}^2T_{2g} + {}^2T_{1g}$; $e_g^4 \Rightarrow {}^1A_{1g}$; e_g^3 and $e_g^1 \Rightarrow {}^2E_g$; $e_g^2 \Rightarrow {}^3A_{2g} + {}^1E_g + {}^1A_{1g}$

In a tetrahedral field, the *g*-suffixes are dropped.

7.9.5 Origin of Luminescence: Phosphorescence, Fluorescence, Jablonski Diagram

When an electron is excited to a higher energy level (*i.e.* spectroscopic state), the excitation energy may be released in two ways:

- **radiative decay** through photon emission.
- **nonradiative path through the thermal degradation of energy as heat to the surrounding; the lattice vibration or collisional quenching** (*i.e.* **external conversion) is the main cause of nonradiative decay;** this nonradiative path leading to the ground state (generally) is very fast (within 10^{-8} s in gas phase, within $10^{-12} - 10^{-14}$ s in liquid phase).

When the nonradiative decay path is relatively inefficient (*i.e.* lifetime for the nonradiative path is elongated), the radiative path and nonradiative path may compete mutually and the radiative decay path may predominate also. In such cases, the system can show the **luminescence.** For the transition metal complexes, generally, the nonradiative decay path is quite efficient even at low temperature. **This is**

why, the transition metal complexes are not generally luminescent. For the lanthanides, the 4*f*–orbitals are deeply seated in the inner core and the 4*f*–orbitals are hardly influenced by the environment. In such cases, the nonradiative path (*i.e.* through lattice vibration) becomes comparatively less efficient. **In fact, many lanthanide complexes can show the luminescene.**

Chemiluminescence: If the excited state required for luminescence is generated by a chemical reaction not by light absorption, then the term chemiluminescence is used. It is illustrated below through the production of singlet O_2 which is very much reactive towards the organic substrates. Singlet oxygen ($^1\Delta_g$) can be prepared in the following reactions.

$$Cl_2 + H_2O_2 \rightarrow 2Cl^- + 2H^+ + O_2(^1\Delta_g)$$
$$ClO^- + H_2O_2 \rightarrow Cl^- + H_2O + O_2(^1\Delta_g)$$

The mechanisms of the given reactions do not allow the production unpaired electrons, *i.e.* the given reactions do not allow the formation of triplet O_2 ($^3\Sigma_g^-$). Thus the produced singlet oxygen produces **red chemiluminescent glow** when it returns to its ground state, *i.e.* triplet oxygen.

Luminescence leading to fluorescence and phosphorescence: Thus **luminescence** (*i.e.* **radiative decay**) occurs only when the **nonradiative path is quite inefficient.** Depending on the life time of the excited state, luminescence can be of two types:

Luminescence (Inefficient non-radiative path) }
- **Fluorescence** – rapid radiative decay
- **Phosphorescence** – delayed radiative decay.

Thus, fluorescence stops if the exiciting radiation is removed but phosphorescence exists even after the removal of the exciting radiation. **For phosphorescence, lifetime of the excited state must be relatively higher.** In fact, for fluorescence, both the excited and ground state are having the same spin-multiplicity and the return to the ground state from the excited state is a **spin-allowed transition** occurring in the scale of nanoseconds. On the other hand, the radiative decay of phosphorescence occurs through the **spin-forbidden transition** (*i.e.* excited state and ground state differ in spin multiplicity). This is why, the life-time of the excited state is relatively longer for the phosphorescence.

Intersystem crossing (ISC) and internal conversion—two important nonradiative paths of deexcitation: Generally, photoexcitation from the ground state into the excited state occurs through a spin-allowed transition. This initial excited state may be converted into another excited state of different multiplicity through **intersystem crossing (ISC) (a nonradiative path).** This intersystem crossing occurs when the potential energy curves of the two excited states cross. At the point of crossing of the two curves, this conversion occurs through a **radiationless transition. Spin-orbit coupling allows the intersystem crossing.**

Sometimes, there may be two excited states of the same spin multiplicity. Then the higher excited state may be converted into the lower excited state of the same spin-multiplicity through the **internal conversion (IC)** or internal quenching. In internal conversion, the electronic energy is converted into the vibrational energy of the molecule itself. For example, in the case of Cr(III)-complexes, the higher excited state $^4T_{1g}(F)$ may be converted into the lower excited state $^4T_{2g}(F)$ through internal conversion. This excited $^4T_{2g}(F)$ state is converted into the ^2E_g state through **intersystem crossing (ISC)** (a radiationless transition) and the ^2E_g state acts as the energy reservoir.

The **spin-orbit coupling** allows the intersystem crossing (*e.g.* triplet to singlet, quartet to doublet, etc.). The second excited state obtained through the intersystem crossing survives relatively longer because its transition to the ground state is spin-forbidden. **This spin forbidden transition** (*i.e.* delayed transition) **to the ground state gives the phosphorescence.**

In the case of **fluorescence** (a radiative decay), return from the excited state to the ground state occurs through the spin-allowed transition. It is very rapid and it completes within 10^{-8} – 10^{-9}s.

Jablonski diagram describing the fate of excited electronic states: ***Jablonski diagram*** (Fig. 7.9.5.1–2) can explain the phenomena like external and internal conversion, phosphorescence and fluorescence.

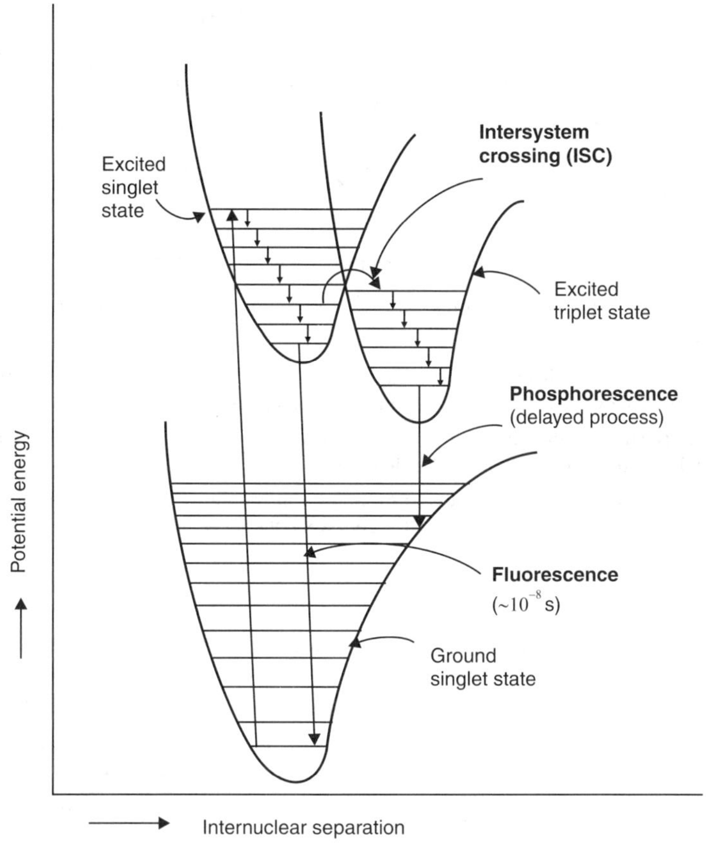

Fig. 7.9.5.1 Light absorption and excitation of molecules leading to fluorescence and phosphorescence. (A simplified version of **Jablonski diagram**).

Photochemistry of the Cr(III) complexes (cf. Sec. 6.9.5) and phosphorescence of the Cr(III) complexes: To understand the **photochemistry of Cr(III) complexes, and the ruby laser action**, the excited states $^4T_{1g}(F)$, $^4T_{2g}(F)$, $^2E_g(G)$ and the ground state $^4A_{2g}$ (Fig. 7.7.3.3) are to be considered (*cf.* Sec. 6.9.5). The excitations, $^4A_{2g}(F) \rightarrow {}^4T_{1g}(F)$, $^4T_{2g}(F)$ followed by **intersystem crossing** (ISC) populates the ^2E_g state. Then, the transition $^2E_g \rightarrow {}^4A_{2g}(F)$ gives the phosphorescence. These aspects have been discussed in Sec 7.14.5. **Photochemistry of $[Ru(bpy)_3]^{2+}$** can also be rationalised in terms of its phosphoresence behaviour (*cf.* Secs. 16.16.1–2, Vol. 3).

In the case of Cr(III), energy difference between the ground state $^4A_{2g}$ ***(high spin t_{2g}^3)*** and the **phosphorescent state** ^2E_g ***(low spin t_{2g}^3)*** does not depend on the crystal field strength. In presence of very weak field ligands, the $^4T_{1g}$ and $^4T_{2g}$ terms may reside below the ^2E_g term and *under this condition, Cr(III) complexes will show the* ***fluorescence not phosphorescence.***

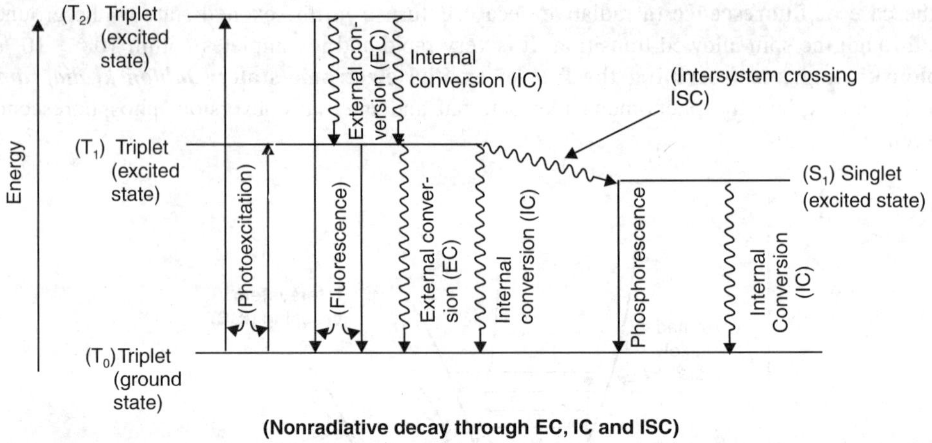

Fig. 7.9.5.2 Schematic representation of simplified *Jablonski diagram.* $T_0 \rightarrow T_1$, T_2 (spin allowed); T_1, $T_2 \rightarrow T_0$ (spin allowed, fluorescence); $S_1 \rightarrow T_0$ (spin forbidden, phosphorescence).

- **External Conversion (EC) or collisional quenching:** Loss of excitation energy occurs as **translational energy** or **heat energy** through the lattice vibration or collision with the surrounding; *It is a nonoradiative decay.*
- **Internal conversion (IC) or internal quenching:** This self quenching does not involve collision and electronic energy is converted into the **vibrational energy** of the molecule iteself.
- **Intersystem crossing (ISC) (radiationless transition):** Spin-orbit coupling allows the conversion of one excited state to another excited state differing in spin-multiplicity. It occurs when the potential energy curves of the two excited states of different spin multiplicities cross.

Conclusion

- Luminescence (*i.e.* radiative decay of the excited state) leading to fluorescence and phosphorescence is important only when the nonradiative decay of the excited state is relatively inefficient.
- Intersystem crossing (ISC) through spin-orbit coupling enhances the population density of an excited state differing in spin-multiplicity with the ground state. Return from this excited state to the ground state through the **spin-forbidden transition gives the phosphorescence.**
- Return from the excited state to the ground state through the **spin-allowed transition gives fluorescence.**
- Fluorescence is an **instantaneous emission** while phosphorescence is a **delayed transition.**

7.10 POSSIBLE SPIN-ALLOWED ELECTRONIC TRANSITIONS IN THE WEAK FIELD OCTAHEDRAL AND TETRAHEDRAL COMPLEXES AND ORGEL DIAGRAM

O_h *vs.* T_d System: The crystal field splitting patterns of the *d*–orbitals in the octahedral and tetrahedral complexes are of opposite nature for a particular d^n configuration (*cf.* Figs. 7.10.1, 2)

d^n *vs.* d^{10-n} systems: The splitting patterns of the terms for d^n and d^{10-n} configurations in an octahedral or tetrahedral field are of opposite nature. This is interpreted in terms of *the hole formalism rule* which states: d^n electronic system ≡ d^n positronic system ≡ d^{10-n} positronic system ≡ d^{10-n} electronic system. It leads to: d^n electronic configuration ≡ $(10 - n)$ holes in the d^{10} electronic system, *i.e.* d^n configuration is **hole equivalent** of d^{10-n} configuration. **But an electron is most stable where a positron is least stable.**

d^n and d^{n+5} systems: In a weak field, d^5–system has a zero orbital angular momentum (*i.e.* $L = 0$). Hence, the orbital angular momentum for the d^{5+n} and d^n systems will be identical. Thus for the d^n and

d^{n+5} systems, L will be the same. Consequently, the ground state terms for the d^n and d^{5+n} systems will be the same but they will differ in S and J values. This is illustrated below:

$$d^1(^2D_{3/2}),\ d^6(^5D_4);\ d^2(^3F_2),\ d^7(^4F_{9/2});\ d^3(^4F_{3/2}),\ d^8(^3F_4);\ d^4(^5D_0),\ d^9(^2D_{5/2})$$

Thus in terms of splitting of the terms in an octahedral or a tetrahedral field, it is follows:

$$d^1 \equiv d^6;\ d^2 \equiv d^7;\ d^3 \equiv d^8;\ d^4 \equiv d^9$$

Considering these factors, we can conclude as follows:

(i) **d^n and d^{10-n} are inversely related.**

(ii) **d^n and d^{n+5} are identical ($n < 5$)**

(iii) **d^{5-n}** (≡ n holes in the half-filled shell, *i.e.* d^5) and **d^{10-n}** (≡ n holes in the full filled shell, *i.e.* d^{10}) **are identical** and **reverse of those for d^n.**

(iv) **d^n (O_h) and d^{10-n} (T_d) are identical.**

(v) **d^n (O_h) and d^n (T_d) are inversely related.**

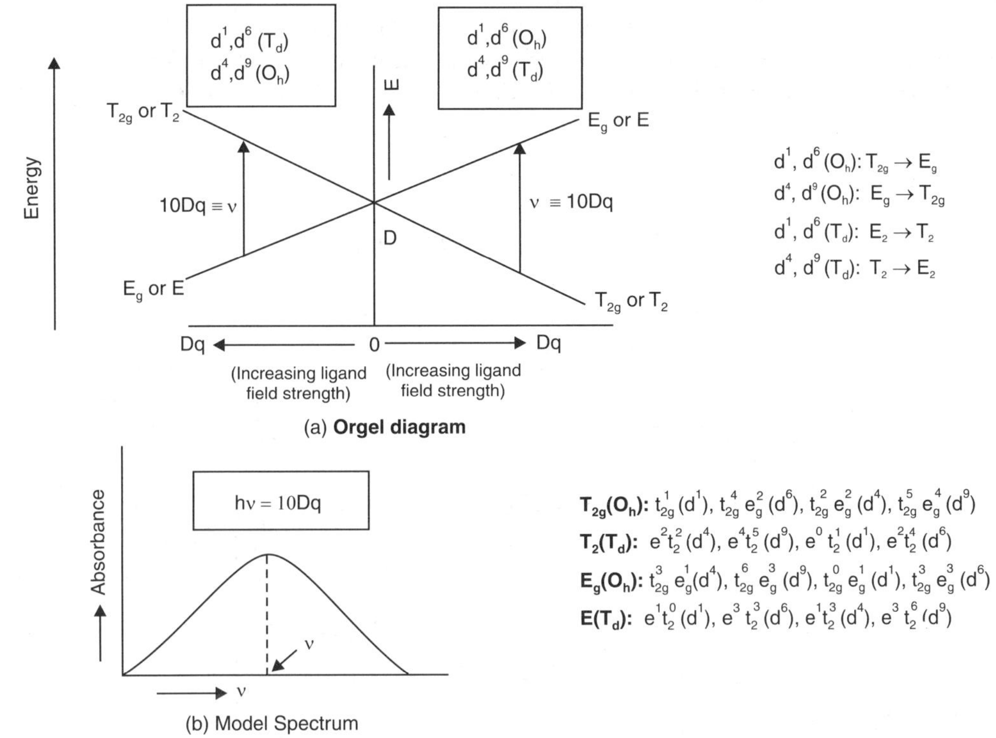

(a) **Orgel diagram**

(b) Model Spectrum

Fig. 7.10.1 Orgel diagram for d^1, d^4, d^6 and d^9 systems (weak field ligands, high spin complexes) and spectral transition in the octahedral (O_h) and tetrahedral (T_d) systems. (**Note:** Here J. Teller effect has been ignored).

The O_h–systems (having the centre of symmetry) will bear the 'g'-subscript while for the T_d–systems (lacking in centre of symmetry), the g–subscript is dropped.

These are illustrated below:

$$\left.\begin{aligned} d^2(O_h) &\equiv \text{inverse of } d^8(O_h) \equiv \text{inverse of } d^2(T_d) \\ &\equiv d^8(T_d) \equiv d^7(O_h) \equiv \text{inverse of } d^3(O_h) \\ &\equiv \text{inverse of } d^7(T_d) \\ &\equiv d^3(T_d). \end{aligned}\right\} \quad (cf.\ \text{Fig. 7.10.2})$$

Similarly, we can write:

$$\left.\begin{aligned} d^1(O_h) &\equiv d^6(O_h) \equiv \text{inverse of } d^4(O_h) \equiv \text{inverse of } d^1(T_d) \\ &\equiv \text{inverse of } d^9(O_h) \\ &\equiv d^4(T_d) \equiv d^9(T_d) \end{aligned}\right\} \quad (\textit{cf.}\ \text{Fig. 7.10.1})$$

Possible spectral transitions for the d^n–configurations are shown in Figs. 7.10.1 and 7.10.2.

(A) Spectral transitions in d^2, d^7, d^3 and d^8 systems: These can be formulated as follows from the Orgel diagram.

$$\mathbf{d^2, d^7\,(O_h)}:\ {}^nT_{1g}(F) \xrightarrow{\nu_1} {}^nT_{2g}(F) \qquad \mathbf{d^3, d^8\,(T_d)}:\ {}^nT_1(F) \xrightarrow{\nu_1} {}^nT_2(F)$$

$${}^nT_{1g}(F) \xrightarrow{\nu_2} {}^nA_{2g}(F) \qquad {}^nT_1(F) \xrightarrow{\nu_2} {}^nA_2(F)$$

$${}^nT_{1g}(F) \xrightarrow{\nu_3} {}^nT_{1g}(P) \qquad {}^nT_1(F) \xrightarrow{\nu_3} {}^nT_1(P)$$

- **Without considering the symmetry interaction energy**, we can write:

$$\nu_1 = 8Dq,\ \nu_2 = 18Dq,\ \nu_3 = 15B' + 6Dq$$

The above energy values are given without considering the **symmetry interaction energy**.

- **By considering the symmetry interaction** $[{}^nT_{1g}(F) - {}^nT_{1g}(P)]$, we can write as follows (*cf.* Fig. 7.10.2).

$$T_{1g}(F) \xrightarrow{\nu_1} T_{2g}(F),\ \nu_1 = 8Dq + x'$$

$$T_{1g}(F) \xrightarrow{\nu_2} A_{2g}(F),\ \nu_2 = 18Dq + x'$$

$$T_{1g}(F) \xrightarrow{\nu_3} T_{1g}(P),\ \nu_3 = 15B' + 6Dq + 2x'$$

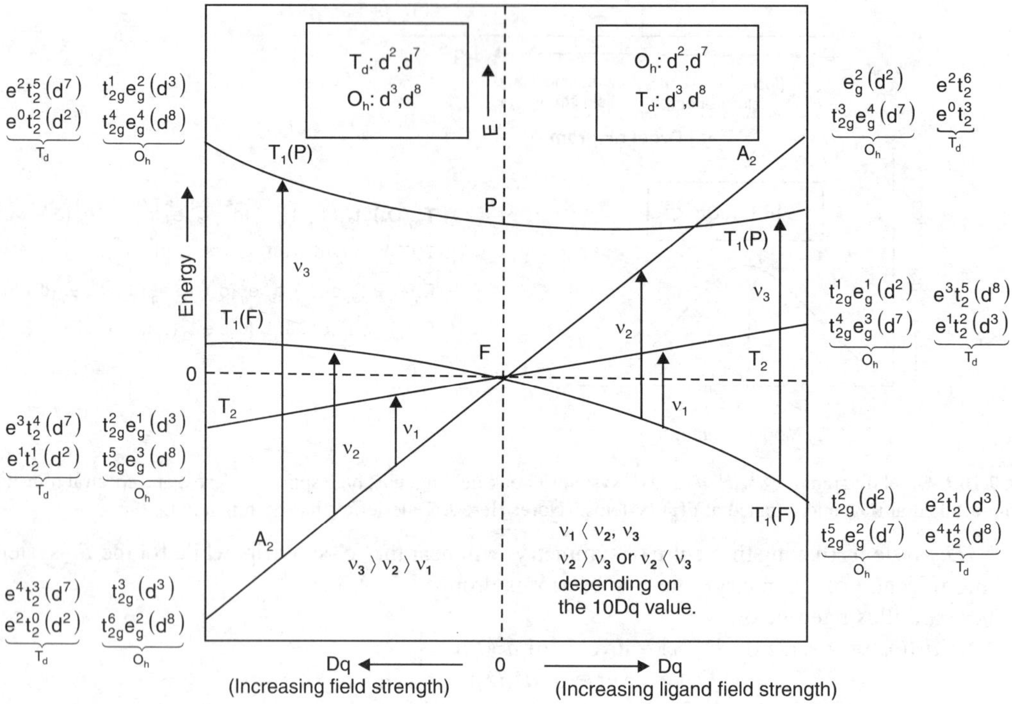

(a)

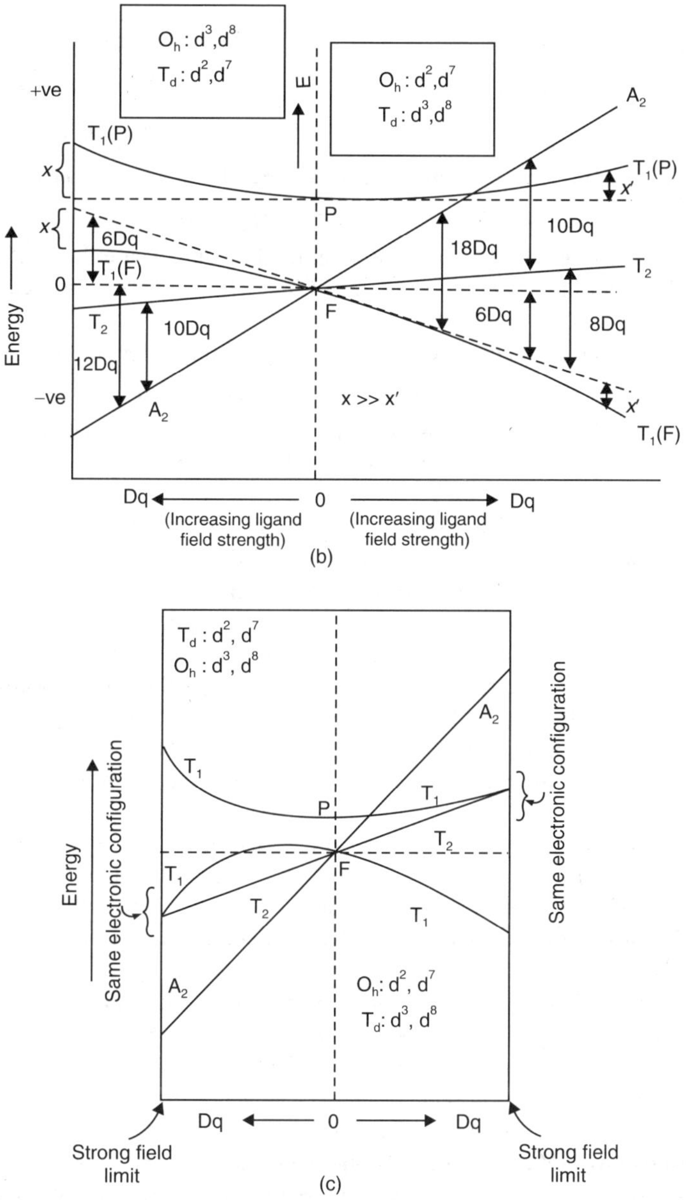

Fig. 7.10.2 Orgel diagram for d^2, d^7, d^3 and d^8 systems in high spin octahedral and tetrahedral crystal fields (considering configurational interaction between $T_1(P)$ and $T_1(F)$ terms). **Note:** In a strong field limit, $T_1(F)$ and $T_2(F)$ states for $d^{2,\,7}(T_d)$, $d^{3,\,8}(O_h)$ systems merge; similarly $T_2(F)$ and $T_1(P)$ states merge for $d^{3,\,8}(T_d)$ and $d^{2,\,7}(O_h)$ systems.

$$\nu_2 - \nu_1 = 10Dq;\ \nu_2 + \nu_3 - 3\nu_1 = 15B'$$

Here x' denotes the *symmetry interaction energy*. $^3T_1$ and $^3A_2$ terms can mutually cross depending on the *Dq*–value. After this crossing point, $\nu_3 \langle \nu_2$ and before this crossing point $\nu_2 \langle \nu_3$.

$$\mathbf{d^3, d^8 (O_h):}\quad {}^nA_{2g}(F) \xrightarrow{\nu_1} {}^nT_{2g}(F) \qquad \mathbf{d^2, d^7 (T_d):}\quad {}^nA_2(F) \xrightarrow{\nu_1} {}^nT_2(F)$$

$$ {}^nA_{2g}(F) \xrightarrow{\nu_2} {}^nT_{1g}(F) \qquad {}^nA_2(F) \xrightarrow{\nu_2} {}^nT_1(F)$$

$$ {}^nA_{2g}(F) \xrightarrow{\nu_3} {}^nT_{1g}(P) \qquad {}^nA_2(F) \xrightarrow{\nu_3} {}^nT_1(P)$$

● **Without considering the symmetry interaction** between the ${}^nT_1(F)$ and ${}^nT_1(P)$ states, we can write:

$$\nu_1 = 10Dq,\ \nu_2 = 18Dq,\ \nu_3 = 12Dq + 15B'$$

These relationships are true only for the very weak field ligands where the configurational interaction is almost absent. According to the **noncrossing rule**, $T_1(P)$ and $T_1(F)$ terms can never cross. This is why, the trend, $\nu_3 \rangle \nu_2 \rangle \nu_1$ is always maintained.

● **In terms of symmetry interaction energy (x),** we have the following relations:

$$\nu_1 = 10Dq,\ \nu_2 = 18Dq - x,\ \ \nu_3 = 12Dq + 15B' + x$$

For a certain value of $10Dq$, $\boldsymbol{x} \rangle\rangle \boldsymbol{x'}$. Thus, the symmetry interaction is relatively more important for the $d^{3,8}$ (O_h) and $d^{2,7}$ (T_d) systems. x and x' increase with the increase of $10Dq$ values. x and x' are almost negligible for the smaller values of $10Dq$.

This is why, for the weak field ligands, the symmetry interaction energy x is not so important.

(B) Interaction between the T_1(F) and T_1(P) states (*cf.* Fig. 7.10.2b): Here it may be pointed out that the $T_{1g}(F)$ state originated from the ground state term F (for free ion)$_3$ cannot cross (*i.e.* **noncrossing rule**) the $T_{1g}(P)$ state originated from the excited state term P of the free ion. Because of their same symmetry, these spectroscopic states mutually interact and the lower energy state is further lowered down while the higher energy state rises more. If it is argued that these two spectroscopic states will mutually intersect then it implies that at the cross-over point, two electrons will have the same symmetry and energy in the same atom or ion. ***But this is not possible according to Pauli's exclusion principle. This is the basis of noncrossing rule according to which the energy states of same symmetry will never cross each other.***

Because of the symmetry interaction between the $T_1(F)$ and $T_1(P)$ states, the energy difference between the $A_2(F)$ and $T_1(F)$ states for d^3, d^8 (O_h) and d^2, d^7 (T_d) is less than $18Dq$. But for d^3, d^8 (T_d) and d^2, $d^7(O_h)$, energy difference between the $A_2(F)$ and $T_1(F)$ states is slightly greater than $18Dq$. But, for the weak field ligands where the symmetry interaction energy is very small, the energy difference is almost equal to $18Dq$.

(C) Interelectronic repulsion parameter (B, B'): In general, the excited ${}^{2s+1}P$ state lies above the ground state ${}^{2s+1}F$ and the energy difference between these two states of free ion is given by $15B$ where B (called Racah parameter) gives the measure of interelectronic repulsion. Generally, in the complexes the degree of interelectronic repulsion reduces and consequently the value of B is reduced by about 20%. In the complexes, the reduced B is denoted by B^1.

(D) Spectral transitions in d^1, d^4, d^6 and d^9 systems: Octahedral and tetrahedral complexes of these electronic configurations are characterised by a single absorption band the corresponding energy is equal to $10Dq$. In fact, it happens so due to the splitting of the D–term (*cf.* Figs. 7.10.1 and 3). However, the J.T. distortion can split the band.

(E) Spectral transitions in d^2, d^3, d^7 and d^8 systems: Octahedral and tetrahedral complexes of these electronic configurations are characterised by three absorption bands.

The lowest band (ν_1) [$A_2 \rightarrow T_2(F)$] for the d^3, d^8 (O_h) and d^2, $d^7(T_d)$ complexes correspond to $10Dq$ (*cf.* Figs. 7.10.2 – 3). For the d^3, d^8 (T_d) and d^2, d^7 (O_h) systems, the energy difference between the lowest band (ν_1) [$T_1 \rightarrow T_2(F)$] and the band (ν_2) [$T_1 \rightarrow A_2$ (F)] corresponds to $10Dq$ (*cf.* Figs.

7.10.2 – 7.10.3). The ν_2–band is the highest energy band if the Dq value allows the crossing of the $A_2(F)$ and $T_1(P)$ states.

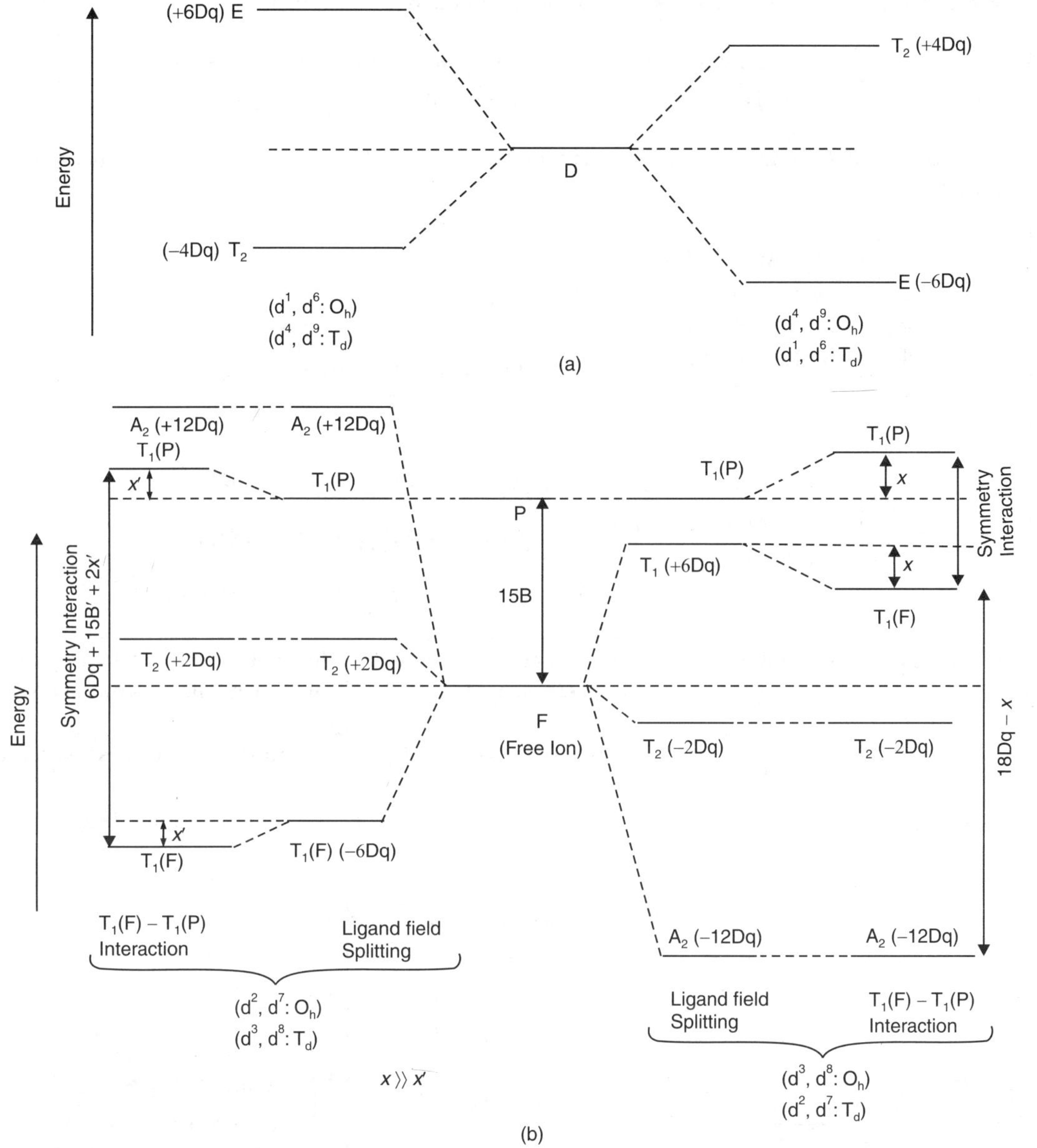

Fig. 7.10.3 Splitting of the D and F terms in the octahedral (O_h) and tetrahedral (T_d) field for the weak field ligands. x or x' denotes the symmetry interaction energy for the repulsive interaction between the $T_1(F)$ and $T_1(P)$ states (*See* Fig. 7.9.4.1 for the energy values of different states).

Note: ● g–subscript to be used for the octahedral systems.

● For the d^2, d^7 (O_h) and d^3, d^8 (T_d) systems, here the crystal field strength is considered to be sufficiently high so that the $A_2(F)$ state can cross the $T_1(P)$ state.

Summary of the ligand field transitions in terms of the Orgel Diagram

- **d^1, d^4, d^6, d^9(O_h, T_d):** One transition (Δ_o or Δ_t) occurs (*cf.* Fig. 7.10.1):

$$d^1, d^6(O_h): T_{2g} \xrightarrow{10Dq_o} E_g \quad \Big| \quad d^4, d^9(O_h): E_g \xrightarrow{10Dq_o} T_{2g}$$
$$d^4, d^9(T_d): T_2 \xrightarrow{10Dq_t} E \quad \Big| \quad d^1, d^6(T_d): E \xrightarrow{10Dq_t} T_2$$

- **d^2, d^3, d^7, d^8(O_h and T_d):** Three transitions are expected (g suffix omitted for T_d) (*cf.* Fig. 7.10.2).

$\mathbf{d^2, d^7(O_h)}$ **and** $\mathbf{d^3, d^8(T_d)}$:
$\nu_1 < \nu_2, \nu_3; \nu_2 - \nu_1 = \Delta_o$ or Δ_t
Relative values of ν_2 and ν_3 depend on $10Dq$ value.
$\nu_2 + \nu_3 - 3\nu_1 = 15B'$

$$\begin{cases} T_{1(g)}(F) \xrightarrow{\nu_1} T_{2(g)}(F), \ \nu_1 \approx 8Dq \\ T_{1(g)}(F) \xrightarrow{\nu_2} A_{2(g)}(F), \ \nu_2 \approx 18Dq \\ T_{1(g)}(F) \xrightarrow{\nu_3} T_{1(g)}(P), \ \nu_3 \approx 6Dq + 15B' \end{cases}$$

$\mathbf{d^3, d^8(O_h)}$ **and** $\mathbf{d^2, d^7(T_d)}$:
$\nu_1 = \Delta_o$ or $\Delta_t \langle \nu_2 \langle \nu_3$

$$\begin{cases} A_{2(g)}(F) \xrightarrow{\nu_1} T_{2(g)}(F), \ \nu_1 \approx 10Dq \\ A_{2(g)}(F) \xrightarrow{\nu_2} T_{1(g)}(F), \ \nu_2 \approx 18Dq \\ A_{2(g)}(F) \xrightarrow{\nu_3} T_{1(g)}(P), \ \nu_3 \approx 12Dq + 15B' \end{cases}$$

(F) Techniques to analyse the electronic spectra of the octahedral and tetrahedral complexes d^n (n = 2, 3, 7, 8) having the ground spectroscopic state A_2(F) or T_1(F): The spectral analysis, involves: band assignments; evaluation of $10Dq$, B' and $\beta\left(=\dfrac{B'}{B_0}\right)$. To determine these spectral parameters, different techniques are known. If all the three ligand field spectral transitions are identified,

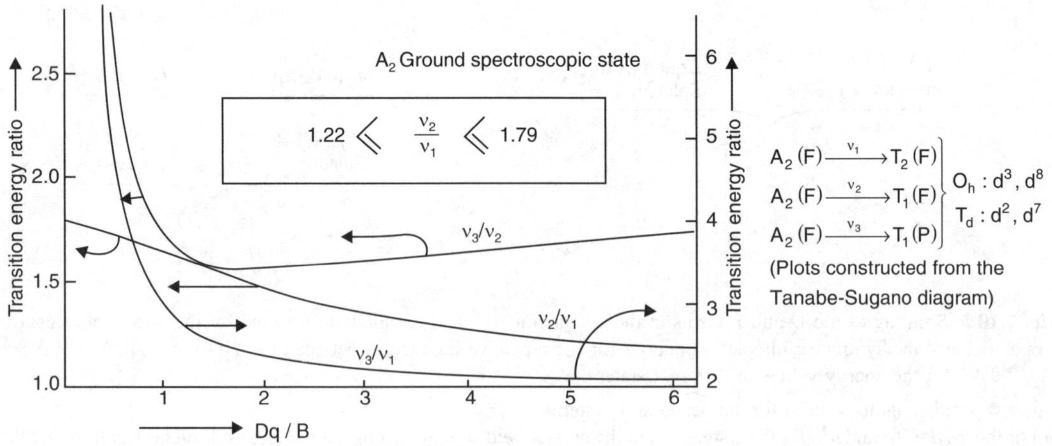

Fig. 7.10.4 (a) Plots of transition energy ratio *vs.* *Dq/B* for the A_2 ground spectroscopic state.

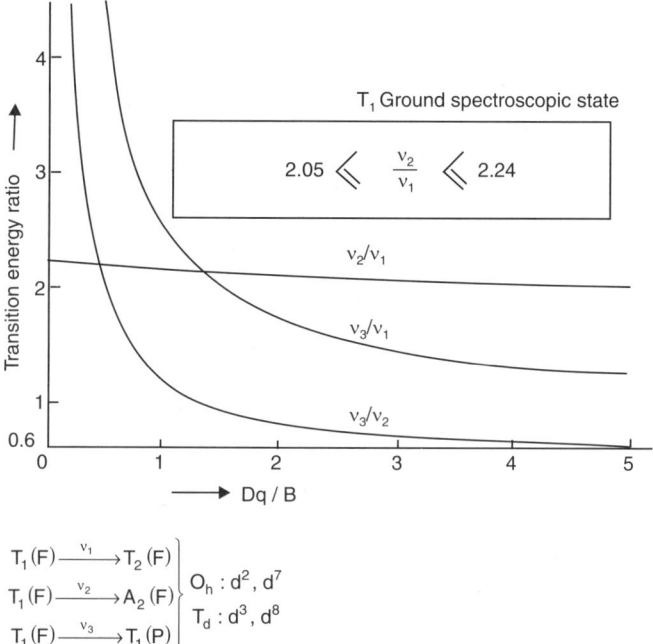

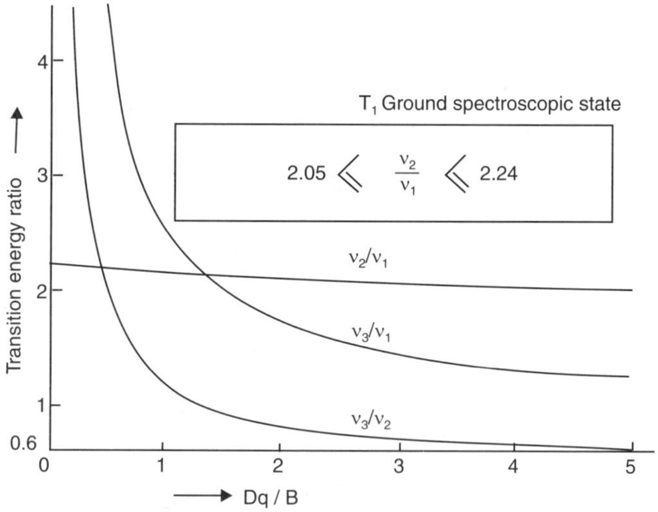

$T_1 (F) \xrightarrow{\nu_1} T_2 (F)$
$T_1 (F) \xrightarrow{\nu_2} A_2 (F)$ $O_h : d^2, d^7$; $T_d : d^3, d^8$
$T_1 (F) \xrightarrow{\nu_3} T_1 (P)$

Note : ν_2/ν_1 is approximately 2.0 regardless of the ligand field strength.
(Plots constructed from the Tanabe-Sugano diagram)

Fig. 7.10.4 (b) Plots of transition energy ratio *vs.* *Dq/B* for the T_1 ground spectroscopic state.

then the spectral analysis becomes straight-forward. When this is not possible, the **transition energy ratios,** *i.e.* ν_3/ν_1, ν_3/ν_2, ν_2/ν_1 are calculated and these are fitted with the standard curves (*cf.* Fig. 7.10.4a, b) of the transition energy ratios *vs.* *Dq/B* derived from the Tanabe-Sugano diagram. These theoretical plots (*i.e.* standard curves) of ν_2/ν_1, ν_3/ν_2 *vs.* *Dq/B* (for the A_2 and T_1 ground terms) are constructed from the Tanabe-Sugano diagram. Standard Tables for the theoretical transition energy ratios, ν_1/B, ν_2/B, ν_3/B and *Dq/B* values are also available (*cf.* A.B.P. Lever, *J. Chem. Edu.*, **45,** 711, 1968; A.B.P. Lever, Inorganic Electronic Spectroscopy, 2nd Edn., 1984; Elsevier).

7.11 SPLITTING OF THE ENERGY LEVELS DUE TO JAHN-TELLER DISTORTION (*cf.* Sec. 3.11.3 – 4 and Sec. 7.14 for details)

Here we shall illustrate the case of octahedral complexes where the J.T. distortion is more important than in the case of tetrahedral complexes. In the octahedral stereochemistry, if any level (t_{2g} or e_g) remains unsymmetrically filled in before or after electron transition, then the corresponding level undergoes splitting. Such a splitting for the d^9 system in an octahedral field is shown in Fig. 7.11.1.

The splitting of energy levels due to the J.T. distortion may broaden or sometimes generate the close spaced bands. *Here it is important to note that there is some freedom about the choice of B_1 and B_2 and they may get interchanged* (*i.e.* the state denoted by B_1 by one author may be represented by B_2 by another author).

The effect of J.T. distortion is well noticed in the high-spin octahedral complexes of Cr^{2+} (d^4), $Ti^{3+}(d^1)$, $Fe^{2+}(d^6)$. These have been discussed separately (*cf.* Sec. 3.11.3 – 4).

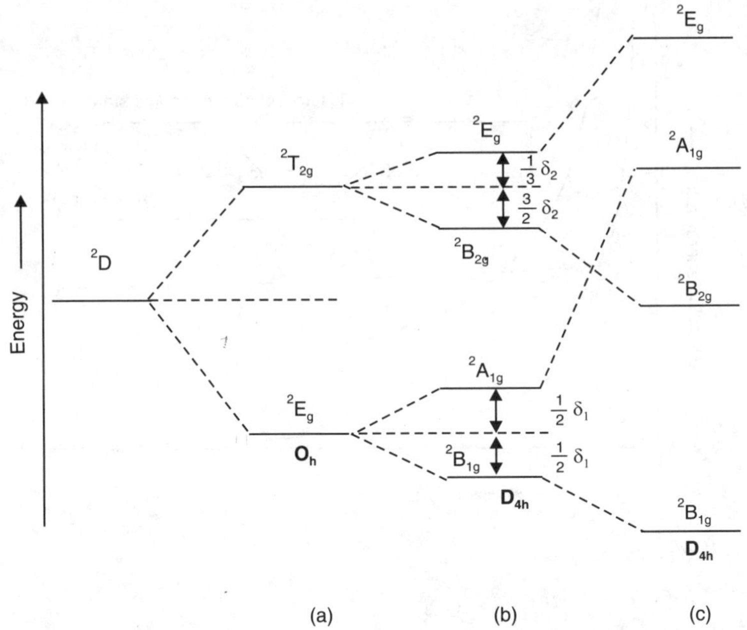

Fig. 7.11.1 (*cf.* Fig. 7.14.2.1 for details): Energy levels for the d^9 configuration in six-coordinate complexes. (a) Perfectly octahedral field (O_h)—without any J.T. distortion (*i.e.* hypothetical case); (b) Medium *z*-out J.T. distortion (*i.e.* medium axial elongation); (c) Strong *z*-out J.T. distortion.

Note: For the six-coordinate d^1–complexes, medium *z*-in J.T. distortion is common. The ordering of energy levels is:

$$^2B_{2g} < ^2E_g << ^2B_{1g} < ^2A_{1g} \text{ } (cf. \text{ Fig. 7.14.1.2 for details})$$

7.12 MODIFIED ORGEL DIAGRAM (INCORPORATING BOTH THE LOW SPIN AND HIGH SPIN STATES): ILLUSTRATION FOR THE d^6 SYSTEM

Orgel diagrams generally represent the spin allowed transitions in the weak field complexes. However, for the d^4, d^5, d^6 and d^7 systems, after a certain value of $10Dq$ (= P), the low spin complexes are formed. In the high-spin complexes, the states of highest spin multiplicity are involved while for the low-spin complexes, the states of lower spin multiplicity are involved. In the simple Orgel diagram (dealing with the states of highest spin multiplicity only), the states of lower spin multiplicities may be included. It is illustrated for the d^6–system (octahedral) in Fig. 7.12.1. Thus the modified Orgel diagram represents both the spin-allowed and spin-forbidden transitions.

The ground state term for the free the ion of d^6 configuration (*e.g.* Co^{3+}) is ^5D which splits into $^5T_{2g}$ and ^5E_g in presence of a weak octahedral field. After the critical $10Dq$ value (= P), it attains the low-spin state ($t_{2g}^6\, e_g^0$). The **first excited state** of the low-spin complex is $t_{2g}^5\, e_g^1$. Here it is assumed that in the excited state, the two unpaired electrons in the t_{2g} and e_g levels remain in an antiparallel condition to give the spin multiplicity 1. It will allow the spin allowed transitions (*cf.* Sec. 7.15.1 for details). The distribution of 5 electrons (≡ one hole) in the d_{xy}, d_{yz} and d_{xz} orbitals of the t_{2g} level can be made in three possible ways. The single electron in the $d_{x^2-y^2}$ and d_{z^2} orbitals of the e_g level can be distributed in two ways. **Thus the excited state $t_{2g}^5\, e_g^1$ can have 6 (= 3 × 2) probable ways of electron distribution.** These 6 probable ways (*i.e.* orbital multiplicity = 6) of electronic configuration will give the $^1T_{1g}$ and

$^1T_{2g}$ states and each of them is having the orbital degeneracy 3. Thus t_{2g}^6 configuration gives the $^1A_{1g}$ level (ground state) while the excited state $t_{2g}^5\,e_g^1$ leads to the $^1T_{1g}$ and $^1T_{2g}$ levels. These can be correlated with the singlet state ^{1}I of the free ion.

$$^1I\,(\text{free ion term}) \xrightarrow[\text{(strong field)}]{O_h\text{–field}} \underbrace{^1A_{1g}}_{\text{ground state}},\ \underbrace{^1T_{1g},\ ^1T_{2g}}_{\text{excited states}} \quad (cf.\ \text{Sec. 7.15.1})$$

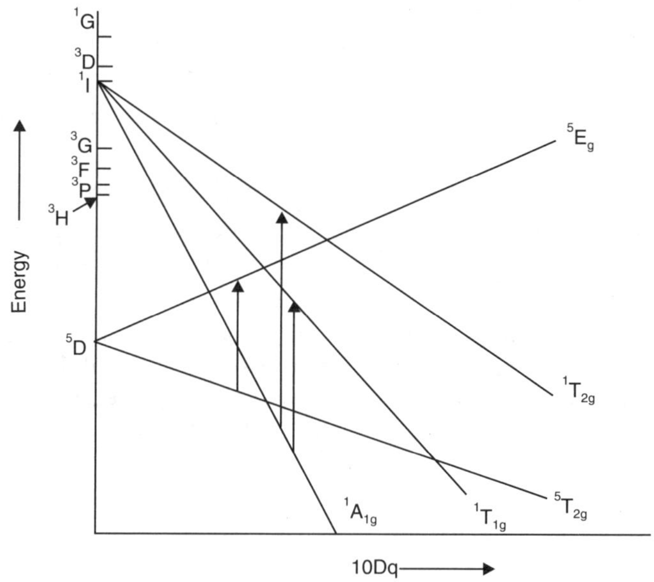

Fig. 7.12.1 Modified Orgel diagram for the d^6 configuration in the octahedral complexes showing the transitions in both the high-spin (*i.e.* lower 10*Dq* value) and low-spin (*i.e.* higher 10*Dq* value) states.

Now in the simple Orgel diagram of d^6 configuration (involving only the ^5D free ion term), the singlet free ion term may be included (Fig. 7.12.1). The diagram shows that up to a certain value of 10*Dq* value ($\langle$ *P*), the ground state is $^5T_{2g}$ while after the critical 10*Dq* value, the ground state is represented by the $^1A_{1g}$ level. By considering the $^5T_{2g}$ level as the ground state, the spin allowed transition is: $^5T_{2g} \rightarrow {}^5E_g$. On the other hand, by considering the $^1A_{1g}$ level as the ground state, there are two possible spin allowed transitions:

$^1A_{1g} \rightarrow {}^1T_{1g}$ and $^1A_{1g} \rightarrow {}^1T_{2g}$. These predictions are experimentally verified.

7.13 LIMITATIONS OF THE ORGEL DIAGRAM: TANABE—SUGANO DIAGRAMS

7.13.1 Limitations of the Orgel Diagram

The simple Orgel diagram deals with the highest spin multiplicity states (*i.e.* high spin complexes) and the spin allowed transitions only. For the high-spin d^5 system, there is no spin allowed transition. For the d^1, d^9, d^4 (h.s.) and d^6 (h.s.) systems, **only one ligand field transition** is expected while for the d^2, d^3, d^7 (h.s.) and d^8 systems, there are only **three spin allowed transitions.** But, very often, the actual spectra cannot be quantitatively interpreted in terms of the only spin-allowed transitions predicted from the simple Orgel diagram because besides the spin allowed transitions, there may be some spin-forbidden transitions of weaker intensity.

- Simple Orgel diagram fails to predict the spectral transitions in the low-spin octahedral complexes.
- Orgel diagram plots the energies of different states (*vs.* crystal field strength) in absolute units. Consequently, quantitatively, **each diagram applies only to a particular metal ion system.**

These are the main limitations of the Orgel diagram. These are majorly removed in the Tanabe-Sugano diagram (*cf.* Y. Tanabe and S. Sugano, *J. Phys. Soc. Jpn.*, **9**, 753–779, 1954).

7.13.2 Characteristics of the Tanabe-Sugano (T-S) Diagram

(i) The ordinate (*i.e.* *y*-axis) represents the energy of the free ion terms in the unit of *B* (Racah parameter). This energy scale may be obtained from the concerned atomic spectra.

(ii) The abscissa or horizontal axis (*i.e.* *x*–axis) represents the crystal field strength (10*Dq* or *Dq*) in the unit of *B*.

(iii) Thus, the *T–S* diagram gives the plot of *E/B vs. Dq/B* or 10*Dq/B* for a particular d^n configuration.

The *T–S* diagram shows the variation of energy of different states with the variation of crystal field strength. Thus it includes the *energy states of both the high spin and low-spin* (if possible, as in the cases of d^4, d^5, d^6, d^7 systems) complexes.

(iv) In the *T–S* diagrams, the abscissa represents the ground state while in the Orgel diagram the energy of the ground state decreases with the increase of the ligand field strength. Thus in the *T–S* diagram, **the lowest term is considered of zero energy regardless of the ligand field strength.** Energies of the higher states are plotted with respect to that of this ground state energy.

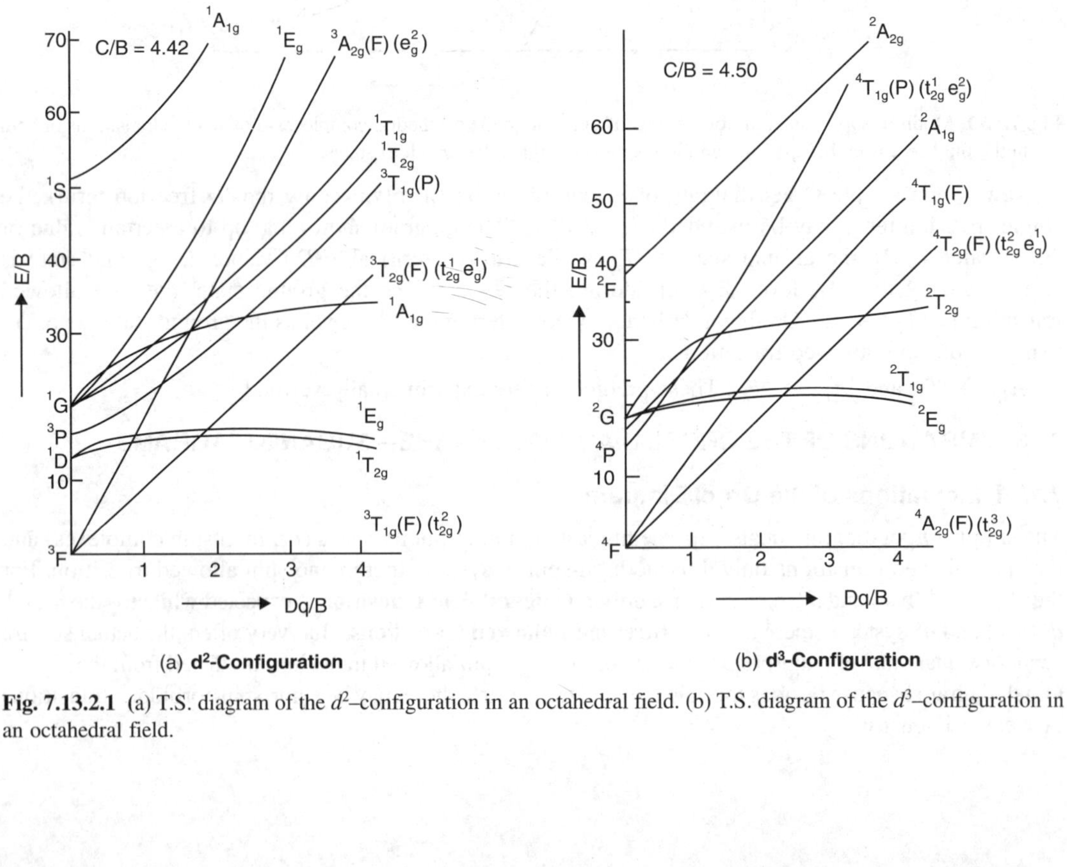

(a) **d²-Configuration** (b) **d³-Configuration**

Fig. 7.13.2.1 (a) T.S. diagram of the d^2–configuration in an octahedral field. (b) T.S. diagram of the d^3–configuration in an octahedral field.

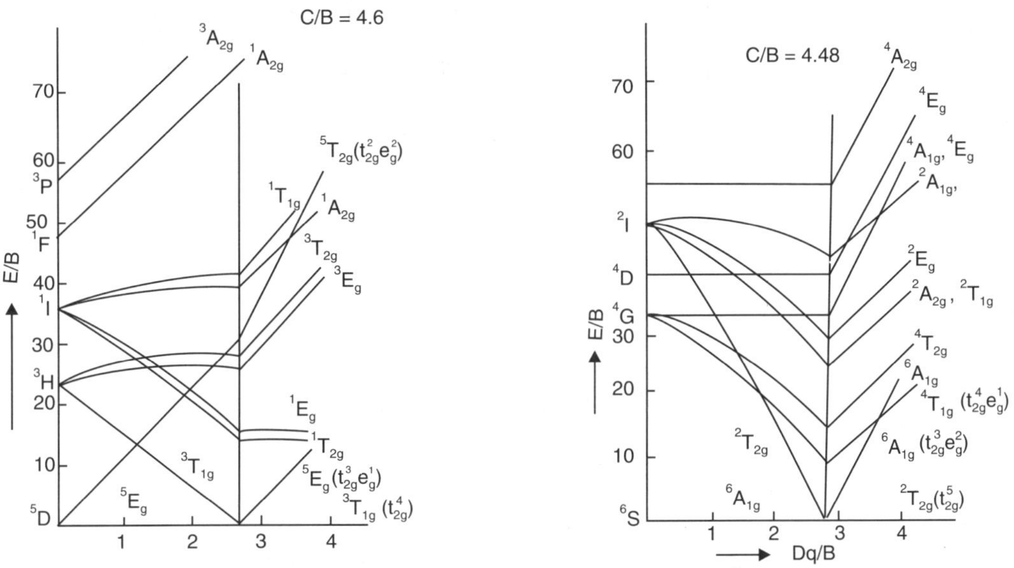

(c) **d⁴-configuration**

(d) **d⁵-configuration**

Fig. 7.13.2.1 (c) T.S. diagram of the d^4–configuration in an octahedral field; (d) T.S. diagram of the d^5–configuration in an octahedral field.

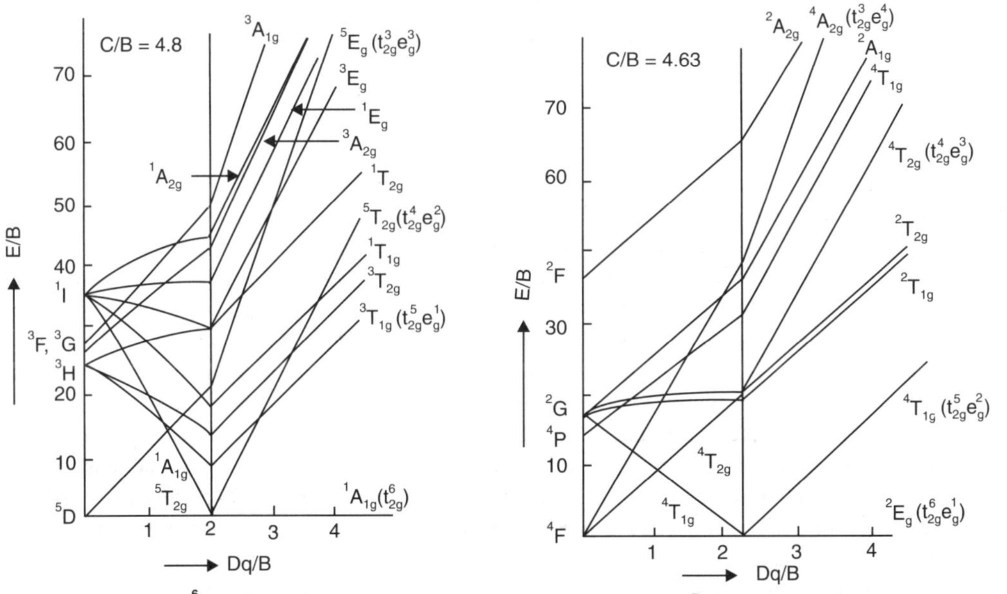

(e) **d⁶-configuration**

(f) **d⁷-configuration**

Fig. 7.13.2.1 (e) T.S. diagram of the d^6–configuration in the octahedral field; (f) T.S. diagram of the d^7–configuration in an octahedral field.

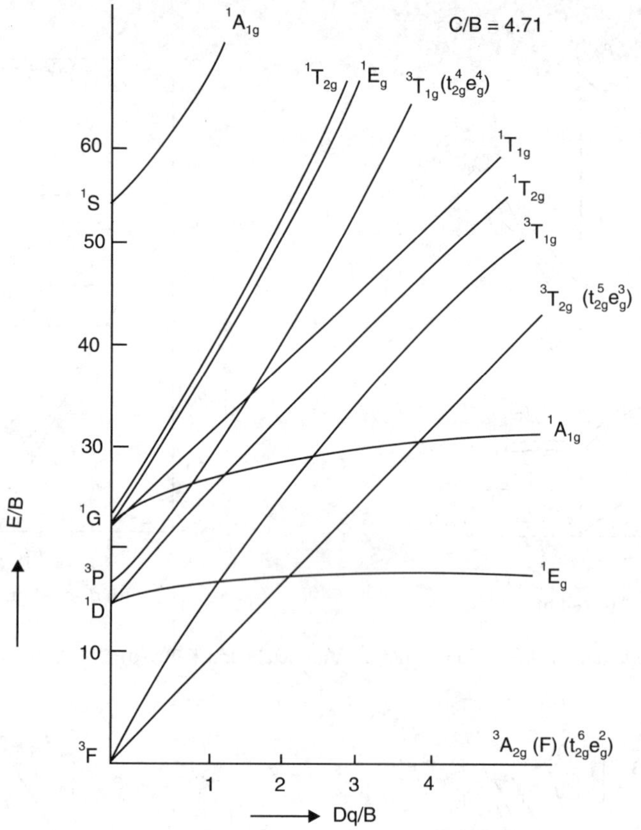

(g) d^8-configuration

Fig. 7.13.2.1 (g) T.S. diagram of the d^8–configuration in an octahedral field.

(v) For the d^4, d^5, d^6 and d^7 systems, at the point of critical $10Dq$ value (= P), there is a vertical line that indicates the cross-over region. The left side of the cross-over region represents the high-spin state while the right side of the cross-over region represents the low-spin state. In the region of high-spin state, the abscissa represents the ground state of the high-spin state while in the region of low-spin state, the abscissa represents the ground state of the low-spin state.

(vi) In the *T–S* diagrams, the energies of the states and crystal field strength are expressed in terms of the Racah parameter *B* which is different for the different isoelectronic systems. The *B* value is shown at the top of the diagram.

In the *T–S* diagrams, as the energies of the states are plotted in the unit of *B*, one diagram can be used for all members of a particular isoelectronic system. ***Thus, the T–S diagrams, unlike the Orgel diagrams, are specific for a particular group of isoelectronic ions of d^n configuration.*** This is the advantage of the *T–S* diagram over the Orgel diagram.

(vii) In the *T–S* diagrams, states of different spin multiplicities are involved and their energies are the functions of Racah parameters *B* and *C*. The energy difference between the states of the same spin multiplicity is the function of *B* only while energy difference between the states of different spin multiplicities is the function of both *B* and *C* (*cf.* Table 7.9.3.1).

This is why, the T – S diagrams are constructed for a given ratio of the C/B value. Thus the systems differing in the *C/B* values cannot be treated with a *T–S* diagram drawn for a particular *C/B* value. Thus, even for a particular d^n configurations, theoretically, a large number of *T–S* diagrams are possible depending on the *C/B* values. **This is the serious limitation of the *T–S* diagram.** However, in reality, it has been found that for most of the transition metal ions, the *C/B* ratio has been found close to 4.0. This aspect is discussed below.

The *B* and *C* values are decreased during the complexation *i.e.* the values of *B* and *C* for the free ions are higher than those found in their complexes. However, the ratio *C/B* remains more or less unchanged *i.e.* $(C/B)_{\text{Free ion}} \approx (C/B)_{\text{Complex}}$. For most of the transition metal ions, the value of *B* is close to 1000 cm^{-1} and $C \approx 4B$ *i.e.* $C/B \approx 4.0$. The *T–S* diagrams are drawn based on this approximation and *one particular T–S diagram can be used for the different metal ions with the same d^n configuration for the various types of ligands.*

(viii) As the *T–S* diagrams involve the states of different spin multiplicities, these can predict both the spin allowed and spin forbidden transitions.

(ix) The *T–S* diagrams can interpret the spectra of the metal complexes of cubic symmetry (both O_h and T_d). The *T–S* diagram of the d^n–complex of O_h symmetry is applicable for the d^{10-n} complex of T_d symmetry.

(x) If the electronic transition occurs between the states which run parallel (*i.e.* having the same slopes) in the *T–S* diagram, ***then the peak becomes sharp because of the absence of peak broadening by the metal-ligand bond symmetric vibration*** (*cf.* Secs. 7.7.1–2). On the other hands, if the slopes of two states differ then the peak will be broadened by the metal-ligand bond vibration.

The *T–S* diagrams are given in Fig. 7.13.2.1. Application of the *T–S* diagram in the spectral analysis has been discussed in Sec. 7.7.3 and will be discussed later.

7.14 LIGAND FIELD SPECTRA OF THE HIGH-SPIN OCTAHEDRAL COMPLEXES OF FIRST TRANSITION SERIES

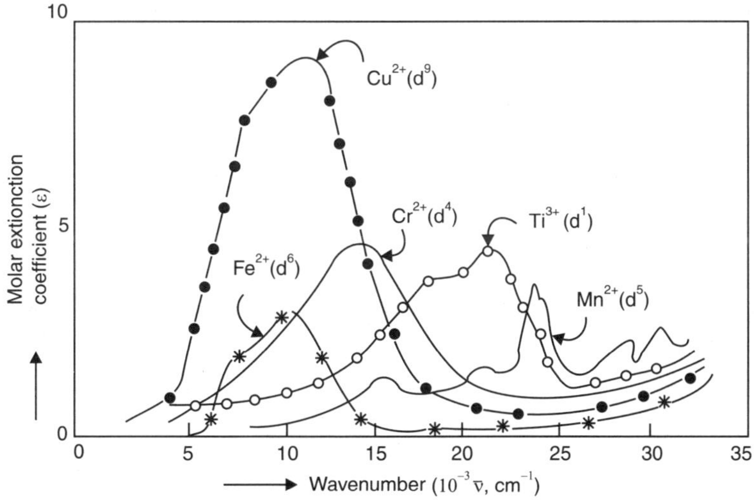

Fig. 7.14.1 (a) Electronic absorption spectra of the **aqua complexes (high-spin)** of some bivalent and trivalent metal ions of 1st transition series.

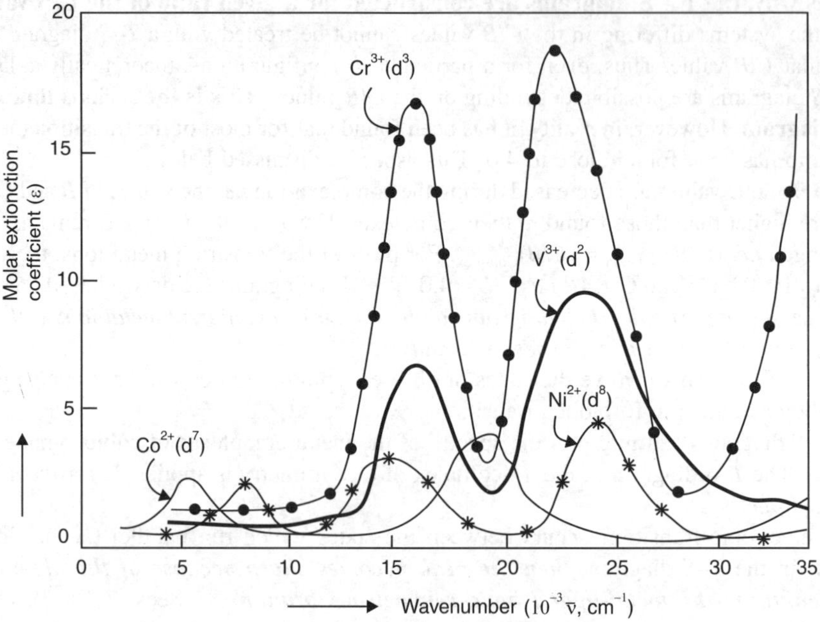

Fig. 7.14.1 (b) Electronic absorption spectra of the **aqua complexes (high-spin)** of some bivalent and trivalent metal ions of 1st transition series.

The aqua metal ions *i.e.* $[M(OH_2)_6]^{n+}$ of 1st transition series are octahedral and their colour originates from their ligand field bands. Their spectra and shown in Fig. 7.14.1.

7.14.1 Octahedral Complexes of $3d^1$ Metal Ions

The representative examples are the complexes of Ti(III) ($3d^1$), V(IV) ($3d^1$), VO^{2+} ($3d^1$).

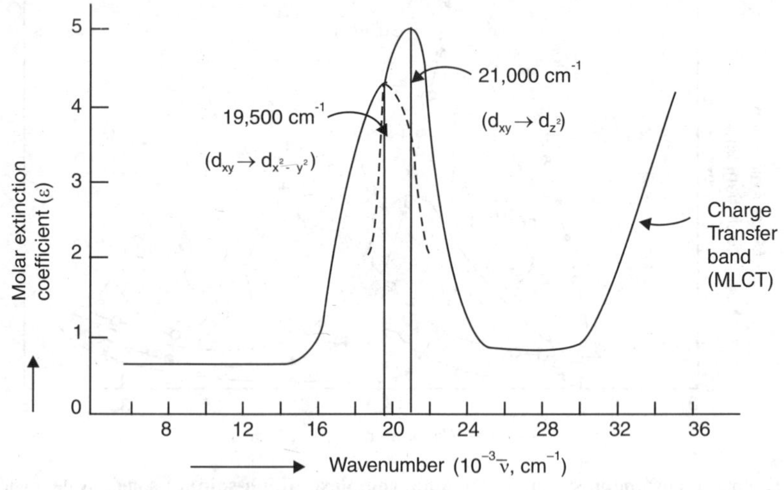

Fig. 7.14.1.1 The absorption spectrum of $[Ti(OH_2)_6]^{3+}$.

(A) Octahedral complexes of Ti(III) (*cf.* Sec. 3.11.4): The aqueous solution of titanous salt *i.e.* $[Ti(OH_2)_6]^{3+}$ looks purple or violet colour having the absorption maximum at ~20,000 cm^{-1} (*i.e.* ~500 nm). This purple coloured solution is obtained by dissolving $TiCl_3$ in dilute HCl. The purple colour of $[Ti(OH_2)_6]^{3+}$ indicates that its broad absorption peak occupies in such a way that most of the red and violet light of the visible spectrum remain unabsorbed, *i.e.* blue and green light are absorbed. *Thus the unabsorbed red-violet colour gives the purple colour of the solution.*

The **characteristic features** of the absorption spectrum of $[Ti(OH_2)_6]^{3+}$ are: **weak, broad, and unsymmetrical.** Thus we are to explain: ***position, poor intensity ($\varepsilon \approx 5$), bandwidth and shape.***

(i) **Position:** Analysis of the spectrum of $[Ti(OH_2)_6]^{3+}$ indicates that the absorption band appears at 21,000 cm^{-1} **with a shoulder** at 19,500 cm^{-1}. The **average maximum** lies at about 20,300 cm^{-1} (*i.e.* the band is centred at about 20,300 $cm^{-1} \approx \frac{1}{2}(21{,}000 + 19{,}500)\,cm^{-1}$).

From the Orgel diagram, the electron transition responsible for the electronic spectrum of $[Ti(OH_2)_6]^{3+}$ is:

$$^2T_{2g} \rightarrow {}^2E_g \text{ i.e. } t_{2g}^1\, e_g^0 \rightarrow t_{2g}^0\, e_g^1$$

The involved energy change is $10Dq$. Thus we can write:

$$10Dq = 20{,}300 \text{ cm}^{-1} = 493 \text{ nm} \approx 243 \text{ kJ mol}^{-1}$$
$$(1000 \text{ cm}^{-1} = 12.0 \text{ kJ mol}^{-1}).$$

(ii) **Poor intensity:** The poor intensity of the peak is characterised by $\varepsilon \approx 4–5$. It is due to the Laporte forbidden *d–d* transition. It is just partially allowed due to the vibronic coupling.

(iii) **Bandwidth and asymmetric shape:** It gives a wide and unsymmetric band. Analysis of the spectrum indicates that the **broad peak actually consists of two overlapping symmetric peaks.** It takes place due to the splitting of the ^2E_g state, a consequence of Jahn-Teller effect.

According to Jahn Teller theorem, the t_{2g}^1(**ground state**) level will undergo splitting and it will definitely suffer the **z–in distortion** rather than the *z*–out distortion to gain the additional cfse $\frac{2}{3}\delta_2$ compared to the undistorted octahedron. The upper e_g state will undergo splitting but it will have no energetic contribution towards the ground state because it remains unoccupied in the ground state ($t_{2g}^1 e_g^0$). But the **excited state** (*i.e.* $t_{2g}^0\, e_g^1$) will suffer the distortion and will have the energetic effect of a larger magnitude compared to that of the ground state. The possible electron transitions are shown in Fig. 7.14.1.2. These are:

$$^2B_{2g} \xrightarrow{\nu_3} {}^2E_g;\ {}^2B_{2g} \xrightarrow{\nu_1} {}^2B_{1g};\ {}^2B_{2g} \xrightarrow{\nu_2} {}^2A_{1g}, \textbf{(in terms of energy state)}$$

$$d_{xy} \xrightarrow{\nu_3} (d_{xz}, d_{xy});\ d_{xy} \xrightarrow{\nu_1} d_{x^2-y^2};\ d_{xy} \xrightarrow{\nu_2} d_{z^2}, \textbf{(in terms of orbitals)}$$

$$b_{2g} \xrightarrow{\nu_3} e_g;\ b_{2g} \xrightarrow{\nu_1} b_{1g};\ b_{2g} \xrightarrow{\nu_2} a_{1g}, \textbf{(in terms of orbitals)}$$

The peaks due to ν_1 and ν_2 (cf. Fig. 7.14.1.2) are closely spaced and they are overlapping to produce a resultant broad peak with a shoulder. The ν_3 transition lies only in the infra-red region and the corresponding energy change (δ_2) is very small (*ca.* 400 – 800 cm^{-1}). In fact, splitting of the E_g level is more significant (*ca.* δ_1 is of the order of 1000 – 3000 cm^{-1}).

The main peak at about 21,000 $cm^{-1}\left(d_{xy} \xrightarrow{\nu_2} d_{z^2} \text{ i.e., } {}^2B_{2g} \longrightarrow {}^2A_{1g}\right)$ and the other peak at about 19,500 $cm^{-1}\left(d_{xy} \xrightarrow{\nu_1} d_{x^2-y^2} \text{ i.e., } {}^2B_{2g} \longrightarrow {}^2B_{1g}\right)$ showing a shoulder overlap to produce a resultant broad peak having the average maximum at about 20,300 cm^{-1}. *The v_1 transition actually involves the 10Dq value of the corresponding hypothetical octahedral complex.* An ideal octahedral complex (which is a hypothetical case for Ti(III)) of a d^1–system should have a single peak. By considering this fact, average peak position of a ν_1($\approx$ 19,500 cm^{-1}) and ν_2 ($\approx$ 21,000 cm^{-1}) has been *conventionally regarded* as the peak position of the **hypothetical octahedral complex,** $[Ti(OH_2)_6]^{3+}$

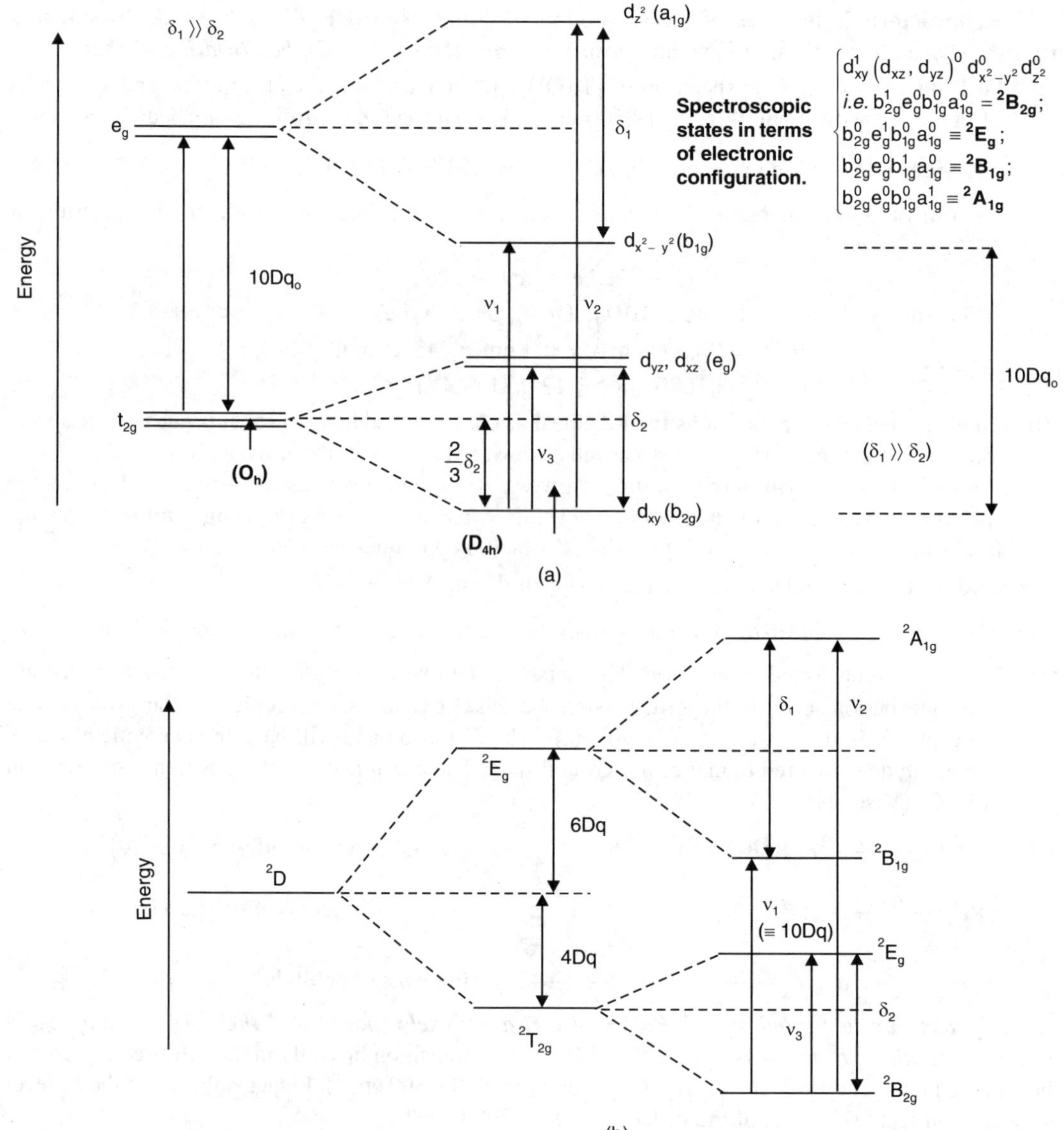

Fig. 7.14.1.2 Possible ways of electron transition in d^1 system due to **z-in Jahn-Teller distortion**

(Note: b_{1g}, b_{2g}, e_g and a_{1g} are used **to describe the orbital** while B_{1g}, B_{2g}, E_g and A_{1g} denote the **spectroscopic states**).

and this peak position has been considered to estimate the 10Dq value. However, this is an **approximation** but it does not affect the analysis seriously because the ν_1 and ν_2 bands are very close to each other. For $[Ti(OH_2)_6]^{3+}$, the average peak position appears at about 20,300 cm^{-1} which is considered as the 10Dq value of the corresponding **hypothetical octahedral complex.** With the increase of the crystal field strength, the 10Dq value increases.

	$[TiCl_6]^{3-}$	$[TiF_6]^{3-}$	$[Ti(OH_2)_6]^{3+}$	$[Ti(CN)_6]^{3-}$	$[Ti(bpy)_3]^{3+}$
10Dq (cm^{-1}):	13,000	18,900	20,300	22,300	25,000

[**Note:** Some authors analyse the peaks of $[Ti(OH_2)_6]^{3+}$ as follows:

absorption band at 20,300 cm^{-1} and shoulder at 18,400 cm^{-1}.

Without the J.T. distortion, a single peak ($^2T_{2g} \rightarrow {}^2E_g$) will appear at $\frac{1}{2}(20,300 + 18,400)\ cm^{-1}$ *i.e.* 19350 cm^{-1} which is considered as the $10Dq$ value of the hypothetical octahedral $[Ti(OH_2)_6]^{3+}$ complex].

Here it is worth mentioning that if the peaks due to ν_1 and ν_2 are widely separated (*i.e.* δ_1 is relatively high) then the two peaks may not overlap and **two distinct peaks will appear.** It happens so in the high-spin $[Co^{III}F_6]^{3-}$ complex which involves the transition:

$$t_{2g}^4\, e_g^2 \rightarrow t_{2g}^3\, e_g^3$$

This aspect will be discussed separately for the d^6–system.

The spectral features of other octahedral Ti(III)–complexes like $[TiCl_6]^{3-}$, $[TiF_6]^{3-}$, $[Ti(urea)]^{3+}$, Ti(III) in Al_2O_3 matrix or in $CoTi(SO_4)_3.12H_2O$ are similar to those of $[Ti(OH_2)_6]^{3+}$.

(B) Possibility of spin-orbit coupling to split the energy levels to explain the broad peak of Ti(III)–complexes: Splitting of the $^2T_{2g}$ term (originated from ^2D) due to spin-orbit coupling has been shown in Fig. 7.7.4.1. The ^2E_g term does not split due to spin-orbit coupling but the $^2T_{2g}$ term splits into two components characterised by $J = \frac{3}{2}$ and $J = \frac{1}{2}$. In the case of Ti(III), the spin-orbit coupling constant is quite small (*cf.* $\lambda \approx +\ 155\ cm^{-1}$) and consequently, the energy difference between the states ($J = 3/2$ and $J = 1/2$, for ground state $J = 3/2$) is $3\lambda/2$ which is approximately 232 cm^{-1}. This value is negligibly small compared to the crystal field splitting parameter $10Dq$ ($\approx 20,000\ cm^{-1}$). **Thus the spin-orbit coupling leading to the splitting of the $^2T_{2g}$ state can hardly explain the peak broadening phenomenon in the case of Ti(III).** This can be explained better by J.T. distortion as discussed above.

Note: Spin-orbit coupling constants of the 3*d*–metal ions are quite negligible compared to the $10Dq_o$ value of the octahedral complexes. Thus the peak broadening by this mechanism is not significant. **But, for the tetrahedral complexes where $10Dq_t$ is relatively small, spin-orbit coupling phenomenon can broaden the peak appreciably.** In fact, splitting of the spectrum of the tetrahedral $[CoCl_4]^{2-}$ complex by spin-orbit coupling is well documented. This aspect will be discussed later.

(C) Charge transfer band in $[Ti(OH_2)_6]^{3+}$: The metal to ligand charge transfer (MLCT) band due to the electron transfer from Ti(III) (which is reducing in character) to oxygen (O) appears at the ultraviolet range (beyond 30,000 cm^{-1}). In that region, there is a strong peak for this MLCT band.

(D) Blue colour of gemstone sapphire due to Ti(III): In the gemstone sapphire, Ti(III) is present in the structure of *corundum* (Al_2O_3). In corundum, Ti^{3+} (r = 81 pm) being larger than Al^{3+} (r = 67 pm) experiences the shorter Ti^{III}—O distance compared to the normal Ti^{III}—O distance as in $[Ti(OH_2)_6]^{3+}$. This is why, Ti(III) experiences the higher crystal field splitting when placed in corundum through the substitution of Al(III). It explains the blue colour of sapphire (*cf.* $[Ti(OH_2)_6]^{3+}$ looks violet where the crystal field splitting is less). The colour of gemstones has been discussed in Sec. 11.11.4, Vol. 2.

(E) Six–coordinate complexes of V(IV) and six-coordinate complexes of oxovanadium, *i.e.* VO^{2+}: Vanadium(IV) or oxovanadium(IV) represents the $3d^1$ system. The electronic spectra of the six-coordinate V(IV) complexes like $[VCl_6]^{2-}$ in $K_2[VCl_6]$ or $[VF_6]^{2-}$ in $(NH_4)_2[VF_6]$ can be analysed as in the case of six-coordinate Ti(III) complexes. For $[VCl_6]^{2-}$ and $[VF_6]^{2-}$, their absorption bands are centred at around 15,400 cm^{-1} and 20,100 cm^{-1} respectively.

The spectra of the octahedral (O_h) and distorted octahedral (D_{4h}, *z*-in distortion) complexes of V(IV), *e.g.* $[VCl_6]^{2-}$ etc. can be explained as in the cases of Ti(III) complexes. The six coordinate complexes of oxovanadium(IV) are of C_{4v} symmetry and their spectra can be explained by using the Fig. 7.14.1.3 (which is similar to Fig. 7.14.1.2a). In the oxovanadium(IV) complexes like $[VOL_5]$, the V$=$O unit remains projected to an axial direction (say *z*-direction). In $[VO(OH_2)_5]^{2+}$, the V—O (for VO^{2+} unit) bond length is 167 pm while the V—$O(OH_2)$ bond length is in the range of 203-230 pm. This supports the **tetragonal compression in $[VOL_5]$**.

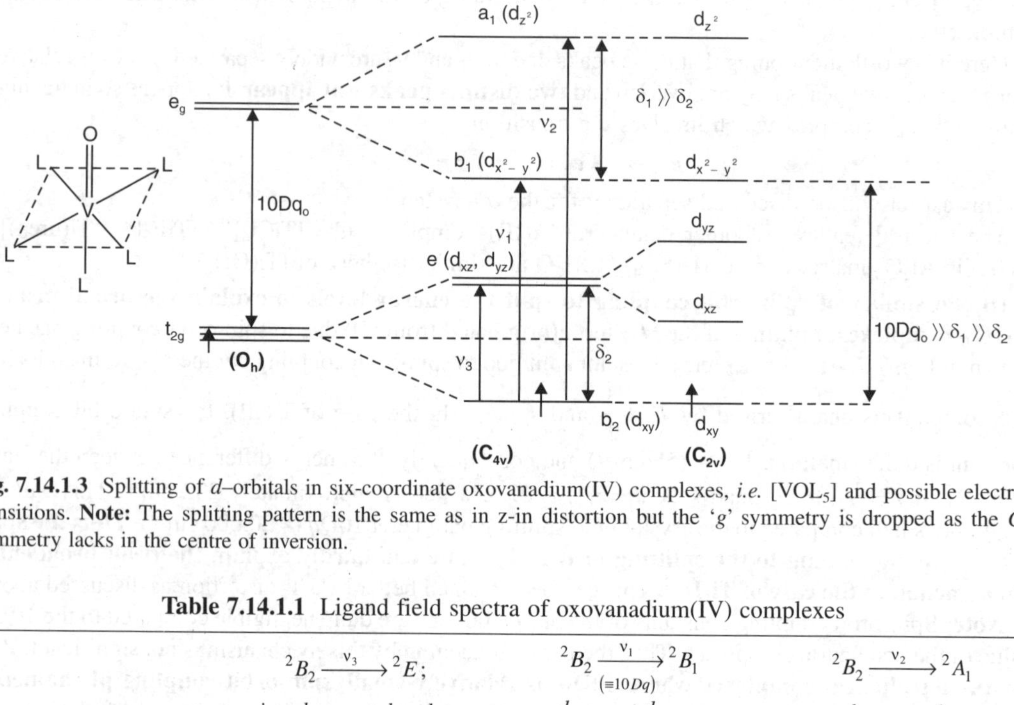

Fig. 7.14.1.3 Splitting of *d*–orbitals in six-coordinate oxovanadium(IV) complexes, *i.e.* $[VOL_5]$ and possible electron transitions. **Note:** The splitting pattern is the same as in z-in distortion but the '*g*' symmetry is dropped as the C_{4v} symmetry lacks in the centre of inversion.

Table 7.14.1.1 Ligand field spectra of oxovanadium(IV) complexes

	$^2B_2 \xrightarrow{\nu_3} {^2E}$; *i.e.*, $d_{xy} \longrightarrow d_{xz}, d_{yz}$ ν_3 (cm^{-1})	$^2B_2 \xrightarrow[(\equiv 10Dq)]{\nu_1} {^2B_1}$ $d_{xy} \longrightarrow d_{x^2-y^2}$ ν_1 (cm^{-1})	$^2B_2 \xrightarrow{\nu_2} {^2A_1}$ $d_{xy} \longrightarrow d_{z^2}$ ν_2 (cm^{-1})
$[VO(OH_2)_5]^{2+}$	13,100	16,000	–
$[VO(NCS)_5]^{3-}$	13,500	17,200	23,900
$[VO(ox)_2]^{2-}$	12,600	16,500	29,400
$[VO(acac)_2]$, (in methanol)	13,000	17,450	25,630

Here it may be noted that the $^2B_2 \rightarrow {^2E}$ transition is **allowed** but the $^2B_2 \rightarrow {^2B_1}$ and $^2B_2 \rightarrow {^2A_1}$ transitions are **forbidden.** However, these can be allowed by vibronic coupling. These two transitions give the **relatively weaker bands.**

Note: • The charge transfer bands of oxovanadium(IV) complexes are similar in both solid state and solution. It indicates the similar fate of the $(V{=}O)^{2+}$ group both in solution and solid state. In fact, **basicity of the oxygen site in the VO^{2+} group is drastically reduced due to π–bonding with V(IV)** and consequently, it is not protonated easily.

The six-coordinate complexes of oxovanadium (*i.e.* vanadyl cation, VO^{2+}) are not octahedral and they experience the **tetragonal compression** (*i.e.* **z–in distortion**) and the complexes like $[VO(OH_2)_5]^{2+}$, $[VO(NCS)_5]^{3-}$ etc. belong to the C_{4v} point group. The orbital splitting in the six-coordinate complexes of VO^{2+} is shown in Fig. 7.14.1.3.

The possible electronic transitions are:

$$^2B_2 \xrightarrow{\nu_3} {}^2E \text{ i.e. } b_2 \xrightarrow{\nu_3} e;\ {}^2B_2 \xrightarrow[(\equiv 10Dq)]{\nu_1} {}^2B_1 \text{ i.e. } b_2 \xrightarrow{\nu_1} b_1;\ {}^2B_2 \xrightarrow{\nu_2} {}^2A_1 \text{ i.e. } b_2 \xrightarrow{\nu_2} a_1$$

The electronic transitions in terms of the energy states and d^n-configurations are correlated for the C_{4v} oxovanadium(IV) complexes as follows:

$$d^1_{xy}\left(d_{xz}, d_{yz}\right)^0 \left(d_{x^2-y^2}\right)^0 d^0_{z^2} \text{ i.e. } b_2^1 e^0 b_1^0 a_1^0 \equiv {}^2B_2 \Big\}\ \textbf{ground state}$$

$$\left.\begin{array}{l} d^0_{xy}\left(d_{xz}, d_{yz}\right)^1 \left(d_{x^2-y^2}\right)^0 d^0_{z^2} \text{ i.e. } b_2^0 e^1 b_1^0 a_1^0 \equiv {}^2E \\ d^0_{xy}\left(d_{xz}, d_{yz}\right)^0 \left(d_{x^2-y^2}\right)^1 d^0_{z^2} \text{ i.e. } b_2^0 e^0 b_1^1 a_1^0 \equiv {}^2B_1 \\ d^0_{xy}\left(d_{xz}, d_{yz}\right)^0 \left(d_{x^2-y^2}\right)^0 d^1_{z^2} \text{ i.e. } b_2^0 e^0 b_1^0 a_1^1 \equiv {}^2A_1 \end{array}\right\}\ \textbf{excited state}$$

$[MoX_5]^{2-}$: The spectral behaviour of $[MoX_5]^{2-}$ is very much similar to that of $[VOX_5]^{3-}$ but the transition $d_{xy} \rightarrow d_{z^2}$ is very often obscured by the CT band for $[MoX_5]^{2-}$.

- The oxovanadium complexes like $[VO(NCS)_5]^{3-}$, $[VO(acac)_2]$, etc. should be written as $[V(NCS)_5O]^{3-}$, $[V(acac)_2O]$, ...respectively (*See* Chapter 1, IUPAC recommendation).

(F) Five-coordinate complexes of oxovanadium(IV): The spectra of five-coordinate complexes like $[VO(ox)_2]^{2-}$, $[VO(acac)_2]$ of C_{4v} symmetry can also be explained in terms of Fig. 7.14.1.3.

(G) Spectral properties of the oxovanadium(IV) complexes of C_{2v} symmetry: Sometimes, the oxovanadium complexes, *e.g.* vanadyl tartarate, show **four bands instead of three bands** (expected from the C_{4v} symmetry). Origin of the four spectral transitions has been explained by considering the **C_{2v} symmetry**. In moving from the C_{4v} symmetry to the C_{2v} symmetry, degeneracy of the *e*-levels (*i.e.* d_{xz} and d_{yz} orbitals) is lifted. Then the possible transitions are:

$$\mathbf{C_{2v}}: d_{xy} \xrightarrow{11\text{ kK}} d_{xz};\ d_{xy} \xrightarrow{17\text{ kK}} d_{yz};\ d_{xy} \xrightarrow{18.8\text{ kK}} d_{x^2-y^2};\ d_{xy} \xrightarrow{25.3\text{ kK}} d_{z^2}$$

(For vanadyl tartarate).

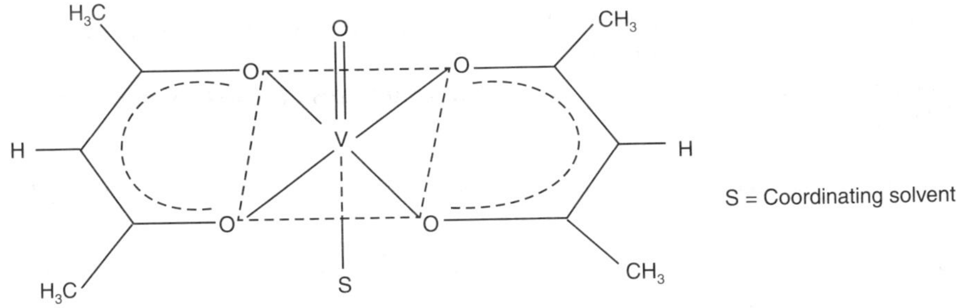

$[VO(acac)_2] \cdot (C_{4v}/C_{2v})$

Blue-green [VO(acac)$_2$]: When dissolved in various organic solvents varying in coordinating power, it shows the spectral bands of different energies. It happens so due to the coordination by the solvent molecule (S) at the *trans*-axial site to the V = O bond to give the 6-coordinate species. Thus for different solvents, the average 10*Dq* value changes and the spectral transitions occur at different energies. In fact, the extents of splitting of the *d*-orbitals in C_{4v} symmetry depend on the average crystal field strength (*cf.* spectral behaviour of [Cu(acac)$_2$] in different solvents, Fig. 7.14.2.5).

Detailed studies (both is solid state and mixed solvents) on **[VO(acac)$_2$]** indicate that it also adopts the C_{2v} symmery and the transition $d_{xy} \rightarrow d_{xz}, d_{yz}$ (*i.e.* $b_2 \rightarrow e$ in C_{4v}) gives two spectral components, *i.e.* two separate transitions as expected for the C_{2v} symmetry. However, some authors have attempted to explain the four peaks (*i.e.* components) in the electronic absorption spectra of [VO(acac)$_2$] in terms of vibrational structure.

7.14.2 Six-Coordinate Complexes of 3d^9–System

(A) Possible spectral transitions in the distorted octahedral complexes of Cu(II): Cu(II) complexes are the representative examples of the d^9–system. Energy level diagram for the spectral transitions in a d^9–system is the inverted energy level diagram of the d^1–system. It is expected so from the **hole formalism rule.** Thus by using the Orgel diagram, for the octahedral complexes of Cu(II), the spectral transition is:

$$^2E_g \rightarrow {}^2T_{2g}, \text{ (cf. for } d^1\text{–system: } {}^2T_{2g} \rightarrow {}^2E_g)$$

In terms of the electronic configuration, the transition is:

$$t_{2g}^6 e_g^3 \rightarrow t_{2g}^5 e_g^4$$

The ground state is **orbitally doubly degenerate.**

$$t_{2g}^6 e_g^3 ({}^2E_g) : t_{2g}^6 \left(d_{x^2-y^2}\right)^2 \left(d_{z^2}\right)^1, t_{2g}^6 \left(d_{x^2-y^2}\right)^1 \left(d_{z^2}\right)^2$$

The excited state is **orbitally triply degenerate.**

$$t_{2g}^5 e_g^4 ({}^2T_{2g}): (d_{xy})^2 (d_{yz})^2 (d_{xz})^1 e_g^4; (d_{xy})^2 (d_{yz})^1 (d_{xz})^2 e_g^4; (d_{xy})^1 (d_{yz})^2 (d_{xz})^2 e_g^4$$

The ground state electronic configuration, $t_{2g}^6 e_g^3$ indicates that the unsymmetrical filling in the e_g level causes the J.T. distortion (*i.e.* tetragonal distortion) at the ground state. Thus it can never have the perfectly octahedral complexes. The six-coordinate complexes of Cu(II) are the distorted octahedrals. The tetragonal distortion can produce both the *z*-out (*i.e.* tetragonal elongation) and *z*–in (*i.e.* tetragonal compression) distortion with an equal probability because both the possibilities lead to the same additional cfse $\frac{1}{2}\delta_1$ (δ_1 = splitting energy of the e_g level) for a d^9–system. But in reality, most of the Cu(II)–complexes are experiencing the *z*–out distortion. The tetragonal compression is known only in some rare cases. One such example is mentioned here. In solid state, **K$_2$[CuF$_4$] shows the *z*–in distortion** (*i.e.* compressed octahedral) around Cu(II).

The *d*–orbital splitting in the **elongated octahedral complexes of Cu(II)** is shown in Fig. 7.14.2.1. The possible transitions are:

$$^2B_{1g} \xrightarrow{\nu_3} {}^2A_{1g}, \text{ i.e. } d_{z^2} \xrightarrow{\nu_3} d_{x^2-y^2} \text{ or, } a_{1g} \xrightarrow{\nu_3} b_{1g}, \text{ (12,000 – 17,000 cm}^{-1})$$

$$^2B_{1g} \xrightarrow{\nu_2} {}^2B_{2g}, \text{ i.e. } d_{xy} \xrightarrow[(\equiv 10Dq)]{\nu_2} d_{x^2-y^2} \text{ or, } b_{2g} \xrightarrow{\nu_2} b_{1g}, \text{ (15,500 – 18,000 cm}^{-1})$$

$$^2B_{1g} \xrightarrow{\nu_1} {}^2E_g, \text{ i.e. } d_{xz}, d_{yz} \xrightarrow{\nu_1} d_{x^2-y^2} \text{ or, } e_g \xrightarrow{\nu_1} b_{1g}, \text{ (17,000 – 20,000 cm}^{-1})$$

The ν_3 transition actually gives the measure of splitting of the e_g level denoted by δ_1.

The extents of splitting of the t_{2g} and e_g levels measured by δ_2 and δ_1 respectively (*cf.* Fig. 7.14.2.1.) due to the J.T. distortion depend on the ligand field strength of the ligands. In most of the cases, the splittings are not too large. Consequently, the peaks due to ν_1, ν_2 and ν_3 very often overlap to give ***a broad resultant peak of unsymmetrical shape.*** In some cases, where the splittings are quite large, separate peaks may appear.

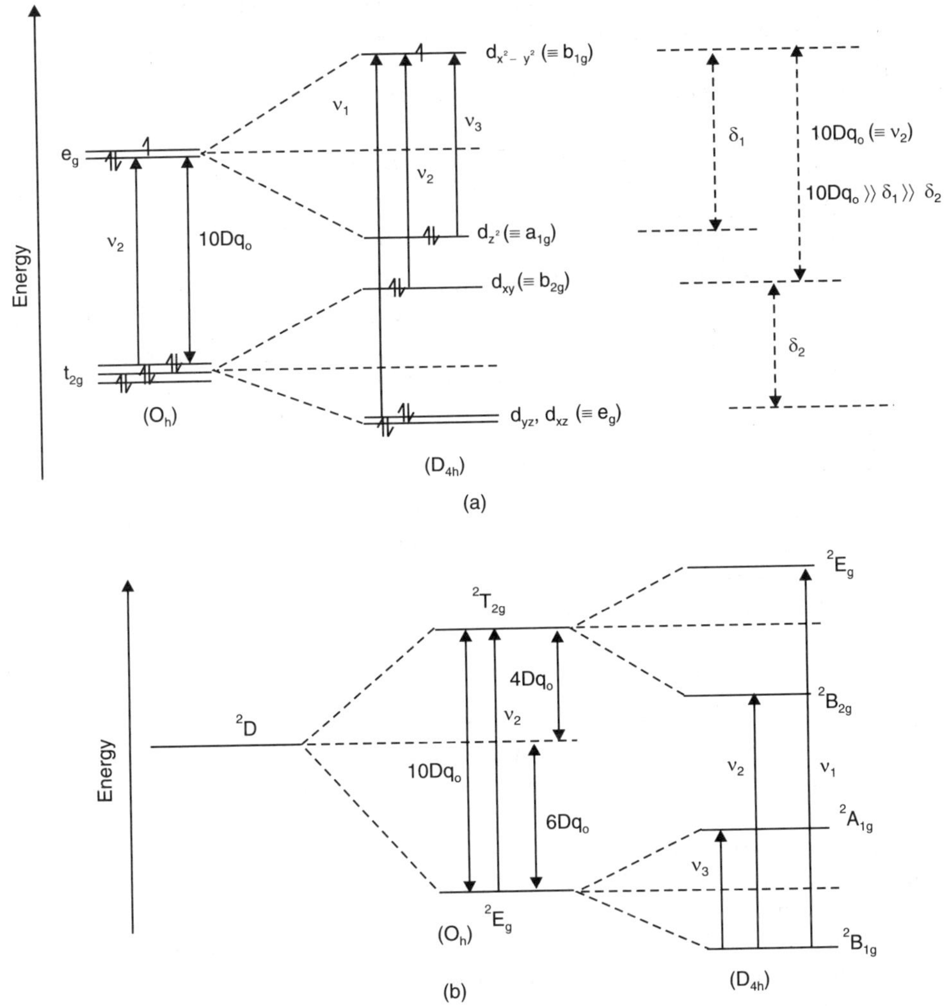

Fig. 7.14.2.1 Splitting of the energy levels due to the medium *z*-out tetragonal distortion in the six-coordinate complexes of Cu(II) (d^9-system). (a) Splitting of the *d*-orbitals; (b) splitting of the spectroscopic states.

Note: The ^2D term splits in the same way both in the C_{4v} and D_{4h} symmetry.

(B) Peak broadening by spin-orbit coupling in Cu(II) complexes: The absorption spectra of the Cu(II)–complexes are broadened due to J.T. distortion discussed above. Besides this, the peak may be further broadened due to the spin-orbit coupling. The T_{2g} level is split (*cf.* Secs. 7.7.4, 8.20.3) due to the spin-orbit coupling to broaden the peak. But it is relatively less important compared to that in tetrahedral complexes where the spin-orbit interaction energies are more or less comparable to the ligand field energies.

(C) Colour of the Cu(II)–complexes: Most of the Cu(II) complexes are **blue or green** in colour. Some complexes look **red or brown**. Such red or brown complexes experience a strong absorption in the near ultraviolet region due to the charge transfer (CT) bands. The tails of the CT bands extend into the blue end of the visible spectrum. It makes the red-brown colour. The possible electronic transitions in the distorted octahedral complexes of Cu(II) are given below.

$$e_g^4 b_{2g}^2 a_{1g}^2 b_{1g}^1 \left(i.e.\ {}^2B_{1g}\right) \xrightarrow{\nu_1} e_g^3 b_{2g}^2 a_{1g}^2 b_{1g}^2 \left(i.e.\ {}^2E_g\right) i.e.\ {}^2B_{1g} \longrightarrow {}^2E_g$$

$$e_g^4 b_{2g}^2 a_{1g}^2 b_{1g}^1 \left(i.e.\ {}^2B_{1g}\right) \xrightarrow{\nu_2} e_g^4 b_{2g}^1 a_{1g}^2 b_{1g}^2 \left(i.e.\ {}^2B_{2g}\right) i.e.\ {}^2B_{1g} \longrightarrow {}^2B_{2g}$$

$$e_g^4 b_{2g}^2 a_{1g}^2 b_{1g}^1 \left(i.e.\ {}^2B_{1g}\right) \xrightarrow{\nu_3} e_g^4 b_{2g}^2 a_{1g}^1 b_{1g}^2 \left(i.e.\ {}^2A_{1g}\right) i.e.\ {}^2B_{1g} \longrightarrow {}^2A_{1g}$$

(D) Spectral properties of $[Cu(OH_2)_6]^{2+}$: It possesses the elongated octahedral structure and splitting of the energy levels occurs as shown in Fig. 7.14.2.1. H_2O is a weak field ligand and consequently splitting of the t_{2g} and e_g levels is not very high and, the three possible spectral transitions give three peaks (ν_1, ν_2 and ν_3) and they overlap to produce a **broad and unsymmetric peak** having the average maximum at 12,600 cm^{-1} *i.e.* ~800 nm ($\varepsilon \approx 11$). In $(NH_4)_2[Cu(OH_2)_6](SO_4)_2$ (solid state), three transitions have been identified,

$d_{z^2} \longrightarrow d_{x^2-y^2}$ (6,400 cm^{-1}); $d_{xy} \longrightarrow d_{x^2-y^2}$ (10,650 cm^{-1}); $d_{xz}, d_{yz} \longrightarrow d_{x^2-y^2}$ (11,550 cm^{-1}).

The crystal structure of $CuSO_4 \cdot 5H_2O$ is quite interesting (*cf.* Sec. 13.1.6, Vol. 3). Here each Cu(II)–centre is coordinated by four water molecules in a square planar geometry and two *trans*-axial positions are occupied by two sulfate O–atoms. These two *trans*–axial bonds are relatively longer as expected from the *z*-out distortion. The fifth water molecule (called **anion water**) is H-bonded with the SO_4–moiety and the water molecule coordinated with the metal centre. Analysis of the polarised crystal spectrum of $CuSO_4 \cdot 5H_2O$ shows three absorptions peaks as expected:

$$10{,}500\ \text{cm}^{-1}\ (\nu_3), 13{,}000\ \text{cm}^{-1}\ (\nu_2)\ \text{and}\ 14{,}500\ \text{cm}^{-1}\ (\nu_1)$$

Note: ***Anhydrous $CuSO_4$ is colourless.*** Crystal field splitting of the *d*-orbitals by the sulfate ligands is reduced to such an extent that the spectral transitions (*d*—*d* transitions) occur **at the infrared region.** It makes it colouress.

(E) Spectral properties of $[Cu(NH_3)_x(OH_2)_y]^{2+}$ (x + y = 6): With the replacement of weak field ligands by the relatively stronger field ligands, the magnitude of crystal field splitting increases and consequently the absorption peaks move towards the lower wavelength, *i.e.* higher frequency. NH_3 is a stronger field ligand than H_2O. Thus, the gradual replacement of H_2O by NH_3 (upto $x = 4$), the absorption maximum shifts towards the higher frequency (*i.e.* towards UV).

$$[Cu(OH_2)_6]^{2+}: \quad \lambda_{max} \approx 800\ \text{nm},\ \varepsilon_{max} \approx 11.0$$

$$[Cu(NH_3)_4(OH_2)_2]^{2+}: \lambda_{max} \approx 590\ \text{nm},\ \varepsilon_{max} \approx 50.0$$

In the mixed ligand complex $[Cu(NH_3)_4(OH_2)_2]^{2+}$, the average ligand field strength is the average of the fields due to water and ammonia.

$$\bar{\nu}_{M[A_xB_{6-x}]} = \frac{x}{6}\bar{\nu}_{[MA_6]} + \frac{(6-x)}{6}\bar{\nu}_{[MB_6]}$$

or, $$10Dq_{M[A_xB_{6-x}]} = \frac{x}{6}10Dq_{[MA_6]} + \frac{6-x}{6}10Dq_{[MB_6]}$$

Here it is interesting to note that **entry of the fifth NH_3 molecule** in $[Cu(NH_3)_4(OH_2)_2]^{2+}$ shifts the peak **towards the longer wavelength**. No evidence for the formation of $[Cu(NH_3)_6]^{2+}$ in aqueous media is reported. On entering the stronger field ligand, the peak is expected to shift towards the lower wavelength but on placing the fifth NH_3 molecule there is a **reversal of spectral shift.** The detailed explanation of the *reversal of spectral shift* will be discussed later in connection with the spectral shift of $[Cu(acac)_2]$ in different solvents.

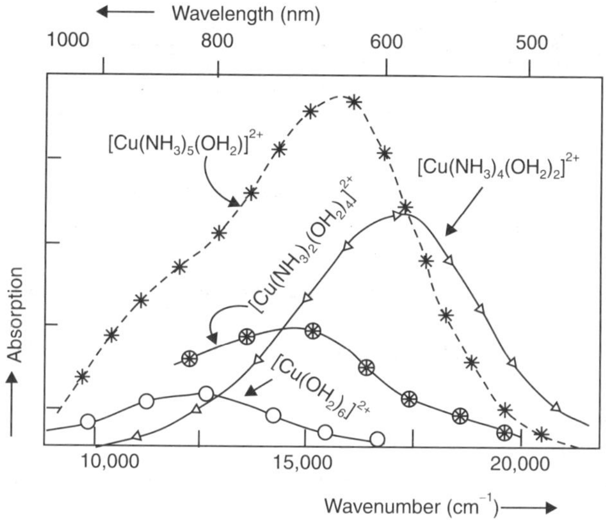

Fig. 7.14.2.2 Comparison of the electronic spectra of $[Cu(OH_2)]^{2+}$ and its different ammine complexes in aqueous solution.

(F) Spectral properties of $[Cu(acac)_2]$ in different solvents: Belford and cowerkers (*cf. J. Chem. Phys.*, **26,** 1165, 1957) studied the spectral properties of *bis*-acetylacetonatocopper(II), *i.e.* $[Cu(acac)_2]$ in different solvents varying in coordinating power to investigate **the effect of the degree of tetragonal distortion on the spectral transitions.** $[Cu(acac)_2]$ adopts the square planar geometry (a limiting situation of *z*–out distortion). When $[Cu(acac)_2]$ is dissolved in different solvents, the solvent molecules (S) may coordinate along the *trans*– axial directions to give a tetragonally distorted octahedral structure in solution (Fig. 7.14.2.3).

Ligation by the solvent molecules in the *trans*–axial positions will convert the s**quare planar $[Cu(acac)_2]$ complex** into a **distorted octahedral $[Cu(acac)_2(S)_2]$ complex**. If the coordinating power of solvent is poor, then the complex mainly remains as a square planar one. Limiting situation of the splitting of the t_{2g} and e_g level due to the *z*–out distortion will give the square planar geometry. **In a poor coordinating solvent (*e.g.* $CHCl_3$), $[Cu(acac)_2]$ remains practically as a square planar complex,** *i.e.* splitting of the t_{2g} and e_g level will be the maximum. On the other hand, in a good coordinating solvent (*e.g.* piperidine), coordination along the *trans*–axial directions by S will give the distorted octahedral geometry where the splittings of the t_{2g} and e_g levels are relatively less compared to those in the poor coordinating solvents.

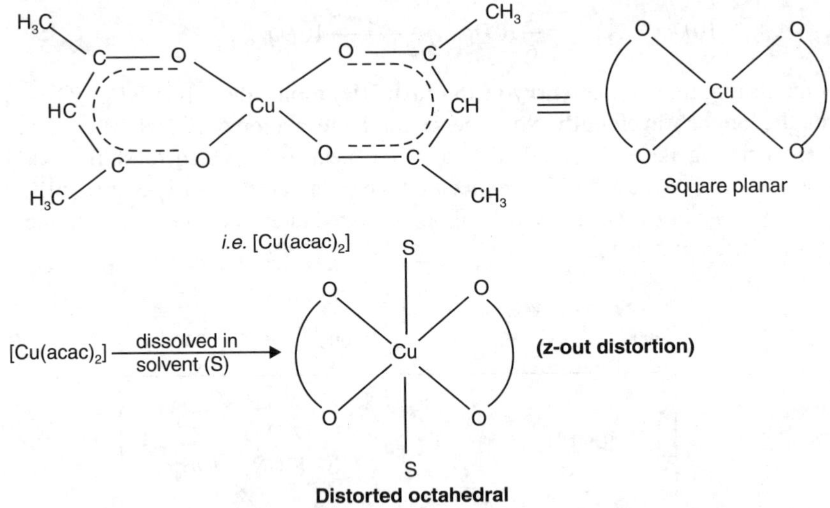

Fig. 7.14.2.3 Structural features of $[Cu(acac)_2]$ in coordinating solvents.

● δ_1, δ_2 (*cf.* Fig. 7.14.2.1):	CH_3Cl_3 ⟩ Dioxane ⟩ Pentanol ⟩ Pyridine ⟩ Piperidine
● Coordinating power and basicity of the solvent (S):	$CHCl_3$ ⟨ Dioxane ⟨ Pentanol ⟨ Pyridine ⟨ Piperidine
● Tendency to adopt the octahedral geometry from the square planar geometry (in different solvents):	$CHCl_3$ ⟨ Dioxane ⟨ Pentanol ⟨ Pyridine ⟨ Piperidine

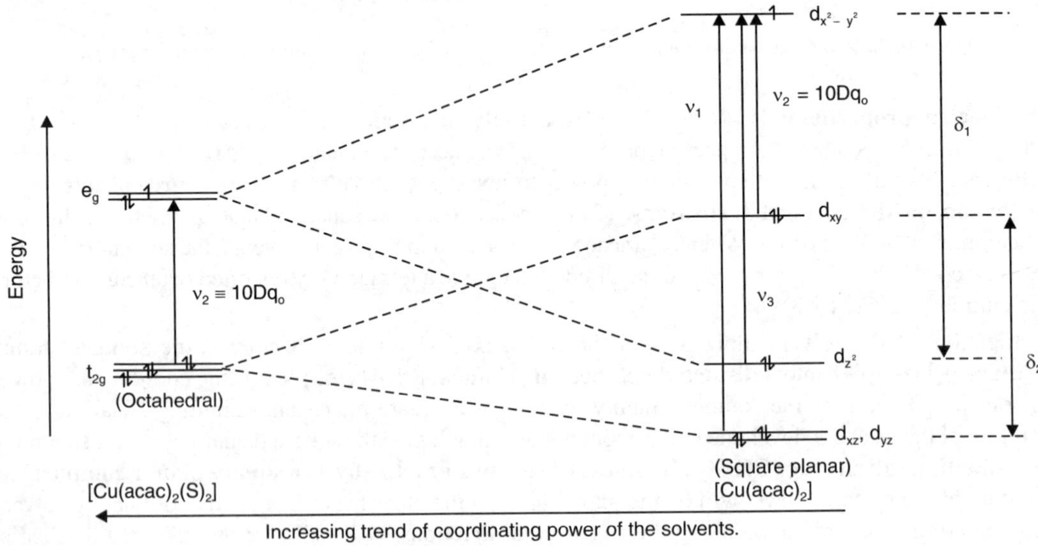

Fig. 7.14.2.4 Gradual transition of the square planar geometry of $[Cu(acac)_2]$ into the octahedral complex $[Cu(acac)_2(S)_2]$ with the increase of coordinating power (*i.e.* basicity) of the solvents (S).

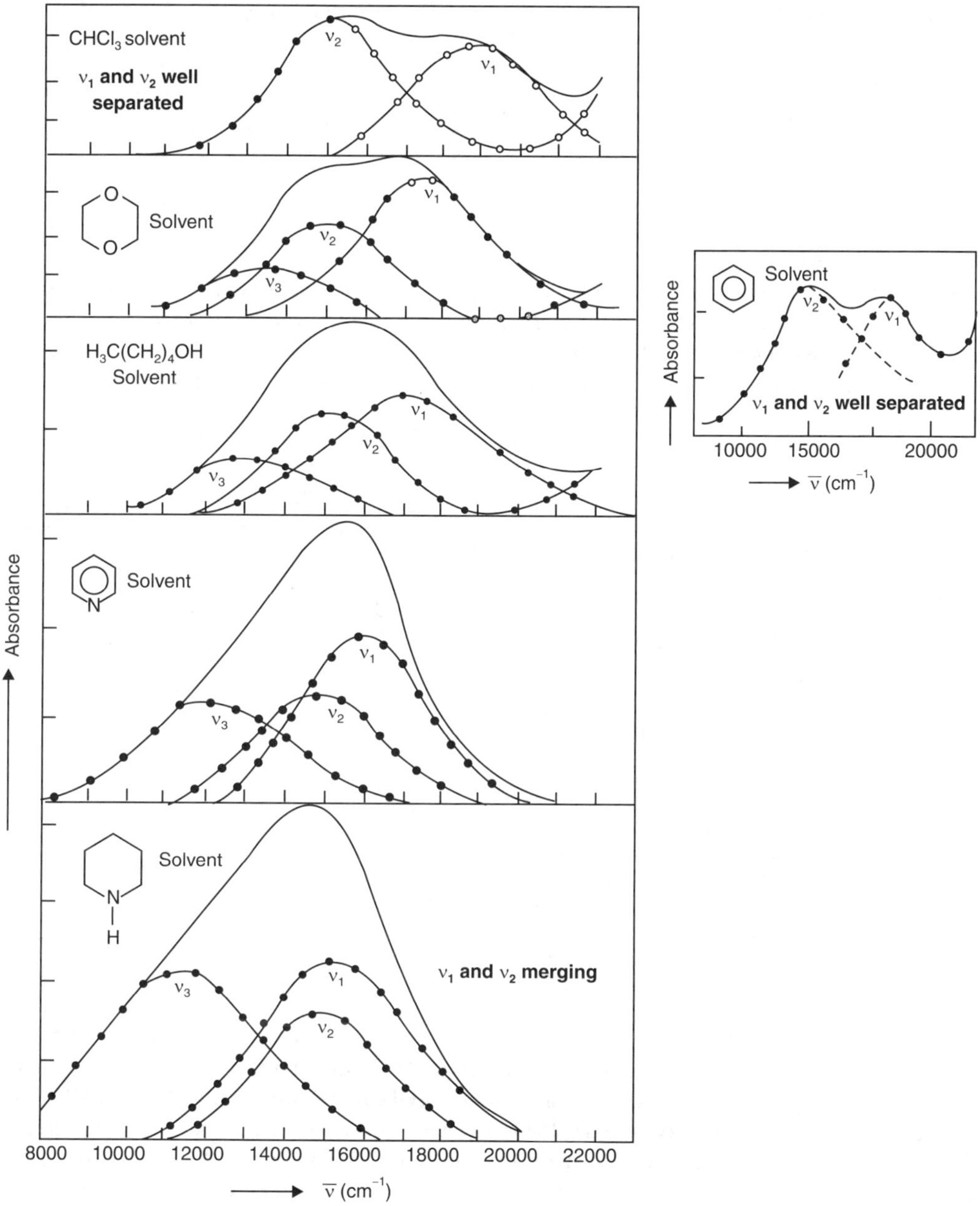

Fig. 7.14.2.5 Spectral behaviour of [Cu(acac)$_2$] in different solvents differing in basicities (*i.e.* coordinating power as a ligand).

The splittings measured by δ_1 and δ_2 (*cf.* Fig. 7.14.2.1a) are minimum in the most octahedral geometry as in the piperidine solvent. With the decrease of δ_1 and δ_2, ν_3 (frequency) decreases towards the lower frequency while ν_1 (frequency) decreases towards ν_2 that remains unchanged. For the ν_2 transition, the involved orbitals possess no z–component (*cf.* effect of J.T. distortion along the $\pm z$–direction).

$$\left.\begin{array}{ll} \nu_3(\equiv \delta_1): & d_{z^2} \longrightarrow d_{x^2-y^2} \\ \nu_2(\equiv 10Dq_o): & d_{xy} \longrightarrow d_{x^2-y^2} \\ \nu_1(\equiv 10Dq_o + \delta_2): & d_{yz}, d_{xz} \longrightarrow d_{x^2-y^2} \\ \multicolumn{2}{l}{\delta_2, \delta_1 \longrightarrow 0,\ \nu_3 \longrightarrow 0;\ \nu_1 \longrightarrow \nu_2} \end{array}\right\} cf.\ \text{Fig. 7.14.2.1a}$$

Thus we can conclude:

With the increase of the basicity (i.e. coordinating power) of the solvents, ν_3 decreases to the lower frequency and ν_1 moves to merge with ν_2 that remains more or less unchanged; in the poor coordinating solvents, the peaks due to ν_1, ν_2 and ν_3 become more manifested.

The above predictions have been experimentally verified (*cf.* Fig. 7.14.2.5 and Table 7.14.2.1).

Table 7.14.2.1 Effect of coordinating power of solvent on the spectra of [Cu(acac)$_2$].

Solvent (Increasing coordinating power ↓)	$d_{xz}, d_{yz} \xrightarrow[(cm^{-1})]{\nu_1} d_{x^2-y^2}$ (Decreasing frequency ↓)	$d_{xy} \xrightarrow[(cm^{-1})]{\nu_2} d_{x^2-y^2}$ (More or less unchanged ↓)	$d_{z^2} \xrightarrow[(cm^{-1})]{\nu_3} d_{x^2-y^2}$ (Decreasing frequency ↓)
Chloroform	18,810	15,190	–
1, 4–Dioxane	17,500	15,100	13,500
n–Pentanol	17,100	15,200	13,000
Pyridine	15,900	14,800	12,100
Piperidine	15,100	14,800	11,300

Chloroform: $CHCl_3$; 1, 4–Dioxane: 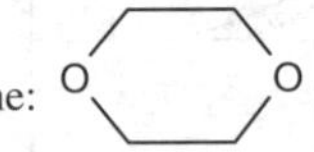; *n*–Pentanol: $H_3C(CH_2)_4OH$; Pyridine: 

Piperidine: 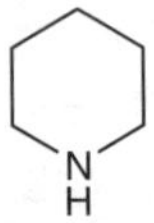

(G) Spectral shift in moving from [Cu(NH$_3$)$_4$(OH$_2$)$_2$]$^{2+}$ to [Cu(NH$_3$)$_5$(OH$_2$)]$^{2+}$: Now let us consider the effect of entry of the fifth NH_3 molecule in [Cu(NH$_3$)$_4$(OH$_2$)]$^{2+}$. *Trans*– [Cu(NH$_3$)$_4$(OH$_2$)$_2$]$^{2+}$ is more or less square planar in geometry (where two H_2O molecules are weakly held at the *trans*-axial directions) while in [Cu(NH$_3$)$_5$(OH$_2$)]$^{2+}$ one axial position is occupied by NH_3 giving rise to a distorted octahedral geometry. It is reasonable because NH_3 is a better base than H_2O (*cf.* [Cu(acac)$_2$] in the solvent, piperidine *vs.* dioxane). Thus [Cu(NH$_3$)$_5$(OH$_2$)]$^{2+}$ is going to have approximately the splitting of *d*–orbitals as in the octahedral geometry while *trans*– [Cu(NH$_3$)$_4$(OH$_2$)]$^{2+}$ is going to have the splitting pattern as in the square planar geometry. The extent of splitting in the square planar geometry is more than that in the octahedral geometry. This is why, when [Cu(NH$_3$)$_4$(OH$_2$)$_2$]$^{2+}$ is converted into [Cu(NH$_3$)$_5$(OH$_2$)]$^{2+}$, **the maximum of absorption is shifted towards the longer wavelength.**

(H) Spectral properties of the square pyramidal (C_{4v}) complexes of Cu(II): Splitting pattern of the *d*–orbitals in the square pyramidal geometries is the same as in the case of *z*–out distortion. In fact, the ^2D term splits in the same way both in C_{4v} (lacking in centre of symmetry) and D_{4h} (having the centre of inversion). Thus the spectral transitions are the same as given in Fig. 7.14.2.1 However, the subscript '*g*' is dropped because the square pyramidal complexes lack in the centre of symmetry. Thus in the **square pyramidal complexes** like $[Cu(NH_3)_5]^{2+}$, $[Cu(HN_3)_4(OH_2)]^{2+}$, etc. the transitions are:

$$\left.\begin{array}{l} d_{z^2} \xrightarrow{\nu_3} d_{x^2-y^2} \text{ i.e. } {}^2B_1 \xrightarrow{\nu_3} {}^2A_1 \\ d_{xy} \xrightarrow{\nu_2} d_{x^2-y^2} \text{ i.e. } {}^2B_1 \xrightarrow{\nu_2} {}^2B_2 \\ d_{xz}, d_{yz} \xrightarrow{\nu_1} d_{x^2-y^2} \text{ i.e. } {}^2B_1 \xrightarrow{\nu_1} {}^2E \end{array}\right\} \begin{array}{l} (9{,}000 - 10{,}000 \text{ cm}^{-1}), \\ (11{,}500 - 16{,}000 \text{ cm}^{-1}), \\ (15{,}000 - 19{,}000 \text{ cm}^{-1}), \end{array}$$

Very often the three bands (ν_1, ν_2 and ν_3) are not detected separately. They can overlap to produce a **broad band.** For $[Cu(NH_3)_5]^{2+}$ in $NH_4[Cu(NH_3)_5]^-(PF_6)_3$ two bands at 15000 cm^{-1} and 11400 cm^{-1} are noted.

(I) Spectral properties of the square planar complexes (D_{4h}) of copper(II): The possible electronic transitions are:

$$\left.\begin{array}{l} {}^2B_{1g} \xrightarrow{\nu_3} {}^2A_{1g} \text{ i.e. } d_{z^2} \xrightarrow{\nu_3} d_{x^2-y^2};\ {}^2B_{1g} \xrightarrow{\nu_2} {}^2B_{2g} \text{ i.e. } d_{xy} \xrightarrow{\nu_2} d_{x^2-y^2} \\ \text{and } {}^2B_{1g} \xrightarrow{\nu_1} {}^2E_g \text{ i.e. } d_{xz}, d_{yz} \xrightarrow{\nu_1} d_{x^2-y^2} \end{array}\right\} \textit{cf. Fig. 7.14.2.1}$$

In the square planar complexes (where there is a limiting situation of *z*-out distortion), the common energy order is:

$$^2B_{1g} \langle {}^2B_{2g} \langle {}^2E_g \langle {}^2A_{1g} \quad \text{i.e. } d_{z^2} \langle d_{xz}, d_{yz} \langle d_{xy} \langle d_{x^2-y^2}$$

The spectral properties of $[Cu(acac)_2]$ have been already discussed. The rose red *bis*(biguanide)copper(II) chloride dihydrate, *i.e.* **$[Cu(bigH)_2]Cl_2 \cdot 2H_2O$ shows a broad spectral band** (Fig. 7.14.2.6) at about 19000 cm^{-1} (which consists of three components: 20,200 cm^{-1}, 18,000 cm^{-1} and

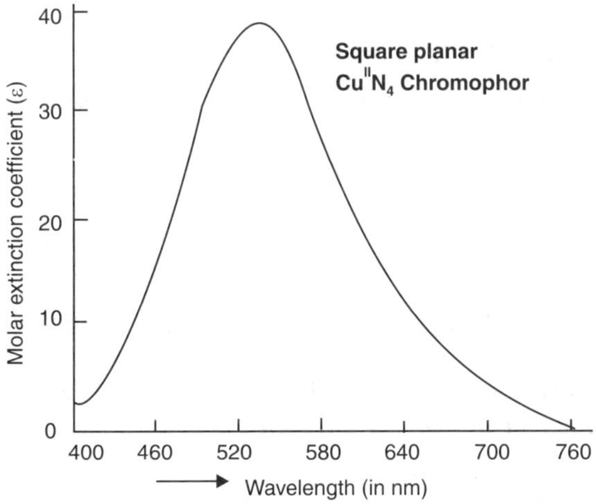

Fig. 7.14.2.6 Absorption sectrum of $[Cu(bigH)_2]^{2+}$ in solution.

17,200 cm^{-1}). In fact, the expected three bands (*i.e.* ν_1, ν_2 and ν_3) overlap to produce a broad band and it happens so for many square planar CuN_4 (D_{4h}) structures. In a lower symmetry like D_{2h} (*e.g.* CuN_2O_2 chromophore), the E_g level further splits into the A_g and B_g components.

(J) Spectral properties of *tris*-chelates, *e.g.* $[Cu(bpy)_3]^{2+}$, $[Cu(phen)_3]^{2+}$ of D_3 symmetry: In moving from the O_h to D_3 symmetry, the $^2T_{2g}$ level splits into the ^2E and $^2A_1$ levels while the ^2E_g state remains as ^2E (without any spliting). Thus the possible transitions are:

$$\boldsymbol{O_h}: {}^2E_g \rightarrow {}^2T_{2g};\ \boldsymbol{D_3}: {}^2E \rightarrow {}^2A_1 \text{ and } {}^2E \rightarrow {}^2E$$

In fact, $[Cu(bpy)_3]^{2+}$ and $[Cu(phen)_3]^{2+}$ show two well defined peaks near 6000 cm^{-1} and 15,000 cm^{-1} (Fig. 7.14.2.7). The higher energy band corresponds to the transition $^2E \rightarrow {}^2E$. This higher energy band shows a doublet structure which may be due to the vibrational transition and/or further splitting of the E-term giving rise to the C_2 symmetry.

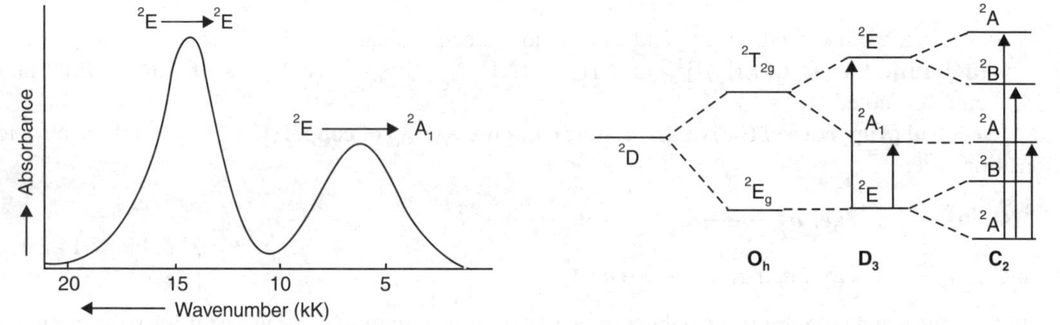

(a) (b)

Fig. 7.14.2.7 (a) Absorption spectrum of trigonally distorted $[Cu(bpy)_3]^{3+}$ (D_3 symmetry); (b) Energy level diagram for $O_h \rightarrow D_3 \rightarrow C_2$ in a d^9 system.

(K) Spectral properties of the tetrahedral (T_d) and distorted (D_{2d}) tetrahedral complexes of Cu(II): These are discussed in Sec. 7.16.

(L) Spectral transitions in the dimeric form of Cu(II)-acetate, *i.e.* $[Cu(CH_3CO_2)_2(OH_2)]_2$: These are discussed in Chapter 8 (*cf.* Fig. 8.31.1.2) in terms of the square pyramidal symmetry around Cu(II).

7.14.3 Six-Coordinate Complexes of 3d²–System

$[V(OH_2)_6]^{3+}$ ($3d^2$) is the representative example of this group. In terms of Orgel diagram (Fig. 7.10.2), the spin allowed transitions are:

$$\nu_1: {}^3T_{1g}(F) \rightarrow {}^3T_{2g}(F);\ \nu_1 = 8Dq + x' \approx 8Dq$$

$$\nu_2: {}^3T_{1g}(F) \rightarrow {}^3A_{2g}(F);\ \nu_2 = 18Dq + x' \approx 18Dq$$

$$\nu_3: {}^3T_{1g}(F) \rightarrow {}^3T_{1g}(P);\ \nu_3 = 15B' + 6Dq + 2x' \approx 15B' + 6Dq$$

$\nu_3 \rangle \nu_2 \rangle \nu_1$ or $\nu_2 \rangle \nu_3 \rangle \nu_1$ depending on the conditions.

The ν_2-transition involves the ***2e transition*** (*cf.* $^3T_{1g}(F) \Rightarrow t_{2g}^2$; $^3A_{2g}(F) \Rightarrow (e_g^2)$) and thus it is a **spin-forbidden transition.**

x' denotes the symmetry interaction energy (*cf.* Figs. 7.10.2; 7.14.3.1) and it can be neglected **for the relatively weak field ligands for the sake of simplicity.** B' denotes the Racah parmeter in the complex and it gives the measure of interelectron repulsion.

The above relations lead to:

$$10Dq = \nu_2 - \nu_1;\ 15B' = (\nu_2 + \nu_3 - 3\nu_1);\ \nu_3 - 2\nu_1 = 15B' - 10Dq$$

Here it is important to mention that the spectroscopic state $^3A_{2g}$ arises from the electronic configuration e_g^2. Thus the ν_2–peak due to the transition, $^3T_{1g} \rightarrow {}^3A_{2g}$ involves **a two-electron excitation** (*i.e.* simultaneous transition of two electrons) and this is a **spin forbidden transition**. Generally, it occurs in the UV–region where the charge transfer band prevails. Because of the strong charge transfer band, the spin forbidden peak (ν_2) of poor intensity very often remains uncharacterised.

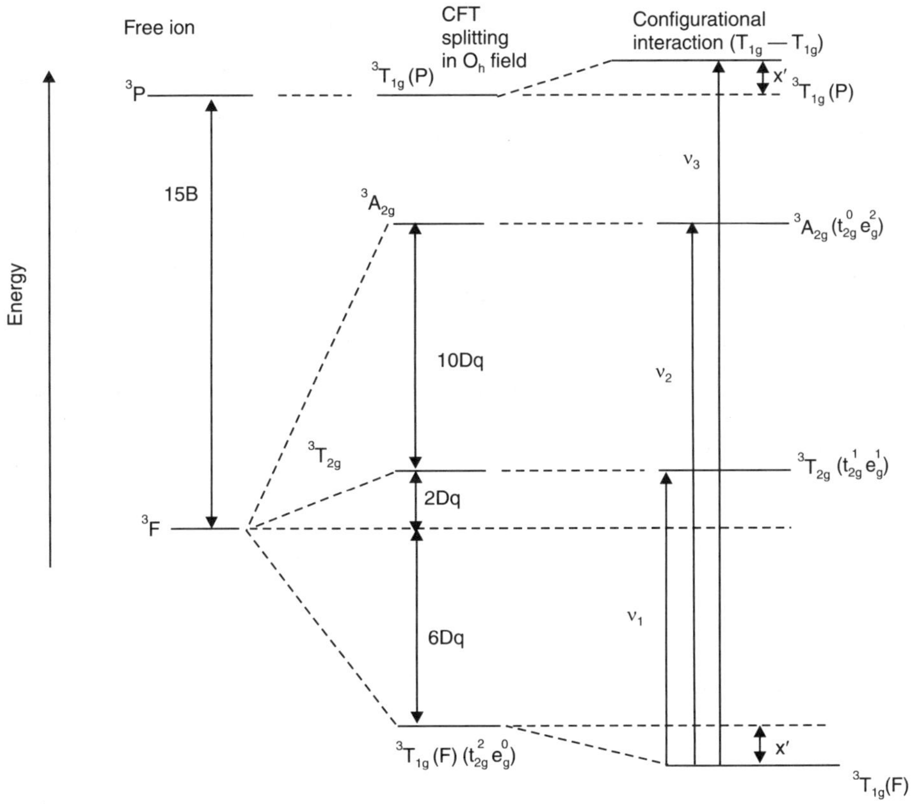

Fig. 7.14.3.1 Energy level diagram for a d^2–octahedral system. ($^3T_{1g}(F) \Rightarrow t_{2g}^{1.8}e_g^{0.2}$ of energy $-6Dq$)

Note: The $^3T_{1g}(P)$ and $^3A_{2g}(F)$ states may intersect depending on the $10Dq$ value. Here it is assumed that $^3T_{1g}(P)$ state lies above the $^3A_{2g}(F)$ state; x' denotes the configurational interaction energy for the $^3T_{1g}(P)$ – $^3T_{1g}(F)$ interaction. (*See* Fig. 7.9.4.1 for energy values of different states and **$-6Dq$ energy of the $^3T_{1g}(F)$ state corresponds to $t_{2g}^{1.8}e_g^{0.2}$).**

A. Spectral behaviour of $[V(OH_2)_6]^{3+}$: It looks green with two strong absorption peaks at 17,200 cm^{-1} ($\varepsilon \approx 6$) and 25, 600 ($\varepsilon \approx 8$) (*cf.* Fig. 7.14.3.2).

Now let us try to assign the peaks with the help of Tanabe-Sugano diagram (*cf.* Figs. 7.13.2.1, 7.14.3.3). Let us assume arbitrarily that the peak at 17,200 cm^{-1} is due to the following transition.

$$^3T_{1g}(F) \xrightarrow[(17{,}200\ \text{cm}^{-1})]{\nu_1} {}^3T_{2g}(F)\ (\textbf{lowest energy band});\ \nu_1 \approx 8Dq = 17{,}200\ \text{cm}^{-1}$$

Then, the following transition should occur at about 38,500 cm^{-1}.

$$^3T_{1g}(F) \xrightarrow{\nu_2} {}^3A_{2g}(F);\ \nu_2\,(\text{expected}) \approx 18Dq = 38{,}500\ \text{cm}^{-1}$$

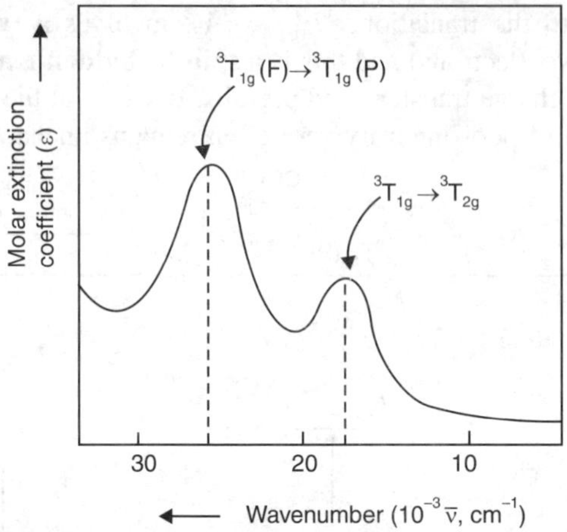

Fig. 7.14.3.2 The electronic absorption spectrum of $[V(OH_2)_6]^{3+}$.

Thus the peak at 25,600 cm^{-1} is not due to the $^3T_{1g} \rightarrow {^3A_{2g}}$ transition, if we consider that the peak due to $^3T_{1g} \rightarrow {^3T_{2g}}$ occurs at 17,200 cm^{-1}. Under the situation, we may assume that the peak at 25,600 cm^{-1} is due to the following transition.

$$^3T_{1g}(F) \xrightarrow[(25{,}600\ cm^{-1})]{\nu_3} {^3T_{1g}}(P);\ \nu_3 \approx 6Dq + 15B' = 25{,}600\ cm^{-1}$$

Note: From Fig. 7.10.4b, it is evident that for the $T_1(F)$ ground state, the transition energy ratio, **ν_2/ν_1 is close to 2 regardless of the ligand field strength**. If ν_2 is taken as 25,600 cm^{-1}, then **the ratio ν_2/ν_1 becomes 1.48 < 2.** It indicates that ν_2 does not correspond to 25,600 cm^{-1}. Thus it is reasonable to assume ν_3 = 25,600 cm^{-1} (*see* Solved Problem 12).

Now let us verify whether these assignments are correct or not. The above assignments lead to:

$$\frac{^3T_{1g} \longrightarrow {^3T_{1g}}(\text{P})}{^3T_{1g} \longrightarrow {^3T_{2g}}} = \frac{\nu_3}{\nu_1} = \frac{25{,}000}{17{,}200} \approx 1.50$$

This ratio of ν_3 and ν_1 can be fitted in the enlarged Tanabe-Sugano diagram (*cf.* Fig. 7.14.3.3) and this ratio fits for the value of $10Dq/B'$ as 28 (*i.e.* ***Dq*/*B′* = 2.8**). This ν_3/ν_1 ratio can also be fitted in the standard curve, ν_3/ν_1 vs. Dq/B' to find Dq/B' (*cf.* Fig. 7.10.4). For this $10Dq/B'$ or Dq/B' value, we find:

$$\frac{\Delta E\left[{^3T_{1g}}(F) \rightarrow {^3T_{2g}}(F)\right]}{B'} = 25.9$$

$$i.e.\quad \frac{\nu_1}{B'} \approx \frac{8Dq}{B'} = 25.9$$

$$\text{or,}\quad \frac{17{,}200}{B'} = 25.9$$

$$\text{or,}\quad B' = 664.1\ cm^{-1}$$

$$\frac{\Delta E\left[{^3T_{1g}}(F) \rightarrow {^3T_{1g}}(P)\right]}{B'} = 38.6$$

$$i.e.\quad \frac{\nu_3}{B'} \approx \frac{15B' + 6Dq}{B'} = 38.6$$

$$\text{or,}\quad \frac{25{,}600}{B'} = 38.6$$

$$\text{or,}\quad B' \approx 663.2$$

Thus the evaluated B' becomes about 664 cm^{-1}(*cf.* B_0 = 860 cm^{-1}). This reduced value of B' compared to that of B_0 (for free ion) is quite reasonable in terms of the ***nephelauxetic effect***. Thus the nephelauxetic ratio ($\beta = B'/B_0$) becomes about 0.80, *i.e.* B is reduced by about 20% in $[V(OH_2)_6]^{3+}$. The $10Dq$ value can be obtained by using the following relationship.

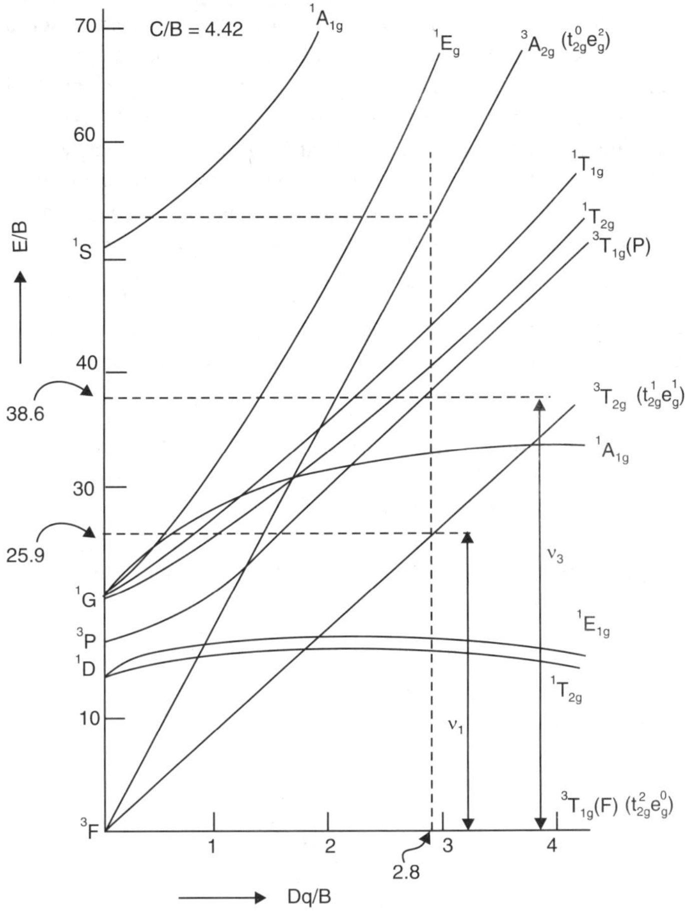

Fig. 7.14.3.3 T.S. diagram for a d^2–octahedral system and analysis of the absorption spectrum of $[V(OH_2)_6]^{3+}$.

$$\frac{10Dq}{B'} = 28 \text{ or, } 10Dq \approx 28 \times 664 \text{ cm}^{-1} = 18,592 \text{ cm}^{-1}$$

For $\dfrac{Dq}{\beta'} = 2.8$, theoretically predicted energy ratios (*cf.* Fig. 7.10.4) are:

$$\frac{\nu_2}{\nu_1} = 2.08 \text{ and } \frac{\nu_3}{\nu_2} = 0.72 \;\; i.e. \;\nu_2 = \nu_1 \times 2.08 = 17,200 \times 2.08 = 35,776 \text{ cm}^{-1}$$

or, $$\nu_2 = \frac{\nu_3}{0.72} = \frac{25,600}{0.72} = 35,555 \text{ cm}^{-1}$$

Thus the redicted ν_2 should appear at about 35,666 cm^{-1}($\approx \frac{1}{2}$ (35,776 + 35,555) cm^{-1}.

Now let us consider the peak due to the transition, $^3T_{1g}(F) \xrightarrow{\nu_2} {}^3A_{2g}(F)$ and it should occur at about 35,666 cm^{-1} from the above assignments. *We have already mentioned that the transition is forbidden because it involves the simultaneous transition of two electrons.* Such a weak ligand field band in the near ultraviolet region is generally obscured by the intense charge transfer bands. Thus, by taking the spectrum of $[V(OH_2)_6]^{3+}$ in solution, it is not possible to verify the prediction of the existence of ν_2 ($\approx 18Dq + x'$) at about 35,666 cm^{-1}. However, in solid state when V(III) is present in the matrix crystal of Al_2O_3 where the charge transfer band appears at still higher energy, a weak band occurs at about 34,500 cm^{-1}. In the solid matrix of Al_2O_3, V(III) centre is octahedrally surrounded by the *O*–atoms. Thus the 10*Dq* of V(III) in the solid matrix of Al_2O_3 is comparable to that of $[V(OH_2)_6]^{3+}$. The observed peak at 34,500 cm^{-1} can reasonably be considered due to $^3T_{1g}(F) \rightarrow {}^3A_{2g}(F)$.

- Thus the assignments, *i.e.* $^3T_{1g}(F) \xrightarrow{17{,}200\ cm^{-1}} {}^3T_{2g}(F)$ and $^3T_{1g}(F) \xrightarrow{25{,}600\ cm^{-1}} {}^3T_{1g}(P)$ are correct. These are supported by the fact: extimated *B* value (= 664 cm^{-1}) is less than that of the free ion ($B' \approx 8\%$ of B_0); the predicted ν_2 value for $^3T_{1g}(F) \rightarrow {}^3A_{2g}(F)$ is experimentally verified.

- If the assignments, $^3T_{1g}(F) \xrightarrow{17{,}200\ cm^{-1}} {}^3T_{1g}(P)$ and $^3T_{1g}(F) \xrightarrow{25{,}600\ cm^{-1}} {}^3A_{2g}(F)$ are done, **then it will lead to *B′* value which is higher than that of free ion** (*i.e.* $B' \rangle B_0$) **and this is not acceptable.** Besides this, it predicts the band for the $^3T_{1g}(F) \rightarrow {}^3T_{2g}(F)$ transition in the infra-red region and this prediction is not experimentally verified.

Note: If ν_1 (17,200 cm^{-1}) is considered to be 8*Dq*, then it leads to 10*Dq* to be 21,500 cm^{-1}. But this value is quite different from that (18, 592 cm^{-1}) obtained from fitting in the Tanabe-Sugano diagram. It indicates the *importance of symmetry interaction energy x′*.

$$\nu_1 = 17{,}200\ cm^{-1} = 8Dq + x';\ \nu_3 = 25{,}600\ cm^{-1} = 15B' + 6Dq + 2x'$$

$$\nu_2 \approx 35{,}666\ cm^{-1}\ \text{(predicted value)}$$

$$= 18Dq + x'$$

$$\nu_2 - \nu_1 = 10Dq = 35{,}666 - 17{,}200 = 18{,}466\ cm^{-1}$$

i.e. $$Dq \approx 1846\ cm^{-1}$$

$$\nu_1 = 17{,}200\ cm^{-1} = 8Dq + x' = 8 \times 1846 + x'$$

i.e. $$x' = 2432\ cm^{-1}$$

$$\nu_3 = 25,600\ cm^{-1} = 15B' + 6Dq + 2x' = 15B' + 6 \times 1846 + 2 \times 2432$$

$$= 15B' + 15{,}940$$

or, $$15B' = 9660 \text{ or, } B' \approx 644\ cm^{-1}$$ **(acceptable value)**

Spectral properties of some other V(III)–complexes are given in Table 7.14.3.1.

Table 7.14.3.1 Characteristic features of the ligand field bands of some complexes of V(III)

Complex	$^3T_{1g}(F) \rightarrow {}^3T_{2g}(F)$ ν_1 (cm^{-1})	$^3T_{1g} \rightarrow {}^3A_{2g}(F)$ ν_2 (cm^{-1})	$^3T_{1g}(F) \rightarrow {}^3T_{1g}(P)$ ν_3 (cm^{-1})	*B′* (cm^{-1})	*10Dq* (cm^{-1})
$[V(OH_2)_6]^{3+}$	17,200	(36000)*	25,200	665	18,600
$[VF_6]^{3-}$	14,800	(31,000)*	23,000	650	16,200
$[V(urea)_6]^{3+}$	16,200	(33,500)*	24,000	610	17,500

* Predicted approximate values. $B_0 = 862\ cm^{-1}$.

B. Energy of the spectroscopic states arising from the ^{3}F–term (for d^2):

(i) ***Without considering the symmetry interaction between $^3T_{1g}(F)$ and $^3T_{1g}(P)$***

$$E[^3T_{1g}(F)] = -6Dq;\ \ E[^3T_{1g}(P)] = 15B'$$

$$E[^3T_{2g}(F)] = 2Dq;\ \ E[^3A_{2g}(F)] = 12Dq$$

(ii) ***After considering the symmetry interaction between $^3T_{1g}(F)$ and $^3T_{1g}(P)$*** (*cf.* Fig. 7.14.3.1)

$$E[^3T_{1g}(F)] = -6Dq - x';\ E[^3T_{1g}(P)] = 15B' + x'$$

$$E[^3T_{2g}] = 2Dq;\ E[^3A_{2g}] = 12Dq$$

where x' is the energy due to the symmetry interaction.

In a very weak field, the symmetry interaction energy (x') is negligibly small and the $^3T_{1g}(F)$ state is stabilised by $6Dq$ compared to the ^3F term of free ion. In the absence of crystal field, the $^3T_{1g}(P)$ state is destabilised by $15B_0$ compared to the ^3F term. In absence of the crystal field, the $^3T_{1g}(F)$ term has the same energy as that of unsplit ^3F term. In absence of symmetry interaction, energy of the $^3T_{1g}(P)$ state is the same as that of ^3P of free ion, *i.e.* energy of the $^3T_{1g}(P)$ state is zero (0) with respect to that of the ^3P term. Thus we can write for $H = H_0 + V_{oct}$:

$$\left.\begin{array}{l}\int {}^3T_{1g}(F)|H_0|{}^3T_{1g}(F)\partial\tau = 0\\ \int {}^3T_{1g}(P)|H_0|{}^3T_{1g}(P)\partial\tau = 15B\end{array}\right\}\text{(No crystal field)}\quad \left.\begin{array}{l}\int {}^3T_{1g}(F)|V_{oct}|{}^3T_{1g}(F)\partial\tau = -6Dq\\ \int {}^3T_{1g}(P)|V_{oct}|{}^3T_{1g}(P)\partial\tau = 0\end{array}\right\}\text{(Crystal field)}$$

$^3T_{1g}(F)$ and $^3T_{1g}(P)$ terms having the same group theoretical representation will experience a symmetry interaction or **configurational interaction** (*i.e.* **noncrossing rule** which states that the same symmetry levels in terms of both spin and orbital cannot cross one another). If the interaction energy is x', then we can write;

$$\int {}^3T_{1g}(F)|V_{oct}|{}^3T_{1g}(P)\partial\tau = x'$$

The relevant secular determinant with $H = H_0 + V_{oct}$ is given by:

$$\begin{array}{ccc} & {}^3T_{1g}(F) & {}^3T_{1g}(P) \\ \begin{array}{c}{}^3T_{1g}(F)\\ {}^3T_{1g}(P)\end{array} & \multicolumn{2}{c}{\begin{vmatrix} -6Dq - E & x' \\ x' & 15B' - E \end{vmatrix} = 0} \end{array}$$

i.e. $\quad (-6Dq - E)\,(15B' - E) = x'^2$ ...(7.14.3.1)

when the limiting strong field ligands are considered, the interelectron repulsion parameter B' becomes negligible (*i.e.* when $10Dq$ or Δ is very large, B' becomes negligible. Under this condition, *i.e.* $E \rangle\rangle 15B'$, we can write:

$$E^2 + 6DqE - x'^2 = 0 \qquad \text{...(7.14.3.2)}$$

For $x' = 4Dq$, the above quadratic equation gives: $E = -8Dq, 2Dq$.

$E = -8Dq$ corresponds to the crystal field energy of t_{2g}^2 configuration and $E = +2Dq$ corresponds to the crystal field energy of $t_{2g}^1e_g^1$ configuration.

Putting $x' = 4Dq$ in Eqn. 7.14.3.1, we can write:

$$(-6Dq - E)(15B' - E) - (4Dq)^2 = 0$$

or, $\quad E^2 + E(6Dq - 15B') - 90DqB' - 16(Dq)^2 = 0$

Solution of the above quadratic equation gives two roots of the state $^3T_{1g}$. These are:

$$E = \frac{1}{2}\left[15B' - 6Dq \pm \left\{(6Dq - 15B)^2 + 4\left(90DqB' + 16(Dq)^2\right)\right\}^{1/2}\right]$$

$$=\frac{1}{2}\left[15B'-6Dq\pm\left\{225B'^2+100(Dq)^2+180DqB'\right\}^{1/2}\right]$$

$$=\frac{1}{2}\left[15B'-\frac{3}{5}\Delta\pm\left\{225B'^2+\Delta^2+18B'\Delta\right\}^{1/2}\right]$$

$$\left(cf.\ ax^2+bx+c=0,\quad x=\frac{-b\pm\sqrt{b^2-4ac}}{2a}\right)$$

Energies of two $^3T_{1g}$ states are given as follows:

$$E\left[{}^3T_{1g}(F)\right]=\frac{1}{2}\left[15B'-\frac{3}{5}\Delta-\left(225B'^2+18B'\Delta+\Delta^2\right)^{1/2}\right],\ \text{(corresponding to } t_{2g}^2e_g^0)$$

$$E\left({}^3T_{1g}(P)\right)=\frac{1}{2}\left[15B'-\frac{3}{5}\Delta+\left(225B'^2+18B'\Delta+\Delta^2\right)^{1/2}\right],\ \text{(corresponding to } t_{2g}^1e_g^1)$$

Energies of the other two states do not involve x' (*i.e.* symmetry interaction energy term). These are:

$$E\left[{}^3T_{2g}(F)\right]=\frac{1}{5}\Delta,\ E\left[{}^3A_{2g}(F)\right]=\frac{6}{5}\Delta$$

● **In the case of limiting weak field ligand** (*i.e.* $\Delta \to 0$), energies of the $^3T_{1g}(F)$, $^3T_{2g}(F)$ and $^3A_{2g}(F)$ states become zero with respect to that of the ^3F term, *i.e.* $^3T_{1g}(F)$, $^3T_{2g}(F)$ and $^3A_{2g}(F)$ merge with ^3F. However, energy of the $^3T_{1g}(P)$ becomes $15B$ with respect to that of the ^3F term.

● **In the case of limiting strong field ligand,** B' can be ignored and we can write:

$$E\left[{}^3T_{1g}(F)\right]=-\frac{1}{2}\times\frac{8}{5}\Delta=-\frac{4}{5}\Delta=-8Dq=\text{energy of } t_{2g}^2$$

$$E\left[{}^3T_{2g}(F)\right]=\frac{1}{5}\Delta=2Dq=\text{energy of } t_{2g}^1e_g^1$$

$$E\left[{}^3T_{1g}(P)\right]=\frac{1}{2}\times\frac{2}{5}\Delta=\frac{1}{5}\Delta=2Dq=\text{energy of } t_{2g}^1e_g^1,\ E\left[{}^3A_{2g}(F)\right]=\frac{6}{5}\Delta=12Dq=\text{energy of } e_g^2$$

Thus the transition energies can be expressed as follows:

$$^3T_{1g}(F)\xrightarrow{\nu_1}{}^3T_{2g}(F),\ \nu_1=\frac{1}{2}\left[\Delta-15B'+\left(225B'^2+18B'\Delta+\Delta^2\right)^{1/2}\right]$$

i.e.
$$\frac{E(\nu_1)}{B'}=\frac{1}{2}\left[\frac{\Delta}{B'}-15+\left(225+18\frac{\Delta}{B'}+\frac{\Delta^2}{B'^2}\right)^{1/2}\right]$$

$$^3T_{1g}(F)\xrightarrow{\nu_3}{}^3T_{1g}(P);\ \nu_3=\left(225B'+18B'\Delta+\Delta^2\right)^{1/2}$$

i.e.
$$\frac{E(\nu_3)}{B'}=\left(225+18\frac{\Delta}{B'}+\frac{\Delta^2}{B'^2}\right)^{1/2}$$

Energy change in the spin-allowed ligand field bands for the high-spin octahedral d^2, d^7 octahedral and d^3, d^8 tetrahedral systems

$$T_1(F) \xrightarrow{\nu_1} T_2(F): \nu_1 = \frac{1}{2}\left[\Delta - 15B' + \left(225B' + 18B'\Delta + \Delta^2\right)^{1/2}\right]$$

$$T_1(F) \xrightarrow{\nu_2} A_2(F): \nu_2 = \frac{1}{2}\left[3\Delta - 15B' + \left(225B' + 18B'\Delta + \Delta^2\right)^{1/2}\right]$$

$$T_1(F) \xrightarrow{\nu_3} T_1(P): \nu_3 = \left(225B' + 18B'\Delta + \Delta^2\right)^{1/2}; \nu_3 - 2\nu_1 = 15B' - 10Dq$$

It is evident from the expressions of $E(\nu_1)/B'$ and $E(\nu_3)/B'$ which are the functions of Δ/B' that for the electronic transitions, the transition energies can be graphically represented by plotting E/B' against Δ/B'. This practice was introduced by *Y.* Tanabe and *S.* Sugano and consequently the diagrams (*i.e.* plot of E/B *vs.* Δ/B) are named as **Tanabe-Sugano diagrams** (*cf.* Sec. 7.13) to honour the scientists.

7.14.4 Six-Coordinate Complexes of $3d^8$–System

Octahedral complexes of Ni(II) are the representative examples of this group. In terms of Orgel diagram (Fig. 7.10.2) and energy level diagram shown in Fig. 7.14.4.1, the spin-allowed ligand field transitions are:

$$\left.\begin{array}{ll} {}^3A_{2g}(F) \longrightarrow {}^3T_{2g}(F), & \nu_1 = 10Dq \\ {}^3A_{2g}(F) \longrightarrow {}^3T_{1g}(F), & \nu_2 = 18Dq - x \approx 18Dq \\ {}^3A_{2g}(F) \longrightarrow {}^3T_{1g}(P), & \nu_3 = 12Dq + 15B' + x \approx 12Dq + 15B' \end{array}\right\} \nu_3 \rangle \nu_2 \rangle \nu_1$$

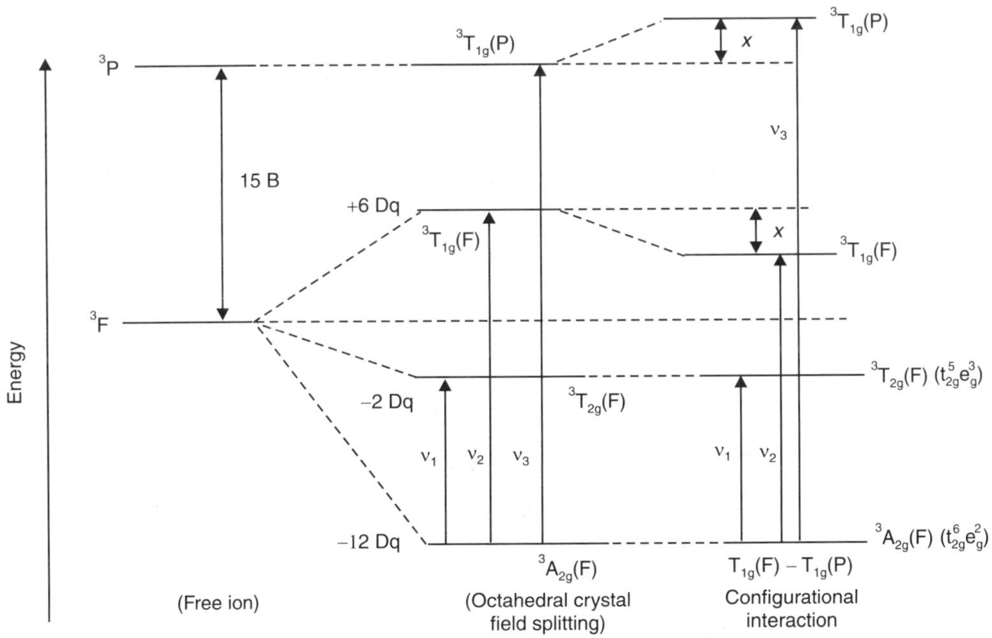

Fig. 7.14.4.1 Splitting of ^3F and ^3P terms of $3d^8$ system in an octahedral field and electronic transitions.

The ν_3-transition involves the 2e-transition (*cf.* $^3A_{2g}(F) \Rightarrow t_{2g}^6e_g^2$; $^3T_{1g}(P) \Rightarrow t_{2g}^4e_g^4$) and thus it is a **spin-forbidden transition.**

(**Note:** Sometimes two spin-forbidden transitions, $^3A_{2g}(F) \rightarrow {}^1E_g(D)$ and $^3A_{2g}(F) \rightarrow {}^1T_{2g}(D)$ are observed (*cf.* Fig. 7.13.2.1)).

Here x denotes the **configurational interaction energy.** Because of the configurational interaction between the terms $^3T_{1g}(F)$ and $^3T_{1g}(P)$, energy of the $^3T_{1g}(F)$ term is lowered by x and energy of the $^3T_{1g}(P)$ term goes up by x.

The configurational interaction between the $^3T_{1g}(F)$ and $^3T_{1g}(P)$ states becomes accountable when the ligands are fairly strong (*see* Fig. 7.9.4.1 for the different energy values).

For the octahedral Ni(II)–complexes, the three spin-allowed ligand field bands appear in the range:

$$\nu_1(8{,}000 - 13{,}000\ cm^{-1}),\ \nu_2(15{,}000 - 19{,}000\ cm^{-1}),\ \nu_3(25{,}000 - 30{,}000\ cm^{-1})$$

Table 7.14.4.1 Position of ligand field bands in some octahedral Ni(II) complexes.

Complex	$^3A_{2g} \xrightarrow{\nu_1} {}^3T_{2g}(F)$ (cm^{-1})	$^3A_{2g} \xrightarrow{\nu_2} {}^3T_{1g}(F)$ (cm^{-1})	$^3A_{2g} \xrightarrow{\nu_3} {}^3T_{1g}(P)$ (cm^{-1})	*10Dq* (cm^{-1})	*B′* (cm^{-1})
$[Ni(OH_2)_6]^{2+}$	8,700	13,800	25,300	8,700	905
$[Ni(NH_3)_6]^{2+}$	10,750	17,500	28,200	10,750	895
$[Ni(en)_3]^{2+}$	11,200	18,350	29,000	11,200	920
$[Ni(bpy)_3]^{2+}$	12,650	19,200	Obscured	12,650	–
$[Ni(DMSO)_6]^{2+}$	7,730	12,970	24,040	7,730	920
$[Ni(DMF)]_6]^{2+}$	8,500	13,600	25,000	8,500	875

DMSO = dimethyl sulfoxide, *DMF* = *N, N*–dimethyl formamide, B_0(for free ion) = 1030 cm^{-1}

The energy change in ν_1 is independent of configurational interaction. It gives the measure of $10Dq$. If all the three ligand field bands (ν_1, ν_2 and ν_3) are characterised then B' can be estimated as follows:

$$15B' = (\nu_2 + \nu_3 - 3\nu_1)$$

In many cases, the ν_3 band gets obscured due to overlapping with the strong charge transfer band. In such cases, B' value can be estimated with the help of Tanabe–Sugano diagram.

Energy of the spectroscopic states are:

$$E\left({}^3T_{2g}\right) = -2Dq;\ E\left({}^3A_{2g}\right) = -12Dq;\ E\left({}^3T_{1g}(F)\right) = 7.5B' + 3Dq$$

$$-\frac{1}{2}\left(225B'^2 + 100Dq^2 - 180DqB'\right)^{1/2};\ E\left({}^3T_{1g}(P)\right) = 7.5B' + 3Dq$$

$$+\frac{1}{2}\left(225B' + 100Dq^2 - 180DqB'\right)^{1/2}$$

$$^3A_{2g} \Rightarrow t_{2g}^6e_g^2;\ {}^3T_{2g} \Rightarrow t_{2g}^5e_g^3;\ {}^3T_{1g}(F) \Rightarrow t_{2g}^5e_g^3;\ {}^3T_{1g}(P) \Rightarrow t_{2g}^4e_g^4 \ (\textit{see}\ \text{Fig. 7.10.2})$$

For the said spin-allowed ligand field transitions, the energy changes are as follows:

$$\nu_1 = E\left({}^3T_{2g}\right) - E\left({}^3A_{2g}\right) = -2Dq + 12Dq = 10Dq$$

$$\nu_2 = E\left[{}^3T_{1g}(F)\right] - E\left({}^3A_{2g}\right) = 15Dq + 7.5B' - \frac{1}{2}\left[100Dq^2 + 225B'^2 - 180B'Dq\right]^{1/2}$$

$$\nu_3 = E\left[{}^3T_{1g}(P)\right] - E\left({}^3A_{2g}\right) = 15Dq + 7.5B' + \frac{1}{2}\left[100Dq^2 + 225B'^2 - 180B'Dq\right]^{1/2}$$

$$Q = \left[100Dq^2 + 225B' - 180B'Dq\right]^{1/2} = \nu_3 - \nu_2$$

$$10Dq = \nu_1 = \frac{1}{3}\left[\nu_2 + \nu_3 - 15B'\right]$$

These relations are valid for the octahedral d^3 and d^8 systems and for the tetrahedral d^2, d^7 systems. From the above relations, it is evident that when ν_3 is not characterised due to overlapping with the charge transfer band, Dq and B' can be estimated from ν_1 and ν_2 and position of the ν_3–band can be predicted.

• **Transition, $^3A_{2g}(F) \rightarrow {}^3T_{1g}(P)$:** The ground state $^3A_{2g}$ represents the electronic configuration $t_{2g}^6e_g^2$ and the excited state $^3T_{1g}$(P) arises from the $t_{2g}^4e_g^4$ the configuration. Thus the said transition involves the promotion of two electrons from the t_{2g} level. ***This is a spin forbidden transition*** (*cf.* $^3T_{1g} \rightarrow {}^3A_{2g}$ in d^2 configuration). Thus it is reasonable to expect that intensity of the band for the $^3A_{2g} \rightarrow {}^3T_{1g}$(P) transition should be poor. ***However intensity of the band is comparable to that of other spin-allowed transition.*** It is believed that mixing with the spin allowed transitions gives the intensity to the spin-forbidden transitions. This phenomenon is referred to as ***intensity borrowing or intensity stealing.***

• **Prediction of the ligating behaviour of NO_3^- in $[Ni(NO_3)_4]^{2-}$ from the spectral characteristics:** Electronic absorption spectrum of $[Ni(NO_3)_4]^{2-}$ represents the characteristics of ligand field spectra of the octahedral complexes. It indicates that some of the NO_3^- groups act as the didentate ligands in $[Ni(NO_3)_4]^{2-}$.

For most of the octahedral Ni(II) complexes, the ratio ν_2/ν_1 lies in the range 1.5 to 1.7 (*cf.* Fig. 7.10.4) and ***it is the characteristic feature of the octahedral complexes of Ni(II).***

(A) Spitting of the ν_2–band [*i.e.* $^3A_{2g} \rightarrow {}^3T_{1g}(F)$] in Ni(II)-octahedral complexes: In some cases as in $[Ni(DMSO)_6]^{2+}$ and $[Ni(OH_2)_6]^{2+}$, the ν_2 band undergoes splitting, *i.e.* the ν_2–band appears as a doublet. Analysis of the bands indicates the overlapping of two peaks. It is believed that there is a spin-forbidden transition $^3A_{2g}(F) \rightarrow {}^1E_g(D)$ in the region of the spin-allowed transition $^3A_{2g}(F) \rightarrow {}^3T_{1g}(F)$ in $[Ni(OH_2)_6]^{2+}$ and $[Ni(DMSO)_6]^{2+}$. In fact, spin-orbit coupling allows the mixing of the singlet state (^1E_g) and the triplet state ($^3T_{1g}$(F)) to some extent when these two states are of comparable energy. These singlet and triplet states are of comparable energy when the crystal field strength leads to $\frac{Dq}{B} \approx 1$ (*cf.* Figs. 7.13.2.1g, 7.14.4.2d). *This condition (i.e.* $\frac{Dq}{B} \approx 1$ *) of spin orbit coupling mixing of the* $^1E_g(D)$ *and* $^3T_{1g}(F)$ *states is attained for the ligands like* H_2O *and* DMSO. For the stronger field ligands like en, the two energy states, *i.e.* $^1E_g(D)$ and $^3T_{1g}$(F) are so widely apart that no significant mixing between the states, occurs. This is why, in $[Ni(OH_2)_6]^{2+}$ and $[Ni(DMSO)_6]^{2+}$, the spin forbidden transition gains some intensity **but in $[Ni(en)_3]^{2+}$, the spin forbidden transition is not at all important**. The said spin-forbidden transition gains some intensity through the interaction with the $^3T_{1g}(F)$ level in the complexes like $[Ni(OH_2)_6]^{2+}$, $[Ni(DMSO)_6]^{2+}$ where $\frac{Dq}{B} \approx 1$. In fact, at low temperature, these two overlapping peaks have been characterised in $[Ni(OH_2)_6]^{2+}$.

$$^3A_{2g}(F) \xrightarrow{14{,}000\ \text{cm}^{-1}} {}^1E_g(D);\ {}^3A_{2g}(F) \xrightarrow{15{,}200\ \text{cm}^{-1}} {}^3T_{1g}(F).$$

In $[Ni(en)_3]^{2+}$ (where $\frac{Dq}{B} \gg 1$), there is no splitting in the ν_2–band and it is also expected so. Here it may be pointed out that some authors have attempted to explain the splitting of the ν_2 band of $[Ni(OH_2)_6]^{2+}$ through the spin-orbit coupling splitting of the $^3T_{1g}(F)$ term (*see* Sec. 7.7.4, 8.20.3).

(B) Assignment of the ligand field bands in $[Ni(OH_2)_6]^{2+}$: Aqueous solution of Ni(II) looks light green in colour (*i.e.* absorption occurs in the red and blue portion of the visible spectrum). The observed strong bands are:

$$8{,}700\ cm^{-1}\ (\varepsilon = 1.6),\ 13{,}800\ cm^{-1}\ (\varepsilon = 2.0),\ 25{,}300\ cm^{-1}\ (\varepsilon = 4.6)$$

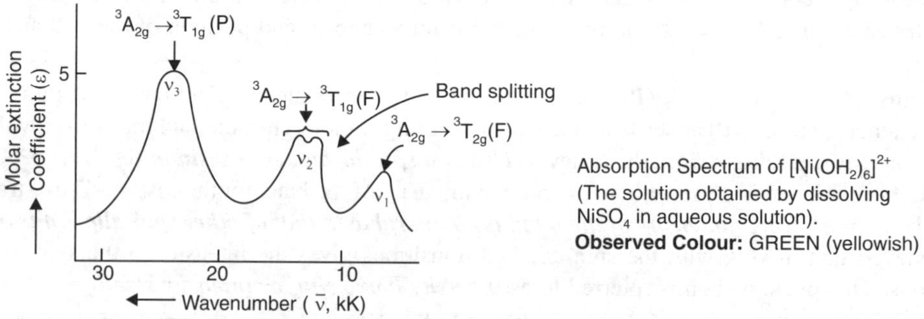

Fig. 7.14.4.2 (a) Electronic absorption spectrum of $[Ni(OH_2)_6]^{2+}$ (*cf.* Fig. 7.1.1).

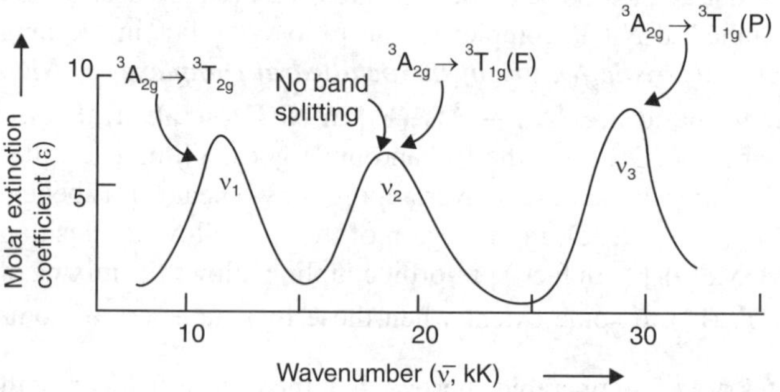

Fig. 7.14.4.2 (b) Electronic absorption spectrum of $[Ni(en)_3]^{2+}$ (*cf.* Fig. 7.1.1).

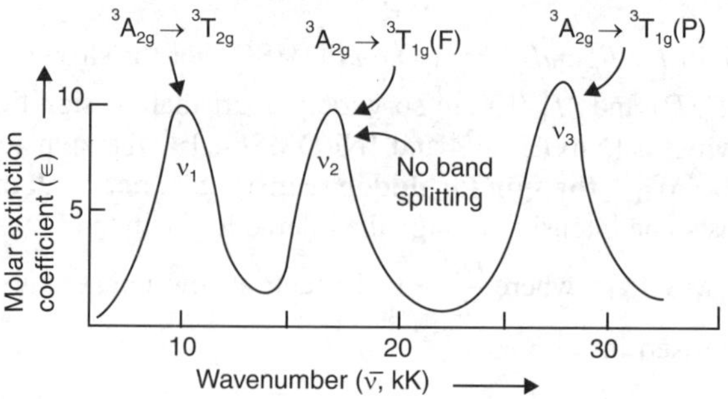

Fig. 7.14.4.2 (c) Electronic absorption spectrum of $[Ni(gly)_3]^-$, gly = glycinate.

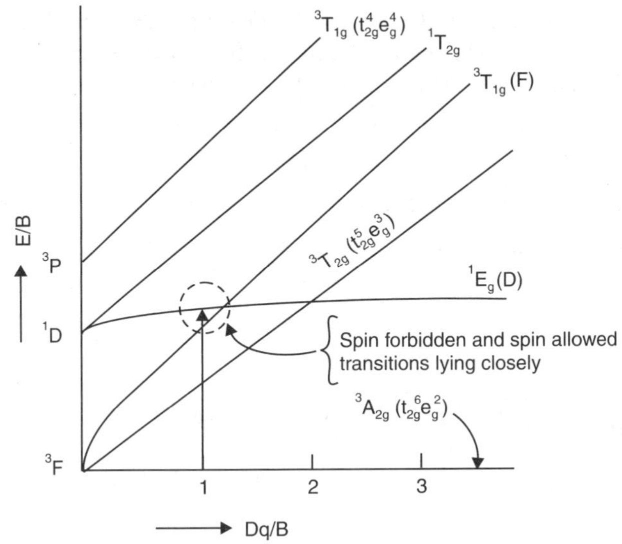

Fig. 7.14.4.2 (d) For d^8 (O_h); condition (*i.e.* $Dq/B \approx 1.0$) for gaining intensity in the spin forbidden transition to ^1E_g through the spin-orbit coupling **mixing of the $^3T_{1g}$ (F) and ^1E_g (D) states**.

The bands assignments are:

$$^3A_{2g}(F) \to {}^3T_{2g}(F),\ \nu_1 = 8{,}700\ \text{cm}^{-1},\ (= 10Dq)$$
$$^3A_{2g}(F) \to {}^3T_{1g}(F),\ \nu_2 = 13{,}800\ \text{cm}^{-1}$$
$$^3A_{2g}(F) \to {}^3T_{1g}(P),\ \nu_3 = 25{,}300\ \text{cm}^{-1}$$

Now let us verify the above assignments by considering the Tanabe-Sugano diagram.

$$\frac{^3A_{2g}(F) \longrightarrow {}^3T_{1g}(P)}{^3A_{2g}(F) \longrightarrow {}^3T_{1g}(F)} = \frac{\nu_3}{\nu_2} = \frac{25{,}300}{13{,}800} = 1.83$$

The above ratio is fitted into the Tanabe-Sugano diagram of d^8–system. The above ν_3/ν_2 ratio corresponds to $Dq/B' = 0.88$ (*cf.* Fig. 7.10.4).

The ν_1 band, *i.e.* $^3A_{2g}(F) \to {}^3T_{2g}(F)$ gives the $10Dq$ value (= 8,700 cm^{-1}).

$$10Dq = 8{,}700\ \text{cm}^{-1}\ \textit{i.e.}\ Dq = 870$$

$$\frac{Dq}{B'} = 0.88\ \text{or,}\ B' = \frac{870}{0.88} \approx 989\ \text{cm}^{-1}$$

For $Dq/B' = 0.88$, we get from the relevant Tanabe-Sugano diagram:

$$\frac{^3A_{2g}(F) \longrightarrow {}^3T_{1g}(P)}{B'} = \frac{\nu_3}{B'} = 26.6$$

i.e.
$$\frac{25{,}300}{B'} = 26.6,\ \text{or}\ B' = \frac{25{,}300}{26.6} = 951\ \text{cm}^{-1}.$$

Taking $\dfrac{Dq}{B'} = 0.88$ and $B' = 951$ cm^{-1}, we get: $Dq = 951 \times 0.88 \approx 837$ cm^{-1}

i.e. $10Dq = 8370\ cm^{-1}$. It predicts ν_1 (= $10Dq$) to appear at 8,370 cm^{-1} (*cf.* experimental value 8,700 cm^{-1}). Thus the prediction is in good agreement with the experimental finding.

By taking all the three ligand field bands, B' can be also calculated.

$$B' = \frac{\nu_2 + \nu_3 - 3\nu_1}{15} = \frac{13,800 + 25,300 - 3 \times 8,700}{15} = 867\ cm^{-1}$$

The B' values obtained in different ways are more or less comparable. The average value is: B'(average) $= \frac{1}{3}(867 + 989 + 951) \approx 936\ cm^{-1}$.

Taking $B_0 = 1030\ cm^{-1}$ and $B' = 936\ cm^{-1}$, we can calculate the nephelauxetic parameter (β) as follows:

$$\beta = \frac{B'}{B_0} = \frac{936}{1030} \approx 0.91, \text{ i.e. } B' \text{ is reduced to about 90\% of } B_0.$$

We can verify whether the assignments are self consistent or not in another way.

Transition energy ratio (experimental value)	*Theoretically calculated value for $\frac{Dq}{B'} = 0.88$ from Fig. 7.10.4*
$\frac{^3A_{2g}(F) \longrightarrow {}^3T_{1g}(F)}{^3A_{2g}(F) \longrightarrow {}^3T_{2g}(F)} = \frac{\nu_2}{\nu_1} = \frac{13,800}{8,700} \approx 1.59$	1.67
$\frac{^3A_{2g}(F) \longrightarrow {}^3T_{1g}(P)}{^3A_{2g}(F) \longrightarrow {}^3T_{2g}(F)} = \frac{\nu_3}{\nu_1} = \frac{25,300}{8,700} \approx 2.9$	3.0

Thus the calculated energy ratios are in good agreement with the experimental values and the B' value determined (< B_0) is also quite reasonable. All these support that the band assignments are correct.

(Note: The ν_2 band of $[Ni(OH_2)_6]^{2+}$ splits into a doublet and it has been already explained by considering the spin-forbidden transition to $^1E_g(D)$ state and spin-orbit coupling splitting of the $^3T_{1g}(F)$ term.)

The $10Dq$ value can be determined from the ν_1 band without any uncertainty. It leads to: $10Dq$ = 8,700 cm^{-1}. If the energy change (without considering the configurational interaction) in the ν_2 band is considered to be $18Dq$ then we get:

$$\nu_2 = 18Dq = 13,800 \text{ i.e. } 10Dq = 7,667\ cm^{-1}.$$

This value is different from the value of $10Dq$ obtained from the ν_1–band. ***It indicates that in $[Ni(H_2O)_6]^{2+}$, the configurational interaction between the $^3T_{1g}(F)$ and $^3T_{1g}(P)$ states is quite important.*** Configurational interaction energy (x) can be calculated as follows:

$\nu_2 = 18Dq - x = 18 \times 870 - x = 13,800$

i.e. $x = 18 \times 870 - 13,800 = 15,660 - 13,800 = 1860\ cm^{-1}$

(C) More rigorous analysis of the spectral parameters of Ni(II)–octahedral complexes: It has been shown that in $[Ni(OH_2)_6]^{2+}$, the $10Dq$ value calculated from the ν_2 (taking as $18Dq$) band is significantly different from the value obtained from the ν_1–band. It indicates the importance of configurational interaction. It leads to:

$$\nu_2 = 18Dq - x$$

Energies of the $^3A_{2g}(F)$ and $^3T_{2g}(F)$ states do not depend on the configurational interaction.

$$E[^3A_{2g}(F)] = -12Dq;\ (t_{2g}^6e_g^2\text{: cfse} = -12Dq)$$

$$E[^3T_{2g}(F)] = -2Dq;\ (t_{2g}^5e_g^3\text{: cfse} = -2Dq)$$

The excited states $^3T_{1g}(F)$ and $^3T_{1g}(P)$ experience the configurational interaction. Their energies are given by the following *second order secular determinant.*

$$\begin{array}{c|cc|} & ^3T_{1g}(F) & ^3T_{1g}(P) \\ ^3T_{1g}(F) & 6Dq-E & 4Dq \\ ^3T_{1g}(P) & 4Dq & p-E \end{array} = 0,$$

$p = 15B'$ = separation energy between the 3F and 3P terms in the complex.

i.e. $(6Dq)p - 16(Dq)^2 - (6Dq)E - pE + E^2 = 0$

Roots of this equation give the energy values of $^3T_{1g}(P)$ and $^3T_{1g}(F)$ states.

If energy of the ^3F state (free ion) is considered to be zero, then energy of the ^3P state is p.

Energy of the $^3T_{1g}(P)$ state is given by:

$$\nu_3 = E[^3T_{1g}(P)] - E[^3A_{2g}(F)]$$
$$= E\,[^3T_{1g}(P)] - (-12Dq)$$

or, $E\,[^3T_{1g}(P)] = \nu_3 - 12Dq$

For **[Ni(*DMSO*)$_6$]$^{2+}$** : $\nu_1 = 7{,}730\ \text{cm}^{-1} = 10Dq$; *i.e.* $Dq = 773\ \text{cm}^{-1}$

$$\nu_2 = 12{,}970\ \text{cm}^{-1}$$
$$\nu_3 = 24{,}040\ \text{cm}^{-1}$$

We get: $E[^3T_{1g}(P)] = \nu_3 - 12Dq = 24{,}040 - 12 \times 773$

$$= 14,\ 764\ \text{cm}^{-1}$$

Now let us use this energy value ($=14{,}764\ \text{cm}^{-1}$) and Dq ($= 773\ \text{cm}^{-1}$) in the following equation.

$$(6Dq)p - 16(Dq)^2 - (6Dq)E - pE + E^2 = 0.$$

or $\quad 6 \times 773p - 16 \times (773)^2 - 6 \times 773 \times 14{,}764 - p \times 14{,}764 + (14{,}764)^2 = 0.$

It leads to:

$$p = 13{,}818 = 15B'$$

i.e. $\quad B' \approx 921\ \text{cm}^{-1}$

By using the values of Dq and p in the following equation, we get the two values of E.

$$(6Dq)p - 16(Dq)^2 - 6Dq(E) - pE + E^2 = 0$$
$$p = 13{,}818\ \text{cm}^{-1},\ Dq = 773\ \text{cm}^{-1}$$

The solutions are: $E_1 = 3,\ 694\ \text{cm}^{-1}$, $E_2 = 14{,}762\ \text{cm}^{-1}$

i.e. $\quad E[^3T_{1g}(P)] = 14{,}762\ \text{cm}^{-1}$,

$$E[^3T_{1g}(F)] = 3{,}694\ \text{cm}^{-1}.$$

Now let us calculate the ν_2 and ν_3 values.

$$\nu_2 = E[^3T_{1g}(F)] - E[^3A_{2g}(F)] = 3{,}694\ \text{cm}^{-1} - (-12 \times Dq)\ \text{cm}^{-1}$$
$$= 3{,}694\ \text{cm}^{-1} + 12 \times 773\ \text{cm}^{-1} = 12{,}970\ \text{cm}^{-1}$$
$$\nu_3 = E[^3T_{1g}\,(P)] - E[^3A_{2g}(F)] = 14{,}762 - (-12 \times 773) = 24{,}038\ \text{cm}^{-1}$$

These calculated values of ν_2 abd ν_3 are in good agreement with the experimental values. Thus, the assignments of the spectral transitions are self-consistent.

(D) Spectral analysis of $[Ni(bpy)_3]^{2+}$ where the third band, *i.e.* ν_3-band is obscured: The two well defined peaks are:

$$\nu_1 = 12{,}650\ \text{cm}^{-1}\ (= 10Dq)$$
$$\nu_2 = 19{,}200\ \text{cm}^{-1}$$

There is a strong band at 26,000 cm^{-1} having a very high extinction coefficient. Thus, the band at 26,000 cm^{-1} is reasonably expected to be a charge transfer band into which the third ligand field band of weak intensity may be merged. However, by considering the positions of ν_1 and ν_2, the ligand field bands can be characterised with the help of Tanabe-Sugano diagram.

$$\frac{{}^3A_{2g}(F) \longrightarrow {}^3T_{1g}(F)}{{}^3A_{2g}(F) \longrightarrow {}^3T_{2g}(F)} = \frac{\nu_2}{\nu_1} = \frac{19{,}200}{12{,}650} \approx 1.52.$$

This ratio fits into the Tanabe-Sugano diagram for $Dq/B' \approx 1.7$ (*cf.* Fig. 7.10.4)

The ν_1 (= $10Dq$) band gives Dq = 1265 cm^{-1}.

$$\frac{Dq}{B'} = 1.7 = \frac{1265}{B'};\quad \text{or}\quad B' = \frac{1265}{1.7} = 744\ \text{cm}^{-1}.$$

i.e.
$$\beta = \frac{B'}{B_0} = \frac{744}{1030} = 0.72.$$

Note: $[Ni(bpy)_3]^{2+}$ belongs to the D_3 point group but in terms of electronic transitions, it effectively adopts the O_h symmetry of the chromophore $[Ni(N)_6]$. A similar situation arises for $[Cr(ox)_3]^{3-}$, $[Cr(en)_3]^{3+}$, etc. (*cf.* Secs. 7.14.5 and 7.15.4 for details).

(E) Spectral properties of the tetragonally distorted high spin Ni(II)-complexes like *trans*-$[Ni^{II}A_4B_2]$ of D_{4h} symmetry: In moving from the O_h to D_{4h} symmetry, the following changes occur.

${}^3A_{2g} \rightarrow {}^3B_{1g}$; ${}^3T_{2g} \rightarrow {}^3E_g + {}^3B_{2g}$; ${}^3T_{1g} \rightarrow {}^3E_g + {}^3A_{2g}$ (*cf.* **Correlation Table** 7.23.2).

Let us consider the *z*-out distortion to give the D_{4h} symmetry. In the limiting situation, it will produce the diamagnetic square planar complex (which is discussed separately in Sec. 7.17). The six coordinate tetragonally distorted Ni(II) cmplexes are paramagnetic and in such cases, generally the energy order of the *d*-orbitals is as follows:

$e_g^4 \langle b_{2g}^2 \langle a_{1g}^1 \langle b_{1g}^1$ or $e_g^4 \langle a_{1g}^2 \langle b_{2g}^1 \langle b_{1g}^1$ (*i.e.* the relative positions of b_{2g} and a_{1g} depend on the degree of *z*-out distortion.

The different spectroscopic states are related with the electronic configurations as follows:

$$e_g^4 b_{2g}^2 a_{1g}^1 b_{1g}^1 \Rightarrow {}^3B_{1g};\ e_g^4 b_{2g}^1 a_{1g}^2 b_{1g}^1 \Rightarrow {}^3A_{2g}\left({}^3T_{1g}(F)\right);\ e_g^3 b_{2g}^2 a_{1g}^2 b_{1g}^1 \Rightarrow {}^3E_g\left({}^3T_{2g}\right).$$

$$e_g^4 b_{2g}^1 a_{1g}^1 b_{1g}^2 \Rightarrow {}^3B_{2g};\ e_g^3 b_{2g}^2 a_{1g}^1 b_{1g}^2 \Rightarrow {}^3E_g\left({}^3T_{1g}(F)\right);\ e_g^3 b_{2g}^3 a_{1g}^2 b_{1g}^2 \Rightarrow {}^3A_g\left({}^3T_{1g}(P)\right);$$

$$e_g^2 b_{2g}^2 a_{1g}^2 b_{1g}^2 \Rightarrow {}^3A_g\left({}^3T_{1g}(P)\right).$$

The possible spin allowed **electronic transitions** in the 6-coordinate Ni(II) **paramagnetic complexes** of D_{4h} symmetry are as follows:

$${}^3B_{1g} \xrightarrow{\nu_1} {}^3E_g^{(a)};\ {}^3B_{1g} \xrightarrow{\nu_2} {}^3B_{2g};\ {}^3B_{1g} \xrightarrow{\nu_3} {}^3A_{2g}(F);\ {}^3B_{1g} \xrightarrow{\nu_4} {}^3E_g^{(b)};$$

$${}^3B_{1g} \xrightarrow{\nu_5} {}^3E_g(P);\ {}^3B_{1g} \xrightarrow{\nu_6} {}^3A_{2g}(P)$$

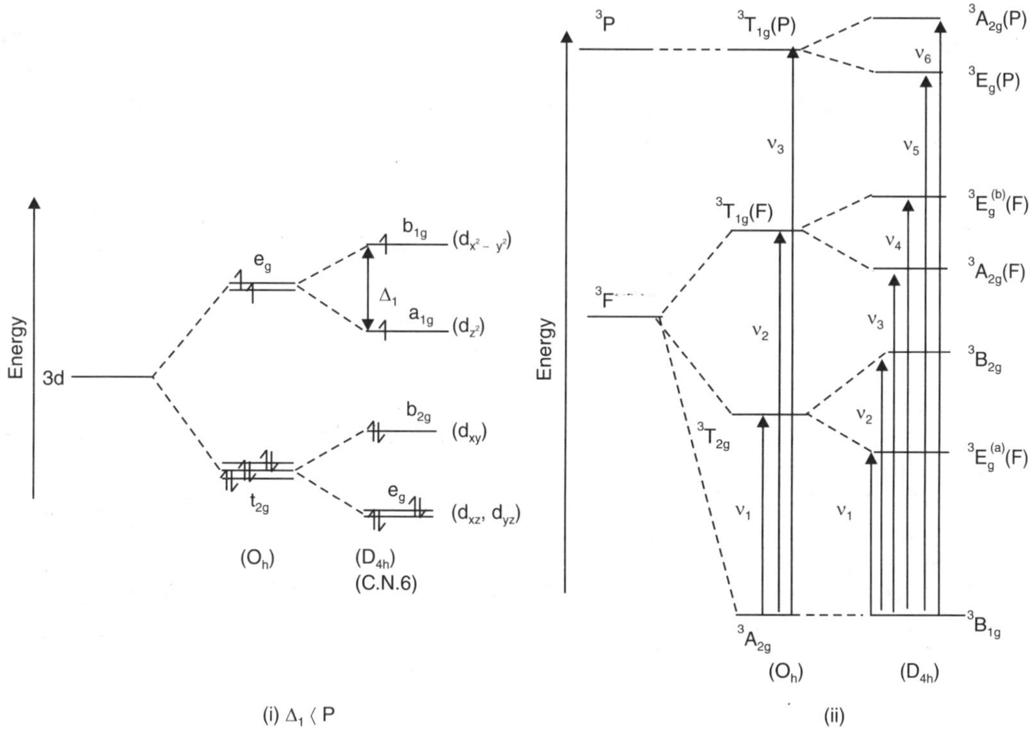

(a) and (b) suffixes are used to denote the different ^3E_g levels coming from ^3F.

Fig. 7.14.4.3 Effect of tetragonal elongation on (i) *d*-orbital splitting, (ii) energy levels for the 6 coordinate Ni(II) complexes. **Note:** Energy order: $^3A_{2g}$ (P) $\rangle$ ^3E_g (P) but the reverse order for ^3E_g and $^3A_{2g}$ generated from the splitting of the $^3T_{1g}$(F) level. It occurs due to the **configurational interaction** between the $^3A_{2g}$(P) and $^3A_{2g}$(F) levels. This interaction minimises the splitting of both the $^3T_{1g}$(P) and $^3T_{1g}$(F) levels. Thus, in moving from $O_h \rightarrow D_{4h}$, **splitting of the $^3T_{2g}$(F) level is only appreciable.** ***This is why, generally in moving from O_h to D_{4h} (C.N. 6), the octahedral ν_2 and ν_3 bands are not affected much but the octahedral ν_1 band spits into two components.***

The splitting of the $^3T_{1g}(P)$ state is very small (*cf.* $^3E_g(P) \sim {}^3A_{2g}(P) \sim 500$ cm^{-1}) and generally **the ν_5 and ν_6 transitions merge to produce a relatively broader peak.** The magnitude of tetragonal splitting in *trans*-[$Ni^{II}A_4B_2$](D_{4h}) depends on the difference of the ligand field strength of A and B. Peak positions of some representative 6-coordinate Ni(II) complexes of D_{4h} symmetry are given below:

[$NiBr_2(py)_4$]: ν_1(8.42 kK); ν_2(11.55 kK); ν_3(14.10 kK); ν_4(16.37 kK); $\nu_{5,6}$(25.9 kK);

[$Ni(NH_3)_4(NO_2)_2$]: ν_1(11.2 kK); ν_2(12.0 kK); ν_3(19.9 kK); ν_4(20.35 kK);

Note: (i) **$O_h \rightarrow D_{4h}$ (C.N. = 6),** *i.e.* **[$Ni^{II}A_6$] → *trans*-[$Ni^{II}A_4B_2$]:** In moving from **O_h to D_{4h}(C.N. 6),** splitting of the energy levels can generate ***more than 3 ligand field bands*** (*cf.* Fig. 17.4.4.3), but in reality, in most of the cases, the ligand field bands can be interpreted in terms of the ν_1, ν_2 and ν_3 transitions of O_h symmetry, *i.e.* splitting of the energy levels is not sufficiently large to give all the separate peaks as expected in D_{4h} (C.N. 6). This occurs due to two reasons: splitting of the T-term is not large specially when the ligand field strengths of A and B in *trans*-[$Ni^{II} A_4B_2$] do not differ widely; both the $^3T_{1g}(F)$ and $^3T_{1g}(P)$ terms yield $^3A_{2g}$ + ^3E_g in D_{4h} and symmetry interaction (*i.e.* **configurational interaction**) between the $^3A_{2g}$(F) and $^3A_{2g}$(P) states; and ^3E_g(F) and ^3E_g(P) states reduce the splitting.

This configurational interaction leads to the energy order $^3E_g \rangle {}^3A_{2g}$ for $^3T_{1g}$(F) while the order is inverted for ^3E_g and $^3A_{2g}$ generated from $^3T_{1g}$(P). **This symmetry interaction also minimises splitting of the $^3T_{1g}$(P) and $^3T_{1g}$(F) terms and splitting of the $^3T_{2g}$(F) term is only accountable.** This is why, in moving from O_h to D_{4h} (C.N. 6), the ν_2 and ν_3 octahedral bands are not greatly affected but the octahedral ν_1 band splits into two components.

$$\left[\mathrm{Ni^{II}}A_6\right](O_h) \longrightarrow trans\text{-}\left[\mathrm{Ni^{II}}A_4B_2\right](D_{4h})$$

$$^3A_{2g} \longrightarrow {}^3T_{2g}(F): \nu_1 \text{ splits into two components}$$

$$\left.\begin{array}{l} ^3A_{2g} \longrightarrow {}^3T_{1g}(F): \nu_2 \\ ^3A_{2g} \longrightarrow {}^3T_{1g}(P): \nu_3 \end{array}\right\} \begin{array}{l}\text{(slightly broadened)} \\ \text{almost unchanged.}\end{array}$$

Conclusion: (i) If the ligand field strengths of A and B do not differ largely, the ligand field bands of both the *cis*- and *trans*- isomers of [$Ni^{II}A_4B_2$] can be interpreted in terms of the spin allowed transitions in the octahedral complexes.

(ii) **$O_h \rightarrow C_{2v}$ (C.N. 6),** ***i.e.*** **[$Ni^{II}A_6$] →*cis*- [$Ni^{II}A_4B_2$]:** The *cis*-complex also experiences the tetragonal field **electronically** as in the *trans*-complex but the order and magnitude of splitting are different. These aspects have been discussed in detail in Secs. 17.4.5 and 7.15.2.

(iii) **$O_h \rightarrow D_{4h}$ (C.N. 4),** ***i.e.*** **[$Ni^{II}A_6$] → [$Ni^{II}A_4$] (square planar):** The limiting situation of *z*-out distortion gives the diamagnetic square planar complex. Its spectral transitions are separately discussed in Sec. 7.17.

7.14.5 Six-Coordinate Complexes of 3d³-System

Complexes of Cr(III) and V(II) are the representative examples of this group. From the Orgel diagram, the spin-allowed transitions are:

$$^4A_{2g}(F) \xrightarrow{\nu_1} {}^4T_{2g}(F);\ \nu_1 = 10Dq \ (\text{independent of configuration interaction})$$

$$^4A_{2g}(F) \xrightarrow{\nu_2} {}^4T_{1g}(F);\ \nu_2 = 18Dq - x \approx 18Dq$$

$$^4A_{2g}(F) \xrightarrow{\nu_3} {}^4T_{1g}(P);\ \nu_3 = 15B' + 12Dq + x$$

$$\approx 15B' + 12Dq$$

ν_3 involves **2*e* transition** $\left(i.e.\ t_{2g}^3 \longrightarrow t_{2g}^1 e_g^2\right)$ and it is a **spin forbidden transition.**

x is the symmetry interaction energy for the configurational interaction between the $^4T_{1g}(F)$ and $^4T_{1g}(P)$ states. However, in a weak field ligand, *x* becomes negligibly small (*cf.* Figs. 7.10.2; 7.14.5.2);

The B' and $10Dq$ value may be calculated as follows:

$$\nu_1 = 10Dq,\ 15B' = (\nu_2 + \nu_3 - 3\nu_1).$$

Calculation of B' by this method is applicable in the cases where the three ligand field bands (ν_1, ν_2 and ν_3) are well characterised. But in most of the cases, the third ligand field band (ν_3) is obscured by the strong charge transfer band. Moreover, the band due to ν_3 is **spin forbidden** because it involves a **two-electron transfer,** *i.e.* $^4A_{2g}(t_{2g}^3 e_g^0) \rightarrow {}^4T_{1g}(P)\ (t_{2g}^1 e_g^2)$. Thus it is a weak band. In the cases like $[CrF_6]^{3-}$ where the charge transfer band lies at still higher energy, all the three ligand field bands are well characterised and B' can be calculated as follows:

$[CrF_6]^{3-}$: ν_1 = 14,900 cm^{-1}, ν_2 = 22,700 cm^{-1}, ν_3 = 34,400 cm^{-1}

It leads to: $10Dq = \nu_1 = 14{,}900\ \text{cm}^{-1}$

$$B' = \frac{34{,}400 + 22{,}700 - 3\times 14{,}900}{15} = 827\ \text{cm}^{-1}.$$

The energy change in the spin allowed ligand field bands for the high-spin octahedral d³ and d⁸ and tetrahedral d² and d⁷ systems

$$^4A_2 \Rightarrow t_2^3 \Rightarrow E = -12Dq;\ ^4T_2\,(\text{F}) \Rightarrow t_2^2e^1 \Rightarrow E = -2Dq,$$

$$^4T_1(F) \Rightarrow t_2^2e^1 \Rightarrow E = 7.5B' + 3Dq - X;\ ^4T_1(P) \Rightarrow t_2^1e^2 \Rightarrow E = 7.5B' + 3Dq + X$$

$$X = \frac{1}{2}\left(225B'^2 + 100Dq^2 - 180DqB'\right)^{1/2}$$

$$A_2(F) \longrightarrow T_2(F),\ \nu_1 = \Delta,$$

$$A_2(F) \longrightarrow T_1(F),\ \nu_2 = 7.5B' + 1.5\Delta - \frac{1}{2}\left[225B'^2 + \Delta^2 - 18B'\Delta\right]^{1/2}$$

$$A_2(F) \longrightarrow T_1(P),\ \nu_3 = 7.5B' + 1.5\Delta + \frac{1}{2}\left[225B'^2 + \Delta^2 - 18B'\Delta\right]^{1/2}$$

$$B' = \frac{\nu_2 + \nu_3 - 3\nu_1}{15} = \frac{2\nu_1^2 + \nu_2^2 - 3\nu_1\nu_2}{(15\nu_2 - 27\nu_1)} = \frac{2\nu_1^2 + \nu_3^2 - 3\nu_1\nu_3}{15\nu_3 - 27\nu_1}$$

(A) Analysis of the spectrum of $[Cr(OH_2)_6]^{3+}$: Aqueous solution of $[Cr(OH_2)_6]^{3+}$ looks light green due to the absorption of yellow and blue parts of the visible spectrum. There are two intense bands at 17,000 cm^{-1} ($\varepsilon \approx 14$) and 24,000 cm^{-1} ($\varepsilon \approx 15$). A weak band at 37,000 cm^{-1} ($\varepsilon \approx 4$) appears as a shoulder of a strong charge transfer band. Besides these, there are two weak bands at about 15,000 cm^{-1} and 22,000 cm^{-1} and these are the **spin forbidden transitions.** These are not shown in Fig. 7.14.5.1.

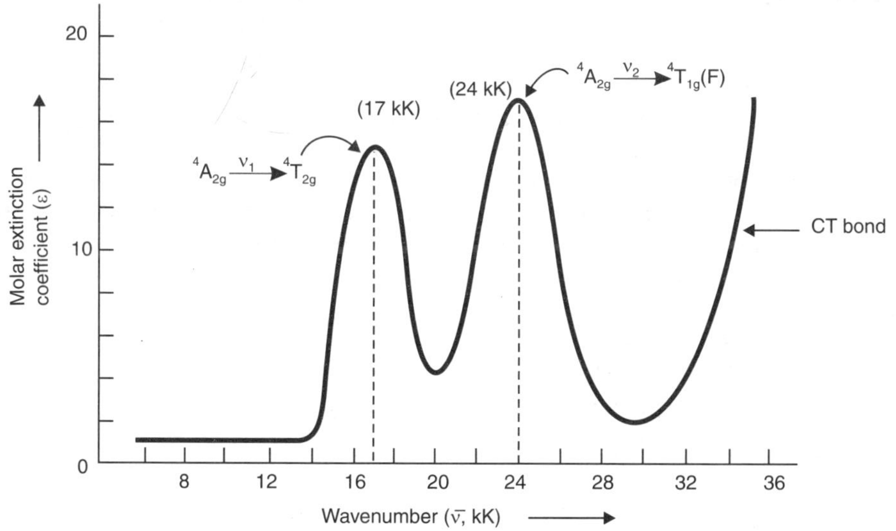

Fig. 7.14.5.1 The electronic absorption spectrum of $[Cr(OH_2)_6]^{3+}$.

To assign the peaks, we can start with the intense peaks at 17,000 cm^{-1} and 24,000 cm^{-1}. Let us assign the transitions as follows:

$$^4A_{2g}(F) \rightarrow {}^4T_{2g}(F);\ \nu_1 = 17{,}000\ cm^{-1}\ (= 10Dq)$$

$$^4A_{2g}(F) \rightarrow {}^4T_{1g}(F);\ \nu_2 = 24{,}000\ cm^{-1}\ (= 18Dq - x).$$

We can verify whether these assignments are correct or not by comparing with the Tanabe-Sugano diagram. The ratio, $\nu_2/\nu_1 \approx 1.41$ fits with the Tanabe-Sugano diagram for the value $Dq/B' = 2.45$ (*cf.* Fig. 7.10.4). This Dq/B' value corresponds to $E(\nu_1)/B' = 24.5$ for the ν_1 transition, *i.e.*

$$\frac{^4A_{2g}(F) \longrightarrow {}^4T_{2g}(F)}{B'} = \frac{E(\nu_1)}{B'} = \frac{17{,}000}{B'} = 24.5$$

i.e. $$B' = \frac{17{,}000}{24.5} \approx 694\ cm^{-1}$$

and $$10Dq = \nu_1 = 17{,}000\ cm^{-1}$$

The B' value may compared with B_0 value (of free ion).

$$B_0 = 1030\ cm^{-1}\ \textit{i.e.}\ \beta\ \text{(nephelauxetic ratio)} = \frac{B'}{B_0} = \frac{694}{1030} = 0.67$$

Thus in $[Cr(OH_2)_6]^{3+}$, B value is reduced by about 33%.

Fitting of the ν_1 and ν_3 bands at $Dq/B' = 2.45$ in the Tanabe-Sugano diagram gives the following result.

$$\frac{^4A_{2g}(F) \longrightarrow {}^4T_{1g}(P)}{^4A_{2g}(F) \longrightarrow {}^4T_{2g}(F)} = \frac{E(\nu_3)}{E(\nu_1)} = 2.2.\ (\textit{cf.}\ \text{Fig. 7.10.4})$$

It predicts that ν_3 should appear at 17,000 × 2.2, *i.e.* 37,400 cm^{-1}. In fact, a weak band at 37,000 cm^{-1} appears. It supports the validity of the assignment of the spectral transitions.

(B) Spectral analysis of $[Cr(OH_2)_6]^{3+}$ in terms of the octahedral field subjected to configurational interaction: It has been already mentioned that due to the configurational interaction between the $^4T_{1g}(F)$ and $^4T_{1g}(P)$ terms, energy of the $^4T_{1g}(P)$ term goes up by x and energy of the $^4T_{1g}(F)$ term goes down by the same amount. The energy level diagram for the d^3–configuration is shown in Fig. 7.14.5.2.

We have already mentioned that the peaks at 17,000 cm^{-1} and 24,000 cm^{-1} are easily detected. If the $^4A_{2g}(F) \rightarrow {}^4T_{2g}(F)$ transition is considered to occur at 17,000 cm^{-1}, then $10Dq$ should be equal to 17,000 cm^{-1}, *i.e.* $Dq = 1700\ cm^{-1}$. Thus the transition, $^4A_{2g}(F) \rightarrow {}^4T_{1g}(F)$, *i.e.* ν_2 (= $18Dq$) should occur at 1700 × 18 cm^{-1} = 30,600 cm^{-1}. This prediction is based on the fact of **no configurational interaction**. But in reality, the peak appears at 24,000 cm^{-1}. This huge discrepency (by about 6000 cm^{-1}) is not due to the experimental error. Rather it indicates that in $[Cr(OH_2)_6]^{3+}$, the **configurational interaction** is quite significant. The ν_1–transition is independent of the configurational interaction, *i.e.* there is no error in the relation, $10Dq = \nu_1 = 17{,}000\ cm^{-1}$. Due to the configurational interaction between the $^4T_{1g}(F)$ and $^4T_{1g}(P)$ terms, we have the condition:

$$^4A_{2g}(F) \rightarrow {}^4T_{1g}(F),\ \nu_2 = 24{,}000\ cm^{-1} = 18Dq - x$$

$$^4A_{2g}(F) \rightarrow {}^4T_{1g}(P),\ \nu_3 = 37{,}000\ cm^{-1} = 15B' + 12Dq + x$$

It leads to:

$$24{,}000 = 18Dq - x = 18 \times 1700 - x$$

i.e. $$x = 6{,}600\ cm^{-1}$$

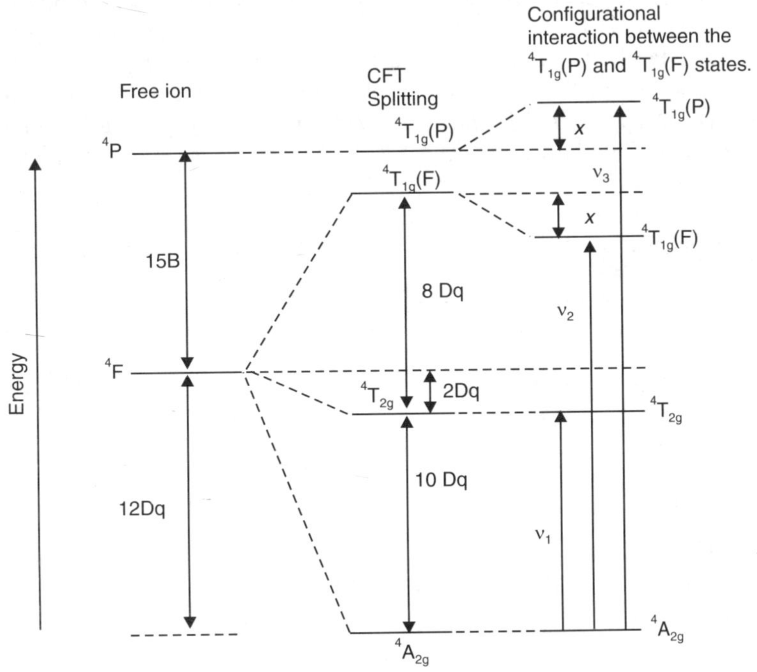

Note: The electronic configurations and crystal field energies of the $^4T_{2g}$ and $^4T_{2g}$ states are: $^4A_{2g}(t_{2g}^3, -12Dq)$, $^4T_{2g}(t_{2g}^2e_g^1, -2Dq)$. Due to the mixing between the $^4T_{1g}(P)$ and $^4T_{1g}(F)$ states, the $^4T_{1g}(F)$ corresponds to $t_{2g}^x e_g^y$ (pDq, say; $x + y = 3$). Considering the energy conservation and orbital degeneray: $1 \times 12\ Dq + 3 \times 2Dq = 3pDq$, *i.e.* $p = 6$; $(-4x + 6y)Dq = pDq = 6Dq$ and $x + y = 3$ lead to $x = 1.2$ and $y = 1.8$. Thus in a weak field, $^4T_{1g}(F)$ corresponds to $t_{2g}^{1.2}\, e_g^{1.8}$ (not $t_{2g}^1 e_g^2$).

Fig. 7.14.5.2 Energy level diagram and spectral transitions (spin-allowed) in the d^3–octahedral system (x denotes the configurational interaction energy for the $^4T_{1g}(P) - {}^4T_{1g}(F)$ interaction) (*See* Fig. 7.9.4.1 and Solved Problem 14 for the energy values of different states).

Assuming $\nu_3 = 37{,}000\ \text{cm}^{-1}$, we can find out the value of B' as follows:

$$37{,}000 = 15B' + 12Dq + x = 15B' + 12 \times 1700 + 6{,}600$$

or, $$30{,}400 - 20{,}400 = 15B'$$

or, $$B' \approx 667\ \text{cm}^{-1}\ (cf.\ B_0 = 1030\ \text{cm}^{-1})$$

i.e. $$\beta = \frac{B'}{B_0} = \frac{667}{1030} = 0.64$$

i.e. B is reduced to about 64% of the free ion value.

(C) O_h *vs.* C_{4v}, *e.g.* $[Cr(NH_3)_6]^{3+}$ (O_h) *vs.* $[Cr(NCS)(NH_3)_5]^{2+}$, $[CrCl(NH_3)_5]^{2+}$ (C_{4V}): Reduction in symmetry in moving from O_h to C_{4v} splits the energy levels as follows (*See* Correlation Table, Table 7.23.2) :

$A_{2g} \rightarrow B_1$; $T_{1g} \rightarrow A_2 + E$; $T_{2g} \rightarrow B_2 + E\}$ $\Rightarrow$ **splitting of the octahedral peaks into the doublets**

If we compare the spectral transitions in $[Cr(NH_3)_6]^{3+}$ and $[Cr(NH_3)_5X]^{2+}$, the degree of splitting of the T_1 and T_2 levels depends on the **difference of the ligand field strength of NH_3 and X^-.** The splitting of the T_{2g} level is, in general, higher than that of the T_{1g} level (*cf.* Sec. 7.14.4). In general, the spectral transitions of $[Cr(NH_3)_5X]^{2+}$ can be explained in terms of the octahedral transitions as splitting of the T-levels in the C_{4v} symmetry is not sufficiently high to develop the separate peaks. For $[Cr(NCS)(NH_3)_5]^{2+}$, the peaks do not split. For $[CrCl(NH_3)_5]^{2+}$, the peak due to the $^4A_{2g} \rightarrow {}^4T_{1g}$(F) transition does not split but the peak due to the $^4A_{2g} \rightarrow {}^4T_{2g}$(F) transition splits into a doublet as expected. In $[Cr(NH_3)_5X]^{2+}$(X = Cl, Br, NCS), the *LMCT bands are shifted towards the longer wavelength* compared to those of $[Cr(NH_3)_6]^{3+}$. The ligand field bands shift *towards the lower frequence,* if X^- is a weaker field ligand than NH_3.

Table 7.14.5.1a Spectral characteristics of the spin-allowed ligand field bands of the d^3-systems in an octahedral field.

Complex	ν_1 (cm^{-1})	ν_2 (cm^{-1})	ν_3 (cm^{-1})	*10Dq* (cm^{-1})	*B′* (cm^{-1})	β
$[Cr(OH_2)_6]^{3+}$	17,000	24,000	37,000	17,000	694	0.67
$[Cr(NH_3)_6]^{3+}$	21,550	28,500	–	21,550	–	–
$[CrCl_6]^{3-}$	13,200	18,700	28,600	13,200	513	0.56
$[Cr(CN)_6]^{3-}$	26,700	32,200	–	26,700	–	–
$[CrF_6]^{3-}$	14,900	22,700	34,400	14,900	827	0.80
$[Cr(ox)_3]^{3-}$	17,400	23,600	(38,100)*	17,400	600	0.58
$[Cr(en)_3]^{3+}$	21,800	28,500	(46,500)*	21,850	–	–
$[V(OH_2)_6]^{2+}$	12,300	18,500	27,900	12,300	633	0.84
$[VCl_6]^{4-}$	7,200	12,000	19,000	7,200	627	0.83
$[V(en)_3]^{2+}$	15,600	21,500	31,700	15,600	420	0.57

* The values are predicted but not experimentally verified. These are masked by the strong charge transfer bands.

Table 7.14.5.1b Comparison of the spectral properties of $[Cr^{III}(NH_3)_5X]^{2+}$ and $[Cr(NH_3)_6]^{3+}$ ($\bar{\nu}$ in cm^{-1}) (*cf.* Figs. 7.14.5.3a, 7.5.1.2).

	$^4A_{2g} \rightarrow {}^4T_{2g}$(F)		$^4A_{2g} \rightarrow {}^4T_{1g}$(F)	
	$^4B_1 \rightarrow {}^4E$	$^4B_1 \rightarrow {}^4B_2$	$^4B_1 \rightarrow {}^4A_2, {}^4E$	LMCT
$[CrCl(NH_3)_5]^{2+}$	19,400	22,100 (shoulder)	26,600 (no peak splitting)	42,000
$[Cr(NCS)(NH_3)_5]^{2+}$	20,370 (no peak splitting)		27,000 (no peak splitting)	
$[Cr(NH_3)_6]^{3+}$	21,500		28,500	50,000 (tail)

$^4A_{2g} \rightarrow {}^4T_{1g}$(P) transition is not characterised.

Spectral characteristics of $[Co(NH_3)_6]^{3+}$ (O_h) and $[CoCl(NH_3)_5]^{2+}$ have been compared and discussed in Sec. 7.15.3 (*cf.* Figs. 7.15.3.1, 2).

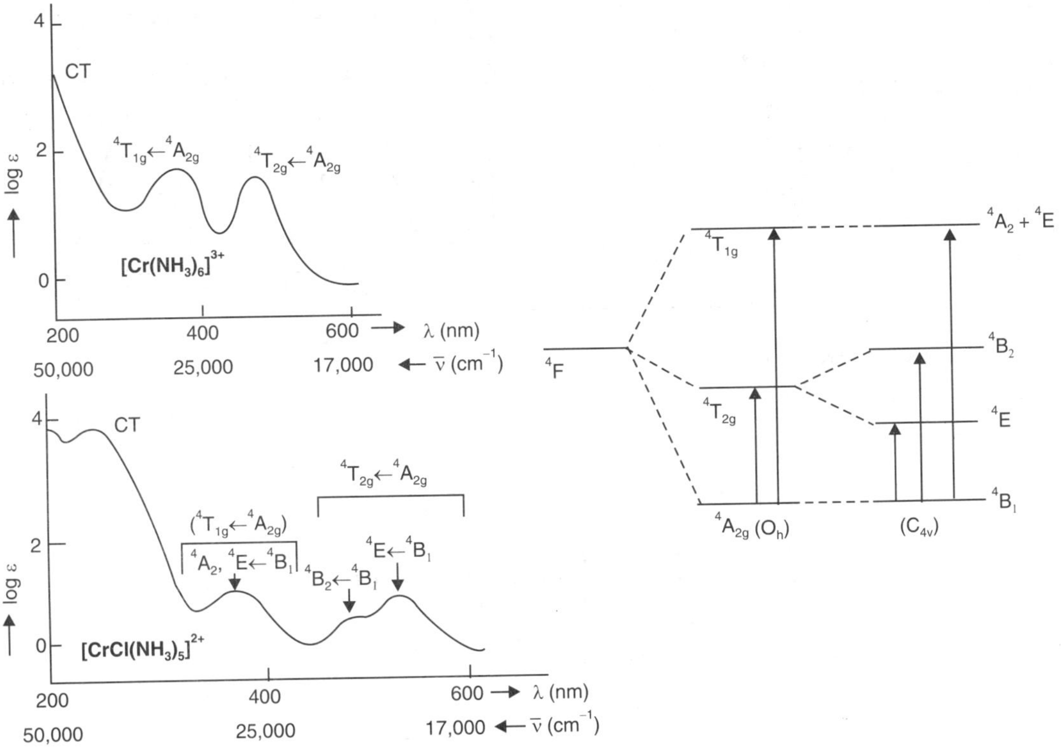

Fig. 7.14.5.3 (a) Spectral transitions of $[Cr(NH_3)_6]^{3+}$ (O_h) and $[CrCl(NH_3)_5]^{2+}$ (C_{4v}). The ligand field transition, ${}^4A_{2g} \rightarrow {}^4T_{1g}(P)$ is masked by the CT band. Splitting of the ${}^4T_{1g}(F)$ state in C_{4v} is not shown.

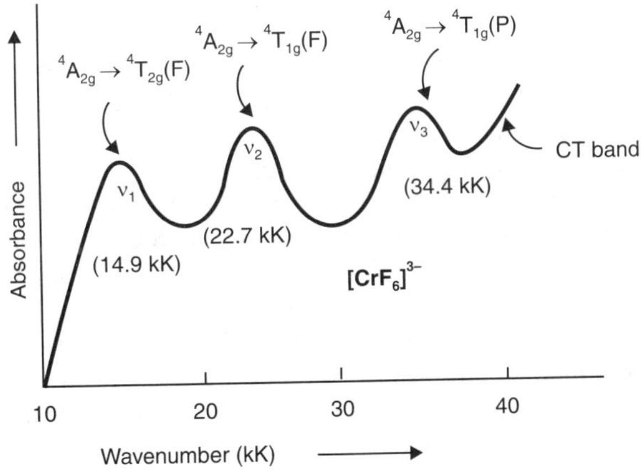

(b)

Fig. 7.14.5.3 (b) Absorption spectram of $[CrF_6]^{3-}$.

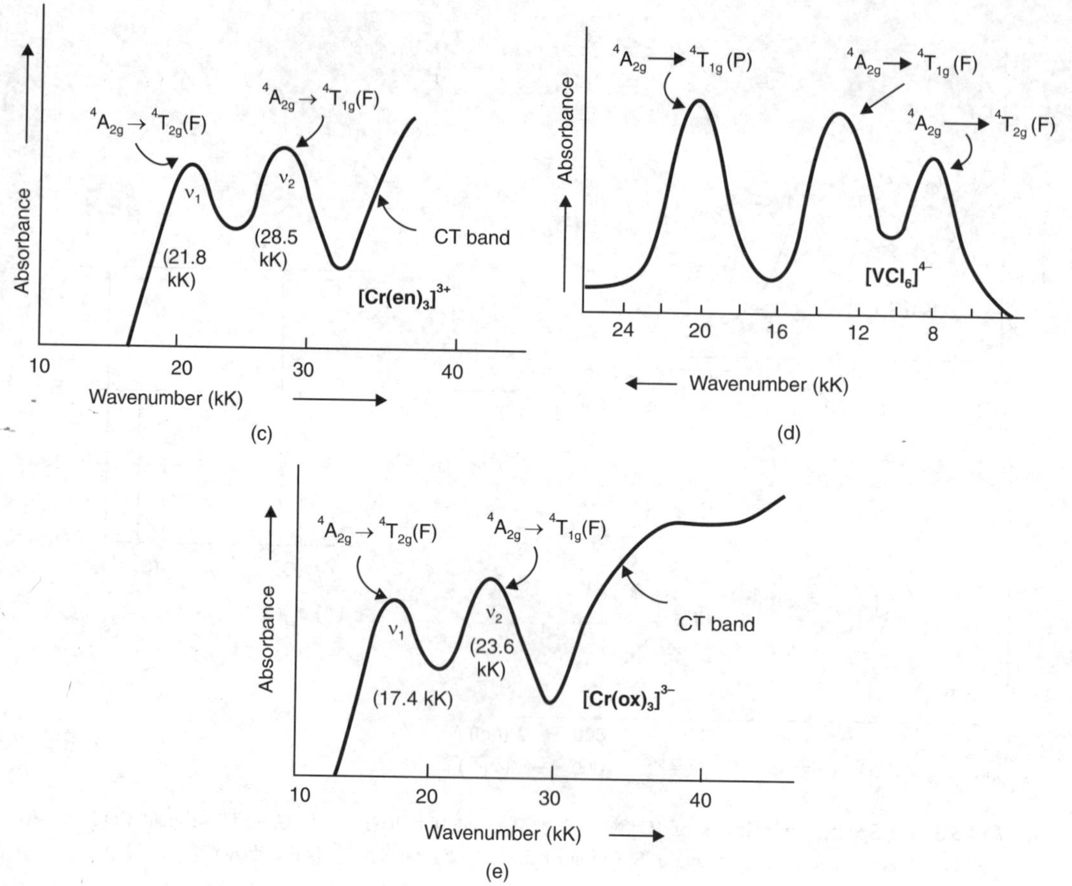

Fig. 7.14.5.3 (c) Absorption spectrum of $[Cr(en)_3]^{3+}$. (d) Absorption spectrum of $[VCl_6]^{4-}$. (e) Absorption spectrum of $[Cr(ox)_3]^{3-}$.

(D) Spectral behaviour of $[Cr(ox)_3]^{3-}$ and $[Cr(en)_3]^{3+}$: For green $[Cr(ox)_3]^{3-}$, the two spin allowed transitions are at 17,400 $cm^{-1}(\nu_1)$ and 23,600 cm^{1-} (ν_2). The 3rd ligand field band (ν_3) is predicted to appear at about 38,100 cm^{-1} but this is not experimentally verified because in that region, there is a strong charge transfer band to mask the ligand field band (ν_3).

For yellow $[Cr(en)_3]^{3+}$, the spin allowed transitions are: 21,850 cm^{-1} (ν_1) and 28,500 cm^{-1} (ν_2). The predicted third ligand field band (ν_3) at about 46,500 cm^{-1} is obscured by the strong charge transfer band. In addition to the spin-allowed transitions, there are two spin-forbidden transitions at 15,100 cm^{-1} and 15,600 cm^{-1}. These will be discussed later on.

Here it is worth mentioning that the said chelated complexes like $[Cr(en)_3]^{3+}$, $[Cr(ox)_3]^{3-}$, etc. are belonging to the **lower symmetry point group D_3.** They are not perfectly octahedral. However, their ligand field bands can be successfully explained by considering the octahedral field as in $[CrL_6]^{3+}$. It indicates that the perturbation is insignificant in these chelated complexes to destroy the ideal octahedral symmetry.

Note: Here it should be mentioned that the *tris*–chelates $[M(A-A)_3]$, *e.g.* $[Cr(en)_3]^{3-}$, $[Cr(ox)_3]^{3-}$ are belonging to the D_3 point group while the complexes like $[M(A)_6]$, *e.g.* $[Cr(NH_3)_6]^{3+}$ are belonging to the O_h point group. D_3 is a lower symmetry point group. The energy levels T_{1g} and T_{2g} of the O_h point group are split in the D_3 point group. The correlation between the O_h and D_3 point groups is as follows (*cf.* **Correlation Table 7.23.2** and Table 7.15.4.2):

O_h	D_3 (lacking in centre of symmetry)	
$^4A_{2g}$	$^4A_2$	(*See* Fig. 7.5.1.2)
$^4T_{1g}$	$^4A_2 + {}^4E$	
$^4T_{2g}$	$^4A_1 + {}^4E$	

Thus the transitions will be modified due to splitting of the T_1 and T_2 terms (*See* Fig. 7.5.1.2).

	O_h	D_3	
(*See* Fig. 7.5.1.2)	$^4A_{2g} \xrightarrow{\nu_1} {}^4T_{2g}$	$^4A_2 \rightarrow {}^4A_1, {}^4E$	(doublet)
	$^4A_{2g} \xrightarrow{\nu_2} {}^4T_{1g}(F)$	$^4A_2 \rightarrow {}^4A_2, {}^4E(F)$	(doublet)
	$^4A_{2g} \xrightarrow{\nu_3} {}^4T_{1g}(P)$	$^4A_2 \rightarrow {}^4A_2, {}^4E(P)$	(doublet)

Thus, all the three bands (ν_1, ν_2, ν_3) of the O_h ligand field should split into the doublets in the D_3 point group. But, in the ligand field spectra of $[Cr(en)_3]^{3+}$ and $[Cr(ox)_3]^{3-}$, no such splitting is noticed. (*cf.* Fig. 7.14.5.3c, e). In other words, the degree of splitting is too small to be reflected in the spectral transitions. ***This indicates that chelation has not effectively changed the effective octahedral symmetry in understanding the ligand field spectra (cf.* Sec. 7.15.4).**

D_3 symmetry (lacking in the centre of inversion) can allow the **d–p mixing** and the electronic transitions are **Laporte allowed**. But the molar extinction coefficients of $[Cr(ox)_3]^{3-}$ and $[Cr(en)_3]^{3+}$ are not remarkably increased compared with those of the O_h-complexes (*cf.* Table 7.15.4.1). Thus effectively they maintain the O_h-symmetry, *i.e.* CrO_6 in $[Cr(ox)_3]^{3-}$ and 'CrN_6' in $[Cr(en)_3]^{3+}$.

● **Group theory treatment of the D_3 point group and spectral transitions in $[Cr^{III}(AA)_3]$ (*cf.* Sec. 7.5.1N):** The dipole moment operator μ_z (along the C_3-axis) transforms as A_2 and μ_x, μ_y ($\perp$ to the C_3-axis) transform as E in a D_3 point group (*cf.* Character Table, Appendix 12A).

The electronic transitions in a noncentrosymmetric system are allowed when the direct product $\psi_{el}^{gd}\mu\psi_{el}^{ex}$ contains the totally symmetric A_1 irreducible representation. The following electronic transitions are examined.

$A_2 \rightarrow A_1$ transition: $A_2 \times \mu_z \times A_1 = A_2 \times A_2 \times A_1 = A_1 \times A_1 = A_1$; *i.e.* allowed transition (||)
$A_2 \rightarrow A_2$ transition: $A_2 \times \mu_z \times A_2 = A_2 \times A_2 \times A_2 = A_2 \times A_1 = A_2$; *i.e.* forbidden transition (||)
$A_2 \rightarrow E$ transition: $A_2 \times \mu_z \times E = A_2 \times A_2 \times E = E \times A_1 = E$; *i.e.* forbidden transition (||)
$A_2 \rightarrow A_1$ transition: $A_2 \times \mu_{x,y} \times A_1 = A_2 \times E \times A_1 = E \times A_2 = E$; *i.e.* forbidden transition ($\perp$)
$A_2 \rightarrow A_2$ transition: $A_2 \times \mu_{x,y} \times A_2 = A_2 \times E \times A_2 = A_1 \times E = E$; *i.e.* forbidden transition ($\perp$)
$A_2 \rightarrow E$ transition: $A_2 \times \mu_{x,y} \times E = E \times E \times A_2 = E \times E = A_1 + A_2 + E$; *i.e.* allowed transition ($\perp$)

It indicates **polarisation** (*i.e.* transitions are allowed only in certain directions) of the electronically allowed transitions in $[Cr^{III}(AA)_3]$ of D_3 symmetry (*see* 7.5.1, N). It may be noted that in *trans*-$[CoCl_2(en)_2]^+$, **vibronic polarisation** (dichroism) occurs. This aspect has been discussed earlier (Sec. 7.5.1, O).

(E) Spectral transitions in the tetragonally distorted Cr(III)–complexes (*cf.* Sec. 7.15.2 for details): Spectral transitions in the perfectly octahedral and tetrahedral complexes can be explained by considering the Orgel or Tanabe-Sugano diagrams. When the complexes are deviated from the cubic symmetry, many energy levels are split to complicate the interpretation. ***This is why, in the complexes of lower symmetry, more spectral transitions become possible.***

It has been already mentioned that the chelated complexes like $[Cr^{III}(LL)_3]$ posses the D_3–symmetry instead of the O_h symmetry. But spectra of such chelated complexes can be explained by considering the energy level diagrams of $[Cr^{III}(L)_6]$ complexes of O_h symmetry. However, the energy level diagrams of *cis/trans*–$[Cr^{III}A_4B_2]$ are quite different from that of the ideal octahedral complex of Cr(III). In fact, *cis*–$[Cr^{III}A_4B_2]$ belongs to the C_{2v} point group and *trans*–$[Cr^{III}A_4B_2]$ belongs to the D_{4h} point group. In moving from the O_h symmetry to the D_{4h} symmetry, many degenerate energy levels are split (*cf.* **Correlation Table** 7.15.4.2). It makes a larger number of spectral transitions possible in the D_{4h} molecules (*cf.* Fig. 7.14.5.5).

● **Splitting of energy levels in *cis*- and *trans*- isomers of $[MA_4B_2]$** (*cf.* Sec. 7.15.2): The *cis*- and *trans*- isomers of $[MA_4B_2]$ belong to the C_{2v} and D_{4h} symmetry point groups respectively but from the view of electronic environment **both isomers, in practice, belong to the tetragonal one (D_{4h})** (*cf.* F. Teixidor and J. Casabo, *J. Chem. Edu.*, **64,** 461-62, 1987). **In reality, both the isomers show the tetragonal splitting of the O_h-symmetry but in the *cis*-isomers, the tetragonal splitting is only half of the splitting generated in the *trans*-isomer.** The order of splitting in the energy levels is of opposite nature for the isomers. Thus for a d^3-system, ***the three band spectrum of an octahedral complex becomes a six band spectrum in D_{4h} symmetry (i.e. tetragonal symmetry).*** However, the expected six bands are rarely seen. Moreover, the band splitting is noted for the *trans*-complexes not for the *cis*-complexes in which the tetragonal splitting is too small to generate the separate peaks.

In ***trans*–$[Cr(en)_2F_2]^+$**, the possible transitions are:

$$\left.\begin{array}{l} {}^4B_{1g} \xrightarrow{\nu_1} {}^4E_g(F);\ {}^4B_{1g} \xrightarrow{\nu_2} {}^4B_{2g};\ {}^4B_{1g} \xrightarrow{\nu_3} {}^4E_g(F); \\ {}^4B_{1g} \xrightarrow{\nu_4} {}^4A_{2g}(F);\ {}^4B_{1g} \xrightarrow{\nu_5} {}^4A_{2g}(P);\ {}^4B_{1g} \xrightarrow{\nu_6} {}^4E_g(P) \end{array}\right\} (cf.\ \text{Fig. 7.14.5.5})$$

$\nu_1 = 18{,}500\ cm^{-1}$, $\nu_2 = 21{,}700\ cm^{-1}$, $\nu_3 = 25{,}300\ cm^{-1}$, $\nu_4 = 29{,}300\ cm^{-1}$, ν_5 and ν_6 are obscured by the strong charge transfer band in the far UV region. The expected values of ν_5 and ν_6 are:

$$\nu_5 = 41{,}000\ cm^{-1},\ \nu_6 = 43{,}750\ cm^{-1}$$

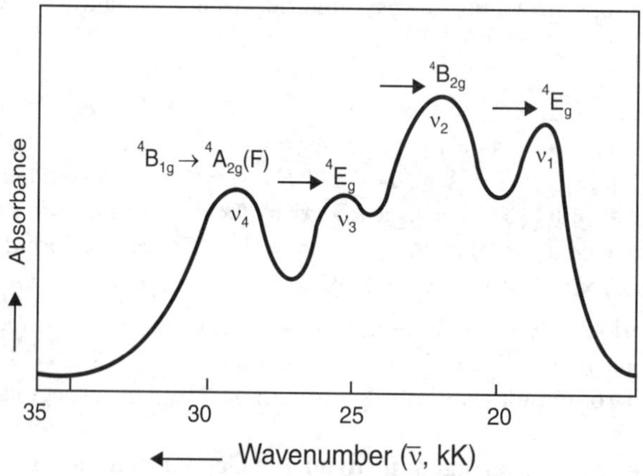

Fig. 7.14.5.4 Absorption spectrum of *trans*–$[Cr(en)_2F_2]^+$, (*cf.* Fig. 7.14.5.5 for the assignment of ν_1, ν_2, ν_3 and ν_4).

● **Comparison of the ligand field spectra of [$Cr^{III}A_4B_2$] (*cis* and *trans*–isomers):** The geometrical isomers *cis*–[$Cr^{III}A_4B_2$] (C_{2v}) and *trans*–[$Cr^{III}A_4B_2$] (D_{4h}) can be distinguished by comparing their ligand field spectra. In such cases, splitting (measured by δ) of the *T* term is important. The splitting energy is determined by the average ligand field strength difference between the A—Cr—A and B—Cr—B axes (for the *trans*-complex) and between the A—Cr—A and A—Cr—B axes (for the *cis*- complex). It gives:

$$\delta(\text{trans}) : \delta(\text{cis}) = 2:1$$

In other words, we can say that this splitting energy (δ) depends on the relative positions of the ligands *A* and *B* in the spectrochemical series.

$$\boldsymbol{\delta(trans) = 2C(\Delta_A \sim \Delta_B),\ \delta(cis) = -C(\Delta_A \sim \Delta_B)}$$

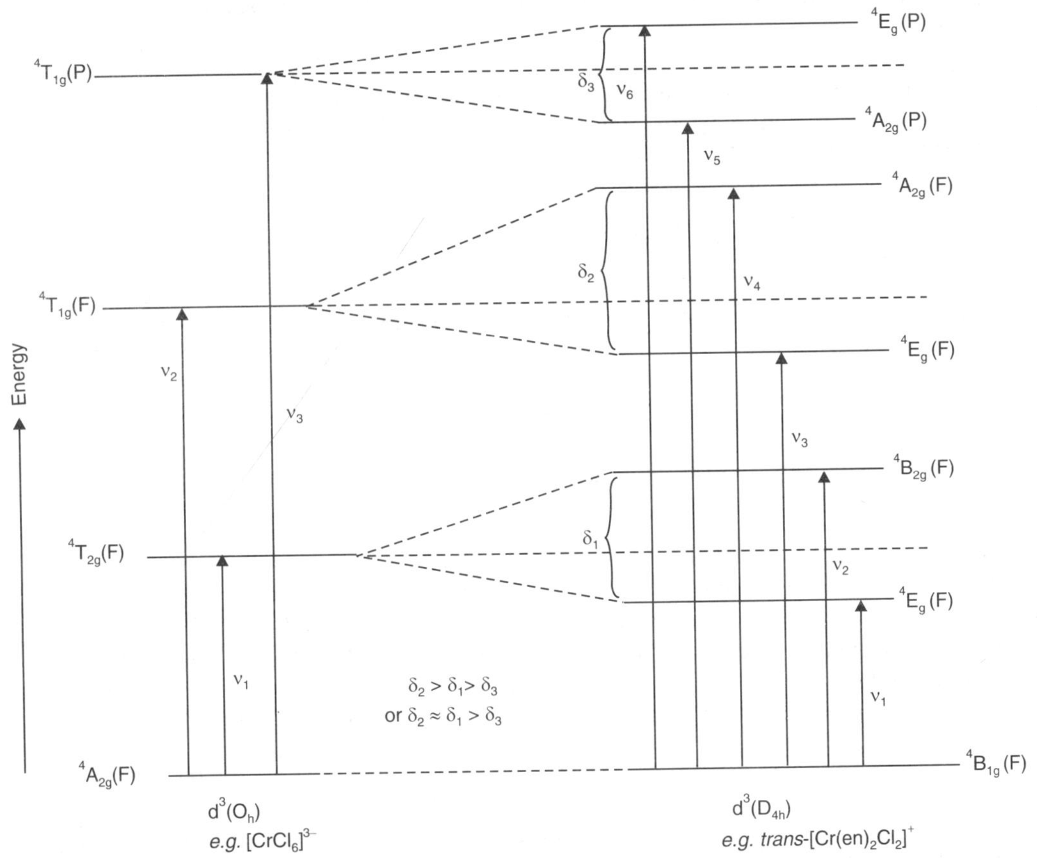

Fig. 7.14.5.5 Comparison of the spin-allowed transitions in the octahedral (O_h) and tetragonal (D_{4h}) complexes of d^3 system. (**Note:** There is a difference in opinion regarding the extent of splitting of the T_{1g} terms and the order of A_{2g} and E_g terms generated from the two T_{1g} terms. These are discussed in Sec. 7.14.4 and Fig. 7.14.4.3).

$\Delta_A \sim \Delta_B$ gives the measure of the difference of the crystal field splitting powers of the ligands *A* and *B*. For the *trans*–isomer, the splitting is twice the splitting occurring in the corresponding *cis*–isomer and the minus sign indicates that the splitting pattern is inverted in the *cis*–isomer compared to that in the *trans*–isomer. **We can expect that for the *trans*–isomer, the peaks due to transition to the $^4T_{1g}$**

and $^4T_{2g}$ levels (in the octahedral complex) will show the doublet character or two separate peaks but in the corresponding *cis*–isomer, this peak splitting is minimised because the *T* level does not split much in the *cis*–complex. Thus, we can distinguish the *cis*– and *trans*– isomers by comparing their ligand field spectra. *It may be mentioned that the* ***peaks are more intense in the cis–complex*** *lacking in the centre of symmetry.* In the ***noncentrosymmetric cis-isomers,*** the electronic transitions are more allowed. *It gives also a better d–p mixing in the cis–complex compared to that in the trans–isomer having the centre of symmetry.*

If the ligands *A* and *B* occupy the positions close to each other in the spectrochemical series, *i.e.* $\Delta_A \sim \Delta_B$ is quite small, then analysis of the ligand field spectra cannot differentiate the *cis*– and *trans*– isomers. In fact, the geometrical isomers of **$[Cr(en)_2F_2]^+$** can be distinguished in this way but the isomers of **$[Cr(NCS)_2(OH_2)_4]^+$** cannot be distinguished in this way. This aspect has been discussed in Sec. 7.15.2a.

Table 7.14.5.2 Comparison of the spectral characteristics of *cis*- and *trans*- isomers of **$[Cr(en)_2F_2]^+$** (*cf.* F. Teixidor et al. *J. Chem. Edu.*, **64,** 461, 1987).

	Transitions in terms of O_h symmetry		
Complex	$^4A_{2g} \xrightarrow[\text{(nm)}]{\nu_1} {}^4T_{2g}(F)$	$^4A_{2g} \xrightarrow[\text{(nm)}]{\nu_2} {}^4T_{1g}(F)$	$^4A_{2g} \xrightarrow[\text{(nm)}]{\nu_3} {}^4T_{1g}(P)$
cis-$[Cr(en)_2F_2]^+$	517	378	235
trans-$[Cr(en)_2F_2]^+$	530[a], 465[b]	397[c], 351[d]	248[e], 229[f]

$$\underbrace{a \Rightarrow {}^4B_{1g} \longrightarrow {}^4E_g,\ b \Rightarrow {}^4B_{1g} \longrightarrow {}^4B_{2g}}_{\text{Components of } \nu_1 \text{ in terms of } D_{4h} \text{ symmetry}};\ \underbrace{c \Rightarrow {}^4B_{1g} \longrightarrow {}^4E_g,\ d \Rightarrow {}^4B_{1g} \longrightarrow {}^4A_{2g}}_{\text{Components of } \nu_2 \text{ in terms of } D_{4h} \text{ symmetry}};$$

$$\underbrace{e \Rightarrow {}^4B_{1g} \longrightarrow {}^4A_{2g},\ f \Rightarrow {}^4B_{1g} \longrightarrow {}^4E_g}_{\text{Components of } \nu_3 \text{ in terms of } D_{4h} \text{ symmetry}}$$

Thus it is evident that though the peaks undergo splitting for the *trans*-isomer but the corresponding peaks of the *cis*-compound undergo no splitting, *i.e.* O_h symmetry is practically maintained for the *cis*-isomer. The ν_1 transition corresponds to the $10Dq$ value (*i.e.* 517 nm ≡ 19,342 cm^{-1}) for the *cis*-isomer). By applying the **rule of average environment,** $10Dq$ value for *cis*-$[Cr(en)_2F_2]^+$ can be calculated as follows:

$$\nu_1 = 14{,}900 \text{ cm}^{-1} \text{ for } [CrF_6]^{3-};\ \nu_1 = 21{,}800 \text{ cm}^{-1} \text{ for } [Cr(en)_3]^{3+}$$

$$\nu_1 \equiv 10Dq \text{ of } cis\text{-}[Cr(en)_2F_2]^+ = \frac{2\times 14{,}900 + 4\times 21{,}800}{6} = 19{,}500 \text{ cm}^{-1}$$

Thus the calculated ν_1 value is in good conformity with the experimental value.

(F) Spin-forbidden transitions in d^3 (*e.g.* Cr^{III})–system: The spin-forbidden transitions (quartet to doublet) in the Cr(III)–complexes are quite important to understand their behaviour in **photochemistry** (*cf.* Chapter 6) and in ***ruby laser*****.** Obviously, these spin-forbidden transitions are of low intensity and these are not important in interpreting the colour and absorption spectra of the Cr(III)–complexes.

In this connection, we are to consider the crystal field splitting of the ^2G term (of free ion) along with the ^4F and ^4P terms. The ^2G term splits as follows:

$$^2G(\text{free ion, } d^3) \xrightarrow{O_h \text{ field}} {}^2E_g,\ {}^2T_{1g},\ {}^2T_{2g},\ {}^2A_{1g}$$

These doublet terms obviously arise from the *spin-pairing* in the t_{2g} level. The $^2E_g(G)$ and $^2T_{1g}(G)$ states are of almost identical energy and thse two states are generally jointly represented by $^2E_g(G)$.

The ^2G term of free ion arises from the **low-spin configuration** of the t_{2g} level (*i.e.*, ⇅ ↑ –) where one of the t_{2g} orbitals is occupied by a pair of electrons and the 3rd electron remains in an unpaired condition in one of the remaining two t_{2g} orbitals. In the octahedral field, the ^2G term splits into 4-states: $^2E_g(G)$ and $^2T_{1g}(G)$ of almost the same energy, and $^2T_{2g}(G)$, $^2A_{1g}(G)$ of higher energy. The energy order is:

$$\underbrace{^2E_g(G),\ ^2T_{1g}(G)}_{(\equiv\, ^2E_g(G))} < {}^2T_{2g}(G) < {}^2A_{1g}(G).$$

The $^2E_g(G)$ and $^2T_{1g}(G)$ states are of almost the same energy and these are jointly represented by $^2E_g(G)$. The energy level diagram for both the quartet and doublet states of a d^3 (octahedral) system are shown in Fig. 7.14.5.6.

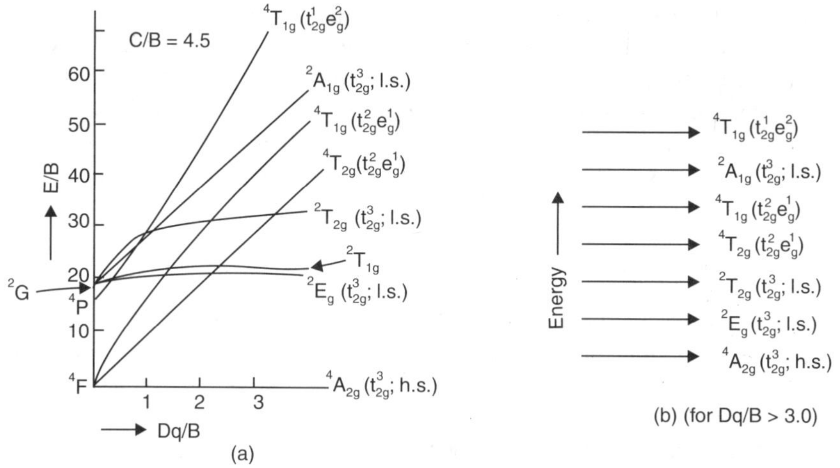

Fig. 7.14.5.6 (a) Tanabe-Sugano diagram of a d^3–octahedral system. (b) Relative energy order of different energy states of a d^3–octahedral system ($^2T_{1g}$, $^2E_g \equiv {}^2E_g$).

The characteristic features of the ^2E_g, $^2T_{1g}$ and $^2T_{2g}$ states are that **they run parallel to the ground state** ($^4A_{2g}$). These doublet states arise due to the rearrangement of electrons in the t_{2g} level. This is why, transition from the $^4A_{2g}$ state to these doublet states requires simply the electron rearrangement in the t_{2g} level and the transition energy does not depend on Δ which gives the measure of $t_{2g} - e_g$ separation. ***The metal-ligand bond distance remains more or less the same for both the ground $^4A_{2g}$ and excited ^2E_g, $^2T_{1g}$ and $^2T_{2g}$ states as the electronic configuration in the e_g level remains unchanged.*** Consequently, peak broadening due to the metal-ligand bond vibration does not occur in such spectral transitions (*cf.* Fig. 7.7.3.1 and Secs. 7.7.2–3). Δ value depends on the metal-ligand distance and a symmetric vibration (called **breathing mode**) (*i.e.* periodic contraction and elongation of the *M* — *L* bond) causes a continuous change of Δ within a limit. This is why, when energy of the two states (between which the spectral transition occurs) depends on Δ in different ways, the peak broadening will occur due to the *M* — *L* bond vibration. The mechanism of peak broadening by the *M* — *L* bond vibration has been discussed in detail in Secs. 7.7.2–3.

The following **spin-forbidden transitions** do not experience any peak broadening phenomenon due to the metal-ligand bond vibration. **These are very sharp peaks.**

$$^4A_{2g} \to {}^2E_g;\ {}^4A_{2g} \to {}^2T_{1g};\ {}^4A_{2g} \to {}^2T_{2g}$$

Photochemistry of the Cr(III) Complexes: Importance of the Excited ^2E_g Level

The excited ^2E_g state plays a crucial role in controlling the photochemistry of the Cr(III) complexes. Photoexcitation gives a significant amount of population density in the ^2E_g state whose life time is relatively higher (*ca.* 10^{-3}–10^{-4} s) because the transition $^2E_g \to {}^4A_{2g}$ is spin-forbidden. Photochemical excitation of the ground $^4A_{2g}$ state to the quartet states will cause some population to the ^2E_g state. Nonradiative deactivation of the higher quartet state will give the population density in the ^2E_g state. This ^2E_g state (low-spin t_{2g}^3 configuration) **bears a vacant t_{2g} orbital capable of receiving a nucleophilic attack and it is the cause of reactivity.** This aspect has been discussed in Chapter 6 in detail.

(G) **Role of Cr(III) in ruby laser: Role of the excited ^2E_g level (*cf.* Sec. 7.9.5):** In a ruby laser, a small amount of Cr(III) is embedded in α–Al_2O_3 lattice (corundum) in which Cr(III) experiences an octahedral field of 6 oxide O–sites. When a ruby crystal is irradiated to the radiation of an appropriate wavelength to excite Cr(III) from the $^4A_{2g}$ ground state to the $^4T_{2g}(F)$ state and partly to the $^4T_{1g}(F)$ state, the $^4T_{2g}(F)$ state deexcites through an **intersystem crossing** (ISC) and comes to the ^2E_g state which possesses a relatively longer half-life time. The ISC process involves the spin-orbit coupling mechanism (Sec. 7.9.5). Thus the paths,

$$^4A_{2g}(F) \xrightarrow{h\nu} {}^4T_{2g}(F) \xrightarrow[\text{(lattice vibration)}]{\text{dexcitation }(ISC)} {}^2E_g(G)$$

$$^4A_{2g}(F) \xrightarrow{h\nu} {}^4T_{1g}(F) \xrightarrow[\text{conversion}]{\text{internal}} {}^4T_{2g}(F) \xrightarrow[\text{(intersystem crossing)}]{\text{Dexcitation}} {}^2E_g(G).$$

increase the population density (*i.e.* **population inversion**) in the excited $^2E_g(G)$ state. The $^4T_{2g}$(F) term corresponds to the $t_{2g}^2e_g^1$ configuration; the $^4T_{1g}(F)$ term also corresponds to $t_{2g}^2e_g^1$ configuration; and the ground state term corresponds to the t_{2g}^3 configuration. It may be noted that the same electronic configuration is attained for both the $^4T_{2g}(F)$ and $^4T_{1g}(F)$ states for the relatively stronger crystal field (Fig. 7.10.2) and it happens so when Cr^{III} is doped in the corundum crystal; this aspect is discussed later. The transition to $^4T_{1g}(F)$ needs the absorption of violet light while the transition to $^4T_{2g}(F)$ needs the absorption of green light. **The absorption of violet and green light makes the ruby red.** The excited $^4T_{1g}(F)$ state is converted into the lower excited state $^4T_{2g}(F)$ through an **internal conversion.** The $^4T_{2g}(F)$ state is close to the ^2E_g state but they are of different spin multiplicities.

The excited $^4T_{2g}(F)$ state is converted through the lattice vibration to another excited state ^2E_g of different spin multiplicity. This conversion is described as the **nonradiative intersystem crossing (ISC)**. In the case of ruby, **the energy difference between the excited states $^4T_{2g}(F)$ and $^2E_g(G)$ matches appropriately the energy of a particular lattice vibrational mode.** This allows the ISC efficiently. Thus, the population density in the ^2E_g level is increased and this excited state acts as the **reservoir of energy.** This excited state can experience a radiative decay to the ground state through a spin-forbidden transition (*i.e.* **phosphorescence, cf. Sec. 7.9.5**). The return from the ^2E_g state to the ground state $^4A_{2g}$ is spin-forbidden. This enhances the life time of the ^2E_g state and at the cost of the exciting radiation, population density in the ^2E_g state may be more than that in the ground state. ***It leads to a population inversion.*** When the $^2E_g(G)$ state comes back to the ground state $^4A_{2g}(F)$, it gives the **monochromatic radiation** (16,000 cm^{-1}, 627 nm) because both the $^2E_g(G)$ and $^4A_{2g}(F)$ run parallel in the Tanabe-Sugano diagram.

Special colours through the isomorphous substitution of Al^{3+} in corundum (Al_2O_3)

The trivalent cations like Al^{3+}, Ti^{3+} can lead to the isomorphous substitution of Al^{3+}, in Al_2O_3. But their sizes slightly differ: $r(Al^{3+}) = 67$ pm; $r(Cr^{3+}) = 75$ pm; $r(Ti^{+3}) = 81$ pm. Thus, Cr^{3+} and Ti^{3+} are larger than Al^{3+} and consequently when Cr^{3+} and Ti^{3+} are placed in the positions of Al^{3+} in corundum, the Cr^{III} — O and Ti^{III} — O distances are shorter than the normal Cr^{III} — O and Ti^{III} — O distances. ***Because of this shorter M — O distance, Cr^{3+} and Ti^{3+} experience a larger crystal field splitting than the normal crystal field splitting in presence of the O^{2-} ligands.*** This is why, Cr^{3+} in corundum gives the unusual *red colour of ruby* (*cf.* Cr_2O_3 green). Similarly Ti^{3+} in corundum gives the unsual *blue colour of gemstone sapphire* (*cf.* $Ti(aq)^{3+}$ violet colour). This explains the ***red colour of ruby** and **blue colour of gemstone sapphire.*** When Cr^{3+} is placed into the beryl, $Al_2Be_3[Si_6O_{18}]$ in the positions of Al^{3+}, the Cr — O distance is longer than the Cr — O distance in ruby. Thus Cr(III) in beryl experiences a weaker crystal field splitting. It explains the ***green colour of gemstone emerald.***

The spin-forbidden transition, $^2E_g \rightarrow {^4A_{2g}}$ gives the **red emission.** Both the ^2E_g (low-spin t_{2g}^3) and $^4A_{2g}$ (high-spin t_{2g}^3) terms correspond to t_{2g}^3 configuration but spin orientations of the electrons in the two states are different. As the transition leads to the rearrangement of electrons in the t_{2g} level, the transition energy is independent of crystal field strength measuring the energy difference between the t_{2g} and e_g level. **In fact, the spin-forbidden transition, $^2E_g \rightarrow {^4A_{2g}}$ in many octahedral Cr(III) complexes gives the phosphorescence of red light emission as in ruby emission.** If the ligands are rigid then life-time of the ^2E_g term may be several microseconds in solution. It happens so in the case of $[Cr(bpy)_3]^{3+}$.

In fact, return to the ground $^4A_{2g}(F)$ state from the $^2E_g(G)$ state is accompanied by *a stimulated emission of monochromatic radiation in phase with the stimulating radiation.* This gives the **laser beam.**

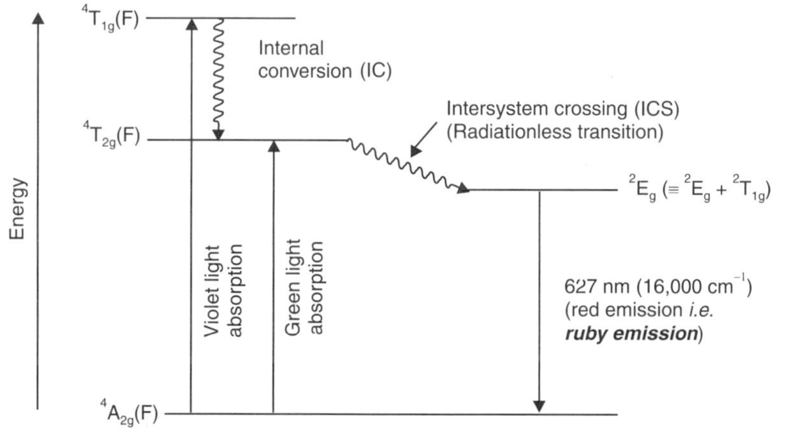

Fig. 7.14.5.7 Role of the ^2E_g level in **ruby emission** for the octahedral Cr(III) complexes in terms of Jablonski diagram. ^2E_g, $^2T_{1g}$ = ^2E_g (*cf.* Fig. 7.14.5.6).

7.14.6 Six-Coordinate High-spin Complexes of 3d^7–Systems

Co(II) complexes are the main examples of this group. In fact, Co(II) can adopt both the high-spin complexes ($t_{2g}^5\, e_g^2$, $^4T_{1g}$ term) and low-spin complexes ($t_{2g}^6\, e_g^1$, ^2E_g term) depending on the relative values of $10Dq$ and pairing energy. In fact, only for the very strong field ligands ($10Dq \rangle$ 15000 cm^{-1}), the low-spin complexes of Co(II) may be formed. However, most of the six-coordinate Co(II) complexes are the high-spin complexes.

For the high-spin d^7–system, Orgel diagram is similar to that of the d^2 system and the ground state is $^4T_{1g}$.

The spin-allowed transitions are:

$$^4T_{1g}(F) \xrightarrow{\nu_1} {}^4T_{2g}(F),\ \nu_1 = 8Dq + x' \approx 8Dq$$

$$^4T_{1g}(F) \xrightarrow{\nu_2} {}^4A_{2g}(F),\ \nu_2 = 18Dq + x' \approx 18Dq$$

$$^4T_{1g}(F) \xrightarrow{\nu_3} {}^4T_{1g}(P),\ \nu_3 = 6Dq + 15B' + 2x' \approx 6Dq + 15B'$$

(x' is quite negligible for the weak field ligands)

$\nu_2, \nu_3 > \nu_1$; ν_2 > or < ν_3; depending on the $10Dq$ values.

$$\nu_2 - \nu_1 = 10Dq,\ 15B' = \nu_3 + \nu_2 - 3\nu_1$$

Here it is worth mentioning the transition (ν_2), $^4T_{1g}(F) \rightarrow {}^4A_{2g}(F)$ denotes the transition, $t_{2g}^5 e_g^2 \rightarrow t_{2g}^3 e_g^4$, *i.e.* a two-electron transition. **This is a spin-forbidden transition.** This is why, it is a weak band.

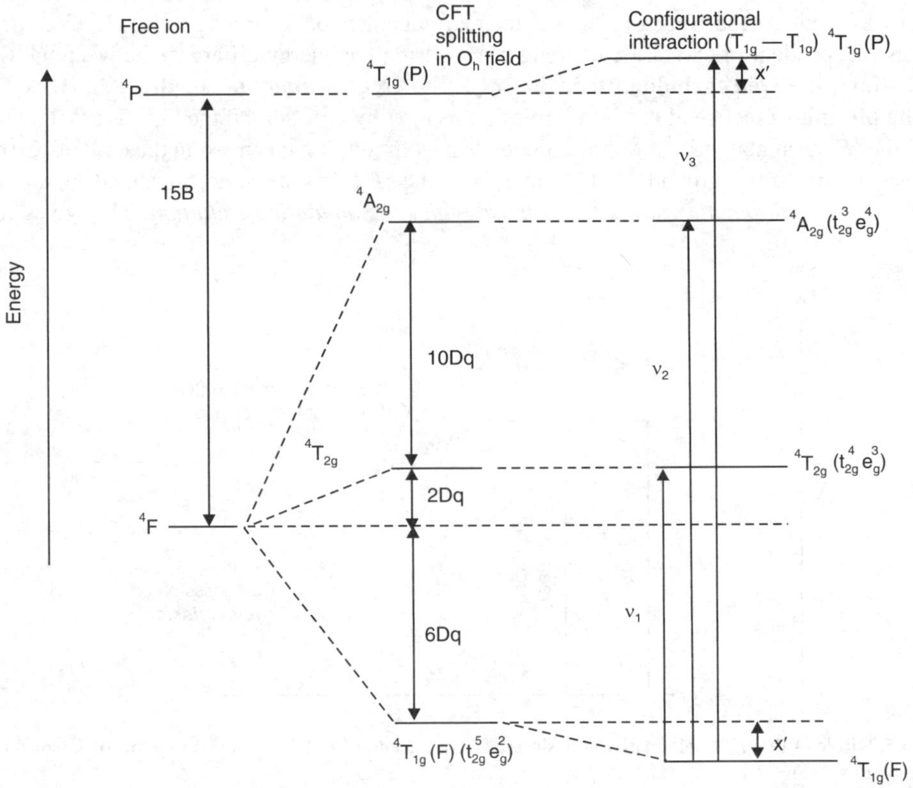

Note: $^4T_{1g}(P)$ and $^4A_{2g}(F)$ states may cross each other at a particular $10Dq$ value. When the $10Dq$ value crosses this particular value, $^4A_{2g}$ becomes of higher energy compared to that $^4T_{1g}(P)$. x' indicates the configurational interaction energy for the interaction between the $^4T_{1g}(P)$ and $^4T_{1g}(F)$ states.

Fig. 7.14.6.1 Energy level diagram of the high-spin d^7–systems. **Note:** The $^4T_{1g}(P)$ and $^4A_{2g}(F)$ states may intersect depending on the Dq values.

(A) **Analysis of the ligand field spectra of $[Co(OH_2)_6]^{2+}$:** The absorption spectrum of light pink coloured $[Co(OH_2)_6]^{2+}$ gives a well defined weak peak at 8,100 cm^{-1} and a broad band at about 20,000 cm^{-1}. **Analysis of the broad band indicates the presence of three overlapping peaks at** 16,000 cm^{-1}, 19,400 cm^{-1} and 21,600 cm^{-1}. The peak assignments are as follows:

$^4T_{1g}(F) \rightarrow {}^4T_{2g}(F)$, $\nu_1 = 8{,}100$ cm^{-1}, ($\varepsilon = 1.3$)

$^4T_{1g}(F) \rightarrow {}^4A_{2g}(F)$, $\nu_2 = 16{,}100$ cm^{-1},

$^4T_{1g}(F) \rightarrow {}^4T_{1g}(P)$, $\nu_3 = 19{,}400$ cm^{-1}, ($\varepsilon = 4.8$)

The peak at 21,600 cm^{-1} is believed to be originated either due to the spin-orbit coupling effect (*i.e.* splitting of the T–term, *i.e.* excited $^4T_{1g}$(P) state) or the spin-forbidden transition to a doublet state.

(Note: Some authors believe that the broad band at about 20,000 cm^{-1} consists of two peaks not three peaks. They suggest the following two transitions:

$^4T_{1g}(F) \rightarrow {}^4A_{2g}(F)$, 19,600 cm^{-1}

$^4T_{1g}(F) \rightarrow {}^4T_{1g}(P)$, 21,600 cm^{-1})

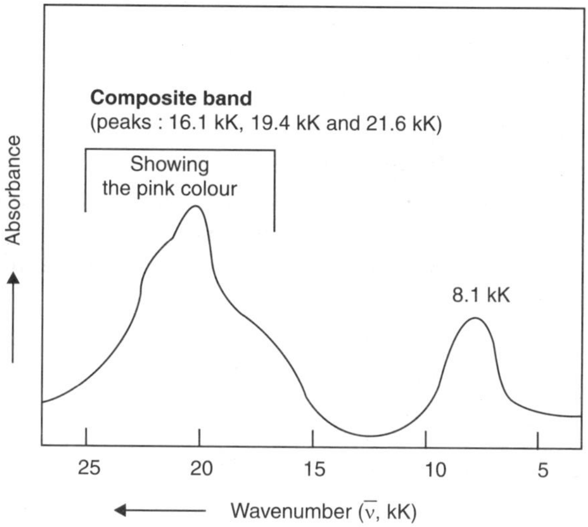

Fig. 7.14.6.2 Electronic absorption spectrum of $[Co(OH_2)_6]^{2+}$.

The said ν_2 (16,100 cm^{-1}) and ν_3 (19,400 cm^{-1}) peaks are closely spaced and they overlap. The ν_2–peak (2*e* transition, **a spin-forbidden transition**) earns the intensity due to mixing with the ν_3-spin-allowed transition (*cf.* **intensity stealing effect**). Close positions of the ν_2 and ν_3 peaks indicate that in $[Co(OH_2)_6]^{2+}$, the two states, *i.e.* $^4A_{2g}(F)$ and $^4T_{1g}(P)$ are in the **cross-over region:**

Now let us try to verify the assignments in terms of the Tanabe-Sugano diagram of the d^7–system.

$$\frac{\nu_3}{\nu_1} = \frac{^4T_{1g}(F) \longrightarrow {}^4T_{1g}(P)}{^4T_{1g}(F) - {}^4T_{2g}(F)} = \frac{19{,}400}{8{,}100} \approx 2.40$$

The above ratio fits in the corresponding Tanabe-Sugano diagram for $Dq/B = 1.13$ (Fig. 7.10.4). This gives the following result.

$$\frac{^4T_{1g}(F) \longrightarrow {}^4T_{1g}(P)}{B'} = \frac{\nu_3}{B'} = 23.54$$ (obtained from the Tanabe-Sugano diagram for $Dq/B = 1.13$).

$$\frac{\nu_3}{B'} = \frac{19{,}400}{B'} = 23.54 \ i.e.\ B' = \frac{19{,}400}{23.54} = 824 \text{ cm}^{-1} \left(cf.\ B_0 = 971 \text{ cm}^{-1}\right)$$

$$\beta = \frac{B'}{B_0} = \frac{824}{971} = 0.85 \ i.e.\ B' \text{ is reduced to about 85\% of the free ion value } (B_0).$$

$$\frac{Dq}{B'} = 1.13 \text{ or, } Dq = B' \times 1.13 = 824 \times 1.13 = 931 \text{ cm}^{-1}$$

i.e. $10Dq = 9{,}310 \text{ cm}^{-1}$

From the experimental data we get: $\frac{\nu_2}{\nu_1} = \frac{16{,}100}{8{,}100} = 1.99; \frac{\nu_3}{\nu_2} = \frac{19{,}400}{16{,}100} = 1.20$

For $\frac{Dq}{B'} = 1.13$, the calculated ratios (from the Tanabe-Sugano curve; *cf.* Fig. 7.10.4) are: $\frac{\nu_2}{\nu_1} = 2.14$, $\frac{\nu_3}{\nu_2} = 1.13$. These are in good agreement with the band assignment data.

We can calculate $10Dq$ as follows:

$$10Dq = \nu_2 - \nu_1 = 16{,}000 - 8{,}100 = 7{,}900 \text{ cm}^{-1}$$

This value is approximately close to 9,310 cm^{-1}. *But the agreement is not very good.* However, the given spectral assignments give a reasonable account of the spectrum.

From the above peak assignments, we can calculate the configurational interaction energy (x') and other parameters as follows:

$$\nu_1 = 8{,}100 \text{ cm}^{-1} = 8Dq + x'$$

$$\nu_2 = 16{,}000 \text{ cm}^{-1} = 18Dq + x'$$

$$\nu_3 = 19{,}400 \text{ cm}^{-1} = 6Dq + 15B' + 2x'$$

$$10Dq = \nu_2 - \nu_1 = (16{,}000 - 8{,}100) \text{ cm}^{-1} = 7{,}900 \text{ cm}^{-1};\ i.e.\ Dq = 790 \text{ cm}^{-1}$$

$$\nu_1 = 8{,}100 \text{ cm}^{-1} = 8Dq + x' = 8 \times 790 + x'$$

i.e. $x' = 1780 \text{ cm}^{-1}$

$$\nu_3 = 19{,}400 \text{ cm}^{-1} = 6 \times 790 + 15B' + 2 \times 1780 = 15B' + 8{,}300$$

or, $15B' = 11{,}100 \text{ cm}^{-1}$, *i.e.* $B' = 740 \text{ cm}^{-1}$

B. **Analysis of the ligand field spectra of $[CoF_6]^{4-}$ in the crystals of $KCoF_3$:** The spectral transitions are:

$$^4T_{1g}(F) \xrightarrow{\nu_1} {}^4T_{2g}(F),\ \nu_1 = 8Dq + x' = 7{,}150 \text{ cm}^{-1}$$

$$^4T_{1g}(F) \xrightarrow{\nu_2} {}^4A_{2g}(F),\ \nu_2 = 18Dq + x' = 15{,}200 \text{ cm}^{-1}$$

$$^4T_{1g}(F) \xrightarrow{\nu_3} {}^4T_{1g}(P),\ \nu_3 = 6Dq + 15B' + 2x' = 19{,}200 \text{ cm}^{-1}$$

Besides, the above three spin-allowed transitions, there are some **spin-forbidden transitions** (quartet to doublet). Analysis of the above results yields:

$$\nu_2 - \nu_1 = 10Dq = (15{,}200 - 7{,}150) \text{ cm}^{-1} = 8{,}050 \text{ cm}^{-1},\ i.e.\ Dq = 805 \text{ cm}^{-1}$$

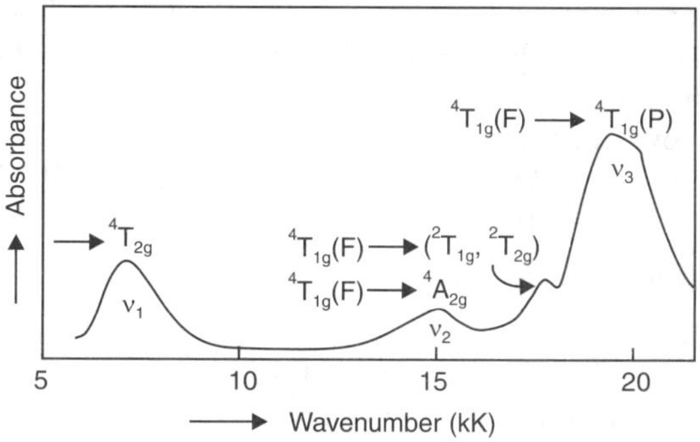

Fig. 7.14.6.3 Absorption spectrum of $[CoF_6]^{4-}$ in solid crystals of $KCoF_3$.

Calculation of $10Dq$ in this way is not affected by configurational interaction.

$\nu_1 = 8Dq + x'$, or, $x' = \nu_1 - 8Dq = (7{,}150 - 8 \times 805)\ \text{cm}^{-1} = 710\ \text{cm}^{-1}$

$$\nu_3 = 15B' + 6Dq + 2x'.\ \text{or}\ B' = \frac{\nu_3 - 6Dq - 2x'}{15} = \frac{(19{,}200 - 6 \times 805 - 2 \times 710)\text{cm}^{-1}}{15} = 863\ \text{cm}^{-1}$$

$$\beta = \frac{B'}{B_0} = \frac{863}{970} = 0.88,\ \textit{i.e. } B \text{ is reduced by } 12\%..$$

Here it is worth mentioning that Co(II) (h.s.) is highly **labile** and the species like $[CoF_6]^{4-}$ cannot be studied in solution.

C. **More rigorous analysis of the spectrum of high-spin octahedral d^7–systems:** Energies of the $^4T_{2g}(F)$ and $^4A_{2g}(F)$ terms do not depend on the configurational interaction.

$$^4T_{2g}(F) \Rightarrow t_{2g}^4 e_g^3;\ ^4A_{2g}(F) \Rightarrow t_{2g}^3 e_g^4;\ ^4T_{1g}(P) \Rightarrow t_{2g}^4 e_g^3;\ ^4T_{1g}(F) \Rightarrow t_{2g}^5 e_g^2$$

$$E\,[^4T_{2g}(F)] = +2Dq,\ E\,[^4A_{2g}(F)] = +12Dq.$$

The terms $^4T_{1g}(F)$ and $^4T_{1g}(P)$ experience the configurational interaction and energy matrix for the two $^4T_{1g}$ terms is given by:

$$\begin{vmatrix} -6Dq - E & 4Dq \\ 40Dq & 15B' - E \end{vmatrix} = 0$$

$$E = \frac{1}{2}\left[15B' - 6Dq \pm \left\{225B'^2 + 100(Dq)^2 + 180DqB'\right\}^{1/2}\right]$$

$$\nu_1\left[^4T_{1g}(F) \longrightarrow {}^4T_{2g}(F)\right] = \frac{1}{2}\left[\Delta - 15B' + \left(225B'^2 + 18B'\Delta + \Delta^2\right)^{1/2}\right]$$

$$\nu_2\left[^4T_{1g}(F) \longrightarrow {}^4A_{2g}(F)\right] = \frac{1}{2}\left[3\Delta - 15B' + \left(225B'^2 + 18B'\Delta + \Delta^2\right)^{1/2}\right]$$

$$\nu_3\left[^4T_{1g}(F) \longrightarrow {}^4T_{1g}(P)\right] = \left[\left(225B'^2 + 18B'\Delta + \Delta^2\right)^{1/2}\right]$$

$$\nu_3 - 2\nu_1 = 15B' - \Delta = 15B' - 10Dq$$

The ν_2–band involves a two electron transition, *i.e.* $t_{2g}^5 e_g^2 \rightarrow t_{2g}^3 e_g^4$ and this spin-forbidden transition is of poor intensity. **This is why, for the spectral characterisation of the high-spin d^7 system, the ν_1 and ν_3 bands are quite important.**

Let us illustrate the above equations to analyse the spectrum of high-spin octahedral complex, [Co(salicylaldehydebenzothiazolecarbohydrazide)$(OH_2)_2$] having $\nu_1 = 9100$ cm^{-1} and $\nu_3 = 18,500$ cm^{-1}.

$$\nu_3 - 2\nu_1 = 15B' - 10Dq$$

or, $18,500 - 2 \times 9,100 = 15B' - 10Dq$;

or, $18,500 - 18,200 = 15B' - 10Dq$ or, $15B' = 10Dq + 300$; or $B' = \dfrac{10Dq + 300}{15}$

$2\nu_1 = 10Dq - 15B' + [225B'^2 + 100(Dq)^2 + 180DqB']^{1/2}$.

Putting the values of ν_1 (= 9,100) and B' we get:

$$32(Dq)^2 + 960Dq - 34,21,600 = 0$$

or, $Dq \approx 1019$ cm^{-1} (taking the positive value of Dq).

$$B' = \frac{10Dq + 300}{15} = \frac{10190 + 300}{15} = \frac{10,490}{15} \approx 699 \text{ cm}^{-1}$$

i.e. $\beta = \dfrac{B'}{B_0} = \dfrac{699}{971} = 0.72.$

7.14.7 Six-Coordinate High-spin Complexes of d^4-System (*cf.* Sec. 3.11.4)

Complexes of Cr(II) and Mn(III) (with the weak field ligands) are the representative examples of this group. The ground state electronic configuration $t_{2g}^3 e_g^1$ indicates that it experiences the *z*-out Jahn-Teller distortion. The free ion term is ^5D which splits into the ^5E_g (lower) and $^5T_{2g}$ (higher) states in a weak octahedral field. If no distortion is considered, then we have the spectral transition:

$$^5E_g \rightarrow {}^5T_{2g}.$$

However, Jahn-Teller splitting of the terms will occur as in the case of Cu^{2+}(d^9-system). The energy level diagram for a d^4 (high-spin) system is given in Fig. 7.14.7.1.

The possible electronic transitions are shown in Fig. 7.14.7.1. These transitions are:

$$^5B_{1g} \rightarrow {}^5A_{1g}\ (\nu_1);\ {}^5B_{1g} \rightarrow {}^5B_{2g}\ (\nu_2);\ {}^5B_{1g} \rightarrow {}^5E_g\ (\nu_3).$$

Table 7.14.7.1 Spectral features of some d^4-(high-spin)-complexes.

Complex	$d_{z^2} \xrightarrow{\nu_1} d_{x^2-y^2}$ $^5B_{1g} \xrightarrow{\nu_1} {}^5A_{1g}$ **(cm^{-1})**	$d_{xy} \xrightarrow{\nu_2} d_{x^2-y^2}$ $^5B_{1g} \xrightarrow[(=10Dq)]{\nu_2} {}^5B_{2g}$ **(cm^{-1})**	$d_{xz}, d_{yz} \xrightarrow{\nu_3} d_{x^2-y^2}$ $^5B_{1g} \xrightarrow{\nu_3} {}^5E_g$ **(cm^{-1})**
$[Cr(OH_2)_6]^{2+}$	9,500	14,000	–
CrF_2 (solid crystal)	11,500	14,700	–
$[MnF_6]^{3-}$	9,000	17,400	19,600
*$[Mn(DMSO)_6]^{3+}$ (O-donor)	14,500	17,000	20,000
$[Mn(OH_2)_6]^{3+}$	–	–	21,000

* There are two types ν_{S-O} ir-stretching frequencies. This aspect has been discussed in Secs. 3.11.3–4 in detail in terms of Jahn-Teller effect.

The blue coloured aqua complex $[Cr(OH_2)_6]^{2+}$ (chromus solution **unstable with respect to oxidation**) shows a peak at about 14,000 cm^{-1} which is generally considered as:

$$^5E_g \xrightarrow{\nu} {}^5T_{2g} \quad (\nu = 14{,}000 \text{ cm}^{-1} = 10Dq).$$

But the finer analysis shows another band at 9,500 cm^{-1}. The peak assignments are:

$$\textbf{[Cr(OH}_2\textbf{)}_6\textbf{]}^{2+}\textbf{:}\ {}^5B_{1g} \xrightarrow{\nu_1} {}^5A_{1g} \ (9{,}500 \text{ cm}^{-1}),$$

$$^5B_{1g} \xrightarrow{\nu_2} {}^5B_{2g} \ (14{,}000 \text{ cm}^{-1}).$$

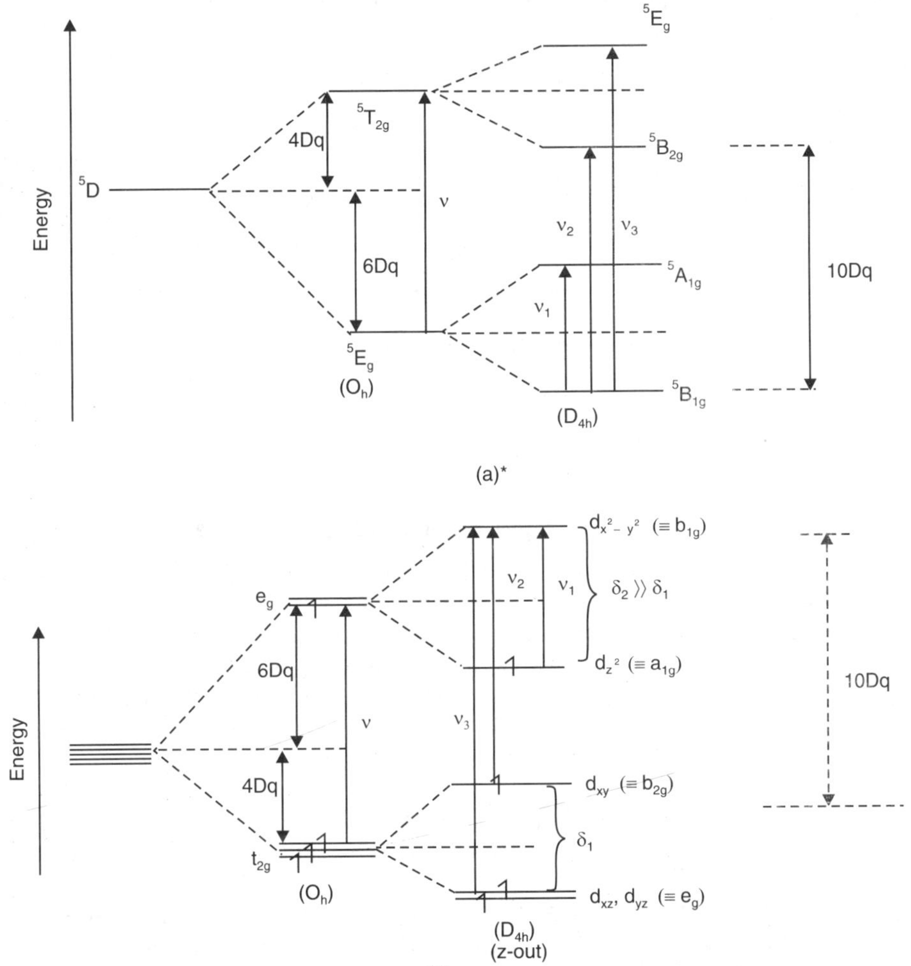

(b)

Relationships between the spectrocopic states and electronic configurations

$e_g^2 b_{2g}^1 a_{1g}^1 b_{1g}^0 \equiv {}^5B_{1g}$; $e_g^2 b_{2g}^1 a_{1g}^0 b_{1g}^1 \equiv {}^5A_{1g}$; $e_g^2 b_{2g}^0 a_{1g}^1 b_{1g}^1 \equiv {}^5B_{2g}$; $e_g^1 b_{2g}^1 a_{1g}^1 b_{1g}^1 \equiv {}^5E_g$

Fig. 7.14.7.1 Energy level diagram in the six-coordinate high-spin complexes of d^4-system. (a) Splitting of the spectroscopic states, (b) Splitting of *d*-orbitals *Further splitting of ^5E_g of D_{4h} into ^5A_g and ^5B_g gives the C_{2h} (rhombic) symmetry.

Probably, ν_2 ($d_{xy} \rightarrow d_{x^2-y^2}$) and ν_3 ($d_{xz}, d_{yz} \rightarrow d_{x^2-y^2}$) peaks are close in positions because splitting of the t_{2g} level is less important, *i.e.* $\delta_2 \rangle\rangle \delta_1$ (*cf.* splitting of the e_g level is more significant). **This is why it is difficult to characterize the ν_2 and ν_3 band separately.** Spectral characteristics of some d^4-systems (high-spin) are given in Table 7.14.7.1.

A regular octahedral geometry for the Cr(II) complexes should show a **single spin allowed transition** (*i.e.* $^5E_g \rightarrow {}^5T_{2g}$) but very often, the Cr(II) complex (high spin) experiences the distortion (giving rise to splitting of the energy levels) to show the multiple peaks which may overlap. However, the spectrum of $[CrCl_6]^{4-}$ ($CrCl_2$ in molten $AlCl_3$ at 230°C) shows a single absorption (as expected for the O_h symmetry) at 10.2 kK without any apparent splitting (Fig. 7.14.7.2).

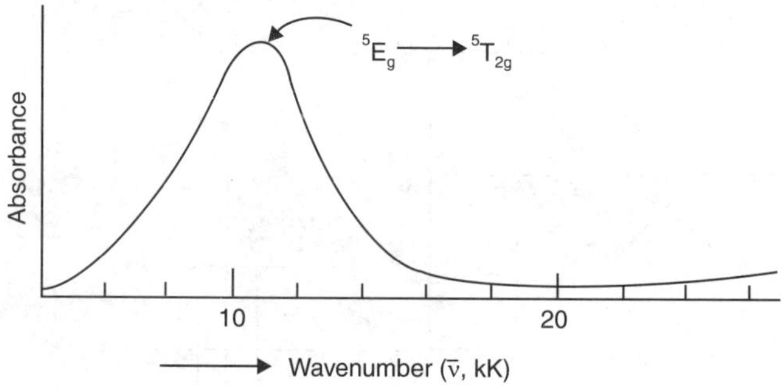

Fig. 7.14.7.2 Absorption spectrum of $[CrCl_6]^{4-}$ ($CrCl_2$ in molten $AlCl_3$ at 230°C and 5.5 alm pressure).

Here it may be mentioned that Cr(II) (d^4 system, h.s.) is highly **labile.** Consequently, the species like $[CrCl_6]^{4-}$ cannot be studied and chacterised in aqueous solution.

7.14.8 Six-Coordinate High-Spin Complexes of 3d⁶-System (*cf.* Sec. 3.11.4)

Fe(II) – complexes are the important examples in this group. **Two high-spin complexes of Co(III), *i.e.* $[CoF_6]^{3-}$ and $[CoF_3(OH_2)_3]$** are also known to belong to this group.

Both the ground state ($t_{2g}^4e_g^2$) and excited state ($t_{2g}^3e_g^3$) are prone to experience the J.T. distortion. Obviously, the excited state is more sensitive to the J.T. distortion because in the excited state, origin of the distortion lies at the e_g-level while in the ground state, origin of the distortion lies at the t_{2g} level.

If the high-spin complexes like $[Fe(OH_2)_6]^{2+}$, $[CoF_6]^{3-}$, etc. are supposed to have the perfectly octahedral geometry, we can predict the following spectral transition.

$$^5T_{2g}\left(t_{2g}^4e_g^2\right) \xrightarrow[(=10Dq)]{\nu} {}^5E_2\left(t_{2g}^3e_g^3\right)$$

However, in terms of Jahn-Teller distortion, there are **three possible spin allowed spectral transitions (Fig. 7.14.8.1)**:

the ν_1 (due to splitting of the t_{2g} level) lies in the ir-range, ν_2 and ν_3 lie in the visible range.

Very often, ν_2 and ν_3 overlap to produce a **broad peak with a shoulder.** But in some cases, the resultant broad peak may show a clear doublet structure as in $[CoF_6]^{3-}$.

$[Fe(OH_2)_6]^{2+}$ (pale green) shows a doublet peak (10,400 cm^{-1} and a shoulder at 8,300 cm^{-1}). It indicates the peak due to the transition, $^5T_{2g} \rightarrow {}^5E_g$ (in the hypothetical octahedral complex) is split into a doublet by about 2000 cm^{-1}. **$[CoF_3(OH_2)_3]$** shows a broad peak at about 13,000 cm^{-1}. For **$[CoF_6]^{3-}$**, the two well defined peaks appear at 11,800 cm^{-1} and 14,400 cm^{-1}. ***In such cases, conventionally 10Dq is determined from the average position of the two peaks.*** Thus, for **$[CoF_6]^{3-}$**, the

$10Dq$ value is taken as $\frac{1}{2}(14,400 + 11,800)\,\text{cm}^{-1} = 13,100\ \text{cm}^{-1}$. It is assumed that the **single ideal octahedral peak** of $[CoF_6]^{3-}$ arises at 13,100 cm^{-1}.

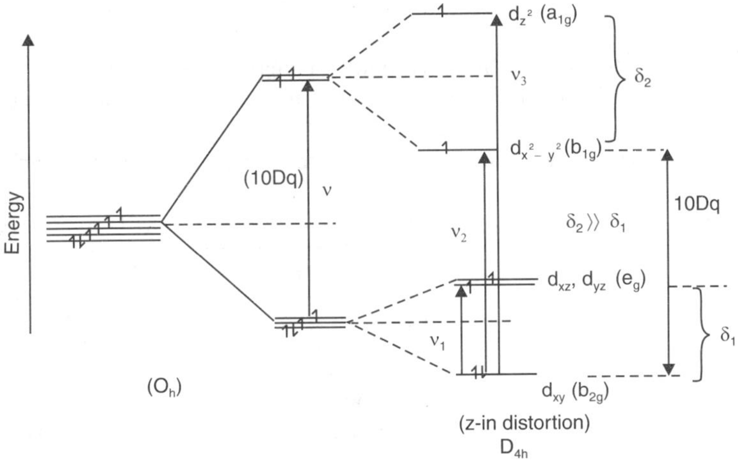

Spectroscopic terms and electronic configurations

$d_{xy}^2 (d_{xz}, d_{yz})^2 d_{x^2-y^2}^1 d_{z^2}^1 \equiv {}^5B_{2g}$;

$d_{xy}^1 (d_{xz}, d_{yz})^3 d_{x^2-y^2}^1 d_{z^2}^1 \equiv {}^5E_g$;

$d_{xy}^1 (d_{xz}, d_{yz})^2 d_{x^2-y^2}^2 d_{z^2}^1 \equiv {}^5B_{1g}$;

$d_{xy}^1 (d_{xz}, d_{yz})^2 d_{x^2-y^2}^1 d_{z^2}^2 \equiv {}^5A_{1g}$

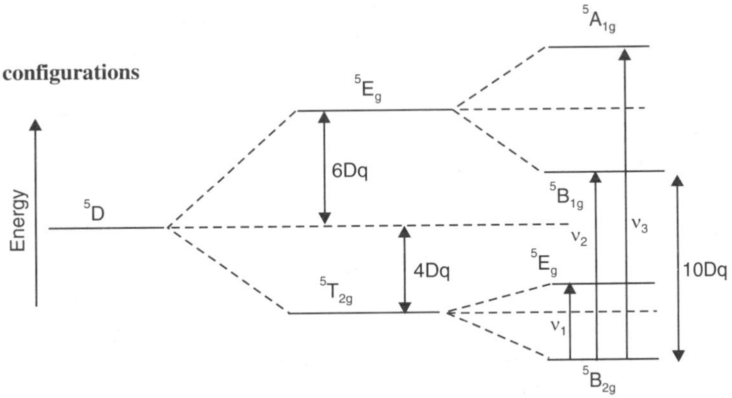

Fig. 7.14.8.1a Possible spectral transitions in high-spin six-coordinate complexes of d^6-system.

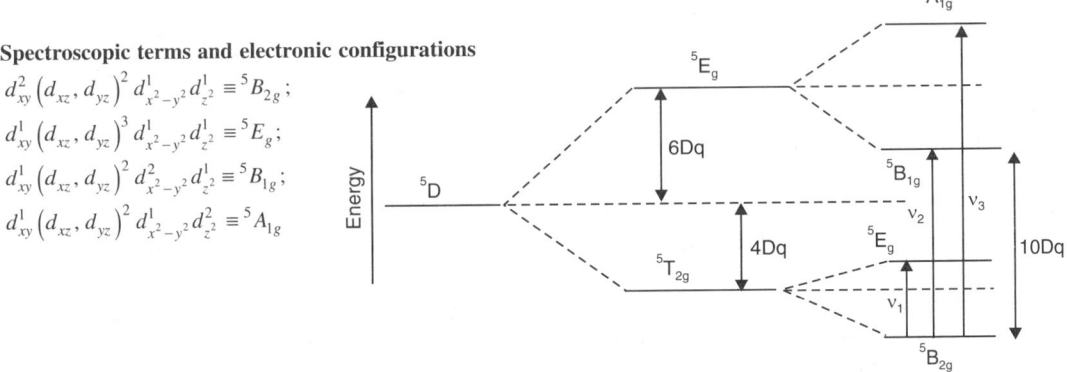

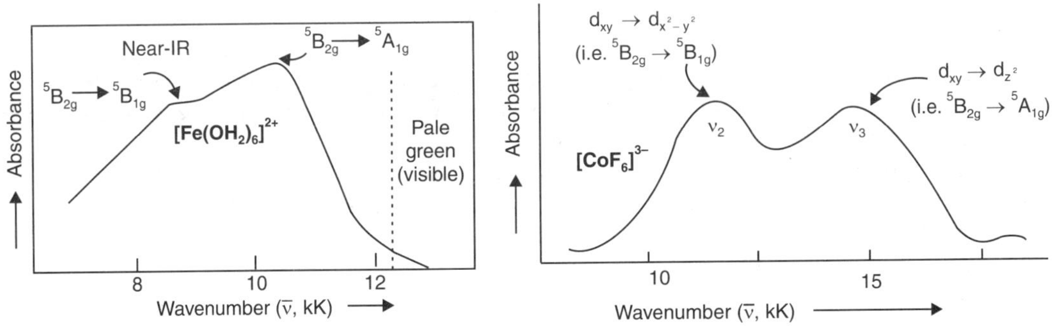

Fig. 7.14.8.1b Absorption spectra of $[Fe(OH_2)_6]^{2+}$ and $[CoF_6]^{3-}$.

Pale green **$[Fe(OH_2)_6]^{2+}$** shows a broad and weak absorption band at about 10,000 cm^{-1} which actually consists of two peaks at 10,400 cm^{-1} and 8,300 cm^{-1} (appearing as a shoulder) (J.T. effect). It leads to the 10Dq value as $\frac{1}{2}(10{,}400 + 8{,}300) \approx 9{,}400\ cm^{-1}$. The case may be compared with the d^1-systems, *e.g.* $[Ti(OH_2)_6]^{3+}$.

However, it may be pointed out that from the splitting pattern of the *d*-orbitals, the ν_2 peak (*i.e.* lower energy peak, *e.g.* 8,300 cm^{-1} for $[Fe(OH_2)_6]^{2+}$, 11,800 cm^{-1} for $[CoF_6]^{3-}$) should represent the 10*Dq* value.

Spectral features of some representative high-spin d^6-complexes are given in Table 7.14.8.1.

7.14.8.1 Spectral features of some high-spin Fe(II) and Co(III) complexes.

Complex	$^5T_{2g} \rightarrow {^5E_{2g}}$ *(cm^{-1}) (with a doublet structure)*	
	$d_{xy} \xrightarrow{\nu_2} d_{x^2-y^2}$ $^5B_{2g} \longrightarrow {^5B_{1g}}$	$d_{xy} \xrightarrow{\nu_3} d_{z^2}$ $^5B_{2g} \longrightarrow {^5A_{1g}}$
$[Fe(OH_2)_6]^{2+}$	8,300	10,400
$[FeCl_2, 4H_2O]$ *i.e.* $[FeCl_2(OH_2)_4]$	6,800	11,400
$[FeCl_2(phen)_2]$	8,470	10,510
$[FeCl_2(py)_4]$	8,550	10,300
$[FeBr_2(py)_4]$	7,750	10,650
$[CoF_6]^{3-}$	11,800	14,400

Note: The octahedral complexes of Fe(II) with the unsaturated amines like bpy, phen, etc. are highly coloured due to **metal to ligand charge transfer band**. These transitions are: $t_{2g} \rightarrow \pi^*$ (of the unsaturated ligand.)

7.14.9 Six-Coordinate High-Spin Complexes of d^5-System

The Mn(II) and Fe(III) complexes are the major examples of this group. The free ion ground state term ^6S generates the $^6A_{1g}$ ground state level. There is no excited state level having the spin multiplicity 6. Thus, ***there is no spin-allowed transition.*** The important excited states bear the spin multiplicity 4 (*i.e.* quartet). Before to discuss the spin-forbidden transitions, we are to consider the characteristic features of the energy level diagram of the high-spin complexes of d^5 system.

- **Free ion-terms:** The important free ion terms are ^6S (ground state), ^4G, ^4D, ^4P and ^4F. Their energy order is:

  $$\underset{(L=0)}{^6S} < \underset{(L=4)}{^4G} < \underset{(L=1)}{^4P} < \underset{(L=2)}{^4D} < \underset{(L=3)}{^4F}$$

  The first excited state (^4G) and the ground state (^6S) are in the expectation of Hund's rule. *But the higher excited states do not follow the energy order as predicted from the Hund's rules* (*see* Sec. 7.9.3).
- **Splitting of the free ion terms in an octahedral field:**
  $^6S \rightarrow {^6A_{1g}}$; $^4G \rightarrow {^4A_{1g}}, {^4E_g}, {^4T_{1g}}, {^4T_{2g}}$; $^4P \rightarrow {^4T_{1g}}$; $^4F \rightarrow {^4A_{2g}}, {^4T_{1g}}, {^4T_{2g}}$; $^4D \rightarrow {^4E_g}, {^4T_{2g}}$.
- **Configurational interaction (*i.e.* symmetrical interaction):**
  (i) $^4T_{1g}(G)$, $^4T_{1g}(P)$, $^4T_{1g}(F)$
  (ii) $^4T_{2g}(G)$, $^4T_{2g}(D)$, $^4T_{2g}(F)$
  (iii) $^4E_g(G)$, $^4E_g(D)$

The $^4T_{1g}$ *levels experience a higher configurational interaction than the* $^4T_{2g}$ *levels* because of the higher symmetry of $^4T_{1g}$. $^4T_{1g}(P)$ *is sandwiched between the upper* $^4T_{1g}(F)$ *and lower* $^4T_{1g}(G)$ *levels.* The interaction between the $^4T_{1g}(P)$ and $^4T_{1g}(G)$ states is relatively more important than the interaction between the $^4T_{1g}(P)$ and $^4T_{1g}(F)$ levels because the $^4T_{1g}(P)$ and $^4T_{1g}(G)$ levels are closer in energy. The repulsive interaction between the $^4T_{1g}(P)$ and $^4T_{1g}(G)$ levels raises the $^4T_{1g}(P)$ level with the Dq value. Then the raised $^4T_{1g}(P)$ level begins to experience a repulsive interaction (*i.e.* lowering of its energy) from the $^4T_{1g}(F)$ level. Thus, at the higher Dq values, the $^4T_{1g}(P)$ level experiences the repulsive interactions of opposite nature from the higher energy $^4T_{1g}(F)$ level and the lower energy $^4T_{1g}(G)$ level. ***This is why, after a certain Dq value,*** $^4T_{1g}(P)$ ***level runs parallel to the ground*** $^6A_{1g}$ ***level. This is why, energy curve of the sandwiched*** $^4T_{1g}(P)$ ***level is of sigmoidal nature.***

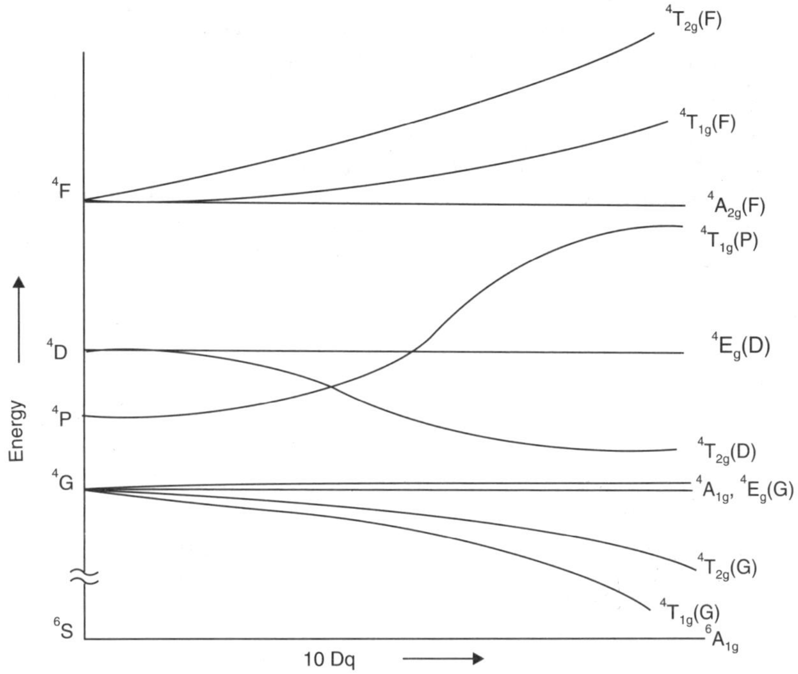

Fig. 7.14.9.1 Energy diagram for d^5-ion in high-spin octahedral complexes.

Because of the same ground (discussed above), the **sandwiched** $^4T_{2g}(D)$ level between the $^4T_{2g}(F)$ (higher energy) and $^4T_{2g}(G)$ (lower energy) levels experiences the **sigmoidal variation.** Initially, the repulsive interaction between the $^4T_{2g}(F)$ and $^4T_{2g}(D)$ states is more important. The configurational interaction between the $^4E_g(G)$ and $^4E_g(D)$ states is not so important. *Here it may be pointed out that the configurational interaction increases with the orbital degeneracy and it follows the order:* $T \rangle E \rangle A$. ***In fact, the two*** E_g ***levels run parallel to the ground state level,*** $^6A_{1g}$.

- **Degenerate character of** $^4A_{1g}(G)$ **and** $^4E_g(G)$**:** These two excited levels are accidentally degenerate. Consequently these two transitions produce a single peak.

$$^6A_{1g} \rightarrow {}^4A_{1g}(G),\ {}^4E_g(G)$$

- **Bandwidth:** Band broadening due to the vibronic coupling depends on the relative slopes of the two levels between which the transition occurs (*cf.* Fig. 7.7.3.1). The four excited energy levels, *i.e.* $^4E_g(G)$, $^4A_{1g}(G)$, $^4E_g(D)$, $^4A_{2g}(F)$ run parallel to the ground state $^6A_{1g}(S)$. In such cases, the relative slopes of the states between which the transition occurs is zero and **the peaks are sharp.**

  $$^6A_{1g} \rightarrow {}^4E_g(G), {}^4A_{1g}(G); {}^6A_{1g} \rightarrow {}^4E_g(D); {}^6A_{1g} \rightarrow {}^4A_{2g}(F).$$

  In other words, the above transition energies are independent of the magnitude of Dq. Here it is worth mentioning that among the above mentioned three transitions, the last two transitions produce the bands which are not as sharp as the band produced in the first transition.

On the other hand, the excited states $^4T_{1g}(G)$ and $^4T_{2g}(G)$ do not run parallel to the ground state $^6A_{1g}$. Consequently, transitions to the $^4T_{1g}(G)$ and $^4T_{2g}(G)$ states produce the **relatively broader peaks.**

In fact, the excited states $^4A_{1g}$ and ^4E_g, and the ground state $^6A_{1g}$ originate from the **same electronic configuration** $t_{2g}^3e_g^2$ but different spin multiplicities. Consequently, equilibrium bond distance remains more or less the same for both the ground and excited states (*i.e.* peak broadening by metal-ligand bond vibration does not occur). But the excited states $^4T_{1g}$ and $^4T_{2g}$ originate from the electronic configuration $t_{2g}^4e_g^1$ which is different from the ground state electronic configuration $t_{2g}^3e_g^2$. Thus, the transitions $^6A_{1g} \rightarrow {}^4T_{1g}$; $^6A_{1g} \rightarrow {}^4T_{2g}$ involve a change in the equilibrium metal-ligand bond distance. It causes **peak broadening** through the metal-ligand bond vibration (*cf.* Secs. 7.7.1-3).

In fact, for high spin Mn(II), some of the transitions involve only the **electronic rearrangement** (in terms of spin orientation) without any change of electron density in the t_{2g} and e_g level. Different possible electronic arrangements of a d^5-system in an octahedral field are shown in Fig. 7.14.9.2.

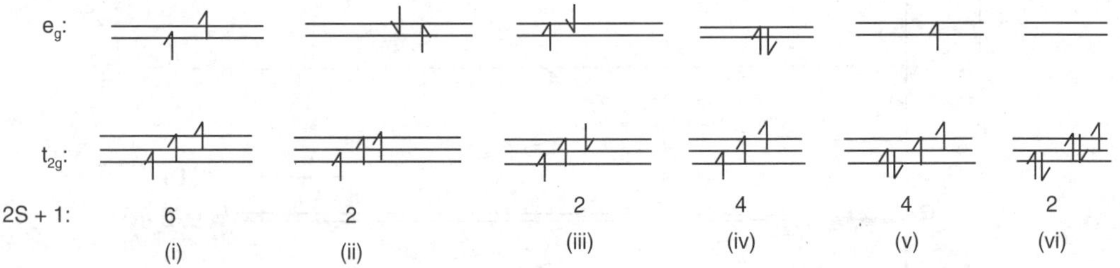

Fig. 7.14.9.2 Different possible electronic arrangements for a d^5-system in an octahedral field.

The spin forbidden transitions, (i) → (ii), (iii), (iv) do not require any electron excitation from the t_{2g} level to the e_g-level and all these states are having the same electronic configuration $t_{2g}^3e_g^2$ with different spin multiplicities. These transitions will produce the **sharp peaks** (*cf.* Secs. 7.7.2-3). The spin forbidden transitions, (i) → (v), (vi) will produce the **broader peaks** because electron density in the e_g-level is different in the states between which the electron transition occurs. In such cases, the metal-ligand bond length is different in the excited and ground electronic states.

The characteristics of ligand field spectra of $[Mn(OH_2)_6]^{2+}$, $[Fe(OH_2)_6]^{3+}$, $[FeF_6]^{3-}$ are given in Table 7.14.9.1.

- **Intensity:** In the high-spin d^5-complexes, the transitions are both Laporte forbidden and spin forbidden. Consequently, the peaks are of **very poor intensity.**
- **Spectral analysis of $[Mn(OH_2)_6]^{2+}$:** The peaks positions given in Table 7.14.9.1 fit in the Tanabe-Sugano diagram for $Dq/B = 1.1$. For the following transitions, we get the result:

$$\frac{^6A_{1g} \longrightarrow {}^4T_{1g}(G)}{B'} = \frac{18,900}{B'} = 24.0 \text{ (from the Tanabe-Sugano diagram, for } Dq/B = 1.1)$$

{ It leads to: $B' = 787.5\ cm^{-1}$
and $Dq = 1.1 \times 787.5 \approx 866\ cm^{-1}$, *i.e.* $10Dq = 8660\ cm^{-1}$

$$\beta = \frac{B'}{B_0} = \frac{787.5}{960} \approx 0.82$$

i.e. *B* is reduced to about 80% of the free ion value.

Table 7.14.9.1 Characteristics of ligand field transitions of some high-spin d^5-complexes

Transitions **(cm^{-1})**	*[Mn(OH$_2$)$_6$]$^{2+}$* **(cm^{-1})**	*[Fe(OH$_2$)$_6$]$^{3+}$* **(cm^{-1})**	*[FeF$_6$]$^{3-}$* **(cm^{-1})**
$^6A_{1g}\rightarrow{}^4T_{1g}(G)$	18,900 (0.013)	12,600	16,600
$^6A_{1g}\rightarrow{}^4T_{2g}(G)$	23,100 (0.009)	18,500	21,000
{ $^6A_{1g}\rightarrow{}^4E_g(G)$	24,970 (0.031)	24,300	26,500
$^6A_{1g}\rightarrow{}^4A_{1g}(G)$	25,300 (0.014)	24,600	–
$^6A_{1g}\rightarrow{}^4T_{2g}(D)$	28,000 (0.018)	28,800	29,800
$^6A_{1g}\rightarrow{}^4E_g(D)$	29,750 (0.013)	–	32,000
$^6A_{1g}\rightarrow{}^4T_{1g}(P)$	32,960 (0.02)	30,200	–
$^6A_{1g}\rightarrow{}^4A_{2g}(F)$	35,400	–	
$^6A_{1g}\rightarrow{}^4T_{1g}(F)$	36,900	–	
$^6A_{1g}\rightarrow{}^4T_{2g}(F)$	40,820	–	

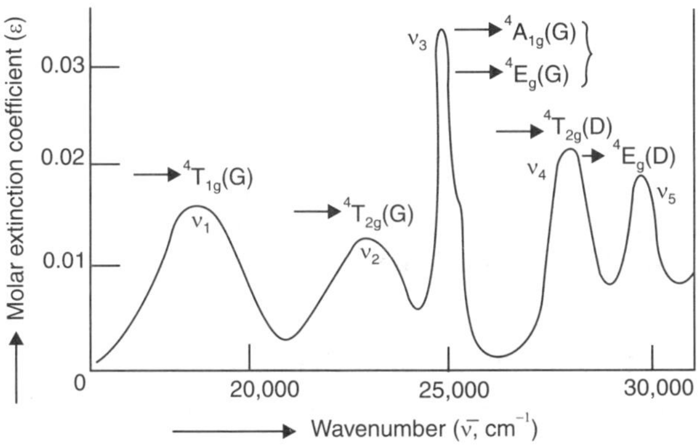

Fig. 7.14.9.2 Electronic absorption spectrum of $[Mn(H_2O)_6]^{2+}$.

Generally, for the high-spin Mn(II) complexes, the three lowest transitions are assigned and then the values of *Dq*, *B* and *C* are calculated therefrom to predict the positions of the higher energy transitions. The lowest three transitions are:

$$^6A_{1g} \xrightarrow{\nu_1} {}^4T_{1g}(G), \quad {}^6A_{1g} \xrightarrow{\nu_2} {}^4T_{2g}(G), \quad {}^6A_{1g} \xrightarrow{\nu_3} {}^4E_g, {}^4A_{1g}(G)$$

For $[Mn(OH_2)_6]^{2+}$, the ν_3-band (24,970 cm^{-1}) involves the energy change $10B' + 5C$, *i.e.* $10B' + 5C = 24{,}970\ cm^{-1}$.

Taking $Dq = 866\ cm^{-1}$ and $B' = 787\ cm^{-1}$ (obtained by fitting the spectral data in the Tanabe-Sugano diagram), we can evaluate *C*.

$10B' + 5C = 24970$; or, $10 \times 787 + 5C = 24{,}970$; or, $C = \dfrac{24{,}970 - 7{,}870}{5} = 3{,}420\ \text{cm}^{-1}$.

For the transition, ${}^6A_{1g} \to {}^4E_g(D)$, the energy change is given by: $17B' + 5C = 17 \times 787 + 5 \times 3420 = 30{,}479\ \text{cm}^{-1}$ (*cf.* experimental value, 29,750 cm^{-1}).

Note: Energy difference between the different states in terms of Racah parameters (A, B, C) or Condon-Shortley parameters (F_0, F_2, F_4) can be obtained from Table 7.9.3.1.

- **Spectral properties of the high-spin Fe(III)-complexes:** In the Fe(III) complexes, because of the ***ligand to metal charge transfer band,*** interpretation of the weak ligand field bands becomes very difficult. Ligand field bands of light violet $[Fe(OH_2)_6]^{3+}$ (in a fairly strong $HClO_4$ media) have been characterized. It is still complicated due to the strong propensity of $[Fe(OH_2)_6]^{3+}$ to hydrolyse to $[Fe(OH)(OH_2)_5]^{2+}$ which looks brown due to the charge transfer band (LMCT) (*cf.* ^-OH group is more polarizable than H_2O). The interpretation is further complicated due to the tendency of $[Fe(OH)(OH_2)_5]^{2+}$ to undergo dimerization.

The ligand field bands of $\mathbf{[Fe(ox)_3]^{3-}}$, $\mathbf{[Fe(acac)_3]}$ can be explained by considering the spin-forbidden transitions of a d^5 (high-spin, O_h) system. But lowering of symmetry (compared to that of an O_h system) in such *tris*-chelates splits the bands. The **strong LMCT bands** obscure some of the ligand field bands. Red $[Fe(acac)_3]$ in acetone solution shows four bands in the infrared and visible regions (molar extinction coefficient in parenthesis):

9,760 cm^{-1} (0.45), 13,160 cm^{-1}(0.70), 23,120 cm^{-1} (3,250), 28,410 cm^{-1} (3,420)

7.15 LIGAND FIELD BANDS IN SPIN PAIRED SIX-COORDINATE COMPLEXES OF FIRST TRANSITION SERIES

Under the condition, $10Dq \rangle P$, the d^4, d^5, d^6 and d^7 systems can produce the low-spin complexes. The possible spectral transitions can be assigned by analyzing the right hand portion of the Tanabe-Sugano diagram.

Very often, the π-bonding ligands (*i.e.* π-acid ligands like CN^-, NO, etc.) are the strong field ligands to induce the spin pairing. Such π-bonding ligands are highly polarizable. This is why, very often the charge transfer bands are extended in the visible range where the ligand field bands are overshadowed by the intense charge transfer bands. In such cases, the interpretation of the ligand field bands becomes complicated.

7.15.1 Ligand Field Spectra of the Low-spin Octahedral d⁶ Systems

The Fe(II) and Co(III) complexes are the representative examples. For the d^6-configuration, the spin-pairing is relatively easier compared to many other d^n-configurations. Both the high cfse in the low spin complexes and relatively less loss of exchange energy in pairing the electrons favour the d^6 system to adopt the low-spin configuration.

The ground state electronic configuration t_{2g}^6 gives the ${}^1A_{1g}$ state. The excited state electronic configuration $t_{2g}^5 e_g^1$ generates the following states (*cf.* Figs. 7.15.1.1, 2 and Sec. 7.12).

$$t_{2g}^5 e_g^1: {}^3T_{1g} + {}^3T_{2g} + {}^1T_{1g} + {}^1T_{2g}$$

The **energy order** is: ${}^3T_{1g} \langle {}^3T_{2g} \langle {}^1T_{1g} \langle {}^1T_{2g}$

The singlet states ${}^1T_{1g}$ and ${}^1T_{2g}$ are developed due to the antiparallel arrangement of the spins of the two unpaired electrons (one in the t_{2g} level and the other in the e_g level).

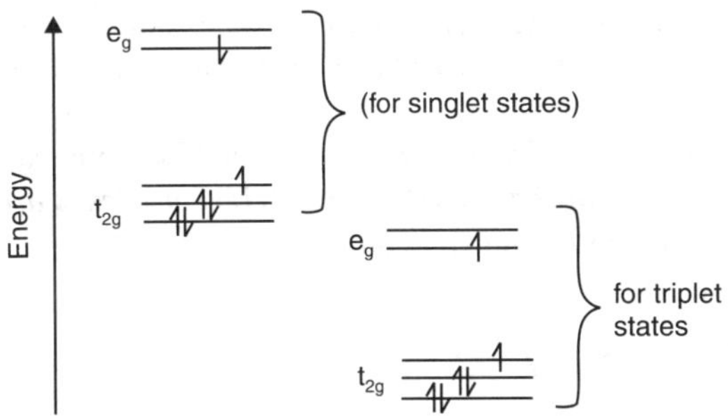

Fig. 7.15.1.1 Singlet and triplet states arising from the $t_{2g}^5e_g^1$ configuration.

- t_{2g}^5 *i.e.* one hole in the d_{xy}, d_{yz} and d_{xz} orbitals: **3 possible ways of distribution.**
- e_g^1 *i.e.* one electron in the $d_{x^2-y^2}$ and d_{z^2} orbitals: **2 possible ways of distribution.**
- $t_{2g}^5e_g^1 : 3 \times 2 = 6$ possible ways of electron distribution, *i.e.* **orbital multiplicity = 6.**
- $^1T_{1g} + {}^1T_{2g}$: total orbital multiplicity = 3 + 3 = 6

Because of the antiparallel arrangement of the spins, the singlet states are of higher energy.

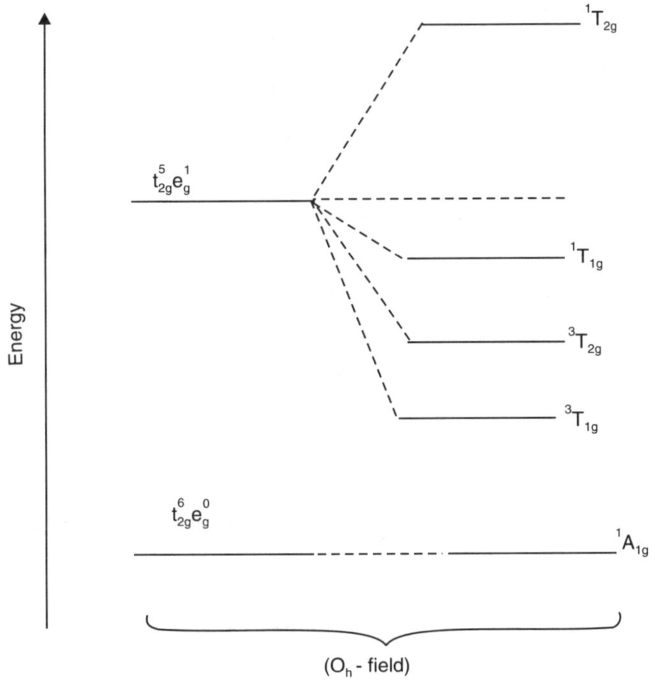

Fig. 7.15.1.2 (a) States arising from the low-spin d^6-configuration in an octahedral fields.

From Fig. 7.15.1.2, the possible electronic transitions giving rise to the ligand field bands in the low-spin octahedral complexes of d^6-configuration are as follows:

$^1A_{1g} \rightarrow {}^3T_{1g}$; $^1A_{1g} \rightarrow {}^3T_{2g}$ } spin-forbidden transitions.

$^1A_{1g} \rightarrow {}^1T_{1g}$; $^1A_{1g} \rightarrow {}^1T_{2g}$ } spin-allowed transition.

Obviously, transitions to the triplet states give the peaks of poor intensity and these spin-forbidden transitions occur at a lower energy. Energies of the $^3T_{2g}$ and $^1T_{1g}$ states are close to each other. This is why, the strong band arising from the spin-allowed $^1A_{1g} \rightarrow {}^1T_{1g}$ transition very often obscures the spin-forbidden transition $^1A_{1g} \rightarrow {}^3T_{2g}$. In fact, characterization and analysis of the said weak peaks due to the spin-forbidden transitions are not of much importance. On the other hand, analysis of the spin-allowed transitions is quite important. However, in some cases, all the four peaks have been characterized (*cf.* Table 7.15.1.1).

By considering the relatively stronger spin allowed transitions in the low-spin octahedral complexes of Co(III), it may be pointed out that ***the spectral behaviours of the Co(III) and Cr(III) complexes are quite similar. The Cr(III) complexes also record two strong ligand field bands.***

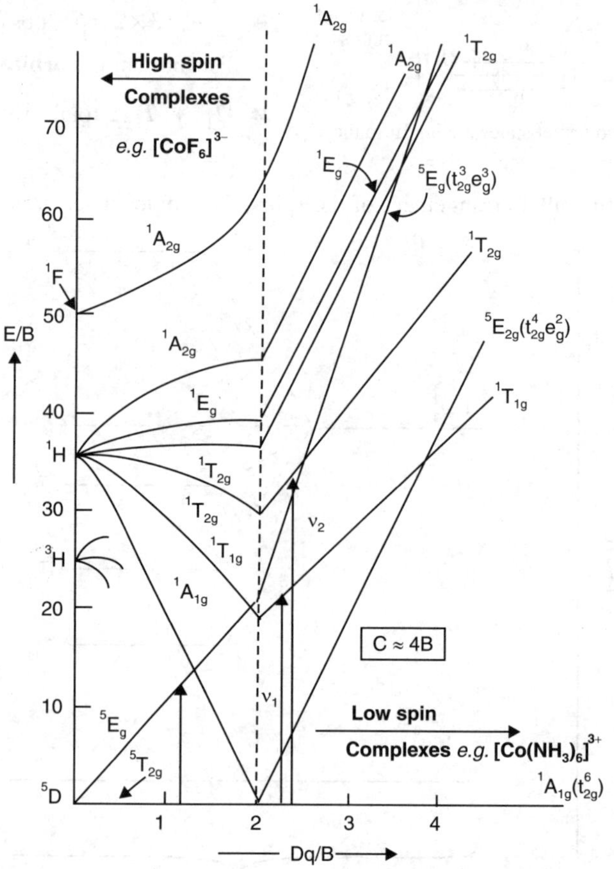

Fig. 7.15.1.2 (b) Tanabe-Sugano diagram for the d^6-octahedral complexes showing the spin-allowed transitions.

Octahedral Cr(III) complexes

$$^4A_{2g} \xrightarrow{\nu_1} {}^4T_{2g}(F) \quad \text{and} \quad ^4A_{2g} \xrightarrow{\nu_2} {}^4T_{1g}(F)$$

(The high energy 3rd ligand field and $^4A_{2g} \xrightarrow{\nu_3} {}^4T_{1g}(P)$ is very often obscured by the strong charge transfer band).

Octahedral low spin Co(III) complexes

$$^1A_{1g} \xrightarrow{\nu_1} {}^1T_{1g}$$

$$^1A_{1g} \xrightarrow{\nu_2} {}^1T_{2g}$$

For the octahedral low spin Co(III)-complexes, the energy changes in the ν_1 and ν_2 transitions are given by:

$$\left.\begin{aligned}\nu_1 &= 10Dq - 4B' + 86B'^2 / 10Dq = \Delta - 4B' + 86B'^2/\Delta \\ \nu_2 &= 10Dq + 12B' + 2B'^2 / 10Dq = \Delta + 12B' + 2B'^2/\Delta\end{aligned}\right\}(\text{Taking, } C \approx 4B')$$

In the low spin complexes, the B'-parameter (measuring the interelectronic repulsion) is relatively less important and the terms containing B'^2 can be neglected. This approximation leads to:

$$\nu_1 \approx 10Dq - 4B' \text{ and } \nu_2 = 10Dq + 12B'$$

i.e. $$\nu_2 - \nu_1 \approx 16B' \text{ and, } 10Dq \approx \nu_1 + 4B' \approx \nu_2 - 12B'$$

Let us analyze the spectral results of $\mathbf{[Co(en)_3]^{3+}}$ by using the approximate relations.

$$\nu_1 = 21{,}400 \text{ cm}^{-1},\ \nu_2 = 29{,}400 \text{ cm}^{-1}$$

$$16B' \approx \nu_2 - \nu_1 = 29{,}400 - 21{,}400 = 8{,}000,\ i.e.\ B' \approx 500 \text{ cm}^{-1} = 0.5 \text{ kK}$$

$$10Dq \approx \nu_1 + 4B' = 21{,}400 + 4 \times 500 = 23{,}400 \text{ cm}^{-1} = 23.4 \text{ kK}$$

Here it may be mentioned that the said spectral analysis can also be done by using the Tanabe-Sugano diagram. It is illustrated below.

The ratio of ν_2 and ν_1 is given by:

$$\frac{{}^1A_{1g} \longrightarrow {}^1T_{2g}}{{}^1A_{1g} \longrightarrow {}^1T_{1g}} = \frac{\nu_2}{\nu_1} = \frac{29{,}400}{21{,}400} \approx 1.37$$

The above mentioned ratio can be fitted into the Tanabe-Sugano diagram by sliding a ruler along the abscissa. This fitting is achieved for $Dq/B = 4.0$. For this value of Dq/B, ν_1/B' becomes 38 (obtained from the diagram).

Table 7.15.1.1 Characteristic features of ligand field spectra of low-spin octahedral complexes of d^6-system (For $[M(CO)_6]$, the ligand field bands appear as the **shoulders of the strong MLCT bands**).

Complex	$^1A_{1g} \rightarrow {}^3T_{1g}$ (cm^{-1})	$^1A_{1g} \rightarrow {}^3T_{2g}$ (cm^{-1})	$^1A_{1g} \rightarrow {}^1T_{1g}$ (cm^{-1})	$^1A_{1g} \rightarrow {}^1T_{2g}$ (cm^{-1})
$[Co(en)_3]^{3+}$	13,700	17,500	21,400	29,400
$[Co(NH_3)_6]^{3+}$	13,000	–	21,100	29,500
$[Co(OH_2)_6]^{3+}$	8,000	12,500	16,500	24,600
$[Fe(CN)_6]^{4-}$	23,700	–	31,000	37,000
$[Fe(phen)_3]^{2+}$	8,950	9,250	12,600	14,700
$[RhF_6]^{3-}$	–	–	21,300	27,800
$[Cr(CO)_6]$	–	–	29,500	31,500
$[Mo(CO)_6]$	–	–	30,150	31,950

$$\frac{{}^1A_{1g} \longrightarrow {}^1T_{1g}}{B'} = \frac{\nu_1}{B'} \approx \frac{21{,}400}{B'} = 38,\ i.e.\ B' \approx 563 \text{ cm}^{-1} = 0.563 \text{ kK}$$

$$\frac{Dq}{B'} = 4.0 \text{ or, } Dq = B' \times 4 = 2{,}252 \text{ cm}^{-1}\ (i.e.\ 10Dq = 22{,}520 \text{ cm}^{-1} = 22.52 \text{ kK})$$

These $10Dq$ and B' values are close to the values obtained from the approximate relations.

$$\beta = \frac{B'}{B_0} = \frac{563}{1100} \approx 0.51 \text{ i.e. } B \text{ value is reduced by about 50\%.}$$

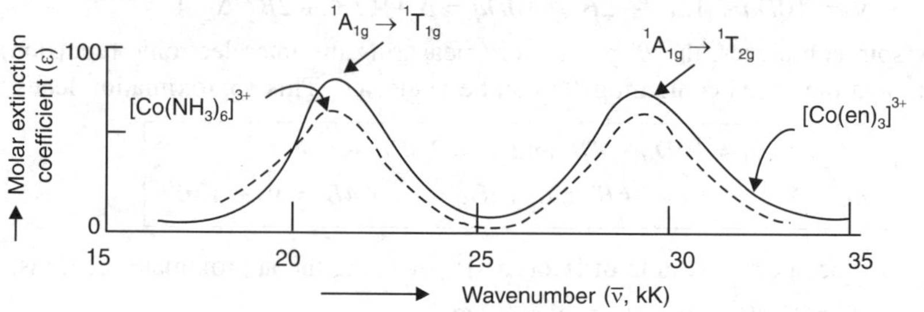

Fig. 7.15.1.3 Electronic absorption spectra of $[Co(NH_3)_6]^{3+}$ (dotted line) and $[Co(en)_3]^{3+}$ (solid line).

● $\mathbf{[Co(en)_3]^{3+}}$ ***vs.*** $\mathbf{[Co(ox)_3]^{3-}}$**:** The spectra of **yellow** $[Co(en)_3]^{3+}$ and **green** $[Co(ox)_3]^{3-}$ complexes are compared in Fig. 7.15.1.4. Both are of D_3 symmetry but their spectral behaviours indicate that they maintain effectively the O_h symmetry, *i.e.* $[M(N)_6]$ and $[M(O)_6]$. The two spin allowed transitions are: $^1A_{1g} \xrightarrow{\nu_1} {}^1T_{1g}$ and $^1A_{1g} \xrightarrow{\nu_2} {}^1T_{2g}$. For $[Co(ox)_3]^{3-}$, the spin allowed transition (ν_2) at the higher energy is close to the strong CT band. The **better reducible character of the oxalate ligand** leads the LMCT band to appear in the lower energy (*cf.* ν_{LMCT} : $[Co(en)_3]^{3+} \gg [Co(ox)_3]^{3-}$). Thus the Laporte selection rule is relaxed better in $[Co(ox)_3]^{3-}$ to give the *higher intensity* probably due to the **intensity stealing phenomenon** from the nearby allowed LMCT band. *The slope of the* $^1T_{2g}$ *state is higher than that of the* $^1T_{1g}$ state (*cf.* Fig. 7.15.1.2b). Thus separation of the two peaks (*i.e.* $\Delta\nu = \nu_2 - \nu_1$) will increase with the ligand field strength. *This is why, these two peaks are separated more in* $[Co(en)_3]^{3+}$ (*cf.* crystal field strength: en > ox^{2-}).

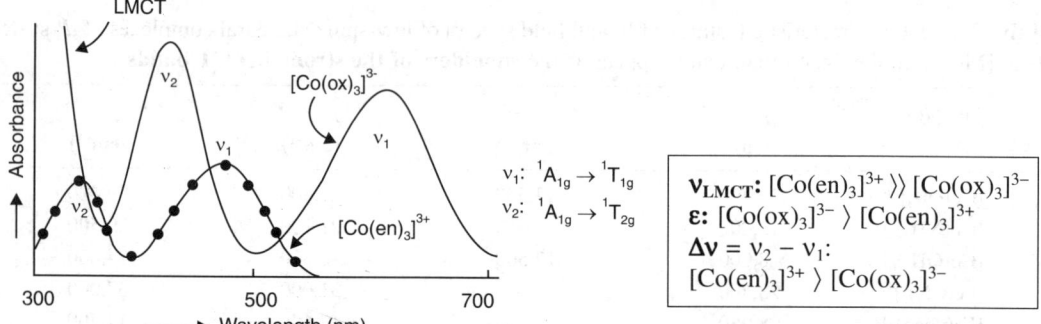

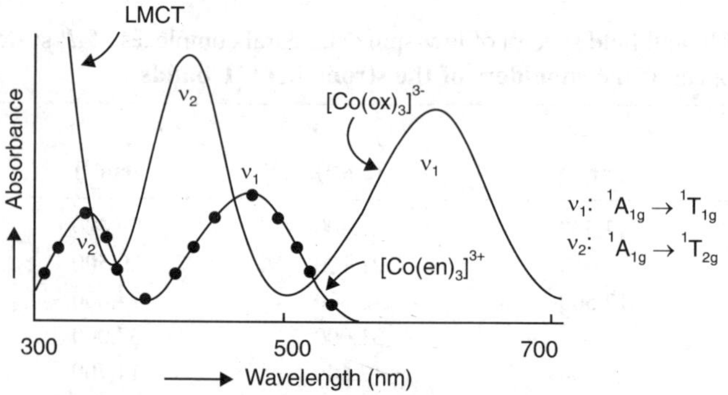

ν_{LMCT}: $[Co(en)_3]^{3+} \gg [Co(ox)_3]^{3-}$ ε: $[Co(ox)_3]^{3-} > [Co(en)_3]^{3+}$ $\Delta\nu = \nu_2 - \nu_1$: $[Co(en)_3]^{3+} > [Co(ox)_3]^{3-}$

Fig. 7.15.1.4 Comparison of the absorption spectra of $[Co(ox)_3]^{3-}$ and $[Co(en)_3]^{3+}$. **Note:** Higher intensity for the $[Co(ox)_3]^{3-}$ complex may arise from the allowed charge transfer band (*cf.* **intensity stealing effect**).

7.15.2 Ligand Field Spectra of Low-Symmetry Six-coordinate Complexes of Co(III). (*cf.* Sec. 7.14.5 E for Cr(III) Complexes and Sec. 7.14.4 for Ni(II) Complexes)

(A) Splitting of the octahedral spectroscopic states in tetragonal and rhombic symmetry: When the ligands are nonequivalent, this situation will arise. Among the six-coordinate complexes, the three types of field symmetries are quite important. These are:

cubic (*i.e.* perfectly octahedral), **tetragonal** and **rhombic.**

These are classified in terms of μ which is the sum of charge (q) in the directions of x, y and z axes along which the ligands are placed. If the μ–s along the respective axes are denoted by μ_x, μ_y, and μ_z, then we can write:

cubic and octahedral: $\mu_x = \mu_y = \mu_z$, *e.g.* $[Co(NH_3)_6]^{3+}$
tetragonal: $\mu_x = \mu_y \neq \mu_z$, *e.g.* $[CoCl_2(NH_3)_4]^+$
rhombic: $\mu_x \neq \mu_y \neq \mu_z$, *e.g.* *mer*-$[CoCl_3(NH_3)_3]$

Table 7.15.2.1 Symmetries of some common types of six-coordinate complexes.

Symmetry	***Complex***
Cubic	$[MA_6]$ (O_h); 1,2,3-$[MA_3B_3]$ (*fac* or *cis*)(C_{3v});
($\mu_x = \mu_y = \mu_z$)	
Tetragonal	
($\mu_x = \mu_y \neq \mu_z$)	
C_{2v}	*cis*-$[MA_4B_2]$
D_{4h}	*trans*-$[MA_4B_2]$
Rhombic	1,2,6-$[MA_3B_3]$ (*mer* or *trans*) (C_{2v})
($\mu_x \neq \mu_y \neq \mu_z$)	

In moving from the octahedral to tetragonal (D_{4h}) field symmetry, the T-state is split into two states A_2 and E or B_2 and E., *i.e.* $T_{1g}(O_h) \rightarrow A_{2g}, E_g(D_{4h})$ and $T_{2g} \rightarrow B_{2g}, E_g(D_{4h})$ (*cf.* Correlation Table 7.15.4.2). Again the E-state is split into B_1 and B_2 in moving from the tetragonal to rhombic field symmetry. These are illustrated in Fig. 7.15.2.1. ***In terms of electronic environment, both the cis- and trans- isomers of [MA₄B₂] may be considered to have the tetragonal symmetry.*** This aspect has been discussed later.

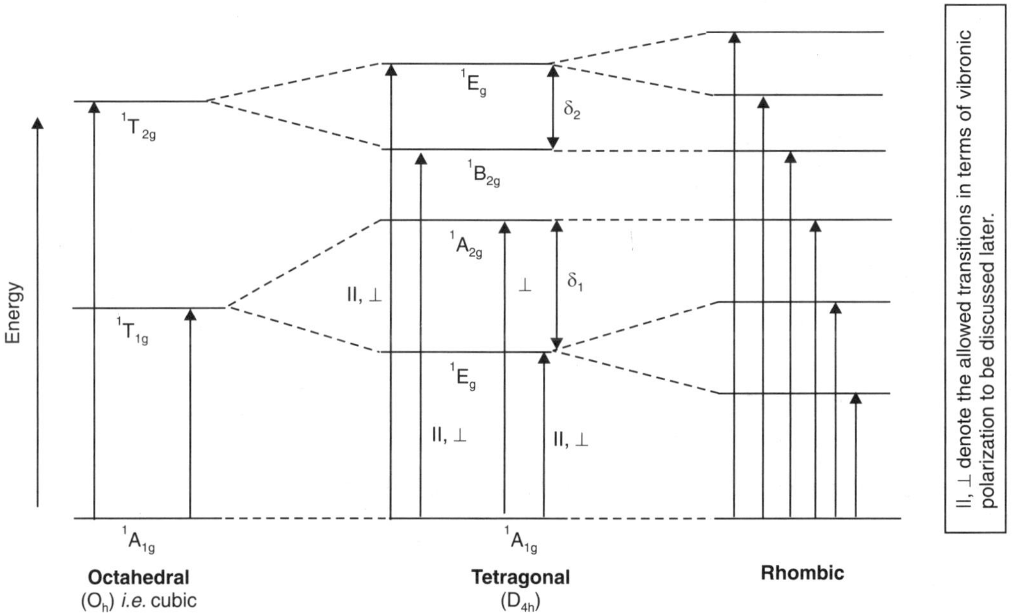

Fig. 7.15.2.1 Energy level diagrams for the spin-allowed spectral transitions in the octahedral, tetragonal and rhombic field symmetries of the low-spin Co(III) complexes.

(Note: Some authors represent the energy order of the ^1E_g and $^1A_{2g}$ levels generated from the splitting of $^1T_{1g}$ as: $^1E_g \rangle\, ^1A_{2g}$. However, polarisation (*i.e.* dichroism) studies on *trans*-$[CoCl_2(en)_2]^+$ (D_{4h}) indicate that $^1A_{2g}$ is the higher energy state (*cf.* S. Yamada et. al. *Bull. Chem. Soc. Japan*, **25**, 127, 1952; **28**, 222, 1955; C.J. Ballhausen et. al. *J. Inorg. Nucl, Chem.*, **3,** 178, 1956). This aspect will be discussed later.

Note: ● $\delta_1 \rangle\rangle \delta_2$ ● For the D_{4h} tetragonal symmetry (*e.g. trans*-$[MA_4B_2]$) the ordering is $^1E_g \langle\, ^1A_{2g}$ while for C_{2v} tetragonal symmetry (*e.g. cis*- $[MA_4B_2]$) the ordering is $^1A_2 \langle\, ^1E$. Splitting of the $^1T_{1g}$ level in the *trans*-complex is almost double to that of the corresponding *cis*-complex (*cf.* Fig. 7.15.2.3). In an octahedral complex, there are **two possible transitions;** in a tetragonal complex, there are **four possible transitions** while in a rhombic system, there are **six possible transitions**).

From the energy level diagram given in Fig. 7.15.2.1, it is evident that for the low-spin complexes of Co(III), we can expect the following transitions (*if the splitting is sufficiently large to develop the separate peaks*).

Cubic: two peaks; **Tetragonal:** four peaks; **Rhombic:** six peaks.

However, in most of the cases, splitting of the $^1T_{2g}$ level is too small to be detected. Similarly, splitting of the ^1E_g level in a rhombic field is again found to be too small to be detected. Considering this fact, we can expect the following peaks:

Cubic: two peaks; **Tetragonal:** three peaks; **Rhombic:** three peaks.

(B) ***Characterization of the cis- and trans- isomers of low-spin $[Co^{III}A_4B_2]$ from their ligand field spectra:*** Both the *cis- and trans*-isomers are expected to give the three ligand field bands. **Splitting of the $^1T_{2g}$ level for the both isomers is too small to be detected, but splitting of the $^1T_{1g}$ level is more for the *trans*- isomer compared to that of the corresponding *cis*-isomer.** These can be understood in terms of the μ-values. Here we shall consider the splitting (δ_1) of the T_{1g} term.

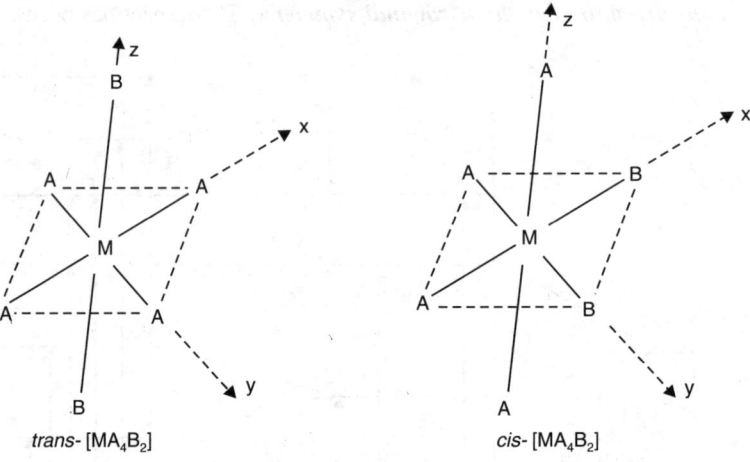

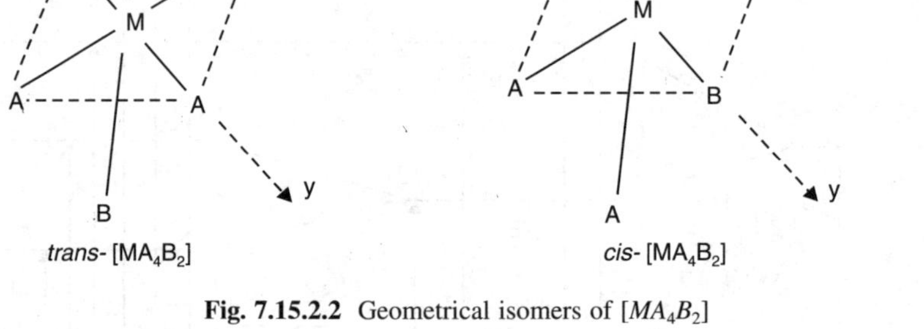

Fig. 7.15.2.2 Geometrical isomers of $[MA_4B_2]$

The **tetragonality parameter** $\mu_x - \mu_z$ gives the measure of splitting of the T-term. It is obtained as follows:

Trans-$[MA_4B_2]$

$\mu_x = \mu_y = 2q_A$

$\mu_z = 2q_B$

i.e. $\mu_x - \mu_z = 2(q_A - q_B)$

and $\delta_1(trans) \propto 2(q_A - q_B)$

Cis-$[MA_4B_2]$

$\mu_x = \mu_y = q_A + q_B$

$\mu_z = 2q_A$

i.e. $\mu_x - \mu_z = -\,(q_A - q_B)$

and $\delta_1(cis) \propto -\,(q_A - q_B)$

From the relative values of the **tetragonality parameter** $\mu_x - \mu_z$, *it is evident that for the trans-complex, the T_{1g} term undergoes splitting twice that of the cis-complex.* The minus sign indicates the ordering of the ^1E_g and $^1A_{2g}$ levels originated due to splitting of the $^1T_{1g}$ level is inverted in the *cis*-complex compared to that of the *trans*- complex. In fact, the energy order of ^1E_g and $^1A_{2g}$ terms follows the sequence:

$$trans\text{-}[MA_4B_2]: {}^1E_g \langle {}^1A_{2g};\ cis\text{-}[MA_4B_2]: {}^1E \rangle {}^1A_2 \text{ and, } \delta_1\ (trans): \delta_1\ (cis) = 2:1$$

The magnitude of splitting (δ_1) of $^1T_{1g}$ term can be expressed in terms of positions of the ligands *A* and *B* in the spectrochemical series. The splitting energy is determined by the average ligand field strength difference between the A—Co—A and B—Co—B axes (for the *trans*-complex) and between the A—Co—A and A—Co—B axes (for the *cis*-complex). It gives:

$$\boldsymbol{\delta_1(trans) = 2C(\Delta_A \sim \Delta_B),\ \delta_1(cis) = -C(\Delta_A \sim \Delta_B).}$$

Here *C* is a proportionality constant and Δ_A and Δ_B denote the crystal field splitting power of the ligands *A* and *B* respectively. **Thus the $\Delta_A \sim \Delta_B$ parameter represents their difference in position in the spectrochemical series.**

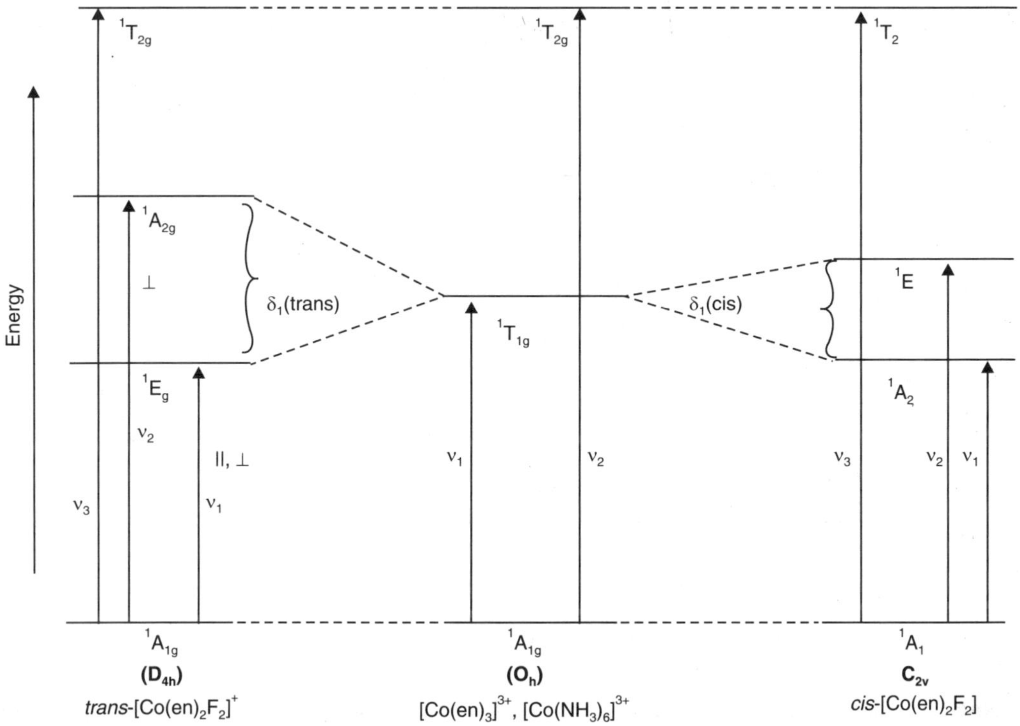

Fig. 7.15.2.3 The spin allowed transitions in the low spin $[Co^{III}A_6]$ (O_h) and *cis-trans-* isomers of $[Co^{III}A_4B_2]$ type complexes: $\delta_1(trans) \approx 2\delta_1(cis)$; **Splitting of the $^1T_{2g}$ is negligibly small and it is ignored.**

Note: ● Though the *cis-* and *trans-* isomers belong to the C_{2v} and D_{4h} symmetry point groups respectively, **from the view of electronic environment around the metal centre, both the isomers experience the tetragonal splitting only differing in the magnitude of splitting and energy order of the splitted levels** (*cf.* F. Teixidor *et al, J. Chem. Edu.,* **64,** 461, 1987).

● Some authors indicate the energy order as ${}^1E_g \rangle {}^1A_{2g}$ (for the *trans*-isomer) but the **polarisation studies** on *trans*-$[CoCl_2(en)_2]^+$ indicate the energy order: ${}^1A_{2g} \rangle {}^1E_g$ (for the *trans*-isomer) and this practice is followed in this book.

Now let us consider the relative positions of the peaks due to the following transitions.

$$(cf.\ \text{Fig. 7.15.2.3})\begin{cases} {}^1A_1 \xrightarrow{\nu_1} {}^1A_2;\ {}^1A_1 \xrightarrow{\nu_2} {}^1E,\ {}^1A_1 \xrightarrow{\nu_3} {}^1T_2\ (\textbf{for cis - isomer}) \\ {}^1A_{1g} \xrightarrow{\nu_1} {}^1E_g;\ {}^1A_{1g} \xrightarrow{\nu_2} {}^1A_{2g},\ {}^1A_{1g} \xrightarrow{\nu_3} {}^1T_{2g}\ (\textbf{for trans - isomer}) \end{cases}$$

The difference, $\nu_1 \sim \nu_2$ in the *trans*-isomer is twice that of the corresponding *cis*-complex, *i.e.*

$$\frac{(\nu_1 \sim \nu_2)_{\text{trans}}}{(\nu_1 \sim \nu_2)_{\text{cis}}} \approx 2.0.$$

In reality, in the *cis*-complexes, *e.g.* *cis*-$[Co(en)_2F_2]^+$, *cis*-$[Co(C_2H_5COO)_2(en)_2]^+$, the positions of ν_1 and ν_2 are so close that two separate peaks are not observed, *i.e.* they merge or overlap with each other to give a relatively broad peak. On the other hand, for their corresponding *trans-isomers*, the transitions due to ν_1 and ν_2 are well separated to be characterized as the separate peaks. This is why, the *trans*-isomer gives three peaks (*cf.* Fig. 7.15.2.4a).

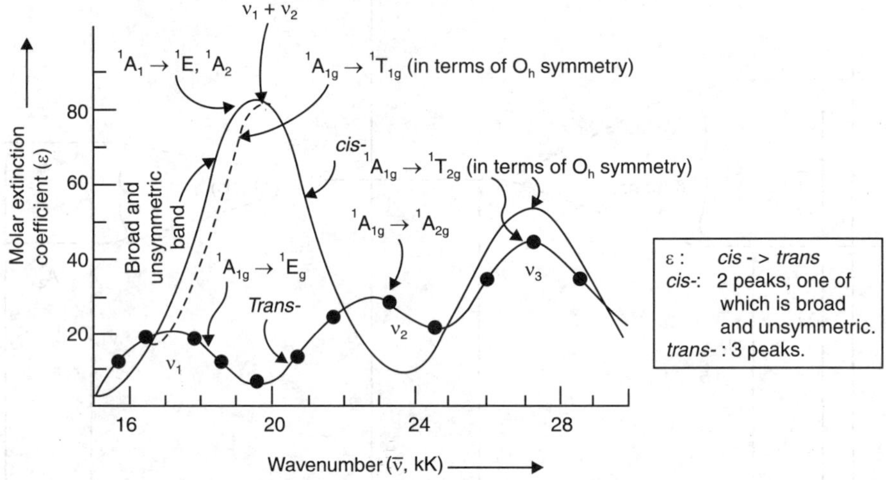

Fig. 7.15.2.4 (a) Electronic absorption spectra of *cis*- and *trans*- isomers of $[Co(en)_2F_2]^+$ (ν_1, ν_2, ν_3 in terms of tetragonal symmetry).

(**Note:** The dotted line in Fig. 7.15.2.4a for the *cis*-isomer indicates the position of the symmetrical band for the ${}^1A_{1g} \rightarrow {}^1T_{1g}$ transition without any splitting of the ${}^1T_{1g}$ term (assuming the O_h symmetry).

Positions of the ligand field bands in the *cis*- and *trans*- isomers of $[Co(en)_2F_2]^+$ are given below:
(***cis*-$[Co(en)_2F_2]^+$:** $\underbrace{\nu_1, \nu_2}$ (19,000 cm^{-1}, ε = 85), ν_3 (27,500 cm^{-1}, ε = 50)

Merged to give a broad and asymmetric peak.

***trans*- $[Co(en)_2F_2]^+$:** ν_1 (17,500 cm^{-1}, ε = 20), ν_2 (22,500 cm^{-1}, ε = 30), ν_3 (27,800 cm^{-1}, ε = 50).

Here it may be mentioned that the ligand field spectra can distinguish the geometrical isomers of $[Co^{III}A_4B_2]$ only when the positions of *A* and *B* in the spectrochemical series are widely apart. Because, the extent of splitting of the ${}^1T_{1g}$ level depends on the relative crystal field splitting powers of *A* and *B*.

Here it is interesting to note that **the intensity of the peaks of the *cis*-$[Co^{III}A_4B_2]$ complexes is much higher than those of the *trans*-$[Co^{III}A_4B_2]$ complexes.** The *cis*-complexes lack in centre of symmetry. Consequently, the *d-p* mixing is more efficient in relaxing the Laporte selection rule in the *cis*-complex compared to that in the *trans*- isomer which is possessing the centre of symmetry. Besides this, in the ***noncentrosymmetric cis- complex,*** some of the electronic transitions should be allowed by their own right because in such cases the Laporte selection rule is not applicable. However, the molar extinction coefficient values (ε) do not differ widely for the *cis*- and *trans*- complexes and the ε-values are also more or less comparable with those of the octahedral complexes (*cf.* for $[Co(NH_3)_6]^{3+}$: ν_1(21.2 kK, ε = 56); ν_2(29.55 kK, ε = 46)). These observations suggest that in the given complexes, ***effective octahedral symmetry is maintained.***

In $[Co(en)_2F_2]^+$, the parameter $\Delta_{en} \sim \Delta_F$ giving the measure of difference in crystal field splitting power of the ligands en and F^- is significantly high. Consequently, their *cis*- and *trans*- isomers can be distinguished by comparing their ligand field spectra. For the *cis*- complex, the octahedral ν_1 and ν_2 transitions do not experience any splitting but in the *trans*-complex, the ν_1 transition gives two separate peaks as expected for the D_{4h} symmetry (*cf.* Fig. 7.15.2.4a). ***In the cis-complex, the octahedral ν_1-band is only broadened to have an asymmetric shape. However, even in the trans-complex, the octahedral ν_2 band does not experience any splitting.***

F^- and H_2O are very close in their positions in the spectrochemical series (*i.e.* $\Delta_{en} - \Delta_F \approx \Delta_{en} - \Delta_{H_2O}$) and the geometrical isomers of $[Co(en)_2(OH_2)_2]^{3+}$ can also be distinguished by comparing their ligand field spectra as in the case of $[Co(en)_2F_2]^+$. *Trans*- $[Co(en)_2(OH_2)_2]^{3+}$ shows one absorption band with two components at 18,200 cm^{-1} and 22,500 cm^{-1} and another absorption band at 29,000 cm^{-1}. The positions of H_2O and NCS^- in spectrochemical series are very close and in fact, the geometrical isomers of $[Cr(NCS)_2(OH_2)_4]^+$ cannot be distinguished in this way (*cf.* Sec. 7.14.5D). Here it is worth mentioning that the crystal field splitting power of Co(III) is greater than that of Cr(III) (*cf.* $10Dq = fg$, g = 18.2 for Co^{3+}; g = 17.4 for Cr^{3+}). It causes more splitting of the T-term in the Co(III) complexes compared to that in the Cr(III)-complexes for the same ligands.

(C) Characterization of the *fac-mer* isomers of $[MA_3B_3]$: It has been already mentioned that the *fac*-isomer (*i.e. cis*-or 1,2,3-$[MA_3B_3]$) experiences the octahedral ligand field splitting of *d*-orbitals ($\mu_x = \mu_y = \mu_z$). In an octahedral ligand field, *fac*-$[Co^{III}A_3B_3]$ experiences two ligand field transitions, *i.e.*

$$ {}^1A_{1g} \xrightarrow{\nu_1} {}^1T_{1g} \text{ and } {}^1A_{1g} \xrightarrow{\nu_2} {}^1T_{2g} $$

On the other hand, in the rhombic ligand field ($\mu_x \neq \mu_y \neq \mu_z$) of *mer*-$[Co^{III}A_3B_3]$, both the octahedral ν_1 and ν_2 peaks are expected to split into triplet components because of the splitting of T_{1g} and T_{2g} levels into three energy levels (*i.e.* each *T*-term splits into 3 levels, *cf.* Fig. 7.15.2.1). In reality, splitting of the T_2-term is negligibly small. Consequently, the ν_2-band remains more or less the same for both the isomers. Splitting of the T_{1g} term in a rhombic field is relatively higher, but the splitting energy may not be sufficiently high to generate the three different peaks through the splitting of the octahedral ν_1-band. **However very often, the ν_1-band may be broadened in a rhombic field** (compared to that of the octahedral field).

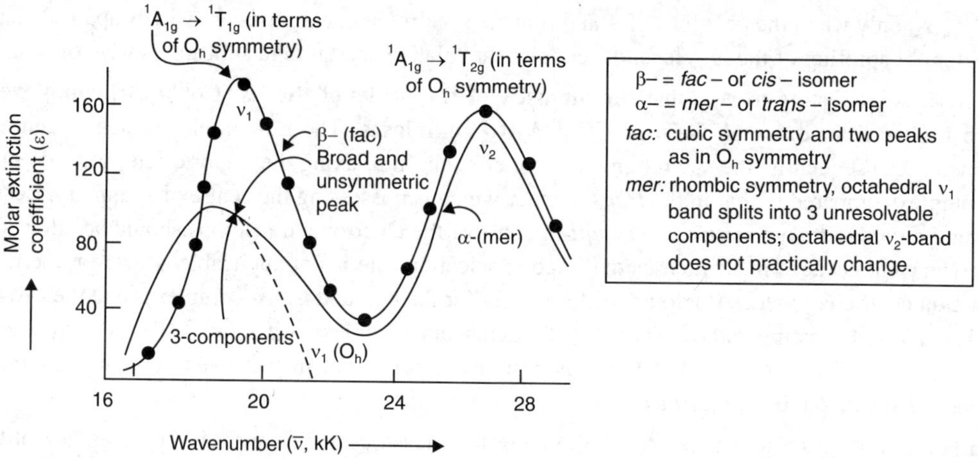

Fig. 7.15.2.4 (b) Electronic absorption spectra of α- and β isomers of $[Co(gly)_3]$, *i.e.* $[Co(N,O)_3]$.

(Note: The broken line in Fig. 7.15.2.4b for the α-isomer indicates the position of the the symmetrical band for the $^1A_{1g} \rightarrow {}^1T_{1g}$ transition (assuming the O_h symmetry) without any splitting of the $^1T_{1g}$ term.)

Now let us consider to compare the ligand field spectra (Fig. 7.15.2.4b) of α (*violet, dihydrate*) and β (*red, monohydrate*) isomers of $[Co(glycinate)_3]$, *i.e.* $[Co(N,O)_3]$. The α-isomer is actually *mer-* or 1,2,6- or, *trans*-isomer while the β-isomer is *fac-* or, 1,2,3- or *cis*-isomer. ***Thus the α-isomer experiences the rhombic ligand field splitting while the β-isomer experiences the octahedral ligand field splitting.*** In fact, for the α-isomer, the lower energy band (*i.e.* ν_1-band) is broadened with a shoulder (*i.e.* broad and unsymmetric band) while the corresponding band for the β-isomer is symmetrical. However, the higher energy band (*i.e.* ν_2-band) is having more or less of the same shape for both the α- and β- isomers.

Here it is worth mentioning that characterization of the geometrical isomers by considering the band splitting property in their ligand field spectra is well documented for ***the inert complexes of Cr(III), Co(III), Rh(III), Ir(III), Ru(III),*** etc. But the technique is not applicable for the labile complexes which are expected to experience the rearrangement and solvolysis quickly.

(D) Vibronic polarisation in ***trans*-$[CoCl_2(en)_2]^+$ (D_{4h}) and characterisation of energy states derived from the splitting of the $^1T_{1g}$ level (*cf.* Figs. 7.15.2.1, 3):** It has been already indicated in Sec. 7.5.1 that *trans*-$[CoCl_2(en)_2]^+$ shows the **dichroism** (*i.e.* vibronic polarisation). In this centrosymmetric system, the forbidden electronic transitions are partially allowed through the vibronic coupling but the transitions experience a **vibronic polarisation** (*see* Sec. 7.5.1 for details) as indicated below:

	$^1A_{1g} \longrightarrow {}^1A_{2g}$	$^1A_{1g} \longrightarrow {}^1E_g$	$^1A_{1g} \longrightarrow {}^1B_{2g}$ (arising from the splitting of $^1T_{2g}$ term).
For μ_z (II to C_4 axis):	Forbidden	Allowed	Allowed
For $\mu_{x,y}$ (⊥ to C_4 axis):	Allowed	Allowed	Allowed

When the spectra are recorded with the polarised light perpendicular (⊥) to the C_4 axis (*i.e.* Cl—Co—Cl), **three peaks are observed** but when the spectra are recorded with the polarised light parallel (II) to the C_4-axis, only **two peaks are noted,** *i.e.* one peak is missing. The group theory treatment predicts that the missing peak is due to the $^1A_{1g} \longrightarrow {}^1A_{2g}$ transition. The **three peaks**

(⊥ polarisation) appear at ~16 kK, ~22 kK and ~27 kK (which merges with the LMCT band; *cf.* Cl^-, a polarisable ligand). For the polarised light (|| to the C_4 axis), the peak at about ~22 kK is mssing and it is definitely due to the $^1A_{1g} \longrightarrow {}^1A_{2g}$ transition. Thus the peaks are reasonably assigned as:

$$^1A_{1g} \longrightarrow {}^1E_g \text{ (~16 kK); } {}^1A_{1g} \longrightarrow {}^1A_{2g} \text{ (~22 kK); } {}^1A_{1g} \longrightarrow {}^1B_{2g} \text{ (~27 kK).}$$

It indicates the energy order as $^1B_{2g} \rangle {}^1A_{2g} \rangle {}^1E_g$. As the splitting in the unsplit $^1T_{2g}$ level is negligibly small, energy of the $^1B_{2g}$ level corresponds to that of the unsplit $^1T_{2g}$ level (approximately). For details of the dichroism studies on *trans*-$[CoCl_2(en)_2]^+$ (*Ref.* S. Yamada et. al. *Bull. Chem. Soc. Japan*, **25**, 127, 1952; **28**, 222, 1955) and spectral analysis/interpretation (*Ref.* C.J. Ballhausen et. al. *J. Inorg. Nucl. Chem.*, **3**, 178, 1956), the readers may consult the original references. Dichroism of *trans*-$[CoCl_2(en)_2]$ is shown in Fig. 7.15.2.4c.

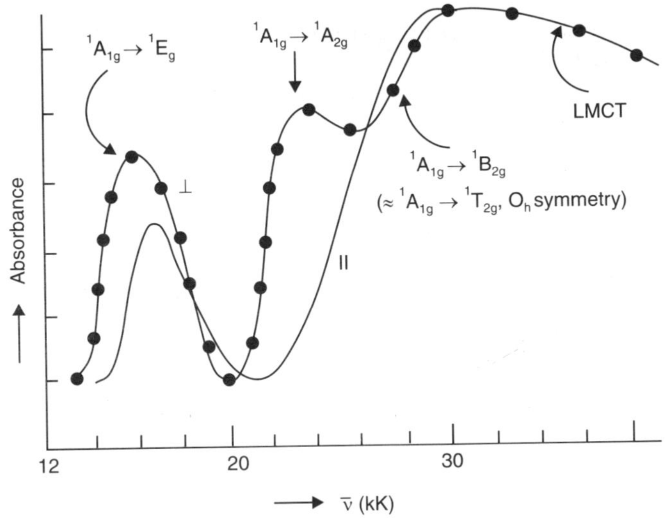

Fig. 7.15.2.4 (c) Dichroism (*i.e.* vibronic polarisation) of *trans*-$[CoCl_2(en)_2]^{2+}$.

7.15.3 Spectral Transitions in Six-Coordinate Low-spin Complexes of Co(III) Belonging to O_h, C_{4v} and D_{4h} Point Group (*cf.* Sec. 7.14.4E for Ni(II) complexes; Sec. 7.14.5E for Cr(III) complexes)

We have already mentioned that lowering of symmetry (with respect to O_h) leads to splitting of the $^1T_{1g}$ and $^1T_{2g}$ terms. But splitting of the $^1T_{2g}$ term is too small to be detected in the ligand field spectra. On the other hand, splitting of the T_{1g} term is quite significant to introduce its effect in the ligand field transition. The representative examples are:

O_h: $[Co(NH_3)_6]^{3+}$; C_{4v}: $[CoCl(NH_3)_5]^{2+}$

D_{4h}: *trans*- $[CoCl_2(NH_3)_4]^+$;

We have already discussed and compared the ligand field spectra of *cis* (C_{2v})- and *trans* (D_{4h})-isomers of $[Co(en)_2F_2]^+$ both of which **electronically maintain the tetragonal symmetry** (*cf.* earlier discussion). In the same way, the ligand field spectra of the *cis*- and *trans*- isomers of $[CoCl_2(NH_3)_4]^+$ can be explained. The *cis*- isomer is violet while the *trans*- isomer is bright green. Spectral characteristics of $[Co^{III}(NH_3)_5X]^{2+}$ (X = F, Cl, Br, I) have been discussed in detail in Sec. 7.21.1 (*cf.* Fig. 7.21.1.1).

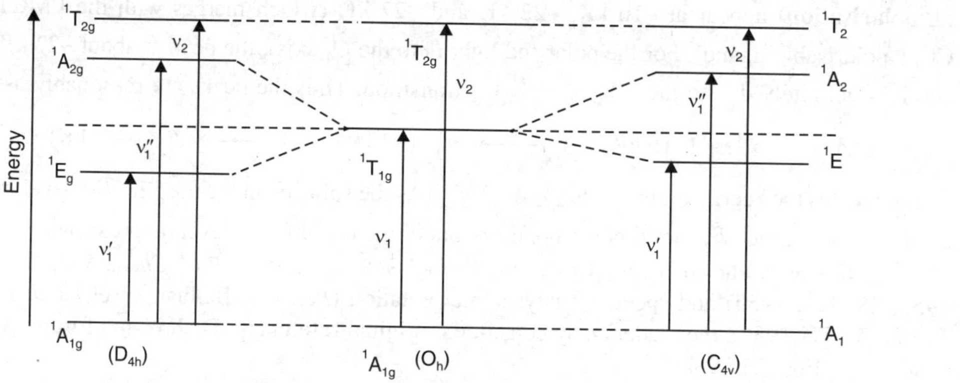

Fig. 7.15.3.1 Electronic transitions in O_h, D_{4h} and C_{4v} point group complexes of Co(III) (*cf.* Correlation Table 7.15.4.2) (Splitting of the T_{2g} level is ignored).

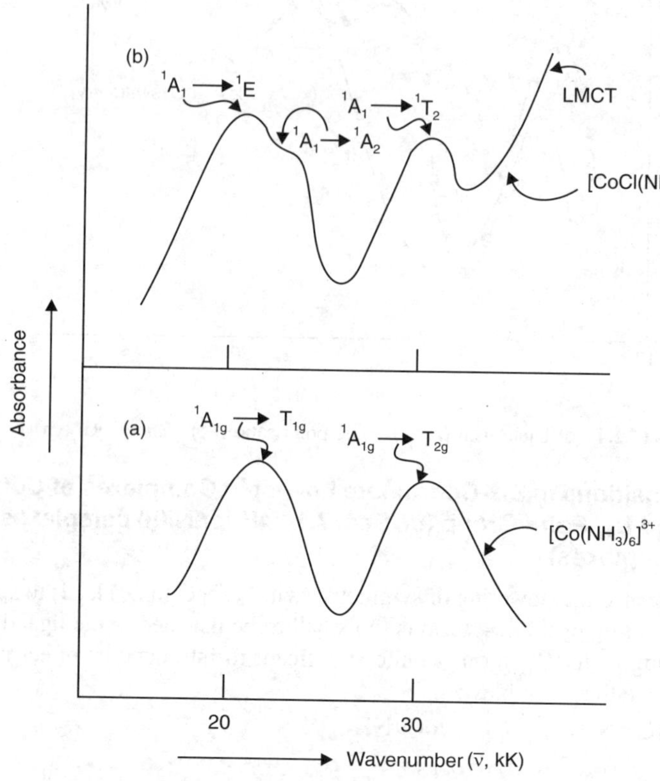

Fig. 7.15.3.2 Comparison of the electronic spectra of $[Co(NH_3)_6]^{3+}(O_h)$ and $[CoCl(NH_3)_5]^{2+}$ (C_{4v}).

In **$[CoCl(NH_3)_5]^{2+}$** (C_{4v} symmetry lacking in centre of symmetry), the $^1T_{1g}$ term of the O_h system splits into the ^1E and $^1A_2$ terms (*cf.* **Correlation Table** 7.15.4.2) but the splitting energy is not sufficiently large to generate two separate peaks (ν_1' and ν_1''). Rather the ν_1 band shows the **doublet character** indicated by the presence of a shoulder (*cf.* Fig. 7.15.3.2). Splitting of the energy levels in $[CoCl(NH_3)_5]^{2+}$ can be understood from that of $[Co(NH_3)_6]^{3+}$. By replacing one NH_3 ligand (stronger field ligand) by a Cl^- ligand (weaker field ligand), we get $[CoCl(NH_3)_5]^{2+}$ (C_{4v}) from $[Co(NH_3)_6]^{3+}$ (O_h). If Cl^- is assumed to occupy one axial position in the z-direction, then the d-orbitals will split as in **z-out distortion,** *i.e.* the orbitals bearing the z-component will be stabilised more. Splitting patterns of the energy levels are of similar type for both *trans*-$[CoCl_2(NH_3)_4]^+$ and $[CoCl(NH_3)_5]^{2+}$ (both experiencing the z-out distortion). However, the T_{1g} state is splitted more in *trans*-$[CoCl_2(NH_3)_4]^+$ (*cf.* effect of two Cl^- *vs.* one Cl^-).

In moving from **$[Co(NH_3)_6]^{3+}$ to $[Co(NH_3)_5X]^{2+}$ (X = F, Cl, Br, I),** the T_{1g} level splits into the components E and A_2 levels and the T_{2g} level does not practically split. The magnitude of splitting of the T_{1g} level depends on the difference of the average ligand field strength between the N—Co^{III}—N and N—Co^{III}—X axes. This difference is maximum for X = I and minimum for X = F. In fact, for X = F, the band does not practically split (*cf.* ligand field strength difference between NH_3 and X^- is minimum for X = F) and for X = Cl, Br only a shoulder (towards the higher energy) appears. For, X = I, practically all the ligand field bands are masked by the LMCT bands. These aspects are discussed in detail in Sec. 7.21.1.

7.15.4 Comparison of Ligand Field Spectra in the Co(III) and Cr(III) Complexes of O_h and D_3 symmetry (*cf.* Sec. 7.14.5 for group theoretical treatment)

Let us illustrate with $[M(NH_3)_6]^{3+}$(O_h) and $[M(en)_3]^{3+}$ (D_3) complexes for M = Co, Cr. In an O_h symmetry, we have the following ligand field transitions.

$$\textbf{Cr(III):}\ {}^4A_{2g} \xrightarrow{\nu_1} {}^4T_{2g};\ {}^4A_{2g} \xrightarrow{\nu_2} {}^4T_{1g}(F);\ {}^4A_{2g} \xrightarrow{\nu_3} {}^4T_{1g}(P)$$

(The high energy ν_3 band is very often obscured by the strong CT band).

$$\textbf{Co(III):}\ {}^1A_{1g} \xrightarrow{\nu_1} {}^1T_{1g};\ {}^1A_{1g} \xrightarrow{\nu_2} {}^1T_{2g}$$

The T_{1g} and T_{2g} terms of an O_h symmetry are split in a D_3 symmetry (*cf.* Correlation Table 7.15.4.2):

$$T_{1g}(O_h) \rightarrow A_2 + E;\quad T_{2g}(O_h) \rightarrow A_1 + E$$
$$A_{1g}(O_h) \rightarrow A_1;\quad A_{2g}(O_h) \rightarrow A_2$$

Thus the ν_1-peak of an O_h system should split into two peaks in a D_3 symmetry.

For Cr(III): $$\underbrace{{}^4A_{2g} \xrightarrow{\nu_1} {}^4T_{2g}}_{O_h};\ \underbrace{{}^4A_2 \xrightarrow{\nu_1'} {}^4A_1;\ {}^4A_2 \xrightarrow{\nu_1''} {}^4E}_{D_3}$$

For Co(III): $$\underbrace{{}^1A_{1g} \xrightarrow{\nu_1} {}^1T_{1g}}_{O_h};\ \underbrace{{}^1A_1 \xrightarrow{\nu_1'} {}^1A_2;\ {}^1A_1 \xrightarrow{\nu_1''} {}^1E}_{D_3}$$

Similarly the ν_2 band of an O_h-symmetry should also split into two components in a D_3 symmetry. Comparison of the ligand field spectra of $[M(NH_3)_6]^{3+}$ (O_h) and $[M(en)_3]^{3+}$(D_3) indicates that in both the cases, the two bands are symmetrical and **no splitting into the doublets is indicated in** $[M(en)_3]^{3+}$. Thus we can consider that in both types of complexes, the **effective octahedral symmetry, $[M(N)_6]^{3+}$ is maintained.** The chelation has not effectively changed the effective octahedral symmetry in understanding the ligand field spectra. It is also supported by the fact that the ε-values are not dramatically increased in moving from the O_h to D_3 symmetry.

Table 7.15.4.1 Comparison of ligand spectra data of $[M(NH_3)_6]^{3+}(O_h)$ and $[M(en)_3]^{3+}(D_3)$ complexes.

Complexes	$^4A_{2g} \xrightarrow[(cm^{-1})]{\nu_1} {^4T_{2g}}$		$^4A_{2g} \xrightarrow[(cm^{-1})]{\nu_2} {^4T_{1g}(F)}$	
$[Cr(NH_3)_6]^{3+}$	21,550	(ε = 30)	28,500	(ε = 30)
$[Cr(en)_3]^{3+}$	21,850	(ε = 75)	28,450	(ε = 60)
	$^1A_{1g} \xrightarrow[(cm^{-1})]{\nu_1} {^1T_{1g}}$		$^1A_{1g} \xrightarrow[(cm^{-1})]{\nu_2} {^1T_{2g}}$	
$[Co(NH_3)_6]^{3+}$	21,200	(ε = 55)	29,550	(ε = 45)
$[Co(en)_3]^{3+}$	21,500	(ε = 90)	29,600	(ε = 80)

Relative intensities of the peaks of $[M(NH_3)_6]^{3+}$ and $[M(en)_3]^{3+}$ complexes: It is evident that the ε-values of the peaks of $[M(en)_3]^{3+}$ complexes are relatively higher compared to those of the $[M(NH_3)_6]^{3+}$ complexes. **It is due to the fact that in a D_3 symmetry, relaxation of Laporte selection rule is more efficient than that in an O_h symmetry.** In the D_3 point group, there is no centre of symmetry. Thus for the complexes of D_3 symmetry, Laporte selection rule is not applicable and in fact, some of the electronic transitions are allowed by their own rights. However, some of these electronic transitions experience the polarisation (*i.e.* **dichroic properties**). These aspects have been illustrated in terms of group theory treatment (*cf.* Secs. 7.14.5, 7.5.1). In the D_3 point group, the orbitals p_x; p_y; d_{xz}, d_{yz}; $d_{x^2-y^2}$, d_{z^2} belong to the *E*-species. In other words, these *p* and *d* orbitals are of the same symmetry and they can undergo *d-p* mixing. On the other hand, in an O_h point group, the asymmetric vibration can only allow the *d-p* mixing momentarily. Thus, the vibronic coupling is the main pathway to relax the Laporte selection rule in the O_h complexes. **In fact, in general, the lowering of symmetry favours the relaxation of Laporte selection rule** (*cf.* T_d *vs.* O_h; D_3 *vs.* O_h; D_{4h} *vs.* C_{2v}). Table 7.15.4.1 indicates that in moving for the O_h symmetry to the D_3 symmetry, the molar extinction coefficients are increased to some extent but the increase is not significantly high (*cf.* ε values: *d–d* transition *vs.* CT band). Thus it is reasonable to conclude that in both the cases, the effective O_h symmetry is maintained at least roughly.

Table 7.15.4.2 Correlation Table for O_h group.

O_h	T_d	D_{4h}	C_{4v}	C_{2v}	D_3^*
A_{1g}	A_1	A_{1g}	A_1	A_1	A_1
A_{2g}	A_2	B_{1g}	B_1	A_2	A_2
E_g	E	$A_{1g} + B_{1g}$	$A_1 + B_1$	$A_1 + A_2$	E
T_{1g}	T_1	$A_{2g} + E_g$	$A_2 + E$	$A_2 + B_1 + B_2$	$A_2 + E$
T_{2g}	T_2	$B_{2g} + E_g$	$B_2 + E$	$A_1 + B_1 + B_2$	$A_1 + E$

(*For D_{3d} (centrosymmetric), same result with the *g*-subscripts.)

7.15.5 Ligand Field Spectra of the Low-Spin Complexes of other d^n-Configurations

d^4-system: The low spin complex $[Mn(CN)_6]^{3-}$ (t_{2g}^4) shows the following two spin allowed ligand field transitions.

$$^3T_{1g} \longrightarrow {^3E_g};\ ^3T_{1g} \longrightarrow {^3A_{1g}}\ (\textit{see}\ \text{Tanabe-Sugano diagram})$$

Theoretically the possible transitions are: $^3T_{1g}\rightarrow{^3E_g}$; $^3T_{1g}\rightarrow{^3T_{2g}}$; $^3T_{1g}\rightarrow{^3A_{1g}}$; $^3T_{1g}\rightarrow{^3A_{2g}}$

$[Mn(CN)_6]^{3-}$ in solution shows a band at about 31,000 cm^{-1} with a shoulder. However, in solid state, the spectrum becomes a complex one due to the spin forbidden transitions along with the spin allowed transitions.

d^5-system: The low spin complex, **$[Fe(CN)_6]^{3-}$ (t_{2g}^5)** experiences the following spin-allowed transitions.

$${}^2T_{2g}\rightarrow{}^2A_{2g},\ {}^2T_{1g};\ {}^2T_{2g}\rightarrow{}^2E_g;\ {}^2T_{2g}\rightarrow{}^2A_{1g}.\ (cf.\ \text{Tanabe-Sugano diagram})$$

In reality, two bands are noted for $K_3[Fe(CN)_6]$.

$${}^2T_{2g}\xrightarrow[30{,}250\ cm^{-1}]{}{}^2A_{2g},\ {}^2T_{1g};\ {}^2T_{2g}\xrightarrow[36{,}700\ cm^{-1}]{}{}^2E_g$$

d^6-system: The low-spin $[Fe(CN)_6]^{4-}$(t_{2g}^6) records two spin-allowed transitions:

$${}^1A_{1g}\xrightarrow[31{,}000\ cm^{-1}]{}{}^1T_{1g};\ {}^1A_{1g}\xrightarrow[37{,}000\ cm^{-1}]{}{}^1T_{2g}\ ;\ (cf.\ \text{Table 7.15.1.1})$$

d^7 system: The low spin ($t_{2g}^6e_g^1$) octahedral complex gives the spin allowed transition ${}^2E_g \rightarrow {}^2T_{1g}$ (*cf.* Tanabe-Sugano diagram). But the complexes (*e.g.* $[Co(NO_2)_6]^{4-}$ (yellow), $[NiF_6]^{3-}$ (violet), etc.) invariably experience the J.T. distortion (generally *z*-out distortion). Depending on the nature of distortion, the complex may adopt the D_{4h} or C_{4v} symmetry. The electron distribution in the ground state is:

$$(d_{xy}, d_{yz})^4 < d_{xy}^2 < d_{z^2}^1 < d_{x^2-y^2}^0$$

Several electronic transitions are possible (*cf.* low-spin square planar complexes of Co(II)).

7.16 LIGAND FIELD SPECTRAL TRANSITIONS IN THE TETRAHEDRAL COMPLEXES OF FIRST TRANSITION SERIES

● **High intensity:** The tetrahedral complexes lacking in centre of symmetry allows some of **the electronic transitions by their own right.** Besides this, the noncentrosymmetric tetrahedral complex can also allow the ***d-p*** **mixing**. Thus, the Laporte selection rule can be more effectively broken down and **the ligand field bands gain more intensity** (*cf.* Sec. 7.5.1 for details). Sometimes, compared to the octahedral complexes, it may be difficult to differentiate the ligand field bands from the charge transfer bands, because the ligand field bands are also quite strong. In the tetrahedral complexes, the transitions should be interpreted **in terms of MOT.** In the T_d-complexes, the *e*-set (*i.e.* $d_{x^2-y^2}$ and d_{z^2}) is basically constituted by the **nonbonding metal d-orbitals** while the t_2-set is constituted by the overlap of the metal *d*-orbitals (*i.e.* d_{xy}, d_{yz}, d_{xz}) and ligand orbitals (*i.e.* *p* or hybrid orbitals), thus the **$e \rightarrow t_2$ is not a pure d–d transition.**

● **Spin-orbit coupling splitting: Peak splitting or peak broadening by spin-orbit coupling interaction is quite important in understanding the ligand field spectra of the tetrahedral complexes (*cf.* Secs. 7.7.4, 8.20.3):** The spin-orbit coupling leads to splitting of the *T*-term and the splitting energy is comparable to that of the tetrahedral splitting energy (Δ_t). Effect of spin-orbit coupling is relatively less important in interpreting the ligand field spectra of the octahedral complexes of 1st transition series because the octahedral splitting energy is much higher than the spin-orbit coupling splitting energy.

● **Orgel diagram:** The spin-allowed transitions can be easily predicted from the Orgel Diagram (Figs. 7.10.1,2). In reality, the spin-paired tetrahedral complexes are rarely known.

The octahedral energy diagram of d^{10-n} configuration is applicable for the tetrahedral energy diagram for d^n configuration. In fact, for the sake of simplicity, the Tanabe-Sugano diagrams of high-spin octahedral complexes of d^{10-n} configuration are used to interpret the ligand field transitions in the tetrahedral complexes of d^n-configuration.

(1) *d^1 and d^6 systems*: The spin allowed transition is:

$$^2E \rightarrow {}^2T_2\ (d^1\text{-system});\ ^5E \rightarrow {}^5T_2\ (d^6\text{-system}).$$

$[VCl_4]$ (d^1-configuration) shows the ligand absorption band at 9,000 cm^{-1} ($\varepsilon \sim 100$) for the $^2E \rightarrow {}^2T_2$ transition. This absorption band can be resolved **into three components** (6,600; 7,850 and 9,250 cm^{-1}). These may arise in different possible ways: ***vibrational fine structure, splitting of the excited state by spin-orbit coupling or Jahn-Teller effect.*** However, theoretical calculations support the Jahn-Teller distortion. The complex also shows the strong LMCT band (Cl → V). Some tetrahedral complexes like $[V(NR_2)_4]$ (R = Me, Et) may adopt the distorted tetrahedral structure of D_{2d} symmetry. It may be noted that in moving from the T_d to D_{2d} symmetry, E splits into A_1 and B_1; and T_2 splits into B_2 and E. For the d^1 system, B_1 is the ground state for the D_{2d} symmetry (*cf.* Fig. 7.16.1). The transitions are:

D_{2d} (d^1 system): $^2B_1 \rightarrow {}^2E$; $^2B_1 \rightarrow {}^2B_2$ (orbitally forbidden; Solved Problem 15); **$[FeCl_4]^{2-}$, $[FeBr_4]^{2-}$**, etc. are the representative examples of tetrahedral complexes of d^6-system. The spin allowed transition is:

$$^5E \xrightarrow{\nu} {}^5T_2,\ \nu = 4{,}000\ \text{cm}^{-1}\ ([FeCl_4]^{2-})$$
$$= 3{,}000\ \text{cm}^{-1}\ ([FeBr_4]^{2-})$$
$$= 2{,}500\ \text{cm}^{-1}\ ([FeI_4]^{2-})$$

For **$[FeL_2X_2]$ (C_{2v}, rhombic symmetry)**, the low symmetry ligand field splits the E and T_2 terms as follows:

$$E(T_d) \rightarrow A_1 + A_2\ (C_{2v})$$
$$T_2(T_d) \rightarrow A_1 + B_1 + B_2\ (C_{2v})$$

In practice, in such cases only **two broad bands** are observed. It indicates that the splitting of the peaks simply broaden the main peaks.

(2) *d^4 and d^9 systems*: The spin allowed transition is:

$$^5T_2 \rightarrow {}^5E\ (\text{for } d^4 \text{ system});\ ^2T_2 \rightarrow {}^2E\ (\text{for } d^9 \text{ system}).$$

$[CrCl_4]^{2-}$, $[CrBr_4]^{2-}$ are the representative examples of the tetrahedral complexes of d^4-system. They show the absorption band at about 10,000 cm^{-1} for $^5T_2 \rightarrow {}^5E$ transition.

Perfectly tetrahedral complexes of Cu(II) (d^9-system) should record **only one absorption band** the transition $^2T_2 \rightarrow {}^2E$. The bands at ~8,000 cm^{-1} and ~9,000 cm^{-1} for **$[CuBr_4]^{2-}$ and $[CuCl_4]^{2-}$** respectively have been considered for the $^2T_2 \rightarrow {}^2E$ transitions (assuming the tetrahedral geometry).

Undistorted tetrahedral Cu(II) complex ($e^4t_2^5$) is unexpeced because it will experience the **Jahn-Teller distortion** (*cf.* t_2 level is unsymmetrically filled in). Cu(II) has a tendency to adopt the square planar geometry and in such cases, the tetrahedral geometry experiences a compression (*cf.* Table 3.11.3.3) along one of the S_4-axis (*i.e.* flattening of the tetrahedron along a C_2-axis leading towards a square planar geometry). This distortion will generate a **structure of D_{2d} symmetry** (which is an intermediate state of the two limiting geometries, *i.e.* tetrahedron and square planar). Splitting of the energy levels and d-orbitals in the T_d and D_{2d} symmetry are shown in Fig. 7.16.1 (*cf.* Fig. 3.11.2.1).

The possible electronic transitions are:

$$\mathbf{T_d}:\ ^2T_2 \rightarrow {}^2E;\ i.e.\ (e \rightarrow t_2)$$

D_{2d} (*cf.* Fig. 7.16.1): $^2B_2 \xrightarrow{\nu_3} {}^2E$; $^2B_2 \xrightarrow{\nu_2} {}^2B_1$ (**orbitally forbidden,** *see* Solved Problem 15);

$$^2B_2 \xrightarrow{\nu_1} {}^2A_1$$

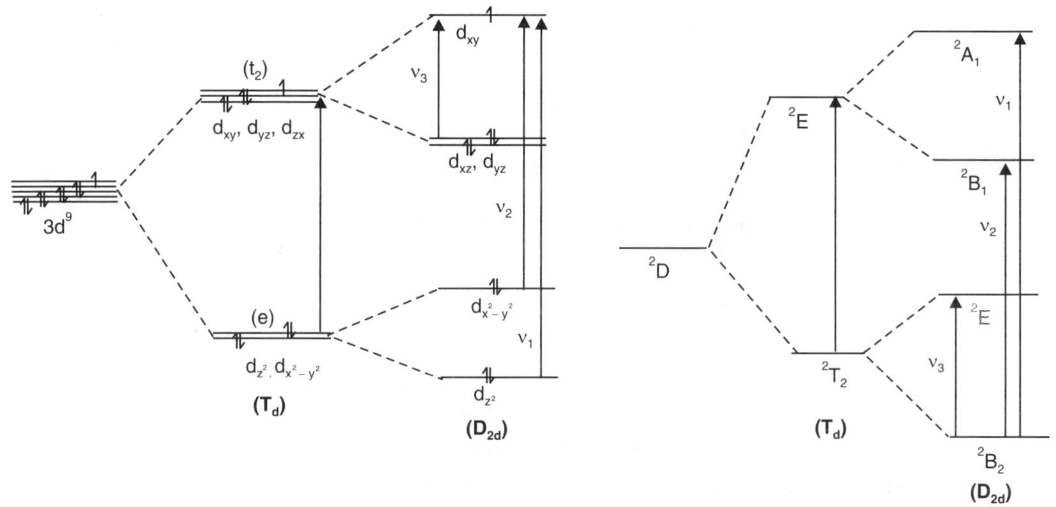

Fig. 7.16.1 Splitting of the *d*-orbitals and energy levels in T_d and D_{2d} (compressed tetrahedron) symmetry for Cu(II) (d^9 system).

In terms of the *d*-orbitals, the possible transitions in the D_{2d} symmetry are:

$$d_{z^2}^2 d_{x^2-y^2}^2 \left(d_{xz}d_{yz}\right)^4 d_{xy}^1 \left(i.e.\ {}^2B_2\right) \xrightarrow[\left({}^2B_2 \longrightarrow {}^2A_1\right)]{\nu_1} d_{z^2}^1 d_{x^2-y^2}^2 \left(d_{xz}d_{yz}\right)^4 d_{xy}^2 \left(i.e.\ {}^2A_1\right)$$

$$d_{z^2}^2 d_{x^2-y^2}^2 \left(d_{xz}d_{yz}\right)^4 d_{xy}^1 \left(i.e.\ {}^2B_2\right) \xrightarrow[\left({}^2B_2 \longrightarrow {}^2B_1\right)]{\nu_2} d_{z^2}^2 d_{x^2-y^2}^1 \left(d_{xz}d_{yz}\right)^4 d_{xy}^2 \left(i.e.\ {}^2B_1\right)$$

$$d_{z^2}^2 d_{x^2-y^2}^2 \left(d_{xz}d_{yz}\right)^4 d_{xy}^1 \left(i.e.\ {}^2B_2\right) \xrightarrow[\left({}^2B_2 \longrightarrow {}^2E\right)]{\nu_3} d_{z^2}^2 d_{x^2-y^2}^2 \left(d_{xz}d_{yz}\right)^3 d_{xy}^2 \left(i.e.\ {}^2E\right)$$

It may be noted that the ν_1 transition is *z*-polarised and the ν_3 transition is *x*, *y*-polarised. The ν_2-transition is orbitally forbidden. In short, the transitions in D_{2d} symmetry can be represented as follows:

$$d_{z^2} \xrightarrow{\nu_1} d_{xy};\ d_{x^2-y^2} \xrightarrow{\nu_2} d_{xy};\ d_{xz}, d_{yz} \xrightarrow{\nu_3} d_{xy}$$

In **$Cs_2[CuCl_4]$,** transitions for $[CuCl_4]^{2-}$ (in D_{2d} symmetry) are:

$$\nu_1 = 9050\ \text{cm}^{-1};\ \nu_2 = 7900\ \text{cm}^{-1};\ \nu_3 \approx 5{,}000\ \text{cm}^{-1}$$

The finer resolution indicates that the ν_3-band consists of two components (5,500 cm^{-1} and 4,800 cm^{-1}). It is suggested that further distortion leads to D_{2d} to C_s symmetry through the splitting of the d_{xz} and d_{yz} orbitals. This splitting leads to the following transitions.

$$d_{xz} \rightarrow d_{xy}\ (5{,}500\ \text{cm}^{-1});\ d_{yz} \rightarrow d_{xy}\ (4{,}800\ \text{cm}^{-1}).$$

There is another view of **spin-orbit coupling effect** ($\lambda = -830\ cm^{-1}$) to split the ^2E level. This can explain the possible four transitions in $[CuCl_4]^{2-}$. The orbitally **forbidden transition, $^2B_2 \rightarrow {}^2B_1$** gains the intensity through the mixing with one component of the ^2E state (which splits through the spin-orbit coupling). Thus the spectral features of $[CuCl_4]^{2-}$ may be explained by the spin-orbit coupling and/or further distortion of D_{2d} to C_s symmetry.

(3) *d^2 and d^7 systems*: From the Orgel diagram (Fig. 7.10.2), we have the following three spin-allowed transitions.

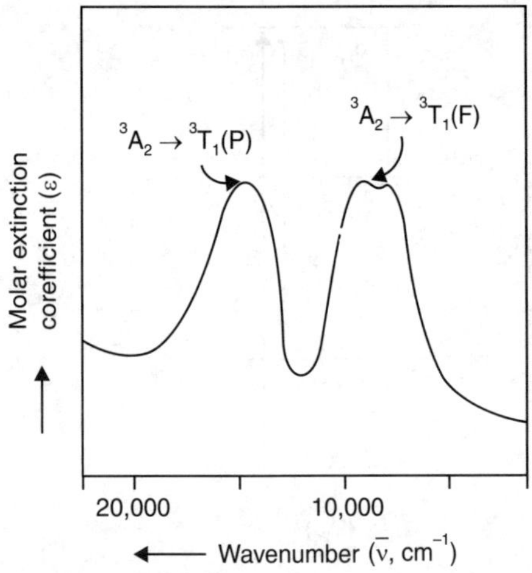

Fig. 7.16.2 Electronic absorption spectrum of $[VCl_4]^-$ (a d^2-tetrahedral complex).

$$^xA_2(F) \xrightarrow{\nu_1} {}^xT_2(F)(\textbf{orbitally forbidden});\ {}^xA_2(F) \xrightarrow{\nu_2} {}^xT_1(F) \text{ and } {}^xA_2(F) \xrightarrow{\nu_3} {}^xT_1(P).$$
$$(x = 3 \text{ for } d^2; = 4 \text{ for } d^7)$$

Without considering any configurational interaction,

$$\nu_1 = 10Dq,\ \nu_2 = 18Dq,\ \nu_3 = 12Dq + 15B'$$

Due to the configurational interaction between the $T_1(P)$ and $T_1(F)$ states, they follow the **noncrossing rule** and the energy order is: $\nu_3 \rangle \nu_2 \rangle \nu_1$.

Tetrahedral complexes of d^2-system (*e.g.* **$[VCl_4]^-$**) are rarely known. On the other hand, tetrahedral complexes of Co(II) (d^7-system), have been studied in more detail to assign their spectral transitions.

(A) $[VCl_4]^-$: Here, one band consists of three components (8,000 cm^{-1}, 9,500 cm^{-1} and 10,500 cm^{-1}) while the high energy band appears at 15,000 cm^{-1}. Splitting of the T-term through the spin-orbit coupling is to be considered to interpret the spectra.

(B) $[CoCl_4]^{2-}$: The spectrum of **$[CoCl_4]^{2-}$** in aqueous solution indicates two bands:

$$5{,}800\ cm^{-1}\ (\varepsilon = 64),\ 15{,}000\ cm^{-1}\ (\varepsilon = 520).$$

The bands have been assigned as follows:

$$^4A_2(F) \xrightarrow[(5{,}800\ cm^{-1})]{\nu_2} {}^4T_1(F);\ {}^4A_2(F) \xrightarrow[(\sim 15{,}000\ cm^{-1})]{\nu_3} {}^4T_1(P)$$

If the band at 5,800 cm^{-1} is considered for the transition, $^4A_2(F) \rightarrow {}^4T_2(F)$, *i.e.* ν_1, then the $10Dq$ value becomes 5,800 cm^{-1} ***which is too high for the tetrahedral complexes of Co(II)***. In fact, this value cannot be fitted in the energy diagram. This is illustrated below:

if ν_1 = 5,800 cm^{-1}, and ν_2 = 15,000 cm^{-1}, then $\frac{\nu_2}{\nu_1}$ becomes 2.58.

From Fig. 7.10.4, it is evident that for the A_2-ground spectroscopic term, the $\frac{\nu_2}{\nu_1}$ ratio lies within the range 1.8 – 1.22. Thus the band assignments, ν_1 = 5,800 cm^{-1} and ν_2 = 15000 cm^{-1} are not theoretically acceptable.

In the energy diagram, the ν_3/ν_2 ratio $\left(\frac{15{,}000}{5{,}800} = 2.58\right)$ can be fitted for Dq/B' = 0.45 and ν_2/B' = 7.9.

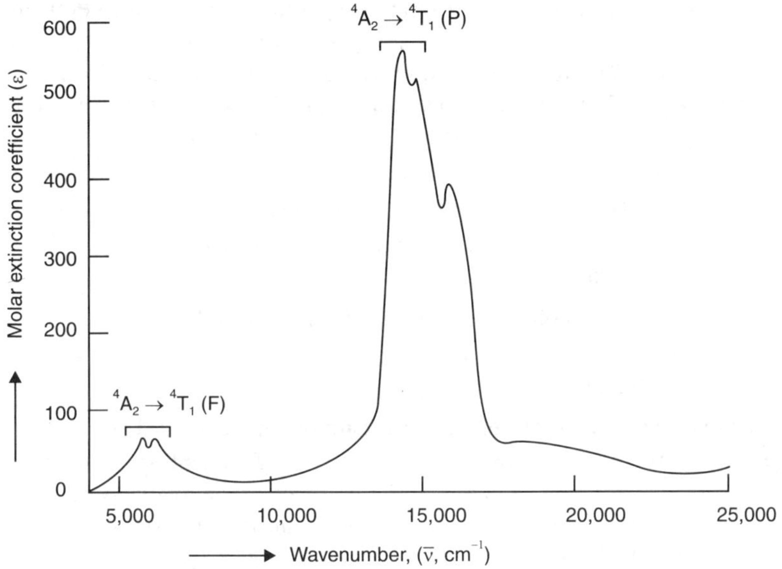

Fig. 7.16.3 Electronic absorption spectrum of $[CoCl_4]^{2-}$ (a d^7-tetrahedral complex).

It leads to:

$$\frac{5,800}{B'} = 7.9 \text{ or, } B' = 734 \text{ cm}^{-1}$$

and
$$Dq = B' \times 0.45 \approx 330 \text{ cm}^{-1}$$
$$\text{or, } 10Dq = 3,300 \text{ cm}^{-1}$$

This B' value is about 70% of the free ion value (*cf.* $B_0 = 1033$ cm^{-1}). It indicates the presence of an extensive ***covalent interaction*** in the complex, $[CoCl_4]^{2-}$.

We can verify the band assignments from Fig. 7.10.4.

$$\left.\frac{\nu_3}{\nu_1} = \frac{15,000}{3,300} = 4.54; \frac{\nu_2}{\nu_1} = \frac{5,800}{3,300} = 1.75\right\}(\text{Experimental values})$$

For $\dfrac{Dq}{B'} = 0.45$, Fig. 7.10.4 theoretically gives the values 4.59 and 1.74 for ν_3/ν_1 and ν_2/ν_1 respectively. These calculated values are in good agreement with the experimental values.

Thus the transition, $^4A_2(F) \rightarrow {}^4T_2(F)$, *i.e.* ν_1, is expected to appear at 3,300 cm^{-1}. In fact, studies in the infra-red region indicates the predicted band to appear at about 3,500 cm^{-1}.

The ν_1 (= $10Dq$) band at 3,500 cm^{-1} indicates:

$$10Dq = 3,500 \text{ cm}^{-1}.$$

Thus, the calculated positions of ν_2 becomes as follows:

$$\nu_2 = 18Dq = 18 \times 350 \text{ cm}^{-1} = 6,300 \text{ cm}^{-1} \text{ (without configurational interaction)}$$

This calculated value is significantly different from the experimental value (= 5,800 cm^{-1}). Similarly, if ν_3 (= $12Dq + 15B_0$) is calculated, it becomes:

$$\nu_3 = 12 \times 350 + 15,500 = 19,700 \text{ cm}^{-1}; \text{ taking } 15B_0 = 15,500 \text{ cm}^{-1}.$$

This calculated value differs significantly from the experimental value (= 15,000 cm^{-1}). Apparently it is due to the reduction of B-value in the complex (*i.e.* $15B$ value is dropped to about 10,800 cm^{-1} which is about 70% of $15B_0$). ***The above discrepancy can be explained by considering the configuration interaction between the $^4T_1(F)$ and $^4T_1(P)$ levels and the reduced B-value.*** The modified relations are:

$$\nu_1 = 10Dq,\ \nu_2 = 18Dq - x,\ \nu_3 = 12Dq + 15B' + x$$

(x = configurational interaction energy).

Taking $10Dq\ (= \nu_1) = 3{,}500\ cm^{-1}$ (experimental value), ν_2 becomes:

$$\nu_2 = 18 \times 350 - x = 6{,}300 - x = 5{,}800\ cm^{-1}$$

i.e. $x = 500\ cm^{-1}$

$$\nu_3 = 15{,}000\ cm^{-1} = 12 \times 350 + 15B' + 500$$

or, $15B' = 15{,}000 - 4{,}700 = 10{,}300\ cm^{-1}$

or, $B' \approx 687\ cm^{-1}$

(C) Important characteristic features of the spectrum of blue coloured $[CoCl_4]^{2-}$

(i) **Allowedness of the spectral transitions:** The $A_2 \rightarrow T_1$ transition is **orbitally allowed** (*cf.* $A_2 \times \mu \times T_1 = A_2 \times T_2 \times T_1 = A_1 + E + T_1 + T_2$; presence of the totally symmetric representation A_1 in the direct product result makes the electronic transition allowed by its own right) but the $A_2 \rightarrow T_2$ transition is **orbitally forbidden** (*cf.* $A_2 \times \mu \times T_2 = A_2 \times T_2 \times T_2 = A_2 + E + T_1 + T_2$, absence of A_1 species in the direct product result makes the transition orbitally forbidden). However, the $A_2 \rightarrow T_2$ transition may gain some intensity through the vibronic coupling. In fact, the $A_2 \rightarrow T_1$ transition gives the more intense band (*see* Fig. 7.16.3). This aspect has been discussed in detail in Sec. 7.5.1.

(ii) **High intensity of the high energy band (ν_3):** The visible spectrum of $CoCl_4^{2-}$ is dominated by the highest energy band (at ~ 15,000 cm^{-1}) and intensity of this peak is very high ($\varepsilon \approx 500$). On the other hand, for $[Co(OH_2)_6]^{2+}$, intensity of the spectrum is very low and it looks pale pink. The **high intensity** of the absorption bands in the tetrahedral complexes is the common feature, because some of the electronic transitions are allowed by their own right in the tetrahedral complexes. Some forbidden transitions may be allowed through vibronic coupling. Besides these, for the noncentrosymmetric tetrahedral complexes, **d-p mixing allows the relaxation** of Laporte Selection rule better. All these aspects have been discussed in detail in Sec. 7.5.1.

(iii) **Relatively lower intensity of the lower energy band ν_2:** Here it is worth mentioning that, the transition, $^4A_2(F) \rightarrow {}^4T_1(F)$ *i.e.* ν_2 band is relatively weak in intensity ($\varepsilon \approx 60$).

(iv) **ν_1-band in the infra-red region:** The $^4A_2(F) \rightarrow {}^4T_2$(F) orbitally forbidden transition generally occurs in the infra-red region and this band (*i.e.* ν_1) is very often obscured by ligand field bands due to the vibrational transitions. ***Thus spectrum of the tetrahedral Co(II)-complexes is mainly characterized by the ν_2 and ν_3 ligand field transitions.***

(v) **Bandwidth and fine-structure components in the ν_2 and ν_3 bands:** For $[CoCl_4]^{2-}$, both the ν_2 *i.e.* $^4A_2(F) \rightarrow {}^4T_1(F)$ and ν_3, *i.e.* $^4A_2(F) \rightarrow {}^4T_1(P)$ bands appear with some **structure** (*i.e.* splitting). **Spin-orbit coupling can split the T-terms** (*cf.* Secs. 7.7.4 and 8.20.3). The fine-structure in the bands are believed to be caused due to spin-orbit coupling. Splitting of the T-terms may be also caused by **lowering of symmetry,** *e.g.* $T_d \rightarrow D_{2d} \rightarrow C_s$, *cf.* Fig. 7.16.1). The spin-orbit coupling also partially allows the **spin-forbidden quartet → doublet transitions.** Some of these spin-forbidden transitions occur in the range of spin-allowed transitions. These factors like *vibronic coupling, symmetry lowering, spin-orbit coupling, spin-forbidden transitions are responsible to give* the **fine structure components** of the ν_2 and ν_3 bands.

(vi) **Complex envelopes of the spectral bands:** The visible transitions in the tetrahedral complexes, generally have the **complex envelopes** because of the presence of several spin-forbidden transitions (to the doublet states) occurring in the same region of spin-allowed transitions. These spin-forbidden transitions earn some intensity through the *spin-orbit coupling allowing the mixing of quartet and doublet states and intensity stealing from the nearby spin allowed transitions*. The spin-forbidden transitions in $[CoCl_4]^{2-}$ also produce the bands near 20,000 cm^{-1}. Intensities of these are bands almost comparable to those of the spin-allowed bands of the octahedral complexes.

In fact, bandwidths of the ν_2 and ν_3 bands are in the range of about 1500-3000 cm^{-1}. For the computational purpose, ν_2 and ν_3 are assumed to represent the **centre of gravity of the broad bands.**

Peak broadening through the spin-orbit coupling in T_d *vs.* O_h Complexes: Peak splitting or peak broadening effects by spin-orbit coupling is not so important for the octahedral complex of 1st transition series because the energy separation caused by crystal field splitting in the O_h systems is much larger than the energy separation caused by spin-orbit coupling (*cf.* Sec. 7.7.4).

Positions of the ligand field bands of tetrahalidocobalt(II) complexes: Ligand field strength $Cl^- > Br^- > I^-$

	ν_2 ($\approx 18Dq$) (cm^{-1})	ν_3 ($\approx 12Dq + 15B'$)
$[CoCl_4]^{2-}$	5,800	15,000
$[CoBr_4]^{2-}$	5,300	13,700
$[CoI_4]^{2-}$	5,000	12,500

(D) Characteristics of the ligand field bands of $[Co(NCS)_4]^{2-}$: $\nu_2 \approx 18Dq \approx 7{,}780$ cm^{-1}, $\nu_3 \approx 12Dq + 15B' \approx 16{,}250$ cm^{-1}. The ν_1 transition is not definitely 7,780 cm^{-1} because it will lead to $10Dq = 7{,}780$ cm^{-1} which is too large for the tetrahedral complex of Co(II). Thus it is quite reasonable to consider:

$$\nu_2 = 7{,}780 \text{ cm}^{-1} \text{ and } \nu_3 = 16{,}250 \text{ cm}^{-1}$$

Without considering the symmetry interaction we have:

$$\nu_2 = 18Dq = 7{,}780 \text{ cm}^{-1} \; i.e. \; Dq \approx 432 \text{ cm}^{-1} \text{ or } 10Dq = 4320 \text{ cm}^{-1}$$

$$\nu_3 = 12Dq + 15B' = 16{,}250 \text{ cm}^{-1} \; i.e. \; B' = \frac{16{,}250 - 12 \times 432}{15} \text{ cm}^{-1} = 738 \text{ cm}^{-1}$$

More refined values of B' and $10Dq$ can be obtained by using the Tanabe-Sugano diagram or Fig. 7.10.4

$$\frac{\nu_3}{\nu_2} = \frac{^4A_2(F) \rightarrow {}^4T_1(P)}{^4A_2(F) \rightarrow {}^4T_1(F)} = \frac{16{,}250}{7{,}780} = 2.1$$

This ratio leads to the following values (obtained from Figs. 7.10.4 and T.S. diagram):

$$\frac{Dq}{B'} = 0.65, \; \frac{\nu_3}{B'} = 23.38, \; \frac{\nu_2}{\nu_1} = 1.71 \text{ and } \frac{\nu_3}{\nu_1} = 3.6$$

$$\frac{\nu_3}{B'} = \frac{16{,}250}{B'} = 23.38 \; i.e. \; B' = \frac{16{,}250}{23.38} = 695 \text{ cm}^{-1}$$

$$\frac{Dq}{B'} = 0.65 \; i.e. \; Dq = 695 \times 0.65 \approx 452 \text{ cm}^{-1} \; i.e. \; \nu_1 = 10Dq = 4520 \text{ cm}^{-1}$$

$$\frac{\nu_2}{\nu_1} = 1.71 \text{ i.e. } \frac{7{,}780}{\nu_1} = 1.71 \text{ or } \nu_1 = 4550 \text{ cm}^{-1}$$

$\nu_2 = 18Dq - x = 18 \times 452 - x = 7{,}780 \text{ cm}^{-1}$ *i.e.* $x = 18 \times 452 - 7{,}780 = 8{,}136 - 7{,}780 = 356 \text{ cm}^{-1}$

x(configurational interaction energy) = 356 cm^{-1}

(E) More rigorous energy expression for the transitions in the tetrahedral d^7-complexes: Here it may be mentioned that the detailed calculations of ν_1, ν_2 and ν_3 bands can be done as follows:

$$\nu_1 = \Delta$$

$$\nu_2 = 1.5\Delta + 7.5B' - \frac{1}{2}\left[225B'^2 + \Delta^2 - 18B'\Delta\right]^{1/2} = 1.5\Delta + 7.5B' - Q$$

$$\nu_3 = 1.5\Delta + 7.5B' + \frac{1}{2}\left[225B'^2 + \Delta^2 - 18B'\Delta\right]^{1/2} = 1.5\Delta + 7.5B' + Q$$

$$\Delta = \frac{1}{3}[\nu_2 + \nu_3 - 15B']$$

$$Q = \frac{1}{2}[\nu_3 - \nu_2]$$

$$4Q^2 = \Delta^2 - 18\Delta B' + 225B'$$

Let us illustrate the application of the above relations for a particular tetrahedral complex of Co(II):

(F) Bis(kojato)cobalt(II): One band at 7,140 cm^{-1}; a doublet peak (18,180 cm^{-1} and 19,600 cm^{-1}). *For the Co(II) - tetrahedral complexes, the ν_1 band lies in the infra-red region.* Thus it is reasonable to assume: $\nu_2 = 7{,}140 \text{ cm}^{-1}$, ν_3 (split by spin-orbit coupling) as the mean of 18,180 cm^{-1} and 19,600 cm^{-1}, *i.e.* $\nu_3 = \frac{1}{2}(18{,}180 + 19{,}600) = 18{,}890 \text{ cm}^{-1}$

$$\Delta = \frac{1}{3}[\nu_2 + \nu_3 - 15B'] = \frac{1}{3}[7{,}140 + 18{,}890 - 15B']$$

$$= 8676.7 - 5B'$$

$$Q = \frac{1}{2}[\nu_3 - \nu_2] = \frac{1}{2}[18{,}890 - 7{,}140] = 5875 \text{ cm}^{-1}$$

$$4Q^2 = \Delta^2 - 18\Delta B' + 225B'^2$$

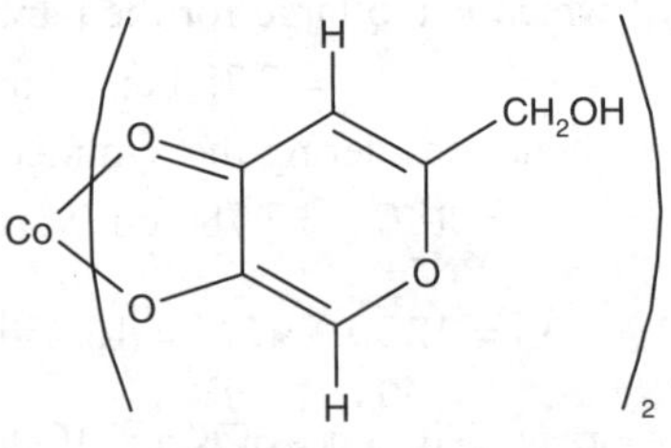

Complex of kojic acid

Using, $\Delta = 8676.7 - 5B'$ and $Q = 5875 \text{ cm}^{-1}$, the acceptable solution of the equation leads to $B' = 916 \text{ cm}^{-1}$ (*cf.* other root is negative which is not acceptable).

Thus, $\Delta = 8676.7 - 5 \times 916 \approx 4100 \text{ cm}^{-1}$.

(4) *d*^{5}*-system*: For the d^5 system ($\equiv d^{10-5}$), the same energy level diagram (Fig. 7.14.9.1) is applicable for the octahedral, tetrahedral or cubic fields. **$[MnCl_4]^{2-}$** and **$[MnBr_4]^{2-}$** are the representative examples in this group. The ground state is $^6A_1$. There is no excited state having the spin multiplicity 6. ***Thus there are only spin-forbidden transitions.***

These are: $^6A_1 \xrightarrow{\nu_1} {}^4T_1(G)$; $^6A_1 \xrightarrow{\nu_2} {}^4T_2(G)$; $^6A_1 \xrightarrow{\nu_3} {}^4E(G)$; $^6A_1 \xrightarrow{\nu_4} {}^4A_1(G)$, etc. The ν_3 and ν_4 transitions are merged *i.e.* $^6A_1 \to {}^4A_1(G)$, $^4E(G)$. These transitions are independent of 10Dq and consequently these are **sharp** (*cf.* Fig. 7.7.3.1). It may be noted that in the high-spin Mn(II)-octahedral complexes, the transitions are also *spin-forbidden*. But compared to the octahedral complexes, in general, the spectral bands of the tetrahedral complexes are about 200 times stronger in intensity (*cf.* the effect of *d–p* mixing and allowedness of the electronic transitions by their own rights).

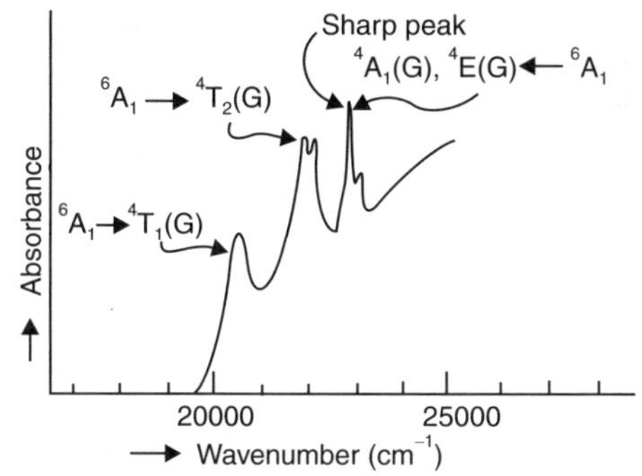

Fig. 7.16.5 Electronic absorption spectrum of $[MnCl_4]^{2-}$ in $Cs_2[MnCl_4]$.

$[FeCl_4]^-$ is also an example of tetrahedral complex of d^5-system. The strong charge transfer bands in $[FeCl_4]^-$ complicate the interpretation of the ligand field bands. Three bands have been characterized.

(5) *d^3 and d^8-system*: The possible transitions are:

$$\left.\begin{aligned} &{}^xT_1(F) \xrightarrow{\nu_1} {}^xT_2(F), \ \nu_1 = 8Dq + x' \approx 8Dq \\ &{}^xT_1(F) \xrightarrow{\nu_2} {}^xA_2(F), \ \nu_2 = 18Dq + x' \approx 18Dq \\ &{}^xT_1(F) \xrightarrow{\nu_3} {}^xT_1(P), \ \nu_3 = 15B' + 6Dq + 2x' \approx 15B' + 6Dq \end{aligned}\right\} \left(x = 4 \text{ for } d^3,\ x = 3 \text{ for } d^8\right)$$

$$i.e.\ \nu_2 - \nu_1 = 10Dq;\ \nu_3 - 2\nu_1 = 15B' - 10Dq \text{ and } 15B' = \nu_2 + \nu_3 - 3\nu_1$$

Here x' gives the measure of configurational energy due to the **symmetry interaction** between $^xT_1(F)$ and $^xT_1(P)$. More rigorously, we can write:

$$\nu_1[{}^xT_1(F) \rightarrow {}^xT_2] = \frac{1}{2}\left[10Dq - 15B' + \left\{225B'^2 + 100(Dq)^2 + 180DqB'\right\}^{1/2}\right]$$

$$\nu_2[{}^xT_1(F) \rightarrow {}^xA_2(\text{F})] = \frac{1}{2}\left[30Dq - 15B' + \left\{225B'^2 + 100(Dq)^2 + 180DqB'\right\}^{1/2}\right]$$

$$\nu_3[{}^xT_1(F) \rightarrow {}^xT_1(\text{P})] = \left[225B'^2 + 100(Dq)^2 + 180DqB'\right]^{1/2}$$

The ν_2 transition represents a **two-electron transition**. Thus it is a **spin forbidden transition** and consequently its intensity is relatively less.

The energy levels $^xA_2(F)$ and $^xT_1(P)$ can cross each other. Thus, the relative positions of ν_2 and ν_3 depend on the crystal field strength. At the lower crystal field strength, $\nu_2 \langle \nu_3$ while at the higher crystal field strength, $\nu_3 \langle \nu_2$. In most of the tetrahedral complexes (where the crystal field splitting is relatively smaller), $\nu_2 \langle \nu_3$ condition is maintained.

Tetrahedral complexes of Ni(II) are the common examples of the d^8-system. The ν_1-band $[^3T_1(F) \rightarrow {}^3T_2(F)]$ generally lies in the *infra-red region* (3,000–5,000 cm^{-1}). The ν_2 band $[^3T_1(F) \rightarrow {}^3A_2(F)]$ generally lies in the near infra-red region (6,500 to 9,000 cm^{-1}). This two-electron transition band is a forbidden transition and it is weak. The ν_3-band $[^3T_1(F) \rightarrow {}^3T_1(P)]$ lies in the *visible range*

(12,000–17,000 cm^{-1}, ε ≈ 100–200). Some **spin forbidden transitions** to the components of ^1D and ^1G term have been noted. These are very weak bands.

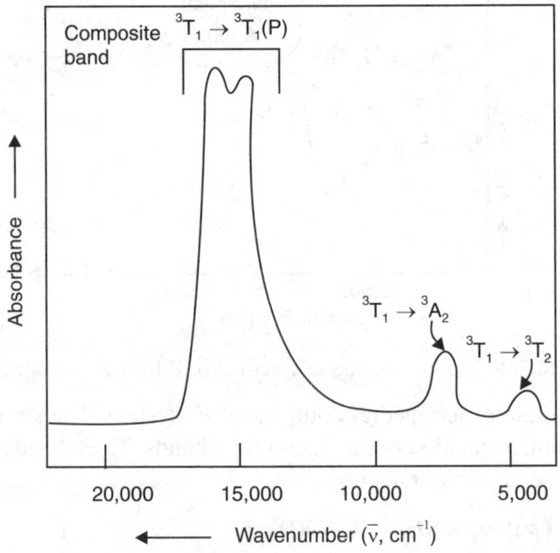

Fig. 7.16.6 Electronic absorption spectrum of $[NiCl_4]^{2-}$ (a d^8-tetrahedral complex).

$[NiCl_4]^{2-}$ shows two bands at 7,800 cm^{-1} and at about 16,000 cm^{-1}. In fact, the band at 16,000 cm^{-1} represents a **multiplet band (*i.e.* splitting pattern)**. The band splitting may be due to **splitting of the T-term due to spin-orbit coupling and/or Jahn-Teller distortion.** The centre of gravity of this multiplet band lies at 16,000 cm^{-1}. The assigned transitions are:

$$^3T_1(F) \xrightarrow{\nu_2} {}^3A_2(F),\ \nu_2 = 7{,}800\ \text{cm}^{-1}\ (\varepsilon \approx 12)$$

$$^3T_1(F) \xrightarrow{\nu_3} {}^3T_1(P),\ \nu_3 = 16{,}000\ \text{cm}^{-1}\ (\varepsilon \approx 70)$$

The band (at 7,800 cm^{-1}) cannot be assigned for ν_1 *i.e.* $^3T_1(F) \rightarrow {}^3T_2(F)$. If it is considered for ν_1, then it leads to: $\nu_1 = 8Dq$ *i.e.* 7,800 cm^{-1} = $8Dq$ or $10Dq$ = 9,750 cm^{-1} which is too high for the tetrahedral $[NiCl_4]^{2-}$ complex (*cf.* $10Dq$ for octahedral $[Ni(OH_2)_6]^{2+}$ is 8,500 cm^{-1}). The ratio ν_3/ν_2 $= \dfrac{16{,}000}{7{,}800} = 2.05$ fits with the following value:

$$Dq/B' = 0.5 \text{ (obtained from Fig. 7.10.4)}$$

and	$\dfrac{^3T_1(F) \longrightarrow {}^3A_2(F)}{B'} = \dfrac{\nu_2}{B'} = 9.2$	$\dfrac{^3T_1(F) \rightarrow {}^3T_1(P)}{B'} = \dfrac{\nu_3}{B'} = 18.44$
or,	$\dfrac{7{,}800}{B'} = 9.2;$ or $B' \approx 848$ cm^{-1}	or, $\dfrac{16{,}000}{18.44} = B' \approx 868$ cm^{-1}
and	$Dq = B' \times 0.5 = 848 \times 0.5 = 424$	and $Dq = B' \times 0.5 = 868 \times 0.5 = 434$ cm^{-1}
i.e.	$10Dq = 4{,}240$ cm^{-1}	*i.e.* $10Dq = 4340$ cm^{-1}

Taking, $\nu_2 \approx 18Dq$ (without considering the configurational interaction energy), we get:

$$7{,}800\ \text{cm}^{-1} = 18Dq \text{ or } 10Dq \approx 4{,}330\ \text{cm}^{-1}$$

and $\nu_3 \approx 6Dq + 15B'$; or, $16{,}000 = 6 \times 433 + 15B'$

or, $15B' = 13{,}402$

or, $B' = 893.5\ \text{cm}^{-1}$

The results (without considering the configurational interaction energy) are fairly in good agreement with the results obtained by the more refined method *i.e.* analysis by using the Tanabe-Sugano diagram and transition energy ratio values (*i.e.* Fig. 7.10.4). It indicates that the symmetry interaction energy (x') is not very much significant and in fact, it happens so for the weaker crystal field splitting (*cf.* Fig. 7.10.2).

Taking, $Dq \approx 425\ \text{cm}^{-1}$, the ν_1 (= $8Dq$) band is predicted to appear at about 3,400 cm^{-1} (*cf.* experimentally found at 4,080 cm^{-1}).

For **$[NiBr_4]^{2-}$**, the spectral bands appear at: $\nu_2 = 7{,}000\ \text{cm}^{-1}$ ($\varepsilon = 30$), $\nu_3 = 15{,}200\ \text{cm}^{-1}$ ($\varepsilon \approx 200\ \text{cm}^{-1}$).

7.17 SPECTRAL TRANSITIONS IN THE SQUARE PLANAR COMPLEXES: SPECTRAL CHARACTERISTICS OF THE OCTAHEDRAL, TETRAHEDRAL AND SQUARE PLANAR COMPLEXES OF NICKEL(II)

7.17.1 Ligand Field Transitions in the Square Planar Complexes of Ni(II) and other d^8-Systems

We have already discussed the spin-allowed transitions in the octahedral and tetrahedral complexes of Ni(II). Now we shall discuss the spectral transition in the square planar complexes of Ni(II).

In moving from the O_h to square planar geometry (D_{4h}), the following changes occur:

$$\left.\begin{array}{l} A_{1g} \to A_{1g};\ A_{2g} \to B_{1g};\ E_g \to A_{1g} + B_{1g} \\ T_{1g} \to A_{2g} + E_g;\ T_{2g} \to B_{2g} + E_g \end{array}\right\} (cf.\ \text{Correlation Table 7.15.4.2})$$

The limiting situation of z-out distortion leads to the square planar geometry (D_{4h}). Positions of different d-orbitals in the square planar geometry is a matter of discussion. However, the uppermost orbital is $d_{x^2-y^2}$ (b_{1g}). The possible ways of d-orbital splitting in a square planar geometry are shown in Fig. 7.17.1.1.

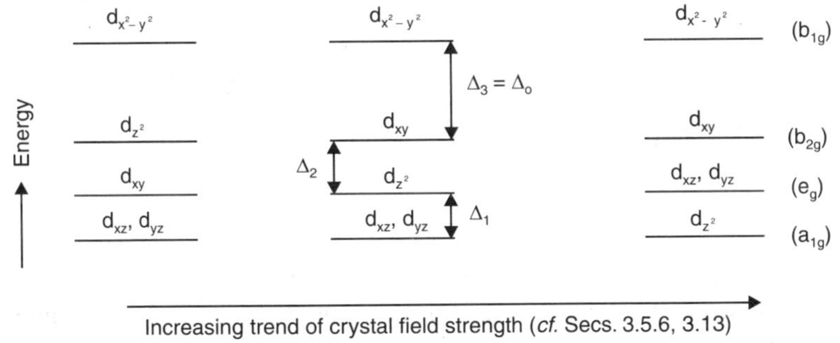

Fig. 7.17.1.1 Relative position (not in scale) of the d-orbitals in the square planar complexes of d^8-system depending on the crystal field strength. **Note:** In all situations for moving from the O_h to D_{4h} symmetry, energy separation between the d_{xy} and $d_{x^2-y^2}$ orbitals will remain the same (= Δ_o or $10Dq_o$).

From Fig. 7.17.1.1, it is evident that the $d_{x^2-y^2}$ orbital is of the highest energy while there is some uncertainty in the relative positions of the remaining four d-orbitals. For the diamagnetic Ni(II)

(d^8 system) complexes, the lower four d-orbitals are filled in and these four d-orbitals are so close in energy that it may be difficult to identify individually the three possible transitions to the topmost vacant $d_{x^2-y^2}$ orbital. *The possible transitions may overlap to produce a single absorption band.*

Here it is reasonable to point out that in most of the square planar complexes of Ni(II), the energy difference (Δ_1) between a_{1g} (d_{z^2}) and e_g (d_{xz}, d_{yz}) is quite small compared to the energy difference (Δ_2) between $a_{1g}\left(d_{z^2}\right)$ and b_{2g} (d_{xy}); and the energy difference (Δ_3) between b_{2g} (d_{xy}) and $b_{1g}\left(d_{x^2-y^2}\right)$. Thus we can write:

$$\Delta_3(\approx \Delta_o \text{ for the hypothetical octahedral complex}) \rangle\ \Delta_2 \rangle\rangle\ \Delta_1$$

For most of the square planar complexes of Co(II), Ni(II), Cu(II), the following relationship is maintained (*cf.* Secs. 3.5.6, 3.13).

$$\Delta_3\ (\approx \Delta_o),\ \Delta_2 \approx \frac{2}{3}\Delta_3,\ \Delta_1 \approx \frac{1}{12}\Delta_3$$

But, for **[Ni(CN)$_4$]$^{2-}$**, the following relationship is found from the spectroscopic evidence.

$$\Delta_3\ (\approx \Delta_o),\ \Delta_2 \approx \frac{2}{5}\Delta_3,\ \Delta_1 \approx \frac{1}{38}\Delta_3$$

For the tetragonal (D_{4h}) Ni(II) (d^8) complexes, theoretically two possibilities may arise: $e_g^4\, a_{1g}^2\, b_{2g}^2$ (*i.e.* **diamagnetic; $^1A_{1g}$ state**) when $\Delta_3 > P$; and $e_g^4\, a_{1g}^2\, b_{2g}^1\, b_{1g}^1$ (*i.e.* **paramagnetic, $^3A_{2g}$ state**) when $P > \Delta_3$. But in most of the cases, the ground state is characterised by the singlet (diamagnetic state) $^1A_{1g}$ sate for the d^8 system. For the **paramagnetic 6-coordinate Ni(II) complexes of D$_{4h}$ symmetry,** the possible electronic transitions are shown in Fig. 7.14.4.3.

Now let us consider all the **possible electronic transitions in the diamagnetic square planar complexes of Ni(II)** (*cf.* Fig. 7.17.1.2).

$$e_g^4a_{1g}^2b_{2g}^2b_{1g}^0 \text{ i.e. } {}^1A_{1g} \xrightarrow[({}^1A_{1g}\rightarrow{}^1A_{2g})]{\nu_1} e_g^4a_{1g}^2\,b_{2g}^1b_{1g}^1 \text{ i.e. } {}^1A_{2g};\ \nu_1 \approx \Delta_3\ (\approx \Delta_o)$$

$$e_g^4a_{1g}^2b_{2g}^2b_{1g}^0 \text{ i.e. } {}^1A_{1g} \xrightarrow[({}^1A_{1g}\rightarrow{}^1B_{1g})]{\nu_2} e_g^4a_{1g}^1b_{2g}^2b_{1g}^1 \text{ i.e. } {}^1B_{1g};\ \nu_2 \approx \Delta_3 + \Delta_2$$

$$e_g^4a_{1g}^2b_{2g}^2b_{1g}^0 \text{ i.e. } {}^1A_{1g} \xrightarrow[({}^1A_{1g}\rightarrow{}^1E_g)]{\nu_3} e_g^3a_{1g}^2\,b_{2g}^2b_{1g}^1 \text{ i.e. } {}^1E_g;\ \nu_3 \approx \Delta_3 + \Delta_2 + \Delta_1 \approx \Delta_3 + \Delta_2 = \nu_2$$

(Δ_1 is quite small)

The said electronic transitions **in terms of the orbitals** can be simply represented as follows:

$$e_g\left(i.e.\ d_{xz}, d_{yz}\right)\xrightarrow{\nu_3} b_{1g}\left(i.e.\ d_{x^2-y^2}\right);\ a_{1g}\left(i.e.\ d_{z^2}\right)\xrightarrow{\nu_2} b_{1g}\left(i.e.\ d_{x^2-y^2}\right);$$

$$b_{2g}\left(i.e.\ d_{xy}\right)\xrightarrow{\nu_1} b_{1g}\left(i.e.\ d_{x^2-y^2}\right);\ \nu_3 \approx \nu_2 > \nu_1\left(=\Delta_0\right)$$

Thus it is difficult to characterize ν_2 and ν_3 individually. They are expected to overlap. The ν_2 transitions are expected to occur at a higher energy compared to ν_1. Thus, it is quite expected that ν_2 and ν_3 may be overlapped by the strong charge transfer band. If Δ_2 is quite small compared to Δ_3, then it is difficult to characterize the ν_1 and ν_2 transitions individually.

The possible electronic transitions in the **tetragonally distorted octahedral (D_{4h}) complexes of Cu(II)** (d^9-system) have been discussed earlier (*cf.* Sec. 7.14.2). The possible transitions are (*cf.* Fig. 7.14.2.1):

$$b_{2g}^2\,b_{1g}^1 \xrightarrow[(=\Delta_o)]{\nu_1} b_{2g}^1\ b_{1g}^2;\ a_{1g}^2b_{1g}^1 \xrightarrow{\nu_2} a_{1g}^1b_{1g}^2;\ e_g^4b_{1g}^1 \xrightarrow{\nu_3} e_g^3b_{1g}^2.$$

The relative positions of ν_2 and ν_3 depend on the degree of tetragonal distortion (*see* Fig. 7.14.2.1).

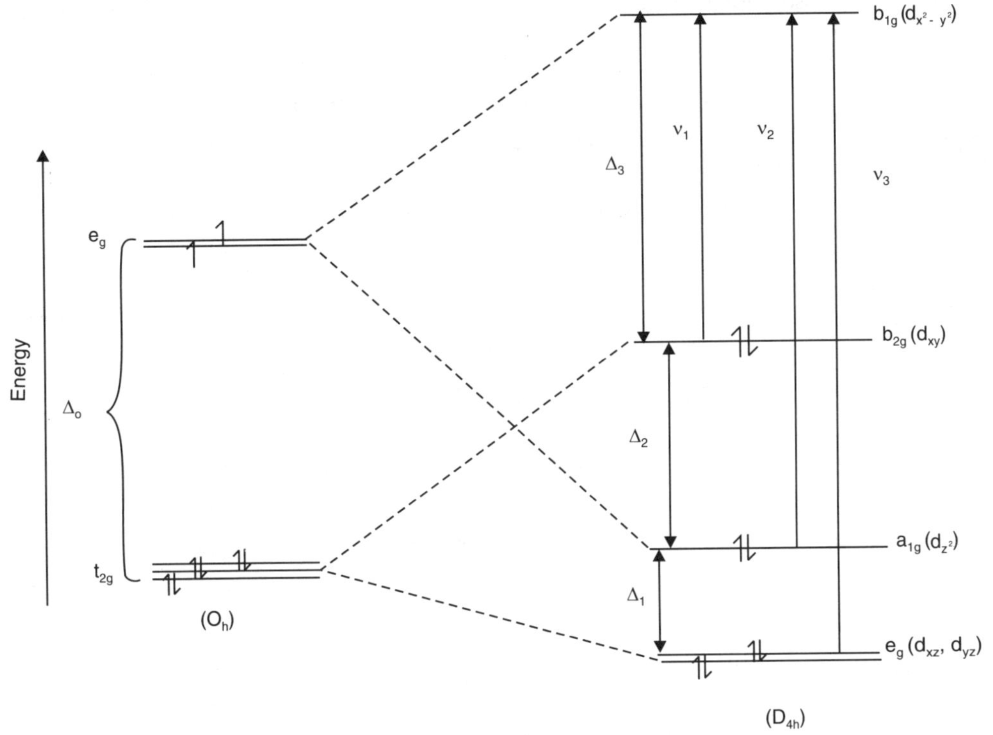

Fig. 7.17.1.2 Possible electron transitions (*i.e.* ligand field bands) in the square planar complexes of Ni(II).

Conventionally, for the square planar complexes of Ni(II), the band is assigned to the $b_{2g}(d_{xy}) \rightarrow b_{1g}\left(d_{x^2-y^2}\right)$ transition. This transition energy is the Δ_o of the corresponding hypothetical octahedral geometry. **Most of the square planar complexes are usually red or orange (sometimes green)** and the intense spectral band appears in the range 15,000 cm^{-1}–25,000 cm^{-1} ($\varepsilon \approx$ 50–500). This transition is generally assigned as:

$$b_{2g}\left(d_{xy}\right) \longrightarrow b_{1g}\left(d_{x^2-y^2}\right) \text{ i.e. } {}^1A_{1g} \longrightarrow {}^1A_{2g}$$

There is also a **second more intense** band appearing in the range 25,000–35,000 cm^{-1}. This is a charge transfer band. For example, in **[Ni(CN)$_4$]$^{2-}$**, the *d–d* transitions making shoulders occur at 22,500 cm^{-1} and 30,500 cm^{-1} and the MLCT bands have been assigned as follows:

$$d_{xy} \xrightarrow{32{,}000 \text{ cm}^{-1}} \pi^*(\text{CN});\ d_{z^2} \xrightarrow{35{,}000 \text{ cm}^{-1}} \pi^*(\text{CN});\ \left(d_{xz}, d_{yz}\right) \xrightarrow{37{,}600 \text{ cm}^{-1}} \pi^*(\text{CN}).$$

Some diamagnetic square planar complexes of Ni(II) are: **[Ni(dmgH)$_2$],** *i.e.* bis(dimethylglyoximato)-nickel(II) (red; absorption at 21,000 cm^{-1}, 24,390 cm^{-1} and 27,000 cm^{-1}); **[Ni(bigH)$_2$]Cl$_2$,** *i.e.* bis(biguanide)nickel(II) chloride (orange); **[Ni(CN)$_4$]$^{2-}$,** *i.e.* tetracyanidonickelate(II) (yellow).

For the square planar biguanide complexes of Ni(II), Pd(II), the absorption maxima are as follows:

[Ni(bigH)$_2$]Cl$_2$	**[Pd(bigH)$_2$]Cl$_2$**	**[Pt(bigH)$_2$]Cl$_2$**
22,472 cm^{-1} (445 nm)	33,400 cm^{-1} (300 nm)	45,500 cm^{-1} (220 nm)

Positions of the absorption maxima (*i.e.* ν: $Pt^{II} \rangle Pd^{II} \rangle Ni^{II}$) is in conformity with the splitting power of Pt^{II}, Pd^{II} and Ni^{II}.

Ni(II)-Complexes: Ligand field transitions

Octahedral	*Tetrahedral*	*Square planar*
$^3A_{2g}(F) \xrightarrow[(\approx 10Dq_o)]{\nu_1} {}^3T_{2g}(F)$	$^3T_1(F) \xrightarrow[(\approx 8Dq_t)]{\nu_1} {}^3T_2(F)$	$^1A_{1g} \xrightarrow[(\sim 10Dq_o)]{\nu} {}^1A_{2g}$ *i.e.* $d_{xy} \longrightarrow d_{x^2-y^2}$
$^3A_{2g}(F) \xrightarrow[(\approx 18Dq_o)]{\nu_2} {}^3T_{1g}(F)$	$^3T_1(F) \xrightarrow[(\approx 18Dq_t)]{\nu_2} {}^3A_2(F)$	*$10Dq_o$ for the hypothetical O_h complex.
$^3A_{2g}(F) \xrightarrow[(\approx 15B'+12Dq_o)]{\nu_3} {}^3T_{1g}(P)$	$^3T_1(F) \xrightarrow[(\approx 15B'+6Dq_t)]{\nu_3} {}^3T_1(P)$	
ν_1 (8,000–14,000)	ν_1 (infra-red region)	ν (15,000–25,000 cm^{-1})
ν_2 (12,000–20,000)	ν_2 (7,000–9,000 cm^{-1})	
ν_3 (20,000–30,000)	ν_3 (14,000–16,000 cm^{-1})	$\varepsilon \approx$ 50–500
$\varepsilon \approx$ 1–5	$\varepsilon \approx$ 20 for ν_2,	**Colour:** red, orange
	$\varepsilon \approx$ 100–200 for ν_3	
Colour: blue-green	**Colour:** brown, green	

Intensity of the spectral bands of square planar complexes: Square planar geometry is centrosymmetric. According to Laporte Selection rule, the $g \rightarrow g$ transition is forbidden. Thus, the ligand field bands of the square planar complexes are expected to be of poor intensity. But in reality, these bands are quite strong. It arises due to the following facts:

(i) extensive mixing of the metal *d*-orbitals and ligand-orbitals, specially the π-bonding interaction.

(ii) the ligand field bands generally occur at high energy because of the high crystal field splitting in the square planar complexes; consequently, the ligand field bands very often lie close to the strong charge transfer bands and the ligand fields bands gain the intensity through the **intensity stealing phenomenon.**

The characteristics of electronic spectra of some representative square planar complexes of Ni(II) are given below.

	$\bar{\nu}$ (in cm^{-1})		
	$^1A_{1g} \rightarrow {}^1A_{2g}$ *i.e.* $d_{xy} \rightarrow d_{x^2-y^2}$	$^1A_{1g} \rightarrow {}^1B_{1g}$ *i.e.* $d_{z^2} \rightarrow d_{x^2-y^2}$	$^1A_{1g} \rightarrow {}^1E_g$ *i.e.* $d_{xz}, d_{yz} \rightarrow d_{x^2-y^2}$
[Ni(dmgH)$_2$]:	21,000	24,390	27,000
[Ni(dtc)$_2$]:	17,000	21,000	23,910

dtc ⇒ dithiocarbamate.

Conditions Favouring the Square Planar Complex Formation for a d^8-System

For Ni(II), only the strong field ligands favour the formation of square planar complex. But, sometimes, the weak field *S*-donor ligands can form the square planar complex. The *S*-donor sites can use the vacant *d*-orbitals to withdraw the metal *d*-electrons through π-bonding. Due to this **pronounced nephelauxetic effect,** the pairing energy (P) is reduced. Consequently, the splitting energy of the e_g orbital due to the *z*-out distortion can exceed the pairing energy easily. It gives the spin-paired, diamagnetic square planar complexes of Ni(II). This aspect has been discussed in detail in Chapter 3. For the heavier congeners like Pd(II), Pt(II), even for the relatively weak field ligands, the conditions for square planar complex formation are satisfied because of their higher crystal field splitting power.

In **$[PdCl_4]^{2-}$**, the ligand field band appears at about 23,000 cm^{-1} (ε ≈ 200). Another high energy band at about 30,000 cm^{-1} (ε ≈ 400) is overlapped by the shoulder of a charge transfer band at about 36,000 cm^{-1} (ε ≈ 3,000). The ligand field bands in $[PdCl_4]^{2-}$ may be assigned as follows:

$(a_{1g}^2 \rightarrow a_{1g}^1\, b_{1g}^1$, *i.e.* ${}^1A_{1g} \rightarrow {}^1B_{1g}$: 23,000 cm^{-1}) and $(a_{1g}^2 e_g^4 \rightarrow e_g^3 a_{1g}^2 b_{1g}^1$, *i.e.* ${}^1A_{1g} \rightarrow {}^1E_g)$.

However, some authors have assigned the peak at 23,000 cm^{-1} due to the ${}^1A_{1g} \rightarrow {}^1A_{2g}$ $\left(i.e.\ d_{xy} \rightarrow d_{x^2-y^2}\right)$ transition and the shoulder at 30,000 cm^{-1} due to the higher energy ligand field bands merging with the CT band.

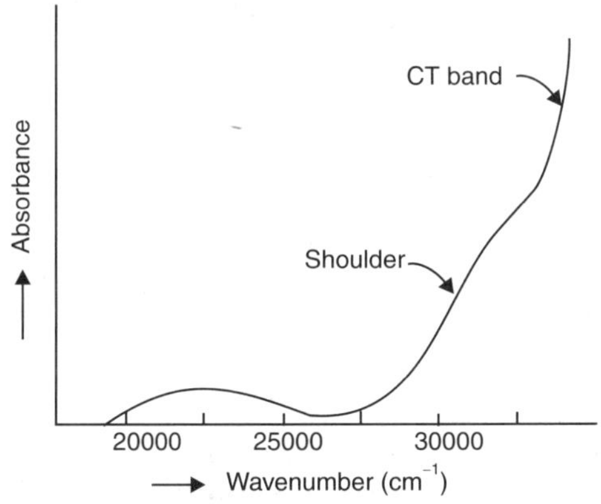

Fig. 7.17.1.3 Absorption spectra of $[PdCl_4]^{2-}$ in dilute aqueous solution of HCl.

In general for the **square planar complexes of d^8-systems,** *e.g.* Ni(II), Pd(II), Pt(II), Cu(III), Ag(III), the **spin allowed ligand field transitions** are:

$$\left.\begin{array}{l} d_{xy}\left(b_{2g}\right) \longrightarrow d_{x^2-y^2}\left(b_{1g}\right)\ i.e.\ {}^1A_{1g} \xrightarrow{\nu_1} {}^1A_{2g},\ \nu_1 \approx \Delta_3 \approx \Delta_o \\ d_{z^2}\left(a_{1g}\right) \longrightarrow d_{x^2-y^2}\left(b_{1g}\right)\ i.e.\ {}^1A_{1g} \xrightarrow{\nu_2} {}^1B_{1g},\ \nu_2 \approx \Delta_3 + \Delta_2 \\ d_{xz}, d_{yz}\left(e_g\right) \longrightarrow d_{x^2-y^2}\left(b_{1g}\right)\ i.e.\ {}^1A_{1g} \xrightarrow{\nu_3} {}^1E_g,\ \nu_3 \approx \Delta_3 + \Delta_2 + \Delta_1 \end{array}\right\} (cf.\ \text{Fig. 7.17.1.2})$$

In addition to these, the weak **spin forbidden ligand field transitions** are:

${}^1A_{1g} \rightarrow {}^3A_{2g}$; ${}^1A_{1g} \rightarrow {}^3B_{1g}$; ${}^1A_{1g} \rightarrow {}^3E_g$ } (lying in the near infra-red region).

The triplet energy states are as follows (where the unpaired two electrons are with parallel spins):

$$^3A_{2g} \rightarrow e_g^4 a_{1g}^2 b_{2g}^1 b_{1g}^1;\ ^3B_{1g} \Rightarrow e_g^4 a_{1g}^1 b_{2g}^2 b_{1g}^1;\ ^3E_g \Rightarrow e_g^3 a_{1g}^2 b_{2g}^2 b_{1g}^1$$

7.17.2 Coordination Equilibria in Ni(II)-Complexes and Thermochromism (*cf.* Secs. 8.26.1-4)

The coordination equilibria are known in many Ni(II)-complexes.

$$\text{octahedral} \rightleftharpoons \text{planar} \rightleftharpoons \text{tetrahedral}$$

Existence of the equilibria is established by considering the magnetic properties of the complexes and ligand field spectra. These are illustrated in the following examples.

The temperature dependent coordination equilibria are the examples of **thermochromism** experiencing the colour change with the change of temperature. The octahedral $\rightleftharpoons$ tetrahedral $\rightleftharpoons$ square planar equilibria of the Ni(II) complexes are the classical examples of thermochromism. Some examples of thermochromism in Ni(II) complexes have been discussed here.

(A) N-substituted-salicylaldiminato complexes of Ni(II) have been studied in detail.

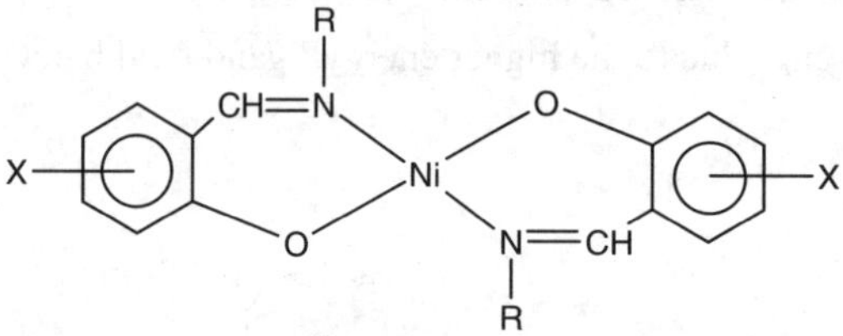

, (M = Ni^{II}, Co^{II}, Cu^{II}),

$$\underset{\nu\approx 16000\ cm^{-1}}{\text{square planar (\textbf{olive green})}} \underset{\text{cooling}}{\overset{\text{heating}}{\rightleftharpoons}} \underset{\nu\approx 7200\ cm^{-1},\ 11200\ cm^{-1}}{\text{tetrahedral (\textbf{brown})}} \text{(thermochromism for Ni}^{II}\text{-complexes)}$$

For the above mentioned complex, existence of the temperature dependent equilibrium (*i.e.* **thermochromism**) square planar $\rightleftharpoons$ tetrahedral has been established. In solution, the Ni(II) complex shows three absorption bands:

$$\sim 7{,}200\ cm^{-1},\ \sim 11{,}200\ cm^{-1},\ \sim 16{,}000\ cm^{-1}$$

With the increase of temperature, intensity of the bands at 7,200 cm^{-1} and 11,200 cm^{-1} increases while the intensity of the band at 16,000 cm^{-1} decreases. **The bands at 7,200 cm^{-1} and 11,200 cm^{-1} are the characteristic bands of the tetrahedral geometry for Ni(II)** and they correspond to the transitions, $^3T_1(F) \rightarrow {}^3A_2(F)$ and $^3T_1(F) \rightarrow {}^3T_1(P)$ respectively. On the other hand, the band at 16,000 cm^{-1} represents the characteristic spectral band of the square planar complexes and this represents the $^1A_{1g} \rightarrow {}^1A_{2g}$ transition, *i.e.* $d_{xy} \rightarrow d_{x^2-y^2}$. The change of intensity of different bands with the increase of temperature indicates that at higher temperature, the tetrahedral geometry is more preferred. *The tetrahedral complexes look brown while the square planar complexes look olive green.*

For the substituted N(sec-alkyl)-salicylaldiminato-Ni(II) complexes, the **bulky R-groups favour the tetrahedral geomtery** (where the steric repulsion at ~ 109° bond angle is relatively less). In fact, with the increase of the size of the R-group, the brown colour predominates and the magnetic moment also increases. It may be noted that the square planar complexes of Ni(II) are diamagnetic while the tetrahedral complexes are paramagnetic.

(B) The tetrahedral $\rightleftharpoons$ square planar equilibrium is well documented for **[Ni(PR$_3$)$_2$X$_2$]**.

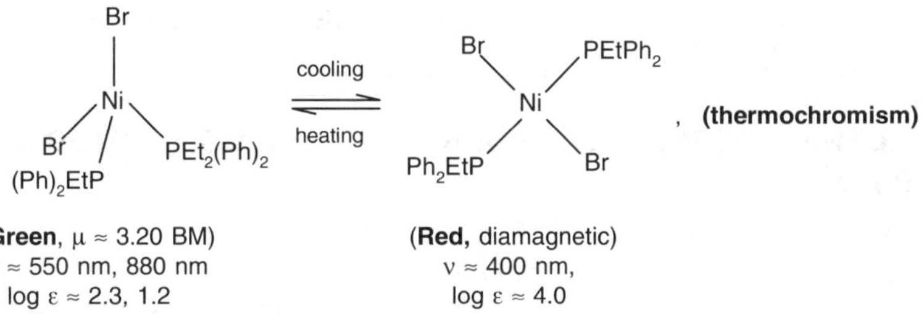

In benzene solution, all the three absorption bands are found. On cooling, the red coloured square planar complex predominates while at a relatively higher temperature towards the room temperature, the green complex predominates.

(C) For the **Ni(II)-N-arylsalicylaldimine complexes,** the equilibrium, tetrahedral (monomeric) $\rightleftharpoons$ planar (monomeric) $\rightleftharpoons$ octahedral (polymeric) has been established.

(D) The planar $\rightleftharpoons$ octahedral equilibrium has been found for *cis*-$\mathbf{[Ni(OH_2)_2(trien)]^{2+}}$, *cis*-$\mathbf{[Ni(en)_2(OH_2)_2]^{2+}}$ and *cis*-$\mathbf{[Ni(OH_2)_2(pn)_2]^{2+}}$. The octahedral *cis*-$\mathbf{[Ni(OH_2)_2(trien)]^{2+}}$ is blue in colour while the planar $\mathbf{[Ni(trien)]^{2+}}$ is yellow in colour.

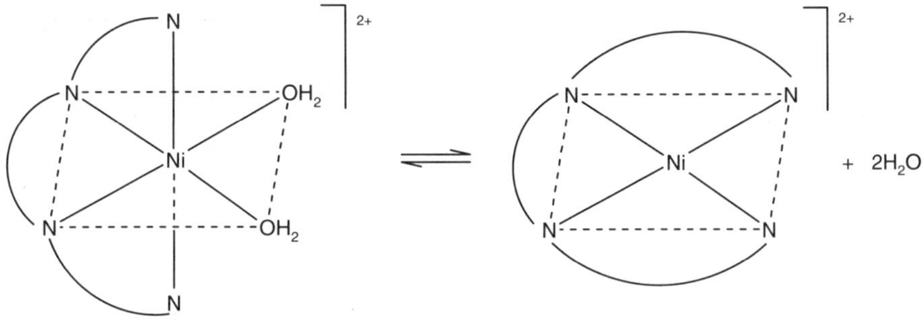

Blue, (Oh) **Yellow**, (sq. planar)

(E) In fact, dissolution of many planar Ni(II)-complexes in the coordinating solvents like water, pyridine, etc. leads to the following **planar $\rightleftharpoons$ octahedral equilibrium.**

$$\underset{(\text{square planar})}{\left[\text{NiL}_4\right]} + 2S \rightleftharpoons \underset{(\text{octahedral})}{\left[\text{NiL}_4(S)_2\right]},\ S = \text{coordinating solvent molecule.}$$

Thus, in solution both the absorption peaks of octahedral geometry and square planar geometry are noted. The equilibrium also leads to an **anomalous magnetic moments.**

(F) In **Lifschitz salts, $\mathbf{[Ni(N\text{–}N)_2]}X_2$** (N–N = substituted en, X = different types of anions), the planar $\rightleftharpoons$ octahedral equilibrium is well documented.

$$\underset{\substack{\textbf{Planar, yellow}\\ \textbf{diamagnetic}}}{\left[\text{Ni}(\text{N}-\text{N})_2\right]X_2} \rightleftharpoons \underset{\substack{\textbf{Octahedral, blue}\\ \textbf{paramagnetic}}}{\left[\text{Ni}(\text{N}-\text{N})_2X_2\right] \text{ or } \left[\text{Ni}(\text{N}-\text{N})_2S_2\right]}$$

(G) $\mathbf{(NR_xH_{4-x})_2[NiCl_4]}$ (x = 1,2,3) experiences the reversible thermocromism.

At a lower temperature, they have the yellow brown or green colour but with the increasing temperature they become blue. This colour change occurs due to structural change, octahedral (through the bridging Cl-ligands) to tetrahedral, *i.e.*

$$\underset{\substack{\text{(Yellow-brown/green)} \\ \text{(Lower temperature)}}}{\text{octahedral}} \underset{\text{cooling}}{\overset{\text{heating}}{\rightleftharpoons}} \underset{\substack{\text{(blue)} \\ \text{(Higher temperature)}}}{\text{tetrahedral,}} \quad \textbf{(thermochromism)}$$

7.18 SPECTRAL TRANSITIONS IN THE SQUARE PLANAR COMPLEXES OF COBALT(II) (d^7)

The well known square planar complexes of Co(II) are:

[Co(salen)], [Co(acacen)], $[Co(bigH)_2]$, etc.

The *d*-orbital splitting pattern leading to the square planar complexes of Co(II) depends on the crystal field strength of the ligands. The $d_{x^2-y^2}$ orbital (b_{1g}) is of the highest energy while energy order of the remaining four *d*-orbitals depends on the ligand field strength. The ground state electronic configuration of Co(II) in the low-spin square planar complex is:

$e_g^4\, a_{1g}^2\, b_{2g}^1\, b_{1g}^0$ (low spin) (when tetragonality is very high)

$e_g^4\, b_{2g}^2\, a_{1g}^1\, b_{1g}^0$ (low spin) (when tetragonality is small)

If the tetragonality is extreme then d_{z^2} (a_{1g}) becomes the orbital of the lowest energy. High-spin square planar complexes of Co(II) with 3 unpaired electrons are theoretically possible. But no such complex is known. The possible electronic transitions are shown in Fig. 7.18.1

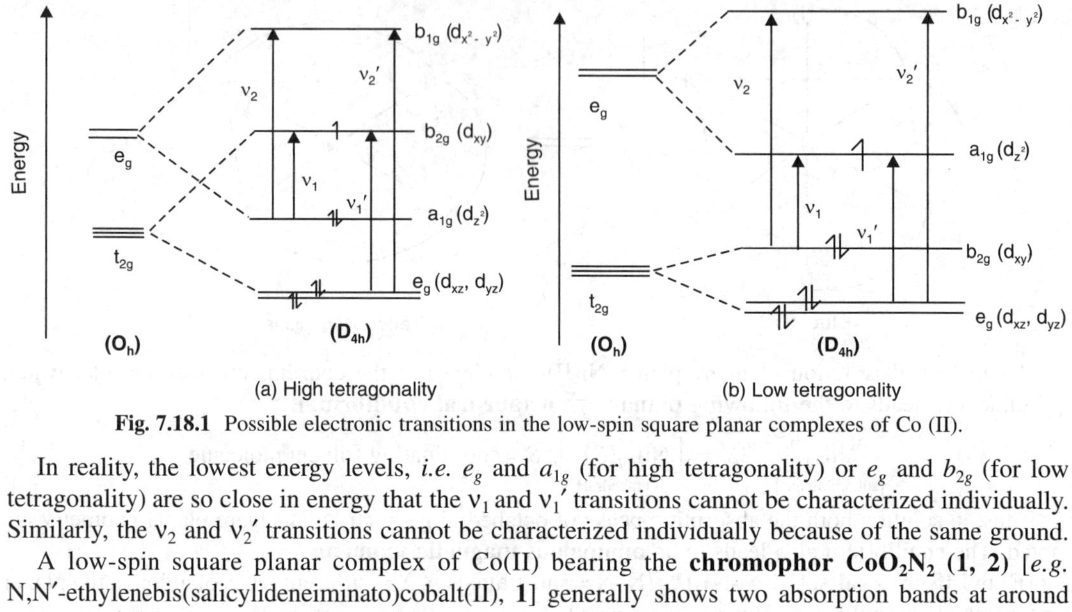

Fig. 7.18.1 Possible electronic transitions in the low-spin square planar complexes of Co (II).

In reality, the lowest energy levels, *i.e.* e_g and a_{1g} (for high tetragonality) or e_g and b_{2g} (for low tetragonality) are so close in energy that the ν_1 and ν_1' transitions cannot be characterized individually. Similarly, the ν_2 and ν_2' transitions cannot be characterized individually because of the same ground.

A low-spin square planar complex of Co(II) bearing the **chromophor CoO_2N_2 (1, 2)** [*e.g.* N,N′-ethylenebis(salicylideneiminato)cobalt(II), **1**] generally shows two absorption bands at around 8,500 cm^{-1} ($\varepsilon \approx 20$), and 20,000 cm^{-1} ($\varepsilon \approx 300$).

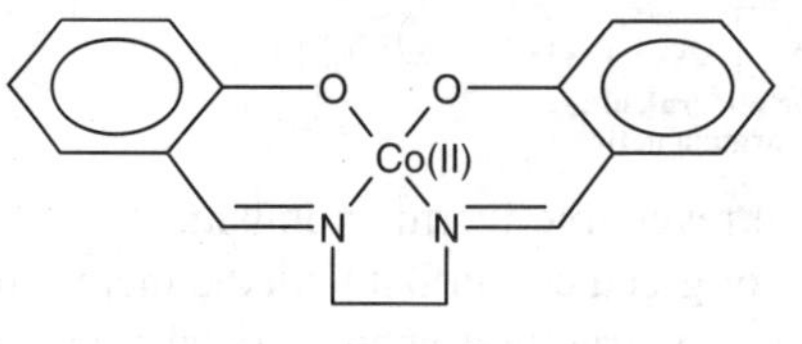

1 [Co(salen)]

N, N′-ethylene-*bis*(salicylideneiminato)cobalt(II)

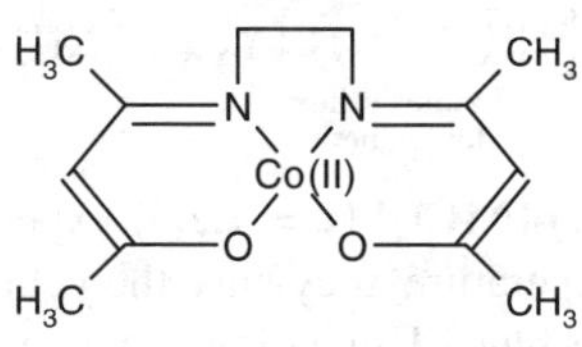

2 [Co(acacen)]

N, N′-ethylene-*bis*(acetylacetoneiminato)cobalt(II)

$$\left.\begin{array}{l}\nu_1, \nu_1' \left(i.e., a_{1g}, e_g \longrightarrow b_{2g}\right): \sim 8{,}500\ \text{cm}^{-1} \\ \nu_2, \nu_2' \left(i.e., a_{1g}, e_g \longrightarrow b_{1g}\right): \sim 20{,}000\ \text{cm}^{-1}\end{array}\right\}\ \text{for the square planar } CoO_2N_2 \text{ chromophore.}$$

7.19 LIGAND FIELD SPECTRA OF THE SECOND AND THIRD TRANSITION SERIES

The heavier congeners generally form the **low-spin complexes** because of their higher crystal field splitting power and reduced spin pairing energy. The heavier congeners also more readily can attain their **higher oxidation states** because the 4*d* and 5*d* electrons are relatively loosely bound compared to the 3*d* electron. It is partly due to the **relatavistic expansion of d-orbitals (*cf.* Ch. 8, Vol. 1).**

Interpretation and characterization of the ligand field bands of the 2nd and 3rd transition series are complicated because of the following reasons.

(i) For the heavier congeners, the **charge transfer bands** appear at a relatively lower energy and these intense charge transfer bands overlap with the ligand field bands which appear at a relatively higher energy (*cf.* 10*Dq* value increases for the heavier congeners). It complicates the characterization of the ligand field bands.

(ii) The **spin-orbit coupling interaction** increases for the heavier congeners and it runs in the sequence: 5*d*-series 〉 4*d*-series 〉 3*d*-series. Thus peak splitting (*i.e* multiplet structure) due to spin-orbit coupling interaction is quite predominant for the heavier congeners. It also complicates the interpretation of the ligand field bands.

In fact, there are only very few cases (*e.g.* Mo^{3+}, Rh^{3+}, Ir^{3+}) where the ligand field bands are sufficiently free from the complicating effect of the charge transfer band and spin-orbit coupling interaction.

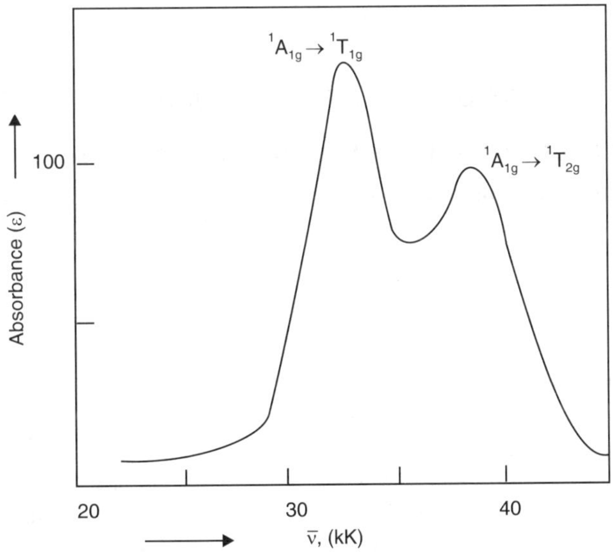

Fig. 7.19.1 Absorption spectra of $[Rh(NH_3)_6]^{3+}$.

$[MoCl_6]^{3-}$ (d^3, *i.e.* t_{2g}^3) in aqueous solution shows two spin-allowed transitions at 19,000 cm^{-1} and 24,000 cm^{-1}. The bands are assigned as:

$$^4A_{2g}(F) \xrightarrow{\nu_1} {}^4T_{2g}(F),\ \nu_1 = 10Dq$$

$$^4A_{2g}(F) \xrightarrow{\nu_2} {}^4T_{1g}(F),\ \nu_2 = 18Dq - x$$

$$^4A_{2g}(F) \xrightarrow{\nu_3} {}^4T_{1g}(P),\ \nu_3 = 12Dq + 15B' + x$$

(x = symmetry interaction energy)

$$\nu_1 = 19{,}000\ \text{cm}^{-1},\ i.e.\ 10Dq = 19{,}000\ \text{cm}^{-1} \text{ and } \nu_2 = 24{,}000\ \text{cm}^{-1}$$

The third band (*i.e.* ν_3) appears at a much higher energy and it is obscured by the overlapping charge transfer band.

The ligand field spectra of **Rh(III) low-spin d^6** (*i.e.* t_{2g}^6) like $[RhCl_6]^{3-}$, $[Rh(NH_3)_6]^{3+}$, etc. can be explained as in the case of low-spin Co(III)-complexes (Sec. 7.15.1). The Rh(III) complexes generally record two absorption peaks ($^1A_{1g} \rightarrow {}^1T_{1g}$ and $^1A_{1g} \rightarrow {}^1T_{2g}$) towards the blue end of the visible spectrum. This is very often accompanied with the charge transfer bands in the blue region. These make the colour of the Rh(III) complexes orange, red, yellow or brown. For $[Rh(NH_3)_6]^{3+}$, the assignments are:

$$^1A_{1g} \xrightarrow{\nu_1} {}^1T_{1g},\ \nu_1 = 32{,}700\ \text{cm}^{-1}$$

$$^1A_{1g} \xrightarrow{\nu_2} {}^1T_{2g},\ \nu_2 = 39{,}100\ \text{cm}^{-1}$$

The ligand field bands in $[RhF_6]^{3-}$ are:

$$^1A_{1g} \xrightarrow{\nu_1} {}^1T_{1g},\ \nu_1 = 21{,}300\ \text{cm}^{-1}$$

$$^1A_{1g} \xrightarrow{\nu_2} {}^1T_{2g},\ \nu_2 = 27{,}800\ \text{cm}^{-1}$$

Thus in terms of band assignments, Rh(III)-octahedral complexes are similar to the low-spin octahedral complexes of Co(III).

In the same way, spectra of the $Ir^{III}(t_{2g}^6)$ complexes can be explained.

7.20 ELECTRONIC SPECTRA OF THE LANTHANIDE AND ACTINIDE COMPLEXES

7.20.1 Characteristics of the Electronic Spectra of the Lanthanides

In general, +3 is the **most common oxidation state** of the lanthanides. The occurance of +2 oxidation state (*e.g.* Eu^{2+}, f^7) and +4 oxidation state (*e.g.* Ce^{4+}, f^0) can be rationalized in terms of half-filled, full-filled and empty *f*-shell.

Except the lanthanides (M^{3+}) having f^0, f^1, f^7, f^{13} and f^{14} configurations all other lanthanides are having their characteristic colours.

Table 7.20.1.1 Colours, electronic configurations and ground state terms of trivalent lanthanides.

M^{3+} ions (f^n)	*Ground state*	*Colours*
$La^{3+}(f^0)$, $Lu^{3+}(f^{14})$	$^1S_0\,(f^0)$, $^1S_0\,(f^{14})$	Colourless (no *f-f* transition)
$Ce^{3+}(f^1)$, $Yb^{3+}(f^{13})$	$^2F_{5/2}\,(f^1)$, $^2F_{7/2}\,(f^{13})$	Colourless (absorption in UV-region)
$Pr^{3+}(f^2)$, $Tm^{3+}(f^{12})$	$^3H_4\,(f^2)$, $^3H_6\,(f^{12})$	Green
$Nd^{3+}(f^3)$, $Er^{3+}(f^{11})$	$^4I_{9/2}\,(f^3)$, $^4I_{15/2}\,(f^{11})$	Lilac (blue-violet)
$Pm^{3+}(f^4)$, $Ho^{3+}(f^{10})$	$^5I_4\,(f^4)$, $^5I_8\,(f^{10})$	Pink, yellow
$Sm^{3+}(f^5)$, $Dy^{3+}(f^9)$	$^6H_{5/2}\,(f^5)$, $^6H_{15/2}\,(f^9)$	Yellow
$Eu^{3+}(f^6)$, $Tb^{3+}(f^8)$	$^7F_0\,(f^6)$, $^7F_6\,(f^8)$	Pale pink
$Gd^{3+}(f^7)$	$^8S_{7/2}\,(f^7)$	Colouless

It is evident that the ***lanthanides with f^n and f^{14-n} configurations are having the same colour.*** Thus the colour sequence in the Lu–Gd series ($f^{14} - f^7$) is repeated in the La–Gd series ($f^0 - f^7$).

f^n	La(III) f^0	Ce(III) f^1	Pr(III) f^2	Nd(III) f^3	Pm(III) f^4	Sm(III) f^5	Eu(III) f^6	Gd(III) f^7
	colourless	**colourless**	**green**	**blue-violet**	**pink**	**yellow**	**pale pink**	**colourless**
f^{14-n}	Lu(III) f^{14}	Yb(III) f^{13}	Tm(III) f^{12}	Er(III) f^{11}	Ho(III) f^{10}	Dy(III) f^9	Tb(III) f^8	Gd(III) f^7

(i) ***f-f* transition in Ln^{3+} producing the sharp (*i.e.* Rydberg type spectra; *cf.* line spectra for atoms) and less intense peaks:** In the lanthanides (Ln^{3+}), the *f-f* transition is forbidden in terms of Laporte selection rule (*cf.* for *d-d* transition). For the transition elements, the *d-d* transition is made partially allowed through the vibronic coupling **which is a crystal field effect** (*i.e.* asymmetric stretching of the metal-ligand bond). For Ln^{3+}, the 4*f*-orbital is deeply seated and the 4*f*-electrons are screened well by the filled shells of higher principal quantum number. **Because of this fact, the (*n* – 2)*f* orbitals, in general, are hardly exposed to the ligands or crystal field, and consequently the *f-f* transitions are hardly influenced by the ligands or crystal field effect.** The vibronic coupling arising from the metal-ligand bond vibration partially allows the electronic transition and at the same time, it broadens the peak. **In absence of this effect, the peaks arising from the *f-f* transitions are sharp** (*i.e.* **line like Rydberg spectra) and of low intensity.**

Here it is to be mentioned that in the lanthanides, the **spin-orbit coupling splitting is more important than the crystal-field splitting.** In general, the crystal field effect on the deeply seated (*n* – 2)*f* orbitals is negligibly small and in fact, in the lanthanides, the crystal field splitting can split the various spectroscopic states by about 100 cm^{-1} only. **Thus the crystal field splitting cannot appreciably split or broaden the peak.** Absence of vibronic coupling (*i.e.* metal-ligand bond vibration) and very weak crystal field effect are responsible in making the peaks (due to *f-f* transition) very sharp in the lanthanides. In other words, **the metal-ligand bond distance is hardly affected due to the f–f transition** (*i.e.* metal-ligand bond distance remains more or less the same for the ground and excited electronic state) and consequently the peaks become sharp (*cf.* Fig. 7.7.3.1; Secs. 7.7.2-3). **The narrow** (*i.e.* **sharp) absorption bands of the lanthanides are very often used to calibrate the wavelength in different instruments.**

Transition metal ions (d-d transitions): Vibronic coupling makes the peaks stronger and broader. Peak broadening by spin-orbit coupling is important for the tetrahedral complexes but not for the octahedral complexes. ***Colour (i.e. electronic transition energies) of a particular metal ion depends on the nature of the ligands.***

Lanthanides (f-f transitions): Absence of vibronic coupling and poor crystal field effect (compared to spin orbit coupling splitting) make the peaks sharper. Due to the absence of vibronic coupling, it is difficult to relax the orbital selectron rule, *i.e.* Laporte selection rule and **the peaks become weak.** In fact, other mechanisms like electric dipole, magnetic dipole, electric quarrupole transitions can only partially break down the Laporte selection rule. Because of the no effect of ligand field on the *f–f* transitions, ***colour of a particular metal ion does not depend on the nature of the ligands.***

Here it may be mentioned that through the 4*f* ⟶ 4*f* transitions are forbidden, the 4*f* ⟶ 5*d* transitions are allowed. **The peaks arising from the 4*f* ⟶ 5*d* transitions are more intense and relatively broader.** The *d*-orbitals are more exposed to the ligands and the metal-ligand bond vibration can broaden the peaks as usual. In such cases, the *f-f* transitions may earn some intensity from the *f-d* allowed transitions through the **intensity stealing mechanism.**

(ii) **Minimum effect of crystal field on the *f-f* transition:** The crystal effect has no significant effect on the *f-f* transitions. External fields can split the states arising from the f^n configurations to the magnitude of **100 cm^{-1}** (*cf.* spin-orbit coupling splitting energy is quite high; spin-orbit coupling constant is of the order of **1000 cm^{-1} for Ln^{3+}**). Thus the crystal field effect which depends on the nature of ligands is of minimum importance for the *f-f* transition. **Thus the colour and absorption spectra of a particular Ln^{3+} ion are more or less the same for the various types of ligands. In other words, each Ln^{3+} ion has its own characteristic colour.** On the other hand, for the transition metals, the nature of the electronic transitions largely depends on the nature of the ligands and a particular transition metal ion shows different types of colours and absorption spectra for the different types of ligands.

(iii) **Spectroscopic states generated through the splitting of energy states by spin-orbit coupling interaction and a large number of transitions in the higher energy end:** For the lanthanides, the spin-orbit coupling constant **(λ) is fairly high** (~1000 cm^{-1}) and the $\vec{L}$ and $\vec{S}$ vectors couple to produce the $\vec{J}$ vector. In other words, J is a good quantum number for Ln^{3+}. Because of **the high spin-orbit coupling constant,** the energy difference among the different spectroscopic states generated through the splitting of the energy states by spin orbit coupling is very high (*cf.* ***for the transition metal ions, the spectroscopic states are generated by crystal field effect***).

For the Ln^{3+} ions, generally, the energy difference between the ground state and first excited state (J values differ by one unit between these levels) is sufficiently higher than the thermal energy k_BT ($\approx$ 200 cm^{-1}). However, for Eu(III) and Sm(III), the spin-orbit coupling constants are fairly small (*cf.* $\lambda \approx 230$ cm^{-1} for Eu^{3+}, $\lambda \approx 240$ cm^{-1} for Sm^{3+}). Thus for Sm^{3+} and Eu^{3+}, the energy separation between the ground state and excited state is relatively small.

$$\boldsymbol{Sm^{3+}\ (f^5)}: E\ (^6H_{7/2}, \text{ first excited state}) - E\ (^6H_{5/2}, \text{ ground state})$$

$$= \frac{7}{2} \times 240 = 840 \text{ cm}^{-1}$$

$$\boldsymbol{Eu^{3+}\ (f^6)}: E(^7E_1) - E(^7E_0) = 1 \times 230 = 230 \text{ cm}^{-1}$$

Energy difference (ΔE) between the successive states differing in J values (*i.e.* J and $J + 1$) by one unit can be obtained by Lande internal rule: $\Delta E = (J + 1)\lambda$; $J = |L + S|$ to $|L - S|$

$$E_J = \frac{1}{2}\lambda\left[J(J+1) - L(L+1) - S(S+1)\right]$$

By considering the λ-values, it is expected that except for Sm^{3+} and Eu^{3+}, the excited states remain virtually unpopulated at ordinary temperature under the condition,

$$\Delta E_{J,J+1} \rangle\rangle k_BT.$$

(iv) **Energy states and selection rules for the *f-f* transitions:** For a particular Ln^{3+} ion of f^n configuration, the *f-f* transition occurs from one state of a particular J value to the excited state of another J value [between the J-states $\Delta J \le 2l$, $l = 3$]. As the energy separation is fairly large, **the transitions generally occur in the high energy end of the visible spectrum.**

For an *f*-electron, $l = 3$, $m_l = 3,2,1,0,-1,-2,-3$. **Thus a large number of transitions are possible and it gives a larger number of peaks.** The spectral transitions are governed by the following **selection rule:**

$$\Delta l = \pm 1,\ \Delta S = 0,\ \Delta L \le 2l\ (l = 3),\ \Delta J \le 2l$$

Application of the selection is illustrated for Dy^{3+}.

$$^6H_{15/2} \longrightarrow {}^6F_{1/2} \quad \text{(forbidden transition; } \Delta J = 7 > 2 \times 3\text{)}$$

$$^6H_{15/2} \longrightarrow {}^6F_{3/2} \quad \text{(allowed transition; } \Delta J = 6 = 2 \times 3\text{)}$$

If $J = 0$ for the initial state or final state, then the restriction for ΔJ is as follows:

$$\Delta J = 2,4 \text{ or } 6$$

Thus for Eu^{3+}, the transitions, $^7F_0 \rightarrow {}^7F_3$ ($\Delta J = 3$), $^7F_0 \rightarrow {}^7F_5$ ($\Delta J = 5$) are forbidden.

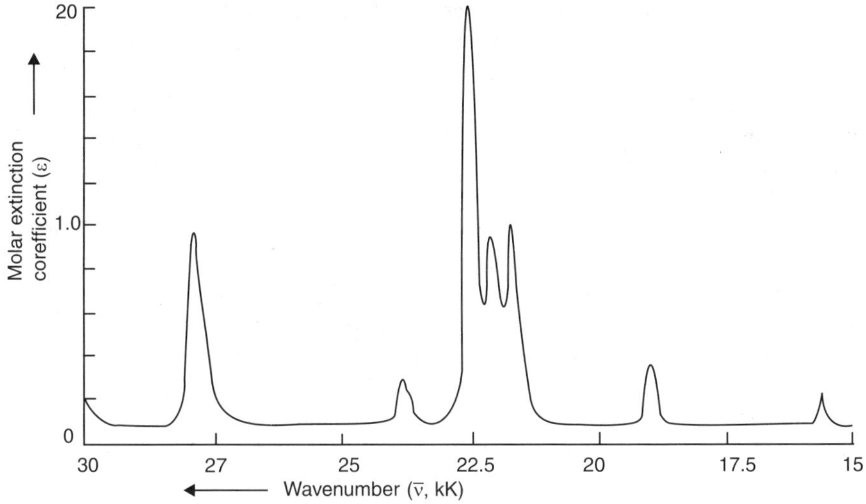

Fig. 7.20.1.1 Electronic absorption spectra of **Ho^{3+}** in dilute acidic solution.

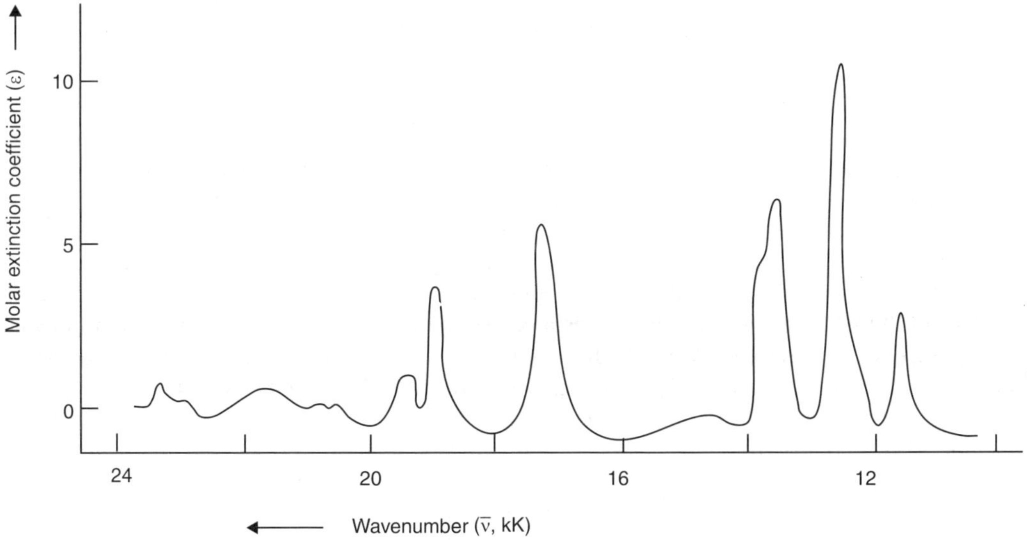

Fig. 7.20.1.2 Electronic absorption spectra of **Nd^{3+}** in aqueous acidic medium.

The crystal field effect can lift some of the $(2J+1)$ fold degeneracy of the states of f^n configuration but the crystal field splitting can cause only fine structure in some of the bands of lanthanide complexes ***because the spin-orbit coupling effect is more important than the crystal field splitting effect.***

Hypersensative transitions in the trivalent lanthanides

It has been already mentioned that the absorption spectra of lanthanides (Ln^{3+}) are more or less insensitive to the ligand field environment. **Thus colour of the aqua-ions**, *i.e.* **$[Ln(aq)]^{3+}$ remains unchanged in presence of the complexing ligands.** However, there are some transitions obeying the selection rules $|\Delta J| \le 2$, $|\Delta L| \le 2$ very much sensitive to the ligand environment. Such transitions are called **hypersensitive transitions.** Representative examples of hypersensitive transitions are:

Pr^{3+} ($^3H_4 \rightarrow {}^3F_2$), Nd^{3+} ($^4I_{9/2} \rightarrow {}^4G_{5/2}$), Eu^{3+}($^5F_0 \rightarrow {}^5D_2$), Ho^{3+}($^5I_8 \rightarrow {}^3H_6$, $^5I_8 \rightarrow {}^5G_6$).

The mechanism of hypersensitivity of these transitions will not be discussed here. This lies beyond the scope of this book.

(v) **Luminescence (*cf.* Sec. 7.9.5):** It has been already mentioned that the ligand field environment does not have any strong influence on the *f-f* transition. The excited *f*-electron can weakly interact with the environment. ***Thus the nonradiative path (i.e. thermal degradation of energy as heat in the surroundings) is quite inefficient to bring back the excited electron to the ground state*.** **Thus the nonradiative lifetime of the excited state is quite long and luminescence is observed in many cases.**

In the lanthanides, there are many excited states of different multiplicities. These excited states are not normally populated but in presence of some exciting radiation, population density in such excited states can be enhanced in absence of any efficient nonradiative deexcitation. **Thus population inversion state** can be easily achieved and stimulated emission can produce the **laser beams** in such cases.

Based on this property, the oxides of some lanthanides (*e.g.* Ho, Tb, Eu) are used in **oxide phosphores for colour TV screens and similar devices.**

Pr^{3+}, Eu^{3+}, Yb^{3+} as the NMR shift reagents (*See* Chapter 12)

Several highly paramagnetic lanthanides (Ln^{3+}) are used as the NMR shift reagents. Some organic molecules possess very complicated NMR spectra and the analysis is very difficult. When such organic molecules are coordinated to the highly paramagnetic lanthanides, the large magnetic moment of these ions can cause splitting and spreading of the NMR signals of the coordinated organic moieties. This technique helps to analyze the NMR signals (*cf.* Sec. 12.2.13).

By making the use of phosphorescence behaviour of the lanthanide ions, it has been possible to replace the ruby lasers and phosphors in commercial applications. **Lanthanide oxide phosphors are almost ubiquitously used in colour televisions.** Isomorphous substitution of Ca^{2+} ions in many **metalloproteins and metalloenzymes** by the lanthanide ions has been quite useful for characterization and detection by making the use of phosphorescence property of the lanthanide ions.

(vi) **Charge transfer (CT) bands:** The charge transfer bands are **relatively less important** in the lanthanides (Ln^{3+}) compared to those in the transition metal complexes. *This is due to the fact that the ligand orbitals can hardly interact with the 4f orbitals of the lanthanides.* LMCT (ligand to metal CT) bands are only important in the stable divalent lanthanides (*e.g.* Eu^{2+}, Sm^{2+}, Yb^{2+}, Tm^{3+}) and tetravalent cerium (*i.e.* Ce^{4+}). Thus the yellow colour of Ce^{4+} and blood red colour of Sm^{2+} are due to the charge transfer bands.

Spectral Properties of the Lanthanides in terms of the Special Properties of the 4*f*-Orbitals: A Summary

The 4*f* orbitals are more deeply seated compared to the 5*s*, 5*p* and valence 6*s* orbitals, *i.e.* $<r_{max}>_{4f} \langle\langle <r_{max}>_{5s, 5p, 6s}$. This is why, the ligands can hardly interact with the 4*f* orbitals. It imparts the following properties:

- Spectral properties and colours are more or less **independent of the crystal field effects.**
- Electronically excited states (*i.e.* excited *f*-electrons) do not easily deexcite by the nonradiative paths, *i.e.* through lattice vibration; it imparts the properties of **luminescence.**
- Inefficient vibronic coupling (*i.e.* inefficient coupling with the metal-ligand vibration) makes the spectral **peaks sharp** (*i.e.* Rydberg type spectra, or line-like spectra of atoms) and of **poor intensity** (*i.e.* low ε value).
- In absence of crystal field splitting, the electronic states can be described by the **free-ion states.** The **spectroscopic states are generated by spin orbit coupling splitting of the energy states** (*cf.* crystal field splitting generates the spectrospcopic states for transition metal ions).
- No interaction of the ligand orbitals with the deeply seated 4*f*-orbitals makes the **CT-bands less important.**

(vii) **Colourless $La^{3+}(4f^0)$ and $Lu^{3+}(4f^{14})$:** Their ground state term is ^1S. No *f-f* transition is possible and the CT bands are absent. These make them colourless.

(viii) **Very poor intensity in the absorption peaks of Gd(III) $(4f^7)$:** The ground state term is $^8S_{7/2}$ ($L = 0$, $S = 7/2$, $J = 7/2$). There is no excited state having the same spin multiplicity. Thus, in Gd^{3+}, the *f-f* transitions are orbitally forbidden as well as spin forbidden (*cf.* Mn^{2+}). This is why, intensity of the absorption peaks of Gd^{3+} is very weak.

(ix) **Colourless Ce^{3+} $(4f^1)$ and $Yb^{3+}(4f^{13})$:** They cannot absorb in the visible or near UV region. In such cases, the **spin-orbit coupling constant (λ) is very high** and consequently the spin-orbit coupling splitting energy is very high and no *f-f* transition in the visible or near UV region is possible. However, they show very strong absorption bands in the UV region due to the $4f \rightarrow 5d$ transitions. These $f \rightarrow d$ transitions are not orbitally forbidden and peaks are relatively stronger.

7.20.2 Characteristics of Electronic Spectra of the Actinides

(i) **Two types of actinides:** The electronic spectra of the trivalent actinides can be broadly classified into two groups.

(a) **Heavier actinides (starting from Am^{3+})** show the electronic absorption spectra that resemble those of the lanthanides.

(b) **Lighter actinides (upto Pu^{3+})** show the electronic absorption spectra that approximately resemble those of the transition metals.

(ii) ***f-f* transition:** Just like the lanthanides (*i.e.* members of the 4*f*-series), the actinides (*i.e.* members of the 5*f* series) experience the $5f \rightarrow 5f$ transitions to produce the electronic absorption spectra.

Compared to the 4*f*-orbitals, the 5*f*-orbitals are more exposed towards the ligand environment (*i.e.* 4*f* orbitals are more deeply seated). **Thus the crystal field effect is more important to effect the *f-f* transition for the actinides compared to that for the lanthanides.**

This crystal field effect leading to the splitting of the energy levels, distortion of the symmetry, vibronic coupling and metal-ligand bond vibration becomes relatively more important for the actinides and consequently this effect will **broaden the peak and the peaks will get more intensity.**

Thus the peaks of the absorption spectra of the actinides are expected to be more intense and broader compared to those of the lanthanides.

For the lighter members of the actinides for which the effective nuclear charge on the metal ion is relatively less, the 5*f*-orbitals are relatively more exposed towards the environment compared to those of the heavier members of the series. **Thus the ligand field effects leading to peak broadening and peak intensifying are more important for the lighter members (*e.g.*** Th, U, Pu, etc.) and their spectral characteristics resemble to those of the transition metal ions (for which the *d*-orbitals experience the crystal field effect heavily) to some extent. In fact, in many other aspects like participation of the 5*f*-orbitals in bonding, attainment of higher oxidation states and variable oxidation states, etc. the lighter actinide members resemble to those of the transition metal ions. ***In fact, because of this ground, at the early days, the lighter members of the actinides were mistakenly considered as the members of the transition series.***

In conclusion, for the heavier members of the actinide series, the 5*f*-orbitals are more deeply seated compared to those of the earlier members. **Thus in this respect, the 5*f* orbitals of the heavier members of the series behave spectroscopically just like the 4*f* orbitals of the lanthanides.** It leads to:

- peaks are sharp (*i.e.* Rydberg type line spectra) and of poor intensity for the heavier members of the actinides where crystal field effect on the 5*f*-orbitals is insignificant; thus their spectra qualitatively resemble those of the lanthanides for which the **deeply seated 4*f* orbitals do not practically experience the ligand field effect;** however, these peaks are about 10 times more intense than those of the lanthanides; it indicates that even for the heavier actinides, the 5*f*-orbitals experience the ligand field effect to some extent.
- peaks are broad and of higher intensity for the lighter members of the actinides where the crystal field effect on the 5*f*-orbitals is more important.

(iii) **5f → 6d transitions:** These electronic transitions are not orbitally forbidden. These peaks are relatively broader and of higher intensity as expected. These 5*f* → 6*d* transitions occur at the relatively lower energies compared to those of the 4*f* → 5*d* transitions of the lanthanides (*cf.* **relativistice expansion** of the *d* and *f* orbitals of the heavier congeners).

(iv) **Charge transfer bands:** Charge transfer bands are more important (specially for the lighter members of the actinides) in the actinides than in the lanthanides because the 5*f* orbitals are exposed better than the 4*f* orbitals towards the ligands.

7.20.3 Comparison of the Absorption Spectra of the Transition Metal Ions and Inner-transition Metal Ions (*cf.* Sec. 8.23, Vol. 1)

(i) ***d-d vs. f-f* Transitions:** For the transition metal ions, the ligand field bands arises from the *d* → *d* transitions while for inner transition series members, the bands arise from the *f* → *f* transitions.

(ii) **Shape and intensity:** Both the *d-d* and *f-f* transition are **orbitally forbidden** (*i.e.* Laporte forbidden transitions). Crystal field effect introducing the lower symmetry and metal-ligand bond vibration (*i.e.* vibronic coupling) are the important phenomena to relax the Laporte selection rule. These are important for the transition metals where the $(n-1)$ *d*-orbitals are exposed towards the ligand environment but these are not important for the inner-transition series where the $(n-2)f$ orbitals are deeply seated and they are hardly affected by the ligands (Fig. 7.20.3.1). The peak broadening by the metal-ligand bond vibration and crystal field effect are important for the metal ions of transition series. ***Because of these reasons, the peaks due to the d-d transitions are relatively more intense and broader compared to those of the inner transition series.*** In fact, the weak spectra of the metal ions of the inner transition series, in general, resemble the **Rydberg type line spectra of the atoms.**

(iii) **Spin-orbit coupling splitting *vs.* crystal field splitting:** The deeply seated $(n-2)f$ orbitals are poorly influenced by the ligand field. Thus, the crystal field splitting of the energy states is less important for the inner transition series. On the other hand, crystal field splitting of the energy states arising from the d^n-configuration of the transition metal ions is highly significant.

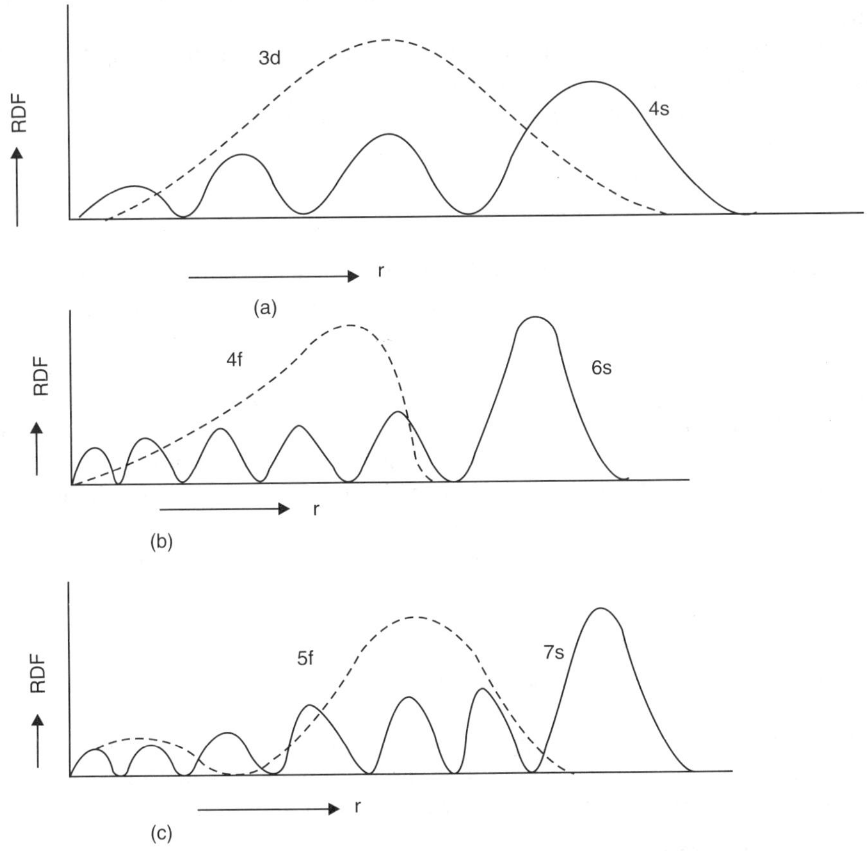

Fig. 7.20.3.1 Qualitative representation of the RDF curves for (a) 3d *vs.* 4s, (b) 4f *vs.* 6s; (c) 5f *vs.* 7s. RDF = $4\pi r^2[R_{n,l}(r)]^2$, r = distance from the nucleous.

Spin-orbit coupling constants for the inner transition series members are of much higher magnitude compared to those of the transition series.

In conclusion, the spectroscopic states arising from the splitting of states by spin-orbit coupling interaction are important to explain their electronic absorption spectra for the inner transition series. On the other hand, the spectroscopic states arising from the crystal field effect are important to rationalize the electronic spectra of the transition series.

- Free ion term $\xrightarrow{\text{Crystal field splitting}}$ spectroscopic states **(transition metal ions)**.
- Free ion term $\xrightarrow{\text{Spin-orbit coupling splitting}}$ spectroscopic states **(inner transition metal ions)**.

(iv) **Energy and number of transitions for the metal ions of the transition series and inner transition series:** For the inner transition series, the spin-orbit coupling interaction splits the energy states, and the ***energy difference among the spectroscopic states differing in the J values (i.e. $\Delta J \leq 2l$, $l = 3$) are very high.*** In fact, the electronic transition for the *f*-series generally occur in the high energy end of the visible spectrum. For the transition series, splitting of the energy states occurs by the crystal field and ***the energy difference among the spectroscopic states is relatively smaller.*** In fact, for the transition metal ions, the electronic transition energy covers the whole visible spectrum.

For the transition metal ions, the number of spin allowed transitions among the ***spectroscopic states generated by the crystal field*** splitting is very few (generally not more than > 3). On the other hand, for the inner-transition metal ions, the ***spectroscopic states are generated by spin-orbit coupling*** and for the *f*-orbitals, $l = 3$, $m_l = +3, +2, +1, 0, -1, -2, -3$, the *f-f* transition can occur in many more possible ways under the conditions, $\Delta L \leq 2l$ ($l = 3$), $\Delta J \leq 2l$. In fact, in general, the number of spin allowed spectral transitions for the inner transition series is more than those found for the transition series.

(v) **Colour dependence on the nature of the ligands:** Crystal field splitting is of no much significance in the case of inner transition series. This is why, the energy of electron transition for a particular metal ion of *f*-series does not depend on the nature of the ligands. Thus colour of a particular metal ion of the inner transition series is **its own characteristic property** (*i.e.* ***colour does not depend on the nature of the ligands***). Crystal field splitting of the spectroscopic states is very much important for the transition metal ions and the energy of electronic transition depends on the ligand field strength for a particular metal ion. **This is why, colour of a particular transition metal ion largely depends on the nature of the ligand** (*i.e.* crystal field splitting power of the ligand) and symmetry of the complex determined to some extent by the ligands.

(vi) **Luminescence:** Deexcitation of the excited states through the nonradiative path (*i.e.* thermal degradation of energy as heat in the surroundings) is less important for the inner transition series because the $(n-2)f$-orbitals are hardly affected by the environment. Thus the lifetime of the excited state, *i.e.* excited *f*-electron, is relatively longer for the inner transition series and luminescence is expected for such members. Luminescence is only rarely found in the transition metal ions.

Note: The above comparison is strictly applicable for the transition metal ions *vs.* lanthanides and heavier actinides. The lighter actinides behave spectroscopically like the transition metals at least to some extent. The reason behind this has been already discussed in Sec. 7.20.2.

7.21 CHARGE TRANSFER (CT) TRANSITIONS

We have already discussed the *d-d* transitions in the transition metal complexes and *f-f* transitions in the lanthanides and actinides. In some complexes, metal ion and the ligand may have the **complimentary redox properties** (*i.e.* metal centre is oxidizing and ligand moiety is oxidizable or vice-versa). In such cases, electron transfer can occur from one component to another component of the same complex. These two components of a complex are: metal and ligand. If the electron transfer occurs from metal to ligand, *i.e.* $M\text{–}L \xrightarrow{h\nu} M^+L^-$, then the absorption band developed is described as **metal to ligand charge transfer (MLCT) band.** On the other hand, if the electron transfer occurs from the ligand to metal, *i.e.* $M\text{–}L \xrightarrow{h\nu} M^-L^+$, then it produces the **ligand to metal charge transfer band (LMCT).** Obviously, in such charge transfer transitions, **a temporary redox reaction** takes place and the absorption bands developed in this way are also described as the **redox bands.** In MLCT, the metal centre is oxidized while the ligand moiety is reduced. **Thus, MLCT occurs when the metal centre is oxidizable and the ligand moiety is reducible.** In LMCT, the ligand is oxidized and the metal is reduced. **Thus LMCT occurs when the metal centre is oxidizing and the ligand moiety is oxidizable.**

Here it is important to note that the CT band may also appear in the d^{10} (*e.g.* HgI_2, HgS) and d^0 (*e.g.* MnO_4^-, CrO_4^{2-}) systems also. It may be noted that the d^0 and d^{10} systems cannot give any ligand field band.

Thus to develop a CT band or redox band, there must be an electron donor (*D*) and an electron acceptor (A). The energy of the CT band depends on the ionization energy (IE_D) of the donor and electron affinity (EA_A) of the acceptor.

$$\Delta E = h\nu = IE_D - EA_A - C, \ C \text{ is a constant}$$

The origin of the CT band can be simply explained by considering the MO orbitals formed in the DA (donor-acceptor) complex (Fig. 7.21.1).

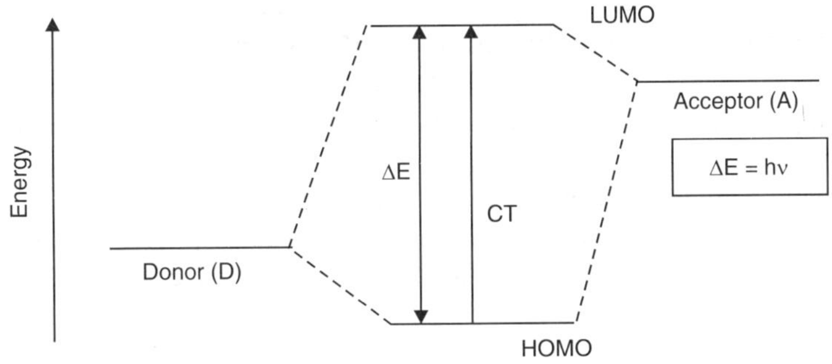

Fig. 7.21.1 Development of a CT band in terms of MOs.

The CT spectra can be best explained in terms of **molecular orbital theory** (MOT) not in terms of **crystal field theory** (CFT). In the transition metal complexes, the molecular orbitals are formed by the combination of the metal *d*-orbitals and ligand orbitals. Because of the energy difference between the metal and ligand orbitals, the metal and ligand orbital character are not uniformly distributed among the molecular orbitals. When an electron is transferred from a MO enriched in ligand orbital character to a MO enriched with the metal orbital character, it is referred to as $L \rightarrow M$, *i.e.* LMCT transition.

Similarly, $M \rightarrow L$, *i.e.* MLCT transition can be assigned for the electron transition from a MO enriched with the metal orbital character to a MO enriched with the ligand orbital character.

Here it is worth mentioning that compared to the ligand field bands arising from the *d-d* transition, **the charge transfer bands are more intense.** The MOs bear the character of both *d*-orbital (coming from the metal centre) and *p*-orbital or *s-p* hybrid orbital (coming from the ligand). Thus, in the CT bands, the $p \rightarrow d$ or $d \rightarrow p$ transition component prevails along with the $p \rightarrow p$ and $d \rightarrow d$ transition components. The $p \rightarrow p$ and $d \rightarrow d$ transitions are not orbitally allowed but the $p \rightarrow d$ or $d \rightarrow p$ transition component is orbitally allowed. The CT bands are also the spin-allowed transitions. Thus the CT bands are both orbitally and spin allowed. **This is why, the CT bands are highly intense.** The intense colours developed due to the CT bands find many important applications in analytical chemistry.

7.21.1 LMCT Bands

The $L \rightarrow M$ charge transfer bands are very much common. In this transition, the ligand is oxidized and the metal centre is reduced. This CT is favoured when the metal centre is oxidizing and the ligand moiety is oxidizable. **Thus with the increase of the oxidizing power of the metal centre and oxidizability of the ligand, the charge transfer band will move towards the longer wavelength (*i.e.* red shift).** The LMCT spectra are common when the ligands have the filled π-orbitals (*e.g.* heavier halides and chalcogenides) and the metal centre with a relatively high ionization energy. The transition and post transition metals with the high oxidation states are the suitable metal centres for LMCT. In such cases, the electrons from the filled orbitals of relatively higher energy of the ligands will move towards the low energy orbitals of the metal.

Examples of LMCT spectra: The halidocomplexes like $[MX_6]$, $[MX_4]$; Fe^{III}-SCN complex; Fe^{III}-phenol complex; Fe^{III}-hydroxamate complex; V^{V}-hydroxamate complex; $[Co^{III}(NH_3)_5X]$ (X = Cl, Br, I); $[CrO_4]^{2-}$, *i.e.* K_2CrO_4; $[MnO_4]^-$, *i.e.* $KMnO_4$, etc. shows the LMCT bands.

The **ease** with which a **LMCT** in MX_6 (*i.e.* hexahalido complexes) occurs runs in the sequence:

$Rh^{4+} \rangle Ru^{4+} \rangle Cu^{2+} \rangle Os^{4+} \rangle Fe^{3+} \rangle Ru^{3+} \rangle Pd^{4+} < Pd^{2+} \sim Pt^{4+} \sim Rh^{3+} \rangle Pt^{2+} \rangle Ti^{4+} \sim Ir^{3+}$.

and, $I^- \rangle Br^- \rangle Cl^- \rangle F^-$.

Favourable conditions to shift the LMCT band towards the higher wavelength region

• Oxidizing metal centre; • lighter congener of more oxidising property (*e.g.* $Fe^{3+} \rangle Ru^{3+} \rangle Os^{+3}$; $MnO_4^- \rangle TcO_4^- \rangle ReO_4^-$; etc.); • high ionization energy of the metal centre; • high polarizing power of the metal centre; • high polarizablity of the ligand; • higher reducing power of the ligands; • filled π-orbitals of the ligand and vacant *d*-orbitals of the metals; • HOMO (enriched with the ligand orbital character) of high energy and LUMO (enriched with the metal orbital character) of low energy.

- **$[Co^{III}(NH_3)_5X]^{2+}$:** Charge transfer spectra of $[Co(NH_3)_5X]^{2+}$ (X = F, Cl, Br, I) are quite interesting. For X = F, the LMCT band lies in the far UV-region as in the case of $[Co(NH_3)_6]^{3+}$. For X = Cl, Br, I, the intense LMCT band overlaps with the higher energy weaker ligand field (*i.e.* *d-d* transition) band (*i.e.* $^1A_{1g} \rightarrow {}^1T_{2g}$) and the corresponding ligand field band gains the intensity from the CT band through the **intensity stealing mechanism.**

Table 7.21.1.1 Spectral characteristics of $[Co^{III}(NH_3)_5X]^{2+}$ and $[Co(NH_3)_6]^{3+}$ (*cf.* Figs. 7.15.3.1-2).

Complex	O_h: $^1A_{1g} \rightarrow {}^1T_{1g}$		$^1A_{1g} \rightarrow {}^1T_{2g}$	*LMCT bands*
	C_{4v}: $^1A_1 \rightarrow {}^1E$	$^1A_1 \rightarrow {}^1A_2$	$^1A_1 \rightarrow {}^1T_2$	
	$\bar{\nu}$ (cm^{-1})		$\bar{\nu}$ (cm^{-1})	$\bar{\nu}$ (cm^{-1})
$[Co(NH_3)_6]^{3+}$	21,200 (O_h)		29,550 (O_h)	–
$[CoF(NH_3)_5]^{2+}$	19,450	21,470	28,270	
$[CoCl(NH_3)_5]^{2+}$	18,720	21,350 (sh)	27,500	37,100, ~45,000
$[CoBr(NH_3)_5]^{2+}$	18,230	21,950 (sh)	Masked by CT band	32,000, ~39,000
$[CoI(NH_3)_5]^{2+}$	17,250	Masked by CT band	Masked by CT band	

$[Co(NH_3)_5X]^{2+}$ belongs to the C_{4v} point group while $[Co(NH_3)_6]^{3+}$ belongs to the O_h point group. However, the ligand field transitions in $[Co(NH_3)_5X]^{2+}$ can be roughly explained in terms of the octahedral

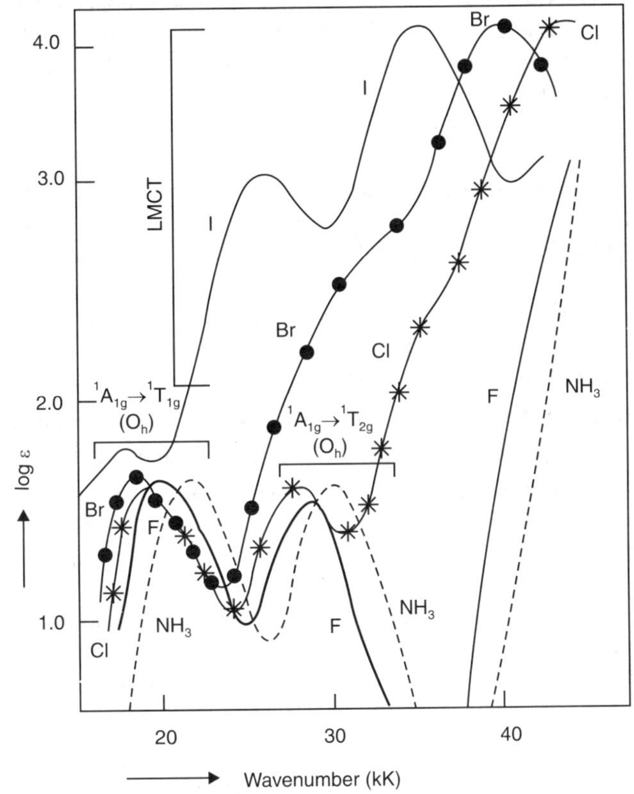

Fig. 7.21.1.1 Absorption spectra of $[Co^{III}(NH_3)_5(X)]$ (X = F^-, Cl^-, Br^-, I^-, NH_3)

transitions (*see* Sec. 7.15.3 for discussion). In moving from $[Co(NH_3)_6]^{3+}$ to $[Co(NH_3)_5X]^{2+}$, the T_{1g} state splits into the E and A_2 components but the T_{2g} state practically does not split (*cf.* Fig. 7.15.3.1). *The magnitude of splitting of the T_{1g} state depends on the difference of average ligand field strength between the N—Co^{III}—N and N—Co^{III}—X axes.* In terms of crystal field strength of the ligands (*cf.* H_3N 〉 F^- 〉 Cl^- Br^- 〉 I^- ligand fied strength sequence), this difference is maximum for $[CoI(NH_3)_5]^{2+}$ and minimum for $[CoF(NH_3)_5]^{2+}$. In fact, the ligand field band related to the T_{1g} level practically does not split for X = F and only a shoulder appears on the higher energy side for X = Br and Cl; for X = I, these expected separate peaks are masked by the LMCT band. The peak related to the T_{2g} level does not split but it gets masked by the LMCT band for X = Br, I.

The LMCT band should shift towards the longer wavelength (*i.e.* **red shift**) in the sequence: X = I 〉 Br 〉 Cl 〉 F and it happens so. For the **IPCT (ion-pair charge transfer)** band in **$[Co(NH_3)_6]^{3+} \cdot X^-$**, the similar spectral shift (*i.e.* **red shift** for the more polarisable anion) is also noted (*cf.* Fig. 7.21.3.1). For X = F and NH_3, the LMCT shoulder only appears at $\bar{\nu} \geq 40{,}000$ cm^{-1}. For X = I, practically all the ligand field bands are masked by the LMCT bands; for X = F, Cl, the ligand field bands are not masked; for X = Br, the higher energy ligand field band (*i.e.* related to the $^1T_{2g}$ level) is only masked by the LMCT band. In for X = Cl and Br, the lowest energy LMCT band appears as a shoulder at about 37,100 cm^{-1} and 32,000 cm^{-1} respectively. The next higher energy LMCT band appears distinctly with log $\varepsilon \approx 4.0$ for X = Cl and Br. For X = F, the lowest energy LMCT band shows a shoulder at 40,000 cm^{-1}.

In the same way, the **LMCT band in $[Cr^{III}(NH_3)_5X]^{2+}$** can be explained. For X = Cl, reduction in symmetry from O_h to C_{4v} does not much split the octahedral peaks but in general, the ligand field bands shift towards the longer wavelength (compared to those of $[Cr(NH_3)_6]^{2+}$) because Cl^- is a weaker field ligand than NH_3. The interesting observations is that for $[Cr(NH_3)_6]^{3+}$, a tail of the CT band appears at 50,000 cm^{-1} (*i.e.* 200 nm) while in $[CrCl(NH_3)_5]^{2+}$ the LMCT ($Cl^- \rightarrow Cr^{III}$) band appears at 42,000 cm^{-1} (*i.e.* 240 nm).

(B) Oxidizing power of the metal centre and position of the LMCT band: The colours of the following oxyanions with the d^0 metal centre are:

$[VO_4]^{3-}$ **(colourless)**, $[CrO_4]^{2-}$ **(yellow)**, $[MnO_4]^-$ **(pink)**
(V^{5+}) (Cr^{6+}) (Mn^{7+})

Among the said oxyanions, $[MnO_4]^-$ is the most oxidizing (*i.e.* highest E^0 value, 1.52 V) and the LMCT band appears at the lowest energy while $[VO_4]^{3-}$ is the poorest oxidizing agent for which the LMCT band appears at the highest energy (in the ultraviolet region).

(C) $\left[MnO_4\right]^-$ *vs.* $\left[ReO_4\right]^-$; $\left[CrO_4\right]^{2-}$ *vs.* $\left[MoO_4\right]^{2-}$, $\left[WO_4\right]^{2-}$ **:** Now let us consider the positions of the LMCT bands in $[MnO_4]^-$ (**strongly pink coloured**) and $[ReO_4]^-$ (**colourless**). **We know that the higher oxidation state is more stabilized for the heavier congeners** (*cf. relativistic effect and expansion of d-orbital*; Vol. 1). This is why, Re(VII) is more stable than Mn(VII). In other words, $[MnO_4]^-$ is more oxidizing than ReO_4^-. This is why, the LMCT band lies in the visible range (565 nm) for $[MnO_4]^-$ while the corresponding LMCT band of $[ReO_4]^-$ lies in the ultraviolet region. It explains the difference in colours of $[ReO_4]^-$ and $[MnO_4]^-$. Similarly, we can explain the following observation.

$[MoO_4]^{2-}$, $[WO_4]^{2-}$ **(colourless**, LMCT band at ultraviolet region),
$[CrO_4]^{2-}$ **(yellow colour**; LMCT band at visible region).

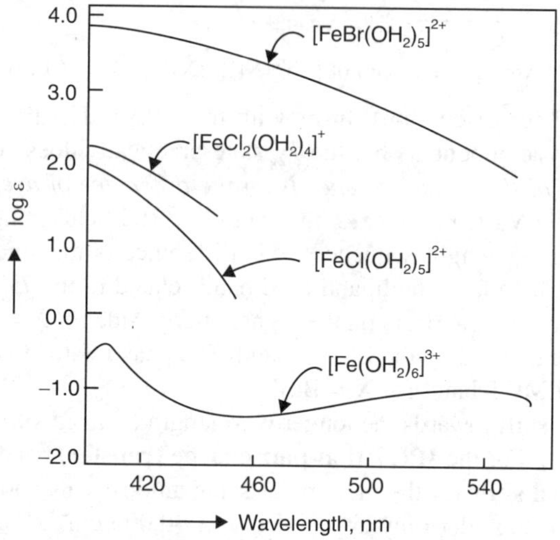

Fig. 7.21.1.2 Comparison of the electronic absorption spectra of different Fe(III) species in aqueous solution. $[Fe(OH_2)_6]^{3+}$ is almost colourless (forbidden *d–d* transition, both Laporte forbidden and spin forbidden). The halido-complexes are coloured due to the LMCT bands.

- **Fe(III)-aqua and halido complexes:** The strong absorption band (*i.e.* LMCT band) of Fe(III) complexes moves towards the longer wavelength when the ligand becomes more polarizable and oxidizable. This is why, $[Fe(aq)]^{3+}$ is colourless, but $[Fe^{III}(OH)_x(OH_2)_{6-x}]$ (*i.e.* hydrolyzed species), $[Fe(aq)X]^{2+}$ (X = Cl, Br) are strongly coloured.

- **Fe^{III}–SCN^- complex:** From the filled π-orbitals of the ligand, electron transfer occurs to the t_{2g}-level of Fe(III). This gives the intense red colour and this colour formation reaction can be utilized for the qualitative and quantitative analysis of Fe(III).

(E) Colour of d^0 and d^{10} system: The yellow colour of $[CrO_4]^{2-}$, pink colour of $[MnO_4]^-$, yellow colour of As_2S_5, orange colour of Sb_2S_3, the red colour of HgI_2, yellow colour of PbI_2, orange-red colour of BiI_3, etc. can be explained in terms of LMCT.

(F) Inorganic pigments: Colouration of some common inorganic pigments can be explained in terms of LMCT transfer.

CdS (bright yellow): $p_\pi(S) \rightarrow 5s(Cd)$

HgS (deep red, vermilion): $p_\pi(S) \rightarrow 6s(Hg)$

PbO (yellow): $p_\pi(O) \rightarrow 6s(Pb)$.

CrO_4^{2-} (yellow, as $PbCrO_4$): $p_\pi(O) \rightarrow 3d(Cr)$.

As_2S_3 (yellow): $p_\pi(S) \rightarrow 4s, 4p(As)$

Fe_2O_3 (brown): $p_\pi(O) \rightarrow 3d(Fe)$

Sb_2O_5 (yellow, as $Pb_3(SbO_4)_2$): $p_\pi(O) \rightarrow 5s, 5p(Sb)$.

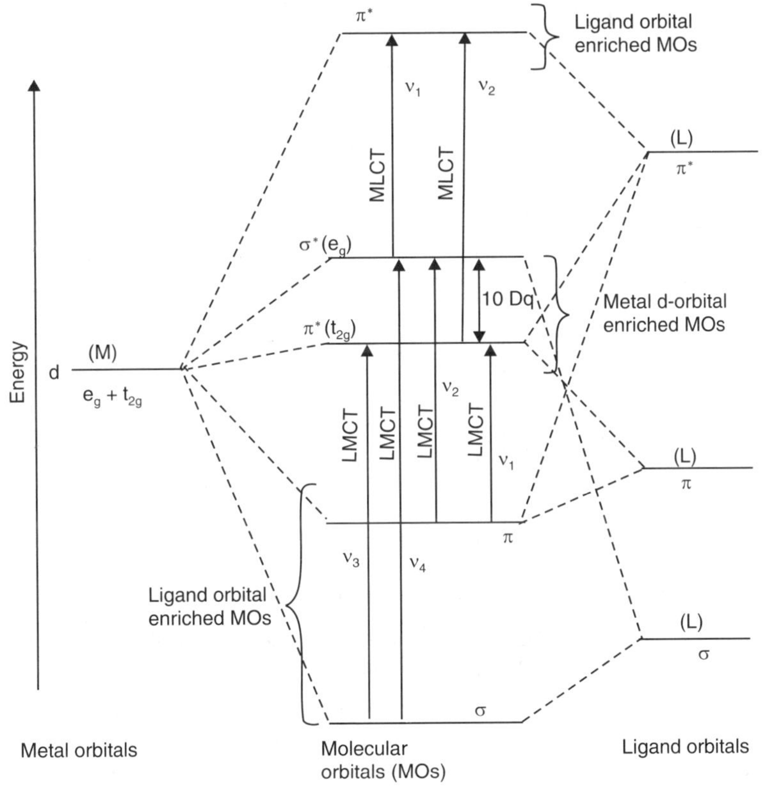

Fig. 7.21.1.3 (a) Simplified MO diagram for showing the LMCT and MLCT transitions in the octahedral complex.

(G) LMCT inducing the instability of compounds: If the LMCT is highly favoured (*i.e.* $\Delta E \langle$ 10,000 cm^{-1}, ΔE = energy difference between the HOMO predominantly enriched with the ligand orbital character and the LUMO enriched with the character of metal orbital), then the complete electron transfer from the ligand to the metal will establish a permanent redox decomposition of the complex. **It indicates that if the CT band arises at a sufficiently long wavelength then a very small amount of energy will be necessary to allow the permanent redox decomposition of the complex.** If we consider the halido-complexes of Fe(III), the CT band will arise in the ultraviolet region for the fluorido-complex while for the other halido-complexes, the CT band will move to the visible range (*i.e.* towards the higher wavelength). For the iodo-complex, the electron transfer process is so favoured that the hypothetical $[FeI_4]^-$ complex undergoes a redox decomposition giving rise to Fe(II) and I_2. A similar situation arises for the halidocomplexes of Cu(II). In fact, CuI_2 or $[CuI_4]^{2-}$ undergoes a redox decomposition giving rise to CuI and I_2 while $[CuCl_4]^{2-}$ and $[CuBr_4]^{2-}$ are dark coloured and stable. The redox decomposition of $[Co(OH_2)_6]^{3+}$ can also be explained in this way.

$$Fe^{3+} + I^- \text{ (i.e. } Fe^{III}\text{–I)} \rightarrow Fe^{2+} + I_2$$

$$Cu^{2+} + I^- \text{ (i.e. } Cu^{II}\text{–I)} \rightarrow CuI + I_2$$

$$[Co(OH_2)_6]^{3+} \rightarrow [Co(OH_2)_6]^{2+} + O_2$$

Here it may be noted that the thermodynamically unstable $[Co(OH_2)_6]^{3+}$ can be characterized well because of its **kinetic stability.**

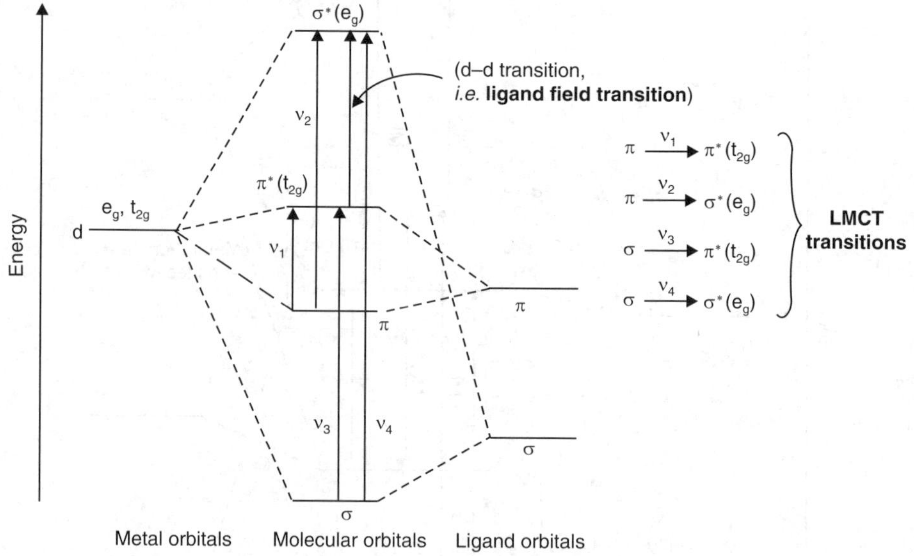

Fig. 7.21.1.3 (b) Simplified MO energy diagram describing ligand to metal charge transfer (LMCT) transitions ν_1, ν_2, ν_3 and ν_4 in the octahedral complexes like $[MX_6]^{n-}$ (X^- = halide).

Note: By considering the *s*- and *p*- orbitals on the halides, the π-MOs transform as $t_{1g} + t_{2g} + t_{1u} + t_{2u}$ and the σ-MOs transform as $a_{1g} + e_g + t_{1u}$. Obviously, the $g \rightarrow g$ transitions, *e.g.* $t_{1g}(\pi)$, $t_{2g}(\pi) \rightarrow t_{2g}(\pi^*)$ are the **forbidden transitions** and $t_{1u}(\pi)$, $t_{2u}(\pi) \rightarrow t_{2g}(\pi^*)$; $t_{1u}(\sigma) \rightarrow e_g(\sigma^*)$, etc. transitions of the $u \rightarrow g$ type are the **allowed transitions**. It may be mentioned that the *p*(ligand) → *d*(metal) component of the LMCT band is **orbitally allowed**.

(H) Molecular orbital picture for LMCT in the octahedral complexes: A simplified MO energy diagram is shown in Fig. 7.21.1.3a, b for the octahedral complex $[ML_6]$. This simplified MO energy diagram has been extracted from the complete MO energy diagram of $[ML_6]$ (*cf.* Fig. 3.17.2.5).

The possible LMCT transitions are:

$$\pi \xrightarrow{\nu_1} \pi^*(t_{2g});\ \pi \xrightarrow{\nu_2} \sigma^*(e_g);\ \sigma \xrightarrow{\nu_3} \pi^*(t_{2g}) \text{ and } \sigma \xrightarrow{\nu_4} \sigma^*(e_g^*)$$

The ν_1-transition gives the lowest energy band. In reality, the ν_1 and ν_2 LMCT bands are only important. The high energy ν_3 and ν_4 LMCT bands are rarely observed and these may appear in the far UV region. In fact, the broad and weak ν_3 band is not observed. Only in few cases, the ν_4 band is noticed. The important expectations are discussed below.

(i) **LMCT ($u \rightarrow g$):** The LMCTs (ν_1, ν_2, ν_3 and ν_4) arise for the transitions from the MOs predominantly enriched with the ligand orbital character of *u*-symmetry to the MOs (*i.e.* $t_{2g} = \pi^*$ and $e_g = \sigma^*$) enriched with the character of metal *d*-orbitals of *g*-symmetry. **If the t_{2g} level is completely filled in (*i.e.* t_{2g}^6), the lowest energy band (ν_1) remains absent and if both the t_{2g} and e_g levels are filled in (*i.e.* $t_{2g}^6 e_g^4$) then both ν_1 and ν_2 remain absent** and the high energy bands due to ν_3 and ν_4 are also not possible. In reality, for the d^{10} configuration, the tetrahedral complexes are preferred over the octahedral complexes and for such systems, the LMCTs can be explained by considering the MO energy diagram of the tetrahedral complexes.

(ii) **Narrowness of the ν_1 - band:** The π^*-MOs ($\equiv t_{2g}$ set of metal) are mainly nonbonding in character and consequently, the metal-ligand bond length does not remarkably change due to the excitation of electron to these MOs. In such cases, energy change for the electronic transition does not change remarkably due to the metal-ligand bond vibration (*cf.* Secs. 7.7.2-3). This is why, bandwidth broadening by the metal-ligand bond vibration is not of much importance for this lowest energy transition (*i.e.* ν_1 band). This band is quite sharp *i.e.* narrow and the bandwidth is within the range 400–1000 cm^{-1} (*cf.* bandwidth, 2000–4000 cm^{-1} for the ν_2-band). For the ν_2-band, the electron transition occurs to the σ^*-MOs. The σ^* (*i.e* e_g) MOs are antibonding in character and excitation of electron to these MOs leads to increase the bond length, *i.e.* $r_e^{ex} \rangle r_e^{gd}$ and *this is why, the ν_2-band experiences peak broadening due to the metal-ligand bond vibration* (*cf.* Secs. 7.7.2-3).

(iii) **Absence of the ν_1 band ($\pi \rightarrow \pi^*$):** This lowest energy band is absent for the systems where π^* (*i.e.* t_{2g}) level is completely filled in as in the low-spin d^6-complexes and d^8-complexes. In such cases, the ν_2-band (*i.e.* $\pi \rightarrow \sigma^*$) is quite important. In fact, this is the lowest energy LMCT band in $[PtX_6]^{2-}$.

Let us consider the positions of the lowest energy LMCT bands in $[Fe(CN)_6]^{4-}$ ($t_{2g}^6 e_g^0$) and $[Fe(CN)_6]^{3-}$ ($t_{2g}^5 e_g^0$). The lowest energy band (*i.e.* ν_1) is possible for $[Fe(CN)_6]^{3-}$ (where t_{2g}-level is remaining partially vacant) but ν_1 is absent in $[Fe(CN)_6]^{4-}$ for which ν_2 is the lowest energy band. **This is why, $[Fe(CN)_6]^{3-}$ shows the first LMCT band at a relatively longer wavelength compared to $[Fe(CN)_6]^{4-}$.** A similar explanation holds good for $[IrBr_6]^{2-}$ (t_{2g}^5) and $[IrBr_6]^{3-}$ (t_{2g}^6).

(iv) **Large bandwidth of the ν_2 band (*i.e.* $\pi \rightarrow \sigma^*$):** It has been mentioned that in the low-spin d^6 systems and d^8-systems, this is the lowest energy LMCT band. Though the π^*-MO (*i.e.* t_{2g}) is predominantly nonbonding but the σ^*–MO (*i.e.* e_g) is antibonding. Consequently, excitation of electron to the σ^*-MO elongates the metal-ligand bond (*i.e.* $r_e^{ex} > r_e^{gd}$). Thus the energy difference between these MOs (*i.e.* π-MO and σ^*-MO between between which the electron transition occurs) depends on the metal-ligand bond vibration. It broadens the peak (*cf.* Secs. 7.7.2-3). In fact, the

ν_2 band is broader than the ν_1-band (*cf.* band-width: for ν_2 band, 2000–4000 cm^{-1}; for ν_1 band 400–1000 cm^{-1}).

(v) **Positions of the LMCT bands:** The required energy for the LMCT transition largely depends on the oxidizing property of the metal centre and oxidizability of the ligand. The more oxidizing property indicates roughly its higher electron affinity, *i.e.* the corresponding metal orbitals are of lower energy. On the other hand, for the more oxidizable ligands, the ionization energy of the ligands is relatively less, *i.e.* the corresponding ligand orbital is of higher energy. Thus we can conclude:

- With the increase of the oxidizing power of the metal centre, the LUMO (predominant in the character of metal orbitals) comes down (energetically) and with the increase of the oxidizability of the ligand, the HOMO (predominant in the character of ligand orbitals) goes up (energetically). Consequently, the energy difference between the HOMO and LUMO decreases and the LMCT band moves towards the longer wavelength (*i.e.* **red shift**). These expectations are illustrated in the examples given in Table 7.21.1.2.

The shifting of both ν_1 and ν_2 towards the longer wavelength in moving from the chlorido complex to the bromido complex occur due to the more oxidizability of Br^- compared to that of $[Cl^-]$.]

$$\nu : [TiCl_6]^{2-} \rangle [TiBr_6]^{2-};\ [OsCl_6]^{2-} \rangle [OsBr_6]^{2-} \rangle [OsI_6]^{2-}.$$

Table 7.21.1.2 Positions of the LMCT bands in some representative hexahalido complexes.

Complex		$\pi \xrightarrow{\nu_1} \pi^*(t_{2g})$ **(cm^{-1})**	$\pi \xrightarrow{\nu_2} \sigma^*(e_g)$ **(cm^{-1})**
$[TiCl_6]^{2-}$	d^0	31,800	42,500
$[TiBr_6]^{2-}$	d^0	25,200	36,500
$[ZrCl_6]^{2-}$	d^0	42,500	–
$[ZrBr_6]^{2-}$	d^0	39,000	–
$[VCl_6]^{2-}$	d^1	21,400	–
$[VBr_6]^{2-}$	d^1	(unstable)	

It is noteworthy that though $[VCl_6]^{2-}$ is stable, $[VBr_6]^{2-}$ is unstable because of its spontaneous redox decomposition. In other words, the LMCT band of $[VBr_6]^{2-}$ is shifted to so longer wavelength that it cannot survive from its spontaneous redox decomposition.

The position of ν_1-band for the following chlorido complexes varies as follows:

$$\nu_1\ (cm^{-1}): [ZrCl_6]^{2-} \rangle [TiCl_6]^{2-} \rangle [VCl_6]^{2-}$$

This can be explained by considering the oxidizing power of the M(IV) centre.

Oxidizing power: V(IV) ⟩ Ti(IV) ⟩ Zr(IV)

Stability of +4 oxidation state: Zr(IV) ⟩ Ti(IV) ⟩ V(IV) (*i.e.* higher oxidation state is stabilized better for the heavier congeners).

(I) Molecular orbital picture of LMCT in the tetrahedral complexes: For this purpose, we need to consider the MO energy diagram of a tetrahedral complex with both σ- and π-bonding (Fig. 7.21.1.4).

The lowest 4-σ-bonding MOs ($a_1 + t_2$) can be predominantly enriched with the character of the ligand orbitals. They can accommodate 8-electrons. The next higher 8 π-MOs (*i.e.* $t_2 + e + t_1$; t_1 is basically a nonbonding one) can accommodate maximum 16 electrons. These 8 π-MOs are predominantly enriched with the character of ligand π orbitals. **The next higher MOs are $e^*(\pi^*)$ which are basically the metal $d_{x^2-y^2}$ and d_{z^2} orbitals.** These are predominantly enriched with the character of the metal *d*-orbitals and sometimes these are regarded as the **nonbonding orbitals.** The next higher t_2^* (σ^*, π^*) MOs are also predominantly enriched with the characters of metal d_{xy}, d_{yz} and d_{xz} orbitals. The energy difference between e^* and t_2^* gives the measure of Δ_t. The electronic configurations of some representative tetrahedral complexes are given below where each ligand or ligating site provides 6*p* electrons for the total electron count.

$[VCl_4](d^1)$: $[\sigma_b(L)]^8$ $[\pi_{b,\,nb}(L)]^{16}$ $[e^*(\pi^*)]^1$ $[t_2^*(\sigma^*,\pi^*)]^0$ $[a_1^*(\sigma^*)]^0$ $[t_2^*(\sigma^*,\pi^*)]^0$

$[NiCl_4]^{2-}(d^8)$: $[\sigma_b(L)]^8$ $[\pi_{b,\,nb}(L)]^{16}$ $[e^*(\pi^*)]^4$ $[t_2^*(\sigma^*,\pi^*)]^4$ $[a_1^*(\sigma^*)]^0$ $[t_2^*(\sigma^*,\pi^*)]^0$

$[CoCl_4]^{2-}(d^7)$: $[\sigma_b(L)]^8$ $[\pi_{b,\,nb}(L)]^{16}$ $[e^*(\pi^*)]^4$ $[t_2^*(\sigma^*,\pi^*)]^3$ $[a_1^*(\sigma^*)]^0$ $[t_2^*(\sigma^*,\pi^*)]^0$

$[HgX_4]^{2-}(d^{10})$: $[\sigma_b(L)]^8$ $[\pi_{b,\,nb}(L)]^{16}$ $[e^*(\pi^*)]^4$ $[t_2^*(\sigma^*,\pi^*)]^6$ $[a_1^*(\sigma^*)]^0$ $[t_2^*(\sigma^*,\pi^*)]^0$

$[MnO_4]^-$, $[CrO_4]^{2-}$ (d^0): $[\sigma_b(L)]^8$ $[\pi_{b,\,nb}(L)]^{16}$ $[e^*(\pi^*)]^0$ $[t_2^*(\sigma^*,\pi^*)]^0$

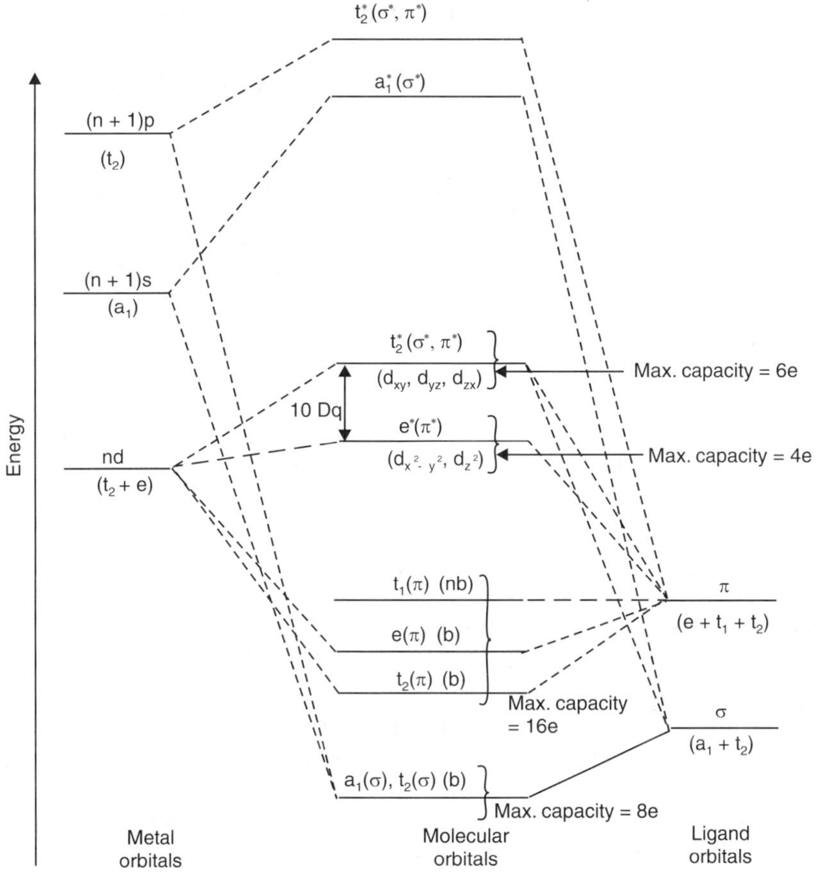

Fig. 7.21.1.4 Molecular orbital energy diagram of a representative tetrahedral complex.

Note: The MO energy diagram given in Fig. 7.21.1.4 is mainly applicable for the tetrahedral complexes of the lower oxidation states like +2, +3. There is a doubt whether the same diagram is applicable for the oxyanions like $[MnO_4]^-$, $[MnO_4]^{2-}$, $[CrO_4]^{2-}$, $[MoO_4]^{2-}$, etc. where the metal centre is in very high oxidation states.

The LMCT transitions occur from the $\pi_{b,\,nb}$ (*L*) *MO*s (which are predominantly enriched with the characters of ligand orbitals) to the higher vacant *MO*s (which are predominantly enriched with the character of metal orbitals). These are shown in Fig. 7.21.15.

Obviously, the ν_1 and ν_2 transitions (*cf.* Fig. 7.21.1.5) are not possible for the d^{10} systems; ν_1 transition is not possible for the d^7 and d^8 configurations; all the transitions are possible for the d^{0-6} configurations. It is understandable by considering the high-spin nature of the tetrahedral complexes (*i.e.* $P \rangle 10Dq_t$). Thus for the d^0 to d^6 systems, vacancies at both the e^* and t_2^* levels remain present to allow all the possible LMCT transitions (*cf.* $d^5 \Rightarrow [e^*]^2 [t_2^*]^3$; $d^6 \Rightarrow [e^*]^3[t_2^*]^3$; $d^7 \Rightarrow [e^*]^4[t_2^*]^3$, etc.). The diagram can also explain the origin of LMCT bands in d^0 (*e.g.* $[MnO_4]^-$) and d^{10} (*e.g.* $[HgX_4]^{2-}$) systems.

For $\mathbf{[NiX_4]^{2-}}$ and $\mathbf{[CoX_4]^{2-}}$ [X = Cl, Br, I), the lowest energy trasition is ν_2 (*cf.* Fig. 7.21.1.5). More correctly, this lowest energy transition represents the $t_1(\pi, nb) \rightarrow t_2(\sigma^*, \pi^*)$ transition (*cf.* Fig. 7.21.1.4). For CdS, HgS, etc. (*i.e.* d^{10} system), the ν_3 transition (*cf.* Fig. 7.21.1.5) is the lowest energy transition which is responsible for their characteristic colour.

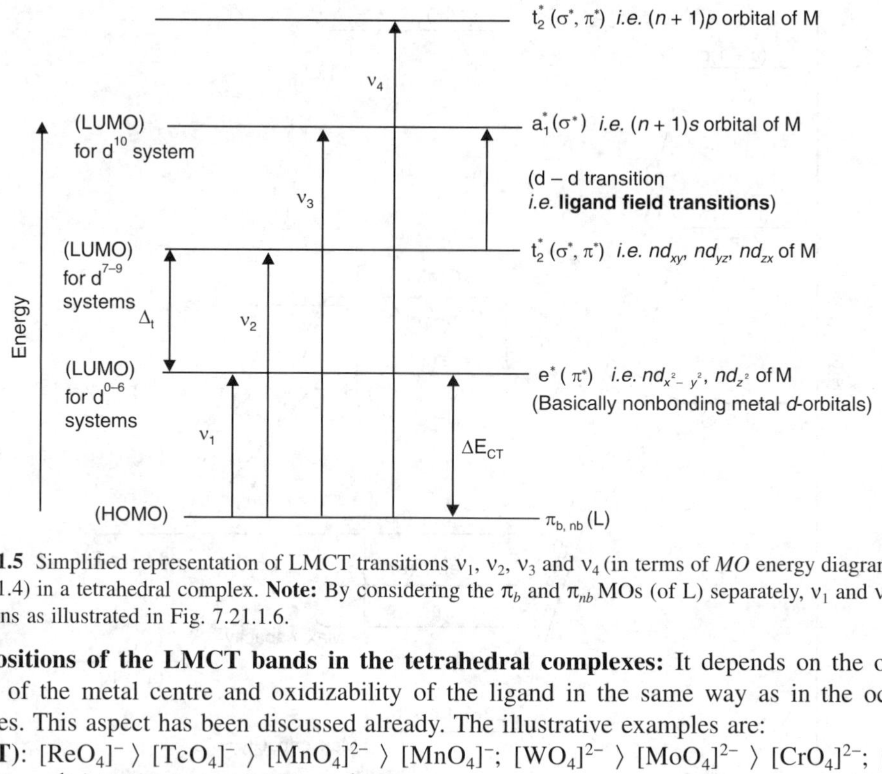

Fig. 7.21.1.5 Simplified representation of LMCT transitions ν_1, ν_2, ν_3 and ν_4 (in terms of *MO* energy diagram given in Fig. 7.21.1.4) in a tetrahedral complex. **Note:** By considering the π_b and π_{nb} MOs (of L) separately, ν_1 and ν_2 produce 4 transitions as illustrated in Fig. 7.21.1.6.

(J) Positions of the LMCT bands in the tetrahedral complexes: It depends on the oxidizing property of the metal centre and oxidizability of the ligand in the same way as in the octahedral complexes. This aspect has been discussed already. The illustrative examples are:
ν(LMCT): $[ReO_4]^- \rangle [TcO_4]^- \rangle [MnO_4]^{2-} \rangle [MnO_4]^-$; $[WO_4]^{2-} \rangle [MoO_4]^{2-} \rangle [CrO_4]^{2-}$; $[CrO_4]^{2-}$ (~ 27,000 cm^{-1}) $\rangle$ $[MnO_4]^-$ (~ 18,000 cm^{-1});
$[HgCl_4]^{2-}$ (43,700 cm^{-1}) $\rangle$ $[HgBr_4]^{2-}$ (40,000 cm^{-1}) $\rangle$ $[HgI_4]^{2-}$ (31,000 cm^{-1})

(K) MLCT bands in $\mathbf{[CrO_4]^{2-}}$ and $\mathbf{[MnO_4]^-}$: In $[CrO_4]^{2-}$, the intense LMCT bands appear at about 26,800 cm^{-1} (= 373 nm) and 36,500 cm^{-1} (= 274 nm) while in $[MnO_4]^-$ the corresponding bands appear at about 17,700 cm^{-1}(= 565 nm) and 29,500 cm^{-1} (= 339 nm). The strong purple colour of $[MnO_4]^-$ is due to ν_1 while the other transitions occur in the ultraviolet region. It is suggested that the lowest energy bands (at 17,700 cm^{-1} and 26,800 cm^{-1}) appear due to the excitation of non-bonding 2*p*-oxygen electron to the vacant e-orbitals (*i.e.* $d_{x^2-y^2}$ and d_{z^2}) of the $3d^0$ metal. There is a difference in opinion regarding the next higher energy band (ν_2). It may be due to $\pi_b(O) \rightarrow e$ or $\pi_{nb}(O) \rightarrow t_2^*$. The possible ways leading to the LMCT bands in $[MnO_4]^-$ and $[CrO_4]^{2-}$ are shown in Fig. 7.21.1.6.

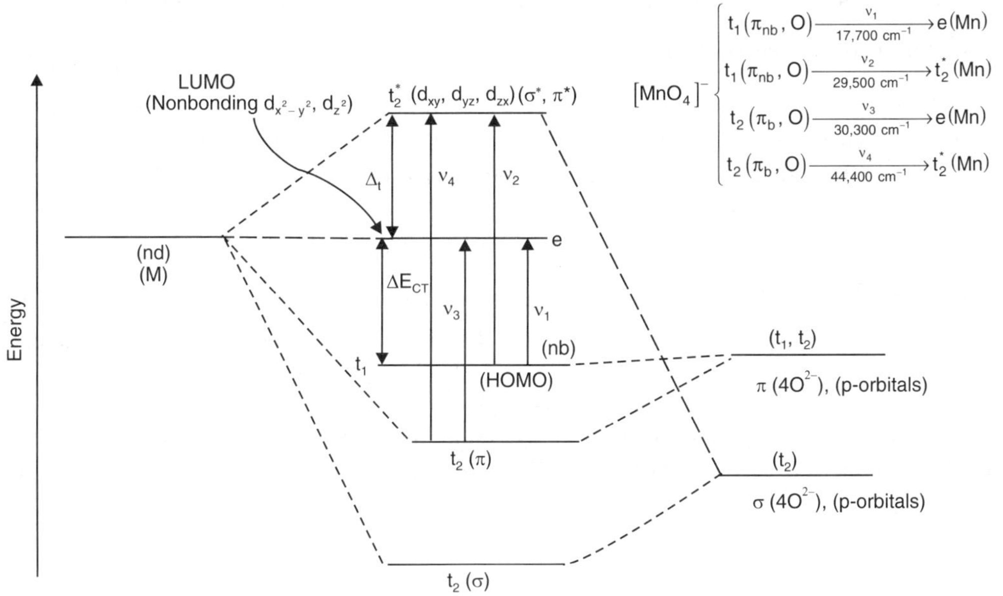

Fig. 7.21.1.6 A simplified partial MO energy diagram (through a partial modification in the diagram given in Fig. 7.21.1.4) of $[MnO_4]^-$ or $[CrO_4]^{2-}$ to show the LMCT transitions.

(L) Simplified MOT for the isoelectronic $[MO_4]^{x-}$ species: E^0 value, η (hardness) value and λ_{max} for the LMCT Band: For the tetrahedral isoelectronic species like $[VO_4]^{3-}$, $[CrO_4]^{2-}$ and $[MnO_4]^-$, energy of the orbitals of the central atom follows the order: $4p \rangle 4s \rangle 3d$. Thus, in terms of energy, compared to the $4p$ and $4s$ orbitals, the $3d$ orbitals of the central metal atom match better with the $2p$ orbitals of the O-atoms. But, the $4s$ and $4p$ orbitals of the central atom having a **greater spatial extent** can interact better with the $2p$ orbitals of oxygen. Thus, approximately, the central atom may be considered to use the $4s$ and $4p$ orbitals (or **$4sp^3$ hybrid orbitals**) to interact (*i.e.* σ-interaction) with the oxygen $2p$-orbitals (σ-symmetry) and the contracted $3d$ orbitals of the central atom remain practically **as the nonbonding ones to act as the LUMOs.** It is reasonable to assume that for such high oxidation states, the $(n-1)d$ orbitals are too contracted (compared to the ns and np orbitals) to participate in bonding through a good overlapping interaction. The filled in nonbonding $2p$ orbitals (π-symmetry) of oxygen atoms act as the **HOMOs.** Thus the lowest energy charge transfer band (LMCT) originates from the following transition.

HOMO (nonbonding $2p_\pi$-orbitals of oxygen) $\rightarrow$ LUMO (nonbonding $3d$ orbitals of M).

Thus HOMO is the same for all the isoelectronic species $[MO_4]^{x-}$ but energy of the LUMO varies as follows:

Energy (LUMO or $3d$-orbital): V(V) $\rangle$ Cr(VI) $\rangle$ Mn(VII)

This energy order is because of two reasons. As the **positive oxidation state** of the central metal atom increases, energy of the $3d$ orbital decreases (*cf.* **contraction of *d*-orbital**). Secondly, in a period, in moving from left to right, energy of the d-orbital decreases because with the **increase of nuclear**

charge the *d*-orbitals are attracted more to lower their energies. Thus, ΔE_{CT} (*i.e.* HOMO-LUMO energy gap) varies as:

$$\Delta E_{CT} \text{ or } \nu_{CT} : [VO_4]^{3-}\text{(colourless)} \rangle [CrO_4]^{2-}\text{(yellow)} \rangle [MnO_4]^{-}\text{(purple)}$$

● **Hardness (η) and ν_{CT} (*i.e.* ΔE_{CT}):** Hardness of a molecular species is determined by the energy separation between the HOMO and LUMO. It is given by: $\eta = \frac{1}{2}(E_{LUMO} - E_{HOMO}) = \frac{1}{2}\Delta E_{CT}$ *i.e.* $\Delta E_{CT} = 2\eta$, (*cf.* Sec. 14.15. 3. Vol. 3). Obviously, η or ν_{CT} runs as: $[MnO_4]^{-} \langle [CrO_4]^{2-} \langle [VO_4]^{3-}$.

● **E^0 and ν_{CT}:** The species $[MO_4]^{x-}$ to act as an oxidizing agent will have to use the LUMO to accpet the electron. Obviosuly, the LUMO of higher energy is more reluctant to accept the electron. It explains the order of oxidizing power.

$$\textbf{E}^0 \textbf{ or oxidizing power and } \lambda_{CT}\textbf{:}\ VO_4^{3-} \langle [CrO_4]^{2-} \langle [MnO_4]^{-}$$

● **$[ReO_4]^{-}$ vs. $[MnO_4]^{-}$ (ν_{CT} and E^0):** In the same way, we can explain the higher oxidizing power of $[MnO_4]^{-}$ (purple) compared to that of $[ReO_4]^{-}$ (CT band appearing at the ultraviolet region). For $[MnO_4]^{-}$, the 3*d* orbital of Mn is the LUMO while for $[ReO_4]^{-}$, the 5*d* orbital of 'Re' is the LUMO, *i.e.* LUMO of $[ReO_4]^{-}$ is of relatively higher energy.

Note: Tendency of polymerization of oxyanions $[MO_4]^{x-}$: This can be explained in terms of feasibility of the $p_\pi(O) \rightarrow d_\pi$ (M) bonding. This aspect has been discussed in Sec. 10.7.5.1 (Vol. 2).

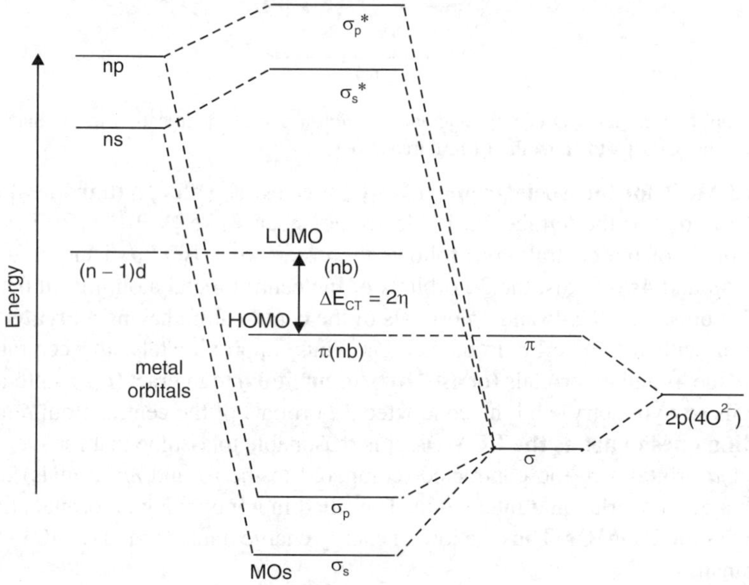

Fig. 7.21.1.7 Simplified MO energy diagram for the tetrahedral species $[MO_4]^{n-}$ *e.g.* $[VO_4]^{3-}$, $[CrO_4]^{2-}$, $[MnO_4]^{-}$, $[ReO_4]^{-}$, etc. assuming the approximate **sp^3 hybridisation of the central element. Note:** The 2*s*-orbitals of O^{2-} sites are so deeply seated (*i.e.* of very low energy) that they do not participate in bonding.

7.21.2 MLCT Bands

When the ligand possesses the low-lying vacant orbitals and the metal centres bear the filled orbitals of relatively higher energy, such a transition occurs. In this transition, the metal centre is oxidized while the ligand is reduced. Thus, for the more reducing metal centre and reducible ligands, the MLCT band shifts towards the longer wavelength.

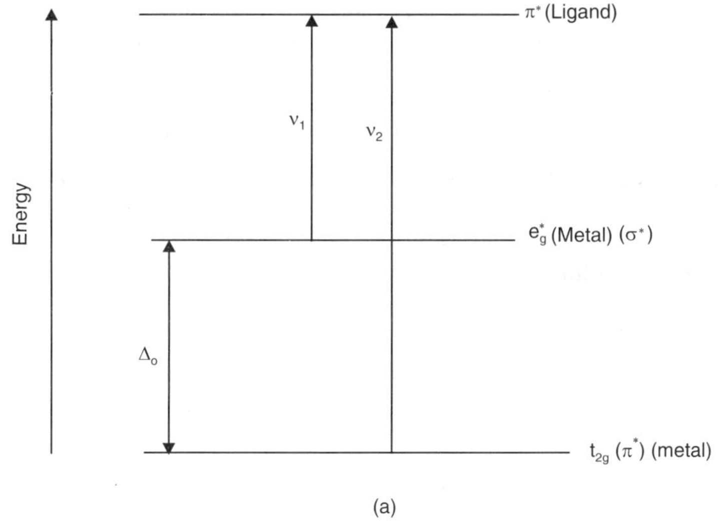

Fig. 7.21.2.1 Simple MO energy diagram to represent the origin of MLCT bands in the octahedral complexes. **Note:** $\nu_2 - \nu_1 = \Delta_o$.

This type of CT band predominates for the relatively lower valent metal centres and heteroaromatic N-bases as the ligands (*e.g.* py, bpy, phen, etc.) and other π-acid ligands like CN^-, CO, etc. In fact, the electron is transferred from the metal orbital to the vacant antibonding orbitals of the ligands like bpy, phen, etc. To observe the MLCT bands of such complexes having the said π-acid ligands, the MLCT bands must appear at lower energy than the **internal $\pi \rightarrow \pi^*$ transition** of the ligands themselves. Otherwises, these intense ligand bands will mask the MLCT bands.

The MLCT transitions in the octahedral complexes can be explained by considering the simplified MO energy diagram (Fig. 7.21.2.1) which is extracted from the complete MO-energy diagram (*cf.* 7.21.1.3) of $[ML_6]$ where L possesses the vacant π^*-MO.

Examples of MLCT bands

- **$[M(CN)_4]^{2-}$ (M = Ni, Pd, Pt), $[Au(CN)_4]^-$:** In these square planar complexes, the MLCT bands appear due to the transitions to the vacant π*-MO of CN^-. The possible transitions are:

  $$d_{xy} \longrightarrow \pi^*;\ d_{z^2} \longrightarrow \pi^*;\ d_{xz}, d_{yz} \longrightarrow \pi^*$$

  For $[Ni(CN)_4]^{2-}$, these MLCT bands appear at 32,300 cm^{-1}, 35,200 cm^{-1} and 37,600 cm^{-1}.
- **Fe(II)-bpy/phen complexes,** *i.e.* **$[Fe(L)_3]^{2+}$ (L = bpy, phen):** The red colour of these complexes arises for the ν_2 band (*i.e.* $t_{2g}^6 \rightarrow t_{2g}^5\pi^{*1}$) at about 19,600 cm^{-1}, *i.e.* 510 nm ($\varepsilon \approx 10^3$)
- **$[Ni(bpy)_3]^{2+}$ and $[Ni(phen)_3]^{2+}$:** For these complexes, the MLCT band appears at much higher energy (about 27,000 cm^{-1}) because Ni(II) is poorly oxidisable (*cf.* Fe(II) is more readily oxidisable).
- **Fe(III)-bpy/phen system:** It is obvious that in the **Fe(III)-bpy/phen complex** (faint blue colour), the MLCT band should occur at higher energy (*i.e.* shorter wavelength) because oxidation of Fe(III) is more difficult than the oxidation of Fe(II).
- **Cu(I)-phen system:** The intense colour of **Cu(I)-phen** complex is due to the MLCT transition.

- **Ir(III)-py complex:** The MLCT band appears in the region 30,000-34,000 cm^{-1} for the $t_{2g}^6 \rightarrow t_{2g}^5\pi^{*1}$ transition. (*i.e.* from the filled t_{2g} orbital of iridium to the vacant π^* of py).
- **[Cr(CO)$_6$] and [Mo(CO)$_6$]:** The MLCT bands (~35,000 cm^{-1}, ~44,000 cm^{-1}) appear for the electron transition from the filled t_{2g}^6 level of Cr or Mo to the vacant π*-MO of CO. The intense MLCT transitions are:

$$t_{2g}^6\left(i.e.\ {}^1A_{1g}\right)\xrightarrow{\nu_1^{CT}} t_{2g}^5\pi^{*1}\left(t_{1u}\right);\ t_{2g}^6 \xrightarrow{\nu_2^{CT}} t_{2g}^5\pi^{*1}\left(t_{2u}\right).$$

The weak ligand field bands ($d-d$ transitions) appear as **the shoulders on the strong CT bands.**

$${}^1A_{1g}\xrightarrow{\nu_1^{d-d}}{}^1T_{1g};\ {}^1A_{1g}\xrightarrow{\nu_2^{d-d}}{}^1T_{2g}.$$

The peak positions are as follows:

	ν_1^{CT} (cm^{-1})	ν_2^{CT} (cm^{-1})	ν_1^{d-d} (cm^{-1})	ν_2^{d-d} (cm^{-1})
[Cr(CO)$_6$]:	35,700	44,800	29,500	31,500
[Mo(CO)$_6$]:	34,600	43,860	30,150	31,950

7.21.3 Charge Transfer Bands in the Ion-Paris: IPCT (Ion-Pair Charge Transfer) Bands

If the cationic complex is oxidizing and the counter anion residing outside the coordination sphere is reducing then a CT transition may occur from the reducing anion to the oxidizing cation (*cf.* Fig. 7.21.3.1; Sec. 12.5.4) (*cf.* E.L. King *et al., J. Phys. Chem.,* **63**, 755, 1959; M.G. Evans, *et al., Trans. Faraday Soc.,* **49**, 363, 1953). It happens so in [Co(NH$_3$)$_6$]I$_3$. Such **intermolecular CT transitions** between the cation and anion are known in many suitable ion-pairs. Intensity and λ_{max} of such peaks depend on the outer sphere formation constant (K$_{OS}$) of the ion pairs.

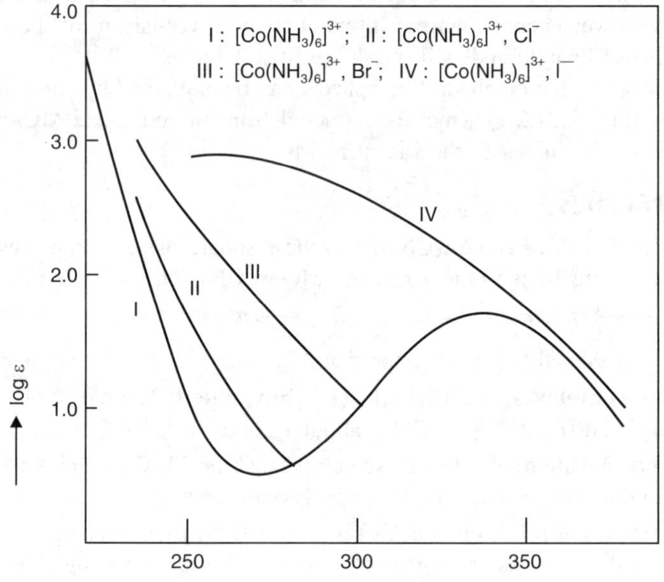

Fig. 7.21.3.1 Shifts of IPCT (*i.e.* ion-pair charge transfer) bands in [Co(NH$_3$)$_6$]$^{3+}$, X$^-$ (X = Cl, Br, I).

From the measurement of the intensity of the **IPCT bands,** K_{OS} can be estimated. The IPCT transition can be represented as follows:

$$\left[M(NH_3)_6\right]^{3+} \bullet X^-(\text{Ion-pair}) \xrightarrow[\text{IPCT}]{h\nu} \left[M(NH_3)_6\right]^{2+} \bullet X = \text{Cl, Br, I; M = Co, Ir}$$

For the more oxidisable and polarisable anion (X^-), the intense IPCT band shifts towards the longer wavelength (*i.e.* **red shift**) *i.e.* ν: $Cl^- \rangle Br^- \rangle I^-$, (*cf.* Fig. 7.21.3.1). In fact, **$[Ir(NH_3)_6]^{3+}$ is itself colourless but it looks yellow coloured in presence of I^-.** Ion-pair formation does not affect the *d-d* transitions.

7.21.4 Metal to Metal Charge Transfer (MMCT) Bands

This aspect has been separately discussed in Sec. 7.22.1.

7.21.5 Complication in Band Classification between the Ligand Field Bands (*i.e.* d-d transition) and Charge Transfer Bands

When the charge transfer band appears at the lower energy (specially in the visible range), it becomes difficult to distinguish between the CT and *d-d* bands. In fact, when a CT band overlaps with a *d-d* band, the *d-d* band also becomes intense through the **intensity stealing effect.** The problem is illustrated for the assignments of bands in $[Ir^{III}(NH_3)_5X]^{n+}$ (X = H_2O, NH_3, Cl, Br) (*cf.* Table 7.21.5.1).

Table 7.21.5.1 Assignments of spectral transitions in $[Ir^{III}(NH_3)_5X]^{n+}$.

	X = NH₃	*H₂O*	*Cl⁻*	*Br⁻*
$^1A_{1g} \xrightarrow{\nu_2} {}^1T_{1g}$ (in cm^{-1})	39,800 (ε = 90)	38,800 (ε = 85)	35,000 (ε = 72)	33,000 (ε = 100)
$^1A_{1g} \xrightarrow{\nu_2} {}^1T_{2g}$ (in cm^{-1})	46,800 (ε = 160)	47,000 (ε = 130)	44,100 (ε =333)	43,500 (ε = 800)
$\nu_2 - \nu_1$ (in cm^{-1})	7,000	8,200	9,100	10,500
$\beta\left(=\frac{B'}{B_0}\right)$	0.71	0.84	0.95	1.12

From Table 7.21.5.1, it is evident that the band assignments in $[Ir(NH_3)_6]^{3+}$ and $[Ir(NH_3)_5(OH_2)]^{3+}$ do not face any complication but the assignments for ν_2 in the corresponding chlorido and bromido complexes are called in question because of the following ground.

(i) So high extinction coefficient is unlikely for a *d-d* transition.

(ii) The energy difference between the $^1T_{1g}$ and $^1T_{2g}$ states measured by ($\nu_2 - \nu_1$) is increased in moving from the chlorido- to bromido-complexes; **this is unusual.**

(iii) The β value ≥1 is unexpected.

In fact, the second band (ν_2) is affected by the low energy LMCT band. This complicates the interpretation. A similar situation arises for $[Co^{III}(NH_3)_5X]^{2+}$ (X = Br, I) (*cf.* Table 7.21.1.1 and Fig. 7.21.1.1).

In the complexes, if the ligand orbitals and metal orbitals are experiencing an extensive mixing to produce the **delocalized MOs (as in many tetrahedral complexes),** then it is not justified to assign the bands as the ligand field bands or charge transfer bands. The bands should be assigned in terms of the MOs involved in the transition.

In the tetrahedral complexes (which are noncentrosymmetric), the *e*-set $\left(i.e.\ d_{x^2-y^2} \text{ and } d_{z^2}\right)$ is basically the nonbonding metal orbitals while the t_2-set is constituted by the overlap of metal *d*-orbitals (*i.e.* d_{xy}, d_{yz} and d_{zx}) with the ligand orbitals (*e.g.* *p*-orbitals or hybrid orbitals). ***Thus in the tetrahedral complexes, the e*** $\rightarrow t_2$ ***transitions should not considered as the pure d-d transitions (i.e. ligand field bands). Such transitions should be described in terms of the MOs involved.***

7.21.6 Comparison Between the d-d Bands, *i.e.* Ligand Field Bands and Charge Transfer (CT) Bands

The basic differences are discussed below.

- **Extinction coefficient measuring the intensity of the peak:** The *d-d* transition may be spin-allowed but it is Laporte forbidden. Consequently, its intensity is relatively less. On the other hand, the CT bands are the fully allowed transitions and the intensity is very high.
- **d^n configuration:** For the d^0 and d^{10} systems, there will be no *d-d* transition and it occurs only when the *d*-orbital is partially filled in. On the other hand, the CT bands may exist even for the d^0 (*e.g.* $[CrO_4]^{2-}$, $[MnO_4]^-$) and d^{10} (*e.g.* HgI_2) systems.
- **Relation with the redox reaction:** A CT band describes a temporary redox reaction. In a limiting situation (when the required energy for the CT band is sufficiently low), the CT process leads to a permanent redox reaction leading to the decomposition of the complexes.
  There is no relation between the redox reaction and *d-d* transition.
- **Involved orbitals:** In a *d-d* transition, the electron transition occurs from one *d*-orbital to another *d*-orbital of the same metal. On the other hand, in a CT transition, the electron transfer occurs either from the metal orbital to the ligand orbital or vice-versa.
- **Position of the bands:** In a CT band, position of the band depends on the oxidizing power of the electron acceptor component and oxidizability of the electron donor component. In a *d-d* transition, position of the band depends on the ligand field splitting of the spectroscopic terms. It depends on the crystal field strength of the ligand, stereochemistry of the complex and the inherent *d*-orbital splitting power of the metal ion under consideration.

In the MLCT transitions, the band shifts to the longer wavelength when the metal centre (*i.e.* donor) becomes more oxidizable and the ligand (*i.e.* acceptor) becomes more reducible. In the LMCT transition, the band shifts towards the longer wavelength when the metal centre (*i.e.* acceptor) becomes more reducible (*i.e.* more oxidizing) and the ligand (*i.e.* donor) becomes more oxidizable.

7.22 INTERVALENCE ELECTRON TRANSFER BANDS IN THE INTERVALENCE COMPOUNDS: METAL TO METAL CHARGE TRANSFER (MMCT) TRANSITIONS

In the inter-valence or mixed-valence compounds, metal ions of different oxidation states are linked through some bridging ligands. Such compounds are generally highly coloured due to the electron transition from the lower valent metal centre to the higher valent metal centre. Some representative examples are discussed here

(i) **Prussian blue** (*cf.* Sec. 1.12): It is generally represented by $KFe^{III}[Fe^{II}(CN)_6]$ obtained by mixing a ferric salt with potassium ferrocyanide, $K_4[Fe(CN)_6]$. Prussian blue gives the deep blue colour but individually $K_4[Fe(CN)_6]$, simple ferric salt and $K_3[Fe(CN)_6]$ are devoid of any such deep colour. It is suggested that prussian blue is a polymeric compound in which each Fe(III)-centre is octahedrally surrounded by the N-ends of cyanides (CN^-) coordinated to Fe(II) through the

C-ends. Thus both the Fe(III)- and Fe(II)-centre maintain the octahedral coordination. The bridging unit is:

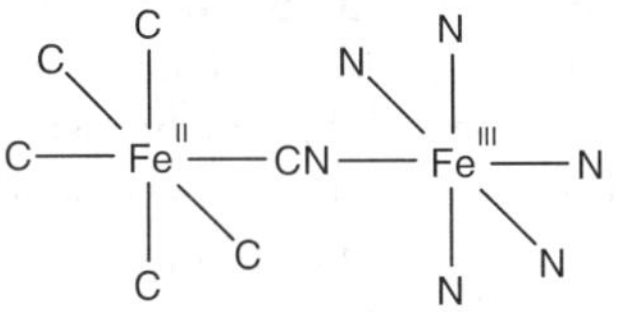

; $[Fe^{II}(CN)_6]$: $t_{2g}^6 e_g^0$ (low-spin)
$[Fe^{III}(NC)_6]$: $t_{2g}^3 e_g^2$ (high-spin)

In Prussian blue, the absorption bands appear at 15,000 cm^{-1} and 25,000 cm^{-1}. These are assigned as:

from t_{2g} orbital of $Fe^{II}(t_{2g}^6)$ to t_{2g} orbital of Fe(III)($t_{2g}^3e_g^2$): 15,000 cm^{-1}

from t_{2g} orbital of $Fe^{II}(t_{2g}^6)$ to e_g orbital of Fe(III)($t_{2g}^3e_g^2$): 25,000 cm^{-1}

This electron transfer from Fe(II) to Fe(III) is mediated through the π-orbital of the bridging 'CN' ligand. It leads to temporary reduction of Fe(III) to Fe(II) and temporary oxidation of Fe(II) to Fe(III). Thus the electron transfer produces the charge transfer bands in the visible region and the compound is intensely coloured.

In such mixed valence compounds experiencing the intravalence electron transfer (IT), the electron is delocalized over the metal centres through the bridging ligand to produce the equivalent metal centres. Thus in Prussian blue, the situation is as follows:

$$\text{Fe(II)} \overset{e}{\cdots\cdots\cdots\cdots} \text{Fe(III)}$$

The colour of Prussian or Turnbull blue is due to the electron transition from Fe(II) to Fe(III) through the π-orbitals of the bridging ligand. It is supported by the fact that if in the polymeric compound both the metal centres are in the same oxidation states (*e.g.*—Fe^{II}—C≡N—Fe^{II}— or —Fe^{III}—CN—Fe^{III}—) then no colour is developed.

Turnbull blue is obtained by mixing a simple Fe(II) salt with $K_3[Fe(CN)_6]$. It is also a polymeric compound and its structure can also be realized in the same framework of Prussian blue. In Turnbull blue, the colour arises from the same IT.

(**Note:** For structural details of prussian and turnbull blue *see* Sec. 1.12.).

(ii) **Creutz-Taube ion (*cf.* Sec. 1.13):** In this complex, Ru(II) and Ru(III) centres are linked through the bridging ligand pyrazine.

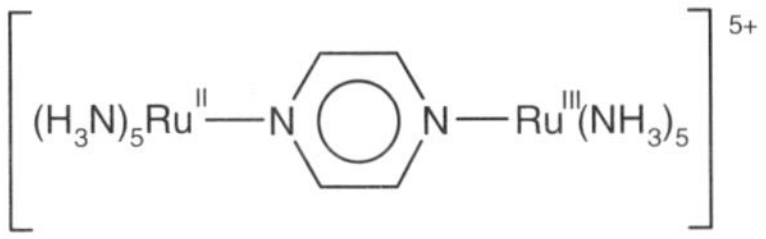

Here the electron transition from Ru(II) to Ru(III) occurs through the π-orbital of the bridging ligand pyrazine. This IT band appears at 1570 nm (*i.e.* in the near infrared region). This transition is schematically represented as:

$$[2, 3] \xrightarrow{h\nu} [3, 2],$$ 2 and 3 refer the oxidation states of ruthenium.

(iii) **Halide bridged mixed valence compounds:** Many compounds having the chain, —X—M^{II}—X—M^{IV}—X (X = Cl, Br; M = Pt, Pd) show the intense intervalence absorption of the polarized light along the chain. The examples are: $[Pt^{II}Br_2(NH_3)_2][Pt^{IV}Br_4(NH_3)_2]$, $[PdCl_2(NH_3)_2][PdCl_4(NH_3)_2]$; etc. **Wolfram's red salt** $Pt(EtNH_2)_4Cl_3.2H_2O$ (Average oxidation state = +3) also belongs to this group. Here the octahedral $[PtCl_2(EtNH_2)_4]^{2+}$ and planar $[Pt(EtNH_2)_4]^{2+}$ ions are linked in chains through the bridging chlorido ligands.

(iv) **Other examples of metal to metal charge transfer (MMCT) bands: Molybdenum blue** is a mixed oxide (MO_3, M_2O_5) having both Mo^V and Mo^{VI}. Its intense blue colour is due to the charge transfer between the metal centres of different oxidation states. Colour of **tungsten blue oxides** is probably due to such metal-metal charge transfer (MMCT). Metal to metal charge transfer also occurs in **black gold**, $Cs_2Au^IAu^{III}Cl_6$ where the electron transfer occurs from the $AuCl_2^-$ moiety to the $Au^{III}Cl_4^-$ moiety. The MMCT transition may occur also between the two different metals as in $[L_5Co^{III}—NC—Fe^{II}(CN)_5]^-$.

$\delta_b \rightarrow \delta^*$ Transition: Origin of colour in $[Re_2Cl_8]^{2-}$, $[Mo_2Cl_8]^{4-}$

In terms of MOT, the **quadruple bond** in $[Re_2Cl_8]^{2-}$ and $[Mo_2Cl_8]^{4-}$ is given by: $\sigma_b^2\pi_b^4\delta_b^2$. The energy difference between the δ_b and δ^*-MO is very small and the $\delta_b \rightarrow \delta^*$ transition gives the **intense red colour** in $[Mo_2Cl_8]^{4-}$ and the **blue colour** in $[Re_2Cl_8]^{4-}$. These are discussed in detail in Sec. 9.19, Vol. 2.

Colour of Magnus Green Salt $[Pt(NH_3)_4][PtCl_4]$

The constituent ions are having different colours: **$[Pt(NH_3)_4]^{2+}$ is colourless; $[PtCl_4]^{2-}$ is red.** The green colour of the salt $[Pt(NH_3)_4][PtCl_4]$ in **solid crystal** is due to the molecular interaction that leads to the **nonadditive colour.** In Magnus's green salt, the square planar cations $[Pt(NH_3)_4]^{2+}$ and square planar anions $[PtCl_4]^{2-}$ stack over one another to form a long –Pt–Pt– chain. Along the –Pt–Pt– chain, the Pt–Pt distance is 325 pm. This short Pt–Pt distance indicates the Pt–Pt interaction through the Pt–*d* orbitals. This **intermetallic interaction** is believed to shift the *d-d* absorption region of $[PtCl_4]^{2-}$ from green to red. Absorption of red light gives the green colour observed. This proposition is supported by its **dichoric behaviour:** intense absorption of the polarized light along the chain, and enhanced electrical conductivity along the chain. The said Pt–Pt interaction (*i.e.* stacking of square planar units) is prevented by the presence of bulky ligands. In fact, $[Pt(EtNH_2)_4][PtCl_4]$ shows a **pink colour which is sum of the colours of the constituent ions.** Here, the Pt–Pt interaction is absent. Here, it is worth mentioning that in the mixed valence compounds, the intervalence charge transfer bands also give the **nonadditive colour**. This aspect has been already discussed.

7.23 SOLVED PROBLEMS ON ELECTRONIC SPECTRA OF TRANSITION METAL COMPLEXES

Some Important Relations

d^2, d^7 (O_h) (High spin)	${}^xT_1(F) \rightarrow {}^xT_2(F)$, $\nu_1 = 8Dq + x'$ ${}^xT_1(F) \rightarrow {}^xA_2(F)$, $\nu_2 = 18Dq + x'$ ${}^xT_1(F) \rightarrow {}^xT_1(P)$, $\nu_3 = 6Dq + 15B' + 2x'$ (x' = configurational interaction energy) $10Dq = \nu_2 - \nu_1$; $10Dq = 2\nu_1 - \nu_3 + 15B'$; $15B' = \nu_2 + \nu_3 - 3\nu_1$
d^3, d^8 (T_d)	$\nu_1 = \frac{1}{2}[10Dq - 15B' + Q]$

$$\nu_2 = \frac{1}{2}[30Dq - 15B' + Q]$$

$\nu_3 = Q$

where $Q = [225B'^2 + 100Dq^2 + 180DqB']^{1/2}$

d^2, d^7 (T_d)

d^3, d^8 (O_h)

(High spin)

${}^xA_2(F) \rightarrow {}^xT_2(F)$, $\nu_1 = 10Dq$

${}^xA_2(F) \rightarrow {}^xT_1(F)$, $\nu_2 = 18Dq - x$

${}^xA_2(F) \rightarrow {}^xT_1(P)$, $\nu_3 = 12Dq + 15B' + x$

(x = configurational interaction energy)

$10Dq = \nu_1$, $15B' = \nu_2 + \nu_3 - 3\nu_1$

$$B' = \frac{2\nu_1^2 + \nu_2^2 - 3\nu_1\nu_2}{15\nu_2 - 27\nu_1}$$

$$= \frac{2\nu_1^2 + \nu_3^2 - 3\nu_1\nu_3}{15\nu_3 - 27\nu_1}$$

$\nu_1 = 10Dq$

$$\nu_2 = \frac{1}{2}[15B' + 30Dq - Q]$$

$$\nu_3 = \frac{1}{2}[15B' + 30Dq + Q]$$

$Q = [(15B' - 10Dq)^2 + 120DqB']^{1/2}$

$= [225B'^2 + 100Dq^2 - 180DqB']^{1/2}$

$= [\nu_3 - \nu_2]$

$$10Dq = \frac{1}{3}[\nu_2 + \nu_3 - 15B']$$

If ν_1 is not known, then the above expressions for ν_2 and ν_3 in terms of Q are sufficient to analyze the result.

d^6 (low spin) Octahedral complex

$${}^1A_{1g} \xrightarrow{\nu_1} {}^1T_{1g}, \nu_1 = 10Dq - 4B' + \frac{86B'^2}{10Dq} \approx 10Dq - 4B'$$

$${}^1A_{1g} \xrightarrow{\nu_2} {}^1T_{2g}, \nu_2 = 10Dq + 12B' + \frac{2B'^2}{10Dq} \approx 10Dq + 12B'$$

$10Dq = f_{\text{ligand}} \times g_{\text{metal}} = fg$;

$$\beta = \frac{B'}{B_0} = 1 - hk$$

Table 7.23.1 Rules of direct product.

For IRs

$A \times A = A$	$B \times A = B$	$E \times A = E$	$T \times A = T$
$A \times B = B$	$B \times B = A$	$E \times B = E$	$T \times B = T$
$A \times E = E$	$B \times E = E$	$E \times E = *$	$T \times E = T_1 + T_2$
$A \times T = T$	$B \times T = T$	$E \times T = T_1 + T_2$	$T \times T = *$

***For T × T**

$T_1 \times T_1 = T_2 \times T_2 = A_1 + E + T_1 + T_2$

$T_1 \times T_2 = T_2 \times T_1 = A_2 + E + T_1 + T_2$

* For ***E* × *E*:** The product depends on the nature of point group

For subscripts and superscripts of IRs

	Primes	**Subscripts**
$g \times g = g$	$' \times ' = '$	$1 \times 1 = 1$
$g \times g \times u = u$	$' \times '' = ''$	$1 \times 2 = 2$
$u \times g = u$	$'' \times ' = ''$	$2 \times 1 = 2$
$u \times u = g$	$'' \times '' = '$	$2 \times 2 = 1$

For O, C_{3v} and T_d: $E_1 \times E_2 = E_2 \times E_1 = B_1 + B_2 + E_1$; $E_1 \times E_1 = E_2 \times E_2 = A_1 + A_2 + E_2$

For C_{4v}, D_{2d} and D_4: $E \times E = A_1 + A_2 + B_1 + B_2$

Direct product Table for O point group (*e.g.* O_h, T_d)

	A_1	A_2	E	T_1	T_2
A_1	A_1	A_2	E	T_1	T_2
A_2	A_2	A_1	E	T_2	T_1
E	E	E	$A_1 + A_2 + E$	$T_1 + T_2$	$T_1 + T_2$
T_1	T_1	T_2	$T_1 + T_2$	$A_1 + E + T_1 + T_2$	$A_2 + E + T_1 + T_2$
T_2	T_2	T_1	$T_1 + T_2$	$A_2 + E + T_1 + T_2$	$A_1 + E + T_1 + T_2$

Examples:

$\mathbf{D_{2d}}: E_g \times E_u = A_{1u} + A_{2u} + E_u$; $\mathbf{C_{3v}}: A_1 \times A_2 = A_2$; $E \times E = A_1 + A_2 + E$;

$\mathbf{T_d}: T_1 \times T_2 = A_2 + E + T_1 + T_2$; $\mathbf{D_{3h}}: E' \times E'' = A_1'' + A_2'' + E''$;

$\mathbf{D_{4h}}: E_g \times B_{2g} = E_g$; $E_g \times E_u = A_{1u} + A_{2u} + B_{1u} + B_{2u}$; $\mathbf{O_h}: E_g \times T_{1u} = T_{1u} + T_{2u}$;

$\mathbf{D_{3d}}: E_g \times E_g = A_{1g} + A_{2g} + E_g$; $\mathbf{T_d}: T_1 \times T_1 = A_1 + E + T_1 + T_2$

Table 7.23.2 Correlation Table for the O_h Point Group.

O_h	O	T_d	D_{4h}	D_{2d}	C_{4v}	D_3	D_{3d}	C_{2v}	C_{2h}
A_{1g}	A_1	A_1	A_{1g}	A_1	A_1	A_1	A_{1g}	A_1	A_g
A_{2g}	A_2	A_2	B_{1g}	B_1	B_1	A_2	A_{2g}	A_2	B_g
E_g	E	E	$A_{1g} + B_{1g}$	$A_1 + B_1$	$A_1 + B_1$	E	E_g	$A_1 + A_2$	$A_g + B_g$
T_{1g}	T_1	T_1	$A_{2g} + E_g$	$A_2 + E$	$A_2 + E$	$A_2 + E$	$A_{2g} + E_g$	$A_2 + B_1 + B_2$	$A_g + 2B_g$
T_{2g}	T_2	T_2	$B_{2g} + E_g$	$B_2 + E$	$B_2 + E$	$A_1 + E$	$A_{1g} + E_g$	$A_1 + B_1 + B_2$	$2A_g + B_g$
A_{1u}	A_1	A_2	A_{1u}	B_1	A_2	A_1	A_{1u}	A_2	A_u
A_{2u}	A_2	A_1	B_{1u}	A_1	B_2	A_2	A_{2u}	A_1	B_u
E_u	E	E	$A_{1u} + B_{1u}$	$A_1 + B_1$	$A_2 + B_2$	E	E_u	$A_1 + A_2$	$A_u + B_u$
T_{1u}	T_1	T_2	$A_{2u} + E_u$	$B_2 + E$	$A_1 + E$	$A_2 + E$	$A_{2u} + E_u$	$A_1 + B_1 + B_2$	$A_u + 2B_u$
T_{2u}	T_2	T_1	$B_{2u} + E_u$	$A_2 + E$	$B_1 + E$	$A_1 + E$	$A_{1u} + E_u$	$A_2 + B_1 + B_2$	$2A_u + B_u$

Table 7.23.3 Normal modes of vibration (*see* Appendix 12B).

O_h([ML_6]): $A_{1g} + E_g + T_{2g} + 2T_{1u} + T_{2u}$; **$T_d$([$ML_4$]):** $A_1 + E + 2T_2$
D_{4h}([ML_4]): $A_{1g} + B_{1g} + B_{2g} + A_{2u} + B_{2u} + 2E_u$; **$D_3$:** $3A_1 + 2A_2 + 5E$
D_{2d}([ML_4]): $2A_1 + B_1 + 2B_2 + 2E$; **D_{4h}([$ML_4L'_2$]):** $2A_{1g} + B_{1g} + B_{2g} + E_g + 2A_{2u} + B_{1u} + 3E_u$
C_{4v}([ML_5]): $3A_1 + 2B_1 + B_2 + 3E$; **C_{2v}([$ML_2L'_2$]):** $4A_1 + A_2 + 2B_1 + 2B_2$

SOLVED NUMERICAL PROBLEMS

Problem 1: Aqueous solution $[Ti(OH_2)_6]^{3+}$ shows a broad absorption peak having the centre at 20,300 cm^{-1}. Calculate the cfse.

Solution: Ti(III) is a d^1 system. The possible electronic transition (without considering the J.T. distortion) is:

$$t_{2g}^1 e_g^0 \xrightarrow[(=10Dq)]{h\nu} t_{2g} e_g^1,\ \bar{\nu} = 20{,}300\ \text{cm}^{-1}$$
$$= 20{,}300 \times 100\ \text{m}^{-1}$$

$$10Dq = h\nu = h\frac{c}{\lambda} = hc\bar{\nu}$$
$$= (6.62 \times 10^{-34}\ \text{J s}) \times (2.99 \times 10^8\ \text{m s}^{-1}) \times 20{,}300 \times 100\ \text{m}^{-1}$$
$$= 4.03 \times 10^{-19}\ \text{J / ion}$$
$$= \frac{4.03\times10^{-19}\times6.022\times10^{23}}{1000}\ \text{kJ/mole}$$
$$\approx 243\ \text{kJ / mole } i.e.\ Dq \approx 24.3\ \text{kJ mol}^{-1}$$

cfse = $4Dq = 4 \times 24.3$ kJ mole = 97.2 kJ mol^{-1}

Problem 2: (a) Calculate the $10Dq$ value of $[TiL_6]^{3+}$ from the following data (L = CH_3COCH_3): $[TiCl_3L_3]$: $10Dq$ = 15,400 cm^{-1}; $[TiCl_6]^{3-}$: $10Dq$ = 12,750 cm^{-1}
(b) Calculate $10Dq$ for $[Co(en)_2F_2]^+$
Given: $10Dq$ for $[Co(en)_3]^{3+}$ = 23,000 cm^{-1}
$10Dq$ for $[CoF_6]^{3-}$ = 13,000 cm^{-1}

Solution: (a) According to Jorgensen's rule of average environment, we have:

$$\bar{\nu}([MA_xB_{6-x}]) = \frac{x}{6}\bar{\nu}([MA_6]) + \frac{6-x}{6}\bar{\nu}([MB_6])$$

or,
$$10Dq([MA_xB_{6-x}]) = \frac{x}{6}\times10Dq([MA_6]) + \frac{6-x}{6}10Dq([MB_6]).$$

In the present case: $10Dq([TiCl_3L_3]) = \frac{3}{6}\times10Dq([TiCl_6]^{3-}) + \frac{3}{6}\times10Dq([TiL_6]^{3+})$

or,
$$15{,}400 = \frac{1}{2}\left[12{,}750 + 10Dq([TiL_6]^{3+})\right]$$

or,
$$10Dq([TiL_6]^{3+}) = 15{,}400 \times 2 - 12{,}750 = 18{,}000\ \text{cm}^{-1}$$

(b) $10Dq\ [Co(en)_2F_2]^+) = \frac{1}{6}[4\times23{,}000 + 2\times13{,}000] = 20{,}000\ \text{cm}^{-1}$

Problem 3: The absorption spectrum of $[Ni(OH_2)_6]^{2+}$ shows peaks at 9,000 cm^{-1}, 14,000 cm^{-1} and 25,000 cm^{-1}. Empirical ligand (H_2O) parameter $h = 1$ and the empirical metal (Ni^{2+}) parameter $k = 0.12$. Comment on the band assignments and calculate the $10Dq$ value and configurational interaction energy (x). (B_0 = 1000 cm^{-1}).

Solutions: The possible ligand field transitions are:

$$^3A_{2g}(F) \rightarrow {}^3T_{2g}(F),\ \nu_1 = 10Dq = 9{,}000\ cm^{-1}$$

$$^3A_{2g}(F) \rightarrow {}^3T_{1g}(F),\ \nu_2 = 18Dq - x = 14{,}000\ cm^{-1}$$

$$^3A_{2g}(F) \rightarrow {}^3T_1(P),\ \nu_3 = 12Dq + 15B' + x = 25{,}000\ cm^{-1}$$

Thus, the lowest energy band gives the $10Dq$ value, *i.e.* $10Dq = 9{,}000\ cm^{-1}$ and $Dq = 900\ cm^{-1}$

$$\nu_2 - \nu_1 = (18Dq - x) - (10Dq) = 8Dq - x$$

or, $$14{,}000 - 9{,}000 = 8 \times 900 - x$$

or, $$5{,}000 = 7{,}200 - x;\ \text{or},\ x = 2{,}200\ cm^{-1}$$

i.e. configurational interaction energy = 2,200 cm^{-1}

$$15B' = \nu_2 + \nu_3 - 3\nu_1 = 14{,}000 + 25{,}000 - 3 \times 9{,}000$$
$$= 39{,}000 - 27{,}000$$
$$= 12{,}000\ cm^{-1}$$

or, $$B' = \frac{1}{15} \times 12{,}000 = 800\ cm^{-1}$$

$$\text{Nephelauxetic parameter}\,(\beta) = \frac{B_{\text{complex}}}{B_{\text{free ion}}} = \frac{B'}{B_0} = 1 - hk$$

i.e. $$\beta = 1 - 1 \times 0.12 = 0.88$$

$$\frac{B'}{B_0} = 0.88;\ \text{or},\ B' = B_0 \times 0.88 = 1000 \times 0.88 = 880\ cm^{-1}$$

Thus the B' value calculated is in good agreement with B' value obtained from the spectral assignments.

Problem 4: The absorption spectrum of $[Co(NH_3)_6]^{3+}$ shows two strong peaks at 21,000 cm^{-1} and 29,500 cm^{-1}. Find the $10Dq$ value and B'. Compare with the $10Dq$ value obtained from $10Dq = fg$ ($f = 1.25$, $g = 18{,}200\ cm^{-1}$).

Solution: The spin allowed transitions in low-spin $[Co(NH_3)_6]^{3+}$ are:

$$^1A_{1g} \xrightarrow{\nu_1} {}^1T_{1g};\ \nu_1 \approx 10Dq - 4B' = 21{,}000\ cm^{-1}$$

$$^1A_{1g} \xrightarrow{\nu_2} {}^1T_{2g};\ \nu_2 \approx 10Dq + 12B' = 29{,}500\ cm^{-1}$$

$$\nu_2 - \nu_1 = (10Dq + 12B') - (10Dq - 4B') = 16B'$$
$$= 29{,}500 - 21{,}000$$
$$= 8{,}500\ cm^{-1}$$

i.e. $B' = 531\ cm^{-1}$

$$10Dq = \nu_1 + 4B' = 21{,}000 + 4 \times 531 = 23{,}124\ cm^{-1}$$

Note: For more correct results, the following relations are to be used:

$$\nu_1 = 10Dq - 4B' + \frac{86B'^2}{10Dq}$$

$$\nu_2 = 10Dq + 12B' + \frac{2B'^2}{10Dq}$$

- $10Dq = f_{\text{ligand}} \times g_{\text{metal}} = f_{NH_3} \times g_{Co^{3+}} = 1.25 \times 18{,}200$
  $= 22{,}750 \text{ cm}^{-1}$

Thus, the calculated $10Dq$ value is approximately close to the value obtained from the spectral band assignment.

Problem 5: The spectral characteristics of high spin $[CoF_6]^{4-}$ (d^7 system) are:

$^4T_{1g} \xrightarrow{\nu_1} {}^4T_{2g}$ ($\nu_1 = 7{,}150 \text{ cm}^{-1}$)

$^4T_{1g} \xrightarrow{\nu_2} {}^4A_{2g}$ ($\nu_2 = 15{,}200 \text{ cm}^{-1}$)

$^4T_{1g} \xrightarrow{\nu_3} {}^4T_{1g}(P)$ ($\nu_3 = 19{,}200 \text{ cm}^{-1}$)

Find $10Dq$, β and B' and configurational interaction energy (x'); Given: $B_0 = 970 \text{ cm}^{-1}$

Solution: $\nu_1 = 8Dq + x'$; $\nu_2 = 18Dq + x'$; $\nu_3 = 15B' + 6Dq + 2x'$

i.e. $\nu_2 - \nu_1 = 10Dq = 15{,}200 - 7{,}150 = 8{,}050 \text{ cm}^{-1}$

$$15B' = \nu_2 + \nu_3 - 3\nu_1 = 15{,}200 + 19{,}200 - 3 \times 7{,}150$$
$$= 34{,}400 - 21{,}450 = 12{,}950$$

or, $B' = 863 \text{ cm}^{-1}$

$$\nu_1 = 8Dq + x' = 8 \times 805 + x' = 6{,}440 + x'$$
$$x' = \nu_1 - 6{,}440 = 7{,}150 - 6{,}440$$
$$= 710 \text{ cm}^{-1}$$

$$\beta = \frac{B_{\text{complex}}}{B_{\text{free ion}}} = \frac{B'}{B_0} = \frac{863}{970} \approx 0.88.$$

Problem 6: The peaks of some Cr(III) complexes are as follows:

(i) $[Cr(OH_2)_6]^{3+}$: 17.4 kK, 24.6 kK, 37.8 kK and a very weak band at 15.0 kK

(ii) $[Cr(NCS)_6]^{3-}$: 16 kK, 17.7 kK, 23.8 kK, 32.4 kK.

Interpret the findings.

Solution: Cr(III) in O_h crystal field shows the following ligand field transitions.

$$\left.\begin{array}{ll} ^4A_{2g}(F) \rightarrow {}^4T_{2g}(F), & \nu_1 = 17.4 \text{ kK} \\ ^4A_{2g}(F) \rightarrow {}^4T_{1g}(F), & \nu_2 = 24.6 \text{ kK} \\ ^4A_{2g}(F) \rightarrow {}^4T_{1g}(P), & \nu_3 = 37.8 \text{ kK} \end{array}\right\} \begin{array}{l} \nu_3 \rangle \nu_2 \rangle \nu_1 \\ \nu_1 = 10Dq \end{array}$$

In $[Cr(OH_2)_6]^{3+}$, the weak band at 15.0 kK is due to a spin-forbidden transition (*cf.* Tanabe-Sugano diagram):

$$^4A_{2g} \rightarrow {}^2T_{2g}, {}^2E_g, \nu_4 = 15.0 \text{ kK}$$

These doublet states reside below $^4T_{2g}(F)$. This is why

$$\nu_4 \langle \nu_1$$

The poor intensity is due to the spin-forbidden transition.

In $[Cr(NCS)_6]^{3-}$, the ligand field peaks are:

$$\nu_1 = 16 \text{ kK}, \nu_2 = 17.7 \text{ kK and } \nu_3 = 23.8 \text{ kK}.$$

The high energy intense band at 32.4 kK is a CT band.

Problem 7: $[VO(acac)_2]$ shows three special bands at: 13 kK, 17.45 kK, 25.65 kK. Interpret the results.

Solution: VO^{2+} represents $3d^1$ where there is a strong covalent interaction with the oxo-ligand. It gives the tetragonally compressed $[VO(acac)_2]$ complex of C_{4v} symmetry. The orbital splitting (*z*-in J.T. distortion) is:

$$d_{xy}^1 < d_{xz}, d_{yz} < d_{x^2-y^2} < d_{z^2} \text{ (i.e. } z\text{-in distortion)}$$

The possible electronic transitions are (*cf.* Figs. 7.14.1.2, 3):

$$d_{xy} \xrightarrow{\nu_1} d_{xz}, d_{yz};\ d_{xy} \xrightarrow{\nu_2} d_{x^2-y^2};\ d_{xy} \xrightarrow{\nu_3} d_{z^2}$$

$$\nu_1 = 13 \text{ kK};\ \nu_2 \ (= 10Dq) = 17.45 \text{ kK};\ \nu_3 = 25.65 \text{ kK}.$$

Problem 8: A tetrahedral complex of Co(II) exhibits three ligand field bands. The lowest energy band in the visible range is 7,780 cm^{-1}. Can you predict the positions of other peaks? (Given: $\beta = 0.76$, $B_0 = 970$ cm^{-1})

Solution: The possible transitions are:

$$\left.\begin{array}{l} ^4A_2 \rightarrow {^4T_2(F)},\ \nu_1 \approx 10Dq \\ ^4A_2 \rightarrow {^4T_1(F)},\ \nu_2 \approx 18Dq \\ ^4A_2 \rightarrow {^4T_1(P)},\ \nu_3 \approx 12Dq + 15B' \end{array}\right\} \begin{array}{l}\text{without considering} \\ \text{the symmetry} \\ \text{interaction energy}\end{array}$$

The absorption band at 7,780 cm^{-1} is not the ν_1 band. It would lead to $10Dq \approx 7.78$ kK which will be too high for a tetrahedral complex of Co(II). In fact, for the tetrahedral complexes of Co(II), the ν_1 band generally lies in the infrared region. Thus, the lowest energy band in the visible range is expected to be the ν_2 band, *i.e.*

$$\nu_2 \approx 18Dq = 7{,}780 \text{ cm}^{-1} \text{ i.e. } Dq \approx 432 \text{ cm}^{-1}$$

Thus, the ν_3 band should appear at:

$$\nu_3 = 12Dq + 15B'.$$

We have: $$\beta = \frac{B'}{B_0} = 0.76 = \frac{B'}{970} \quad i.e.\ B' \approx 737 \text{ cm}^{-1}$$

i.e. $$\nu_3 = 12 \times 432 + 15 \times 737 = 5{,}184 + 11{,}055 = 16{,}239 \text{ cm}^{-1}$$

Note: Taking $\beta \approx 1$ *i.e.* $B' \approx B_0 = 970$ cm^{-1}, $\nu_3 = 12Dq + 15B_0 = 12 \times 432 + 15 \times 970 = 19{,}734$ cm^{-1}

Problem 9. $[VCl_4]$ shows one absorption peak at 9,000 cm^{-1}. Predict the $10Dq$ for $[VCl_6]^{2-}$.

Solution: V(IV) is a $3d^1$ system. In the tetrahedral VCl_4 complex, the single peak arises from the following transition.

$$^2E \xrightarrow{\nu} {^2T_2},\ \nu = \Delta_t = 10Dq = 9000 \text{ cm}^{-1}$$

$$\Delta_o = 10Dq_o \approx \frac{9}{4} 10Dq_t = \frac{9}{4} \times 9000 = 20{,}250 \text{ cm}^{-1}$$

Note: The calculation done above has not considered the J.T distortion. The values of Δ_t and Δ_o appear too high for V(IV).

Problem 10: $[CoCl_4]^{2-}$ shows two spectral bands in the visible range at 5,800 cm^{-1} and 15,000 cm^{-1}. Comment on the $15B'$ parameter. (Given : $15B_0$ = 15,500 cm^{-1})

Solution: The ligand field transitions for the tetrahedral $[CoCl_4]^{2-}$ complex are:

$$^4A_2(F) \xrightarrow{\nu_1} {}^4T_2(F);\ {}^4A_2(F) \xrightarrow{\nu_2} {}^4T_1(F);\ {}^4A_2(F) \xrightarrow{\nu_3} {}^4T_1(P)$$

$$\nu_1 = 10Dq,\ \nu_2 \approx 18Dq,\ \nu_3 \approx 12Dq + 15B'$$

(without the symmetry interaction parameter)

If ν_1 = 5,800 cm^{-1} is considered, then $10Dq$ becomes 5,800 cm^{-1} which is too high for the tetrahedral $[CoCl_4]^{2-}$ complex. Thus it is reasonable to consider: ν_2 = 5,800 $cm^{-1} \approx 18Dq$;

i.e. $$Dq = \frac{5,800}{18} \approx 322\ cm^{-1}.$$

Thus, $\nu_3 \approx 12 \times 322 + 15B' = 3,864 + 15B' = 15,000\ cm^{-1}$

i.e. $15B' = 11,136\ cm^{-1}$

It indicates: $15B_0 = 15,500\ cm^{-1} \rangle\ 15B' = 11,136\ cm^{-1}$ *i.e.* nephelauxetic effect is quite significant.

$$\beta = \frac{15B'}{15B_0} = \frac{11,136}{15,500} \approx 0.72 \text{ i.e. } B' \text{ is } 72\% \text{ of } B_0 \text{ value.}$$

(**Note:** When only ν_2 and ν_3 bands are known for more correct analysis, the procedure to be followed is illustrated in the following example).

Problem 11: A tetrahedral complex of Co(II) gives the two absorption bands in the visible range at 7,140 cm^{-1} and a doublet peak (18,180 cm^{-1}, 19,600 cm^{-1}). Determine its spectral parameters, *i.e.* $10Dq$, β, and assignment of all the possible ligand field bands. (Given: $15B_0$ = 15,500 cm^{-1})

Solution: The possible spin-allowed transitions in the tetrahedral Co(II) complexes are:

$$^4A_2(F) \xrightarrow{\nu_1} {}^4T_2(F);\ {}^4A_2(F) \xrightarrow{\nu_2} {}^4T_1(F);\ {}^4A_2(F) \xrightarrow{\nu_3} {}^4T_1(P)$$

$$\nu_1 = 10Dq,\ \nu_2 = 18Dq - x,\ \nu_3 = 12Dq + 15B' + x$$

[x = configurational interaction energy arising from the symmetry interaction between $^4T_1(F)$ and $^4T_1(P)$]

If the band at 7,140 cm^{-1} is considered to be due to ν_1 then the $10Dq$ (= 7,140 cm^{-1}) becomes too high for the tetrahedral Co(II) complexes. Thus it is reasonable to assign the bands as follows:

$$\nu_2 = 18Dq - x = 7,140\ cm^{-1};\ \nu_3 = 12Dq + 15B' + x = \frac{1}{2}(18,180 + 19,600) = 18,890\ cm^{-1}$$

The doublet character of the ν_3-band is due to the spin-orbit coupling splitting of the T_1 term. Thus, the centre of the doublets is taken as the ν_3 band, *i.e.*

$$\nu_3 = \frac{1}{2}(18,180 + 19,600) = 18,890\ cm^{-1}$$

If the values of all the three bands, *i.e.* ν_1, ν_2 and ν_3 are known, the above relations are sufficient for the spectral analysis. But in absence of the ν_1-band value, we are to use the following relations:

$$\nu_2 = 1.5\Delta_t + 7.5B' - p,\ \nu_3 = 1.5\Delta_t + 7.5B' + p$$

$$p = \frac{1}{2}\left[225B'^2 + \Delta_t^2 - 18B'\Delta_t\right]^{1/2} = \frac{1}{2}\left[\nu_3 - \nu_2\right]$$

$$\nu_1 = \Delta_t = \frac{1}{3}\left[\nu_2 + \nu_3 - 15B'\right];$$

It leads to:

$$\Delta_t = \frac{1}{3}\left[\nu_2 + \nu_3 - 15B'\right] = \frac{1}{3}\left[7{,}140 + 18{,}890 - 15B'\right]$$

$$= 8676.7 - 5B'$$

$$p = \frac{1}{2}\left[\nu_3 - \nu_2\right] = \frac{1}{2}\left[18{,}890 - 7{,}140\right] = 5{,}875 \text{ cm}^{-1}$$

$$4p^2 = 225B'^2 - 18B'\Delta_t + \Delta_t^2$$

Using $p = 5{,}875$ cm^{-1} and $\Delta_t = 8676.7 - 5B'$, we get the acceptable solution: $B' = 916$ cm^{-1} and the the negative root for B' is not acceptable.

Thus, $\Delta_t = \nu_1 = 8676.7 - 5 \times 916 = 4100$ cm^{-1}

i.e. $10Dq = 4100$ cm^{-1}; $Dq = 410$ cm^{-1}

$\nu_2 = 18Dq - x = 18 \times 410 - x = 7{,}380 - x = 7{,}140$ cm^{-1}

i.e. $x = 240$ cm^{-1}

$15B_0 = 15{,}500$ cm^{-1} *i.e.* $B_0 \approx 1033$

It leads to: $\beta = \dfrac{B_{\text{complex}}}{B_{\text{free ion}}} = \dfrac{B'}{B_0} = \dfrac{916}{1033} \approx 0.88.$

Note: If the configurational interaction energy is ignored, then we can write:

$\nu_2 \approx 18Dq = 7{,}140$ cm^{-1} *i.e.* $Dq \approx 397$ cm^{-1}

$\nu_3 = 18{,}890$ cm^{-1} $= 12Dq + 15B' = 12 \times 397 + 15B' = 4764 + 15B'$

or, $15B' = 18{,}890 - 4764 = 14{,}126$ cm^{-1} *i.e.* $B' \approx 942$ cm^{-1}

If leads to: $\beta = \dfrac{B'}{B_0} = \dfrac{942}{1033} \approx 0.9$

Problem 12: $[V(OH_2)_6]^{3+}$ gives two spectral bands at 17.2 kK and 25.6 kK. If these are supposed to be the first two ligand field bands. Calculate the position of the third ligand field band by using the Tanabe-Sugano diagram (given in Fig. 7.23.2).

Solution. The ligand field transitions for the d^2-system are:

$^3T_{1g}(F) \xrightarrow{\nu_1} {}^3T_{2g}(F)$, $\nu_1 = 17.2$ kK, **(lowest energy band)**

$^3T_{1g}(F) \xrightarrow{\nu_2} {}^3A_{2g}(F)$,

$^3T_{1g}(F) \xrightarrow{\nu_3} {}^3T_{1g}(P)$,

Between ν_2 and ν_3, one transition corresponds to 25.6 kK. For the $d^2(O_h)$ system, we have:

$\nu_1 = 8Dq + x' \approx 8Dq$; $\nu_2 = 18Dq + x' \approx 18Dq$; $\nu_3 = 6Dq + 15B' + 2x' \approx 6Dq + 15B'$

Taking, $\nu_1 \approx 8Dq = 17{,}200$ cm^{-1}, *i.e.* $Dq = 2{,}150$ cm^{-1}, ν_2 ($\approx 18Dq$) should appear at 38,700 cm^{-1}. Thus, the band at 25,600 cm^{-1} is not definitely due to the ν_2 transition (*i.e.* $^3T_{1g} \rightarrow {}^3A_{2g}$). Hence, it is reasonable to assume ν_3 to correspond to 25,600 cm^{-1}, *i.e.*

$$^3T_{1g}(F) \xrightarrow{\nu_3} {}^3T_{1g}(P),\ \nu_3 = 25.6 \text{ kK}.$$

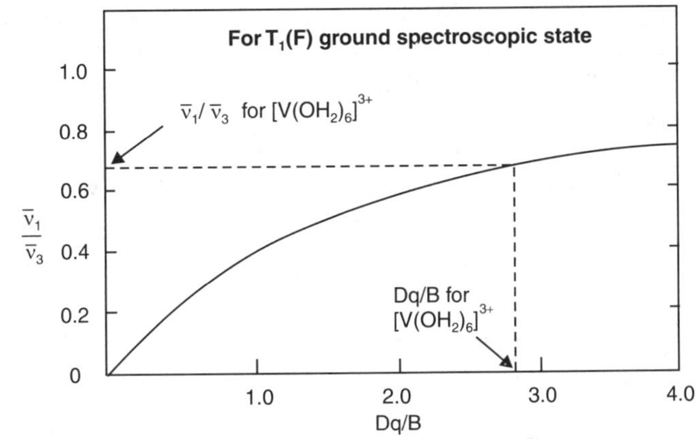

Fig. 7.23.1 Variation of the $\bar{\nu}_1/\bar{\nu}_3$ ratio with *Dq/B* for the d^2-octahedral complexes.

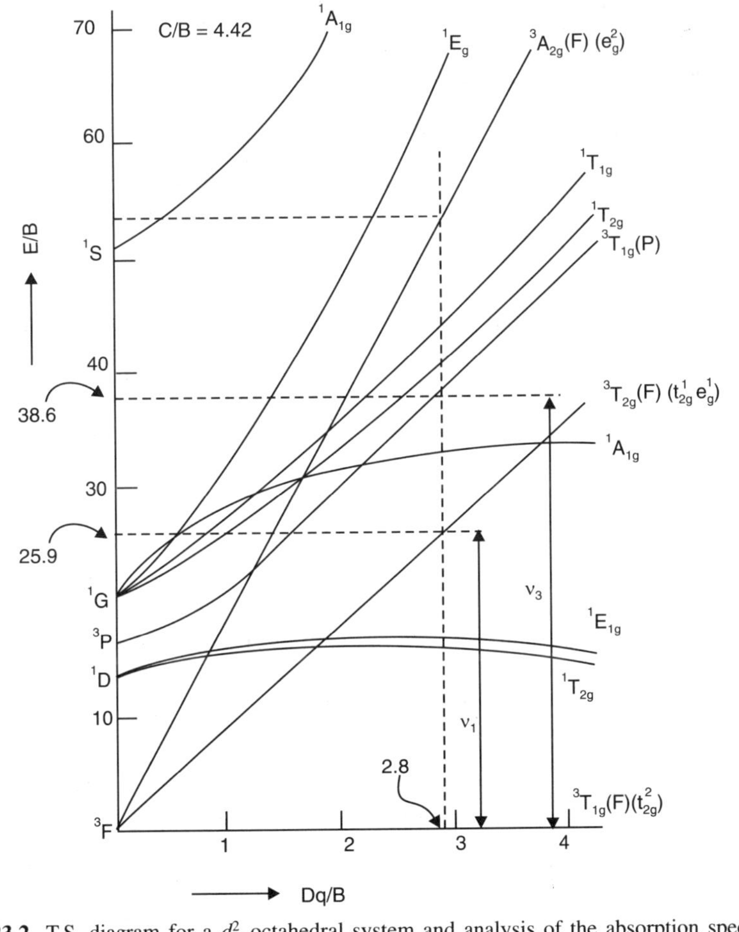

Fig. 7.23.2 T.S. diagram for a d^2–octahedral system and analysis of the absorption spectrum of $[V(OH_2)_6]^{3+}$ · ν_1: $^3T_{1g}(F) \rightarrow {}^3T_{2g}(F)$; ν_2: $^3T_{1g}(F) \rightarrow {}^3A_{2g}(F)$; ν_3: $^3T_{1g}(F) \rightarrow {}^3T_{1g}(P)$.

Now let us verify our prediction from the Tanabe-Sugano diagram and the plot of ν_1/ν_3 *vs.* Dq/B' for the $T_{1g}(F)$ ground spectroscopic state (Figs. 7.23.1 and 7.10.4).

From the Tanabe-Sugano diagram for the d^2-system, the ratio ν_1/ν_3 may be calculated for the different Dq/B' values. This plot is shown in Fig. 7.23.1.

Note: For the T_1 ground spectrocopic state, the transition energy ratio, ν_2/ν_1 is close to 2 (but not less than 2) regardless of the strength of ligand field (*cf.* Fig. 7.10.4b). If ν_2 is taken as 25.6 kK, then ν_2/ν_1 becomes 1.48 ⟨ 2. Thus ν_2 cannot be 25.6 kK.

$$\nu_1\text{: } {}^3T_{1g}(F) \rightarrow {}^3T_{2g}(F);$$

$$\nu_3\text{: } {}^3T_{1g}(F) \rightarrow {}^3T_{1g}(P);$$

In the present case, the ratio $\bar{\nu}_1/\bar{\nu}_3$ becomes 0.672 for which Dq/B' becomes 2.8. For this ratio of Dq/B', E/B' for ${}^3T_{2g}(F)$ becomes 26 and E/B becomes 38 for ${}^3T_{1g}(P)$.

i.e. $$\frac{17,200}{B'} = \frac{\nu_1}{B'} = 26.0 \text{ or, } B' \approx 662 \text{ cm}^{-1}$$

and $$\frac{25,600}{B'} = \frac{\nu_3}{B'} = 38 \text{ or, } B' \approx 674 \text{ cm}^{-1}$$

average value of $B' = 668 \text{ cm}^{-1}$

E/B for ${}^3A_{2g}$(F) at $Dq/B = 2.8$ becomes 52.5

i.e. $$\frac{\nu_2}{B'} = 52.5 \text{ or, } \nu_2 = 52.5 \times 668 = 35,070 \text{ cm}^{-1}$$

(**Note:** Accuracy of the results depends on the accuracy of the Tanabe-Sugano diagram used. Here an approximate Tanabe-Sugano diagram has been used. This is why, B' value becomes different for ${}^3T_{2g}(F)$ and ${}^3T_{1g}(P)$. Here ν_2 gives the approximate value.)

Problem 13: A simplified Tanabe-Sugano diagram for d^3 configuration is given in Fig. 7.23.3. For $[Cr(OH_2)_6]^{3+}$, $10Dq = 17,000 \text{ cm}^{-1}$, and $B' = 680 \text{ cm}^{-1}$. Predict the positions of the ligand field bands.

(**Note:** $10Dq$ and B' values may be obtained by using the relations. $10Dq = f \times g$, $\beta = \frac{B'}{B_0} = 1 - hk$, values of f, g, B_0, h and k may be obtained from literature.)

Solution. The ratio $\frac{Dq}{B'} = \frac{1700}{680} = 2.5$

The possible spin-allowed transitions are:

$${}^4A_{2g}(F) \xrightarrow{\nu_1} {}^4T_{2g}(F);\ {}^4A_{2g}(F) \xrightarrow{\nu_2} {}^4T_{1g}(F) \text{ and } {}^4A_{2g}(F) \xrightarrow{\nu_3} {}^4T_{1g}(P).$$

A vertical line is drawn for the value of $Dq/B' = 2.5$ at the x-axis. This vertical line cuts the energy state ${}^4T_{2g}(F)$ at $y = 24.5$, the energy state ${}^4T_{1g}(F)$ at $y = 34.5$ and the energy state ${}^4T_{1g}(P)$ at $y = 53.0$. The y-axis is plotted in terms of E/B. Thus we have:

$E(\nu_1) = 24.5 \times B' = 24.5 \times 680 = 16,660 \text{ cm}^{-1}$

$E(\nu_2) = 34.5 \times B' = 34.5 \times 680 = 23,460 \text{ cm}^{-1}$

$E(\nu_3) = 53.0 \times B' = 53.0 \times 680 = 36,040 \text{ cm}^{-1}$

The actual experimental values are:

$\nu_1 = 17,000 \text{ cm}^{-1}$, $\nu_2 = 24,000 \text{ cm}^{-1}$, $\nu_3 = 37,000 \text{ cm}^{-1}$

Thus the values calculated from the Tanabe-Sugano diagram are in good agreement with the experimental result.

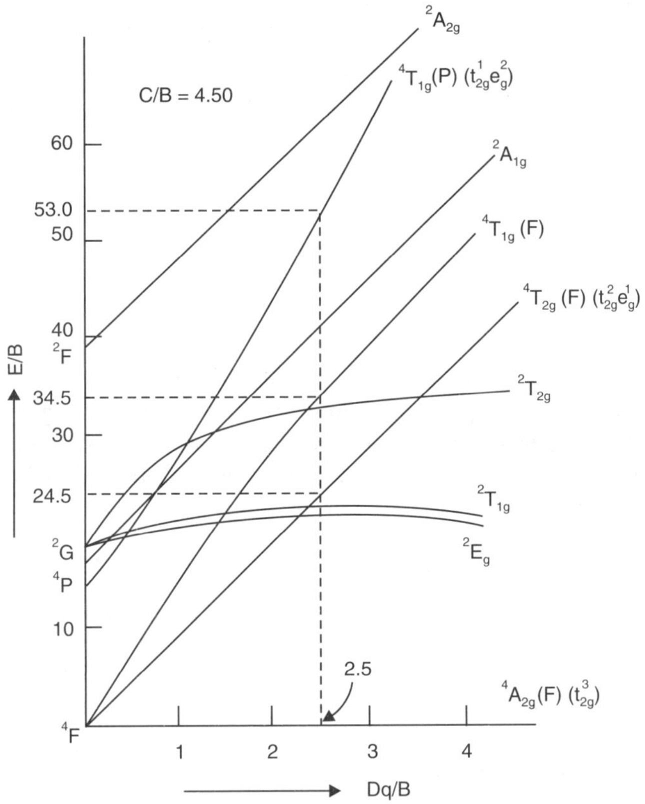

Fig. 7.23.3 Tanabe-Sugano diagram for the d^3-octahedral complex.

(**Note:** In practice, the ratios ν_1/ν_2, ν_2/ν_3, ν_2/ν_3 are determined from the experimental peaks and these are fitted in the T.S. diagram or in Fig. 7.10.4 to determine the $10Dq$ and B' values.

For this purpose, the ratios $\nu_1/\nu_2, \nu_2/\nu_3$, ν_2/ν_3 are calculated for the different Dq/B' values from the *T-S* diagram and these ratios are plotted against the Dq/B' values (*cf.* Fig. 7.10.4). Then the experimental ratio is fitted in the curve to determine Dq'/B. This is illustrated in the Problem 12.)

Problem 14: Calculate energy of the states due to the crystal field splitting of the ^2D term (d^1 configuration), ^3F term (d^2-configuration) and ^4F term (d^3-configuration) in a weak crystal field. Comment on the ground state electronic configuration for the d^2 configuration.

Solution: (i) The ^2D term splits into the $^2T_{2g}$ and ^2E_g levels. The $^2T_{2g}$ level corresponds to the t_{2g}^1 configuration (energy = – $4Dq$ with respect to that of the unsplit ^2D term) and ^2E_g corresponds to the e_g^1 configuration (energy = $+6Dq$ with respect to that of the unsplit ^2D term).

$$t_{2g}^1 : \underbrace{(d_{xy})^1(d_{yz})(d_{xz}), (d_{xy})(d_{yz})^1(d_{xz}), (d_{xy})(d_{yz})(d_{xz})^1}_{^2T_{2g}\,(\text{Orbital degeneracy}=3)}$$

$$e_g^1 : \underbrace{(d_{x^2-y^2})^1(d_{z^2}), (d_{x^2-y^2})(d_{z^2})^1}_{^2E_g\,(\text{orbital degeneracy}=2)}$$

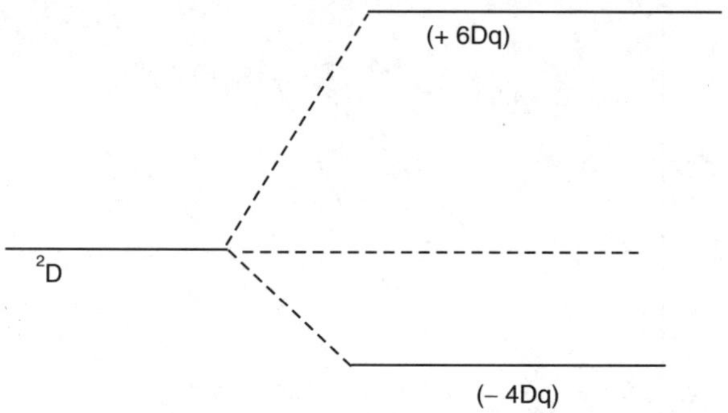

Fig. 7.23.4 Splitting of the ^2D term.

(ii) ^3F (for d^2) splits into $^3T_{1g}$, $^3T_{2g}$ and $^3A_{2g}$. From the Tanabe-Sugano diagram, it is seen that each of the $^3T_{2g}$ and $^3A_{2g}$ terms appears only once but the $^3T_{1g}$ term appears twice, *i.e.* $^3T_{1g}(P)$ and $^3T_{1g}(F)$. $^3T_{2g}$ and $^3A_{2g}$ are the excited states and they correspond to the $t_{2g}^1e_g^1$ and $t_{2g}^0e_g^2$ configurations respectively.

The crystal field energies of the states $^3T_{2g}$ and $^3A_{2g}$ states are $+2Dq$ and $+12Dq$ respectively with respect to that of unsplit ^3F.

Configuration	State	
$t_{2g}^1e_g^1$	$^3T_{2g}(F)$	$^3T_{2g}$(F) corresponds to $(d_{xy})^1\,(d_{z^2})^1$,
e_g^2	$^3A_{2g}(F)$	$(d_{xz})^1(d_{x^2-y^2})^1$, $(d_{yz})^1\,(d_{x^2-y^2})^1$

The ground state t_{2g}^2 configuration works for the **strong ligand field** but in a **weak crystal field** there will be a mixing between the $^3T_{1g}$(F) and the $^3T_{1g}(P)$ terms where $^3T_{1g}(P)$ represents the $t_{2g}^1e_g^1$ configuration like $(d_{xy})^1\,(d_{x^2-y^2})^1$, $(d_{xz})^1\,(d_{z^2})^1$, $(d_{yz})^1\,(d_{z^2})^1$. Thus in a weak crystal field, due to the mixing of the $^3T_{1g}(P)$ and $^3T_{1g}(F)$ states, the ground state is to be represented by $t_{2g}^xe_g^y$ where $x + y = 2$ where x is less than 2).

Splitting of the ^3F term is shown in Fig. 7.23.5.

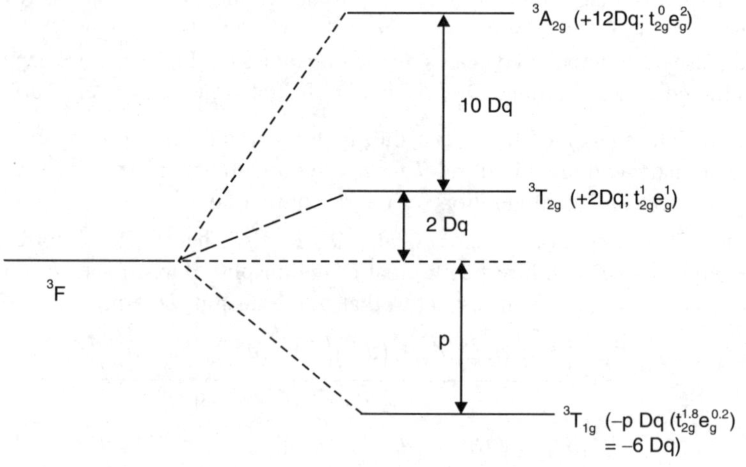

Fig. 7.23.5 Splitting of ^3F term in a weak crystal field.

To maintain the Barrycentre, during splitting of the ^3F term, we have:

$-3p + 3 \times (2Dq) + 1 \times (12Dq) = 0$; orbital degeneracy for the T and A terms are 3 and 1 respectively.

or $p = 6Dq$

Energy of the ground state $t_{2g}^x e_g^y = -4xDq + 6yDq = -6Dq$ and $x + y = 2$

Solution of these two equations give: $x = 1.8$, $y = 0.2$ *i.e.* $t_{2g}^{1.8} e_g^{0.2}$

Note: In a strong field where the crystal field effect is more than electron repulsion effect, the ground state is t_{2g}^2 (energy $= -8Dq$), but in a weak field, the ground state configuration is $t_{2g}^{1.8} e_g^{0.2}$ (energy $= -6Dq$). In a weak field, the electron repulsion is relatively more important. This is why, the electron repulsion forces some electron density to move into the e_g level.

d³ and ⁴F state: By giving the same argument as in the d^2 system, we can write:

$^4A_{2g}(F)(t_{2g}^3, -12Dq)$, $^4T_{2g}(F)$ $(t_{2g}^2 e_g^1, -2Dq)$ and $^4T_{1g}(F)$ $(t_{2g}^x e_g^y; pDq$ say, $x + y = 3)$. Here mixing of the $^4T_{1g}(F)$ and $^4T_{1g}(P)$ states in a weak crystal field has been considered. Applying the principle of energy conversion and orbital degeneracy we can write (*cf.* Fig. 7.14.5.2):

$1 \times 12\,Dq + 3 \times 2\,Dq = 3\,pDq$ *i.e.* $p = 6$; $(-4x + 6y)Dq = pDq = 6Dq$ and $x + y = 3$ lead to $x = 1.2$, $y = 1.8$. Thus in a weak fied, $^4T_{1g}(F)$ corresponds to $t_{2g}^{1.2}\, e_g^{1.8}$ not $t_{2g}^1\, e_g^2$.

Problem 15: In a noncentrosymmetric system, all the electronic transitions are not necessarily allowed. Illustrate for the complexes of T_d and D_{2d} symmetry.

Solution. (a) **$[CoCl_4]^{2-}$ (T_d):** Let us illustrate for the tetrahedral complex $[CoCl_4]^{2-}$ (d^7-system) where the spin allowed electronic transitions are:

$$^4A_2(F) \longrightarrow {}^4T_2(F);\ {}^4A_2(F) \longrightarrow {}^4T_1(F);\ {}^4A_2(F) \longrightarrow {}^4T_1(P).$$

The character table of T_d indicates that the coordinates x, y and z form the basis for the T_2 representation. It indicates that the dipole moment operator transforms as T_2.

- **$A_2 \to T_1$ transition:** $A_2 \times \mu \times T_1 = A_2 \times T_2 \times T_1 = A_2 \times (A_2 + E + T_1 + T_2) = A_1 + E + T_1 + T_2$

The triple product result contains the **totally symmetric** A_1 representation and consequently the *transition is allowed*.

- **$A_2 \to T_2$ transition:** $A_2 \times \mu \times T_2 = A_2 \times T_2 \times T_2 = A_2 \times (A_1 + E + T_1 + T_2) = A_2 + E + T_1 + T_2$

There is no **totally symmetric** A_1 term in the triple product result and thus the **transition is forbidden**. However, coupling of any one of the vibrations of A_2, E, T_1 or T_2 will make the transition allowed.

For the T_d point group, the normal modes of vibration are: $A_1 + E + 2T_2$. Thus coupling of E or T_2 vibration will make the $A_2 \to T_2$ transition allowed. (*cf.* A_1 is present in the direct products: $E \times E = A_1 + A_2 + E$; $T_2 \times T_2 = A_1 + E + T_1 + T_2$).

(b) **$[CuCl_4]^{2-}$(D_{2d}):** Compressed T_d leads to D_{2d} symmetry where the tetrahedral T_2 and E term spit as follows:

T_d	D_{2d}
T_2	$B_2 + E$
E	$B_1 + A_1$

Thus the possible transitions in $[CuCl_4]^{2-}$ (D_{2d}) of d^9 system are:

$$^2B_2 \longrightarrow {}^2E;\ {}^2B_2 \longrightarrow {}^2B_1;\ {}^2B_2 \longrightarrow {}^2A_1$$

For the D_{2d} point group, the dipole moment operators transform as $B_2(\|)$ and $E\ (\perp)$.

Transition	*Triple product*	*Polarisation*
$B_2 \longrightarrow E$	$B_2 \times B_2 \times E = B_2 \times E = E$ $B_2 \times E \times E = E \times E = A_1 + A_2 + B_1 + B_2$	Forbidden (‖) Allowed (⊥)
$B_2 \longrightarrow B_1$	$B_2 \times B_2 \times B_1 = A_2 \times B_2 = B_1$ $B_2 \times E \times B_1 = A_2 \times E = E$	Forbidden (‖) Forbidden (⊥)
$B_2 \longrightarrow A_1$	$B_2 \times B_2 \times A_1 = B_2 \times B_2 = A_1$ $B_2 \times E \times A_1 = B_2 \times E = E$	Allowed (‖) Forbidden (⊥)

Presence of the **totally symmetric A_1** representation in the triple product gives the **allowdness** while the absence of A_1 in the triple product indicates the **forbiddeness** in the respective electronic transitions. It indicates that the $^2B_2 \rightarrow {}^2B_1$ is orbitally forbidden and the other two transitions experience the electronic polarisation.

EXERCISE 7

A. General Type Questions

1. What is Lambert-Beer law? What are its limitations?
2. What are the different types of electronic spectra?
3. What are the orbital and spin selection rules in electronic transitions? Give the quantum mechanical basis of these rules.
   - For the $t_{2g} \rightarrow e_g$ transition, sometimes, more than one peak is obtained. How can you explain this?
4. What are the possible mechanisms to relax the Laporte selection rule?
5. What are the effects of (i) lowering of symmetry and (ii) temperature on the intensity of electronic spectra.
6. What is the effect of spin-orbit coupling splitting of terms on band intensity and band-width?
7. What are the important factors to determine the band-width of electronic spectra?
8. How can you explain the asymmetric shape of the spectral bands?
9. What do you mean by isosbestic point? What is the significance of this point?
10. What do you mean by Condon-Shortley and Racah parameters?
11. Explain the origin of luminescence (with fluorescence and phosphorescence) in terms of Jablonski diagram.
12. Discuss the luminescence of Cr(III)-compounds.
13. Discuss the effect of J.T. distortion on electronic spectra.
14. Predict the spin-allowed spectral transitions in terms of Orgel diagram.
15. Discuss the effect of configurational interaction between the $^3T_1(P)$ and $^3T_1(F)$ terms on the energies of spectral transitions of transition metal ions.
16. Calculate the energies of the states arising from the splitting of ^3F term of a d^2 configuration. Comment on the energies of the $^3T_{1g}$ state in a strong field and in a weak field. In a weak crystal field, the ground state electronic configuration t_{2g}^2 is not correct—comment. (**Hints:** *See* problem 14)
17. What are the limitations of Orgel diagram? Discuss the characteristic features of Tanabe-Sugano diagram.

18. How can you predict the spectral band-width from the Tanabe-Sugano diagram?
19. Generally, the spin-forbidden transitions give the sharp peaks—explain.
20. Give the energies of the spin-allowed transitions in the octahedral and tetrahedral geometries of $d^{1,2,3,4,8,9}$ and high-spin $d^{4,5,6,7}$ systems.
21. Discuss the structural and spectral spectral features of $[Ti(OH_2)_6]^{3+}$ with special reference to intensity, band shape and band-width. How can you determine $10Dq_o$ for $[Ti(OH_2)_6]^{3+}$?
22. Discuss the structural and spectral characteristics of six-coordinate complexes of oxovanadium(IV). How do you determine the $10Dq_o$ value?
23. Compare the spectral behaviour of $[Cu(OH_2)_6]^{2+}$ and $[Cu(NH_3)_x(OH_2)_y]^{2+}$ ($x + y = 6$).
24. Discuss the change of spectral properties of $[Cu(acac)_2]$ in different solvents differing in coordinating power.
25. Discuss the spectral properties of $[V(OH_2)_6]^{3+}$.
26. Discuss the origin of ruby laser action.
27. Discuss the spectral features of $[Cr(NH_3)_6]^{3+}$ and $[Cr(en)_3]^{3+}$. What is the effect of lowering of symmetry O_h to D_3 on the spectral transitions?
28. Discuss the spectra of $[Cr(OH_2)_6]^{2+}$, $[Mn(OH_2)_6]^{2+}$, $[Fe(OH_2)_6]^{2+}$, $[CoF_6]^{3-}$, $[Cr(ox)_3]^{3-}$, $[Co(OH_2)_6]^{2+}$, $[CoCl_4]^{2-}$, $[NiCl_4]^{2-}$, $[Ni(NH_3)_6]^{2+}$, $[Ni(CN)_4]^{2-}$, $[Ni(dmgH)_2]$, $[Co(NH_3)_6]^{3+}$, $[CoF_2(NH_3)_4]^+$ (both *cis*- and *trans* isomers), $[CoCl_3(NH_3)_3]$ (both fac- and mer- isomer), $[Co(gly)_3]$ (both *cis*- and *trans*- isomer), $[Co(salen)]$, $[Co(acacen)]$, $[CuCl_4]^{2-}$, $[FeCl_4]^-$, $[Ni(NH_3)_4(NO_2)_2]$, $[Cu(bpy)_3]^{2+}$.
29. Compare the spectral properties of (i) $[Cr(en)_3]^{3+}$ and $[Co(en)_3]^{3+}$; (ii) *cis*- and *trans*- isomers of $[Co(en)_2F_2]^+$; (iii) *cis*- and *trans*- isomers of $[Cr(en)_2F_2]^+$; (iv) $[Co(NH_3)_6]^{3+}$ and $[CoCl(NH_3)_5]^{2+}$.
30. Tetrahedral complexes of the 1st transition series metal ions are generally more intensely coloured than the octahedral complexes - explain.
31. Compare the spectral properties of the octahedral, tetrahedral and square planar complexes of (i) Ni(II), (ii) Cu(II) and (iii) Co(II).
32. Illustrate the coordination equilibria and thermochromism in the Ni(II) complexes.
33. Comment on the ligand field spectra of the 2nd and 3rd transition series metal ions.
34. Discuss the general characteristic spectral features of the lanthanide and actinide ions.
35. Compare the electronic spectral properties (i) between the $4f$ and $5f$ series ions, (ii) f-series and d-series ions.
36. Explain the chemistry behind the use of lanthanides as the NMR shift reagents and phosphors in colour TV.
37. Discuss the factors favouring the CT band towards the longer wavelength in LMCT and MLCT.
38. Discuss the CT-transitions in LMCT and MLCT in terms of molecular orbital diagram.
39. Illustrate the intravalance electron transfer and MMCT (metal to metal electron transfer).
40. Explain the origin of colour in prussian blue, Magnus green salt, Creutz-Taube ion, $[Mo_2Cl_8]^{4-}$ and $[Re_2Cl_8]^{2-}$.
41. Compare the characteristic features of the CT and ligand field bands.
42. Compare the spectral features of (i) $[Co(NH_3)_5X]^{2+}$ (X = F, Cl, Br, I), (ii) $[Ir^{III}(NH_3)_5X]$ (X = NH_3, H_2O, Cl, F).
43. $[Co(en)_3]^{3+}$ shows two peaks at 21.55 kK and 29.6 kK. Calculate the $10Dq$ and B' value.
44. $[CrF_6]^{3-}$ shows the peaks at 34.8 kK, 22.4 kK and 14.9 kK. Calculate $10Dq$, B' and configurational interaction energy.
45. $[V(OH_2)_6]^{3+}$ shows the ligand field bands at 17.2 kK, 25.7 kK and 36.0 kK. Calculate $10Dq$, B' and configurational interaction energy.

46. $[Co(NCS)_4]^{2-}$ gives peaks at 4.32 kK, 7.8 kK and 16.25 kK. Calculate $10Dq$, B' and configurational interaction energy.
    If the lowest energy band remains uncharacterized, then how do you calculate the values?
47. Discuss the change in spectral transitions due to the lowering of symmetry in the following cases.
    (i) Cu^{II} ($O_h \rightarrow D_{4h}$; C.N. 6);
    (ii) Ni^{II}($O_h \rightarrow D_{4h}$ (C.N. 6) $\rightarrow D_{4h}$ (C.N. 4)).
    (iii) Co^{III}(l.s.) ($O_h \rightarrow D_{4h}$; $O_h \rightarrow C_{4v}$); ($O_h \rightarrow C_{2v}$);
    (iv) Cu^{II}($T_d \rightarrow D_{2d}$).

B. Justify the following statements

1. The electron transitions, $g \rightarrow g$, $u \rightarrow u$ are quantum mechanically forbidden while the $g \rightarrow u$ transitions are quantum mechanically allowed.
   - Presence of centre of symmetry generally disallows the electronic transitions orbitally but the absence of centre of symmetry (as in T_d point group) does not necessarily allow all the possible electronic transitions.
2. For an electron transition, $\Delta S = 0$.This is supported quantum mechanically.
3. A 2e-transition is quantum mechanically forbidden.
   - 1e transition, $t_{2g} \rightarrow e_g$: d^1 system -one peak; d^2-system -two to three peaks; d^3-system -2 to 3 peaks. Apparently, $t_{2g} \rightarrow e_g$ transition should give one peak ($h\nu = \Delta_o$).
4. Franck-Condon vertical transition (0–0) is quite likely for the lanthanides experiencing the $f \rightarrow f$ transitions but for the transition metal complexes, 0–0 transition is the rarest possibility. (**Hints:** Difference in the internuclear distances between the ground and excited state)
   - For the transition metal complexes, vertical transition generally occurs to the higher vibrational levels of the excited electronic state.
5. Vibronic coupling is the main pathway for the relaxation of Laporte Selection rule for the centrosymmetric complexes while *d-p* mixing is the main pathway to to give the intensity of the ligand field bands in the noncentrosymmetric complexes.
6. $T_{1g} \rightarrow T_{2g}$ transition is only allowed when this electronic transition is coupled with the T_{1u} and T_{2u} sets of metal-ligand vibration. (Use the direct product table)
   - For an O_h-system, all the electronic transitions are orbitally forbidden, but in a T_d-system all the electronic transitions are not orbitally forbidden.
7. Vibronic coupling in the electronic transition, $^3T_{1g}(F) \rightarrow {}^3T_{2g}(F)$ (in octahedral system) should give a broad peak with 10 components or subpeaks.
   (**Hints:** Use the results of direct product)
8. In an octahedral system, the electronic transition is possible only when a g-type vibration of the ground state is coupled with the u-type vibration of the excited state and vice-versa.
9. Vibronic coupling is applicable for both the centrosymmetric and noncentrosymmetric complexes while *d-p* mixing is mainly applicable for the noncentrosymmetric systems.
10. Ligand field bands in the tetrahedral complexes are more intense than those in the octahedral complexes.
11. In the tetrahedral complexes, *d-p* mixing is the main pathway to relax the orbital selection rule.
    - In the tetrahedral complexes of 1st transition series, not the metal $(n-1)d$ and np orbital mixing but the metal $(n-1)d$ orbital and ligand p-orbital mixing is the more probable pathway.
12. Very often, the ligand field bands of the octahedral complexes become weaker with the lowering of temperature.

13. $[Co^{III}(NH_3)_5X]$ gives the more intense colour than $[Co^{III}(NH_3)_6]^{3+}$;
   - $[Co^{III}(en)_3]^{3+}$ gives the more intense colour than $[Co^{III}(NH_3)_6]^{3+}$;
   - *cis*-$[Co^{III}(NH_3)_4X_2]$ gives the more intense colour than *trans*-$[Co^{III}(NH_3)_4X_2]$.
14. Lowering of symmetry relaxes the orbital selection rule better.
   - Orbital selection rule is relaxed better in the C_{4v}, D_{4h} and D_3 point groups than in the O_h point group.
   - For the D_{4h}, D_3 and D_{2d} symmetry, electronic transitions are polarised.
   - Polarised electronic transitions cannot be studied in solution phase but in solid crystalline phase.
   - Tetrahedral (T_d) complexes are noncentrosymmetric but all the electronic transitions are not orbitally allowed.
15. A spin-forbidden transition lying close to a spin-allowed transition becomes relatively stronger.
16. A ligand field band lying close to the charge transfer band becomes relatively stronger.
17. Very often the spectral bands are symmetrical.
18. Metal-ligand bond vibration can broaden the peak.
   - The spin-allowed transition peaks are broad while the spin-forbidden transition peaks are sharp.
   - Metal-ligand bond vibration does not always broaden the peak.
19. In a d^2-system, the transition, ${}^3T_{1g}(F) \rightarrow {}^3A_{2g}(F)$ gives a very weak band.
20. In a d^2-system, the band due to the ${}^3T_{1g}(F) \rightarrow {}^3A_{2g}(F)$ transition is broader than the band due to the transition, ${}^3T_{1g}(F) \rightarrow {}^3T_{2g}(F)$.
   - In a d^2-system, transition from the triplet state to the singlet state gives the sharp peaks.
21. In the octahedral Cr(III)-system, the spin-forbidden transitions like, ${}^4A_{2g}(F) \rightarrow {}^2E_g(G)$, ${}^2T_{2g}(G)$ give the sharp peaks while the spin-allowed transitions, *i.e.* ${}^4A_{2g}(F) \rightarrow {}^4T_{1g}(F)$, ${}^4T_{1g}(F)$, ${}^4T_{1g}(P)$ give the broad peaks.
22. Spin-orbit coupling can broaden the ligand field bands. But this is more important in the tetrahedral complexes than in the octahedral complexes of 1st transition series.
23. J.T distortion can broaden the ligand field bands but sometimes it may lead to new bands due to splitting of the bands.
24. Crystallographic data cannot distinguish between the dynamic and static J.T. distortion but spectral analysis can distinguish the both types of J.T. distortion.
25. At low temperature, ligand field peaks are symmetrical but at higher temperature, the peaks become asymmetrical (*i.e.* tail towards the lower frequency).
   - If the condition of Franck-Condon vertical transition is satisfied, the $0 \rightarrow 0$ vibrational transition gives the most intense transition for an allowed electronic transition while the $0 \rightarrow 1$ transition gives the most intense transition for the electronic transition allowed through the vibronic coupling.
   - With the increase of temperature, the peaks generally become broader.
   - With the lowering of temperature, ε_{max} (at λ_{max}) of a peak increases for both the allowed electronic transitions and vibronically allowed electronic transitions, but the intensity measured by the total oscillator strength (f) remains unchanged for an allowed electronic transition while f- decreases for a vibronically allowed electronic transition with the lowering of temperature.
   - If the metal-ligand bond length remains unchanged (*e.g.* $f \rightarrow f$ transition, spin forbidden transition, etc.) during the electronic excitation, the peak becomes sharp but if the bond length in the excited state is longer then the peak becomes broader.

26. Existence of an isosbestic point indicates the presence of an equilibrium between the two absorbing species.
27. The d^n and d^{10-n} configurations are characterized by the same Russel-Saunders Term.
    - The positions of energy states of d^n configuration are inverted in the case of d^{10-n} configuration.
28. The degree of interelectronic repulsion can be measured either by Condon-Shortley parameters (F_0, F_2, F_4) or Racah parameters (A, B, C).
    - For a particular d^n-configuration, the energy difference between the two states of the same spin multiplicity can be expressed by only one Racah parameter (B) while the energy difference between the states of different spin multiplicities can be expressed in terms of two Racah parameters (B, C).
29. During splitting of a free ion term in the crystal field, orbital multiplicity is retained.
30. For the d^2-configuration: high-spin t_{2g}^2 configuration is denoted by a T-term;
    - $t_{2g}^1e_g^1$ excited state leads two T-terms of different energy.
    - $t_{2g}^0e_g^2$ excited state is represented by an A-term.
    - ^3F term splits in a crystal field into $^3A_2$, $^3T_1$ and $^3T_2$ to maintain the orbital degeneracy 7.
    - ^3P term converts in a crystal field into $^3T_1$ to maintain the orbital degeneracy 3.
    - In a weak field, for a d^2 system, the ground state $^3T_{1g}(F)$ does not represent the $t_{2g}^2e_g^0$ configuration. It represents the $t_{2g}^{1.8}e_g^{0.2}$ configuration. However, in a strong field, the ground state represents the t_{2g}^2 configuration.
31. Luminescence (*i.e.* radiative decay through fluorescence or phosphorescence) occurs when the nonradiative paths of decay are relatively inefficient.
    - In chemiluminescence, electronically excited state is produced at the cost of chemical energy.
    - Reaction of Cl_2 or ClO^- with H_2O_2 produces a red chemiluminescent glow.
32. Fluorescence involves a spin-allowed transition while phosphorescence involves a spin-forbidden transition.
    - Intersystem crossing (ISC) – radiationless transition between the two excited electronic states of different multiplicity is an essential condition for phosphorescence.
    - Jablonski diagram explains the luminescence.
33. Luminescence is well known for the lanthanide ions (Ln^{3+}) but it rarely occurs in transition metal ions.
34. Cr(III) can show the luminescence.
    - Existence of $^2E_g(G)$ state close to the $^4T_{2g}(F)$ state is responsible for the luminescence property of Cr(III) in the octahedral geometry.
    - Depending on the crystal field strength, Cr(III) can show both the fluorescence and phosphorescence.
    - Participation of $^2E_g(G)$ state – an excited state controls the photochemical reactions of Cr(III).
    - In $[Cr^{III}(A)_6]$, the electronic transitions are orbitally forbidden but in $[Cr^{III}(A-A)_3]$, some of the electronic transitions are allowed and they show the electronic polarisation.
    - Very often, the spectral transitions in $[Cr^{III}(A-A)_3]$ can be explained in terms of the transitions of Cr^{III}-octahedral complexes.
35. Photochemistry of $[Ru(bpy)_3]^{2+}$ can be rationalized in terms of its luminescence property.
36. Configurational interaction between the $^3T_1(P)$ and $^3T_1(F)$ states is relatively less important in the weak field ligands.
    - Configurational interaction energy is quite important to understand the spectral transition energies for the d^2, d^7 (O_h) and d^3, d^8 (T_d) systems while the configurational interaction energy is relatively less important for the d^2, d^7 (T_d) and d^3, d^8 (O_h) systems for a particular ligand.

- For a d^3 (O_h) system, $10Dq$ value calculated from the two lowest energy bands as $\nu_1 = 10Dq$, $\nu_2 = 18Dq$ are not self consistent.

37. ${}^xA_2(F) \xrightarrow{\nu_1} {}^xT_2(F)$; ${}^xA_2(F) \xrightarrow{\nu_2} {}^xT_1(F)$ and ${}^xA_2(F) \xrightarrow{\nu_3} {}^xT_1(P)$.
The energy order is: $\nu_3 \rangle \nu_2 \rangle \nu_1$ *i.e.* ν_3 is the highest energy band.
    - ${}^xT_1(F) \xrightarrow{\nu_1} {}^xT_2(F)$; ${}^xT_1(F) \xrightarrow{\nu_2} {}^xA_2(F)$; ${}^xT_1(F) \xrightarrow{\nu_3} {}^xT_1(P)$.
    ν_1 is always the lowest energy band but the highest energy band may be either ν_2 or ν_3 depending on the condition (*i.e.* ligand field strength).
38. Orgel diagram can only show the spin-allowed transitions in the high-spin complexes while Tanabe-Sugano diagram can show both the spin allowed and spin forbidden transitions in both the high-spin and low-spin complexes.
39. Orgel diagram cannot predict the relative broadness or sharpness of the peaks but Tanabe-Sugano diagram can predict this.
40. In constructing a Tanabe-Sugano diagram for a d^n-configuration, the ratio of C/B must be mentioned.
    - Different metal ions having the same d^n-configuration should have the different Tanabe-Sugano diagrams but in reality, the same Tanabe-Sugano diagram works for the different but isoelectronic metal ions.
41. $[Ti(OH_2)_6]^{3+}$ shows a weak but broad peak (at 21,000 cm^{-1}) with a shoulder at 19,500 cm^{-1}.
    - Δ_o for $[Ti(OH_2)_6]^{3+}$ is regarded as 20,300 cm^{-1}
    - Large band-width and asymmetric shape of the peak in $[Ti(OH_2)_6]^{3+}$ can be rationalized in terms of J.T. distortion but not in terms of spin-orbit coupling splitting of the T-term.
42. J.T. distortion cannot split the ligand field band into a doublet in $[TiF_6]^{3-}$ but in $[CoF_6]^{3-}$.
43. Six coordinate complexes of V(IV), *e.g.* $[VF_6]^{2-}$ and VO^{2+} *e.g.* $[VO(NCS)_5]^{3-}$ are different in terms of the spectral features: complexes of V(IV) are characterized with a broad band while complexes of VO^{2+} show three different peaks.
    - 5 coordinate complexes of oxovanadium(IV) like $[VO(acac)_2]$ show 4 electronic transitions.
    - Spectral properties of both the 6- and 5- coordinate complexes of oxovanadium(IV) can be explained in the same way.
    - Sometimes, the oxovanadium(IV) complexes show 3 to 4 spectral bands. (**Hints:** C_{4v}, C_{2v} symmetry).
44. For a d^9 configuration, the ground state is orbitally doubly degenerate (*i.e.* ${}^2E_g$) and the excited state is orbitally triply degenrate (${}^2T_{2g}$).
45. Most of the Cu(II) complexes are blue or green but some of them may be red or brown.
46. Analysis of the polarized crystal spectrum of $CuSO_4.5H_2O$ shows three absorption peaks at 10,500 cm^{-1}, 13,000 cm^{-1} and 14,500 cm^{-1}.
47. λ_{max} for $[Cu(OH_2)_6]^{2+}$ is 800 nm while it is shifted to higher energy for $[Cu(NH_3)_4(OH_2)_2]^{2+}$ (λ_{max} = 590 nm). But in moving from $[Cu(NH_3)_4(OH_2)_2]^{2+}$ to $[Cu(NH_3)_5(OH_2)]^{2+}$, λ_{max} shifts towards the longer wavelength.
    - In spite of several possible spectral transitions in the square planar complexes of Cu(II), $[Cu(bigH)_2]^{2+}$ shows only a broad band.
    - $[Cu(bpy)_3]^{2+}$ or $[Cu(phen)_3]^{2+}$ shows two well defined peaks.
48. Ligand field bands of $[Cu(acac)_2]$ are of different types in different solvents having the different coordinating properties.
    - In the poor coordinating solvents, splitting of the spectral peaks of $[Cu(acac)_2]$ is prominent.
49. Square pyramidal (C_{4v}) complexes of Cu(II) are characterized by three ligand field bands.

50. $[V(OH_2)_6]^{3+}$ shows two spectral bands in the visible range. The 3rd peak due to the ${}^3T_{1g}(F) \rightarrow {}^3A_{2g}(F)$ transition is missing.
    - In the six-coordinate complexes of V(III), the intensity of the peak due to ${}^3T_{1g}(F) \rightarrow {}^3A_{2g}(F)$ is very weak.
51. In the Cr(III) octahedral complexes, the third ligand field band due to ${}^4A_{2g}(F) \rightarrow {}^4T_{1g}(P)$ is not always characterized.
    - In the d^3 and d^8 octahedral complexes, $10Dq$ value is given by the lowest energy ligand field band (*i.e.* $\nu_1 = 10Dq$). But, $10Dq$ calculated from the next higher energy band assuming $\nu_2 = 18Dq$ is not correct.
    - In the d^3, d^8 (O_h) systems, if all the three ligand field bands are characterized, then $10Dq$, B' and configurational interaction energy can be calculated easily.
52. $[Cr(OH_2)_6]^{3+}$ shows the following spectral features: two intense bands at 17 kK, and 24 kK; a weak band at 37 kK as a shoulder of a strong CT band; and two weak bands at 15 kK and 22 kK.
53. $[Cr(en)_3]^{3+}$ belongs to the D_3 point group while $[Cr(NH_3)_6]^{3+}$ belongs to the O_h point group. But their spectral characteristics are more or less the same.
    - In moving from the O_h to the D_3 symmetry of Cr(III) complexes, the three spectral ligand field bands should be three doublet peaks.
    - In the $[Cr(ox)_3]^{3-}$ and $[Cr(en)_3]^{3+}$, the third ligand field band cannot easily characterized.
    - Colour of $[Cr(en)_3]^{3+}$ is more intense than that of $[Cr(NH_3)_6]^{3+}$.
    - $[Cr(ox)_3]^{3-}$ is more intensely coloured than $[Cr(en)_3]^{3+}$
54. The *trans*-$[Cr(en)_2F_2]^+$ complex should have the 6 ligand field bands.
    - Nature of the ligand field bands can distinguish the *cis*- and *trans*- isomers of $[Cr(en)_2F_2]^+$. But the ligand field bands cannot distinguish the geometrical isomers of $[Cr(NCS)_2(OH_2)_4]^+$.
    - *Cis*-$[Cr^{III}A_4B_2]$ complexes are more intense in colour than their *trans* isomers.
55. Spin forbidden transition of Cr(III) in the O_h field can explain the action of ruby laser.
    - Ruby (obtained through the isomorphous substitution of Al^{3+} by Cr^{3+} in corundum) looks red while Cr_2O_3 is green.
    - Gem-stone emerald (obtained through the isomorphous substitution of Al^{3+} by Cr^{3+} in beryl) looks green.
    - Ruby red phosphorescence emission does not depend on the nature of ligands.
    - Position of the ${}^2E_g(G)$ level with respect to that of the ${}^4T_{2g}(F)$ is crucial to explain the action of a ruby laser.
    - Many octahedral complexes of Cr(III) give the phosphorescence of ruby emission (*i.e.* red light emission).
56. $[Ti(OH_2)_6]^{3+}$ looks purple but the gemstone sapphire (obtained through the isomorphous substitution of Al^{3+} by Ti^{3+}) looks blue.
57. Most of the Ni(II)-octahedral complexes are characterized by the presence of three ligand field bands.
    - Spectral transitions in $[Ni^{II}A_4B_2]$ (D_{4h} symmetry) can be roughly explained in terms of the ligand field transitions of the octahedral complex of Ni(II).
    - In $[Ni^{II}A_4B_2]$ (D_{4h} symmetry), very often the lowest energy ligand field band shows a doublet character.
    - The octahedral, tetrahedral and square planar geometries of Ni(II) can be distinguished by considering their spectral and magnetic properties.

- In the Ni(II) octahedral complexes, the peak due to the transition, $^3A_{2g}(F) \rightarrow {}^3T_{1g}$ is expected to be weak because it is a spin forbidden transition. But intensity of the band is not so weak compared to that of the other spin-allowed ligand field bands.
(**Hints:** Spin forbidden transition gains the intensity through the intensity stealing phenomenon.)

58. Nature of the ligand field bands in $[Ni(NO_3)_4]^{2-}$ indicates its octahedral structure (*i.e.* some of NO_3^- groups act as the didendate ligands.
59. In $[Ni(OH_2)_6]^{2+}$, the band due to the $^3A_{2g}(F) \rightarrow {}^3T_{1g}(F)$ transition is a doublet. The same thing happens for $[Ni(dmso)_6]^{2+}$).
    - Always, the transition $^3A_{2g}(F) \rightarrow {}^3T_{1g}(F)$ does not give the doublet peak for the O_h Ni(II) complexes.
60. The 3rd ligand field band in $[Ni(bpy)_3]^{2+}$ is missing.
61. The absorption spectrum of $[Co(OH_2)_6]^{2+}$ shows the following features:
One peak at 8.1 kK; a broad peak at about 20 kK constituted by three overlapping bands of 16 kK, 19.4 kK and 21.6 kK.
    - In $[Co(OH_2)_6]^{2+}$, the peak from the spin-forbidden transition, $^4T_{1g}(F) \rightarrow {}^4A_{2g}(F)$ is not so weak.
    - In $[Co(OH_2)_6]^{2+}$, the two energy states $^4A_{2g}(F)$ and $^4T_{1g}(P)$ are in the cross-over region.
62. For the spectral analysis of the high-spin complexes of Co(II), instead of three ligand field bands, two ligand field bands are very often well characterized.
63. Absorption spectrum of $[Cr(OH_2)_6]^{3+}$ shows two bands at 9.5 kK and 14 kK.
64. $[Mn(DMSO)_6]^{3+}$ shows three ligand field transitions at 14.5 kK, 17 kK and 20 kK.
    - In $[Mn(DMSO)_6]^{3+}$, there are two types of ν_{S-O} stretching frequencies.
65. $[CrF_6]^{4-}$ (in CrF_2 crystal) shows two peaks while $[MnF_6]^{3-}$ show three peaks but both Cr(II) and Mn(III) represent the d^4 system.
66. $[CoF_6]^{3-}$ shows a broad peak with a doublet splitting.
67. $[Fe(OH_2)_6]^{2+}$ shows a doublet peak.
68. The colours of $[Mn(OH_2)_6]^{2+}$ and $[Fe(OH_2)_6]^{3+}$ are very faint.
69. In the high-spin d^5-system, some of the ligand field bands are very much sharp.
70. In the d^5-system, $^4T_{1g}(P)$ and $^4T_{2g}(D)$ states are of sigmoidal nature.
71. $[Fe(OH_2)_6]^{3+}$ in aq. $HClO_4$ media is almost colourless but it becomes coloured in aq. HCl media.
    - $[Fe(OH_2)_6]^{3+}$ is colourless but its hydrolyzed species is brown coloured.
72. Low-spin t_{2g}^6 gives $^1A_{1g}$ state but the excited state $t_{2g}^5e_g^1$ gives the states, $^3T_{1g}$, $^3T_{2g}$, $^1T_{1g}$ and $^1T_{2g}$.
73. The low-spin Co(III) complexes show two spin-allowed ligand field bands and two spin-forbidden ligand field bands.
    - The spin-forbidden transitions are of lower energy while the spin-allowed transitions are of higher energy.
74. The low-spin Co(III) complexes and Cr(III) complexes show the similar spectral transitions.
75. For the low-spin Co(III) complexes, energy difference between the spin-allowed ligand field bands is approximately $16B'$.
76. Cubic (*i.e.* octahedral), tetragonal and rhombic symmetries of the low-spin Co(III) complexes can be distinguished by considering their ligand field spectra.
    - Cubic: two peaks; tetragonal: three peaks; rhombic: three peaks.
77. *cis*-$[CoCl_2(NH_3)_4]^+$ and *trans*-$[CoCl_2(NH_3)_4]^+$ can be distinguished by their ligand field spectra.
78. *fac-mer* isomers of $[CoCl_3(NH_3)_3]$ can be distinguished by their absorption spectra.
79. *trans*-$[Co(en)_2F_2]^+$ gives three spectral bands while *cis*-$[Co(en)_2F_2]^+$ gives two peaks.

- Spectral features of the geometrical isomers of $[Co(en)_2F_2]^+$ are:
  cis-: 19 kK (ε = 85), 27.5 kK (ε = 50)
  trans-: 17.5 kK (ε = 20), 22.5 kK (ε = 30), 27.8 kK (ε = 50).

80. The *cis- trans* isomers of $[Co^{III}A_4B_2]$ (low-spin) may not be always distinguished by their absorption spectra.
    - The low spin *trans*-$[Co^{III}A_4B_2]$ complexes show the vibronic polarisation in terms of spectral behaviour.
    - Polarisation studies (on spectral transitions) can be done by using the single crystal in solid state but not in solution.
    - They can be distinguished only when the ligands *A* and *B* are positioned widely apart in the spectrochemical series of the ligands.
81. The isomers of $[Co(gly)_3]$ can be distinguished by their absorption spectra.
82. The ligand field spectra of $[Co(NH_3)_6]^{3+}$, $[CoCl(NH_3)_5]^{2+}$ and $[CoCl_2(NH_3)_4]^+$ are of different types.
83. Though $[Co(NH_3)_6]^{3+}$ and $[Co(en)_3]^{3+}$ belong to the different point groups, their absorption spectra are of similar type. Only they differ in intensity slightly.
    - Electronic spectra of $[Co(ox)_3]^{3-}$ and $[Co(en)_3]^{3+}$ are very much comparable but their intensities and peak separations are different.
84. $K_4[Fe(CN)_6]$ gives two ligand field peaks at 31 kK and 37 kK.
85. In $[CuCl_4]^{2-}$, $[CuBr_4]^{2-}$, the ligand field band is broad.
    - Spectral behaviour of $[CuCl_4]^{2-}$ in $Cs_2[CuCl_4]$ can be best explained in terms of D_{2d} symmetry of $[CuCl_4]^{2-}$
    - In the tetrahedral complexes, $e \rightarrow t_2$ transition does not really represent the $d - d$ transition.
86. Tetrahedral complexes of Co(II) generally show two peaks in the visible range and one peak is more intense.
87. Pale pink colour of Co(II) aqueous solution becomes dark blue-green in presence of HCl.
88. $[CoCl_4]^{2-}$ gives a complex envelope of spectral bands.
89. The peaks of $[CoCl_4]^{2-}$ in the visible range are broad and having the fine structure components.
90. Almost colourless aqueous solution of Mn(II) becomes slightly intensified on addition of HCl.
91. Both the spin-forbidden transitions and peak splitting by spin-orbit coupling are responsible to broaden the peaks to produce a complex envelope of peaks for the tetrahedral complexes of Co(II).
92. The octahedral, tetrahedral and square planar complexes of Co(II) can be distinguished by comparing their electronic spectra and magnetic moments.
93. Tetrahedral complexes of Ni(II) are generally characterized by one intense band in the visible range.
    - For the tetrahedral complexes of Ni(II), one ligand field band lies in the infra-red region, one weak band lies in the near-infra-red region and one strong band lies in the visible range.
    - For $[NiCl_4]^{2-}$, the strong band at 15 kK shows the fine-structure components.
94. Most of the square planar complexes of Ni(II) are red or orange and the intense spectral band appears in the range 15 kK–25 kK and there is another more intense band that lies in the range 25 kK–35 kK.
95. Octahedral complexes of Ni(II) are generally blue-green, tetrahedral complexes of Ni(II) are brown, green while their square planar complexes are orange, red.
96. Spectral bands of the square planar complexes of Ni(II) are generally highly intense.
    - Square planar complexes of d^8 system are expected to give three ligand field bands.

97. Thermochromism is well documented among the Ni(II) complexes.
98. Ni(II) generally adopts the square planar geometry in presence of very strong field ligands but sometimes the weak field S-donor ligands can favour the formation of square planar complexes of Ni(II).
99. In $[PdCl_4]^{2-}$, there are two ligand field bands at about 23 kK and 30 kK. The higher energy band is more intense.
100. N-substituted salicylaldiminato complexes of Ni(II) in solution gives three absorption bands at: ~7.2 kK, ~ 11.2 kK and ~ 16 kK. With the increase of temperature of the solution, intensity of the band at 16 kK decreases while intensity of the other two bands increases.
101. At a lower temperature, $(NR_xH_{4-x})[NiCl_4]$ (x = 1,2,3) appears yellow-brown or green but with increasing the temperature, it becomes blue.
102. Lifschitz salts, *i.e.* $[Ni(N–N)_2]X_2$ (N–N = en or substituted en) are yellow in the noncoordinating solvents and blue in the coordinating solvents.
103. Solution of $[Ni(PR_3)_2X_2]$ shows the absorption peaks at 400 nm, 550 nm and 880 nm. On cooling, intensity of the peak at 400 nm increases while intensity of the other two peaks decreases.
104. In the noncoordinating solvents, $[Ni(trien)]^{2+}$ looks yellow while it becomes blue in the coordinating solvents.
105. [Co(salen)] and [Co(acacen)] complexes are characterized by the two spectral peaks at ~ 8,500 cm^{-1} and ~ 20,000 cm^{-1}.
106. Compared to the 1st transition series, it is more difficult to interpret the ligand field bands for the 2nd and 3rd transition series.
107. $[MoCl_6]^{3-}$ shows two absorptions peaks at 19,000 cm^{-1} and 24,000 cm^{-1} in the visible range. The 3rd ligand field band is probably obscured by the CT bands.
108. The orange-yellow-red complexes of Rh(III) are characterized by two absorption peaks.
109. The absorption peaks of the lanthanides are generally sharp (Rydberg type) and weak.
    - Except the lanthanides (M^{3+}) having f^0, f^1, f^7 and f^{13} and f^{14} configurations, all other lanthanides are coloured.
    - The colour of a particular lanthanide ion does not depend on the nature of the ligands.
    - For the lanthanide ions, a large number of absorption peaks are noticed in the higher energy end.
    - Luminescence property is common for the lanthanide ions while it is rarely observed for the transition metal ions.
    - Absorption peaks of Gd(III) are of very poor intensity.
    - For the lanthanide ions, spin-orbit coupling splitting of the energy states is more important than crystal field splitting of the energy states. For the 3*d*-metal ions, the reverse is true.
    - In the commercial uses, lanthanide compounds are replacing the ruby lasers and phosphors
    - Ce(IV) is a f^0-system, but its complexes are generally coloured.
    - CT bands are relatively less important for the lanthanide ions.
110. Compared to the lanthanides, the absorption peaks are more intense and broader for the actinides.
    - In terms of the spectroscopic properties, the actinides can be broadly classified into two groups.
    - Early members of the actinides are very much similar to the transition metal ions rather than to the lanthanides in terms of chemical properties (*e.g.* variable oxidation states) and spectral properties.
    - For the lanthanides, the $0 \rightarrow 0$ transition (*i.e.* $\upsilon = 0 \rightarrow \upsilon' = 0$) gives the most intense transition while this transition is forbidden (in general) for the transition metal complexes.
    - The heavier actinides are similar to the lanthanides in terms of their spectral properties.

111. In terms of spectral properties, prussian blue and turnbull blue are identical.
- The product obtained through the mixing of Fe(II)-salts and $K_4[Fe(CN)_6]$ or mixing of Fe(III)-salts and $K_3[Fe(CN)_6]$ are colourless.
- Creutz-Taube ion shows an intense band in the near infrared region.
- $[Pt(NH_3)_4]^{2+}$ is colourless and $[PtCl_4]^{2-}$ is red, but Magnus green salt constituted by these ions is green.
- $[Pt(NH_3)_4][PtCl_4]$ (*i.e.* Magnus green salt) shows a colour which is different from the sum of the colours of the constituent ions. But $[Pt(EtNH_2)_4][PtCl_4]$ shows a colour which is the sum of the colours of the constituent ions.
- Molybdenum blue or tungsten blue shows the colour due to metal to metal charge transfer (MMCT)
- $[Re_2Cl_8]^{2-}$ and $[Mo_2Cl_8]^{4-}$ are intensely coloured.

112. The charge transfer bands are more intense than the ligand field bands (*i.e.* *d-d* transitions).

113. Fe^{III}–SCN^-, Fe^{III}–phenol, Fe^{III}–hydroxamate, V^V–hydroxamate complexes are highly coloured.
- In $[Co^{III}(NH_3)_5X]^{2+}$ (X = F, Cl, Br, I), for X = F, the ligand field bands are not influenced by the CT band but for X = Br, I, the high energy ligand field band is overlapped by the CT band.
- In $[Co^{III}(NH_3)_5X]^{2+}$ (X = F, Cl, Br, I) intensity of the ligand field band is maximum for X = I.
- In moving from $[Co(NH_3)_6]^{3+}$ (O_h) to $[Co(NH_3)_5X]^{2+}$ (C_{4v}), the octahedral ligand field bands undergo splitting and this peak splitting is maximum for X = I and minimum for X = F.
- In $[Co(NH_3)_5X]^{2+}$, for X = I, all the ligand field bands are practically influenced by the LMCT band; for X = Cl, Br, only one ligand field band is affected.
- $[VO_4]^{3-}$ is colourless, $[CrO_4]^{2-}$ is yellow while $[MnO_4]^-$ is pink.
- $[ReO_4]^-$ is colourless but $[MnO_4]^-$ is coloured.
- $[CrO_4]^{2-}$ is coloured by $[MoO_4]^{2-}$, $[WO_4]^{2-}$ are colourless.
- The oxidising power sequence, $[MnO_4]^- \rangle [CrO_4]^{2-} \rangle [VO_4]^{3-}$; $[MnO_4]^- \rangle [ReO_4]^-$; $[CrO_4]^{2-} \rangle [MoO_4]^{2-} \rangle [WO_4]^{2-}$ can be explained in terms of the relative positions of their CT bands.
- $[CrO_4]^{2-}$ is harder than $[MnO_4]^-$ and the hardness is related with the ν_{CT} values.
- Colour of $[Fe(aq)]^{3+}$ depends on the pH of solution.
- Compounds of d^0 and d^{10} systems may be coloured.

114. Iodido complex of Cu(II) is unstable while the other halido complexes of Cu(II) are highly coloured.
- Iodido complex of Fe(III) is unstable while the other halido complexes of Fe(III) are dark coloured.
- Aqua-complex of Co(III) is unstable.
- $[Fe(CN)_6]^{3-}$ shows the first LMCT band at a relatively longer wavelength compared to $[Fe(CN)_6]^{4-}$.
- The CT band of $[IrBr_6]^{2-}$ is of lower energy compared to that of $[IrBr_6]^{3-}$.
- $[VCl_6]^{2-}$ is stable but $[VBr_6]^{2-}$ is unstable.
- The energy of the CT bands: $[OsCl_6]^{2-} \rangle [OsBr_6]^{2-} \rangle [OsI_6]^{2-}$; $[TiCl_6]^{2-} \rangle [TiBr_6]^{2-}$
- $[TiX_6]^{2-}$ (X = Cl, Br) shows two CT bands. The lower energy band is relatively narrower and the band-width of the higher energy band is relatively larger.
- Energy of the CT bands follows the sequence:
  $[ZrCl_6]^{2-} \rangle [TiCl_6]^{2-} \rangle [VCl_6]^{2-}$; $[HgCl_4]^{2-} \rangle [HgBr_4]^{2-} \rangle [HgI_4]^{2-}$; $[ReO_4]^- \rangle [TcO_4]^- \rangle [MnO_4]^-$

115. Fe(II)- bpy or phen complex is deep red coloured while the corresponding Fe(III)-complex shows a faint blue colour.

- Cu(I)-bpy/phen complex shows an intense colour.
- $[Cr(CO)_6]$ shows two MLCT bands.

116. For the complex, $[Ir^{III}(NH_3)_5X]$ (X = NH_3, H_2O, Cl^- and Br^-), there is no problem in assigning the ligand field bands for X = NH_3 and H_2O but the task is complicated for X = Cl and Br.
    - For X = Cl, Br, the ligand field peaks are more intense than for X = NH_3 and H_2O.
117. For the tetrahedral complexes, the ligand field transitions represented by $e \rightarrow t_2$ is not a correct description, but for the octahedral complexes, the representation, $t_{2g} \rightarrow e_g$ is correct.
    - Colour of the complex cation $[M(NH_3)_6]^{3+}$ (M = Co, Ir) in solution may depend on the nature of anions present (**Hints:** IPCT band).
    - $[Ir(NH_3)_6]^{3+}$ is colourless but it looks yellow coloured in presence of I^-.
118. In the tetrahedral complexes, for the d^{0-6} system, 4 CT bands can arise; for the $d^{7,8}$ systems, 3 CT bands can arise while for the d^{10} system only one CT band can arise.
    - In $[Cr(CO)_6]$ and $[Mo(CO)_6]$, the strong CT bands mask the expected two ligand field bands.
    - In $[Fe^{II}(L)_3]$. The MLCT band appears at about 19,500 cm^{-1} while in $[Ni^{II}(L)_3]$, the MLCT band appears at much higher energy (L = bpy, phen).
    - In $[M(CN)_4]^{2-}$ (M = Ni, Pd, Pt), there are three MLCT bands.
    - $[Fe(CN)_6]^{3-}$ shows two MLCT bands while $[Fe(CN)_6]^{4-}$ shows one MLCT band. The CT band in $[Fe(CN)_6]^{3-}$ appears at a relatively lower energy compared to that in $[Fe(CN)_6]^{4-}$.
    - Polarisation studies on the electronic transitions for *trans*-$[CoCl_2(en)_2]^+$ can identify the energy order of the ${}^1E_g$ and ${}^1A_{2g}$ levels (generated from the ${}^1T_{1g}$ level of O_h symmetry).
    - The *cis*- and *trans*- isomers of $[CoCl_2(en)_2]^+$ belong to the different point groups, *i.e.* C_{2v} and D_{4h} point groups (by considering the local symmetry) respectively but their spectral transitions can be explained by considering the tetragonal symmetry in both cases.
    - For $[Co^{III}A_4B_2]$ (low spin), the peak splitting is more, if the ligand field strengths of the ligands *A* and *B* differ widely.
    - For $[Co^{III}A_4B_2]$ (low spin), the peak splitting is more for the *trans*-complex compared to that in the *cis*-complex.
    - For $[Co^{III}A_5B]$ (low spin), the octahedral peak splitting becomes prominent if the position of *A* and *B* in the spectrochemical series are widely apart. (*cf.* $[CoF(NH_3)_5]^{2+}$ *vs.* $[CoI(NH_3)_5]^{2+}$.

APPENDICES

APPENDIX I

Units and Conversion Factors

Table 1: Basic physical quantities in SI units

Physical quantity	*Symbol for quantity*	*Name of unit*	*Symbol for unit*
Length	l	metre	m
Mass	m	kilogram	kg
Time	t	second	s
Electric current	I	ampere	A
Temperature	T	kelvin	K
Amount of substance	n	mole	mol
Luminous intensity	I_v	candela	cd

Table 2: Derived physical quantities in SI units

Physical quantity	*Name of SI unit*	*Symbol and definition of SI unit*	*Named after*
Force, weight	newton (N)	$N = kg\ m\ s^{-2} = J\ m^{-1}$	Isaac Newton (1642–1727)
Work, energy, quantity of heat	joule (J)	$J = N\ m$	James Prescott Joule (1818–1889)
Power	watt (W)	$W = J\ s^{-1}$	James Watt (1737–1819)
Pressure	pascal (Pa)	$Pa = N\ m^{-2}$	B. Pascal (1623–1662)
Electrical charge	coulomb (C)	$C = A\ s$	Charles Auguste de Coulomb (1736–1806)
Electrical potential	volt (V)	$V = kg\ m^2\ s^{-3}\ A^{-1}$ $= J\ A^{-1}\ s^{-1}$ $= J\ C^{-1} = W\ A^{-1}$	Allesandro Volta (1745–1827)
Electric capacitance	farad (F)	$F = C\ V^{-1}$	Michael Faraday (1791–1867)
Electric resistance	ohm (Ω)	$\Omega = V\ A^{-1}$	Georg Simon Ohm (1787–1854)
Electric conductance	siemens (S)	$S = \Omega^{-1}$	W. von Siemens (1816–1892)
Magnetic flux	weber (Wb)	$Wb = V\ s$	W. Weber (1804–1891)
Magnetic flux density	tesla (T)	$T = Wb\ m^{-2}$	Nikola Tesla (1856–1943)
Inductance	henry (H)	$H = V\ s\ A^{-1}$	Joseph Henry (1799–1878)
Frequency	hertz (Hz)	$Hz = s^{-1}$	Heinrich Hertz (1857–1894)
Radioactivity	becquerel (Bq)	$Bq = s^{-1}$	A. Henri Becquerel (1852–1908)
Absorbed dose of radiation	gray (Gy)	$Gy = m^2\ s^{-2}$ $= J\ kg^{-1}$	Louis H. Gray (1905–1965)
Area (A)	square metre	m^2	
Volume (V)	cubic metre	m^3	
Density (ρ)	kilogram per cubic metre	$kg\ m^{-3}$	
Velocity (u, v, w, c)	metre per second	$m\ s^{-1}$	
Acceleration	metre per square second	$m\ s^{-2}$	
*Concentration (c)	mole per cubic metre	$mol\ m^{-3}$	

*for concentration (c) in mol per litre, it is given by mol dm^{-3}

Table 3: SI prefixes

Fraction	*Prefix*	*Symbol*	*Multiple*	*Prefix*	*Symbol*
10^{-1}	deci	d	10	deka	da
10^{-2}	centi	c	10^{2}	hecto	h
10^{-3}	milli	m	10^{3}	kilo	k
10^{-6}	micro	μ	10^{6}	mega	M
10^{-9}	nano	n	10^{9}	giga	G
10^{-12}	pico	p	10^{12}	tera	T
10^{-15}	femto	f	10^{15}	peta	P

Table 4: CGS and SI units for some common physical quantities

Physical quantity	*CGS units*		*SI Units*	
	Name	*Symbol*	*Name*	*Symbol*
Length	centimetre	cm	metre	m
	Angstrom (10^{-8} cm)	Å		
Mass	gram	g	kilogram	kg
Time	second	sec	second	s
Temperature	celsius	°C	kelvin	K
	kelvin	°K		
Energy	calorie	cal	joule	J
	kilocalorie	kcal	kilojoule	kJ
	litre-atmosphere ergs	lit-atm erg		

* tonne (t) = 10^3 kg, lb = 453.59 g, Btu = 1.055×10^{10} erg = 252 cal

Table 5: Conversion of CGS units to SI units

Physical quantity	*Relation between the units*
Length	Angstrom (Å) = 10^{-10} m = 10^{-1} nm = 10^2 pm
	micron (μ) = 10^{-6} m
Volume	llitre (L) = dm^3
Force	dyne = 10^{-5} N
Energy*	erg = 10^{-7} J
	cal = 4.185 J
	eV = 1.602×10^{-19} J; eV/molecule = 96.484 kJ mol^{-1}
Pressure	atm = 1.013×10^6 dyne cm^{-2} = 101.33 kN m^{-2}
	atm = 760 torr
	mm Hg or torr = 133.32 N m^{-2}
	bar (10^6 dynes/cm^2) = 10^5 N m^{-2}
Viscosity	poise = 10^{-1} kg m^{-1} s^{-1}
Magnetic flux density	gauses = 10^{-4} T

* 1 cm^{-1} (wave number) = 2.86 cal mol^{-1}, reciprocal centimetre is known as **Kayser (K)** 1 kK (kilo Kayser) = 1000 cm^{-1}, 1 eV per molecule = 23.06 kcal mol^{-1}, 1 erg = 2.39×10^{-11} kcal. 1 eV = (charge of an electron) × 1 V = 1.6×10^{-19} C × V = 1.6×10^{-19} J,

$$1 \text{ eV/molecule} = \frac{1.6 \times 10^{-19} \times 6.02 \times 10^{23}}{1000} \text{ kJ/mole} = 96.488 \text{ kJ mol}^{-1}.$$

$$1 \text{ kK molecule}^{-1} = 10^3 \text{ cm}^{-1} \text{ molecule}^{-1}$$

$$= hc \times 10^3 \text{ cm}^{-1} \text{ molecule}^{-1} \left(cf.\, E = h\nu = \frac{hc}{\lambda} \right)$$

$$= 6.62 \times 10^{-34} \text{ J s} \times 2.99 \times 10^8 \text{ m s}^{-1} \times 10^5 \text{ m}^{-1} \text{ molecule}^{-1}$$

$$= 1.985 \times 10^{-20} \text{ J molecule} = 11.96 \text{ kJ mol}^{-1}$$

Table 6: Atomic Units (*see* text, Vol. 1, Chapter 1, Sec. 1.8.2)

APPENDIX II

Some Physical and Chemical Constants

Table 1: Values of some physical and chemical constants

Constant	*CGS and other units*	*SI units*
Acceleration due to gravity (g)	980.665 cm sec^{-2}	9.80665 m s^{-2}
Avogadro's number (N_A)	6.02217×10^{23} molecules $mole^{-1}$	6.02217×10^{23} mol^{-1}
Bohr magneton (μ_B)	9.274×10^{-21} erg $gauss^{-1}$	9.274×10^{-24} J T^{-1}
Nuclear Magneton (μ_N)	5.04×10^{-24} erg G^{-1}	5.04×10^{-27} J T^{-1}
Bohr radius (a_o)	0.52918 Å	5.2918×10^{-11} m
Boltzmann constant (k_B or k)	1.3806×10^{-16} erg (degree K)$^{-1}$	1.3806×10^{-23} J K^{-1}
Debye	10^{-18} esu cm	3.3356×10^{-30} C m
Electronic charge (e)	4.80298×10^{-10} esu	1.60216×10^{-19} C
Electronic rest mass (m_e)	9.10953×10^{-28} g	9.10953×10^{-31} kg
Faraday (F)	96487 coulomb $equiv^{-1}$	9.6487×10^{4} C mol^{-1}
Gas constant (R)	8.3144×10^{7} erg (degree K)$^{-1}$ $mole^{-1}$	8.3144 J K^{-1} mol^{-1}
	8.31441 Joules (degree K)$^{-1}$ $mole^{-1}$	8.3144 N m K^{-1} mol^{-1}
	82.053×10^{-3} litre-atm (degree K)$^{-1}$ $mole^{-1}$	8.3144 Pa m^3 K^{-1} mol^{-1}
	1.987 cal (degree K)$^{-1}$ $mole^{-1}$	
Permitivity of vacuum (ε_o)		8.854188×10^{-12} C V^{-1} m^{-1}
		(C V^{-1} m^{-1} = C^2 N^{-1} m^{-2})
Permeability of vacuum (μ_o)		$4\pi \times 10^{-7}$ H m^{-1}
Planck's constant (h)	6.62618×10^{-27} erg sec	6.62618×10^{-34} J s
Rest mass of proton (m_p)	1.672648×10^{-24} g	1.672648×10^{-27} kg
Rest mass of neutron (m_n)	1.674954×10^{-24} g	1.674954×10^{-27} kg
Atomic mass unit (u)	1.660565×10^{-24} g	1.660565×10^{-27} kg
(1 u = 10^{-3} kg mol^{-1}/N_A)		(≡ 931.48 MeV)
Vacuum speed of light (c)	2.99793×10^{10} cm sec^{-1}	2.99793×10^{8} m s^{-1}
Standard atmospheric pressure	76 cm Hg	101.325 kPa
	760 mm Hg (or torr)	1.01325 bar
	1.1032×10^{6} dyne cm^{-2}	
Hartree energy (E_h)		4.3597×10^{-18} J
Zero of the Celsius scale	0°C	273.15 K

APPENDIX III
Wavelength and Colours

Table 1: Spectral wavelength and colours

Range of the wavelength absorbed (nm)	*Colour of the radiation absorbed*	*Complementary colour observed*
400–450	Violet	Green-yellow
450–470	Indigo	Yellow
470–490	Blue	Orange
490–500	Blue-green	Red
500–560	Green	Purple
560–575	Green-yellow	Violet
575–590	Yellow	Indigo
590–640	Orange	Blue
640–730	Red	Blue-green

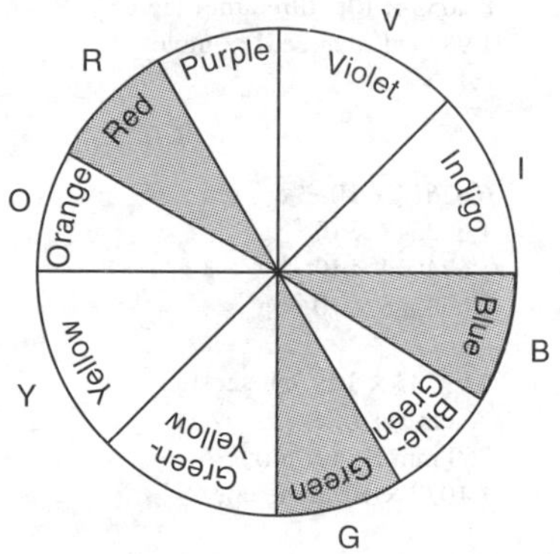

Artist's Colour Wheel: Colours and complementary colours.

Colours that stand opposite each other on the colour wheel are called as the **complimentary colours.** The amount of a colour in a particular image can be decreased by increasing its complimentary colour and vice-versa.

RGB colour model: In this model, **red (R), green (G), and blue (B)** are considered as the **three primary colours** that can generate all the colours of the visible spectrum when mixed together in different proportions. Mixing of these three primary colours in equal parts gives **white** and complete absence of these three colours results in **black**. Computer monitor devices use these **three additive primaries** to produce the different colour shades.

APPENDIX IV

Names, symbols, atomic numbers and atomic weights* of the elements

Element	*Symbol*	*Atomic number*	*Atomic weight*	*Element*	*Symbol*	*Atomic number*	*Atomic weight*
Actinium	*Ac*	89	227.03	Lead	*Pb*	82	207.19
Aluminium	*Al*	13	26.98	Lithium	*Li*	3	6.94
Americium	*Am*	95	(243)	Lutetium	*Lu*	71	174.97
Antimony	*Sb*	51	121.75	Magnesium	*Mg*	12	24.31
Argon	*Ar*	18	39.95	Manganese	*Mn*	25	54.94
Arsenic	*As*	33	74.92	Mendelevium	*Md*	101	(258)
Astatine	*At*	85	(210)	Mercury	*Hg*	80	200.59
Barium	*Ba*	56	137.34	Molybdenum	*Mo*	42	95.94
Barkelium	*Bk*	97	(247)	Neodymium	*Nd*	60	144.24
Beryllium	*Be*	4	9.01	Neon	*Ne*	10	20.18
Bismuth	*Bi*	83	208.98	Neptunium	*Np*	93	237.05
Boron	*B*	5	10.81	Nickel	*Ni*	28	58.71
Bromine	*Br*	35	79.91	Niobium	*Nb*	41	92.91
Cadmium	*Cd*	48	112.40	Nitrogen	*N*	7	14.01
Calcium	*Ca*	20	40.08	Nobelium	*No*	102	(259)
Californium	*Cf*	98	(251)	Osmium	*Os*	76	190.2
Carbon	*C*	6	12.01	Oxygen	*O*	8	16.00
Cerium	*Ce*	58	140.12	Palladium	*Pd*	46	106.4
Cesium	*Cs*	55	132.91	Phosphorus	*P*	15	30.97
Chlorine	*Cl*	17	35.45	Platinum	*Pt*	78	195.09
Chromium	*Cr*	24	52.01	Plutonium	*Pu*	94	(244)
Cobalt	*Co*	27	58.93	Polonium	*Po*	84	(209)
Copper	*Cu*	29	63.54	Potassium	*K*	19	39.10
Curium	*Cm*	96	(247)	Praeseodymium	*Pr*	59	140.91
Dysprosium	*Dy*	66	162.50	Promethium	*Pm*	61	(145)
Einsteinium	*Es*	99	(252)	Protactinium	*Pa*	91	231.04
Erbium	*Er*	68	167.26	Radium	*Ra*	88	226.03
Europium	*Eu*	63	151.96	Radon	*Rn*	86	(222)
Fermium	*Fm*	100	(257)	Rhenium	*Re*	75	186.2
Fluorine	*F*	9	19.00	Rhodium	*Rh*	45	102.91
Francium	*Fr*	87	(223)	Rubidium	*Rb*	37	85.47
Gadolinium	*Gd*	64	157.25	Ruthenium	*Ru*	44	101.07
Gallium	*Ga*	31	69.72	Samarium	*Sm*	62	150.35
Germanium	*Ge*	32	72.59	Scandium	*Sc*	21	44.96
Gold	*Au*	79	196.97	Selenium	*Se*	34	78.96
Hafnium	*Hf*	72	178.49	Silicon	*Si*	14	28.09
Helium	*He*	2	4.00	Silver	*Ag*	47	107.87
Holmium	*Ho*	67	164.93	Sodium	*Na*	11	22.99
Hydrogen	*H*	1	1.008	Strontium	*Sr*	38	87.62
Indium	*In*	49	114.82	Sulfur	*S*	16	32.06
Iodine	*I*	53	126.90	Tantalum	*Ta*	73	180.95
Iridium	*Ir*	77	192.2	Technetium	*Tc*	43	(99)
Iron	*Fe*	26	55.85	Tellurium	*Te*	52	127.60
Krypton	*Kr*	36	83.80	Terbium	*Tb*	65	158.92
Lanthanum	*La*	57	138.91	Thallium	*Tl*	81	204.37
Lawrencium*	*Lr*	103	(260)	Thorium	*Th*	90	232.04

(Contd.)

* Previously it was represented by *Lw*. Now as per IUPAC recommendation, it is represented by *Lr*.

Element	*Symbol*	*Atomic number*	*Atomic weight*	*Element*	*Symbol*	*Atomic number*	*Atomic weight*
Thulium	*Tm*	69	168.93	Xenon	*Xe*	54	131.30
Tin	*Sn*	50	118.69	Ytterbium	*Yb*	70	173.04
Titanium	*Ti*	22	47.90	Yttrium	*Y*	39	88.91
Tungsten	*W*	74	183.85	Zinc	*Zn*	30	65.37
Uranium	*U*	92	238.03	Zirconium	*Zr*	40	91.22
Vanadium	*V*	23	50.94				

* Values are based on the presently accepted ***physical scale*** of ^{12}C = 12.00000 as the standard. The ratio of these atomic weights to those on older ***chemical scale*** assuming the value 16.00000 for the atomic weight of natural oxygen (composition : ^{16}O, 99.757%; ^{17}O, 0.039%; ^{18}O, 0.204%) is 1.000279. The values given within the parentheses are the mass numbers of the most commonly known isotopes.

The true atomic weight of natural oxygen is given by : 16 × 0.99757 + 17 × 0.00039 + 18 × 0.00204 = 16.00447.

$$\text{It gives : } \frac{\text{Physical atomic weight}}{\text{Chemical atomic weight}} = \frac{16.00447}{16.00000} = 1.000279$$

Presently the term ***atomic weight*** is replaced by ***atomic mass*** or ***relative atomic mass*** (RAM). These are expressed in terms of *amu* or *u* or *Da* (Dalton). These aspects have been discussed in Sec. 5.2.3.

** For the ***translawrencium*** elements with $Z > 103$, the currently used names (cf. Chemistry of the Elements by N.N. Greenwood and A. Earnshaw and IUPAC, 1997) are : $_{104}Rf$ (rutherfordium), $_{105}Db$ (dubnium), $_{106}Sg$ (seaborgium), $_{107}Bh$ (bohrium), $_{108}Hs$ (hassium), $_{109}Mt$ (meitnerium). However, other names to these elements have also been suggested from time to time. These are : Z = 104 (kurchatovium, *Ku*; dubnium, *Db*), Z = 105 (hahnium, *Ha*, neilsbohrium, *Ns*, joliotium, *Jl*), Z = 106 (rutherfordium, *Rf*), Z = 107 (neilsbohrium, *Ns*), Z = 108 (hahnium, *Hn*). The IUPAC (1977) recommendation, i.e. a ***hybrid Latin-Greek numerical method*** (cf. Sec. 8.6), is free from such confusions. The elements with Z = 110, 111, 112, 114, 116, 118 have been characterised.

APPENDIX V

Some Useful Mathematical Relationship

(A) Differentials

(*u* and *v* are the functions of *x*; *a* and *m* are constants).

(i) $\frac{dx}{dx} = 1$ (ii) $\frac{d}{dx}(au) = a\frac{du}{dx}$ (iii) $\frac{d}{dx}(u+v) = \frac{du}{dx} + \frac{dv}{dx}$

(iv) $\frac{d}{dx}x^m = m\,x^{m-1}$ (v) $\frac{d}{dx}\ln x = \frac{1}{x}$ (vi) $\frac{d\ln u}{dx} = \frac{1}{u}\frac{du}{dx}$

(vii) $\frac{d}{dx}(e^x) = e^x$ (viii) $\frac{d}{dx}(e^u) = e^u\frac{du}{dx}$ (ix) $\frac{d}{dx}(uv) = u\frac{dv}{dx} + v\frac{du}{dx}$

(x) $\frac{d}{dx}\left(\frac{u}{v}\right) = \frac{v\left(\frac{du}{dx}\right) - u\left(\frac{dv}{dx}\right)}{v^2}$ (xi) $\frac{d}{dx}(\sin x) = \cos x$ (xii) $\frac{d}{dx}(\cos x) = -\sin x$

(xiii) $\frac{d}{dx}(\tan x) = \sec^2 x$ (xiv) $\frac{d}{dx}(\sin u) = \cos u\frac{du}{dx}$ (xv) $\frac{d}{dx}(\cos u) = -\sin u\frac{du}{dx}$

(B) Integrals

(*u* and *v* are the functions of *x*; *a* and *m* are constants; for the indefinite integrals, there should be an addition of *arbitrary constant of integration*).

(i) $\int dx = x$

(ii) $\int audx = a\int udx$

(iii) $\int (u+v)dx = \int udx + \int vdx$

(iv) $\int x^m \, dx = \frac{x^{m+1}}{m+1}, (m \neq -1)$

(v) $\int \frac{dx}{x} = \ln x$

(vi) $\int e^x dx = e^x$

(vii) $\int e^{ax} \, dx = \frac{1}{a} e^{ax}$

(viii) $\int \ln ax \, dx = x \ln ax - x$

(ix) $\int \sin x \, dx = -\cos x$

(x) $\int \cos x \, dx = \sin x$

(xi) $\int uvdx = u \int vdx - \int \left[\frac{du}{dx} \int vdx \right] dx$, (by parts)

(xii) $\int udv = uv - \int vdu$

(xiii) $\int_0^\infty x^n e^{-ax} = \frac{n!}{a^{n+1}}$, ($n$ positive integer)

(C) Trigonometric relationship

(i) $\sin(90^\circ - \theta) = \cos\theta$

(ii) $\cos(90^\circ - \theta) = \sin\theta$

(iii) $\exp(\pm im\theta) = \cos(m\theta) \pm i \sin(m\theta)$, (***Demovier's theorem***)

$\cos(m\theta) = \frac{1}{2}[\exp(im\theta) + \exp(-im\theta)]$ $\quad$ $\sin(m\theta) = \frac{1}{2i}[\exp(im\theta) - \exp(-im\theta)]$

(iv) $\sin^2\theta + \cos^2\theta = 1$

(v) $\sin 2\theta = 2\sin\theta\cos\theta$

(vi) $\cos 2\theta = \cos^2\theta - \sin^2\theta = 2\cos^2\theta - 1 = 1 - 2\sin^2\theta$

(vii) $\sin(\alpha \pm \beta) = \sin\alpha\cos\beta \pm \cos\alpha\sin\beta$

(viii) $\cos(\alpha \pm \beta) = \cos\alpha\cos\beta \mp \sin\alpha\sin\beta$

(ix) $\sin\alpha \pm \sin\beta = 2\sin\frac{1}{2}(\alpha \pm \beta)\cos\frac{1}{2}(\alpha \mp \beta)$

(x) $\cos\alpha + \cos\beta = 2\cos\frac{1}{2}(\alpha + \beta)\cos\frac{1}{2}(\alpha - \beta)$

(xi) $\cos\alpha - \cos\beta = -2\sin\frac{1}{2}(\alpha + \beta)\sin\frac{1}{2}(\alpha - \beta)$

(xii) $\sin\alpha\sin\beta = \frac{1}{2}\cos(\alpha - \beta) - \frac{1}{2}\cos(\alpha + \beta)$

(xiii) $\cos\alpha\cos\beta = \frac{1}{2}\sin(\alpha - \beta) + \frac{1}{2}\cos(\alpha + \beta)$

(xiv) $\sin\alpha\cos\beta = \frac{1}{2}\sin(\alpha + \beta) + \frac{1}{2}\sin(\alpha - \beta)$

(D) Some Useful Expansions ($x^2 \leq 1$)

$(1+x)^{-1} = 1 - x + x^2 - x^3 + ...$ $\quad$ $(1-x)^{-1} = 1 + x + x^2 + x^3 + ...$ $\quad$ $(1-x)^{-2} = 1 + 2x + 3x^2 + 4x^3 + ...$

$\ln(1+x) = x - \frac{x^2}{2} + \frac{x^3}{3} - \frac{x^4}{4} + ...$

$\ln(1-x) = -x - \frac{x^2}{2} - \frac{x^3}{3} - \frac{x^4}{4} ....$

$\exp(x)$ or $e^x = 1 + x + \frac{x^2}{2!} + \frac{x^3}{3!} + \frac{x^4}{4!} + ...$, (for all x)

$\exp(-x)$ or $e^{-x} = 1 - x + \frac{x^2}{2!} - \frac{x^3}{3!} + ...$

$$\sin x = x - \frac{x^3}{3!} + \frac{x^5}{5!} - \frac{x^7}{7!} + ..., \text{ (for all } x)$$

$$\cos x = 1 - \frac{x^2}{2!} + \frac{x^4}{4!} - \frac{x^6}{6!} + ..., \text{ (for all } x)$$

$$\ln x! = x \ln x - x \text{ or } x! = \frac{x^x}{e^x} \textbf{ (Stirling approximation, } 1730)$$
$$(\text{for } x > ca. 10)$$

(E) Progressions

(i) **Arithmatic progression :** $a, a + d, a + 2d, ...$

n-th term $(T_n) = a + (n - 1)d;$

Sum of first n terms $(S_n) = \frac{n}{2}(T_1 + T_n) = \frac{n}{2}\{2a + (n-1)d\}$

Sum of first n-natural numbers $= 1 + 2 + 3 + ... + n = \frac{n(1+n)}{2}$

(ii) **Geometric progression :** $a, ar, ar^2, ...$

n-th term $(T_n) = ar^{n-1}$; Sum of first n terms $(S_n) = \frac{a(1-r^n)}{(1-r)}$ when $r < 1$

$S_n = \frac{a(r^n - 1)}{(r-1)}$ when $r > 1$; sum to infinity $(S_\infty) = \frac{a}{1-r}$ when $|r| < 1$

(iii) **Useful results :**

$$\left.\begin{aligned} \sum_{x=1}^{n} x^2 &= \frac{n}{6}(n+1)(2n+1) \\ \sum_{x=1}^{n} x^3 &= \left[\frac{n}{2}(n+1)\right]^2 \end{aligned}\right\} \text{For first } n \text{ natural numbers, } 1, 2, 3, ..., n$$

(F) Permutations and combinations

$${}^nP_r = \frac{n!}{(n-r)!};\ {}^nP_n = \frac{n!}{0!} = \frac{n!}{1} = n!;\ {}^nC_r = \frac{n!}{r!(n-r)!};\ {}^nP_r = r!\,{}^nC_r;\ {}^nC_r = {}^nC_{n-r};\ {}^nC_n = {}^nC_0 = 1$$

$$n! = 1 \times 2 \times 3 \times ...... \times (n-2) \times (n-1) \times n;\ 0! = 1$$

(G) Logarithms

$\log(mn) = \log m + \log n;\ \log\left(\frac{m}{n}\right) = \log m - \log n,$
$\log(m^n) = n \log m;\ \log 1 = 0;\ \log_a a = 1;\ \ln(x) = 2.3 \log x$

Natural logarithm (base e); common logarithm (base 10).

(H) Mathematical Constants

$\pi = 3.14159265.....,\ e = 2.71828....,\ \ln x = 2.30258 ... \log_{10}{}^x$

APPENDIX VI
Books Consulted

Akhmetov, N.: General and Inorganic Chemistry
Akitt, J.W.: NMR and Chemistry
Anantharaman, R.: Fundamentals of Quantum Chemistry
Aruldhas, G.: Molecular Structure and Spectroscopy
Ballhausen, C.J.: Introduction to Ligand Field Theory
Banerjea, D.: Inorganic Chemistry—Elements and Compounds
Banerjea, D.: Coordination Chemistry
Banerjee, D.: Inorganic Chemistry (Principles)
Banerjea, D.: Fundamental Principles of Inorganic Chemistry
Banwell, C.N.: Fundamentals of Molecular Spectroscopy
Berrow, G.M.: Introduction to Molecular Spectroscopy
Basolo, F. and Pearson, R.G.: Mechanisms of Inorganic Reactions
Bertini, I., Gray, H.B., Lippard, S.J. and Valentine, J.S.: Bioinorganic Chemistry
Carlin, R.L.: Magnetochemistry
Chanda, M.: Atomic Structure and Chemical Bonding
Chatwal, G.R. and Anand, S.K.: Spectroscopy
Connelly, G., Damhus, T. (Editors): Nomenclature of Inorganic Chemistry: IUPAC Recommendations 2005.
Cotton, F.A. and Wilkinson, G.: Advanced Inorganic Chemistry
Cotton, F.A. and Wilkinson, G.: Basic Inorganic Chemistry
Cotton, F.A. and Wilkinson, G., Murillo, C.A. and Bochmann, M.: Advanced Inorganic Chemistry
Cotton, F.A.: Chemical Applications of Group Theory
Das, Asim K.: Bioinorganic Chemistry
Das, Asim K.: Inorganic Chemistry: Biological and Environmental Aspects
Das, Asim K.: Environmental Chemistry with Green Chemistry
Das, A.K.: A Textbook on Medicinal Aspects of Bioinorganic Chemistry
Day (Jr), D.A. and Underwood, A.L.: Quantitative Analysis
Day, M.C. and Selbin, J.: Theoretical Inorganic Chemistry
Dorain, P.B.: Symmetry in Inorganic Chemistry
Douglas, B.E. McDaniel, D.H. and Alexander, J.J.: Concepts and Models of Inorganic Chemistry
Drago, R.S,: Physical Methods in Chemistry
Drago, R.S.: Physical Methods for Chemists
Dyer, J.R.: Applications of Absorption Spectroscopy of Organic Compounds
Dutta, P.K.: General and Inorganic Chemistry
Dutta, R.L. and Syamal, A.: Elements of Magnetochemistry
Dutta, R.L.: Inorganic Chemistry (Part. I and II)
Earnshaw, A.: Introduction to Magnetochemistry
Ebsworth E.A.V., Rankin, D.W.H. and Cradock, S.: Structural Methods in Inorganic Chemistry:
Emeleus, H.J. and Anderson, J.S.: Modern Aspects of Inorganic Chemistry
Figgis, B.N.: Introduction to Ligand Fields
Glasstone, S.: Textbook of Physical Chemistry
Griffith J.S.: The Theory of Transition Metal Ions
Greenwood, N.N. and Earnshaw, A.: Chemistry of the Elements
Huheey, J.E.: Inorganic Chemistry—Principles of Structure and Reactivity
Huheey, J.E., Keiter, E.A., Keiter, R.L. and O.K. Mehdi: Inorganic Chemistry—Principles of Structure and Reactivity
Jolly, W.L.: Principles of Inorganic Chemistry
Jorgensen, C.K.: Absorption spectra and Chemical Bonding in Complexes
Kalshi, P.S.: Spectroscopy of Organic Compounds
Kapoor, K.L.: A Textbook of Physical Chemistry (Vol. 1-4)
Kettle, S.F.A.: Coordination Compounds
Langford, C.H. and Gray, H.B.: Ligand Substitution Processes
Laidler, K.J. and Meiser, J.H.: Physical Chemistry
Lee, J.D.: Concise Inorganic Chemistry
Lever, A.B.P.: Inorganic Electronic Spectroscopy
Lippard, S.J. and Berg, J.M.: Principles of Bioinorganic Chemistry
Mackay, K.M. and Mackay, R.A.: Introduction of Modern Inorganic Chemistry
Mahan, B.H.: University Chemistry
Miessler, G.L. and Tarr, D.A.: Inorganic Chemistry
Mohan, J.: Organic Spectroscopy: Principles and Applications.
Mukherjee, G.N. & Das, A.: Elements of Bioinorganic Chemistry
Nakamoto, K.: Infrared Spectra of Inorganic and Coordination Compounds
Orchin, M. and Jaffe, H.H.: The Importance of Antibonding Orbitals
Pavia, D.L., Lampman, G.M. and Kriz, G.S. : Introduction to Spectroscopy
Potterfied, W.W.: Inorganic Chemistry: A Unified Approach
Purcell, K.F. and Kotz, J.C.: Inorganic Chemistry
Rakshit, P.C.: Physical Chemistry
Rao, C.N.R.: Chemical Applications of Infrared Spectroscopy
Rao, C.N.R.: Ultraviolet and Visible Spectroscopy
Ray, R.K.: Electronic Spectra of Transition Metal Complexes
Reddy, K.V.: Symmetry and Spectroscopy of Molecules
Shriver, D.F., Atkins, P.W. and Langford, C.H.: Inorganic Chemistry
Sarkar, R.: General and Inorganic Chemistry (Part I and II)
Satyanaryana, D.N.: Electronic Absorption Spectroscopy and Related Techniques
Satyanarayana, D.N.: Vibrational Spectroscopy – Theory and Applications
Sharpe, A.G.: Inorganic Chemistry
Skoog, D.A., West, D.M., Holler, F.J. and Crouch, S.R.: Fundamental of Analytical Chemistry
Solomon, E.I. and Lever, A.B.P.: Inorganic Electronic Structure and Spectroscopy
Steed, J.W. and Atwood, J.L.: Supramolecular Chemistry
Topping, J.: Errors of Observation and Their Treatment
Vogel, A.I.: A Textbook of Quantitative Inorganic Analysis
Wilkins, R.G.: The study of Kinetics and Mechanism of Reactions of Transition Metal Complexes
Wulfsberg, G.: Inorganic Chemistry

Acknowledgement

The above listed references and sources have been freely consulted to borrow their views and ideas. The present authors are indebted to all these authors to an endless extent. The present author expresses his heartiest thanks and grateful acknowledgement to all of them.

Subject Index

Reader's Note

Reader's Note

Reader's Note

Reader's Note